TUNNELS AND UNDERGROUND CITIES: ENGINEERING AND INNOVATION MEET ARCHAEOLOGY, ARCHITECTURE AND ART

PROCEEDINGS OF THE WTC2019 ITA-AITES WORLD TUNNEL CONGRESS, NAPLES, ITALY, 3-9 MAY, 2019

Tunnels and Underground Cities: Engineering and Innovation meet Archaeology, Architecture and Art

Volume 7: Long and deep tunnels

Editors

Daniele Peila
Politecnico di Torino, Italy

Giulia Viggiani
University of Cambridge, UK
Università di Roma "Tor Vergata", Italy

Tarcisio Celestino
University of Sao Paulo, Brasil

CRC Press
Taylor & Francis Group
Boca Raton London New York

CRC Press is an imprint of the
Taylor & Francis Group, an **informa** business

A BALKEMA BOOK

Cover illustration:

View of Naples gulf

CRC Press/Balkema is an imprint of the Taylor & Francis Group, an informa business

© 2020 Taylor & Francis Group, London, UK

Typeset by Integra Software Services Pvt. Ltd., Pondicherry, India

Published by: CRC Press/Balkema
Schipholweg 107C, 2316XC Leiden, The Netherlands
e-mail: Pub.NL@taylorandfrancis.com
www.crcpress.com – www.taylorandfrancis.com

ISBN: 978-0-367-46872-9 (Hbk)
ISBN: 978-1-003-03164-2 (eBook)

*Tunnels and Underground Cities: Engineering and Innovation meet Archaeology,
Architecture and Art, Volume 7: Long and deep tunnels – Peila, Viggiani & Celestino (Eds)
© 2020 Taylor & Francis Group, London, ISBN 978-0-367-46872-9*

Table of contents

Tunnels and Underground Cities: Engineering and Innovation meet Archaeology, Architecture and Art, Volume 7: Long and deep tunnels – Peila, Viggiani & Celestino (Eds)
© 2020 Taylor & Francis Group, London, ISBN 978-0-367-46872-9

Preface

The World Tunnel Congress 2019 and the 45th General Assembly of the International Tunnelling and Underground Space Association (ITA), will be held in Naples, Italy next May.

The Italian Tunnelling Society is honored and proud to host this outstanding event of the international tunnelling community.

Hopefully hundreds of experts, engineers, architects, geologists, consultants, contractors, designers, clients, suppliers, manufacturers will come and meet together in Naples to share knowledge, experience and business, enjoying the atmosphere of culture, technology and good living of this historic city, full of marvelous natural, artistic and historical treasures together with new innovative and high standard underground infrastructures.

The city of Naples was the inspirational venue of this conference, starting from the title Tunnels and Underground cities: engineering and innovation meet Archaeology, Architecture and Art.

Naples is a cradle of underground works with an extended network of Greek and Roman tunnels and underground cavities dated to the fourth century BC, but also a vibrant and innovative city boasting a modern and efficient underground transit system, whose stations represent one of the most interesting Italian experiments on the permanent insertion of contemporary artwork in the urban context.

All this has inspired and deeply enriched the scientific contributions received from authors coming from over 50 different countries.

We have entrusted the WTC2019 proceedings to an editorial board of 3 professors skilled in the field of tunneling, engineering, geotechnics and geomechanics of soil and rocks, well known at international level. They have relied on a Scientific Committee made up of 11 Topic Coordinators and more than 100 national and international experts: they have reviewed more than 1.000 abstracts and 750 papers, to end up with the publication of about 670 papers, inserted in this WTC2019 proceedings.

According to the Scientific Board statement we believe these proceedings can be a valuable text in the development of the art and science of engineering and construction of underground works even with reference to the subject matters "Archaeology, Architecture and Art" proposed by the innovative title of the congress, which have "contaminated" and enriched many proceedings' papers.

Andrea Pigorini   Renato Casale
SIG President   *Chairman of the Organizing Committee WTC2019*

Acknowledgements

REVIEWERS

The Editors wish to express their gratitude to the eleven Topic Coordinators: Lorenzo Brino, Giovanna Cassani, Alessandra De Cesaris, Pietro Jarre, Donato Ludovici, Vittorio Manassero, Matthias Neuenschwander, Moreno Pescara, Enrico Maria Pizzarotti, Tatiana Rotonda, Alessandra Sciotti and all the Scientific Committee members for their effort and valuable time.

SPONSORS

The WTC2019 Organizing Committee and the Editors wish to express their gratitude to the congress sponsors for their help and support.

Tunnels and Underground Cities: Engineering and Innovation meet Archaeology,
Architecture and Art, Volume 7: Long and deep tunnels – Peila, Viggiani & Celestino (Eds)
© 2020 Taylor & Francis Group, London, ISBN 978-0-367-46872-9

WTC 2019 Congress Organization

HONORARY ADVISORY PANEL

Pietro Lunardi, President WTC2001 Milan
Sebastiano Pelizza, ITA Past President 1996-1998
Bruno Pigorini, President WTC1986 Florence

INTERNATIONAL STEERING COMMITTEE

Giuseppe Lunardi, Italy (Coordinator)
Tarcisio Celestino, Brazil (ITA President)
Soren Eskesen, Denmark (ITA Past President)
Alexandre Gomes, Chile (ITA Vice President)
Ruth Haug, Norway (ITA Vice President)
Eric Leca, France (ITA Vice President)
Jenny Yan, China (ITA Vice President)
Felix Amberg, Switzerland
Lars Barbendererder, Germany
Arnold Dix, Australia
Randall Essex, USA
Pekka Nieminen, Finland
Dr Ooi Teik Aun, Malaysia
Chung-Sik Yoo, Korea
Davorin Kolic, Croatia
Olivier Vion, France
Miguel Fernandez-Bollo, Spain (AETOS)
Yann Leblais, France (AFTES)
Johan Mignon, Belgium (ABTUS)
Xavier Roulet, Switzerland (STS)
Joao Bilé Serra, Portugal (CPT)
Martin Bosshard, Switzerland
Luzi R. Gruber, Switzerland

EXECUTIVE COMMITTEE

Renato Casale (Organizing Committee President)
Andrea Pigorini, (SIG President)
Olivier Vion (ITA Executive Director)
Francesco Bellone
Anna Bortolussi
Massimiliano Bringiotti
Ignazio Carbone
Antonello De Risi
Anna Forciniti
Giuseppe M. Gaspari

Giuseppe Lunardi
Daniele Martinelli
Giuseppe Molisso
Daniele Peila
Enrico Maria Pizzarotti
Marco Ranieri

ORGANIZING COMMITTEE

Enrico Luigi Arini
Joseph Attias
Margherita Bellone
Claude Berenguier
Filippo Bonasso
Massimo Concilia
Matteo d'Aloja
Enrico Dal Negro
Gianluca Dati
Giovanni Giacomin
Aniello A. Giamundo
Mario Giovanni Lampiano
Pompeo Levanto
Mario Lodigiani
Maurizio Marchionni
Davide Mardegan
Paolo Mazzalai
Gian Luca Menchini
Alessandro Micheli
Cesare Salvadori
Stelvio Santarelli
Andrea Sciotti
Alberto Selleri
Patrizio Torta
Daniele Vanni

SCIENTIFIC COMMITTEE

Daniele Peila, Italy (Chair)
Giulia Viggiani, Italy (Chair)
Tarcisio Celestino, Brazil (Chair)
Lorenzo Brino, Italy
Giovanna Cassani, Italy
Alessandra De Cesaris, Italy
Pietro Jarre, Italy
Donato Ludovici, Italy
Vittorio Manassero, Italy
Matthias Neuenschwander, Switzerland
Moreno Pescara, Italy
Enrico Maria Pizzarotti, Italy
Tatiana Rotonda, Italy
Alessandra Sciotti, Italy
Han Admiraal, The Netherlands
Luisa Alfieri, Italy

Georgios Anagnostou, Switzerland
Andre Assis, Brazil
Stefano Aversa, Italy
Jonathan Baber, USA
Monica Barbero, Italy
Carlo Bardani, Italy
Mikhail Belenkiy, Russia
Paolo Berry, Italy
Adam Bezuijen, Belgium
Nhu Bilgin, Turkey
Emilio Bilotta, Italy
Nikolai Bobylev, United Kingdom
Romano Borchiellini, Italy
Martin Bosshard, Switzerland
Francesca Bozzano, Italy
Wout Broere, The Netherlands

Domenico Calcaterra, Italy
Carlo Callari, Italy
Luigi Callisto, Italy
Elena Chiriotti, France
Massimo Coli, Italy
Franco Cucchi, Italy
Paolo Cucino, Italy
Stefano De Caro, Italy
Bart De Pauw, Belgium
Michel Deffayet, France
Nicola Della Valle, Spain
Riccardo Dell'Osso, Italy
Claudio Di Prisco, Italy
Arnold Dix, Australia
Amanda Elioff, USA
Carolina Ercolani, Italy
Adriano Fava, Italy
Sebastiano Foti, Italy
Piergiuseppe Froldi, Italy
Brian Fulcher, USA
Stefano Fuoco, Italy
Robert Galler, Austria
Piergiorgio Grasso, Italy
Alessandro Graziani, Italy
Lamberto Griffini, Italy
Eivind Grov, Norway
Zhu Hehua, China
Georgios Kalamaras, Italy
Jurij Karlovsek, Australia
Donald Lamont, United Kingdom
Albino Lembo Fazio, Italy
Roland Leucker, Germany
Stefano Lo Russo, Italy
Sindre Log, USA
Robert Mair, United Kingdom
Alessandro Mandolini, Italy
Francesco Marchese, Italy
Paul Marinos, Greece
Daniele Martinelli, Italy
Antonello Martino, Italy

Alberto Meda, Italy
Davide Merlini, Switzerland
Alessandro Micheli, Italy
Salvatore Miliziano, Italy
Mike Mooney, USA
Alberto Morino, Italy
Martin Muncke, Austria
Nasri Munfah, USA
Bjørn Nilsen, Norway
Fabio Oliva, Italy
Anna Osello, Italy
Alessandro Pagliaroli, Italy
Mario Patrucco, Italy
Francesco Peduto, Italy
Giorgio Piaggio, Chile
Giovanni Plizzari, Italy
Sebastiano Rampello, Italy
Jan Rohed, Norway
Jamal Rostami, USA
Henry Russell, USA
Giampiero Russo, Italy
Gabriele Scarascia Mugnozza, Italy
Claudio Scavia, Italy
Ken Schotte, Belgium
Gerard Seingre, Switzerland
Alberto Selleri, Italy
Anna Siemińska Lewandowska, Poland
Achille Sorlini, Italy
Ray Sterling, USA
Markus Thewes, Germany
Jean-François Thimus, Belgium
Paolo Tommasi, Italy
Daniele Vanni, Italy
Francesco Venza, Italy
Luca Verrucci, Italy
Mario Virano, Italy
Harald Wagner, Thailand
Bai Yun, China
Jian Zhao, Australia
Raffaele Zurlo, Italy

Long and deep tunnels

Tunnels and Underground Cities: Engineering and Innovation meet Archaeology,
Architecture and Art, Volume 7: Long and deep tunnels – Peila, Viggiani & Celestino (Eds)
© 2020 Taylor & Francis Group, London, ISBN 978-0-367-46872-9

Long and deep single shield TBM in very complicated geology under the Alps. Saint Martin La Porte

C. Acquista & G. Giacomin
Ghella S.p.A., Rome, Italy

F. Martin & G. Comin
Spie Batignolles Genie Civil, Saint Martin La Port, France

ABSTRACT: The paper describes the extreme geological conditions encountered during the excavation of the first stretch of the Mont-Cenis base tunnel in the new high speed railway Lyon-Turin, SMP4 site, between Saint Martin La Porte and La Praz in France. The total length of the tunnel is approximately 8700 m realized using the mechanized boring through a single shield hard rock TBM. The geology is characterized by the presence of various geological formations, fault zones with high coal content, high fractured rock mass and constant lack of a vertical and stable face. This heterogeneity has affected in a severe way the TBM advance. The aim of the paper is to highlight all TBM requirements foreseen by the contract and all needed improvements applied to overpassed or cope with different geological accidents. All installed equipment on the TBM was designed with special devices assuring to manage particular exceptional conditions. Eventually, the TBM performance is detailed referring to a series of milestones where a series of technical improvements were introduced to regain a normal excavation trend.

1   FRAMEWORK OF THE PROJECT

Saint-Martin-La-Porte 4 is the first lot of the new high speed railway link Lyon-Turin that includes the excavation of the first 1,5+9 km of the future double 57 km long tubes Mont Cenis base tunnel between France and Italy. The project is a key-element of the new European transport network RTE-E.

The Client is the public promoter TELT sas, the construction supervision is entrusted to Egis/Alpina and the tender was awarded by the temporary grouping of companies formed by 6 contractors: Spie batignolles Genie Civil, Eiffage Génie Civil, Ghella SpA, CMC di Ravenna, Cogeis SpA and Sotrabas.

The contract is divided in four parts: three starting from the 2,33 km long Saint-Martin-La-Porte access tunnel (1 and 3A-3B in conventional method and 2 in mechanized method) and one developing from the 2,47 km long La Praz access tunnel in conventional method (4).

2   GEOLOGY

The overall geology of the project is enclosed in two principal formations: "Zone Subbrianconnaise" [ZSB] and "Zone Houillere Brianconnaise" [ZHB]. From east to west, so from Saint-Martin-La-Porte to La Praz, ZSB gradually changes in the ZHB passing from an overlapping tectonic contact commonly named "Front Houiller" [FH]. A short geological description of these encountered formations is detailed below:

– ZSB => marly limestone, massive limestone and calcschists
– FH => anhydrites, gypsums and cornelians (even dolomites as symptom of the geological contact)

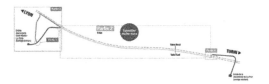

Figure 1. Saint-Martin-La-Porte site.

Figure 2. Geological profiles of the SMP4 part 2 site (left: till to PK14+000, right: from PK14+000 to the La Praz access tunnel).

– ZHB => divided from east to west in 4 geological unit: "Encombres", "Brequin-Orelle", "La Praz" et "Forneaux", substantially from arenaceous and black shales with coal levels to sandstones and conglomerates. Singular geological events as coal faults and shales with a coal fraction up to 15 % were been foreseen in the preliminary project.

Part 2 is bored inside the ZHB passing from "Brequin-Orelle" unit to "La Praz" one. Part 2 starts from the TBM assembly cavern at PK11+793 to its disassembly cavern at PK20+588, so a 9 km long tunnel with a nominal excavation diameter of 11,25 m. Part 2 is an exploratory tunnel necessarily obliged causing the lack of vertical surface drills (approximately one probe/ km) justified by the high overburden, between 800 and 1100, and the complicated environmental conditions for its access.

From the geological synthesis remains several uncertainties along the tunnel alignment concerning:

– Discretization of faults and coal levels
– Presence of mixed faced characterized by intensive fracturing and heterogencity
– Variability of the inclination of the schistosity compared with the theoretical face plan
– Presence of high hydraulic loads and localization of water in-lets.

The geognostic aim of the SMP4 site is correlated to the mandatory improvements that will extended to all future TBMs designs in order to more easily face any geological singularity already encountered.

All synthesis reports, geological and hydrogeological, including the geotechnical model are not considered as contractual scope. This paper is focalized on the SMP4 part 2.

3 CONTRACTUAL PRESCRIPTIONS AND TBM SPECIFICATIONS

As contractual basic prescription, the TBM design had to be suitable to all potential geological risks such as convergences, mixed faces, presence of instable blocs and faults in poor materials (coal levels). In the following the most significant contractual prescriptions [CP] are coupled with the contractor's technical choices [TC] for the chosen TBM.

3.1 Exploratory probing

[CP] => Shield designed with top, bottom and face openings (as shown in figure 3 left) to permit the realization of exploratory destructive surveys or face treatments (such as injections of bentonite, resin…) in fractured rock masses. TBM mandatorily equipped with two twin drilling machines mounted on a structure separate from the erector ring, capable to perform

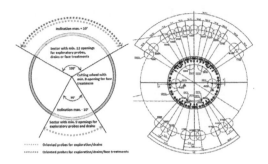

Figure 3.   Positionning of drill openings from the contract (left) to the TBM design (right).

100 m up to 200 m long boreholes with a minimum diameter of 76 mm. Admissible overlap of probing is from 20 to 25 m.

[TC] => top, bottom and face openings as shown in figure 3 right. 2 drilling machines installed on a drilling ring placed on the erector bridge in front of the erector. The main drilling tools normally used are: 108x82 mm casing pipes 2 or 1 or 0,8 m long, T38 M/F bars 3 m or 2,4 long and bottom drill bit of 76 mm. Depending on registered TBM parameters the probing is scheduled in advance to respect a minimum overlapping of 10 m. The overlapping of probes was reduced from 25 to 10 m in order to increase their frequency and perform shorter ones (between 30 to 50 m) to prevent the over-deviation problems. A normal average net advance speed at the drill rig is 30 to 50 m/h but the normal drilling time to 10-15 m/h even considering all preparatory works.

[CP] => TBM back-up equipped with: a coring machine and a bolting machine. The first one to investigate radially and vertically, while the second one to reinforce radially the segmental lining. Radial cores for annular void checking 1,5 m long at a distance of 30 m behind the shield and vertical ones 15 m deep carried out each approx.. 200m for rock sampling and gas measurements.

[TC] => a special mobile ring is designed on the TBM gantry 1 for either the coring machine or the bolting one. During normal conditions, except for particular events, the coring machine is exploitable on the TBM and the bolting one is stored on the site yard. Both machines are able to drill following a rigorous pattern. The coring is carried out through the wire-line method with NQ pipes and a 1,5 m long double corer and with a net cores diameter of 47,6 mm. A down vertical coring of min. 4,5 m long is performed each filling of the belt storage (approx. each 250 m). The bolting machine was employed one time during the overpassing of the first fault at ring 200 to drill a series of self-drilling bolts of 32 mm exploited to inject a filling cement mixture.

3.2   Overcutting, thrust force, unlocking torque and annular gap detector

[CP] => a minimum radial overcutting equal to 150 mm using interchangeable disc cutters, cutting wheel radial displacement and reamer cutting tools. The suggested nominal diameter of the cutting wheel was 11140 mm with worn discs. So with the overcutting: 11290 mm.

Figure 4.   3D view of the probing phase and in-situ performing of the survey with the left drilling machine.

Table 1. TBM parameters: thrust force and cutter head torque.

	[CP] – minimum accepted value	[TC]
Maximum thrust force	180000 kN	123800 kN at 350 bar => mode normal
		185700 kN at 525 bar => exceptional mode
Cutting wheel contact force	25000 kN	32000 to 10000 kN depending on the resulting center of gravity of 3 cutting wheel displacement sectors
Torque during excavation (normal mode)	9000 kNm	9000 kNm at 5 rpm
		25228 kNm at 1,76 rpm
Unlocking torque	35000 kNm	29050 kNm at 1,76 rpm => exceptional mode
		35000 kNm at 0,5 rpm => unlocking mode

[TC] => the available overcutting is a combination of cutting wheel eccentricity, shimming of cutting tools, introduction of new cutting tools on free tracks passing from 19" to 20". Considering worn disc cutters, the nominal diameter of 11210 could be enlarged to 11410 mm passing from an intermediate stage at 11310 mm. The vertical displacement of the cutting wheel is adjustable from 20 mm to 130 mm respect to shield horizontal axis.

Concerning the thrust force and the cutter head torque see table 1.

[CP] => TBM equipped with devices to comprehend over-excavation profiles

[TC] => 7 hydraulic cylinders with a stroke from 0 to 400 mm on the front and intermediate shield. These one are activated during each ring building or standstill to measure the annular gap between shield and real excavation section.

3.3 Backfilling: from mortar to bi-component injection

[CP] => the mortar mix design for filling the annular void has to avoid any movements of the segmental ring and to assure an homogeneous contact between the excavation profile and the extrados of the lining. In presence of soils involving high thrust forces, a complete injection of the annular void, normally demanded at a distance of 15 m from the tail skin, have to be continuously finalized at 360° during the TBM advance assuring the complete filling of ring N-1.

[TC] => in the technical tender offer the first proposition was the using of two mortar mix designs respectively applied on the basis of the TBM thrust parameters. These mix designs obliged the use of lime filler, sand and gravels resulting in the constraint to transport the mortar in tanks to the TBM. Operationally, the first mix design assured the filling of the annular void at 120° at N-1 and at 360° at N-10, while the second one obliged the installation at N-1 of a special ring equipped with a bullflex. This last one is a segmental expansive injectable joint installed directly at the segments extrados on the site yard.

This offer approved solution was on-going replaced by the bi-component system. The bi-component mix design permits to transport their components (A and B) inside pipelines assuring a long-term stability. Operationally, the filling of the annular void is assured at 180° at N-1 and at 360° at N-5. Actually on site, the A-component is made by a particular weight controlled mixture of water, bentonite, cement 52,5 R, slag and retarding agent. The B-component, a liquid activator alkaline base solution, added with A-component with a variable B/A ratio between 5,5 and 7% permits the creation of a gel in the correspondingly range between 8 and 11 seconds. The viscosity at the Marsh cone is variable between 37 and 40 seconds. The bleeding at 3h and 24h are respectively 1 and 5 %. The short-term compressive strengths at 1h and 3h are respectively 0,3 and 0,8 MPa. The long-term compressive strengths at 24h and 28d are respectively at 1,6 and 2,5 MPa.

In figure 5 the transition from the mortar solution to the bi-component one is presented.

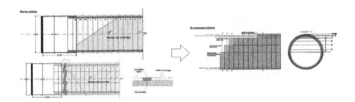

Figure 5.   Mortar solution at technical offer stage and bi-component one applied during TBM advance.

3.4   *Back-up design (G1 as multi-purpose gantry)*

[CP] => a minimum free radial gap from 0,5 to 0,8 m between the segmental lining and the back-up geometry in order to install a steel rib support such as HEB type to limit the risk of convergence. For the same reasons, belt conveyor structures and pipes brackets had to be manufactured to permit a minimum free radial gap of 30 cm from the intrados of the concrete rings.

[TC] => intrados diameter of the segmental lining equal to 9900 mm, back-up size inside the equivalent diameter of 9000 mm (radially resulting in 30 cm for reinforcements and 15 cm for back-up steering) and free available gap, behind TBM passage, equal to 30 cm where steel ribs HEB 240 or 180 were foreseen to be installed by special devices on the TBM:

– Inside the shield: segment erector with an hydraulic clamping tool (bolted on the vacuum plate)
– At 35÷40 m behind the shield: exploiting the gantry 1 with a dedicated steel rib erector.

In figure 6 is shown the TBM equipment to put in place the proposed steel reinforcement.

3.5   *Exceptional mode*

[CP] => the transfer from normal conditions to exceptional ones, causing a locking of the TBM, could occur if a massive coal layer is visible at the face, if a squeezing and high fractured soil is encountered or if a collapse of huge blocs at the face or directly on the shield happened. In this case the Client have to formalize it permitting to the site to start all preparatory works to pass in the TBM exceptional mode.

[TC] => besides all devices previously detailed (steel rib erector, coring/bolting ring…) the TBM was even conceived to void the building of the invert element directly above the second-last ring (note: normally the invert element is placed above the ring N-13, so approx. 21 m behind the tail skin). In order to perform this operation a series of preliminary steps have to be completed:

– Dismantling of the segment feeder
– Installing of the extension beam for the invert hoist
– Installing of the twin crutching cylinders (necessary to permit the cantilevered translation of the invert hoist)
– Installing, alternatively with the ring building, of the clamping tool for steel rib elements

Figure 6.   Clamping tool on the vacuum plate on the erector and steel rib erector on the TBM gantry 1.

Figure 7.   Invert element moved forward by the special hoist and segments storage above the inverts.

The invert hoist normally is in parking position and exclusively in the exceptional mode is used.
During exceptional conditions, from 8 to 12 working hours are averagely needed for a complete cycle of 1,5 m long. In figure 7 the TBM is configured in exceptional mode.

4   CHRONOLOGY OF ENCOUTERED EVENTS DURING EXCAVATION

Four main relevant standstills, from the beginning of the TBM excavation at the end of August 2016, have influenced the theoretical scheduled planning of the TBM advance in SMP4 part 2:

1. Crossing of the 1st geological accident: from December 6, 2016 to May 12, 2017 between rings 197 and 279, so approx. 5 months of standstill where 123 m were been bored
2. 1st cutting wheel exceptional refurbishment: from November 4, 2017 to January 26, 2018 at the ring 1390
3. Crossing of the 2nd geological accident: from April 17, 2018 to May 23, 2018 between rings 2207 and 2266, so approx. 1 month of standstill where 88 m were been bored
4. 2nd cutting wheel exceptional refurbishment: from October 11, 2018 to November 12, 2018

1 and 3 were been defined as singular events [S] => S1 et S3, while 2 and 4 were evaluated as a consequence of the continuous events [C] => C2 and C4.

4.1   *Singular events [S]: faults with high coal fraction content and water in-let*

S1 was a 10-15 m long fault marking a sudden transition to a crushed zone of unstructured rock. It consists of black and charcoal shales with an abrupt orientation changing of the stratification from sub-horizontal to vertical. Immediately, S1 involved significant water in-lets, huge over-excavations and the cutting wheel locking. Between rings 202 and 212, totally 27800 ton were mucked out in 11 rings. So, a quantity/ring equal to 2530 ton was averagely extracted with a pick of 5150 ton on ring 207. Several tests of cutting wheel rotation reached the max. values of the unlocking torque 35 MNm.

S3 was a 3 m long fault characterized by schistose micaceous sandstone with an estimated coal content of 15 %. This zone was high fractured without cohesion and with a large presence of fine material. No changing in the orientation of the stratification was observed: sub-vertical. Between rings 2207 and 2211, totally 16100 ton were extracted in 5 rings. So, a quantity/ring equal to 3220 ton was averagely extracted with a pick of 5155 ton on ring 2210.

4.2   *Continuous events [C]: mixed face and fractured rock mass*

Up to today, all along the TBM bored alignment, a series of constant phenomena were observed:

– Systematic void between the face and the cutter head variable between 50 cm and 3,5 m
– Heterogeneity of the face characterized by an high fractured level causing the excavation of precut rock blocks and the over-stressed of cutter head buckets

Two unscheduled and extraordinary interventions were performed to face the abnormal damages highlighted: C2 and C4.

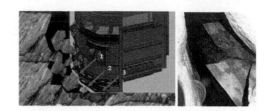

Figure 8. 3D view of precut blocks at the face bottom and typical conditions of a mixed/blocky rock mass.

5 IMPROVING INTERVENTIONS APPLIED FOR S OR C EVENTS

5.1 *Injections: grout mixture and bi-component resin*

Different types of face treatments were performed during the standstills for S1 and S3:

- Partial filling of the cutter head arms with bi-component expansive resin to limit the material in-flow (coupled with polystyrene blocks to reduce the resin consumption). Approx. 8600 kg in 2 phases.
- 1 m long fiber glass bars drilled through disc cages and injected with bi-component expansive resin. Approx. 5200 kg.
- 8 m long casing pipes, both inclined and horizontal, equipped with an injection packer to create a proof resin barrier at the face of the cutting wheel. Approx. 10800 kg.
- 12 m long casing pipes, both inclined and horizontal, equipped with an injection packer to improve the cohesion of fault ground by the injection of a bi-component chemical liquid binder – injections performed with a high pressure pump. Approx. 39500 kg.
- 3-stages areolar grouting through probing pipes through casing pipes and using an high pressure pump. Approx. 24 m3 in 3 days of a mix formed by water, bentonite, cement, plasticizer and retarding agent.

5.2 *Complementary activities in S1 and S3*

During the standstills S1 and S3 a series of particular operations were accomplished:

- Installation of 22 steel rib HEB 180 and 30 steel rib HEB 240 between rings 187 and 212
- Installation of 15 steel rib HEB 180 and 11 steel rib HEB 240 between rings 2207 and 2220
- Realization of self-drilling bolts by pneumatic hammers to reinforce the segmental lining close to the shield
- Realization of multiple expansive resin barriers surrounding the last rings to completely blocked the segmental lining through bi-component injection
- Production/injection at the face and at the extrados of the shield of water based lubrication liquids with a natural polymer, with bentonite and with a foaming agent

5.3 *Continuous updating of the cutting wheel opening ratio*

In general, to definitely cope with fine material in-flows and face instability, the opening ratio of the cutting wheel was reduced from 8 to 4,5 %. As first action during S1, original scrapers rows plugged with thin plates were replaced with additional plates on each arm, new grill bars and peripheral buckets were backward reinforced by steel triangular gussets.

5.4 *Cutting wheel closing with special bolted steel plates*

Up to S3, the cutting wheel normal closing procedure was made with the filling of the arms with expansive resin and polystyrene blocks. From this moment, together with the new design of buckets (see chapter 5.5) the supplier Palmieri proposed the possibility to replace the

Table 2. History of the cutting wheel open ration changing.

Event	Opening ratio	Comment
Start	8,0 %	
S1	5,0 %	4 new plates/arm on 16 arms
	4,5 %	5th plate/arm added on 8 arms
C2	5,0 %	4 plates/arm on 16 arms (gouging of the 5th plate/arm)
S3	≪ 5,0 %	As in C2 but 8 on 16 arms completely closed with see figure 10
C4	6,5 %	As in S3 but with gouging of the 3rd and 4th plate on 4 arms

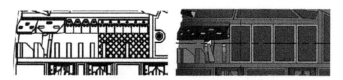

Figure 9. From left to right cutting wheel closing with opening ration from 8 to 5 %.

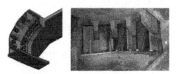

Figure 10. Closing of the arms with special bolted steel plates.

bucket tools with a bolted closing plates. This permitted to avoid the uncertainty of the compressive trend of the injected expansive resin that probably it was the main cause of the progressive augmentation of the torque (due to the muck collection at the end of the arms).

5.5 Cutting wheel adjustment and refurbishment

The standstill for C2 resulted to prevent additional damages to the cutting wheel structure. Following approx. 2 km of boring, at ring 1390, the original peripheral hardox resistant plates were highly worn out up to 15 mm. Besides, the bucket original design was seriously compromised.

In order to find a definitive solution for C2, supplementary more resistant wear plates were welding outside the diameter 7,2 m and buckets (supports and tools) on all arms were replaced with tough ones changing the tool-support bolting way and fixation system.

These exceptional maintenance operations were preceeded by the construction of a welding chamber sizing 6 x 5 x 1,35 m (see figure 12). This one was excavated by the use of emulsion explosive cartridges drilling the face with pneumatic hammers and coring portable machines.

The access to the face was possible through progressive rotations of the cutting wheel.

After C2, from the TBM restart, regular ultrasonic measurements of the on-going wear of the new installed plates were carried out. In September 2018, the TBM overcame 5000 m of advance and a sudden worsening of the wear status of the backward circumference of the cutting wheel, as even the wear of the facing plates at the openings of each arm were noticed (see figure 13).

These facts involved another standstill (C4) to create again a smaller welding chamber sizing 4 x 2,5 x 1 m. In the following main technical actions performed are briefly listed:

– Opening of the cutting wheel from 5 to 6,5 % as briefly described in table 2
– Reinforcement of the peripheral face of the cutting wheel with heavy welded studs
– Reinforcement of the peripheral edge of the cutting wheel with supplementary wear resistant plates.

Figure 11a. 3D face plan of the cutting wheel refurbished, wear status of the cutting wheel external part and new bucket design applied.

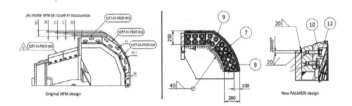

Figure 11b. 3D face plan of the cutting wheel refurbished, wear status of the cutting wheel external part and new bucket design applied.

Figure 12. As-built of the welding chamber and personnel at work for cutter head wear reconstruction.

Figure 13. Observed wear at ring 3350 and application example of heavy studs on a cutter head.

6 TBM PERFORMANCE

In figure 14 the TBM daily performance is shown. T1, T2, T3 and T4 are four time period where TBM continuously excavates, while S1, C2, S3 and C4 are the standstills previously described where the TBM was stopped either for a singular geological event or for extraordinary maintenance operations. In table 3 is pointed out the real TBM productions and, as even in figure 14, is evident the benefits of all technical improvements applied resulting in a regular raising up of productions.

After S3, following the TBM restart, a series of innovative technologies were tested to verify the authenticity of employed methods such as optical televiewing of exploratory probes in advance. Besides, a precautionary stop decisional process was implemented and shared between shift bosses, pilots, TBM engineer and Client. This last one permitted to define a

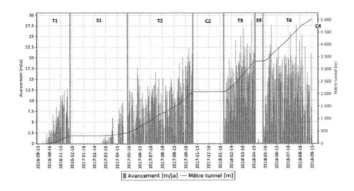

Figure 14. TBM daily performance from September 2016 to September 2018 with detail of standstills.

Table 3. Detail of the daily and monthly TBM production in correspondent advance time periods.

Advance time period	Daily production [m]		Monthly production [m]	
	Average	Maximum	Average	Maximum
T1	5,4	12,8	100,0	190,6
T2	10,0	22,5	238,7	406,6
T3	15,5	29,4	307,4	540,0
T4	14,5	27,7	391,6	477,1

series of thresholds in TBM parameters in order to stop the excavation and start a new exploratory phase at the face.

7 CONCLUSIONS

Meanwhile this writing, the TBM is boring at its full capacity with an average advance speed of 16 m/day. Moreover, the daily advance record was performed on January 12, 2019 with 24 m bored. The breakthrough in the disassembly cavern of La Praz is scheduled at the end of July 2019 in order to start all dismantling operation of TBM and its equipment. At the end of January 2019, 2600 m have to be still bored with the high risk to encounter high water inflows, hard and abrasive rock mass (as even local faults and at least one geological contact).

This project is a clear example of how much relevantly the uncertainty of geological knowledge could influence the advance of a tunnel boring machine.

ACKNOWLEDGEMENTS

Authors would like to acknowledge the Client TELT, the site supervision Egis/Alpina and all construction companies involved in this project: Spie Batignolles Genie Civil, Eiffage Genie Civil, Ghella, CMC and Cogeis. Special thanks have to be addressed to the TBM's manufacturer NFM Technologies and to the belt conveyor's constructor ROWA for their technical support on-field. Furthermore, other suppliers have even played a key role on site assuring a constant on-field technical assistance: Techni-Metal-System, VMT, Mapei, Palmieri Group, Clariant, Ecocem, Sema Ventube, Lorenzetto Loris Srl, Siap, Buzzi Unicem and Condat.

Tunnels and Underground Cities: Engineering and Innovation meet Archaeology,
Architecture and Art, Volume 7: Long and deep tunnels – Peila, Viggiani & Celestino (Eds)
© 2020 Taylor & Francis Group, London, ISBN 978-0-367-46872-9

Experimental setup for studying tunnels in squeezing ground conditions

K. Arora, M. Gutierrez & A. Hedayat
Civil and Environmental Engineering, Colorado School of Mines, Golden, USA

ABSTRACT: Squeezing ground conditions in tunnels are often associated with rock mineralogy, strength, ductility/brittleness, excavation sequence, and magnitude of in situ stresses. Numerous methodologies and empirical correlations have been proposed in the past to determine the level of ground squeezing conditions in tunnels. Most of the correlations are problem-specific and limited in scope. In this work, a fundamental study of tunnel squeezing is carried out using an experimental approach to simulate tunnel boring machine (TBM) excavation in squeezing ground conditions. The experimental setup employs a cubical specimen of a soft rock/soil/synthetic material with each dimensions of 30 cm long. The specimen is subjected to a true triaxial state of stress with different magnitudes of principal stresses and stress levels corresponding to realistic in situ conditions. A miniature TBM is used to excavate a tunnel into the host rock (specimen) while the rock is subjected to true-triaxial state of stress. Embedded extensometers and strain gages glued on the surface of the tunnel liner are used to monitor tunnel response during construction. This paper presents the details of the experimental setup.

1 INTRODUCTION

The problem of squeezing involves time-dependent large deformation in tunnels, which may or may not terminate during construction and arises due to high in-situ stress around the tunnel and problematic geological and geotechnical properties (Barla 1995). Squeezing ground has intrigued engineers over the years in completing underground construction and resulted in the loss of time and money. Wiesmann (1912) first studied the effect of squeezing and since then it has been studied by many researchers including Singh et al. (1992), Aydan et al. (1996), Dube (1993), Barton & Grimnstad (1994), Goel et al. (1995), Yassaghi & Salari-Rad (2005), Gutierrez & Xia (2009), Khanlari et al. (2012), Dwivedi et al. (2012), and Shamsoddin Saeed & Maarefvand (2014).

The squeezing phenomenon in tunnels is still poorly understood (Kovari (1998), and Barla (2000)). Squeezing is associated with high overburden (H), rock mass quality (Q), the uniaxial compressive strength of intact rock (σ_c) and rock mass (σ_{cm}), competency ratio ($\sigma_c/\gamma H$) and tangential stress-strain response around the tunnel. Several definitions have been proposed based on combination of the abovementioned properties (Terzaghi (1946), O'Rourke (1984), Singh (1988), Aydan et al. (1996), Gioda & Cividini (1996), Kovari (1998), Barla (2000)).

To study squeezing problem, it is very important to get a better understanding of the trend in stresses and deformation that will develop around tunnels prior to the excavation. Over the decade's tunnel engineers have been dependent on the empirical methods with limited field data (Schmidt, 1974; Attwell, 1978; O'Reilly & New, 1982; Mair et al. 1993). Various empirical and semi-empirical developed over the years are problem-specific and often contradicts with one other.

Physical models to study two and three-dimensional behavior of the ground in response to tunneling were proposed by many researchers. A comprehensive review of such techniques for tunneling in the soft ground is provided by Meguid et al. (2008).

Figure 1 illustrates different physical models developed all over the world to study tunneling on soft ground and Table 1 lists all the advantages and disadvantages of the developed models.

In this study, a scaled physical model is proposed to study the problem of squeezing in tunnels. Experiments are carried out on a cubical specimen of a soft rock/soil/synthetic material with dimension of 30 cm on each side. The specimen is subjected to compressive true-triaxial

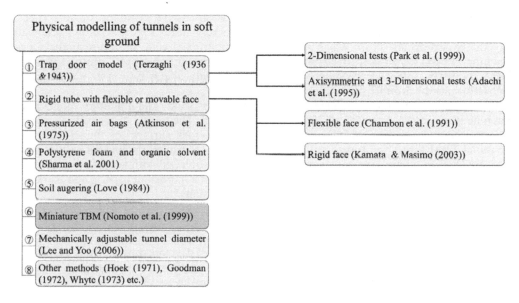

Figure 1. The various physical model developed to study tunneling in soft ground

Table1. Advantage and disadvantages of various physical models developed to study tunneling in soft ground conditions (modified from Meguid et al. 2008).

Method	Applications/advantages	Limitations
Trapdoor	• 2D and 3D tests can be performed under 1g and centrifuge.	• An approximate estimate of stresses and deformation. • Not the actual tunnel process.
Rigid tube with the flexible face	• Can study failure mechanism and face stability under 1g and centrifuge condition.	• No estimate of surface settlement behind the face.
Pressurized airbags	• Studies tunnel stability and induced ground motion around tunnels in 2D as well in 3D under 1g and centrifuge conditions.	• Can be used for unlined tunnels and does not simulate tunnel advance.
Polystyrene foam and organic solvent	• Simulated tunnel advance process under centrifuge conditions.	• Doesn't give accurate results in underwater conditions.
Soil augering	• Tunnels advance process in 1g condition.	• Can be used for cohesive soils only. • Not mechanized for a centrifuge.
Miniature TBM	• Simulated tunneling under centrifuge condition.	• Only up to 25g gravitational acceleration can be applied.
Mechanically adjustable tunnel diameter	• Easy to operate 2D tunnel excavation process.	• It is manually controlled under the 1g condition for 2D models only.

stress state with $\sigma_1 > \sigma_2 > \sigma_3$. A miniature tunnel boring machine (TBM) is designed to simulate excavation similar to real in-situ tunneling. Monitoring is done using acoustic emission (AE), and strain gages that are installed in the TBM and embedded in the cubical specimen. The correlation developed from the experimental results may contribute significantly to better understanding of the tunnel squeezing in rocks. The uniqueness of this experiment is highlighted by the fact that tunnel is excavated in a specimen loaded in all three directions and with magnitudes load in the order of real field stresses.

2 EXPERIMENTAL SETUP

The objective of this experiments is to study the squeezing behavior of soft rock in response to tunnel excavation under true-triaxial stress state. Figure 2 shows schematically the experimental setup, which includes the true-triaxial cell, miniature tunnel boring machine (TBM), servo-controlled pumps and 115V constant speed AC motor for driving TBM, synthetic soft rock specimen (mudstone) and data acquisition system for monitoring deformations around the tunnel.

The mudstone specimen is loaded in true-triaxial stress state. The miniature tunnel boring machine is mounted on the top lid of the true-triaxial cell. The top lid of the cell has a 76 mm diameter circular opening which provides access to the rock surface for tunnel excavation. Required TBM thrust and torque are provided by servo-controlled pumps and electric motor, respectively. For continuous data acquisition, the cell is equipped with acoustic emission (AE) sensors and strain gages embedded in the specimen. All the important aspects and working of this experimental setup are discussed in the following sections.

2.1 *True-triaxial cell*

This experimental incorporates the use of true-triaxial cell developed by Frash et al. (2015) to study the enhanced geothermal system (EGS) at laboratory scale. The apparatus is capable of applying three independently controlled principal stresses up to 13 MPa to a 30x30x30 cm^3 rock specimen.

The apparatus was designed with a mixed flexible bladder (flat jack) and passive confinement system, shown in Figure 3. Each principal stress is applied via one active flat jack per

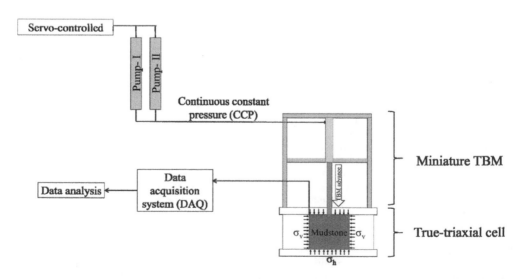

Figure 2.   Schematic diagram of the proposed experimental setup to study TBM excavation in squeezing ground conditions.

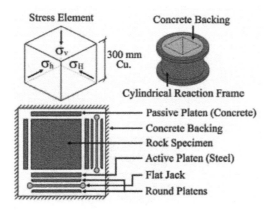

Figure 3. True-triaxial cell at Colorado School of Mines (modified from Frash et al. (2015)).

principal axis when using the typical configuration for the apparatus. Specimen faces directly loaded by the flat jacks and the opposing reaction faces supported by the frame are hereby referred to as active and passive faces, respectively. The steel top lid was furnished with a 63 mm diameter port to pass electrical sensor wires and hydraulic tubing for internal sensors and jacks.

The reaction ring of the central body was constructed from A36 structural steel with the yield strength of 250 MPa. The lids were constructed from A514 steel with the yield strength of 700 MPa to reduce thickness requirements following stress design criteria. Sufficient lid thickness was provided to permit drilling multiple non-intersecting 10 mm holes through the lid while maintaining a safety factor greater than 2.0. The lid of the cell was effectively considered to be a sacrificial component. This lid is modified to conduct the TBM excavation in squeezing ground conditions.

Active face stresses are provided by flat jacks (350 mm diameter circular Freyssinet®) and an assembly of two 300 mm diameter round steel platens and one 300 mm square steel platen. Each flat jack is pressurized via an independent hand pump with active digital pressure monitoring. Using separate pumps bypasses pressure control issues which occur in single pump systems with manifold valves. The square platen's inward edges were beveled to mitigate binding with adjacent platens. A 25 mm thickness was specified for the square platens referencing elastic stress-deflection analysis and limit yield criterion for stress transmission to the specimen corners. The square platen also provided a protective housing for AE sensors. This design decision improved AE data quality by ensuring good face-to-face sensor contact with the specimen, reducing noise transmission through the sensor housing and reducing assembly time. A typical alternative AE sensor installation method in similar true-triaxial devices involves cutting shallow holes into the specimen with consequentially increased sensor alignment difficulty, increased assembly time, decreased stress uniformity and likely reduced AE measurement quality (Frash et al., 2015).

2.2 Miniature tunnel boring machine (TBM)

The miniature TBM, shown in Figure 2, is an integral part of the experimental setup. The miniature TBM is designed keeping in mind different thrust and torque requirement at a different level of field stress. Figure 4 presents the miniature TBM designed and fabricated at the Colorado School of Mines. One of the important parts of this miniature TBM is the 100 kN and 200 mm stroke cylindrical hydraulic jack. The jack is controlled by a pair of D-series Teledyne ISCO pumps. The pump maintains the continuous constant pressure (CCP) at the jack and hence, constant thrust at the face of the TBM.

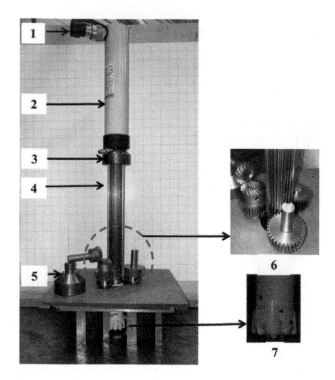

1. Feed in hydraulic oil using D-series ICSO pumps
2. Hydraulic cylinder (10 ton & 8" stroke)
3. Thrust bearing and water/compressed air supply assembly
4. 1 feet long Geared shaft (OD = 1.63")
5. 115V, Single phase AC geared motor (actual motor not available yet)
6. Bevel and planetary geared assembly for shaft rotation
7. 2" diameter, miniature TBM cutter head

Figure 4.   Miniature tunnel boring machine (TBM) designed and fabricated at Colorado School of Mines.

The plunger of hydraulic jack is connected to the rotatory shaft of the miniature TBM through a thrust bearing which allows relative rotation between the rotatory shaft and plunger of the jack. The rotatory shaft is mainly a 300 mm long spur gear with pitch (number of teeths per 25mm pitch diameter) is 16, pitch diameter of 38 mm and the pressure angle of 20°. The shaft (long spur) is connected to a 50 mm button type drill bit which provides the drag action to the rock surface similar to the soft ground TBM.

The required torque is provided by bevel and planetary gear assembly which is driven by a 115V single phase alternating current (AC) constant speed (rpm) AC motor. The main drive gear is the pair of bevel gear having pitch 16, pitch diameter of 38 mm and the pressure angle of 20°. A pair of bevel gear converts the rotation along the horizontal axis into the rotation along the vertical axis.

The vertical bevel gear is axially connected to a spur gear having pitch 16, pitch diameter of 50 mm and the pressure angle of 20°. This spur gear is coupled with the shaft (long spur) of the miniature TBM and provides the required torque to the cutter head at a constant rpm. The longspur is also coupled with two more spur gears (pitch 16, pitch diameter of 50 mm and the pressure angle 20°) which prevents the lateral deformation of the shaft of the miniature TBM (See Figure 4). Therefore, the required thrust and torque for driving this miniature TBM are provided by a pair of the servo-controlled pump at CCP and constant speed AC motor. The muck produced due to the tunnel excavation will be clear off from the excavation face using compressed air at regular intervals.

The whole assembly is supported by a reaction frame designed in such a way that the vertical deflection of miniature TBM at the maximum thrust level will be less than 1 mm. This reaction frame is mounted on the top of the lid of the true-triaxial cell and shown in the schematic diagram of the experimental setup (See Figure 2).

The tunnel advancement rate is measured by a pair of servo-controlled pumps which is continuously monitored. The torque applied by the cutter head is measured by the continuously monitored power output (as power if torque times rpm). Hence, this design of miniature TBM provides all the flexibility in terms of the application of thrust and torque, and at the same time continuously monitors all the essential operating parameters which are monitored in the field as well.

2.3 Synthetic mudstone specimen

A synthetic mudstone specimen is prepared in the laboratory using the methodology proposed by Johnston & Choi (1986). They prepared a synthetic soft rock to eliminate experimental scatter and high variable performance for laboratory model studies. They found the model material to be homogenous and isotropic with physical and mechanical properties well aligned to that of natural mudstone.

In this work, artificial mudstone is prepared in the laboratory by mixing type I/II cement (include cement provider name here), clay which is chemically hydrous aluminum silicate and water in the correct proportion. After performing various trials with different mix proportions, it was found that upon mixing equal proportions of cement, clay, and water with some superplasticizer, a highly workable, consistent and homogenous mix is obtained. The correct proportion of superplasticizer depends on the type and grade of the superplasticizer used. In this case, superplasticizer used is MasterGlenium 7920 supplied by BASF. Table 2 shows the quantity of ingredients required for preparing 1 m^3 of mudstone.

The mix is poured in cylindrical and cubical molds. A cylindrical specimen of 51 mm diameter and 102 mm length were prepared to conduct uniaxial compression tests (UCT) and conventional triaxial tests at a confining pressure ranging from 1 MPa to 10 MPa (which is also the range of confining stress in the true-triaxial test).

Some preliminary compressive strength tests are performed on the five cylindrical specimens of mudstone. Average unconfined compressive strength (UCS) is observed as 6 MPa with a standard deviation of 1 MPa.

A cubical specimen of 30 x 30 x 30 cm^3 is prepared to perform laboratory scaled simulation of TBM excavation in mudstone at stresses equivalent to the stress in the field. Based on the deformation monitoring around the excavation, squeezing behavior of the mudstone will be studied.

2.4 Instrumentation and monitoring

The data acquisition system (DAQ) used to monitor and control the laboratory performed TBM excavation in squeezing ground conditions record all the essential operating parameters as is standard for the field application while also implementing some additional elements to take advantage of improving accessibility that an experimental setup allows. The following section presents the details of instrumentations and monitoring in the proposed experimental setup.

2.4.1 Instrumentation for monitoring TBM output

At the very basic level, as already discussed, the thrust of the TBM will be controlled by the two syringe pumps under continuous constant pressure (CCP) mode. This mean TBM will

Table 2. Mix proportions for one cubic meters of synthetic mudstone.

Ingredient (units)	Quantity for 1m^3 of mudstone
Cement (kg)	592
Clay (kg)	592
Water (kg)	592
Super plasticizer (ml)	24

apply constant thrust at the face of the excavation. The advance rate of the TBM can be back-calculated flow rate (Q) and time (t) data of the servo-controlled pumps. The torque provided by the TBM will be continuously monitored by a separate DAQ which will record the power output from the AC motor running at constant revolutions per minutes (rpm).

2.4.2 *Acoustic emission (AE) to monitor damage*

AE events will be monitored using Physical Acoustic Corporation (PAC) AE monitoring system with six WSα sensor and three PCI-2 cards mounted in the Micro-II chassis. Six AE sensors are used because in case we do not get high quality data from two or three sensor, there should be high quality data from the minimum number of sensors required for geophysical characterization.

As shown in Figure 5, six AE sensors are typically placed on test specimen surfaces and attached directly to the specimen faces using a thin layer of vacuum gel for coupling, to attain direct AE measurement with minimal reflection, surface interference, sensor orientation error, or attenuation effects. The use of six sensors also enables the application of moment-tensor analysis to classify recorded AE events according to location and failure mode, which is either tensile dominated, shear dominated, or mixed-mode.

2.4.3 *Embedded strain gauges for monitoring strain around the excavation*

To monitor the strains around the excavation, various strain gauges will be embedded in a cubical specimen of mudstone. Due to the very high water content of the synthetic mudstone, there will be improper adhesion between strain gages and mudstone. Hence, strain gauges can't be directly embedded in the synthetic mudstone. A multiple point borehole extensometer (MPBEx) will be prepared using a thin flexible material like Teflon and an array of strain gauges as shown in Figure 6.

Since the thin Teflon sheet is more flexible than the mudstone, the embedded strain gauges will show the deformation of the more competent member i.e. mudstone. However, this methodology should be validated by performing a conventional UCT on a cylindrical specimen of

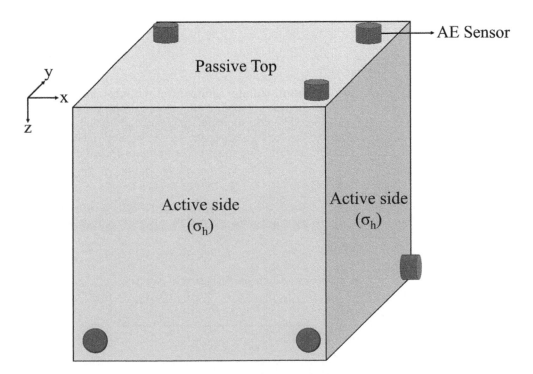

Figure 5. Six AE sensor position for true-triaxial testing on a cubical specimen.

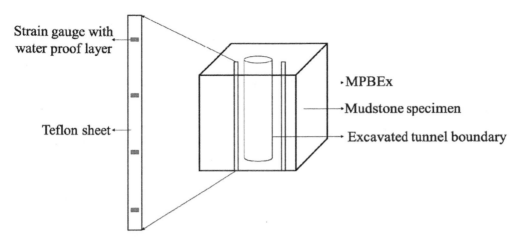

Figure 6.   An array of strain gages (MPBEx) embedded in the synthetic mudstone cubical specimen.

synthetic mudstone with embedded axial strain gauges and measuring axial deformation using a linear voltage differential transducer (LVDT). The axial strain measurements from the embedded strain gauge and LVDT should be comparable.

2.4.4 *Instrumentation in the tunnel liner*

After excavating a two inches diameter tunnel, a thin flexible cylindrical liner will be installed in the excavated tunnel. Elastic properties (Young's modulus and Poisson's ratio) of this liner material will be predetermined by conducting appropriate tests. Four strain gauges, equally spaced in the longitudinal direction, will be glued on the inner surface of the liner. The thin annulus gap between the liner and excavated tunnel will be filled with quick set epoxy with known elastic properties. Using the hoop stress and continuity equation at the tunnel and liner interface, tunnel convergence will be continuously monitored.

3   ANALYSIS AND SYNTHESIS OF THE EXPERIMENTAL RESULTS

Figure 7 shows the flowchart for the post-processing of the test results. Data from the continuous monitoring of the miniature TBM, embedded strain gauges (MPBEx), tunnel liner strain gauges and AE sensors will be closely monitored. Continuous monitoring of the miniature TBM will give the cutter head thrust, advance rate, speed (rpm) and torque with time. In this setup the cutter head thrust, and rpm will be constant throughout the experiment.

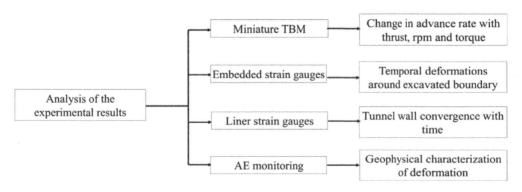

Figure 7.   Flowchart for the analysis of the experimental results.

Monitoring of embedded strain gauges (MPBEx) will provide the strain around the excavated boundary during and after the excavation at multiple points. On other hand, the strain gauges installed in the liner will provide the tunnel wall convergence at multiple points after the excavation.

Post-processing the data from the continuous monitoring using six AE sensors, installed at the surface of the cubical specimen, will allow for the classification of recorded AE events according to location and failure mode.

The deformations and AE events will be continuously monitored even after the TBM excavation till the time no significant change is observed. Results from each test obtained during the excavation stage and the characterization of the samples after testing will be carefully analyzed and synthesized. New methods to predict tunnel squeezing will be formulated or existing ones will be validated and improved.

After the completion of each test, the loaded samples will be sliced into several sections along the tunnel longitudinal axis. Each section will then be imaged and analyzed in terms of deformations, failure and plastic zone formation around the excavation.

4 CONCLUSIONS

A novel experimental setup is proposed to simulate the TBM excavation in squeezing ground condition. This experimental setup is capable of monitoring tunnel advance and tunnel stability at a field stress level. Stress level up to 13 MPa stress can be independently applied in each of the three direction (which is approximately equal to 600 m of overburden pressure). The setup can simulate lined as well as unlined tunnels and will provide a fair estimate of stresses and deformation around the tunnel. This experimental setup also allows to take advantages of some additional features such as slicing of the specimen into thin sections to study the extent of the plastic zone around the tunnel.

ACKNOWLEDGEMENTS

The authors wish to gratefully acknowledge the support from the University Transportation Center for Underground Transportation Infrastructure (UTC-UTI) at the Colorado School of Mines for funding this research under Grant No. 69A3551747118 from the U.S. Department of Transportation (DOT).

REFERENCES

Adachi, T., Tamura, T., Kimura, K. & Nishimura, T., 1995. Axial symmetric trap door tests on sand and cohesion soil, In: *Proceedings of the 30th Japan National Conference on Geotechnical Engineering*: 1973–1976.

Atkinson, J.H., Brown, E.T. & Potts, M., 1975. Collapse of shallow unlined tunnels in dense sand. *Tunnels and Tunnelling* 3, 81–87.

Attwell, P.B., 1978. Ground movements caused by tunneling in soil. *Proceedings of the Large Ground Movements and Structures Conference*: 812–948. Cardiff, Pentish Press, London.

Aydan O, Akagi T. & Kawamoto T., 1996. The squeezing potential of rock around tunnels: theory and prediction with examples taken from Japan, *Rock Mechanics and Rock Engineering* 29:125–143.

Barla G., 1995. Squeezing rocks in tunnels. *ISRM News J* 2(3/4):44–49

Barla G., 2000. Tunnelling under squeezing rock conditions. *Department of structural and geotechnical engineering*, Politecnico.

Barton N. & Grimstad E., 1994. The Q-system following twenty years of application in NMT support selection 12(6):428–436. Felsbau.

Chambon, P., Corte, J.F. & Garnier, J., 1991. Face stability of shallow tunnels in granular soils. *Proceedings of an International Conference on Centrifuge*: 99–105. A.A. Balkema, Rotterdam.

Dube A.K., 1993. Squeezing under high stress conditions, assessment and prevention of failure phenomena in rock engineering. *In:* Pasamehmetoglu AK *et al (eds) Mine*.

Dwivedi R.D., Singh M., Viladkar M.N. & Goel R.K., 2013. Prediction of tunnel deformation in squeezing grounds. *Engineering Geology*, 161:55–64.

Frash, L. P., Gutierrez, M., Hampton, J., & Hood, J., 2015. Laboratory simulation of binary and triple well EGS in large granite blocks using AE events for drilling guidance. *Geothermics*, 55:1–15.

Gioda G. & Cividini A. 1996. Numerical methods for the analysis of tunnel performance in squeezing rocks. *Rock Mechanics and Rock Engineering*, 29(4):171–193.

Goel R.K., Jethwa J.L. & Paithakan A.G., 1995. Tunneling through the young Himalayas a case history of the Maneri–Uttarkashi power tunnel. *Engineering Geology*: 39(1–2):31–44.

Goodman, R.E., 1972. Geological investigations to evaluate stability, In: Geotechnical Practice for Stability in Open Pit Mining. *Proceedings of the Second International Conference on Stability in Open Pit Mining*: 125–132. Vancouver, Canada.

Gutierrez, M., & Xia, C. C., 2009. Squeezing potential of tunnels in clays and clay shales from normalized undrained shear strength, unconfined compressive strength and seismic velocity.

Hoek, E., 1971. Rock Engineering (Inaugural Lecture). *Imperial College*, University of London.

Johnston, I.W. and Choi, S.K., 1986. A synthetic soft rock for laboratory model studies. *Geotechnique*, 36(2): 251–263.

Kamata, H. & Masimo, H., 2003. Centrifuge model test of tunnel face reinforcement by bolting, *Tunnelling and Underground Space Technology*, 18 (2):205.

Khanlari G., Meybodi R.G. & Mokhtari E., 2012. Engineering geological study of the second part of water supply Karaj to Tehran tunnel with emphasis on squeezing problems. *Engineering Geology*, 145:9–17.

Kovári, K., 1998. Tunneling in Squeezing Rock (Tunnelbau in druckhaftem Gebirge). Tunnel, 5:12–31.

Lee, Y.J. and Yoo, C.S., 2006. Behaviour of a bored tunnel adjacent to a line of loaded piles. *Tunnelling and Underground Space Technology*, 21:3–4.

Love, J.P., 1984, Model testing of geogrid in unpaved roads. *D.Phil. Thesis*. Oxford University, UK.

Mair, R.J., Taylor, R.N. & Bracegirdle, A., 1993. Subsurface settlement profiles above tunnels in clays. *Geotechnique*, 43 (2):315–320.

Meguid, M. A., Saada, O., Nunes, M. A., & Mattar, J., 2008. Physical modeling of tunnels in soft ground: a review. *Tunnelling and Underground Space Technology*, 23(2):185–198.

Nomoto, T., Imamura, S., Hagiwara, T., Kusakabe, O., & Fujii, N., 1999. Shield tunnel construction in centrifuge. *Journal of Geotechnical and Geoenvironmental Engineering*, 125 (4):289–300.

O'Reilly, M.P. & New, B.M., 1982. Settlements above tunnels in the UK – their magnitude and prediction. In: *Proceedings of Tunnelling'82*: 173–181. IMM, London.

Park, S.H., Adachi, T., Kimura, M. & Kishida, K., 1999. Trap door test using aluminum blocks. In: *Proceedings of the 29th Symposium of Rock Mechanics*: 106–111. J.S.C.E.

Schmidt, B., 1974. Prediction of settlements due to tunneling in soil: three case histories. In: *Proceedings of the Second Rapid Excavation and Tunneling Conference*, 2:1179–1199. San Francisco, USA.

Shamsoddin Saeed M. & Maarefvand P., 2014. Engineering geological study of NWCT Tunnel in Iran with emphasis on squeezing problems. *Indian Geotechnical Journal*, 44(3):357–369.

Sharma, J.S., Bolton, M.D. & Boyle, R.E., 2001. A new technique for simulation of tunnel excavation in a centrifuge. Geotechnical Testing Journal, 24(4): 343–349.

Singh B., Jethwa J.L. & Dube A.R., 1992. Correlation between observed support pressure and rock mass quality. Tunneling and Underground Space Technology,7(1):59–74.

Terzaghi, K., 1936. Stress distribution in dry and in saturated sand above a yielding trap-door, In: *Proceedings of the International Conference on Soil Mechanics*, 1:307–311.Harvard University. Press, Cambridge.

Terzaghi, K., 1943, *Theoretical Soil Mechanics*. Wiley, New York.

Tezaghi, K., 1946. Rock Tunneling with Steel Supports (RV Proctor and TL White, eds.). *Youngstown, Ohio*: Comercial Shearing Co.

Whyte R.J.A., 1973. Study of progressive hanging wall caving at Chambishi copper mine in Zambia using the base friction model concept. *M.Sc. Thesis*. Imperial College, University of London.

Wiesmann E., 1912. Mountain pressure. *Switz J Struct*, 60: 7(in German).

Yassaghi A., Salari-Rad H., 2005. Squeezing rock conditions at an igneous contact zone in the Taloun tunnels, Tehran–Shomal freeway, Iran: a case study. *International Journal of Rock Mechanics and Mining Sciences*, 42:95–108.

*Tunnels and Underground Cities: Engineering and Innovation meet Archaeology,
Architecture and Art, Volume 7: Long and deep tunnels – Peila, Viggiani & Celestino (Eds)*
© 2020 Taylor & Francis Group, London, ISBN 978-0-367-46872-9

Brenner Base Tunnel – challenges of gripper TBM application for the 15 km long exploratory tunnel Ahrental in challenging rock mass

K. Bäppler & M. Flora
Herrenknecht AG, Schwanau, Germany

ABSTRACT: The Brenner Base Tunnel will be 64 kilometers long and will be the center-piece of the Scandinavian-Mediterranean TEN-T Corridor from Helsinki to Valetta (Malta). It will comprise two single-track rail tunnels with a service and drainage gallery in between. The primary focus in this paper is on the construction of the 15km long exploratory tunnel Ahrental using a well-adapted TBM concept. Tunnelling started from Austrian Ahrental towards the Italian border in challenging terrain using a 7.93m-diameter Gripper TBM. The geology along the section is characterized as extremely challenging with quartz phyllite and shale and high overburden of up to 1300m. Moreover, numerous fault zones were predicted along the drive with expected loose to friable rock mass. The paper addresses in particular the special TBM design and project experiences with the operation of a Gripper TBM in challenging rock mass.

1 INTRODUCTION

The Gotthard Base Tunnel, which was commissioned in June 2017 as the longest railway tunnel in the world, will cede this title in the near future to another epoch-making structure in the Alps, the Brenner Base Tunnel. The Brenner Base Tunnel is an important link between Munich and Verona and will contribute to sustainable mobility.

Together with the existing Innsbruck bypass in Austria, the Brenner Base Tunnel has a total length of 64 kilometers. This will make the joint venture between Italy and Austria seven kilometers longer than the Gotthard Base Tunnel.

The Brenner Base Tunnel (BBT) crosses through the Alps below the Brenner Pass. It has some similarities with the Gotthard, which is partly attributable to the fact that the owner of the Brenner Base Tunnel (BBT) has been working with experts from Alptransit Gotthard for many years. Like the Gotthard Base Tunnel, the Brenner also consists of two single-track parallel tubes and much of the technology from the Gotthard has been adopted. The experience gained during construction of the Gotthard has been taken into consideration in the safety and rescue concept for the BBT, for things such as the emergency stops. Connecting galleries between the tubes are planned every third of a kilometer. These cross passages serve as escape routes in emergency situations and comply with the highest safety standards in tunnel construction.

A special feature of the Brenner Base Tunnel is the exploratory tunnel that is being bored over its entire length about twelve meters below and in between the main tunnel tubes, which will also serve as a service and drainage gallery during operation. The tunnelling work on the exploratory tunnel will provide information about the nature of the rock mass and thereby minimize construction costs and time.

This publication focuses on the drive for the 15 kilometer long Ahrental exploratory tunnel. Since late September 2015, a Gripper TBM with a diameter of 7.93 meters has been boring the tunnel from Austrian Ahrental toward the Italian border. The paper discusses the technical considerations and approaches for tunnelling in the predicted loose to friable rock mass with squeezing rock zones and summarizes the previous experience of the Gripper TBM deployment.

2 PROJECT OVERVIEW AND REQUIREMENTS FOR THE TBM

Part of the construction lot Tulfes-Pfons envisages the construction of the 15km long exploratory tunnel section using TBM tunnelling between the Ahrental hub and the municipality of Pfons with the aim of developing an optimal tunnelling, material management and safety concept for the construction phase of the main tunnel. According to the technical terms of the contract, the use of an open hard rock TBM was specified for the continuous excavation of the tunnel. One of the challenges, together with the extremely demanding predicted geological conditions, which will be discussed in more detail, was the assembly of the TBM in a cavern 3.5 kilometers deep inside the mountain.

The TBM advance started in the assembly and start-up cavern at km 6+922 and extends to km 22+000 (transition to tunnelling of the exploratory tunnel Wolf). An extension of the drive is being considered based on the advance rates so far. The excavated radius of 3.95m takes into account deformation tolerances of up to 150mm, a shotcrete outer shell of up to 300mm and a possible (subsequent) inner shell of 250mm. Overburden between 420m in the Navistal area up to a maximum of 1300m in the Schröflkogel area are found along the tunnelling route. The section of the BBT thus has a comparatively low overburden height compared to the Gotthard Base Tunnel. The expected geological conditions are extremely demanding for TBM tunnelling. Compared to the Gotthard, relatively soft rocks are encountered here, with predominantly formations of Innsbruck quartz phyllite and Graubünden shale. Numerous fault zones are predicted, which make up about 19% of the total distance of the drive. The soft bedrock (shale and phyllite) is thus characterized by a high degree of difficulty in some sections compared to the Gotthard.

For the characterization of the expected rock mass conditions along the Ahrental section, a subdivision into rock mass behavior types took place. Fundamentally these indicate loose to friable rock mass, shear failure or even squeezing rock behavior.

The following minimum requirements for the TBM were thus defined for the TBM design:

- Radial rock deformations of 150mm with a nominal diameter of 7.9m
- Handling of fault zones and radial displacements of up to 400mm
- Water runoff of up to 10l/s/10m
- Radial rock pressure of 500kN/m^2
- Probe drilling for strata exploration from the working area A1

The tunnel boring machine used for the construction lot must meet very high and complex requirements. In addition to technical performance and safety for the personnel, economic aspects also play a prominent role.

On excavating the cavity in the partially squeezing rock zones, without appropriate countermeasures major, long-lasting deformations of the cavity can develop. The phenomenon occurs primarily in rock types of low strength and high deformability, which are also mainly to be expected along the section. In the following section, the machine concept and technical considerations and approaches for drives in squeezing rock mass with an open hard rock machine are explained in more detail and previous experience summarized.

3 TBM LAYOUT AND PROJECT SPECIFIC DESIGN FEATURES

For the exploratory tunnel, Herrenknecht supplied a 200 meter long and 1800 tonne Gripper TBM with an excavation diameter of 7.93 meters.

In keeping with the predicted geology, the machine was designed for areas of squeezing rock. Because of the squeezing rock mass behavior, the design of the machine took into account the reduction of the excavation diameter (geometric influence) following excavation and possible increasing rock loads in preventing this deformation (load influence).

The cutterhead is equipped with 46 19" discs and, for transport reasons, consists of 3 parts, which were welded together on the jobsite. By means of 3 overcutting tools, a maximum overcut of 100mm in radius can be achieved (maximum excavation diameter is 8130mm). The possibility of a variable excavation diameter is a suitable measure to provide more leeway for rock mass deformations and thus pass through zones of squeezing rock mass.

Open TBMs must have a flexible front shield including an adjustable invert shoe, whose kinematic range covers the envisaged variance of the excavation diameter. The same applies to the gripper unit and the rear support as well as to possible relations to the excavation diameter in the back-up area, such as walking legs. In contrast to shield machines, in which the entire back-up is based on the constant internal diameter of the segments, in open machines the remaining excavation diameter remains relevant for all following operations. Predicted variances upward (expansion) and downward (rock mass deformation) must be taken into account in the initial design. Devices for the installation of primary rock support directly behind the dust shield must be designed for the smallest and largest diameters possible. This is of particular importance for mechanized arch setting equipment.

Thanks to the adjustability of the design of the gripper shield or the front shield on all sides (see the following illustration), the machine axis can be kept concentric to the excavation axis even with an increased diameter.

The gripper shield can be extended to suit the maximum excavation diameter. With regard to the defined minimum requirements for the TBM, the gripper shield is designed for a radial load of 500 kN/m^2.

With a time-dependent deformation behavior of the rock mass with a possible accompanying reduction of the excavated diameter, in addition to the overcut the length of the shield skin is paramount. Compared to a shielded TBM the Gripper TBM is equipped with a flexible gripper shield and has a short front shield length. For the Gripper TBM used here, the front shield including fingers has a length of 4420mm. The kinematics of the front shield segments (see figure) allow variation within the structurally possible radial strokes, including a reduction of the diameter when a predetermined rock pressure is exceeded.

The predictions of the ground conditions likely to be encountered along the exploratory tunnel were classified as difficult in places, making the possibilities of early rock support immediately behind the cutterhead shield very important. The challenge for the machine manufacturer is to optimize the operator safety, degree of mechanization and flexibility with regard to the support methods used under the extremely harsh environmental conditions prevailing in this area of the machine.

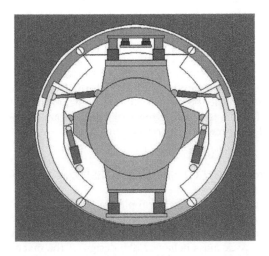

Figure 1.   Flexible gripper shield with adjustable roof shield (green), side wings (yellow) and invert shoe (blue).

Immediately behind the cutterhead shield in the L1 area, two independently moving drills are installed for rock support. The maximum drilling depth per stroke is 3,000mm and by extending the drill rods, longer drilling depths are possible. Also installed in the L1 area is an erector with which the steel ring arches are assembled and installed, and a spraying manipulator with which shotcrete can be applied in the area behind the gripper shield.

The back-up unit consists of 10 trailers and a bridge construction. In back-up 1 is the rock support for the L2 area. It consists of an anchor drilling system with two drill carriages and a spraying manipulator with attached operating platform. On a ring carrier behind the shotcrete manipulator, the drill (setting angle variable between 6–9°) for probing and injection drills is mounted, which can be moved radially 120° in the crown area.

The findings from the driving of the exploratory tunnel, in particular with regard to the encountered nature of the rock mass, are to be used for an optimal tunnelling and safety concept for the TBM tunnelling of the main tunnel tubes. To help with the geological/geotechnical interpretation of the tunnel face, five of the 19 inch disc cutters were equipped with the newly developed DCLM (Disc Cutter Load Monitoring) system. The DCLM system measures the cutting load acting on the disc cutters, which allows conclusions to be drawn about the condition of the tunnel face and thus an optimization of the drive parameters. On the monitored tracks the disc cutter load monitoring system allows detection of changing geological conditions such as fractures, stratifications, etc. in real time. The image for visualizing the DCLM data in the control cabin has been supplemented with an additional display for the proportion of underload and overload. As a result, the machine operator can recognize critical disc cutter overloads during the advance in real time as well as unstable conditions at the tunnel face on the basis of the disc cutter underload. In addition, due to the configuration of certain limit values, the triggering of warning messages as well as further processing via the machine's PLC are easy to implement.

Another five disc cutters are equipped with the DCRM (Disc Cutter Rotation Monitoring) system, which monitors the rotational movement and temperature of the disc cutters, allowing the optimization of tool maintenance intervals.

Additionally, a camera system provides photos of the tunnel face to the control cabin monitors. The cameras are mounted in five positions across the cutterhead at a maximum radial distance apart of 1m. During tunnelling advance the cameras are protected by a cover and when the machine is stopped for maintenance the covers can be opened to begin taking photographs of the tunnel face.

The camera system allows the tunnel face to be photographed during every maintenance shift and the individual images are processed into a 2D image of the tunnel face to assist in the detection of potential tunnel face instabilities such as loose rock falls in the crown area.

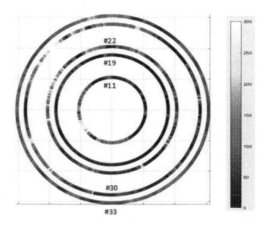

Figure 2.   DCLM, visualization of the load effect at the tunnel face according to color scaling in kN.

Figure 3. Photo-optically recorded image of the tunnel face with five different camera positions on the cutterhead.

4 PROJECT EXPERIENCE TO DATE AND LESSONS LEARNED

After transporting the tunnel boring machine to the jobsite, the components were transported down an access gallery with a 12% gradient to an assembly cavern 3.5km deep inside the mountain where the TBM was assembled. From there, the 15 kilometer advance began in September 2015 from Austrian Ahrental toward the Italian border. After excavating the exploratory tunnel, the Gripper TBM will be pulled back through the excavated tunnel.

Along the first meters of the tunnel the open TBM had to master very complex ground conditions. The challenges the TBM had to face were sometimes extremely difficult rock formations with highly squeezing and swelling rock conditions with extensive invert heave and the encounter of fault zones with sometimes large-volume rock collapses, which were also encountered several times in the course of tunnelling. Tunnel advance rates in these conditions are largely determined by a high degree of rock support work required in the L1 and L2 area. Between the cutterhead and the installation area A1 are approx. 5.5m of tunnelling distance or lead time, which can be used as an "exploration element" for the L1 area to be secured. Due to the multiple collapses that continued to characterize tunnelling even after the first 1000 meters of tunnel, modifications were carried out in the TBM's rock support areas in order to minimize performance losses due to increased rock support requirements.

Rock mass deformations, breakouts and collapses required additional support measures in the L1 area directly behind the gripper shield, including shotcrete for initial rock support and 360° TH arches, which slowed the daily performance of the TBM advance. To date (as of June 2018), seven fault zones were encountered in which a reduction in performance - depending on the thickness of the fault zone - reduced tunnelling speeds to 1-3 m/day due to the high level of rock support required. In these difficult geological zones, in some cases with converging rock mass the available clearance in the L1 area between the bedrock geology and the spray nozzle for applying the shotcrete for initial roof support in the L1 area was sometimes greatly reduced.

In addition, the shotcrete equipment in the L2 area was used much more extensively and required greater availability with reductions in cleaning and maintenance time windows, which in turn had a direct effect on the daily performance. In order to optimize tunnelling performance in these geologically extremely complex areas, a total modification of the shotcrete unit in the L2 area was carried out. The nozzle was designed to allow a larger, adjustable area between the nozzle and the tunnel wall. Also, rebound collection was completely separated from the system in order to be able to dispose of the rebound during spraying operations and perform cleaning work in parallel with the spraying.

Figure 4. Deformed rock support due to prevailing convergences.

In the course of the advance, further system improvements were carried out. The TH arches were replaced with U-profiles and the ring beam erector was greatly modified for arch installation, which resulted in higher performance coupled with greater wear protection. Further optimization of the rock support measures meant that shotcrete in the L1 area was only applied by hand and only for initial support. The logistics were also optimized. The supply of the tunnel boring machine with invert segments, support material, and consumables was carried out with multi-service vehicles (MSV), which - despite the steep slope - transport the material from the surface through the lateral access tunnels directly into the back-up area.

A consideration of the advance rates results in a very satisfactory balance. In solid rock daily top performances of 61m and monthly performances of 825m were achieved, while regular weekly performances of around 200m were standard over many months.

5 CONCLUSION

With the construction of the Brenner Base Tunnel, another epoch-making project is being realized in the Alps. The 15 kilometer long Ahrental exploratory tunnel is part of this major project and is being driven through extremely complex geology. Soft rocks with converging rock mass, unstable tunnel face conditions with accompanying breakouts and collapses at the tunnel face required and continue to require increased rock support efforts and technology adapted to the complex conditions. The use of an open Gripper TBM was based on proven technology. Gripper TBMs have already proven themselves in numerous complex projects in the Alps, including the Lötschberg and the Gotthard tunnels. New in the evolutionary stage of this technology are systems that help to optimize tool maintenance intervals, such as the DCRM system and the DCLM system that makes it possible to draw conclusions about the nature of the tunnel face, thus allowing a kind of tunnel face scanning. Together with the newly developed camera system, in future real-time preliminary exploration will be possible and thus an optimization of the tunnelling parameters along with improvements in terms of performance, safety and quality. The knowledge gained from driving the tunnel is contributing to the development of an optimal tunnelling and safety concept for the construction of the main tunnel.

Tunnels and Underground Cities: Engineering and Innovation meet Archaeology, Architecture and Art, Volume 7: Long and deep tunnels – Peila, Viggiani & Celestino (Eds)

Risk management for the Brenner Base Tunnel

K. Bergmeister
Brenner Base Tunnel BBT SE, Innsbruck, Austria

ABSTRACT: Risk management, along with cost management, is a very important tool in carrying out tunnel and infrastructure projects. Proper risk and opportunities evaluation are especially important in large infrastructure projects. This publication proposes a novel method for risk structuring, based on the models available in literature and including unknown and unidentifiable risks, also known as "black swans". In infrastructure projects, we may also encounter so-called unidentifiable risks and risks which, from a scientific point of view, cannot be directly inferred. The basis of the work is a mathematical relation linking extent to probability of occurrence: It also includes the development of measures to reduce the extent of the damage from unknown risks in large infrastructure projects.

1 INTRODUCTION

1.1 *Risk types*

Large infrastructure projects typically involve long planning and construction phases and a high degree of complexity. According to Flyberg (2009), ex ante underestimations of costs and risks for large infrastructure projects have as their consequences both cost overruns and benefit shortfalls. Large infrastructural projects have a long planning and approval period and complex project interfaces.

Besides known and identifiable risks, which can be analyzed by some deterministic or probabilistic approach, there are also risks that are not only not quantifiable but unidentifiable as well. For this reason, in Austria a specific guideline (ÖGG-Richtlinie) for the estimation of risk according to levels of difficulty has been worked out in 2005 and 2016. This risk estimation is based on experiential values and includes a large area of possible risks. In complex infrastructure projects, we may also encounter so-called unidentifiable risks and risks which, from a scientific point of view, cannot be directly inferred. Such events, were called by Taleb (2009) "The impact of the highly improbable" or "black swans". In the following paper possible management strategies in order to reduce this risk will be presented. From a scientific point of view therefore at least 3 different levels of risk types and assessment methods can be identified.

1.2 *Description of the Brenner Base Tunnel*

The Brenner Base Tunnel is the central part of the Scan-Med corridor which runs for approx. 9,000 km from Helsinki (Finland) to Valletta (Malta). The Brenner Base Tunnel (BBT) will connect Tulfes/Innsbruck, Austria, with Fortezza, Italy. The BBT consists of two main tubes with one railway track each, i.e. trains will travel through them in one-way traffic. In between the two main tunnels, 12 meters below, runs an exploratory tunnel. This exploratory tunnel is being excavated for geological and hydro-geological purposes. The results of this preliminary research will provide detailed insight, and thus it helps to minimize construction costs and risks when realizing the main tunnel tubes. In Innsbruck the BBT will be connected with the

Table 1. Risk types and their methodology.

Risk type	Methodology
Identifiable and quantifiable risks	Probabilistic method supported by a Monte Carlo Simulation
Recognizable but not quantifiable risks	Experience based approach (e.g ÖGG 2016 Austria)
Unforeseen big events (black swan)	Phenomenological approach [Bergmeister, 2013]

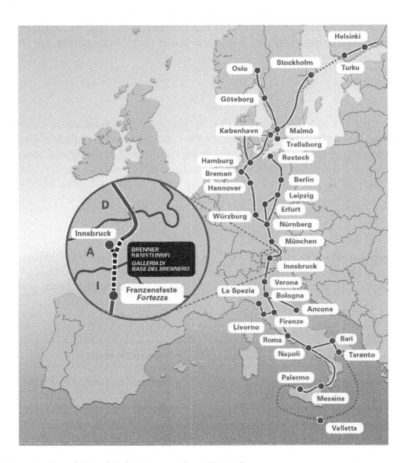

Figure 1. Scan-Med corridor with the Brenner Base Tunnel.

existing underground bypass. Together with the bypass of Innsbruck, the Brenner Base Tunnel will be the longest underground railway line in the world, with a total length of 64 km.

Already since 2010 a sophisticated risk assessment procedure is carried out for the Brenner Base Tunnel project. In Italy on the basis of a national law (D.L. 50/2016), for unforeseen events or risks a maximum of 8% of the total cost of the project can be assumed. In Austria since 2005 a specific guideline describes for every infrastructural project a methodology to forecast the identifiable and quantifiable risks as well as the recognizable but not quantifiable risks.

For the Brenner Base Tunnel as a transnational project, the probability based approach for the identifiable and quantifiable risks has been developed and applied (Bergmeister, 2015). The modelling of risks and opportunities for the Brenner Base Tunnel project is carried periodically, at least one yearly.

2 THE PROBABILISTIC BASED APPROACH FOR THE IDENTIFIABLE AND QUANTIFIABLE RISKS

This assessment of individual opportunities and risks considers the following aspects:

- Risk type,
- Measures for risk prevention,
- Costs for the measures for risk prevention, where the respective maximum, minimum and expected values and the probability of occurrence are all defined.
- Time-related impacts once the opportunity or risk comes to pass, where the respective maximum, minimum and expected values and the probability of occurrence are all defined.
- Uncertainty of the assessment.

2.1 Areas of risk

The assessment of the individual opportunities and risks for the Brenner Base Tunnel takes place in workshops with experts and the project manager of the specific construction lot, taking into account, at least, the following topics aimed at identifying and assessing the individual opportunities and risks. More than 400 independent risks and approx. 20 opportunities have been identified. In the following table 2 the major topics of risks and opportunities are presented.

2.2 Calculation of risk provision

After assessing each individual opportunity and each individual risk, the probability of occurrence "P" is calculated or estimated based on experience. The possible impact "I" and/or the damage that may be the result of a risk occurring is assessed in terms of monetary impact, to

Table 2. Risk and opportunities.

	Macro topics	Some specific issues
Risks		
	Geology, Hydrogeology, Geomechanics, Tunnel construction	Uncertainties, poor rock, gas pockets...
	Natural occurrences, Disposal and Construction sites	Meteorological events, settlements...
	Archaeological risks	
	Construction site logistics	Material supply...
	Spoil	Asbestos...
	Mechanical risks	Loss of machinery...
	Tender procedures	Legal problems
	Contractual risks, claims	
	Authorizations	Changes..
	Labour safety	Additional prescriptions...
	Environment	Noise pollution...
Opportunities		
	Reutilization of excavation material	Aggregates, sale..
	Changes of geology, geomechanical behavior	Improved rock stability
	Improved performance of TBM	Open gripper TBM: excavation time > 20% refereed to total time
		Shielded TBM: excavation time > 25% refereed to total time (Ruepp, 2018)
	Reduced water inlet	Stretches of dry rock
	Reduced vibration transmission	Changed rock behavior...
	Changes in the logistic chain	Concrete manufacturing on site

determine the maximum, minimum and expected values. This gives a monetary amount for risk provision "R".

$$Ri = Pi \times Ii \qquad (1)$$

The time-related impacts of a risk are also assessed with a maximum, a minimum and an expected value. For the probability distributions, a three-point estimation was used (figure 2), based on expert judgement [3].

The three calculated or estimated monetary or time-related values give a statistic distribution. The various steps of the calculation are shown in the flow chart of figure 3.

At least the mean total risk cost Rm and a characteristic (in the sense of statistical confidence) total risk cost Rc need to be estimated (see figure 4). The latter may be viewed as a "safe" value, similar to Value-at-Risk (VaR), widely used in portfolio management. If R denotes the total risk cost and α (typically 0.05) the desired confidence level, then VaR is defined [1] by:

$$P\ [R \geq VaR] = \alpha \qquad (2)$$

Any interdependencies among individual opportunities or risks can be assessed with a correlation matrix. All the risks were considered to have a variable economic impact and none of them was assumed as an "event" or "shock" with a particular probability of occurrence. The dependence was quantified and identified as being positive.

Four scenarios were analysed with regard to the dependence magnitude, namely the cases 0, 0.25, 0.50, 0.75 for Kendall's tau τ, representing independence, weak, moderate and strong dependence.

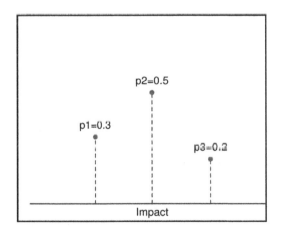

Figure 2.   Risk assessment.

Table 3.   Correlation matrix – example: R1 depend with 0.25 on R2, R1 is independent of R3 and R1 depend with 0.5 on Ri.

	R_1	R_2	R_3	R_i
R_1	1	0,25	0	0,5
R_2	0,25	1	0,75	0
R_3	0	0,75	1	0,5
R_i	0,5	0	0,5	1

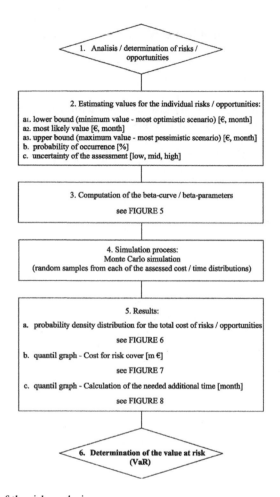

Figure 3. Flow chart of the risk analysis.

By simulating the individual risks with the Monte-Carlo method, statistic distribution curves can be calculated for each technical field or for each construction lot. Considering the assumed probability of occurrence, the distribution of the individual risks were simulated and aggregated with the help of a Monte-Carlo simulation. In the Monte-Carlo simulation, random samples are taken of all opportunities and risks to be taken into consideration. The basis for these samples is the distribution curve. For the Brenner Base Tunnel the beta distribution of the monetary or time-related impacts of the risks or the dependency from other risks have been used. On the basis of various numerical calculations the beta distribution in figure 4 was identified as most appropriate by Tamparopoulos (2012) in his PhD-thesis according.

The various fractile values of the risk costs, Value-at-Risk(%) (VaR) can be determined and calculated. Based on these curves and in accordance with the desired degree of coverage (VaR), the scope of the monetary or time-related risks can be calculated or directly read with a Lorenz curve. This sum curve is a graphical representation of the risk provision expressed in a monetary value (Euros) depending on the probability of risk acceptance. With a Lorenz curve we can see that even a small percentage increase in risk cover (less risk acceptance) can cause significantly higher risk costs. For the Brenner Base Tunnel, a VaR of 50% was assumed and agreed between Italy and Austria. For other infrastructural projects or underground infrastructures a VaR of 75% is widely used. It should be noted that in all cases, the mean (expected) value of the total cost is invariant, since the expectation operator preserves the sums of random variables.

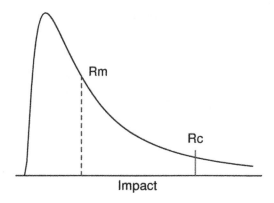

Figure 4.   The beta distribution of the monetary or time-related impact of risks (taken from Tamparo-poulos, 2012).

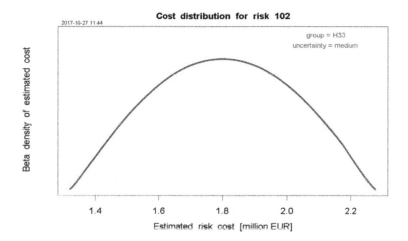

Figure 5.   shows the computation of the beat-curve for the estimated risk costs.

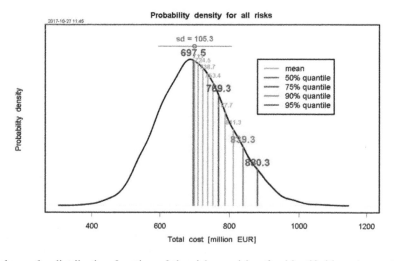

Figure 6.   shows the distribution function of the risk provision for identifiable and quantifiable risks, simulated with the Monte-Carlo method.

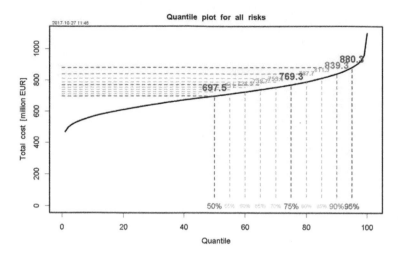

Figure 7. shows the Lorenz curve and/or the sum curve of the risk provision for identifiable and quantifiable risks, simulated with the Monte-Carlo method.

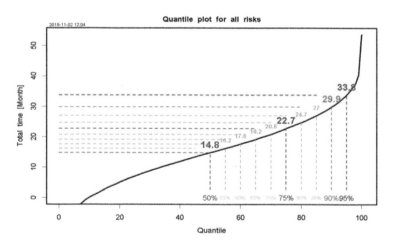

Figure 8. shows the Lorenz curve and/or the sum curve of the risk for additional time needed, simulated with the Monte-Carlo method.

2.3 Calculation of the VaR value for the Brenner Base Tunnel

The total risk provision for the Brenner Base Tunnel with a 50% fractile (VaR50) was calculated as 697.5 mio Euro per 01/01/2017 and, with a 90% fractile (VaR90) as 839.3 mio Euro.

The total amount calculated for the opportunities was after a 5-year optimization period (on the basis of the opportunities carried out in the risk management) ca. 2,6 mio Euro.

3 RECOGNIZABLE BUT NOT QUANTIFIABLE RISKS – THE AUSTRIAN GUIDELINE FOR RISK ASSESSMENT

In Austria, in Switzerland (Ehrbar et al. 2003) and in Germany exists since 2017 a specific guideline for a holistic assessment of all risks (quantifiable and not quantifiable risks).

The first Austrian ÖGG directive of 2005 and the actual guideline 2016 "Costing for transport infrastructure projects" gives estimates of the required risk provision on the basis of

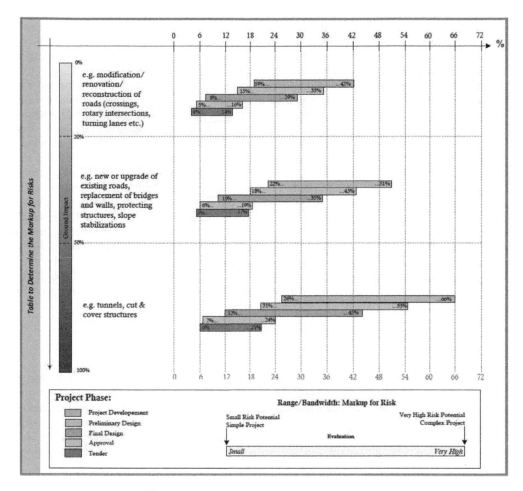

Figure 9. Excerpt from the ÖGG Directive (2016) for the calculation of total risk provision.

many years of experiential data. The economic risk evaluation considers risks in both the promoter's and the project owner's sphere of risk. To calculate the costs for risk provision, certain assumptions are made based on project phase, complexity and construction ground impact.

– Project phase: Project development, preliminary design, final design, approval, tender, realization
– Complexity: The project is categorized as difficult.
– Ground impact: Tunnels, cut&cover structures like the Brenner Base Tunnel are standing for a high ground impact

For the actual situation, depending on the respective status of realization and on the estimate of the complexity of the project Brenner Base Tunnel the risk provision for all risks (quantifiable and not quantifiable risks) amount € 1.215 mio per 01/01/2017.

4 UNKNOWN, EXTREME OPPORTUNITIES AND RISKS

Due to their complexity and long completion horizon, mega-projects are far more likely to be impacted by unexpected events than projects with a shorter completion time. Currently, no quantitative models are to be found in literature that calculate risk provision for such extreme events (giga-risks), but such a model would make sense in this case. There are no experiential or theoretical models to estimate these unknown, extreme events. In Bergmeister (2013) drew up a suggested approach to how such extreme events could be structured for mega-projects and, again as an approach, identified and evaluated.

In modelling unknown, extreme opportunities and risks, several independent or inter-dependent (mutually causal) scenarios can be developed on the basis of the phenomenological Delphi scenario method. These include, for example, mudslides caused by natural events or floods due to high water or interruptions in financing resulting from political crises. The phenomena in the early phases of an event are observed and possible processes are shown, which requires the consideration of possible catalysts (phenomena that suddenly accelerate a process). The scenarios can be developed by experienced experts, supported by computer models. The question is, of course, which scenarios hold the greatest risk or the greatest opportunity. Economic science solves this problem using so-called operational research. Interesting goals are defined as partial goals and their importance is summarized in a function using weighting factors. This simplifies the problem and solves the target function. For the constant monitoring of certain phenomena or scenarios, so-called key indicators must be set. These are important indications that an event is imminent and require the adoption of corresponding measures.

For example, as regards the Brenner Base Tunnel, air quality conditions on disposal sites, possible deformations of the disposal site shape and water resources are the subject of constant monitoring.

5 CONCLUSION

It is important for all infrastructure projects and especially for large projects such as the Brenner Base Tunnel to carry out risk management and periodically update the assessment of opportunities and risks. For the Brenner Base Tunnel, the identifiable and quantifiable opportunities and risks are determined with a probabilistic method. In addition to this, risk provision is also assessed for recognizable but not quantifiable risks using the ÖGG Directive. Risks that are a decisive factor for costs and timing must be prevented or at least reduced by specific measures.

REFERENCES

Bergmeister, K. (2013): Holistisches Chancen- und Risikenmanagement von Megaprojekten. PhD-tesis. University of Bratislava-Vienna.
Bergmeister, K. (2015): Anwendung des holistischen Chancen-Risiken-Managements beim Brenner Basistunnel. In: *Tunnelbau Kompendium der Tunnelbautechnologie*. Deutsche Gesellschaft für Geotechnik. Wiley Online Library
Ehrbar, H.; Kellenberger, J.: Risk Management during the Construction of the Gotthard Base Tunnel. *Proc. Int. Symp. GeoTechnical Measurements and Modelling (GTMM)*, Karlsruhe, 2003
Flyvberg, B.; Bruzelius, N.; Rothengatter, W. (2006): Megaprojects and Risk. An Anatomy of Ambition. Cambridge University Press. ISBN 0 521 00946 4
ÖGG-Richtlinie (2005, 2016): Kostenermittlung für Projekte der Verkehrsinfrastruktur unter Berücksichtigung relevanter Projektrisiken. Salzburg
Ruepp, A. (2018). Evaluation of excavation methods. Master thesis. BOKU University Vienna
Taleb, N.N. (2009). The black swan: The impact of the highly improbable. London.
Tamparopoulos, A. E. (2012). Cost estimation of large construction projects with dependent risks a study on the Brenner Base Tunnel. Ph.D. – thesis BOKU University Vienna
Tamparopoulos, E.A., Spyridis, P., Bergmeister, K. (2011). Small failure probabilities and copula functions: Preliminary studies on structural reliability analysis. In C. Bérenguer, Grall A., and C. Soares, editors, Advances in Safety, Reliability and Risk Management – *Proceedings of the European Safety and Reliability Conference, ESREL 2011*, pages 1115–1120, Troyes, France.

Tunnels and Underground Cities: Engineering and Innovation meet Archaeology,
Architecture and Art, Volume 7: Long and deep tunnels – Peila, Viggiani & Celestino (Eds)
© 2020 Taylor & Francis Group, London, ISBN 978-0-367-46872-9

The second Gotthard tunnel tube

G. Biaggio & V. Kumpusch
Swiss Federal Roads Office (FEDRO), Bern, Switzerland

ABSTRACT: Featuring a length of 16.9 km, the planned second tunnel tube through the Gotthard will prove to be yet another outstanding feat in the Gotthard region. This contribution provides a short overview of the project, explains the investigation of the fault zones, outlines the excavation concept for piercing these zones and addresses both material management and the reuse of the excavation material.

1 INTRODUCTION

Located on the north-south axis of the A2 motorway, the Gotthard Tunnel connects the cantons of Ticino and Uri between Airolo and Göschenen. The existing motorway tunnel was opened in 1980, with forecasts based on ongoing checks indicating that it will be due for renovation in around 15 to 20 years. As part of the 'Gotthard conservation concept' study that began in 2009, efforts were made to identify and investigate on different feasible options for conservation. These included the prospect of constructing a second tunnel and subsequently renovating the first tunnel, as well as the possibility of enforcing complete closures lasting several years in order to enable the renovation of the existing tunnel. This latter option would have required the diversion of traffic via the pass and/or rail loading of the vehicles. On 27 June 2012, the Swiss Federal Council decided in favour of the construction of a second tunnel tube with subsequent renovation of the existing tube. The project was approved by the citizens, the Council of States and the National Council in a referendum on 28 February 2016.

The chosen solution significantly increases the level of safety in the Gotthard Tunnel and ensures that the most important north-south connection will remain open during the renovation of the existing tunnel tube.

When the project is completed, both tubes will feature single-lane operation with one standard lane and one service lane in each direction (Figure 1). This ensures compliance with the constitutional clauses relating to Alpine protection, which also preclude any attempts to increase the capacity of transit roads in the Alpine regions.

While the first tube was being constructed, there was already a project concept for a second tunnel tube. It was planned to subsequently expand the service and infrastructure tunnel that runs parallel to the existing tunnel to enable the establishment of a second tube. The current project deviates from this concept, especially as a result of the easier construction methods featuring TBM excavation that are available nowadays.

2 PROJECT DESCRITPION

The planned second tunnel tube through the Gotthard has a total length of 16,866 m. It runs at a standard clearance of 40 m from the service and infrastructure tunnel located east of the existing Gotthard Tunnel. The layout of the line runs largely parallel to the existing tube (Figure 2). The clearance is reduced in the portal areas in order to let the axis of the second tube run along the axis of the service and infrastructure tunnel. In Airolo and Göschenen, the

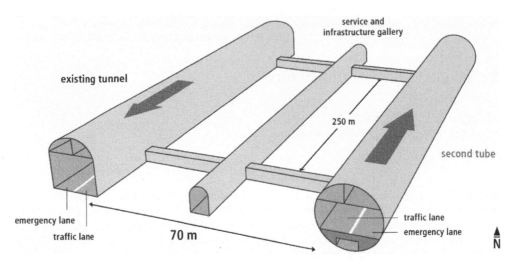

Figure 1.   Overview of the overall system at the completed Gotthard Road Tunnel.

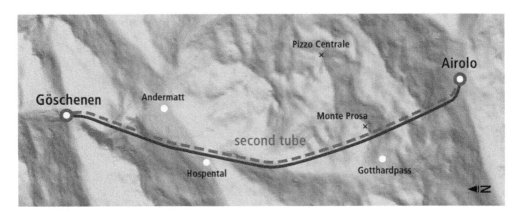

Figure 2.   Overview of the second Gotthard road tunnel.

existing portal structures – which were designed for the inclusion of a second tube already while the first tube was being constructed – and cut-and-cover tunnels are used as tunnel portals. To ensure full use of the service and infrastructure tunnel once the work is completed, the portal areas will be repositioned in advance.

The tunnel tubes are going to be connected at regular intervals by 62 walk-through cross-passages and six vehicle-accessible cross-passages. A majority of the new cross-passages will be arranged in the extension of the existing cross-passages to the service and infrastructure tunnel.

Because of the slab track and the intermediate ceiling, the standard profile (Figure 3) will be divided into three main areas: the driving area (clearance envelope for one standard lane and one service lane with walkways on both sides according to the first tube), the intake and exhaust duct above the intermediate ceiling and two service ducts beneath the slab track (with one service duct intended for the routing of a 380 kV line). The tunnel will be given a double-lining construction (excavation support and inner lining) and an umbrella sealing. The drainage water will be collected and guided to the portals via water lines located in the invert while the water from the road surface will be directed into the wastewater line via slotted channels and siphon shafts.

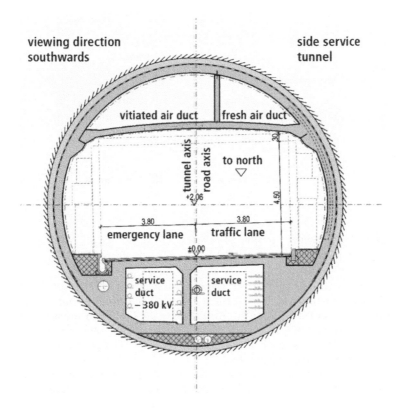

Figure 3. Standard profile of the second Gotthard road tunnel.

Ventilation poses a major challenge when it comes to long road tunnels. When it comes to the second road tube through the Gotthard, a ventilation system will be installed which enables one-way as well as two-way traffic in case of maintenance. The ventilation systems of the two tunnel tubes are independent of one another, with three new underground ventilation control centres and two portal control centres being created for the second tube. Only those ventilation shafts that are already present in the existing underground ventilation control centres and provide a sufficient ventilation cross-section will be used for both tubes. These ventilation shafts will be used both for exhaust air and the intake of fresh air for the traffic area. In case of an emergency, the tunnel will be divided into five ventilation sections. The fumes are to be drawn out of the driving area via exhaust flaps into the intermediate ceiling and blown out through the ventilation shaft by the respective control centre. Jet fans will also be used to control the longitudinal flow in the driving area.

The accompanying measures will also involve the rearrangement of the Airolo motorway connection, including tasks such as the creation of the Galleria di Airolo and corresponding terrain modelling. This cut-and-cover twin-tunnel will be 1086 m long and take the form of a gallery in certain sections on the valley side, making a significant contribution to the eco-logical upgrading of the valley floor.

3 GEOLOGY AND HYDROGEOLOGY

3.1 Geology

Thanks to the excavation of the adjacent first tunnel tube the geology of the second Gotthard road tunnel to be driven through is well known (Figure 4). Approaching from the north, the

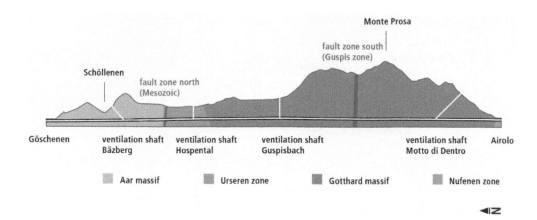

Figure 4.   Geological longitudinal profile.

tunnel crosses the Aar Massif that consists mainly of granites and paragneiss. It then passes through the Urseren zone, which is formed of meta-sediments from the Mesozoic and Permo-Carboniferous periods. The third formation is the Gotthard Massif, largely comprising paragneiss, granite gneiss and granite. The Nufenen zone composed of meta-sediments (from the Mesozoic period) is reached at the Airolo portal. The portal areas themselves feature loose rock sections of varying lengths.

Along the axis of the planned second Gotthard road tube it is expected to encounter two problem zones: these will subsequently be referred to as the 'northern fault zone' (Mesozoic/Permo-Carboniferous) and the 'southern fault zone' (Guspis zone).

Located in the Urseren zone at the transition between the Mesozoic and Permo-Carboniferous sections, the northern fault zone is around 270 m long and largely consists of clay and sericite schists. Limestone, gypsum and dolomite are also expected in this zone. The rock in the northern fault zone shows moderate to very high levels of fault gouging and demonstrates both friable and squeezing rock behaviour.

The southern fault zone is located in the Guspis zone of the Gotthard Massif. Covering a distance of nearly 300 m, it contains a variety of more weathered and fault-gouged and also more compact and firm biotite gneiss and schists, hornblende schists and individual amphibolites. Dyke rocks are also expected to be frequent. As with the northern fault zone, friable and squeezing conditions are also anticipated in this section.

3.2   *Hydrogeology*

The crystalline rocks show average to good permeability, especially along the joint planes. Old crystalline gneiss and schists have a somewhat lower level of permeability than the crystalline rocks. The Alpine metamorphic sedimentary rock located above this, which originates from the Mesozoic and Permo-Carboniferous periods, also features low permeability and indeed acts largely as an impermeable layer.

Despite the high underground water level and increased permeability in the area of the fault zones and the fissured rock, along rock borders and in hydrothermally disintegrated rock, the overall expectations based on the measurements performed in the existing system point to the presence of moderate amounts of water.

Areas bearing more water are particularly clustered in the southern zones near to the portal. The amount of underground water at the portal is forecast to be 30 – 40 l/s in the north and 95 – 110 l/s in the south.

4 EXCAVATION CONCEPT

Given the length of the tunnel, the planned excavation concept involves one TBM drive each approaching from the north and the south portal. Both drives are to proceed at an incline and will feature an excavation diameter of 11.8 m.

The loose rock sections in the portal areas will undergo conventional excavation in advance. Afterwards a further amount of rock (400 m in the north, 750 m in the south) will be excavated by blasting drivage. This is due to the very cramped conditions in the portal areas, which will affect the assembly and launching of the TBMs, as well as additional issues relating to the construction programme and risk-related considerations.

In order to reduce the risks and optimise the overall construction programme, the northern and southern fault zones (totaling approx. 570 m) will be conventionally excavated and secured in advance. Once this has taken place, the two TBMs will be pushed through these areas. The fault zones will be reached via separate access tunnels approaching from the north (approx. 4.4 km) and the south (approx. 5 km), which will be created at an early stage. These tunnels will be excavated using Gripper TBMs with an excavation diameter of 6.0 m.

5 PIERCING OF GEOLOGICAL FAULT ZONES

5.1 *Experience gained from the excavation of the first Gotthard road tunnel*

During the construction of the first Gotthard tube, it was likewise necessary to pass through the two fault zones featuring friable to squeezing rock behaviour.

In the northern fault zone, there were considerable difficulties relating to the excavation of the service and infrastructure tunnel. There were rockfalls that generated so called chimneys, incidents of water ingress that left the rock showing signs of softening and generally low stability times of a few hours. Grouting did not lead to any noticeable improvement in the properties of the rock and the accruing seepage water was only partially displaced.

Based on the experience made while building the service and infrastructure tunnel, the main tunnel was excavated using the so called 'German' method (core method) in the area of the northern fault zone. This involved the sidewall headings and the crown area being excavated first and after completing the tunnel walls the core was removed.

In addition to the loose rock-like conditions, it was also necessary to deal with squeezing rock behaviour as average convergences of approx. 0.4 m were measured. It was possible to establish a link between the occurrence of the convergences and the level of permeability: the deformations occurred early on in the case of high permeability, while deformations were only detected after a couple of hours or days in areas where permeability was low.

Squeezing conditions were also experienced along a nearly 400 m section in the area of the southern fault zone, both in the service and infrastructure tunnel and in the main tube. In light of the experience gained with the excavation of the service and infrastructure tunnel in the Guspis zone, an exploratory heading was driven in advance on the axis of the first tube. Considerable amounts of convergence occurred, especially in foliated sections of the paragneiss, and this could no longer be overcome by means of a simple steel support. For this reason, and also as a result of what happened with the north drive in the Mesozoic/Permo-Carboniferous section, the decision was taken to also excavate the main tube via an intermediate access point using the core method. A high level of rock pressure was detected in the process, with the resulting convergences amounting to as much as 1.5 m in the side walls (Figure 5).

5.2 *Investigation of the fault zones*

Starting from the existing service and infrastructure tunnel, a total of eight sub-horizontal exploratory probe holes each measuring approx. 70 - 75 m were created across both fault zones in 2016. This was done in order to investigate the geological conditions and to extract

Figure 5. Squeezing rock during construction of the Gotthard Road Tunnel.

rock samples for laboratory testing. Three exploratory boreholes were drilled in the Jurassic limestones and clay schists of the Mesozoic layer in the Urseren zone and five boreholes were generated in the Guspis zone. Pressure gauges were installed at two boreholes and dilatometer tests were performed in five boreholes.

In addition, the laboratories of the École polytechnique fédérale de Lausanne (EPFL) and ETH Zurich were commissioned to perform the following rock mechanics tests on rock samples:

- uniaxial compressive strength tests
- Tri-axis tests (with and without consideration of the pore water pressure)
- Split tensile strength test
- Direct shearing tests
- Cerchar abrasivity tests

These were used to determine the strength and deformation properties of the fault-gouged rocks.

5.3 *Excavation and support concept*

Based on the forecast geology and the experience gained from the first Gotthard tube, potential danger patterns were defined for the fault zones and a subsoil model divided into homogeneous areas was created.

Given the poor-quality properties of the rock, it was decided that both fault zones would be excavated by means of conventional excavation.

In the attempt of achieving ring closure as quickly as possible, the planned excavation concept for the fault zones involves full-face excavation. The excavation cross-section ($D_a = 14$–14.5 m) features room for securing, tolerance for movement on the part of the TBM and space for convergences (0.65 m respectively 0.5 m) based on the deformation calculations (see Section 5.4).

Grouted anchor bolts will be deployed in the crown area as an auxiliary construction measure in order to ensure working safety. Steel fibre shotcrete will also be applied to the profile and the face, with face bolts also being installed in the latter (110, l = 8 m, every 12 m). In order to allow the deformations to occur in a controlled manner, the plan calls for a yielding lining (as was used in the Tavetsch intermediate massif of the Gotthard Base Tunnel). This will involve the installation of mesh, steel ring beams with sliding connections (each featuring two interlocking, moveable TH-beams of type TH 44) and a radial bolting system (self-drilling anchors, l = 8–12 m). The ring beams are covered with shotcrete, in which inlays are implemented to keep specific areas open allowing a more controlled deformation (Figure 6). Only once most of the forecast deformations have occurred, the resistance will then be increased – firstly by completely shotcreting the open inlay areas and thus fixing the TH-beams in place, and secondly by ensuring that the excavation support reaches its target thickness of 50 cm through the application of mesh-reinforced shotcrete (Figure 7).

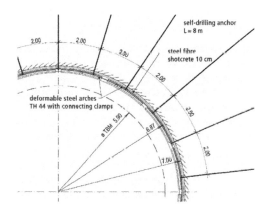

Figure 6.   Yielding lining in area L1.

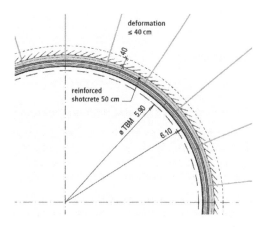

Figure 7.   Yielding lining in area L2.

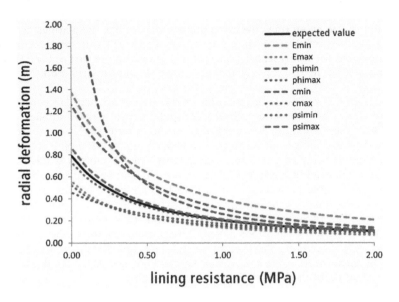

Figure 8.   Characteristic curve for rock according to characteristic values of rock.

5.4   *Deformation calculations and remaining uncertainties*

The convergences in the fault zones were calculated using the characteristic curve procedure. Additionally a FE model was created for the southern fault zone. The calculations were also used to investigate the influence of the wall effect on short fault zones, thereby accounting for the beneficial influence this has on the behaviour of a thin fault zone in the neighbouring unaffected area. Together with the characteristic curve of the excavation, this makes it possible to determine the best time for the installation of the TH-beams and for the closing of the sliding connections so as to enable optimum loading of the excavation with only minor deformations.

Despite the prior knowledge relating to the first Gotthard Road Tunnel tube and good-quality investigation of the rock, the anisotropy of the rock and the wide range of rock characteristics mean that the calculations remain subject to some uncertainties (Figure 8).

In addition, the consideration given to the pore water pressure plays a major role as it significantly influences the short-term and long-term development of the convergences. There is also still an outcrop gap at the northern border of the southern fault zone.

These issues will have to undergo more in-depth analysis in the course of subsequent project planning.

6   MATERIAL MANAGEMENT

6.1   *Reuse concept*

The project has set the following objectives in relation to material management:

–   The excavated material must be utilised as much as technically possible.
–   The environmental impact of material management must be kept to a minimum.
–   The material management concept must be economically acceptable.
–   Material management must not be the ultimate performance-defining factor.

The excavation material is primarily generated from the main drives. Additional excavation material is also amassed from auxiliary structures, access tunnels and caverns. The geology, which has been documented in detail and was further investigated via the exploratory boreholes

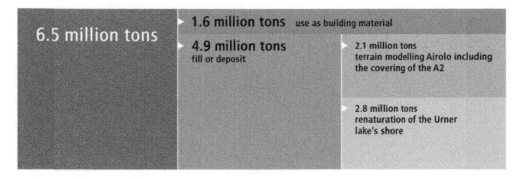

Figure 9. Reuse of excavated material.

drilled in 2016, enables positive forecasts to be made in relation to effective recycling. A total of nearly 6.5 million tons of excavation material will be accumulated (Figure 9), half at the north portal and half at the south portal.

It will be possible to use around 1.6 million tons as aggregates for concrete once this has been processed. Around 4.9 million tons of uncontaminated material that is not suitable for processing or is rejected during material processing will be recycled in the course of filling procedures or renaturation. Lightly contaminated silt and lightly polluted excavation and dismantling material cannot be used for renaturation or terrain modelling and will be disposed in compliance with the law.

6.2 *Material logistics*

Due to the cramped conditions in the vicinity of the portal and the exposed location with respect to natural hazards (especially at the portal area in Göschenen), the material management installations will largely be located in the Airolo area.

The excavation material will be transported from the face to the triage points located outside of the portals by means of a conveyor belt. The material that is accumulated in the north drive and is suitable for processing will be taken by train through the Gotthard Tunnel from the Göschenen portal to Airolo. Once there, both this material and the material from the south drive will be transported via conveyor belt to the processing plants in Stalvedro.

The processed aggregates will then be transported from Stalvedro back to the silos of the concrete plants via conveyor belt and by train (in the case of Göschenen). With the exception of the preliminary lots, it will be possible to cover the full amount of gravel and aggregate required for the concrete through the processing of the A material.

6.3 *Filling and renaturation*

A certain amount of the excess non-polluted excavation material (B material) will be deposited in the immediate surroundings of Airolo. The rearrangement of the Airolo connection features plans for various adjustments to the terrain that are associated with relevant material requirements. These also include the Galleria di Airolo, around 1000 m of which will be covered with excavation material – thereby significantly upgrading the valley floor (Figure 10).

A further approx. 2.8 million tons of excavation material will be used in the canton of Uri for the renaturation of the shallow water zones of the Reuss delta and will therefore serve as embankment protection for the Reuss plain (Figure 11). At the beginning of the 20th century, the point at which the Reuss flowed in Lake Uri consisted of an expansive shallow water zone featuring multiple islands. Following the canalisation of the Reuss and the removal of gravel, which started at the turn of the 20th century, the river landscape ceased to develop in a natural manner. The shorelines moved 200 to 300 m inland in certain areas depending on the water

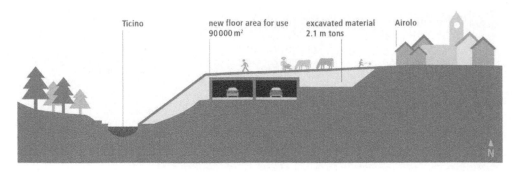

Figure 10. Visualisation of covering of Galleria di Airolo.

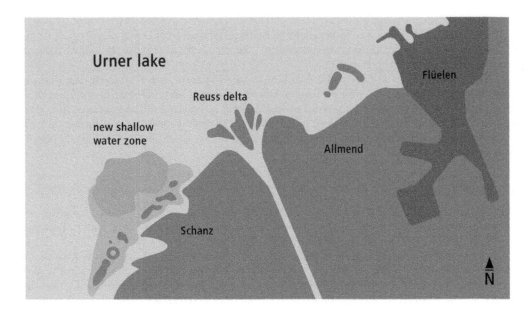

Figure 11. Visualization of renaturation of shallow water zone in Lake Uri.

level, with around 24 ha of reed meadows and fields falling victim to this human intervention. Between 2000 and 2007, a total of 3.3 million tons of excavation material from other tunnel projects was used for the purpose of renaturation in seven stages. The excavation material from the tunnel projects relating to the new Axenstrasse and the second Gotthard road tunnel will make it possible to complete this renaturation process. 'Lakefill III' is a separate project being run by the canton of Uri that was granted approval back in summer 2017.

7 CONCLUSIONS

The building contractor is convinced that the excellent collaboration among all participants has enabled the development of a convincing project. Following public disclosure, it will be possible to quickly turn this project into a reality. Once the construction period of around eight years has been completed, the second Gotthard road tunnel will be opened. This is estimated to take place in 2028, depending on the duration of the approval process. The commissioning of the second Gotthard road tunnel will be immediately followed by the renovation of the existing first tunnel, which will take around three years. After around eleven years of construction work a modern and safe overall system featuring two tunnels – each containing one lane – can be put into operation.

REFERENCES

IG Gottardo Due. 31/08/2017. Construction project. *Secondo tubo San Gottardo, Bozza Rossa*

Swiss Federal Roads Office (FEDRO). 25/10/2017. *Die zweite Gotthard-Strassenröhre. Infodossier zur Genehmigung des Generellen Projekts durch den Bundesrat.*

Cf. https://www.astra.admin.ch/astra/de/home/themen/nationalstrassen/sanierung-gotthard.html

Wild, K. & Amberg, J. 13/10/2017. *Presentation 'Problematik Durchörterung Störzonen'*, STSym autumn workshop

Tunnels and Underground Cities: Engineering and Innovation meet Archaeology,
Architecture and Art, Volume 7: Long and deep tunnels – Peila, Viggiani & Celestino (Eds)
© 2020 Taylor & Francis Group, London, ISBN 978-0-367-46872-9

Dubai Strategic Sewerage Tunnel – challenges to infrastructure at unprecedented depths in the region

D. Brancato, A. Ayoubian, M. Joye & G. Monks
Parsons Corporation

F.A. Al Awadhi & J. Tharamapalan
Dubai Municipality, Dubai, UAE

ABSTRACT: In the rapidly developing Emirate of Dubai, the Dubai Strategic Sewerage Tunnel (DSST) Project will upgrade the existing sewage infrastructure from a pumped system to a gravity-based system. The project will include almost 250 km of link sewer, approximately 82 km of tunnels with inside diameters ranging from 3.5 to 6.5 meters, and two terminal pump station shafts with diameters of 43 meters and 63 meters. The general subsurface conditions are surficial sands underlain by cemented calcareous sand and weak rock, with groundwater near ground surface. Preliminary design was performed for the largest diameter pump station shaft final liner and invert slab, and tunnel liner. Additional analyses include a cylindrical buckling and uplift evaluation of the final shaft structure under computed loads.

1 PROJECT OVERVIEW

Dubai Municipality (DM) has embarked on a major program to expand and improve their existing sewerage system by converting from a pumped system to a deep gravity tunnel system. This Dubai Strategic Sewerage Tunnels (DSST) program will provide the infrastructure to capture and treat flow for the rapidly developing Emirate of Dubai. The program will be expanded in future to suit land use changes. The project will optimize operations for DM by eliminating more

than 100 subsidiary and main pump stations and localized Sewage Treatment Plants (STP) and diverting and centralizing sewage flows to the Al Warsan and Jebel Ali sewage treatment plants (WSTP and JSTP). The DSST project has two main catchments shown in Figure 1. The Deira catchment terminates at WSTP, and the Bur Dubai catchment flows to JSTP, which are called herein DSST-DR and DSST-BD, respectively.

1.1 *Existing Sewerage System*

The existing system includes both subsidiary and main pump stations, with the main pump stations transferring flows by force main directly to the STPs. There are twelve main pump stations in the existing sewerage system. Each of these main pump stations receive flows from subsidiary pump stations and in many instances, from direct gravity sewers. There are over 100 subsidiary pump stations owned and operated by DM. Flows from these subsidiary pump stations are transferred either directly via pumps or by intermediate gravity networks to the main pump stations.

1.2 *Proposed Sewerage System*

The DSST-DR catchment will include approximately 15 km of deep sewer tunnel (3.5 m diameter) ranging in depth from 35 to 80m below ground surface with a 790 mega liter per day

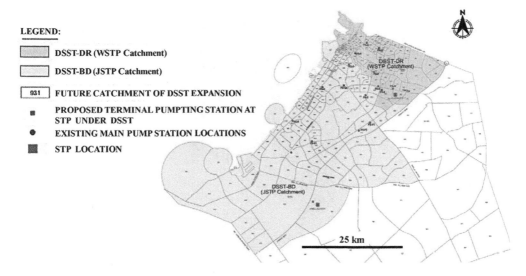

LEGEND:

- DSST-DR (WSTP Catchment)
- DSST-BD (JSTP Catchment)
- **931** FUTURE CATCHMENT OF DSST EXPANSION
- ■ PROPOSED TERMINAL PUMPING STATION AT STP UNDER DSST
- ● EXISTING MAIN PUMP STATION LOCATIONS
- ■ STP LOCATION

25 km

Figure 1. Catchment areas for the Dubai Strategic Sewerage Tunnels for the Al Warsan and Jebel Ali Treatment Plants.

(MLD) ultimate capacity terminal pump station and 55 km of link sewer. The DSST-BD catchment has approximately 60 km of tunnels (ranging in diameter of 3.5 to 6.5 m) and varies from 22 to 108m in depth (Figure 2). DSST-BD has 196 km of link sewers and a 4,270 MLD ultimate capacity terminal pump station. The link sewers will divert existing sewerage networks (i.e. subsidiary and main pumping stations and localized STPs) to the tunnels. The diameters for the terminal pump station shafts are 43 m for the WSTP shaft and 63 m for the JSTP shaft based on preliminary concepts developed for the stations.

In addition to the deep tunnels and the terminal pump station shafts, other ancillary shafts will be constructed. The shaft types include tunnel boring machine (TBM) launch or construction shafts (CS), TBM reception shafts (RS), vortex drop structure shafts (VDS), and maintenance shafts (MS). The elevation view of Figure 2, the primary segment of the DSST-BD tunnel, shows the locations of 27 shafts: five (5) construction shafts, one (1) receiving shaft, fourteen (14) maintenance shafts, and seven (7) vortex drop structure shafts. The vortex drop structure shafts intercept and divert link sewer flows to the main tunnel. Flows intercepted by construction shafts from link sewers or other tunnels can be accommodated by internal structures built within the relatively large diameter shafts. The maintenance shafts provide manhole access to the tunnel and are spaced approximately every 2 to 3 km.

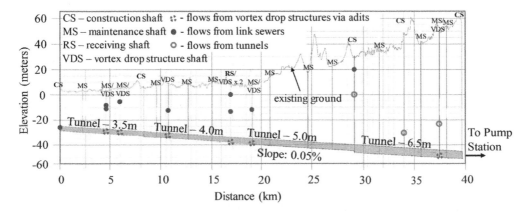

Figure 2. Schematic of the DSST-BD vertical alignment, showing locations of incoming flows.

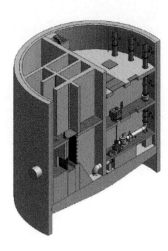

Figure 3.   Terminal Pump Station – Preliminary Concept.

Not including the two terminal pump station shafts, there will be more than 40 intermediate shafts for the DSST project. These shafts are proposed with minimum excavation diameters of 10 m for the maintenance shafts, 12 m for the reception and vortex drop shafts, and 20 m for the construction shafts. Permanent diameters will range from 2 to 4 m for the construction, reception, and maintenance shafts, and between 5 to 10 m for the vortex drop shafts.

The single pump station concept will require terminal pump station shafts to be very large and very deep. Figure 3 shows a section view of a terminal pump station. The alignment was reviewed to identify locations where an intermediate pump station could be located to reduce the depth and size of the pump stations. The evaluation did not indicate a cost-benefit to the intermediate pump station concept. Benefits with respect to depth were offset by land availability and operational considerations. Therefore, the preliminary design is based on the single Terminal Pump Station (TPS) concept. The preliminary pump station designs include wet and dry wells with single-stage vertical mixed flow volute pumps.

2   SUBSURFACE CONDITIONS

The Dubai area is covered with a middle to upper tertiary group of rocks, underlying the recent soil/desert sand or beach sand of varying thickness of 0.5–17m. The coastal sands are largely Holocene, and near the coast consist of mostly calcium carbonate ($CaCO_3$) derived from carbonate sediments, seashells, and coral reefs.

Underlying the Holocene sands is Pleistocene Ghayathi formation. The Ghayathi Formation consists of light brown to reddish brown, poorly cemented calcareous sand, white Calcarenite, calcium carbonate rich sandstone with generally distinct cross bedding. These Calcarenite Sandstones are generally up to 25m thick.

Underlying the Pleistocene Sandstone Conglomerates is the Mio-Pilocene Barzaman formation. The Barzaman formation consists of Pliocene and Miocene buffwhite/creamy white Siltstone, Calcisiltite, and light reddish brown Conglomerate. The thickness of the Barzaman formation is up to 60m.

Underlying the Mio-pliocene Barzaman formation is the Miocene Fars formation. The Fars formation consists of reddish brown/greenish grey Siltstone, Mudstone, Claystone, and colorless to white Gypsum.

Based on the finalized tunnel alignments and by considering the available geotechnical data, the majority of the main tunnel alignment will cross the Barzaman formation. Typical rock core photographs of Barzaman Siltstone/Calcisiltite (left) and Conglomerate (right) are shown in Figure 4. However, the shallower sections of the Deira tunnel north alignment will encounter the Ghayathi formation. A typical rock core photograph of Ghayathi Sandstone is shown in

Figure 4.   Typical Rock Core Photograph of Barzaman Siltstone/Calcisiltite (left) and Conglomerate (right).

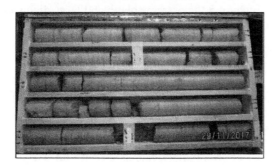

Figure 5.   Typical Rock Core Photograph of Ghayathi Sandstone.

Figure 5 The tunneling operations, along the deepest sections of Bur Dubai tunnels, at the area of Jebel Ali treatment plant, will cross the top layers of Fars formation (mudstone units).

An extensive subsurface investigation for the design of the project was conducted along the tunnel and link sewers alignment. The field investigation included geotechnical borings with downhole seismic velocity measurements, in-situ Packer permeability tests, falling head permeability tests, and installation of groundwater monitoring wells. A wide ranges of laboratory tests were conducted on soil and rock samples collected during subsurface investigation. Soil tests included Atterberg Limit tests, moisture content, hydrometer, and chemical tests. Tests on rock cores included uniaxial compressive strength test with measurement of Young's Modulus, point load index, indirect tensile strength test (Brazillian), total hardness, slake durability test, axial swelling test, Cerchar Abrasivity Index (CAI), petrographic analysis description, petrographic point count analysis, and X-Ray diffraction clay joints.

3   DESIGN CHALLENGES

One of the primary goals of the project is to develop sewer schemes to decommission all pump stations to the new sewerage system. The two terminal pump stations are very large and require managing risk associated with the unique conditions for the shafts. The goals of the sewerage tunnel investment are best satisfied when the system includes the deep shafts and tunnels, but must also satisfy the 100-year design life. This requires a focus on developing a durable design.

4   PERMANENT STRUCTURE DESIGN

The preliminary design for the segmental liner of the sewerage tunnel and the Jebel Ali pump station shaft at the terminus of the Bur Dubai alignment was performed. The tunnel outlet at the Jebel Ali shaft has an invert depth of 110 meters below ground surface, an internal diameter of 6.5 meters, with a liner thickness of 350 mm. The shaft itself has a final depth of about

118 meters, a proposed internal diameter of 63 meters, a final liner thickness of 2.5 meters, and an invert slab varying in thickness of 3.5 to 6.5 meters. Both the shaft liner and shaft invert slab are to be cast-in-place reinforced concrete, and the tunnel liner is proposed as pre-cast concrete segments with steel fibre reinforcement.

The evaluated condition is a long-term case where all anticipated hydrostatic loads, earth loads, and in the case of the shaft liner, dead loads due to the weight of the liner, are fully acting. The tunnel is expected to reach the permanent design condition relatively soon after construction. The shaft, however, is assumed to reach long term condition when the diaphragm wall is completely inundated and the temporary shear keys along the intrados of the diaphragm, necessary for construction, have deteriorated. In the case of the shaft, intermediate conditions should be considered in the final design.

The finite element program, PLAXIS 2D, was used to perform the analysis, though the unique shaft conditions, including the extreme depth and large diameter, warranted independent consideration of buckling and buoyancy resistance.

5   SHAFT DESIGN

For long term loading, the final shaft liner must be designed to take the full hydrostatic water pressure, its self-weight, and any forces applied at the connection of the final liner to the invert slab. A diaphragm wall will be used for excavation support and will also resist the ground pressures in long term as well as the forces transferred from the invert slab due to uplift. These loads can be approximately calculated with an axisymmetric model shown in Figure 6. Separate closed-form analyses are performed to check cylindrical buckling and buoyancy of the structure.

5.1   Finite Element Model

The shaft construction begins with the installation of a diaphragm wall around the shaft perimeter to support the ground during excavation. After construction of diaphragm wall,

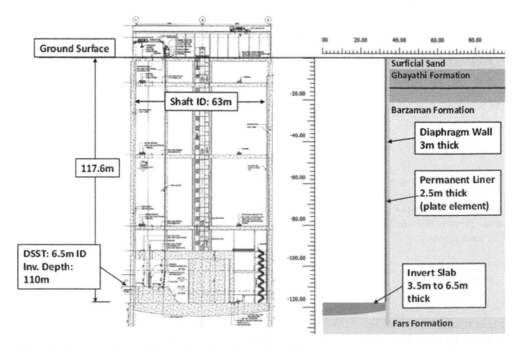

Figure 6.   Elevation View of Proposed Jebel Ali Shaft and Corresponding Plaxis 2D analysis.

sequential excavation within the shaft is performed along with dewatering of the shaft followed by installation of the final liner using a top-down construction method. To temporarily support the permanent liner, shear keys will be constructed at the interface between the final liner and the diaphragm wall. Waterproofing is proposed to be placed between the diaphragm wall and the final liner on the inside face of the diaphragm wall. The following phases were used in the PLAXIS finite element analysis to simulate the construction sequence:

1. Initial Phase: compute the initial ground stresses based on the ground properties and water table depth
2. Install Diaphragm Wall with 5 meter embedment depth below the proposed invert slab
3. Excavate the shaft and dewater by wells placed outside of the shaft
4. Install Invert Slab
5. Cease dewatering: invert slab and diaphragm wall support both earth and water pressure
6. Install Final Liner
7. Allow water pressure to act on the final liner only (this stage of the analysis represents a long term condition)

The final liner force results are presented in Figure 7 and show, despite the high radial force and assuming a 20 cm eccentricity, the pin connection between final liner and invert slab induces large bending moment along the vertical axis of the shaft walls. This is because the invert slab acts as a restraint against the inward elastic movement of the shaft wall. These high stresses for the deep shaft will require special consideration for the connection design and reinforcement at the toe of the liner.

The Jebel Ali Shaft invert slab must resist a water head of approximately 125 meters, or about 1200 kPa. Instead of designing the invert slab as a flexural member, the invert slab can be designed to allow the uplift pressure to arch to the diaphragm wall. This phenomenon was modeled in the axisymmetric finite element model. To ensure no tension could develop in the invert slab when the outside dewatering was ceased, the concrete was modeled as a Mohr-Coulomb material with zero tensile strength. As Figure 8 shows, the principle stresses are compression and arch from the bottom of the invert slab at the center to the top corner, and the minor stresses are showing no tension has developed in the topside of the invert slab. By enabling the arching to develop, the required reinforcement can be significantly reduced.

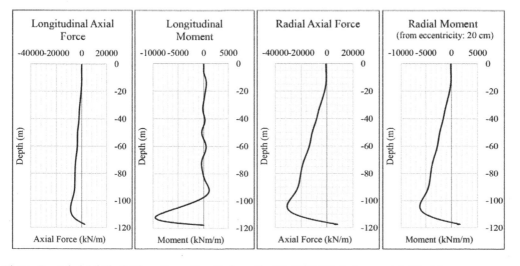

Figure 7.   Jebel Ali Shaft Liner Force Results from the PLAXIS 2D Axisymmetric Model.

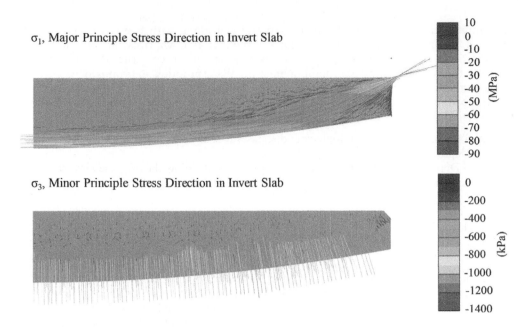

σ₁, Major Principle Stress Direction in Invert Slab

σ₃, Minor Principle Stress Direction in Invert Slab

Figure 8.   Principle stress in Jebel Ali Shaft invert slab (compression is negative).

5.2   *Elastic Buckling and Buoyancy*

For cylinders with thin walls, defined as having a ratio of diameter to wall thickness (D/t ratio) greater than 20, a collapse pressure can be calculated. The Jebel Ali Shaft can be considered as a thin walled cylinder as it has a D/t ratio of 63m/2.5m = 25.2. A conservative estimate of the collapse pressure would be for an infinitely long cylinder with a Poisson's ratio of 0, represented by equation 1 (Sturm, 1936).

$$P_{cr} = \frac{2}{(1 - \nu^2)} E \frac{t^3}{D^3} \qquad (1)$$

At the deepest point, the shaft liner will support about 110 m of water, or about 1.08 MPa. Using equation 1, the critical pressure is 4.75 MPa. The factor of safety for the Jebel Ali Shaft against buckling is calculated to be 4.41.

For buoyancy, it is assumed the water table is at the ground surface, and the uplift resisting forces include the weights of the shaft elements, and the soil friction along the entire depth of the diaphragm wall. The uplift force acts on the bottom of the diaphragm wall and the bottom of the invert slab. Using only shaft weight and soil friction (neglecting adhesion), an adequate factor of safety against uplift was calculated for the Jebel Ali Shaft.

6   TUNNEL DESIGN

The tunnel liner will consist of one-pass, bolted, gasketed pre-cast reinforced concrete segmental liner. During detailed design phase, the analysis of the segmental lining cross-section will consider soil-structure interaction and constructions sequences of the tunnel will be carried out. The tunnel segments must also be designed to resist manufacturing loads, handling, stacking and TBM jacking forces.

For the preliminary tunnel liner design, hand calculations along with finite element analysis using PLAXIS 2D plane strain models were performed. The model consists of four soil layers and a plate and interface element to define the tunnel liner. The construction sequence was

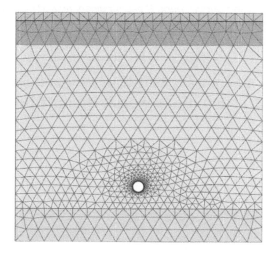

Figure 9.   PLAXIS 2D model for the 6.5 m diameter tunnel.

simulated by allowing for some level of in-situ stress relaxation before the liner was activated in the model. The representative model domain is shown in Figure 9.

7   CONCLUSIONS

The rapidly growing Emirate of Dubai is undertaking the DSST project to transform the existing sewerage system from a pumped system to a fully gravity-based system. The project will replace more than 100 main and subsidiary pump stations with two very deep pump stations constructed at the Al Warsan and Jebel Ali treatment plants. The proposed gravity system will have nearly 250 km of new link sewer, 75 km of deep tunnels (diameter between 3.5 m and 6.5 m), and more than 40 shafts. The design forces on these structures may be quite high, and analysis methodologies must be developed to ensure all temporary and long-term conditions are considered. Some of the most challenging aspects of the project include accounting for the risks associated with the very large terminal pump stations and ensuring the system is operational for more than 100 years.

REFERENCES

Sturm, R.G. 1936. *A Study of the Collapsing Pressure of Thin-Walled Cylinders.* University of Illinois: Urbana, IL

Tunnels and Underground Cities: Engineering and Innovation meet Archaeology, Architecture and Art, Volume 7: Long and deep tunnels – Peila, Viggiani & Celestino (Eds)
© 2020 Taylor & Francis Group, London, ISBN 978-0-367-46872-9

The hugest and more complex belt conveyor system in the longest tunnel under construction in the world: Brenner Base Tunnel

M. Bringiotti
GeoTunnel S.r.l., Genoa, Italy

S. Portner & V. Grasso
Marti Technik AG, Moosseedorf, Switzerland

E.R. Vitale
Ghella S.p.A., Roma, Italy

ABSTRACT: Marti has been working on the Italian side of the Brenner Base Tunnel project since the beginning, starting not only with the bridge which overtakes the Highway, the Isarco river, a National Road, and the Railway line, but also with the supply of the long belt for the Aica-Mules pilot bore. These structures have been designed for those preliminary tunnels as well as for an extensive use in the ongoing project, which is related to the excavation of the main Brenner lines on the Italian side. The operative job sites which the system involves are Mules 1, Genauen 2, Unterplattern and Hinterrigger. These are linked to the already existing plants The system is able to handle contemporarily excavated material produced by n. 3 TBMs and various D&B tunnels (ca. 2.100 t/h capacity) and also brings crushed material to the batching plant installed in the logistic knot cavern, in an Industry 4.0 full integrated optic.

1 INTRODUCTION

Since the beginning of 2017 the BTC consortium, formed by Companies Astaldi, Ghella, PAC, Cogeis and Oberosler, has been involved in the construction of the longest underground railway link in the world: the Brenner Base Tunnel, which forms the central part of the Munich-Verona railway corridor.

The whole project consists of a straight railway tunnel, which reaches a length of about 55 km and connects Fortezza (Italy) to Innsbruck (Austria); next to Innsbruck, the tunnel will interconnect with the existing railway bypass and will therefore reach a total extension of about 64 km.

The tunnel configuration includes two main single-track tubes, which run parallel with a 70m span between each other through most of the track, and linked every 333 m by cross passages (Figure 1).

Between the two main tunnels and driven 12 meters below, an exploratory tunnel will be excavated first. Its main purpose during the construction phase is to provide detailed information about the rock mass Furthermore, its location allows important logistic support during the construction of the main tunnel, for transportation of excavated material as well as that of construction material. During the operations, it will be essentially used for the drainage of the main tunnel.

The excavation process is divided into 2 blocks; the first one will be bored by n. 3 TBMs (n. 2 for the main tunnels and n. 1 for the exploratory tunnel, toward North). The second one will be bored with traditional method, including mainly drill & blast in the competent material and special drilling techniques in the faulty zones, toward South and in some areas toward North.

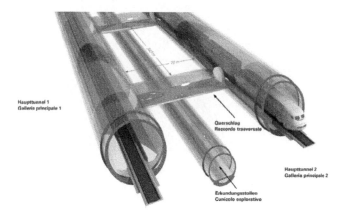

Figure 1.   Brenner Basis Tunnel scheme.

The Company CIPA S.p.A. is currently handling most of the tunnel excavations activities in the lot named "Mules 2–3" with the French drilling partner Robodrill SA These mainly consist of:

- Excavation and lining of the ADIT Tunnel at the Trens Emergency Stop and the Central Tunnel, with a total length of approx. 4,500 m;
- Excavation and lining of the Exploratory Tunnel by traditional method, with a total length of approx. 830 m;
- Excavation and lining of the Main Tunnel toward South, East tube and West tube in single track section, with a total length of approx. 7,320 m;
- Excavation and lining of the Main Tunnel toward South East tube and West tube in double track section, with a total length of approx. 2,590 m;
- Excavation and lining of 19 Connecting Side Tunnels linking the two main tubes, with a total length of approx. 900 m.

All the material excavated in traditional method by the subcontractors Cipa, Europea92 and LSI, runs from the central cavern up to the surface, and sized by crusher, by means of MT belt conveyors. The material is also divided in 3 different geological classes and moved according to needs in 2 different depony areas.

2   GEOLOGICAL CONDITIONS

As anticipated, the Brenner Base Tunnel is the high-speed rail link between Italy and Austria and therefore establishes a connection with North East Europe. It consists in a system of tunnels, which include two single-track tunnels, a service/exploratory tunnel that runs 12 m below them and mostly parallel to the two main tunnels, bypasses between the two main tunnels placed every 333 m, and 3 emergency stop stations located roughly 20 km apart from each other. The bypasses and the emergency stops are the heart of the safety system for the operational phase of this line.

Average Overburden is, between 900 and 1.000 m, with the highest one about 1.800 m at the border between Italy and Austria.

The excavation will be driven through various geological formations forming the eastern Alpine Area. Most of these are metamorphic rocks, consisting of Phyllites (22%), Schist (Carbonate Schist and Phyllite Schist, 41%) and Gneiss of various origin (14%). In addition, there are important amounts of plutonic rock (Brixen Granite and Tonalite, 14%) and rocks with various degrees of metamorphism, such as marble (9%).

Among the tectonic structures in Italy, we find the above mentioned Periadriatic Fault. As mentioned, the unknown characteristics of the rock masses along this stretch determined the

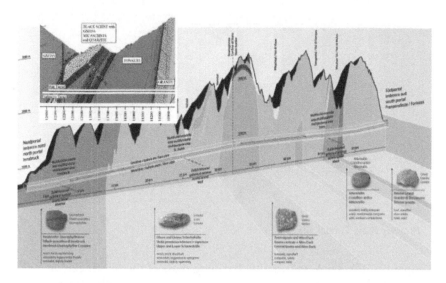

Figure 2. Geological distribution.

necessity of excavating the exploratory tunnel long before the main tunnels, in order to allow the identification of lithological sequences of the various types of rock mass within this heavily tectonized area, as well as a detailed study of their responses to excavation. These analysis were subsequently used to adjust the consolidation and support measures both for the exploratory and for the main tunnels. The excavation of the exploratory tunnel inside the Periadriatic Faults allowed to determine the actual sequence of lithologies within this area.

A geological description is necessary in order to understand the chosen excavation process., This is divided in mechanized tunnelling for the good and competent rock types along the exploratory and the main tunnels toward North, D&B for a limited distance toward South (2–3 km) and in some minor short tunnels as well as several bypasses, and traditional method in the faulty areas. It is also important to understand difficulties related to the excavation in the hard granite as well the related problems linked to the abnormal wear and tear due to high abrasivity.

A summary of the encountered rock sequences is illustrated with Figure 2.

3 THE IDEA

The conveyor belt system allows a straightforward and efficient mean of transportation for material going both in and out from the underground rock crusher, and in and out from the underground concrete batching plants. It evolves while the project is carrying out, and is gradually implemented in accordance to the work's progress.

The conveyor system has been rationalized during the tender phase maintaining its high potential, flexibility and capacity. The differentiation of the two belts in Aica (belt 1 and belt 2) allow each single band to be allocated with a certain type of material; therefore belt 1 is for material A (good material for concrete) and belt 2 is for material B+C (semi-good and not concrete-accepted material). The differentiated configuration avoids alternating different material types with temporary stocking buffer on the same conveyor. However, the system provides the possibility to switch the material types in Aica according to necessities. This kind of realization allows an easier management of the whole system.

Such works may be identified in three major phases. This report is aimed to describe the choices taken within the Construction Project and to list them congruently to the time-related phases.

Without taking into consideration the specific calculations regarding the Executive Work Program's Space-Time Diagram, the 4 major phases may be described as follows:

- Phase 1: excavation of the Exploratory Tunnel (CE) with the conventional excavation method and the TBM assembly chamber; realization of the excavation of the first part of the North Line Tunnels (GLN) by traditional system and excavation of the South Line Tunnels always by traditional method (done by Cipa S.p.A. with its Drilling Partner Robodrill SA);
- Phase 2: continuation of the activities following the previous phase; realization of "in cavern" assembly areas for the two TBMs, which will excavate the Line Tunnels (toward North), excavation prosecution of the South Line Tunnels with drill & blast system (GLS) and realization of the mechanized excavation of the CE Northwards.
- Phase 3: prosecution of the precedent activities, excavation of the North Line by mechanized tunneling (GLN) and subsequent finishing job site works.
- Phase 4: main TBM tunnels secondary final lining. This also represents a complex activity, which will take more than 1 year. A part of this job will be anticipated during TBM excavation, deviating logistic and traffic from one tube to the parallel one.

The phases are integrated and shown in the complex lay out indicated in Figure 3.

All belts generally are connected to the five excavation fronts with the Logistic Knot (N). The system is flexible, but actually in the current phase belts are directly used only for the North bound tunnels (3 TBMs). South bound (conventional excavation) the material is transported by dump truck up to the logistic Knot, where after resizing (by the crushing plant housed underground within the cavern) it is transferred and transported by belt conveyor system.

After that, from Knot N:

- Material type A is carried towards the jobsites Mules and Genauen 2 (depony area reached by a 180m long bridge belt conveyor – 300 ton/h – which crosses the A22 highway, the Isarco river, the National road S.S. 12 and the existing railway line, Figure 4), it is stocked and/or crushed outside and/or transported back to the underground batching plant, in order to provide DB tunnels' primary and secondary lining.

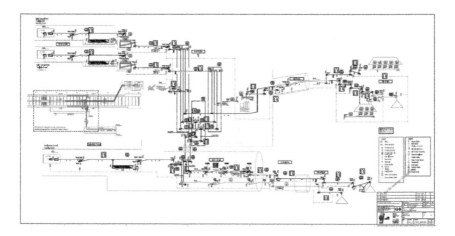

Figure 3.   Belt lay out with cavern zoom (center left).

Figure 4.   Highway A22 transfer belt to Genauen 2 intermediate waste dump.

– Material type A/B+C is transported to Hinterrigger jobsite and is used to produce the precast elements and as pea-gravel for the TBM's back filling. Concrete segments and pea gravel will be then transported back into the tunnel with a dedicated rolling stock through the AICA Tunnel.

4 BELT CONVEYOR SYSTEM DESCRIPTION

Up next is a description of the whole belt conveyor system taking into consideration the three main project's phases thus in order to provide detailed information about each specific installation. In particular it may be detailed by material class handled by each system (specifically type A in case material is high quality, and types B+C when material quality is poorest, material type A may be used to produce "working concrete" and sprayed concrete), its origin and destination, its potential and capacity in ton/h

4.1 *Phase 1*

During Phase 1, the system includes:

– Existing belt no. 1, located at Mules – Aica stretch, with a 500 ton/h capacity.
– Belt no. 2, which has been be implemented on the same stretch and juxtaposed to belt no. 1, thus resulting in a total capacity of 2.100 ton/h, having a capacity of 1600 ton/h.

Figure 5 shows the existing tunnel section and the related clearances of the two belts, as well as the rail system with the booster section where the two runaways become one.

The new conveyor belt system planned for the Aica Tunnel has been designed in order to allow transportation of material toward Hinterrigger in accordance the tunnel's geometries. These geometries grant a tunnel alignment with a reduced curve radius along the initial part of the tunnel. During the tender phase, changes in the tunnel geometries were explicitly forbidden, disqualification being the penalty.

This has been quite an important technical constraint; having the already existing Aica tunnel a very restricted section, contractually forbidden to provide suitable enlargements (niches) in order to host the necessary booster stations, it hasn't been possible to apply standard solutions. MT and BTC designed compact sized boosters in order to fit in the limited spaces available, leaving a sufficient train runaway.

Therefore, the two belts (no. 1 and no. 2) start at the Exploratory Tunnel chamber, arrive at the Aica – Unterplattner jobsite and continue on the outside. Belt no. 2 is designed to transport the incoming material from all excavation fronts (Main Line Tunnels, Exploratory Tunnel, Trens Emergency Stop, FdE Access Tunnel and new Logistic Knot) from the chamber's Exploratory Tunnel toward Aica.

Already existing belt no. 3 connects the Unterplattner jobsite with Hinterrigger, and has a capacity of 500 ton/h. Transition between belt no. 1 and belt no. 3 occurs through an existing transfer chute located in the Unterplattner jobsite.

Belt no. 4 is built in juxtaposition to belt no. 3; its capacity is 1.600 ton/h. Belt no. 4 will be of the same type as belt no. 2 is therefore an extension. The new belt is located on surface

Figure 5. Belts assembly in the Aica-Mules exploratory tunnel with the booster assembly details.

from the Aica portal to the entrance of the Unterplattner tunnel, a part being on an elevated structure and the other sitting on the ground.

The Figure 6 shows the belt conveyor portion standing in front of the Aica-Mules Exploratory Tunnel, together with its general layout out applied in the Unterplattner area.

Figure 7 shows the general layout out applied in the Hinterrigger area with the various distribution points.

During the excavation of the remaining tunnel stretches excavated with conventional method in the Exploratory Tunnel up until the assembly chamber of the TBM, the transportation of the muck is carried by dump trucks until the TBM dismantling chamber. From the existing chamber, material will be loaded on belt conveyors no. 1 and no. 2. Within said tunnel, later belt no. 7 will be assembled.

Belt no. 8 runs through the whole Mules Adit tunnel toward Mules, as mentioned, with a capacity of 600 ton/h. Just a note on CE material; muck coming from the excavation activities within the Exploratory Tunnel will be transported to Hinterrigger by means of conveyor belt system. The plant is able to separate CE material from A and B+C, but it will be moved only to Hinterrigger

4.2 *Aggregate handling within the Logistic Knot*

All belts coming from the GL tunnel discharge material onto the 3 conveyor belts available at the Logistic Knot; which are:

– One for material A towards Mules
– One for material A towards Hinterrigger
– One for material B+C towards Hinterrigger

The two belts towards Hinterrigger merge in order to transport material into the shaft that connects GL and GE tunnels.

The related material will go down from the upper level, and will be unloaded either on belt no. 1 or on belt no. 2, according to its classification and quantity.

Figure 8 (top right) are a close-up representation of belt no. 10 (the picture may be also referred to belt no. 9), which unload on a lower level through the vertical shaft (on the left).

Belt no. 11 connects belt no. 13 with belt no. 8b. related to the transport of material type A, this belt allows to carry material outwards, thanks to a belt plow switch, supplying the Mules jobsite with good quality material thus in order to mix fresh concrete. Belt capacity is about 1.600 ton/h.

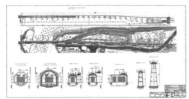

Figure 6.   Unterplattern general lay out.

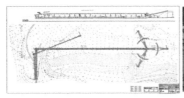

Figure 7.   Hinterrigger general lay out.

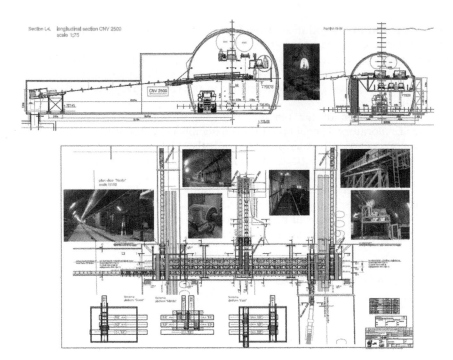

Figure 8.   General logistic knot plan view lay out and details.

Figure 8 shows the general layout out of the logistic knot.

As previously reported, in the caverns, transferring devices are provided for each belt; this is the so-called Switch system. It consists of a belt plow system, transferring material on different belts according to classification. One belt is dedicated to material type A and one to material type B+C. These cross the tunnel connecting the two chambers, reaching the connecting pit area down to the Exploratory Tunnel chamber (km 10+4).

Belt no. 12 connects the East bound Tunnel with the Logistic Knot. In this phase the material can be handled in the Movable Crusher located in the GLE assembling chamber. This material converges in the Logistic Knot and heads to jobsites Hinterrigger or Mules. It comes from the GLE, the GLO and the CE excavations. The belt's capacity is 1.000 ton/h.

Belt no. 13 connects the West bound Tunnel with the Logistic Knot and is implemented during this phase. Its capacity is 1.000 ton/h.

Belt no. 23 allows the handled material (crushed and screened in Mules) to return through the Mules Adit. This material is used to produce concrete with the batching plant available in the Logistic Chamber. This belt's capacity is 350 ton/h. Figure 9 shows the related installation.

Belt no. 24 carries material handled by the Movable Crushing Plant to be transferred on belt no. 12. The material then reaches the Logistic Knot. The belt's capacity is 200 ton/h; during this phase, it receives the excavated material of the GLE, GLO and CE. From there,

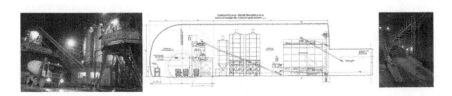

Figure 9.   Batching plant installed in cavern.

material reaches the connecting shaft, goes down to the chamber in the Exploratory Tunnel (km 10+4), and then transported to Hinterrigger (classes A and B + C) or in Mules (class A).

Belt no. 25; as previously mentioned, handles the outgoing material from Mules Adit to the surface. The exiting belt no. 25 reaches the south part of the jobsite area, next to the first transfer chute. Through a specific Switch, from there, the muck can either be discharged on ground, on a temporary stock pile, or it can befurtherly transported by the secondary belts (i.e. branch 1 and branch 0) to a second transfer chute toward Genauen 2. Belt no. 25 has a capacity of 600 ton/h.

The peculiarity of this system is the 2 level belt; in which the upper part is able to move material from the cavern, through Mules Adit to the surface, and the lower part handles crushed and sized aggregates (from the surface close to Mules Adit) to the underground batching plant. It has been specially designed in order to handle the various and different demands, in terms of aggregate size and mucking capacity).

Belt no. 26 handles material to Belt no. 23, allowing material handling back to the underground batching plant. The belt's capacity is about 350 ton/h. It handles the crushed material resized in the crushing plant by means of vibrating dosing transfer chutes.

Figure 10 shows the general plant layout in the Mules site.

Belts no. 27 and no. 28 are in the Hinterrigger jobsite, down from Belt no. 4. They are meant to handle material types B+C; Belt no. 28 is a slewing conveyor, suitably designed for stock piling. The capacity of both is 1.600 ton/h each.

Belt no. 29 handles material from Belt no. 3 and serves, as the mentioned, for type A material stock piling. Belt no. 29 has a maximum capacity of 500 ton/h.

4.3   Phase 2

All of the mentioned belts are installed during Phase 2.

The excavation of the Exploratory Tunnel with conventional method up to the TBM dismantling chamber is conducted during phase 1. Transportation of muck is preferably handled with conveyor belts when the supply time is favorable. In case of delay, the material is handled by means of dump truck. During said phase Belt no. 7 handles the mechanized excavation material to the Logistic Knot, from where it is transported to Hinterrigger as material type A or B+C. Belt no. 7 has a capacity of 500 ton/h.

All the material excavated by D&B is moved to the underground crushing plant, and then transported via belt conveyor on surface (Hinterrigger or Mules).

4.4   Phase 3

In main phase 3 the previously installed belts are continuously extended in accordance to the mechanized excavation advancement rate. together with the belts for the mechanized excavation towards North.

Belt no. 7 is extended consequently to the advancement of the mechanized excavation in the Exploratory Tunnel. The extension belt station is located within the TBM launch chamber (km 1+29). From there, the excavation material is carried to the chamber at km 10+47 with a fixed belt. Muck is transported from the TBM belts in the GLN to 100mc transfer chutes in order to manage potential minor stops without interrupting the excavation activities as well as to level off the flow of material coming from the node.

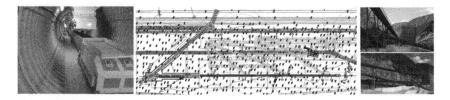

Figure 10.   General plan overview site Mules with external crushing plant and extraction pits.

Belts no. 19 and 20 allow transportation of the excavation material coming from the mechanized excavation of the North-East Tunnels to the Logistic Knot. Material is crushed within the TBM and transported on conveyor belts toward the east TBM assembly chamber (km 49+0) first, and to the Logistic Knot through Belt no. 12 after. An independent system of conveyor belts is planned, and it gradually extendes as the job progresses. Afterwards, from the Logistic Knot it is transferred to Hinterrigger or Genauen 2, according to the material quality and the stock capacity. Both belts have a 1.000 ton/h capability.

Belts no. 20 and 21 allow transportation of excavation material coming from the mechanized excavation of the North-West Tunnels to the Logistic Knot. Material is crushed within the TBM and transported on the conveyor belts toward the East TBM assembly chamber (km 49+0) first, and to the Logistic Knot through belt no. 13 next. An independent system of conveyor belts is planned, and it is gradually extended as the job progresses. Further, from the Logistic Knot it is then transferred to Hinterrigger or Genauen 2, according to the material quality and the stock capacity. Both belts have a 1.000 ton/h capability.

5 THE SYSTEM IN NUMBERS

At the end, the approx. 80 km long system will be installed (integrated with the already existing one, ca. 14 km), with a total installed power of more than 10 MW, which is a really huge number, taking into consideration the complexity of the working sites; see summary table in Table 1.

6 THE SYSTEM'S BRAIN

The various MCCs (Motor Control Centres) are installed in a special container equipped with a suitably dimensioned air conditioning system, which is able to safely handle high working temperatures and potentially dusty atmosphere.

The belt conveyor system in its complexity is equipped with a PLC system, which is able to manage all the functions provided with HDMI design for easy and friendly interface.

Safety devices have been installed in all critical points in order to detect any critical occurrence in advance, and to keep the system working safely in accordance with its quite long working life (at least 4 years).

Table 1. General design data in items [n], length [m] and power [kW].

Areas	Items	N.	Length [m]	Power [kW]
Mules logistic knot	Belts	6	211	206
	Chutes	2		
	Dosing units	2		
	Service platforms	3		
Tunnel North	Belt explorative tunnel	1	16.763	
	N-E belt	2	7.195	
	Main tunnel N-E	1	13.182	
	Main tunnel N-W	1	13.223	
			50.363	5.981
Tunnel South	Belt	1	50	22
Aica - Unterplattner - Hinterrigger	Aica belt	2	10.761	
	Unterplattner - Hinterrigger	1	1.169	
	Hinterrigger	2	499	
	Movable stacker	1	48	
			12.477	2.778
Adit Mules	Belts	3	1.979	673
Site Mules	Belts	6	831	213
TOTAL - new installation			65.911	9.873

Figure 11.   Control rooms and MT's Team in Mules.

Since the handled material is highly wearing, due to the extreme quartz content in the excavated rock (mainly granite), ordinary and extraordinary maintenance is planned according to the various usage coefficients managed by a specifically designed algorithm.

No. 3 control rooms have been foreseen in the critical areas. They are all interconnected between each other. From the job site MT Technicians can control easily the full system (Figure 11).

24/7 Tele-Assistance is granted by Marti Technik with a simple Wi-Fi connection, which link directly to Marti offices in Switzerland.

7   CONCLUSION

Handling such a complex project, is a matter of Team work between highly professionals; in no way can it be considered as simple "procurement and supply" activities. Main target has been the complex design considering the already existing systems within the jobsite. The challenge to ease the work flow of 6 excavation fronts, selecting and conveying muck into classes and needing to feed the underground batching plant with suitable aggregates as well. Partnership between the Main Contractor and the Supplier is the key to success.

ACKNOWLEDGMENT

All the BTC Team must be acknowledged for the hard and professional work done together, starting from feasibility study to the contractual and supply set-up; C. Bernardini, G. Frattini, S. Citarei, M. Ferroni, M. Secondulfo, A. Caffaro, R. Paolini, G. Vozza, A. Cicolani. In MT: R. Damaris, H.J. Meier, P. Rufer and the CEO W. Aebersold!

REFERENCES

Bringiotti, M. 2002. *Frantoi & Vagli: trattato sulla tecnologia delle macchine per la riduzione e classificazione delle rocce*, Febbraio, Edizioni PEI.

Bringiotti, M., Duchateau, JB, Nicastro, D. & Scherwey, P.A. 2009. *Sistemi di smarino via nastro trasportatore - La Marti Technik in Italia e nel progetto del Brennero* Convegno "Le gallerie stradali ed autostradali - Innovazione e tradizione", SIG, Società Italiana Gallerie, Bolzano, Viatec.

Bringiotti, M., Parodi, G.P. & Nicastro D. 2010 *Sistemi di smarino via nastro trasportatore*, Strade & Autostrade, Edicem, Milano, Febbraio.

Fuoco, S., Zurlo R. & Lanconelli M. 2017. *Tunnel deformation limits and interaction with cavity support: The experience inside the exploratory tunnel of the Brenner Base Tunnel*, Proceedings of the World Tunnel Congress – Surface challenges – Underground solutions. Bergen, Norway.

Rehbock, M., Radončić, N., Crapp, R. & Insam, R. 2017. *The Brenner Base Tunnel, Overview and TBM Specifications at the Austrian Side*, Proceedings of the World Tunnel Congress– Surface challenges – Underground solutions. Bergen, Norway.

*Tunnels and Underground Cities: Engineering and Innovation meet Archaeology,
Architecture and Art, Volume 7: Long and deep tunnels – Peila, Viggiani & Celestino (Eds)*
© 2020 Taylor & Francis Group, London, ISBN 978-0-367-46872-9

CERN (HL-LHC): New underground & surface structures at Point 1 & Point 5

A. Canzoneri
Rocksoil S.p.A., Milan, Italy

J. Amiot
Setec, Paris, France

F. Rozemberg
CSD, Lousannes, Switzerland

D. Merlini & F. Gianelli
Pini Swiss Engineers, Lugano, Switzerland

G. Como & F. De Salvo
Lombardi SA, Minusio, Switzerland

C. Helou
Artelia, Paris, France

L.A. Lopez & P. Mattelaer
CERN, Geneve, Switzerland

ABSTRACT: The Large Hadron Collider (LHC) is the most recent and powerful accelerator constructed on the CERN site. The LHC consists of a 27 km circular tunnel, about 100 m underground, with eight sites positioned around the tunnel's circumference. High-Luminosity LHC (HL-LHC) is a new project aiming to upgrade the LHC, at Point 1 (ATLAS in Switzerland) & Point 5 (CMS in France), in order to maintain scientific progress and exploit its full capacity with new underground and surface structures. The paper describes the HL-LHC design developed by the JV ORIGIN in Point 1 represented by Setec (France), Rocksoil (Italy), and CSD Eng. (Switzerland), & JV LAP in Point 5 represented by Lombardi (Switzerland), Pini Swiss (Switzerland), and Artelia (France).The construction works contracts were awarded in March 2018. The execution of the underground works started in April 2018 and is scheduled to be finished by the end of 2021.

1 INTRODUCTION

CERN, the European Organization for Nuclear Research, is an intergovernmental organization with 22 Member States. Its headquarters are in Geneva but its premises are located on both sides of the Swiss-French border. The Large Hadron Collider (LHC) is the flagship of this complex of accelerators. The data collected by this unique instrument in the world has allowed CERN experiments ATLAS and CMS to discover the Higgs boson in 2012. The goal of the HL-LHC project is to upgrade the LHC experiment in order to produce more data, by increasing the number of particles collisions by a factor of 10. It will be operational in 2026.

The project requires new technical infrastructures near each of the two main detectors (ATLAS at Point 1, CMS at Point 5): an additional shaft and cavern, approximately 500 meters

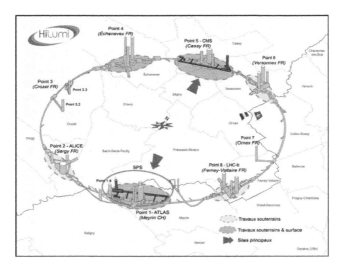

Figure 1. General view of the new works for HL–LHC, in red (https://voisins.cern/en/hl lhc).

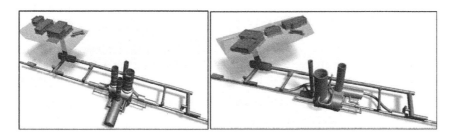

Figure 2. New underground and surface structures (in blue) in Point 1 (left view) and Point 5 (right view).

of tunnels connected to the LHC tunnel, and additional technical buildings at the surface. The design schedule and construction schedule of the civil works are very constrained by the general timeline of the HL-LHC project, and by the high vibration sensitivity of the LHC Machine.
(https://voisins.cern/en/hl-lhc)

During the design phase, it was agreed that most of the underground excavation works would be carried out whilst the LHC machine would not be in operation, i.e. during the "Long Shutdown 2" planned during the years 2019-2020 (enabling works generating vibrations).

The main design and construction phases are:

- 07/2016 – 10/2016: Preliminary Design,
- 11/2016 – 06/2017: Tender Design,
- 07/2017 – 06/2018: Construction Design for underground structures,
- 03/2018: Award of contracts for the works at Point 1 and Point 5 (one different Contractor for each Point),
- 04/2018 – 12/2018: Preparatory works and shafts excavation (deemed possible while the LHC is in operation, with adapted methods if necessary, in case of vibration issues),
- 01/2019 – 08/2020: Excavation of the caverns and tunnels (during the "Long Shutdown 2" of the LHC),
- 09/2021 (Point 1) and 12/2021 (Point 5): contractual milestone for the completion of all underground structures,
- 08/2022: contractual milestone for the completion of the works.

2 SURFACE WORKS AND SITE WORK INSTALLATION

The construction site is located near to the ATLAS site (Point 1) and CMS site (Point 5). Both sites are equipped with the typical installations for underground works like: ventilation, water treatment plant, excavation equipment, site accommodations, workshop, lift installation, etc. The key equipment of the construction site are the gantry crane erected at the top of the shaft and the tunnel excavator. On each Point, several buildings are foreseen on the surface and have been modelled on BIM:

1) **Head Shaft Building**. Steel frame structure with dimensions of 22 m by 35 m housing one cold box for the cryogenic system. The building is about +16.0 m high.
2) **Ventilation Building (SU)**. Reinforced concrete structure with dimensions of nearly 22 m by 30 m. The building is about 13 m high. The SU building will host the necessary cooling and ventilation systems for the HiLumi Underground infrastructures.
3) **Electrical Building (SE)**. The SE building will host the necessary electrical systems for the HiLumi Underground infrastructures. The steel frame superstructure is moment resisting in the transverse direction while the longitudinal stability is ensured with cross bracings.
4) **Cooling Towers (SF)**. Reinforced concrete structure with dimensions of nearly 18 m by 30 m housing 3 cooling towers. The building is about +12.0 m high. The SF building will host the cooling tower structures that are required to extract the heat loads form the machines for the HiLumi Underground infrastructures. The cooling towers exhausts will be built with curved formwork tools.
5) **Compressor Building (SHM)**. Reinforced concrete structure with dimensions of nearly 16 m by 50 m. The building is about +11.0 m high. The SHM building will host the compressors for the cryogenic equipment required for the HiLumi Underground infrastructures.

At the surface the following main site installation works are foreseen:

1) Construction and maintenance of any required temporary roads within the Site Installation areas.
2) Electricity and telecom facilities and connection to the nearest Concessionaire of the Site able to meet the power requirements during the whole period of Works.
3) Connection to the local water supply system that runs adjacent to the Site.
4) Connection to the sewerage system.
5) Treatment of rainwater.
6) Installation and operation of the security system and of the lighting system.
7) Installation and maintenance of the ventilation systems during the underground construction phases.

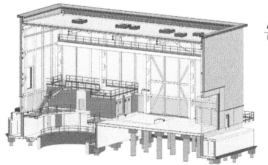
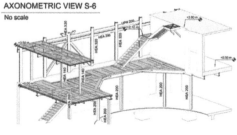

Figure 3.   The Head Shaft Building.

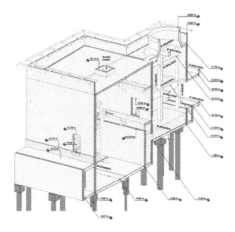

Figure 4. The Cooling towers.

3 UNDERGROUND WORKS

3.1 *Geology*

The Point 1 site lies in an area covered by Würm Quaternary moraine deposits between two outcrops of Molasse. The local geology consists of the Würmienne moraine deposits generally directly overlying Molasse. The Würmienne moraine is a base moraine and consists of compact and consolidated relatively dense sandy silt with variable amounts of sand, clay, 15-30% of 60-100 mm dia. cobbles and hard to very hard gravel. The Molasse (Chattien Inférieur) consists of sub-horizontal bedded lenses of sedimentary rock composed of grain sizes ranging from clay to sand, with a progressive lateral and vertical spatial grading.

At Point 5 the Quaternary moraine is covered by a succession of fluvio-glacial soils that consist of silty and sandy gravel deposits with cobbles and small boulders. It is a heterogeneous, semi-cohesive, compact to very compact sandy (20 35 %), slightly clayey (<10%) gravel layer in abundant (10-15 %) silt paste; light brown, rich in round cobbles and boulders up to few decimetres diameter (max observed diameter ~20 cm, but larger diameters may be expected).

Four main rock types were identified within the Molasse as follows.

1) Sandstone and marly sandstone vary from soft/poorly-cemented (UCS < 20 MPa) to very hard/cemented (UCS >35 MPa). They are composed of 40-70% Quartz, 5-10% Feldspar, 5-20% clay/mica and 5-45% calcareous cement. Stronger rocks are finer grained and more cemented.
2) Sandy marls have these proportions: 20-45% of clay, 20-40% of quartz and 20-30% of calcite.
3) Platy marls (also called marl "fissile"/"feuilletée" in French) can be assimilated to shales. They are composed of 45-60% clay, 15-30% of micro-crystalline quartz, and 20-30% calcareous minerals. The dominant clay is illite (swelling), but smectite (swelling) and chlorite (non-swelling) can also be present at up to 18%.
4) "Grumeleuse" Marl is characterised by numerous closed, polished, discontinuous multidirectional, curved micro-fissures. It can sometimes be referred to as mudstone.

The Molasse is composed of approximately 50% sandstone, 25% of platy marls and marls, and 25% sandy marls. Layers are usually 0.5 to 5 m thick and their stiffness can vary from one layer to another. Available data underline that Molasse is a highly heterogeneous rock mass (Kurzweil, 2004). Sandstone beds can be considered isotropic. The marls are lithologically heterogeneous (succession of weak ductile marls, sandy marls, strong marls, etc. which can be thinly laminated with weak lamination planes) with horizontal and inclined polished fracture surfaces. The upper part (usual range of thickness varying from 1 to 5 m) of the Molasse are usually weathered and softer, with soil-like properties.

At Point 5 two aquifers are individuated within this geological succession:

- The upper aquifer, phreatic, hosted in the fluvio-glacial soils;
- The lower aquifer, artesian, generally hosted in the deep moraine "cailloutis".

The two aquifers do not communicate, being separated by the less pervious layers, though tracking tests carried out in the Point 5 site showed local connections, natural or maybe due to anthropic activities. The geotechnical profile at the location of the shaft (PM17) at Point 1 and (PM57) at Point 5 is presented in the following Figure.

3.2 General description of the underground works

At Point 1 and at Point 5, the new Underground Structures are placed on the inner side of the existing LHC ring (Laigle & Boymond, 2001; Kurzweil, 2004), at an average distance of approx. 50.00 m from the LHC axis. The new HL-LHC Underground Structures will be located approx. 6.00/7.00 m above the level of the existing LHC tunnel crown. The new Underground Structures consist of the following main objects, as shown in the following figure.

Point 1 & Point 5 consist of a new shaft (PM17 & PM57), a service cavern (US17/UW17 & US57/UW57), a power converter gallery (UR15 & UR55), service galleries (UA17, UL17, UA13, UL13 & UA57, UL57, UA53, UL53), and vertical linkage cores to the existing LHC and escape exits (UPR13/UPR17 & UPR53/UPR57).

The shaft connects the surface buildings with the underground facilities (cavern and tunnels). It is approx. 60 m deep with a constant internal diameter of the final lining of 9.80 m. The thickness of the final lining varies from 0.50 m (typical cross section) to 0.80 m (junction with cavern).

The cavern is located at the bottom of the shaft. It is approx. 50 m long, 15 m wide and 11.2 m high (internal dimension). It will house the cryogenic equipment and services.

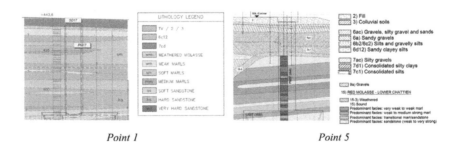

Point 1   Point 5

Figure 5.   Geotechnical profile for the shaft at Point 1 & Point 5.

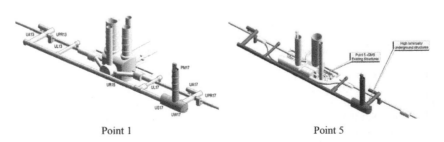

Point 1   Point 5

Figure 6.   General layout underground structures Point 1 and Point 5.

The main tunnel links the cavern with the service tunnels. This tunnel is approximately 300 m long, starting from the cavern. It will mainly house the power converters and the current feed boxes of the superconducting magnets. From this main tunnel, four transversal tunnels begin; these tunnels called "service galleries" are approximately 50 m long and connect the new main tunnel and cavern with the existing LHC tunnel. The UA galleries will house the RF equipment of the crab-cavities and the UL galleries will house the cryogenic distribution system and the superconducting links. The connection with the existing LHC tunnel will be done through two interconnections, called UPR. These connections aim to provide safety escapes for the personnel.

Vertical cores, of approximately 1.0 m in diameter each, are required to let services pass from each of the 4 service tunnels to the existing LHC tunnel. These linkages will allow connecting the existing installed services to the new systems (superconducting magnets, cryogenic distribution lines and RF cavities) that will be located in the HL-LHC. These connections will be executed in a separated working phase after the completion of the civil works.

For the Temporary Support, a design working life of 10 years shall be considered. For the Final Lining and the waterproofing system a design working life of 100 years shall be considered. For the Internal Structures (steel floors, crane railways, internal concrete walls and slabs,...) a design working life of 50 years shall be considered.

3.3 Description of the execution phases

The shafts at both Points have been designed with the following execution steps.

1. Rock excavation (soft ground for the first 25 m at Point 5) performed top down, by steps of approximately 2 m (around 30 steps)
2. Muck removal by wheel loaders and buckets
3. Primary support using reinforced shotcrete and, if needed, rock bolts
4. Laying of waterproofing system
5. Final reinforced concrete lining
6. Construction of the internal structures

The caverns have been designed with the following execution steps.

1. Excavation of the cavern for different portions using rock bolts, shotcrete (and steel ribs for Point 5) for temporary support
2. Muck removal by wheel loaders and buckets
3. Laying of umbrella waterproofing
4. Tunnel invert
5. Drainage network
6. Final reinforced concrete lining
7. Construction of the internal structures

The Tunnels have been designed with the following execution steps.

1. Excavation of the tunnel full face and installation of steel ribs (if needed), rock bolts (Point 5) and shotcrete for temporary support
2. Muck removal by wheel loaders and buckets
3. Laying of umbrella waterproofing
4. Tunnel invert
5. Drainage network
6. Final reinforced (if needed) concrete lining
7. Construction of the internal structures

3.4 Vibrations

The main challenges of the project are a) related to the limitation of the vibrations induced from excavation, which may degrade the operation of the LHC machine and its experiment detectors, and b) the swelling of the rock mass (Glaus & Ingensand 2002). In order to deal with these major issues, several excavation methods, a detailed monitoring system with the adoption of restrictive threshold values, and specific technical solutions have been foreseen in the design.

The excavation of the shaft in rock takes place during the operation of the LHC Machine, before the Long Shutdown Period 2 (LS2), whereas the excavation of the cavern and the galleries is foreseen during the LS2.

In order to manage the vibrations issues, the Engineer has foreseen several excavation methods that may produce different effects on the LHC Machine:

- Method A: Mechanically assisted tunnelling in rock with electrical roadheader;
- Method B: Mechanically assisted tunnelling in rock with rock breaker
- Method C: Excavation with Hydraulic rock splitter inside previously drilled holes
- Method D: bucket excavator.

The vibrations are amplified through two separate channels:

- Amplification through the geological layers from the excavation face to the LHC tunnel;
- Amplification through the support structure of the LHC machine on the LHC tunnel.

The ground displacements induced by the excavation works are expected to be amplified up to 10 times at the LHC beam. In principle, the threshold value is represented by the deformation of the LHC beam, fixed at approx. 1 μm.

During the excavation of the shaft, the vibrations are monitored by the Employer by means of the following instrumentation already installed in the LHC existing tunnels:

- Triaxial orthogonal seismometer Gulap 6T;
- Force balance accelerometer EpiSensor ES-T.

The Employer may stop the excavation works at any time and require a change of excavation method if the project requirements are not complied with. In this case, the Contractor shall quickly dispose of the ongoing excavation installation and set up the new one. In the event that all methods generate vibrations above the acceptable level for the LHC machine, the Contractor shall stop the works until the LHC shutdown. The consequence, in terms of time and money, of the change of excavation method and interruption are ruled, respectively, by the Baseline Schedule and the Bill of Quantities.

4   TENDER AND CONTRACT FOR CIVIL WORKS

The tender and the contract awarding phases have been developed between July 2017, with the dispatch of the bid documentation to the bidders, and April 2018, before the works commencement. The main steps are summarised below:
Phase A: preparatory activities

1. Market Survey: 01/12/2016 - 28/02/2017
   Prior to the Tender Phase, CERN launched a Market Survey in order to assess the interest of pre-selected construction companies in participating in the Invitation to Tender for the civil engineering works
2. Completion of the Tender Design by the Engineer: 03/11/2016 - 31/05/2017

Phase B: bid preparation by the bidders

1. Dispatch of the Invitation to Tender (IT) documents by CERN Procurement: 12/07/2017
2. Bid preparation by the bidders: 14/07/2017 - 27/10/2017

Phase C: bid evaluation
   The bid evaluation process, carried out by CERN and the Engineer, was organised in two successive stages: the first one aimed to evaluate the technical compliance of the bids, through questions and answers with the bidders in order to detect possible non-compliances. Only the bids judged as "technically compliant" were admitted to the next evaluation stage, which concerned the financial evaluation.

1. Technical Bid Evaluation: 31/10/2017 - 05/12/2017
2. Financial Bid Evaluation: 06/12/2017 - 10/01/2018

Phase D: bid discussion: 09/01/2018 - 10/01/2018

The bid discussion took place at the beginning of 2018 with the two companies that, after having passed the technical compliance, offered the best price.

Phase E: contract preparation: 10/01/2018 - 14/02/2018

The contract preparation phase aimed to refine the IT documentation following the results coming from the bid discussion phase. Both construction contracts, for Point 1 and Point 5, have been signed by the parties in March 2018.

The contracts were elaborated by CERN and the Engineer on the basis of the FIDIC Red Book (1999) with the addition of specific contract appendices for the underground works – such as the Baseline Schedule and the Geological Baseline Report – which allow the parties to deal with the geological risks during construction (Watson & Osborne, 2004). In particular, a specific contract sub-clause was prepared in order to define the adjustment of time for completion in relation to the progress of underground excavation.

Time for Completion (Sections) shall be adjusted according to the difference between the encountered and the expected subsurface conditions. The adjustment may reduce or extend the Time for Completion. The expected subsurface conditions are described in the Geotechnical Baseline Report (GBR) in terms of definition of support class and geotechnical baseline condition. The quantities forecast by the Engineer and the performance rates proposed by the contractor arc part of the Baseline Schedule (based on the principle defined in the SIA Norm 118/198). This tool allows the time for completion to be managed and is periodically updated with the quantities remeasured during excavation.

The contract is a remeasurement contract. To this purpose, three different Bills of Quantities (Underground Works, Surface Works and Common Items) have been prepared, based on the Civil Engineering Standard Method of Measurement, Fourth Edition (CESMM4) published by The Institution of Civil Engineers, 2012 and supplemented by all the specific clauses related to this project.

5 CURRENT STATUS OF THE WORK AND CONCLUSIONS

Current status of the works at Point 1 (December 2018)

- Start of the works: April 2018.
- Start of the excavation of the underground works: August 2018.
- Shaft excavation depth: around 50 m.
- Site Works Offices, Ventilation hall, Shaft hall and Workshop hall: completed.

Figure 7.   Point 1 site works in August 2018 – Surface works.

Figure 8.   Point 1 Shaft excavation.

Figure 9.   Point 5 site works at September 2018 – Surface works.

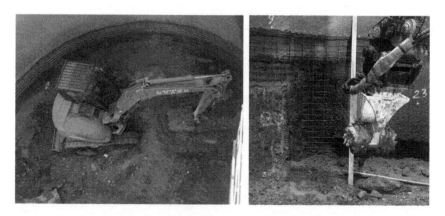

Figure 10.   Point 5 – Shaft excavation (1).

Current status of the works at Point 5 (December 2018)

- Start of the works: April 2018.
- Start of the excavation of the underground works: August 2018.
- Shaft excavation: the section in loose material (fluvio-glacial soils and moraines) is completed and the excavation is currently being carried out in the molasse. The depth is around 60 m and about 10 m still have to be excavated. Up to now, the excavation complies with

Figure 11   Point 5 – Shaft excavation (?).

the advance rate forecast in the design phase and no particular problems related to the monitored vibrations were encountered.
• Site Works Offices and Workshop hall: completed
• Ventilation installation and Shaft hall: completed

The design, from the preliminary design to the construction design, has been successfully provided by two different Engineer for the two different Points, following an ambitious time schedule.

The design for two similar projects for the same Employer was verified from two independent consultant JVs; the technical solutions are similar, but with some specificities. The elaboration of the tender and the contractual documents was coordinated in detail order to have a common basis. At each of the design phases a close cooperation with the Employer and the two consultants represented the key factor aimed to achieve the common goals.

The two Contractors are working according to the Construction Programme and the contractual basis. All technical problems such as vibrations and geotechnical aspects are currently fully under control.

ACKNOWLEDGEMENTS

The authors thank the CERN (Geneve, Switzerland) for their permission to publish the data contained in the present paper.

REFERENCES

Glaus, R. & Ingensand, H. 2002. Tunnel Surveys for New CERN Particle Accelerators. In: *XXII International Congress* Washington, D.C. USA.
International Federation of Consulting Engineers. 2018. FIDIC Contract. Lausanne.
Laigle, F., Boymond, B. 2001. CERN LHC Project - Design and excavation of large span caverns at Point 1 ISRM Regional Symposium, EUROCK 2001, Espoo, Finland.
Kurzweil, H. 2004. The LHC-project of CERN – large caverns in soft rock. A challenge for scientists and engineers. In: *WTC 2004, ITA, Singapore, May 2004.*
Rammer H. 2001. LEP tunnel movements at Point 1 caused by LHC civil engineering. In: *4th ST Workshop Chamonix, France, 30 January - 2 February 2001*
Watson, T., Osborne, J. 2004. Conditions of Contract for Works of Civil Engineering Construction. In: *WTC 2004, ITA, Singapore, May 2004*

Tunnels and Underground Cities: Engineering and Innovation meet Archaeology,
Architecture and Art, Volume 7: Long and deep tunnels – Peila, Viggiani & Celestino (Eds)
© 2020 Taylor & Francis Group, London, ISBN 978-0-367-46872-9

Specificities of the underground structure design of Cigéo. Presentation of the constrains and construction phasing stages of "Phase 1"

G. Champagne De Labriolle & H. Ouffroukh
ARCADIS ESG, Le Plessis Robinson, France

E. Boidy
Tractebel Engineering, Gennevilliers, France

H. Miller
Antea Group, Antony, France

ABSTRACT: The construction of the Cigéo project, future deep geological disposal facility, represents because of its scale, 260 km of underground galleries, a unique case in Civil Engineering. Unlike a classical linear tunnel project, Cigéo is a meshed and complex underground architecture. In this network of underground galleries, many simultaneous excavating faces will have to be deployed while taking into account all the impacts of the different construction sites. Actually, this meshed architecture implies a large number of intersections, requiring the development of a global understanding of the problematics to ensure that the chosen methods are compatible with the flux necessary for each construction site. Thus, the building of the Cigéo project will require all the construction methods of rock tunneling: open shield TBM, deep shafts, conventional drill and blast or hydraulic hammer, and micro-tunneling. This article will focus on the singularity of CIGEO project and the design process based on for the preliminary design stage, to allow optimization and to secure the planning of the project.

1 INTRODUCTION

This article presents the design scheme of Convergences (CVG), the engineering joint venture in charge of the detailed design of the project, to assess the construction methods chosen for the construction of the first phase of Cigéo, named "Phase 1".

CVG is the general contractor of sub-system n°4, focusing on the underground civil structures and of the SUL (Surface-Underground Link). This joint-venture is composed by Tractebel Engineering, Arcadis, Antea Group and Cardet&Huet. The 8 sub-systems are driven by the principal contractor Gaiya working for the client Andra.

2 GENERAL DESCRIPTION OF THE UNDERGROUND ACHITECTURE DURING THE DIFFERENT CONSTRUCTION PHASES

The underground network of Cigéo is spreading over a 12 km² area. It is divided into 3 storage sections depending on the properties of the nuclear waste to be stored (MAVL, HA0/3/7 and HA1/2). Appart from these storage sections, 2 Logistic Areas (Operating Logistic Area – OLA & Construction Logistic Area - CLA) are to be constructed and allow for an efficient operation of the whole industrial center.

Due to the size of the underground network, it cannot be constructed in a single phase. The different operating (storage) phases are divided as shown on the Figure 1 below.

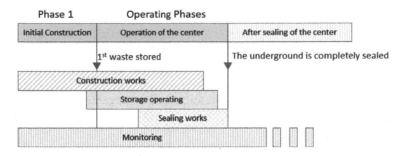

Figure 1. Life cycle of the Cigéo industrial center.

The structures that are required for the storage of the 1st waste (Phase 1) are shown in the Figure 2 above. During the initial construction, 6 access (2 declines and 4 shafts) are available for the expansion underground (boring of the tunnels network). After the beginning of the operating phases, the rest of the underground structures are constructed simultaneously with the storage process. All the works happening during the operating phases are supplied and organized from the CLA underground, linked to the surface by 3 shafts.

The main design rule for the expansion of the network is the full separation of the construction zone and the operating zone.

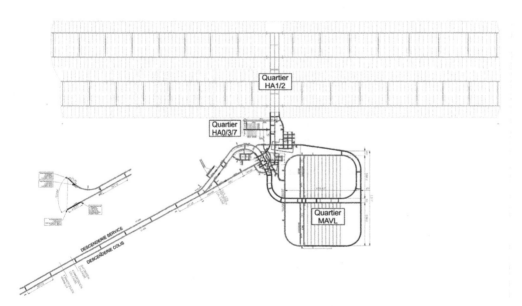

Figure 2. Final architecture of the center (grey), and Phase 1 architecture (black).

3 GEOTECHNICAL AND HYDROGEOLOGICAL CONTEXT

From the ground to the underground level, the main formations are:

- The Tithonian (or Barrois limestone) is a limestone layer with a thickness between 30 and 40 m on TBM launching area. A karstic network is supplied by two different aquifers more or less connected.

- The Kimmeridgian consisting into the marlstone part with a thickness of approximately 100 m and the limestone part with a thickness of 20 m, both layers being aquiferous.
- The Oxfordian, a 260 m thick limestone layer, bearing a non-productive aquifer with water production associated with the porous strata.
- The Callovo-Oxfordian (COx), an argillite layer with a thickness of 160 m and a soft rock behavior. This stratum has no aquifer and very low permeability. It is the layer where the nuclear waste will be disposed.

The first 3 layers are named "cover gound" and have fair geotechnical characteristics (UCS around 50–60 MPa). An elastic behavior of these rock layers during excavation is expected.

The COx has a very different behavior, with a radial convergence around 1% for the short-term (depending on the orientation of the excavation), and a fast and important spalling of the walls if left unsupported. The long-term behavior is favorable if the strains are not blocked, otherwise the stress developing in the concrete lining will require a concrete thickness over 1.00 m in order to fulfil a 100 or 150-years durability.

4 REQUIREMENTS AND CONSTRAINS FOR THE SITE VENTILATION

The design of the site ventilation depends on several constrains and design concepts, which will be summarized in this chapter.

4.1 Requirements

In the case of Cigéo project, the site ventilation must fulfil the following requirements:

- The large range of excavation methods: TBM boring, Drill and Blast, hydraulic hammer;
- The complexity of the architecture (long lines, meshed network, limited inner diameter...). The space available in the section of the tunnels for the ventilation ducts can become a limit to the underground expansion, especially during the phases where only one access is available (blind boring for the shafts and the declines);
- The number of excavation teams working simultaneously can be important;
- The air flow required for each simultaneous excavation face. As a consequence, the ventilation will be operated by blowing, in order to bring the fresh air as close as possible to the excavation face;
- The risk of air recycling must be reduced;
- The operating zone and the construction area must be independent in term of air flows, leading to a full separation of the two-ventilation system.
- The underground fire scenario must be anticipated to allow for evacuation (this topic will not be discussed in this article).

4.2 Regulatory framework

The ventilation during the construction phase must be in conformity with:

- The French Labour Code and the article R4534-44;
- The AFTES national Recommendation;
- The Recommendation n°494 from the French professional organization (Caisse Nationale d'Assurance Maladie des travailleurs salariés);
- The decree regarding the maximum daily working duration on the worksites;

The ventilation design was also inspired from different European recommendations, such as the Swiss « SIA 196 – Ventilation of the underground works » whose principles are similar and complementary to the AFTES recommendation.

The main objective of the ventilation design is to provide a sufficient air flow in the underground galleries. The ventilation system must allow:

- To collect the dust produced by the excavation activities;
- To dilute the gas produced by the thermic engines;
- To ensure the temperature underground is controlled in accordance with the labor regulation, by compensating the heat exhaust of the thermic engines and electric installations.

5 DESCRIPTION OF THE CONSTRUCTION METHODS

5.1 *Declines and TBM bored tunnels*

The range of existing TBM technologies have been assessed in order to come up with a solution that is flexible enough to fulfil all the constrains of the project. The main selective criterions are listed below:

- The Oxfordian low permeability (around 10^{-8} to 10^{-9} m/s) allows for the water table to reach a pressure around 37 bars, but low water inflow that can be handled easily without a confinement of the face → a TBM with face support pressure is not specifically required in this layer.
- The COx has a short-term behavior with fast spalling of the tunnel walls that can be dangerous for the safety if the excavation stays unsupported, even for a short period of time → a hard rock Gripper TBM leaving a long unsupported length is not appropriate. Moreover, the grippers may not be able to lean against the ground as it is a soft rock. In this specific context, a TBM bearing on a segmental ring lining is required. The shield must be as short as possible, et specific technical solutions must be provided to allow for a sufficient annular grouting.
- The route of the TBM includes slopes up to 12% and several curves with curvature radius up to 200 m → a hard rock Gripper TBM or a Double-shield TBM will not be able to fulfil these curves for the excavated diameter (10.40m). Moreover, the geometry of the segmental rings and the conveyor are major design considerations, as they must be adapted to the route that includes long lines and sharp curves. Finally, the backup of the TBM will have to be adapted to both the 12% slope and horizontal parts of the route.
- In the COx, the contact between the ground and water must be reduced at its minimum, even during the mucking phase → All the confinement modes using water supply are hence forbidden. They are indeed inefficient to reduce the extension of the damaged zone due to the high in situ stress of the ground (12 to 16 MPa) compared to the limited confinement allowed by the most recent TBM technologies. An efficient dedusting system must be provided.
- The security and the construction planning are the main concerns of the client ANDRA. Regarding the length of tunnel bored with TBM (between 6 and 10 km depending on the machine), a partial face excavation machine is not adequate compared to a cutting wheel machine that allows higher excavation rates and a better security for the operators especially during the maintenance phases.

Out of these considerations, it appears that choice of CVG to excavate with a full face open TBM is relevant. Regarding the diversity of the constrains in this project, it is also the only technology to fulfill all the requirements.

The combination of TBM and segmental lining chosen is a result of the several feedbacks on large diameter TBM in soft rock, but also the long-term behavior of the COx and the stress induced in the segmental ring. It was hence decided to use the Swiss erection method with 5 main segments and a key-segment always placed on the invert of the tunnel. Approximately 100 km of tunnels were constructed with the specific method, especially in Swiss, Austria but also for a French highway tunnel. The particularity of the method is that the thrust jacks or not required to erect the ring. The key-segment is installed vertically, allowing to have a shorter shield for the TBM. In the cover grounds, i.e. on a 4.2 km distance for each TBM, the concrete segments have a thickness of 40 cm. In the COx, the design of the segments changes:

- On the extrados of the 50 cm thick concrete (C60/75) segments, an additional compressible layer with a 20 cm thickness is added. This material must be able to stick to the concrete and have a behavior similar to compressible elements like HiDCon or TH steel ribs,

meaning that above a given confinement value, a plastic yield is reached and the stress in the material stays limited (around 2 or 3 MPa) when the strain increases. The maximum strain allowed are 50 % of the initial length
- In order to improve the security in this innovative segmental design, the key-segment is also specific as it is also designed to be compressible. The goal is to allow the key-segment to absorb the short-term convergences before being blocked when the concrete invert is casted. The compressible layer can then absorb the long-term convergences.

However, a modification of the TBM will be necessary at the transition between the Oxfordian and the COx.

Regarding all the constrains of the project, other technical choices can be made, but the main requirements for the TBM presented before can ensure the feasibility and fulfill all of the requirements from the client: security of the construction site, construction planning, 150 years durability, and optimization of the excavated diameter.

The excavation rate estimated for the TBM is between 300 and 400 m/month, including the temporary invert allowing for 2 lanes of traffic that are required for the logistic of the works. Both TBM will be launched at the same time and will excavation a total distance of 16 km.

The ventilation concept for the declines and the access galleries of the MAVL section to be constructed by the TBM is the forced ventilation, depicted on the following figure. This ventilation is based on an air duct supplying the fresh air to the excavation face. After the dedusting process, the air exhaust is free in the section of the tunnel until the exit.

5.2 *Deep shafts sinking*

The 5 deep shafts of the project are all large diameter (from 6 to 8.50 effective diameter) and the sinking method is Drill and Blast. Like the declines, they meet the similar geological conditions during their construction. They will also be used to start some excavation works and optimize the planning of the expansion of the 1st phase. It was hence required to achieve an appropriate design in order to secure the construction planning of the 1st past but also to fulfill the operating phase requirements. For more details on the sinking of the shafts refer to the article Constrains, design and construction methods of the Cigéo five deep shafts.

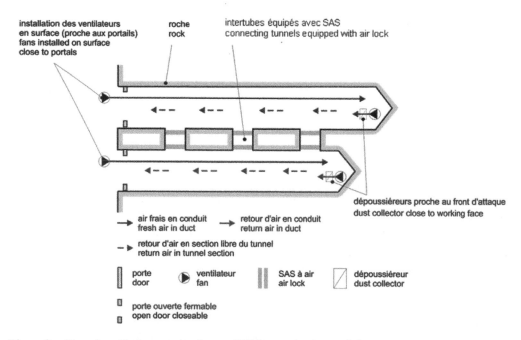

Figure 3.   Forced ventilation concept for two TBM operating in parallel.

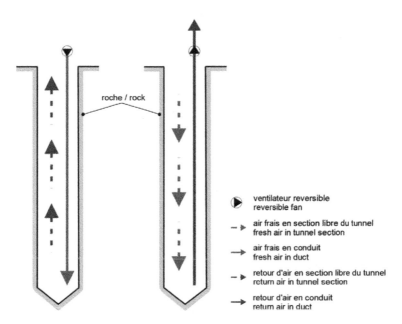

roche / rock

▶ ventilateur reversible
reversible fan

--▶ air frais en section libre du tunnel
fresh air in tunnel section

→ air frais en conduit
fresh air in duct

--▶ retour d'air en section libre du tunnel
return air in tunnel section

→ retour d'air en conduit
return air in duct

Figure 4.   Ventilation concept for shaft sinking.

The ventilation concept associated with the shaft sinking is the reversible ventilation, as shown on Figure 4. The fan will blow air to the excavation face most of the time during all the operations requiring workers at the face. After the blasting, the fan will be reversed, and the smokes created will be exhausted outside of the shaft until the quality of the air is enough to allow for the workers to go back. In this specific case, the air duct must be rigid enough to allow for both ventilations.

5.3   Hydraulic hammer excavation of galleries in the COx

For the tunnels excavated with conventional methods, several constrains were driving the design, and lead to the following choices:

- Water contact with the COx is forbidden → a roadheader machine produces an important quantity of dust and requires a high performance and heavy power dedusting system, especially close to the excavation face.
- The drill and blast excavation is forbidden in order to limit the damages to the rock where the nuclear waste will be stored.
- For durability and normative reasons, the final lining must be circular at the intrados, so are the excavated sections. The excavation height of the galleries is hence around 10 to 12 m → a roadheader machine is not adequate for this kind of geometry and the biggest existing machines can't excavate higher than 7 m so they require a more complex phasing of the excavation with half sections.
- The galleries to be excavated are forming a complex mesh → a roadheader machine is a large size machine with low mobility, and hence not very adequate.

It was decided to use a hydraulic hammer excavator, which is adequate for underground works with its compacity, as it fulfills most of the design constrains. This solution is also more flexible regarding the extension of the works underground in the meshed network. Moreover, this tool has been used for the majority of the tunnels in the Andra Laboratory for similar geological conditions and was mostly adequate.

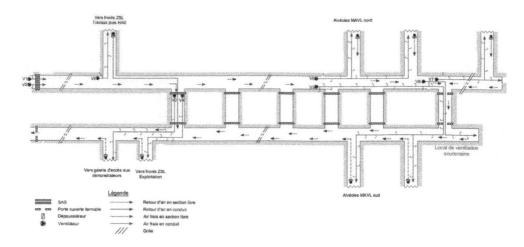

Figure 5.   Ventilation by air circulation concept for the extension of the network.

After confinement of the excavation face with shotcrete and glass fiber bolts, a flexible support system will be installed. It is made of bolts with resin grouting, shotcrete and compressible blocks. The excavation rate is evaluated around 2 ml/day.

In the definitive situation, 3 different cases are encountered:

- The gallery stays in the Construction Area during the whole lifespan of the project: no lining is installed, and the temporary flexible support stays as a final support with a sufficient safety factor including a possibility of maintenance. This allows to reduce the excavated diameter.
- The gallery is in the Operating Area (nuclear conditions): a final concrete lining is casted. This lining has an important thickness (between 1.00 and 1.35 m) and is casted at least 1 year after the excavation of the gallery in order to reduce the stress in the concrete;
- For structural continuity reasons, or in order to ease the final internal structures, a compressible layer can be installed between the final concrete lining and the temporary support. The thickness of the concrete lining can then be reduced (< 70 cm). This solution is studied by CVG because all the lining of the intersections must be compressible (see the following chapter). The lining rate is evaluated around 200 ml/month.

The ventilation during the extension of the underground network from the declines supplies the OLA, the MAVL area and the 4 first storage galleries. The concept used is the ventilation by air circulation. This concept uses one of the declines to supply the fresh air, and the other one as an exhaust. The air circulates in the tunnel section.

Airlocks are provided in the cross passages of the tubes in order to separate the aeraulic conditions of the two declines. The ventilation concept is very adequate to supply important air flow in the case where many simultaneous excavation faces must be supplied.

5.4   Intersections

The meshed network includes a large number of intersections. This chapter will only focus on the intersections between a TBM tunnel and a conventional tunnel in the COx. The construction of these intersections from the TBM gallery is the very complex because of two major issues:

- The diameter of the secondary tunnel is most of the time close to the diameter of the main gallery: a geometrical problem occurs, as the segmental lining must be open in half of its perimeter and hence will not work as a ring.
- The COx long-term behavior is bringing important stress in the intersection lining after 100 or 150 years: even for a secondary tunnel with a diameter lower than the main tunnel, the stress in the remaining segmental rings are too high.

These two issues require for each intersection to remove all the segments over a 24 m distance around the intersection. This represents a total of 1.2 km of segmental rings to be dismantled, including some of them in a slope or in a curve. In order to achieve this, a temporary support similar to the one presented earlier will be installed after the re-excavating a layer of 1.00 m on the whole perimeter in order to allow for a thicker lining. The lining is casted at least 1 year after the excavation of the intersection. Its thickness is important due to the geometric complexity of the intersections.

These specific works on the segmental rings for the intersections require to remove all the equipment in the section (conveyor, pipes, air duct, temporary invert...). Therefore, the intersections can only be constructed when the TBM excavation is finished and all of the equipment have been removed, or when the TBM is not operating (planned break to build the intersection, then restart of the TBM after the equipment are re-installed). These operations are constraining the construction planning.

For this reason, CVG is studying a mechanized solution to remove the segmental rings, according to the feedbacks on similar projects (for example the project MTR Corporation West Island Line 703 Contract in Hong Kong), but also from the understanding of the major lines of the extension of the works. The main goal is to gather existing technologies on a unique machine in order to secure the feasibility of the concept. This solution could allow to industrialize the dismantling operations, and hence increase the safety of the workers and reduce the cost and duration of these operations.

5.5 Microtunnelling

For the high activity waste (HA), the alveoli will be excavated with a method that has been studied and tested in Andra Laboratory. These alveoli are hence co-designed by CVG and Andra, so the construction method is already set. The methods have been studied previously and compared regarding their ability to fulfill the requirements of the project.

The specific solution developed is a mix between the principle of horizontal auger drilling and the guidance system of microtunnelling machines, and is currently being tested in the Laboratory, in similar conditions as Cigéo project.

6 MAJOR ISSUES FOR CONSTRUCTION DEPLOYMENT

The main deployment steps are described below:

- For two years, two TBMs and four shafts are launched simultaneously, in order to connect the underground and the surface and to increase logistic capacity. What is built during this phase provides the main logistics artery for the rest of the construction site: machines, materials, workers, logistic network, ventilation, cooling, etc.

- After the end of the TBM excavation, many intersections have to be built, in order to launch ventilation by air circulation, which is the only solution able to provide enough flow for up to 8 simultaneous excavation faces in conventional methods in each part of the architecture (MAVL, OLA, CLA, HA0). Here, the machines have to combine two opposite constrains: excavate more than 100 m² sections and turn into the intersections. CVG has been in touch with several machine manufacturers to study adaptations.
- Furthermore, the shafts are also used to deploy the construction of the OLA and CLA galleries with smaller machines. The gallery network created thanks to the shafts must be connected quickly to the TBMs' network, to increase the logistic capacity and bring bigger machines to each excavation face.
- Tunnels built with conventional methods are mainly dedicated to long term operation (> 100 years) with quite thick lining (> 1 m concrete) casted at least 1 year after excavation. The final lining of the TBM galleries on the other hand is achieved by the two-layer segmental lining. Therefore, the OLA is on the critical path. CVG studied some optimizations in order

to generalize the two-layer lining to conventional galleries to reduce the concrete thickness and eventually reduce the delay between excavation and concrete cast.

Thus, major issues for construction deployment are focusing on the following questions:

- What is the critical path of the construction planning?
- Which SUL can ensure the transit of the truck flow of each excavation face and lining operations?
- Which tasks are blocking the access?
- Can the required air flow be provided to each excavation face?
- What is the best sequence to build the many intersections with TBM galleries?
- In which galleries the machines can pass simultaneously or make a U-turn?
- Which zone must be delivered in priority for the final equipment/systems installation?

Regarding all the requirements and the complexity of the network, a specific tool has been created to guarantee the respect of the design hypothesis:

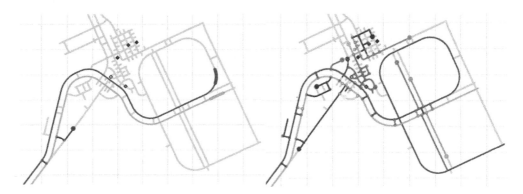

Figure 6.   Construction deployment at two different times: after the end of the first TBM (left), after the end of the second TBM (right).

7   CONCLUSION

The underground Civil works of the « phase 1 » are quite ambitious, because this project is unique and the geomechanic behavior of the COx is a challenge. Moreover, the complexity is also due to the diverse construction methods including some innovative ones. The important number of workers who will operate simultaneously, 24-7, for 8 years is also a challenge.

Thus, in order to reach the wanted advancement rates, every choice which is linked to the whole logistic becomes a unique study because of the complex meshed network to be built.

REFERENCES

Champagne de Labriolle G., Ouffroukh H., Godart S., Arnaud S., *Spécificités de la conception des ouvrages souterrains du projet Cigéo. Présentation des contraintes et du phasage des travaux de la tranche 1.* AFTES Congress, Paris, 2017.

Hagenah B., Cité O., Ouffroukh H., Champagne de Labriolle G., Aparicio Y. *Choix des concepts de ventilation adaptés au creusement d'un réseau maillé souterrain. Application au projet CIGEO.* AFTES Congress, Paris, 2017.

AFTES. 2005. *Recommandation sur la ventilation des ouvrages souterrains en cours de construction.* Tunnels et Ouvrages Souterrains, Arrêté du 14 novembre 1989 relatif à la durée maximale du travail journalier dans les chantiers chauds (CC-1-A, art. 11 § 1)

Société suisse des ingénieurs et des architectes (SIA), *Ventilation des chantiers souterrains,* Zurich, 1998

Drost U., Bettelini, M., Tunnel construction site ventilation and cooling: an integrated flow and heat load solver applied to the lyon-turin high-speed railroad tunnel project, TMI Conference, Turin, 2006

Tunnels and Underground Cities: Engineering and Innovation meet Archaeology,
Architecture and Art, Volume 7: Long and deep tunnels – Peila, Viggiani & Celestino (Eds)
© 2020 Taylor & Francis Group, London, ISBN 978-0-367-46872-9

Engineering challenges of the Snowy 2.0 pumped storage project

B. Chapman, M. Yee & A.R.A. Gomes
SMEC Australia Pty. Ltd., Sydney, Australia

ABSTRACT: The Snowy 2.0 project involves the delivery of a 2000 megawatt pumped storage scheme in south-eastern Australia. The project aims to provide increased storage capacity and security for the national electricity network. The proposed scheme will augment the existing 4100 megawatt Snowy Mountains Hydroelectric Scheme, which is the largest in Australia. The Snowy 2.0 project plans to link the existing Tantangara and Talbingo reservoirs via 28.5 km of waterway tunnels, 2 km of shafts and 16 km of access tunnels. The project will harness the 700 m head difference between the reservoirs within an underground power station housing six 340 megawatt pump-turbines. The project faces several engineering challenges including complex alpine geological and hydrogeological conditions, a remote and environmentally sensitive setting and stringent operational requirements. This paper provides an overview of the project and discusses the main engineering challenges faced during the development of the feasibility study and reference design.

1   INTRODUCTION

The Snowy 2.0 Pumped Storage Project is located 80 km south-west of Canberra, within the Kosciuszko National Park. The project will connect the existing Tantangara and Talbingo Reservoirs through a power waterway and an underground power station with an installed capacity of 2000 MW. There will be no augmentation to the dams or reservoirs. The reservoirs are over 25 km apart and have a gross head difference of approximately 700 m. The major benefit of the proposed scheme compared to other emerging energy storage technologies is the operating duration. At full capacity the scheme could operate for almost 7 days without requiring recharge pumping, for approximately 324 GWh energy production.

2   PROJECT BACKGROUND

With ever increasing energy production by intermittent renewable generation from wind and solar plants and the shift towards a decarbonized energy market, the grid is at risk of both over supply during low demand periods and lack of supply during times of peak demand. The Snowy 2.0 pumped storage project is intended to future-proof the Australian National Electricity Market by generating energy at times of peak demand and storing energy at times of over-supply. By increasing the capacity to dispatch or store energy more flexibly, the national supply network will become more reliable and secure.

2.1   *Snowy Mountains Scheme*

The existing Snowy Mountains Scheme is located in the Alpine region of southern New South Wales. The original scheme was constructed between 1951 and 1974 and consists of nine power stations (including two underground stations) housing 33 turbines with a total generating capacity of 4100 MW, one pumping station, one pump storage facility, sixteen reservoirs and 135 km of interconnected tunnels. The scheme is operated and maintained by Snowy Hydro Limited (SHL), who has recently become wholly owned by the Australian Federal Government.

2.2 Feasibility Study and Reference Design

Augmentation of the Snowy Mountains Scheme to include pumped storage developments was first considered in 1966, with subsequent studies undertaken until the 1990s. One of the schemes proposed was used as the basis for commissioning the preparation of a feasibility study in 2017, which resulted in the feasibility level design of the Snowy 2.0 Pumped Storage Project.

Following the Feasibility Study, the project was further developed into a Reference Design. The purpose of the Reference Design was to provide a basis for contractors to tender, provide cost and program baselines for the contract and enable SHL to arrive at Final Investment Decision (FID). Options proposed during the Feasibility Study were investigated further during the Reference Design stage. Award of contracts is expected to occur in March 2018.

Geotechnical site investigations commenced during the Feasibility Study. However, due to challenges associated with logistics, site access approvals, weather and the length and depth of the alignment, the majority of the data did not become available until during the subsequent Reference Design stage. The Feasibility Study therefore had to be undertaken based on a conceptual ground model developed predominantly from published geological maps, geological field surveys and engineering/geological judgement. Site investigation results have been progressively incorporated to prove and develop the ground model and update the design. This process was ongoing throughout 2018, in conjunction with the planning of additional investigations targeted at outstanding uncertainty and providing inputs for design.

2.3 Detail Design and Construction

Early during the project development, various delivery methods were considered. As an outcome, a modified EPC/Turnkey contract has been adopted for both the civil and mechanical and electrical works. To cater for the inherent variability and uncertainty associated with the subsurface conditions, particular clauses and a risk sharing mechanism were built into the contractual framework to allow for a fair and balanced allocation of geotechnical risks between the Employer and the Contractor in the construction of the project's underground works. The proposed mechanism is based on the recognized principle that risks should be allocated to the party best prepared to manage them. In the case of underground works, this translates to the Employer bearing the risk related to the subsurface conditions and the Contractor bearing the risk of production rates and performance. For this, a comprehensive Geotechnical Baseline Report (GBR) was prepared to set out the anticipated underground conditions under the agreed construction method, which forms part of the civil works contract. To avoid restricting innovation and optimization, Tenderers are given freedom to propose alternative solutions for the design and construction of the scheme.

3 PROJECT DETAILS

3.1 General Arrangement

The project site is located within the northern part of the Snowy Mountains region of New South Wales, Australia. The eastern part of the alignment is located on an elevated plateau, while the western part lies below a well-defined escarpment, with deeply incised valleys forming what is commonly known as the Ravine area. This area is drained by tributaries of the Tumut River and Talbingo Reservoir. The area has a high relief of 500 m to 600 m, with slopes commonly steeper than 30 degrees. Figure 1 shows the scheme horizontal alignment and project area plan.

The scheme developed in the Reference Design consists of intakes located in each reservoir connected by the power waterway. The power waterway consists of a headrace tunnel, approximately 16.9 km long, which trends due west from an intake structure at Tantangara reservoir across the Kiandra plateau to the headrace surge tank located at the escarpment. The power waterway then separates into three vertical pressure shafts which are approximately 540 m high, feeding three pressure tunnels. The pressure tunnels bifurcate into six

Figure 1. Snowy 2.0 alignment and project area plan.

penstock tunnels, each feeding a single unit in the machine hall of the power station. The draft tube tunnels on the downstream side of the machine hall combine back into three collector tunnels which meet at the bottom of the tailrace surge tank. The tailrace surge tank is connected to an intake structure located at the rim of the Talbingo reservoir via the tailrace tunnel which is approximately 7.0 km long.

In total, the scheme comprises approximately 29 km of combined waterway tunnels and 2 km of shafts. The depth of overburden reaches 400 m in the headrace tunnel and 800 m in the tailrace and pressure tunnels. The layout of the major scheme components is summarized in Figure 2.

The diameter of the waterway tunnels was selected based on the expected lining type, the associated roughness, durability and acceptable headlosses. The resulting internal diameter of the headrace and tailrace tunnels was 10.0 m.

Supporting the scheme is a combined length of 16 km of access tunnels and construction adits. The primary access tunnels into the power station are the main access tunnel and the emergency egress, cable and ventilation tunnel, both of which are approximately 3.3 km long. In addition, there is a substantial amount of surface works including access roads, operation and maintenance buildings and services. High voltage power lines currently exist in the area however additional transmission lines and a sub-station are required to distribute and feed Snowy 2.0.

3.2 *Alignment*

The alignment has been continually assessed to determine the optimum scheme arrangement with consideration for topography, geology, hydraulics, constructability, cost and delivery program.

The horizontal alignment is governed by the intake locations and the headrace surge location. The intakes have been located at the edge of the reservoir to minimize waterway length. The headrace surge tank has been located to contain transient surges below ground level, which avoids the requirement for surface structures within the Kosciuszko National Park. A number of alternative horizontal alignments were considered. However, the initial alignment documented in previous studies was found to be close to optimum due to it having the shortest waterway length. The Talbingo Intake was shifted upstream within the reservoir once bathymetry confirmed adequate submergence could be achieved, which reduced the tailrace tunnel length by nearly 800 m.

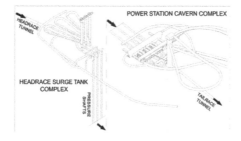

Figure 2. Isometric view of headrace surge tank and power station complex from Reference Design.

The vertical alignment is governed by the intake elevations and headrace surge location. The power station, headrace and tailrace tunnel vertical alignments have been set to meet hydraulic requirements of the system and comply with submergence requirements. Consideration has also been given to the confinement and leakage of the waterway and limiting the extent of steel lining.

One of the main options adopted from the Feasibility Study was the shift of the power station complex downstream whilst maintaining the headrace surge tank location. The power station complex is on the construction critical path given the lengthy activities of cavern excavation, structural erection, mechanical and electrical installation and commissioning. Shifting the power station complex downstream presented the opportunity to shorten the access tunnels and therefore gain early access to the power station complex, leading to a direct program reduction. As the headrace surge tank location is essentially fixed due to topography, the extent to which the power station can shift downstream is constrained by hydraulics and machine operational requirements.

3.3 Geology and Hydrogeology

The project site is located within a green-field area of the Kosciuszko National Park, where only limited information about the geological and hydrogeological conditions was available at the time of the project's inception. While the site is near to the existing Snowy Mountains Scheme, the relevant geological formations have not been majorly intersected by any existing tunnels.

Geotechnical investigations which had commenced during the Feasibility Study provided initial results during the Reference Design stage, allowing a better understanding of the ground conditions and associated geotechnical hazards. The planning of additional investigations is ongoing.

3.3.1 Geological Setting

Regionally, the project site is situated within the south-eastern portion of the Lachlan Orogen (fold belt) of New South Wales, a geological province of old volcanic belts, sedimentary basins and intrusive rocks that have been affected by several episodes of orogenesis and metamorphism.

At a local scale, the alignment passes through several geological formations of highly variable nature, comprising a wide range of rock types and geological structures. The majority of rock units span from the Ordovician to Devonian, with the exception of isolated occurrences of Tertiary basalt. The relevant geological formations and lithologies are summarized in Table 1. A geotechnical longitudinal section for the power waterway alignment is shown in Figure 3.

The project area crosses two major structural zones, including the Tumut Block and the Tantangara Block. The two blocks are bounded by the Long Plain Fault, a major tectonic suture separating the Silurian sedimentary rocks in the west from the Ordovician volcanics in the east. The Tumut Block is anticipated to comprise an open folded syncline in which Devonian sediments and volcanics rest unconformably on Silurian sediments. The Tantangara Block is anticipated to be intercepted by numerous faults, though few have been observed at surface. Folding is well developed throughout the majority of the project area. The project alignment runs perpendicular to the regional structural trends and geological contacts.

Table 1. Geological formations and lithologies.

Underground Structure	Geological Formation	Lithology
Headrace Tunnel	Kellys Plain Volcanics	Dacite
	Tantangara Formation	Meta-Sandstone, Meta-Siltstone
	Temperence Formation	Basaltic Tuff, Chert, Feldspathic Sandstone
	Boggy Plains Suite	Dacite, Diorite
	Gooandra Volcanics	Schist, Rhyolite, Siltstone, Sandstone, Phyllite, Basalt, Gniess, Siltstone, Meta-Siltstone, Dacite
	Ravine Beds East	Meta-Siltstone, Meta-Sandstone
Power Station Complex	Boraig Group	Sandstone, Siltstone, Conglomerate, Volcanics
Tailrace Tunnel	Ravine Beds West	Meta-Sandstone, Meta-Siltstone, Conglomerate

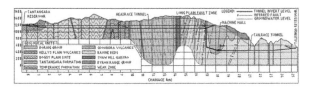

Figure 3.   Snowy 2.0 geotechnical longitudinal section.

Of the geological features identified, the Long Plain Fault is of most engineering significance. The fault can be traced for a distance of more than 200 km and is estimated to comprise
an affected zone which may be over 2 km wide at the intersection with the alignment. This
zone is likely to comprise several individual faults ranging from minor shears to major fractured and brecciated zones of tens of meters thickness.

The fault zone is thought to be associated with a reverse thrust and possible strike-slip
mechanism, generally dipping eastwards at angles of between 45 and 60 degrees. The general
trend is north to north-east, therefore intersecting the project alignment near perpendicularly.
There is poor exposure at surface but the western limit of the fault zone is interpreted to be
near the main escarpment. The majority of the fault zone is therefore expected to intersect the
headrace tunnel, with some associated affected ground expected at the location of the headrace surge shafts and upper portions of the pressure shafts, located immediately to the west.

3.3.2 Hydrogeological Setting

Based on the current information, there appear to be two major groundwater systems along
the alignment separated by the Long Plain Fault. Groundwater on the plateau, east of the
fault appears to have typically high recharge and a relatively shallow flow system, while locations to the west have lower potential recharge and a deeper flow system.

The entire alignment is expected to be contained within fractured rock aquifers which range
from unconfined to confined at depth and are often influenced by steeply dipping structural
controls, which are regionally common. Within the Gooandra Volcanics and Long Plan Fault
zone, an interconnected system is indicated down to headrace tunnel level. A vertical hydraulic
gradient has been detected within the Ravine Beds West unit, indicating less vertical connectivity in the area of the power station complex and high-pressure tunnels. Separate upper and
lower aquifers may exist in this area due to the overlying Boraig Group unit.

Groundwater discharge zones are inferred at four locations along the alignment featuring
major creeks and the two reservoirs. Groundwater recharge zones are inferred at three elevated
hills. Notably, the location of the Long Plain Fault corresponds with a recharge zone and appears
to be a significant hydrogeological feature which is likely influencing groundwater movement

Hydrogeological investigations and modelling indicate that potential tunnel inflow rates
could typically reach between 2 and 5 L/sec/km, with short term inflows likely to be 5 to 10
times higher. The majority of inflows are expected to occur due to a limited number of discrete
structures. There are likely to be several areas of interconnectivity between the surface and
tunnel alignments which may cause high inflows during tunneling.

4   ENGINEERING CHALLENGES

4.1   Complex Geological Conditions

The complex and highly variable alpine geological and hydrogeological setting presents many
challenges for the design and construction of tunnels and other underground works. Tunneling conditions are anticipated to experience a wide range of ground behaviors, comprising
both structurally-controlled and stress-controlled mechanisms, ranging from brittle spalling to
deep-reaching shear failure of weaker rock masses under high stress conditions. In addition,
excavations will frequently face mixed ground conditions, involving regularly transitioning

through different lithologies, faults and weak zones. Based on the understanding of the hydro-geological setting, high groundwater inflow may impact on construction and ground stability if no pre-grouting treatment is carried out in advance of critical areas.

Given the uncertainty and variability of the ground conditions, in addition to the substantial length and depth of the alignment, there has been intrinsic limitations and constraints to the development of the ground model. While the current ground model allows the definition of a reasonable geotechnical baseline, a more complete understanding of the ground conditions will only be obtained once construction is carried out. To cater for this uncertainty, a contractual risk-sharing mechanism has been developed to provide a balanced allocation of risks between the Employer and the Contractor and a fair mechanism of compensation for construction costs and time. The developed GBR provides a baseline of ground types, ground behavior types and tunneling classes along the alignment. The baseline of tunneling classes aims to represent the actual effort required by the Contractor to construct the underground works under the agreed construction method. The intention of the risk sharing mechanism was that appropriate cost and time (production rates) values are assigned and used as a basis for the compensation and re-measurement during construction for the actual encountered ground conditions.

4.2 Long Plain Fault Zone

The Long Plain Fault zone is a defining feature within the project area, presenting several engineering challenges. Significantly, the presence of the fault zone has implications for the selection of suitable scheme arrangements, tunnel construction methods and tunnel lining types, in addition to groundwater inflows, drawdown effects and associated environmental impacts.

Due to the known extent of the Long Plain Fault to the north and south of the project site, a connection of the two reservoirs without intersecting the fault was not feasible. In order to mitigate the impacts of the Long Plain Fault, the selected alignment provides a favorable perpendicular and high elevation intersection with the headrace tunnel, avoiding higher geotechnical risks associated with crossing the fault at greater depth.

Tunneling conditions within the Long Plain Fault are expected to be variable, involving regular transitions between relatively competent material and sheared, fractured and brecciated zones. Initial assessments indicate that ground behaviors in the more complicated zones may involve a combination of stress-controlled and progressive failure mechanisms due to ground loosening. In addition, groundwater interconnectivity between the ground surface and the tunnel horizon has been detected within the Long Plain Fault, thus the potential exists for high water inflows during tunneling. This may affect ground behavior, impede construction progress and potentially result in a drawdown of the groundwater table.

Considering the above, it was recognized early during the project development that any selected tunnel construction method should anticipate forward ground investigation and treatment. It was recognized that the selected method would need to offer flexibility during tunneling and a robust structural final lining during operation, in order to mitigate a major project risk. On this basis, a drill and blast method of construction was proposed in the Reference Design for the extent of the fault zone, combined with a typically reinforced cast in situ final lining. The installation of a steel lining may also be required in certain locations.

4.3 In Situ Stress Conditions

Stress conditions across the project site can be divided into three domains, namely: East, at and west of the Long Plain Fault. Measurements indicate the major principal stresses in the eastern and western domains typically strike in a northerly direction and are approximately perpendicular to the alignment. Stresses in the Long Plain Fault domain rotate to the east, presumably due to faulting effects.

Stress magnitudes within the eastern and western domains are moderate to high, with horizontal to vertical stress ratios typically ranging between 1.5 and 2.0. Complex in situ stress conditions are indicated within and surrounding the Long Plain Fault zone. The complexities in these locations are twofold. At upper elevations, between ground surface and approximately 350 m depth, low confining stresses dominate. Conversely, at lower elevations from 350 m to 900 m depth, high in situ stresses indicate potential overstressing of the rock. These complexities present engineering challenges which vary with depth and in each underground structure.

In the upper elevations, there is potential that the ground has inadequate confining stress to resist hydraulic jacking or fracturing generated by internal static and transient hydraulic pressures. Traditionally, such issues have been resolved through re-alignment of the tunnel and shafts or the installation of costly steel linings. The potentially affected underground structures include the headrace surge shaft, upper sections of the pressure shafts and interconnecting distributor tunnels.

Following an options study, it was recognized that there was little opportunity for realignment to avoid the low confinement zones. Alternative alignments to the north and south both encountered deeply incised valleys or narrow ridges which are suspected to have undergone stress relaxation. Relocating the headrace surge shaft upstream to the east encountered the Long Plain Fault, while downstream to the west encountered inadequate topographic elevation to contain the surge within the ground. Alternative vertical alignments were also considered which would have required the headrace tunnel to be lowered. However, these proved unfavorable for constructability and hydro-mechanical reasons and would require crossing the Long Plain Fault at greater depth, which was considered undesirable. Consequently, the original alignment was maintained, and the affected tunnels and shafts were sized and arranged such that they can be feasibly steel lined, if required. The decision to adopt a steel lining and the length of lining required will be revisited when additional information about the stress conditions has become available.

In the lower elevations, the potential for overstressing indicates stress-controlled ground conditions may be encountered in the lower sections of the pressure shafts, penstocks and power station complex. Given the ground conditions in these locations, it is expected that brittle spalling mechanisms are likely to occur. These ground behaviors have the potential to impede tunnel construction and have implications for the design of excavation support systems. The conditions also have implications for the orientation, sizing and arrangement of the power station complex.

Considering the above, the complex nature of the stress conditions across the project are the subject of ongoing interpretation and additional investigation.

4.4 *Naturally Occurring Asbestos*

During geological field surveys, the presence of naturally occurring, highly asbestiform mineral fibres was identified within the Gooandra Volcanics formation, located on the headrace tunnel alignment to the east of the escarpment. The presence of these fibres has the potential to cause atmospheric contamination if liberated during tunneling works. This presents significant health risks associated with the inhalation of airborne fibrous minerals and may therefore have implications for spoil management and tunnel ventilation systems.

The Reference Design solution proposed the inclusion of an intermediate headrace tunnel construction adit to enable the isolation and management of potentially contaminated tunnel spoil and ventilation exhaust, if the presence of asbestos is confirmed.

4.5 *Environmental Constraints*

The environmentally sensitive project setting, within the Kosciuszko National Park and a heritage listed area, presents significant challenges for all engineering disciplines across the project. The project area is almost entirely undisturbed and contains several threatened species of flora and fauna in addition to a number of sensitive creeks, rivers, bogs and fens. The challenges presented are unique from the original Snowy Mountains Scheme, which was constructed prior to the area being classified as a national park. The environmental setting imposed constraints for the execution of the geotechnical site investigation activities (including drilling) and requirements to minimize surface construction footprints and activities.

The impacts on scheme arrangement, whilst not completely driven by environmental constraints, have resulted in a limited number of tunneling sites within the project area. This, in combination with the large number of underground components and the long project corridor, has resulted in relatively long access tunnels. Other significant risks are associated with the impact on the groundwater system due to tunnel inflows during construction.

Based on hydrogeological studies, tunnel inflows are expected to have the greatest impact in areas of groundwater discharge, where there is potential for drawdown and loss of base flow. This is likely to be exacerbated where connections to the ground surface have been demonstrated such as the Gooandra Volcanics and Long Plain Fault. Lower impacts are expected in areas of recharge.

To manage potential environmental impacts, it is expected that ground treatment through pre-grouting will be required at tunnel sections where large water inflows are anticipated, together with post-grouting if residual water inflows exceed the allowable limits specified in the contract. The requirement for the execution of pre-grouting works is planned to be triggered by pre-specified water inflow levels, as measured during construction.

In order to manage allowable residual water inflow into the tunnel after excavation, three different water leakage/inflow performance criteria or classes have been established. Class 1 applies throughout the alignment as the minimum performance criteria and is the least restrictive. Class 2 is more restrictive and applies in areas of greater environmental sensitivity where recharge is indicated. Class 3 is the most restrictive and applies only to areas known to have sensitive environmental receptors and/or groundwater discharge. Modelling indicates that if inflows are limited to the allowable criteria, there will be minimal or no impact on the existing water system.

The proposed solution aims to find a balance between minimizing potential impacts on the groundwater system, while also minimizing the amount of ground improvement required, and hence impacts on the project cost and schedule.

4.6 Remote Setting

Compounding the complexity of the environmental constraints is the remote project setting. Although not unusually remote for a hydroelectric scheme, the Snowy 2.0 project is not within close proximity to any major towns. This presents logistical challenges for the construction phase, with personnel and materials requiring transportation over long distances. In addition, the main work areas (intakes and access tunnel portals) are relatively spread out, making it difficult to consolidate support services in an efficient manner.

The logistical problems are further complicated by the terrain, particularly around the Ravine area where the power station complex access tunnel portals are located. The existing road network in this area is insufficient to support the construction, operation and maintenance traffic, given the tight radius bends, narrow corridors and basic pavements. With the lower parts of the project located adjacent to the Talbingo Reservoir, over-water transport between a wharf located close to the existing Tumut 3 Pumped Storage Power Station (which has good access provisions) and the Talbingo Intake/Ravine area has been incorporated into the Reference Design of the project.

There is currently no permanent power supply available to most of the work site locations. Therefore, the establishment of construction power networks is also required. Similarly, there are no reliable water supplies at the access tunnel portals which leads to the need to establish new water infrastructure for both construction and operation of the project.

The climatic conditions across the site are variable with the upper reaches of the project being located above the snow line. Wherever possible, locating of permanent and temporary infrastructure within these areas was limited.

4.7 Construction Method Selection

The selection of preferred construction methodologies for the Reference Design solution was based on considerations for constructability, geotechnical risks, costs and construction program. In the first instance, methods were screened for suitability based on a comprehensive risk assessment study. Suitable methods were then ranked based on their cost and program impacts, allowing the preferred method to be identified.

As an outcome, an open type TBM was selected for the eastern half of the headrace tunnel within the Tantangara Formation, due it's flexibility to cater for varying ground conditions. Drill and blast techniques were selected for the headrace tunnel within the Gooandra Volcanics to manage logistics associated with the naturally occurring asbestos. Similarly, drill and blast was selected for tunnels within the Long Plain Fault zone, given the known geotechnical risks associated with driving a TBM through the fault.

For the tailrace tunnel, a double shield TBM was selected based on potential performance benefits. The access tunnels and adits are proposed to be constructed by drill and blast techniques, with most being short drives. The use of TBMs for the longer access tunnels was considered, however the procurement lead times presented program disadvantages.

Given that the shafts are located near the Long Plain Fault, the use of raise boring methods was considered potentially problematic due to the risk of instability of the unsupported excavations during reaming and mucking. Therefore, the shafts are proposed to be constructed by blind-sinking or down-reaming methods, which offer more flexibility to the ground conditions.

The remainder of the underground works are proposed to be constructed with drill and blast methods, given the size, length or shape of the structures.

4.8 Final Lining Design

The decision to line the entire power waterway was made based on the following considerations:

- Fully concrete lined tunnels offer less maintenance requirements
- Head losses in concrete lined tunnels are lower than in unlined tunnels of the same size
- Water loss from the scheme is not expected to be of concern
- Potentially erodible ground is expected which may present durability and stability problems
- High operational hydraulic pressures may have cyclic loading and fatigue implications.

The selection of a preferred final lining type was based on considerations for structural performance, durability, confinement, constructability, hydraulic performance and the maximization of round trip efficiency. Significantly, the minimization of potential groundwater drawdown and loss of base flow was also considered. For the Reference Design solution, a permanent inner cast in situ concrete lining was selected for the majority of the alignment, including the headrace tunnel. A steel lining was applied in limited locations where leakage or confinement issues were identified. A single pass pre-cast segmental lining was selected for the tailrace tunnel.

4.9 Power Station Complex

The main components of the underground power station complex are the machine hall, transformer hall, headrace surge tank and tailrace surge tank. These components are connected via the waterway tunnels, shafts and access tunnels. The complex is located up to 800 m below ground.

The power station complex is the main permanent working area for the scheme and therefore has many specific operation and maintenance requirements which are well established with SHL being operators of the existing scheme. The machine hall will be 33 m wide, 55.4 m tall and 238 m long. It will house the six pump-turbines, motor-generators, main inlet valves and auxiliary balance of plant. At the base of the machine hall will be a drainage gallery and dewatering pit which will service the entire power station complex. The machine hall will be serviced by two 280 tonne overhead travelling cranes.

The transformer hall will be 19 m wide, 28.6 m tall, 295 m long and be located downstream of the machine hall. It will house the six three-phase main transformers, draft tube valves and cooling water equipment. Gas insulated switchgear will also be incorporated into the transformer hall. The machine hall and transformer hall will be connected by two main access

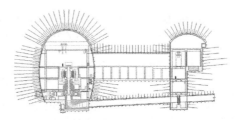

Figure 4.   Typical cross section through the machine hall and transformer hall.

tunnels and six Isolated Phase Busbar (IPB) galleries which will house the electrical equipment required between the motor-generators and the main transformers. A typical cross section through the power station cavern complex is shown in Figure 4.

4.10   *Mechanical and Electrical Design*

The total installed capacity of 2000 MW will make Snowy 2.0 one of the largest pumped-storage schemes in the world. The units proposed are 340 MW reversible Francis pump-turbines with a combination of synchronous (fixed speed) and asynchronous (variable speed) units. The generators are rated at 375 MVA for synchronous units and 425 MVA for asynchronous units.

The required hydraulic performance of the waterway imposes numerous restrictions on both the equipment and the scheme layout. The hydraulic arrangement has considered numerous issues including tunnel roughness, flow velocities, frictional head-losses, tunnel diameter, transient pressures, surge ranges, round-trip efficiency and plant operating scenarios.

The evacuation of the electricity has been closely coordinated with transmission operators (TransGrid) and grid operators (AEMO). The augmentation to the grid around the project is key to achieving the project benefits to the network. Multiple options for connection into the grid were considered with a 500 kV connection being preferred. An underground gas-insulated switchgear facility connected to a surface cable yard was adopted as the preferred arrangement.

The interface between the mechanical and electrical contractor and the civil contractor is complex. For example, the hydraulics of the waterway tunnels will be the responsibility of the civil contractor, however, they impact on the operational requirements of the pump-turbine.

5   CONCLUSIONS

The Snowy 2.0 project involves the delivery of a 2000 MW pumped storage scheme within south-eastern Australia. Following the recent completion of the Feasibility Study and Reference Design, preparations are under way for contracts to be awarded and the construction of the civil works and procurement of the hydro-mechanical plant commencing in 2019.

The project faces a number of significant engineering challenges including complex and variable alpine geological and hydrogeological conditions in a remote environmentally sensitive setting. Other conditions include high overburden and stress conditions, potential presence of naturally occurring asbestos and the crossing of a major regional fault zone. The project is subject to stringent functional and operational requirements, requiring a flexible and a robust scheme design.

These challenges required comprehensive engineering solutions and the consideration of a flexible and robust contractual framework for procurement, engineering and construction.

ACKNOWLEDGEMENTS

The authors wish to acknowledge the support of Snowy Hydro Ltd, SMEC Australia Pty Ltd and the Snowy 2.0 project team in the preparation of this paper.

Tunnels and Underground Cities: Engineering and Innovation meet Archaeology,
Architecture and Art, Volume 7: Long and deep tunnels – Peila, Viggiani & Celestino (Eds)
© 2020 Taylor & Francis Group, London, ISBN 978-0-367-46872-9

Brenner Base Tunnel, construction lot Mules 2–3. Production management and site logistics organization

S. Citarei, M. Secondulfo, D. Buttafoco & J. Debenedetti
B.T.C. Brennero Tunnel Construction, Rome, Italy

F. Amadini
SWS Engineering spa, Trento, Italy

ABSTRACT: Construction of great underground infrastructures and specifically European alpine base tunnels entails a complex, articulated and multidisciplinary setting out of the construction sites, which need to be managed at industrial level in order to meet established time and technical targets. Brenner Base Tunnel represents, in terms of complexity and extension, one of the most significant underground works under construction worldwide. Being a long deep tunnel in rock, the project features high management and environmental complexities. These constraints require detailed analysis and proper system design for successful construction delivery. The temporary joint venture formed by Astaldi-Ghella-Pac-Cogeis was awarded Mules 2 3 construction lot, also based on the improvements proposed during tender phase.

1 INTRODUCTION

1.1 *Project description*

"Mules 2–3" lot is part of the complex Brenner Base Tunnel. Due to its extension of 23 km, it represents the most important contract awarded in Italy in the last years. This lot consists of two single-track tunnels, connected with cross passages at every 333 m. The project goes through different geological formations including the crossing of peculiar areas such as the so-called "Periadriatic Line", which ideally identifies the contact area between the African and European Plates, characterized by various geological intrusions that, in general terms and except for singular episodes, have poor mechanical properties.

The project includes excavation of 40 km of main tunnels, of which about 16 km to be excavated with conventional method and about 24 km with double shielded TBMs, some additional 14.7 km exploratory tunnel, one emergency stop in Trens with a dedicated access tunnel and cross passages located at every 333 m between the main tunnels. Figure 1 shows an overview of the project.

Jobsite logistics is one of the most significant and challenging aspects, due to project size and simultaneous presence of different excavations methods.

These complexities are further exacerbated by the need to set up the jobsite as a closed system, minimizing its impact on the environment and on the surrounding road network.

In particular, the reuse of the excavation material (mainly Bressanone granite) as aggregate for the concrete production and its complete transport through a complex and extended convey or belt system has allowed achieving this goal.

1.2 *Project upgrading during design stage*

During tender stage and following detailed design, several improvements have been proposed and studied, leading to further optimization and upgrading of the project.

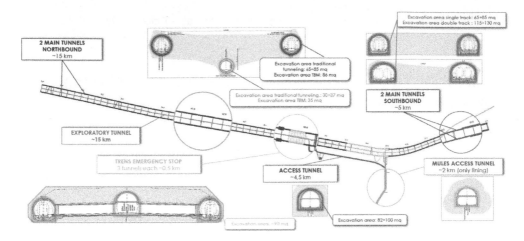

Figure 1. Project overview of construction lot Mules 2–3.

The most important improvements that were introduced are listed below:

– Different excavation method at different chainage in the southbound drive.
– Sooner start of mechanized excavation in the northbound drive.
– Change in ventilation concept, from circulating system to pushing system.
– Excavation material reuse for segmental lining production.
– Segmental lining production and storage within jobsite areas.
– Introduction of additional logistic access between project tunnels in order to increase safety and production.

2   LOGISTIC CONSTRUCTION MANAGEMENT

During the detail design phase, it was decided to improve the site organization as designed at tender stage, by modifying the excavation method of various tunnel stretches, in agreement with the Client.

According to tender design, excavation of the southbound tunnels would be carried out with two TBMs, for a length of about 3.5 km at the end of which conventional method would start for a double-track section of approximately 1.5 km. Excavation of the northbound tunnels would include conventional method for a length of about 3 km, followed by an assembly cavern for the two double shielded TBMs that would complete the remaining 12 km.

During detailed design, a change of excavation method for the southbound tunnels was proposed, from mechanized to conventional. Also, the start of mechanized excavation for the northbound tunnels was anticipated of about 2 km.

These conceptual changes have allowed to enhance separation and independence of the two sites, focusing conventional tunneling in the Southbound drives and mechanized excavation in the Northbound drives.

The most evident improvement can be summarized in the separation of the entire project in two independent sites, each of those with dedicated access.

Mules adit is therefore dedicated only to the transit of wheeled vehicles for conventional excavation, while the Aica tunnel allows rail-only access for the supply of the mechanized excavation.

This separation eliminated interferences between different working processes and therefore improved safety and logistics.

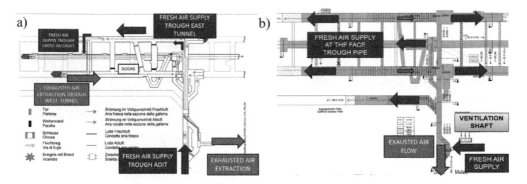

Figure 2. Ventilation scheme during Tender Design (a) and construction phase (b).

3 VENTILATION SYSTEM

3.1 *Tender design solution*

Ventilation is a key point for tunneling because it is needed to ensure the best possible conditions for workers at the faces, particularly in challenging and complex environments such as this one.

The concept of ventilation as foreseen in the tender design has been deeply modified to offer a suitable quality and quantity of air to the front, a greater reduction of dust and in general a healthier working environment.

Tender design was based on a circulation system. Fresh air was naturally supplied trough Mules adit, while the ventilation shaft extracted the exhaust air mechanically.

Projects of such complexity as other base tunnels or underground mines generally adopt this principle. The circulation system is simple, functional but insufficient to guarantee an optimal evacuation of the polluted air from inside the underground complex of the Brenner Base Tunnel.

3.2 *Adopted solution during design stage*

During tender phase, it was proposed to adopt a pushing system, able to supply fresh air directly at the face, pushing exhausted air towards the exit of the tunnel system.

The function of the ventilation unit was therefore reversed from extracting to blowing.

The improvement proposal proved to be a winning choice, appreciated also by the Client for the increase in safety. In fact, in case of an accident the system would completely isolate the affected drive by pressurizing the other tunnels, which thus become safe places for workers.

The ventilation system has been designed to ensure the correct air supply and pressure on the various fronts in all phases of the project. The most critical phases of ventilation coincide with the maximum expansion of the traditional excavation, with regard to the required airflow and with the maximum extension of the mechanized excavation with regards the working pressure.

The electrical consumption of the system is 3200 kW, which is equipped with four 800 kW fans, allowing the system to work at a maximum pressure of 6000 Pa and a capacity of 450 m^3/s.

To improve system management and maintenance, all distribution nodes are made of steel and are equipped with shutter systems that allow the automatic remote adjustment of the airflows to the various tunnels.

4 CONVEYOR BELT SYSTEM

The conveyor belt system has a high degree of general complexity. It must be capable to ensure the selection of the excavated material type, to cover long distances between excavation fronts and deposits and to share the available space with all other services.

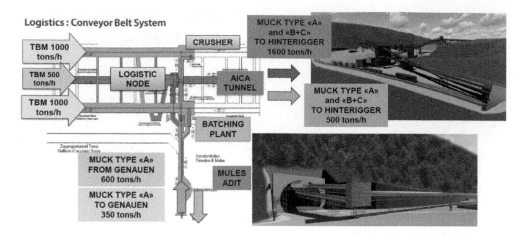

Figure 3.   Flow rate scheme of the conveyor belt system.

4.1 *Excavation material flow*

The flow of excavation materials is completely managed by the system, which conveys the material to the collection point (logistic node) located in the southern caverns.

The muck coming from the mechanized tunnels reaches automatically the logistic node. The muck coming from tunnels excavated with conventional method is transported by truck to the primary crusher at the collection point in the southwest cavern, where it gets reduced in size before being introduced on the conveyor belt system.

The capacity of the belt coming from the main tunnel TBMs is 1000 tons/h while the exploratory tunnel belt has a capacity of 500 tons/h.

From the logistic node, the material can follow two different paths to the spoil areas and concrete plants available on site.

Two conveyor belts, having a respective capacity of 1600 tons/h and 500 tons/h, convey the excavation material through Aica tunnel to the main spoil area located in Hinterigger where the precast segments are also produced.

Muck material type "A" solely, can be sent alternatively to the crushing plant located in Mules (600 tons/h) or to the secondary spoil area in Genauen, that play the role of a buffer deposit.

The crushing plant in Mules prepares and convey (350 tons/h) the different aggregates for the underground concrete plant. In this batching plant all the different types of concrete used for tunnel construction are produced – from sprayed concrete to final lining cast concrete.

5   MUCK MANAGEMENT AND DISPOSAL

Over 6 million cubic meters of spoil will be removed during the work on the Mules 2–3 construction lot.

All material type "A" will be recycled within the construction site itself as aggregate for the production of cast concrete, precast segmental lining in Hinterrigger and for the production of the pea-gravel, eliminating the impact on ordinary traffic.

The proposed solution significantly increases the amount of reused spoil (+62%), reducing the excess quantity that would be destined to sale by approx. 50%, compared to tender design spoil balance.

5.1 *Development plan of Hinterigger disposal area*

The main spoil area of the project is located in the Hinterrigger depression. This site also hosts the production and storage of the segmental lining, which temporarily limits the availability of space.

Prior to disposal, incoming spoil needs to be sorted and checked for polluting content. This procedure requires large surfaces available to handle the following types of material:

- Fertile soil: it requires to be immediately stored because removed initially for deposit preparation. This material has to be available for oxygenation and maintenance up to his final use at the end of the project for re-naturalization of the site.
- Material type "A": it represents the 36% of the total excavated amount. It must be stored for quality control. The material considered suitable is transferred to the crushing plant to be further used in segmental lining concrete production. The exceeding material is to be sold on the market as aggregate.
- Material type "B+C": the material not suitable for concrete production is stored in piles of 10'000 m3 for quality control before its transferred to the final deposit.
- Special waste: the material that fails the qualitative examination for pollutant content exceeding code limits is tagged as special waste and is to be disposed in a designated area which needs to remain accessible at all times.

All of these factors have an evident impact on the management of Hinterrigger site, that requires a complex **development plan** to counteract the growing unavailable surface.

6 SEGMENTAL LINING PRODUCTION AND STORAGE

One of the most important improvements that has been introduced at Mules 2–3 lot is certainly the production of segmental lining directly within construction site area. This has allowed a reduction of traffic on the local road network and consequently less pollution that would have heavily affected the territory, considering that about 200'000 segment will be produced.

Furthermore, the production process is controlled directly by the joint venture and the client, allowing for immediate action and quick problem solving solutions.

The production and storage plan for segmental lining must take into account different sets of segments that are designed for different areas or specific zones such as cross passages.

Total land usage is approximately 37'000 m^2. 15'500 m^2 are occupied by a 21.45 m wide warehouse, the remainder 21'500 m^2 outdoor area is designated for segmental lining storage.

Two concrete mixing plants are located at the South end of the central bay of the warehouse, both equipped with 2 m^3 mixers. Lateral bays host rebar processing and cage production.

All batching plants, aggregate silos and the mixing water are located indoors and can be heated by a dedicated gas power plant if required.

Figure 4.   View of the segmental lining production plant.

Cement and aggregate silos have enough capacity to sustain maximum production for 48 hours. The maximum production within 24 hours is equivalent to 24 complete rings for the main tunnel and 12 complete rings for the exploratory tunnel, for a total production of 865 m3 of concrete per day. The concrete mixing plants have generally been oversized for redundancy. Even in case of maintenance or malfunction, one single plant shall deliver the maximum production.

7 SITE RAILWAY LOGISTICS

7.1 Railway system

A complex system of railway tracks allows the supply of the material to the three TBMs.

Each train allows the transport of two complete rings, including the invert, the annular gap filling material (pea-gravel or expanded clay depending on the geo-mechanical conditions), the bi-component grout for the invert filling, rails and plumbing necessary for the TBM advancement.

Railway profiles feature gradients up to 4.2%, at the very limit of practicability. Considering the weights of the trains, respectively of 380 tons for the main tunnels and 180 tons for the exploratory tunnel, each train requires two locomotives in order to overcome this slope.

The overall railway system requires 110 km of tracks in the maximum extension of the project interconnected by approximately 100 switches to allow maximum capacity of the system even during the numerous planned interruptions such as for maintenance of ventilation and conveyor belts.

Along the railway track, a logistic area is available for the maintenance and stationing of trains, equipped with a workshop operating 24/7.

The railway system will serve maintenance trains, transportation trains for shift workers and specialized personnel such as geologists, surveyors, inspectors and machine engineers, in addition to the ordinary supply trains for TBMs.

7.2 Signaling and traffic management

Rail traffic is managed through a remote signaling and control system, aimed at increasing workers' safety and prioritizing transit of supply trains in order to increase production

Figure 5. Unterplattner logistic site view.

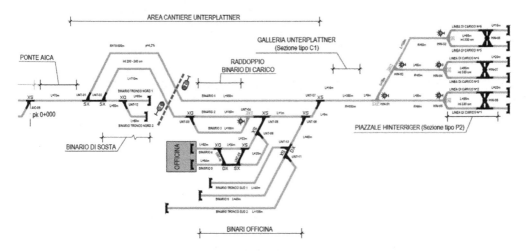

Figure 6.   Railway scheme of Unterplattner and Hinterrigger logistic areas.

capacity. The amount of rolling stock foreseen in the planning phase is likely to increase lest this be a limiting factor to TBM performance.

In the maximum expansion of the project a main control station will have to manage a flow of about 50–60 trains a day. Traffic management is carried out by means of a traffic signaling system controlled directly by the control station, which constantly displays the position of each train on a monitor and plans the route of the trains, moving the electrical switches. This system is being implemented during the exploratory tunnel excavation to become fully operational at the start of the two main TBMs.

The system has been designed to be safe, simple and effective, but fearing the complications that may arise during operation it has been designed to be modular and expandable to increase the ability to remotely manage a greater number of switches and train control, always aimed at ensuring maximum capacity and traffic safety.

8   CONCLUSIONS

As the authors were writing this article, the initial construction phase was almost complete. This phase has been over two years long and was necessary to pave the way for the final excavation configuration that will lead to the North and South boundaries of the lot.

The technological and logistical challenges of this construction site are due to the simultaneous excavation of seven different tunnels with different methodologies. The rationalization of the construction site and the improvements introduced with detailed design have led to an interference reduction and process optimization.

Complications are constantly increasing during construction, especially those related to growing requirements for TBM services, and the progressively diminishing space for storage of precast segmental lining and spoil disposal.

The goal is to complete the work within the expected time-and-space frame by streamlining the complex node of Hinterigger site.

The authors have consciously accepted this challenge, and are looking forward to seeing the continuation of this piece of Italian railway history.

Tunnels and Underground Cities: Engineering and Innovation meet Archaeology,
Architecture and Art, Volume 7: Long and deep tunnels – Peila, Viggiani & Celestino (Eds)
© 2020 Taylor & Francis Group, London, ISBN 978-0-367-46872-9

Challenging mixed face tunneling at India's Sleemanabad Carrier Canal

J. Clark
The Robbins Company, New Delhi, India

ABSTRACT: India's Sleemanabad Carrier Canal is a prime example of just how challenging mixed face conditions can be, although other examples exist. The water transfer tunnel is being bored using a 10 m diameter hybrid-type rock/EPB TBM. However, in 6.5 years of tunneling the machine had only advanced 1,600 m. Commercial issues for the original contractor stalled the project frequently, while ground conditions turned out to be even more difficult than predicted. Low overburden of between 10 and 14 m, combined with mixed face conditions, transition zones and a high water table restricted advance rates. The TBM manufacturer mobilized a team to refurbish the TBM and within a period of 6 weeks a team of 180 people had been deployed to take over all aspects of tunneling and support activities. Production rates improved dramatically as the TBM advanced more than 400 m in four months. This paper will discuss the problems faced and the methodology that enabled good advance rates in highly variable mixed face conditions.

1 INTRODUCTION

The Bargi Diversion Project in Madhya Pradesh, India, is a trans-valley irrigation project consisting of 197 kilometres of canal. It will transfer water from the Narmada river to provide irrigation to 245,000 hectares of land in drought prone areas in Satna, Jabalpur, Katni and Rewa Districts. It will also supply 284 million liters of domestic & industrial water per day to the city of Jabalpur and town of Katni. A section of the scheme comprises a 12 km long tunnel driven by a 10.0 m diameter Robbins hybrid rock/EPB TBM.

2 GEOTECHNICAL CONDITIONS

The geology along the tunnel alignment changes frequently. It consists of compact residual soils, silts, alluvium, highly weathered limestone and dolomite, with stretches of slate, massive crystalline limestone and fresh marble. The strength of the rock varies considerably with UCS values reaching as high as180 MPa. There is a highly permeable boulder horizon, which acts as a conduit for ground water located 2-3 m above the tunnel for the initial 2.7 km of the alignment. The ground water table is above the tunnel for the entire length of the drive. In some instances, the TBM has traversed transition zones between different geological conditions within 1-2 meters of boring, and in other cases the transition zone has extended for over 100 meters. These lengthy transition zones result in difficult boring conditions due to the mixed faced conditions. Figure 1 shows the predicted geological conditions for the initial 3,400 m of the tunnel alignment.

3 TBM SELECTION

The geological reports indicated that approximately 68% (8,160 m) of the tunnel would be driven through residual soils, silts and highly weathered/decomposed rock, with the remaining

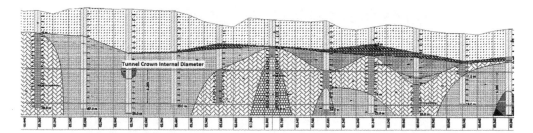

Figure 1. Geological alignment.

Table 1. TBM Technical Specifications.

CUTTERHEAD	
NOMINAL DIAMETER	10.0 m WITH NEW CUTTERS
TYPE	MIXED FACE DESIGN
NUMBER OF DISC CUTTERS/KNIFE BITS	53
CUTTERHEAD DRIVE	
TYPE	VFD ELECTRIC MOTORS (12 X 330KW)
CUTTERHEAD SPEED	0 - 5.4 RPM
CHD TORQUE AT 2.1 RPM	17,615 kNm
CHD TORQUE AT 5.4 RPM	6,966 kNm
MAIN THRUST	
CHD THRUST, EPB	23,731 KN
CHD THRUST, HARD ROCK	14,151 KN
MAX PROPELLING EXTENSION SPEED	100 mm/minute
BELT CONVEYOR	
DRIVE SYSTEM	HYDRAULIC MOTORS
CAPACITY	1,604 CUBIC METER/HR
SCREW CONVEYOR	
SCREW DIAMETER	1,200 mm
DRIVE SYSTEM	HYDRAULIC MOTORS
MAX TORQUE	300 kNm
MAX SPEED	18 RPM

32% (3,840 m) being driven through competent rock. Bearing this in mind, a hybrid rock/EPB machine was considered for the project. The hybrid machine would be configured with the option of switching out the screw conveyor with a TBM belt, and for interchangeability between disc cutters and soft ground tools (see Figure 2).

Approximately 600 m of the anticipated 3,840 m of competent rock was divided into numerous, relatively short stretches. These short stretches would not justify the downtime of one month required to remove the screw conveyor, install the belt conveyor and change from EPB into hard rock mode. However, over 3,200 m of the competent rock was made up of three stretches of 700 m, 1,000 m and 1,500 m. The length of these stretches warranted the downtime required to convert from EPB to hard rock mode; hence, a decision was taken to utilize a hybrid machine for the Sleemanabad project. The main technical specifications of the TBM are shown in Table 1.

4 MACHINE ASSEMBLY & LAUNCH

The machine was assembled at the remote jobsite using Onsite First Time Assembly (OFTA), a method that allows for the machine to be initially assembled on location with testing of critical sub-assemblies in a workshop. The TBM was launched in early March 2011 in EPB mode with a full dress of soft ground tools as the geological information indicated mainly soft ground for the initial 500 m of boring. During the first 200 m of boring the TBM encountered mixed

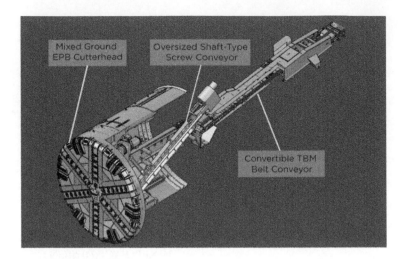

Figure 2.   Sleemanabad TBM features.

geology containing residual soils and hard rock. This was the first of frequent changes in geo-logical conditions that were encountered during the initial 1,600 m of boring. The strength of the rock dictated that the soft ground tools needed to be replaced with disc cutters. After chan-ging out the cutter tools, the machine was boring in rock but still had to be operated in closed/ EPB mode to prevent settlement of the soil above the crown of the tunnel (see Figures 3-4).

Figure 3.   Onsite TBM assembly.

Figure 4.   TBM launch.

5 POST-LAUNCH PROJECT HISTORY

During the following 6.5 years after the launch of the machine the project suffered a multitude of problems both commercial and technical. A comprehensive account of the commercial problems is not in the scope of this paper; nevertheless, it should be noted that they resulted in various minor delays of up to a few weeks due to shortages of spares, segments and consumables etc. They also resulted in the project being completely halted on two separate occasions for ten months and eight months respectively. The main technical problems that the project faced are detailed below.

5.1 Failed Cutterhead Interventions

On numerous occasions cutterhead interventions under hyperbaric conditions had to be either cancelled before they could commence or were aborted during execution due to excessive air losses. A critically long intervention time can lead to erosion blowouts. Small soil particles are blown out of the ground, increasing the pore size and reducing the flow resistance of the air (Babendererde et al. 2014). On the Sleemanabad project the cause of the air losses was air percolating through the permeable material above the crown of the tunnel, especially through the boulder horizon. The problem was exacerbated by the low overburden of only 10-14 m. The standard practice of injection of bentonite into the face to form a cake to prevent, or at least minimize air losses proved to be unsuccessful as the bentonite was pushed up to the surface along with the escaping air.

Due to the failure of bentonite injection it was decided to install safe haven grout blocks via drilling and injection from the surface. Initially this solution proved to be unreliable as there is a shortage of specialist contractors in the region and the sub-contractor employed to carry out the works lacked the necessary experience and equipment. The first attempt at pushing the TBM into a pre-installed grout block to carry out a cutterhead intervention under pressure failed completely due to excessive air loss. This caused a delay of approximately 1 week while waiting for a second grout block to be installed 10 meters ahead of the machine. Unfortunately, after boring forward into the second grout block, the second attempt at an intervention also failed due to excessive air losses. The quality and effectiveness of the safe haven grout blocks improved over time, but they were never close to being 100% effective and were the cause of many further delays.

5.2 Cutterhead Damage

Due to a lack of timely cutterhead interventions the cutterhead suffered wear damage to an extent that required repairs on three separate occasions. Because of the low overburden and mixed geological conditions, it was decided that excavation from the surface down to the TBM would be far easier than creating a subsurface chamber to carry out the repairs. Neither shaft segments nor piling equipment were available at site, but heavy earth moving equipment was being utilized on the canal works. The availability of this equipment made it possible to create an open pit type excavation (see Figure 5).

The residual soils and alluvial material were relatively easy to excavate but their low cohesion called for a multi-level pit with several benches and involved the excavation of around 30,000 m³ of material. One of the problems faced during excavation of the open pits was liquefaction caused by the saturated nature of the soil and vibration of the excavation equipment (see Figure 6).

The three separate operations to repair the cutterhead, including excavation and backfilling of the pits, resulted in a total delay of over eight months.

5.3 Muck spillage

When boring in EPB mode, it is essential that a plug of material can be formed in the screw to maintain face pressure. In order to form the plug an adequate quantity of fines is required in the excavated material. The alignment of the tunnel up to Ch:1600 m contained over 300 m of

Figure 5.   Excavation down to the TBM.

Figure 6.   Liquefaction of the soils.

highly weathered or moderately weathered marble and slate, and 200 m of mixed face conditions where adequate quantities of fines were not present. The absence of fines along with the high water table and permeable geology caused problems in maintaining face pressure. This resulted in a relatively common issue for EPBs in this type of geology: high-pressure water and silt spilling from the transfer point between the screw conveyor and TBM belt conveyor.

A catchment box and pump were available at site (as part of the machine design and supply) but the amount of spillage greatly exceeded its capacity. Polymers and bentonite were injected into the tunnel face, plenum and screw conveyor, but they had minimal effect due to the excessive volume of water. The delays in manually clearing the spillage amounted to almost 80% of working time. Typically a 1.6 m boring stroke would be completed in less than two hours but the remainder of the 10-hour production shift would be taken up clearing away the spillage.

5.4   Sinkholes

A problem related to the failed cutterhead interventions was sinkholes appearing above the machine (see Figure 7). Several interventions were aborted up to 48 hours after they commenced due to progressive increase in air losses. It became standard practice to reduce hyperbaric pressure to a minimum (approx. 0.9 bar) to reduce the amount of air losses; hence, this also extended the

Figure 7.   Sinkhole above the machine.

duration of the intervention. However, this sometimes resulted in water ingress steadily washing fines and silt into the cutterhead chamber and around the extrados of the TBM shields.

The bore diameter of the cutterhead with new cutters is 10.0 m, whereas the outer diameter of the TBM shields is 9.93 m (the differential is required for steering purposes and is standard design practice). The combined length of the TBM shields is 11.2 m which equates to 12.3 m^3 of annular space around the shields. When sufficient face pressure is maintained, the risk of convergence is minimized, but reducing the pressure allowed convergence to occur. The combination of inflow of fines into the chamber and convergence around the annular gap led to cavities forming over the cutterhead and shields and ultimately unravelling up to the surface. The sinkholes generally occurred once the machine had advanced 2-3 rings after the intervention had been completed. The section where the sinkholes occurred is part of the 95% of the tunnel alignment that runs beneath agricultural land, so there was no risk to surface structures. Access to the land over the first 1.5 km of tunnel was relatively straightforward, so earth moving equipment was deployed to backfill the sink holes. Usually after less than 24 hours boring was able to continue.

6   INVOLVEMENT OF THE TBM SUPPLIER

The Robbins Company's initial involvement in the Sleemanabad project was to supply and commission the TBM and tunnel conveyor systems, then to supply key personnel for training the contractor's crews and to troubleshoot any technical problems with the equipment.

In September 2017 after seeing only 1,600 m of tunnel completed in 6.5 years the project owner and senior JV partner held discussions with The Robbins Company, with a view to them taking over the project. Although Robbins had not had a continuous presence at site during the previous 6.5 years they had been involved with the project in various capacities on and off since the launch of the machine. Because of this involvement they were familiar with the difficulties the project had faced and they accepted the challenge. Robbins' new scope of work covered all aspects of production operations including both tunnel and surface works, supply of rails, pipes, cables, consumables, grout, electricity and haulage of excavated material from site. Supply of segments remained in the scope of the senior JV partner. A team was mobilized to begin refurbishment and testing of the TBM in September 2018. By the end of October 2018, the size of the team was increased to 180 people to enable commencement of production operations.

7   PROJECT RE-START

In order to achieve improved production rates a thorough analysis of the cause of delays faced up until that point was carried out. The measures taken to mitigate these problems are listed below.

7.1 Reduction of air losses during interventions

The overall effectiveness and efficiency of cutterhead interventions played a key role in improving overall production on the Sleemanabad project. Historically, creating safe haven grout blocks from the surface had proved to be less than effective in preventing air losses during interventions.

During discussions it was decided that pumping a weak-mix grout solution from inside the TBM offered a better chance of success as the grout pressure would build up from the face of the tunnel, assisting migration of the grout solution into the discontinuities and permeable material above the TBM. Another issue that had to be considered was the fact that the tunnel alignment was moving away from relatively easy access points on the surface. This meant that transportation of drilling and pumping equipment to locations above the tunnel would become much more difficult. Over the course of the following three interventions weak-mix grout was pumped through the mixing chamber of the TBM and into the surrounding geology. On each occasion air losses were minimized, and the interventions were completed without major problems. The grout mix design was modified on each of these three interventions. The final design mix can be seen in Table 2.

Rather than setting a limit for the actual volume of grout to be pumped into the ground, buildup of grout pressure was used to determine when sufficient grout had been pumped. The benchmark used was 0.3 bar above EPB pressure used during boring. Once this pressure was achieved, pumping was stopped for 15-20 minutes to determine if the recorded pressure was a product of back-pressure during pumping. If the pressure dropped by more than 0.2 bar over 20 minutes, grouting recommenced until a steady state pressure was achieved.

7.2 Time consumed with weak-mix grouting methodology

The setting time for the weak mix is between 12-18 hours depending on variables such as the amount of grout pumped, amount of groundwater present, temperature, etc. The time taken to pump the weak mix must also be considered in the overall time calculation. Additional time must also be factored in for cleaning the cutterhead and changing cutters. This is of course a result of having to remove grout rather than merely cleaning soils from the cutterhead spokes, buckets and cutters. Generally, around 20 hours was factored into the overall cutterhead intervention schedule when using weak-mix grout for ground consolidation. It should be noted that 20 hours should only be used in the overall schedule when a decision to pump weak-mix grout is taken prior to an intervention commencing. At this point the mixing chamber is packed with excavated material and the temperature of this material is higher than the ambient temperature on the TBM. The chamber being packed full of muck minimizes the amount of weak mix required and the higher temperature reduces its setting time.

After carrying out several successful interventions using the weak-mix method, the TBM encountered a full face of residual soil. An intervention was attempted without pumping weak-mix grout and it proved to be successful. Unfortunately, the following intervention in very similar ground conditions had to be aborted due to excessive air losses. In order to reduce the air losses and complete this intervention, pumping weak mix grout solution from the TBM was again utilized; however, the addition of weak-mix grout at this point added an extra 18 hours to the standard 20 hours associated with using the weak-mix methodology.

Table 2. Sleemanabad Project, Weak-Mix Grout: Design Mix 1 m³.

Cement	130	kg
Bentonite	50	kg
Water	820	Ltrs
Retarder	2	Ltrs
Sodium Silicate	50	Ltrs

This is because the chamber is more than half empty during an intervention, so time is required to pump an additional 40 + m³ of weak mix to replace the excavated material. Additional setting time is required for the extra grout and for cleaning the cutterhead and cutters as there is less soil in the cutterhead to prevent grout sticking to the cutterhead spokes, buckets and cutter tools.

There is no definitive rule applied at Sleemanabad regarding whether or not to pump weak-grout mix before an intervention because the geology changed too frequently for hard and fast rules. The decision was taken based on an observational approach; however, the overriding philosophy was: "if in doubt pump weak-mix grout". In the long term this philosophy vastly reduced the overall average time spent on each intervention by reducing the amount of interventions that needed to be aborted. It also completely negated the need for installing safe haven grout blocks from the surface.

8   CUTTER TOOL AND CUTTERHEAD DAMAGE

Reducing the amount of cutter changes automatically reduces the amount of downtime taken up for cutterhead interventions, and subsequently improves overall production rates. This was the case at Sleemanabad, where replacing six cutters under hyperbaric interventions could take over nine hours. This is in addition to the time spent cleaning and inspecting the cutterhead and cutter tools, which takes approximately six hours, in addition to the 20 hours associated with pumping weak-grout mix. The total intervention time for replacing six cutters was at least 35 hours, which equates to almost six hours per cutter.

Identifying individual damaged cutters as soon as possible is essential. When one cutter gets blocked and stops rotating, it leads to a higher load on adjacent cutters, with a possibility of a cascading failure (wipe out) of all the cutters in the worst cases (Shanahan 2010). Cascade wipe-outs can result in damage to the cutterhead, and the damage can occur over the course of boring only a couple of strokes. The problem with mixed face excavation in constantly changing geology is that it is extremely difficult to predict cutter wear patterns. Moreover, a couple of damaged cutters do not make a substantial difference to the observed operating parameters of a 10.0 m diameter machine being operated at 24,000 kN of thrust and 12,000 kNm of cutterhead torque. To improve the chances of identifying any change in parameters, Robbins deployed the most experienced operators that were available to this project. Also, a policy of carrying out an intervention if any anomaly in boring parameters was observed was implemented.

Another important factor in reducing cutter consumption was minimizing impact damage to the cutter discs in the mixed face geology. Impact damage occurs when the cutter discs are rotating through relatively soft material before coming into contact with harder material, which can result in radial cracks forming in the cutter discs and the discs breaking away from the hubs. The magnitude of the impact is dependent on speed of cutterhead rotation and depth of penetration per revolution. The higher the percentage of soft material in mixed geology, the higher the risk of damage to the cutter discs.

The closer the cutter is to the periphery of the cutterhead, the higher the risk of damage. Restricting cutter travel speed is the most important factor in avoiding impact damage and the rule of thumb used at Sleemanabad was a maximum cutter travel speed of 30 m/min. If we consider that the gage cutters on a 10.0 meter diameter cutterhead travel a distance of 131.42 m per revolution, then 0.95 rpm is identified as the maximum cutterhead rotation speed in order to remain beneath the maximum cutter travel speed of 30 m/min in mixed face conditions.

The second most important mechanism in reducing impact damage is restricting the depth of cutter penetration at the point of impact between soft and harder material (see Figure 8). Experience on the Sleemanabad project identified this depth to be approximately 8 mm.

Figure 9 shows two mixed face conditions typical of the transition zones that were encountered along the alignment of the Sleemanabad project. In the mixed face conditions shown in A the gage cutters would travel through 90° of soft material before coming into contact with rock. If we look at a penetration rate of 15 mm/revolution (which is achievable in soils and softer rock), the penetration at the point of contact with the harder geology is as follows:

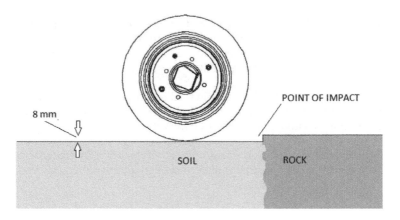

Figure 8. Restricting depth of cutter penetration at point of impact between hard and soft material.

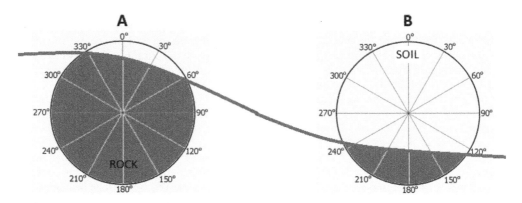

Figure 9. Two mixed face conditions typical of the transition zones at Sleemanabad.

15 mm/360°*90° = 3.8 mm penetration at the point of contact with the harder geology, which is within the acceptable parameters set at site.

In the mixed face conditions shown in B the gage cutters would travel through 240° of soft material before coming into contact with rock. With the same penetration rate of 15 mm/revolution the penetration at the point of contact with the harder geology is as follows:

15 mm/360°*240° = 10 mm penetration at point of contact with the harder geology, which is beyond the acceptable limits set at site. In this case the TBM main thrust pressure would be decreased to reduce the penetration per revolution to approximately 12 mm (12 mm/360°*240° = 8mm).

Face mapping was carried out during each intervention and muck samples taken during each stroke to provide the best possible information to ensure correct operating parameters could be identified. Information on the actual boring parameters of the machine was downloaded from the data logger on a daily basis to confirm that the designated operating parameters were being adhered to.

9 MUCK SPILLAGE

Muck spillage was the most significant cause of delays after failed cutterhead interventions and cutterhead damage. Major spillage often occurred immediately after failed interventions or extended interventions, as well as in areas containing insufficient fines to hold a plug in the

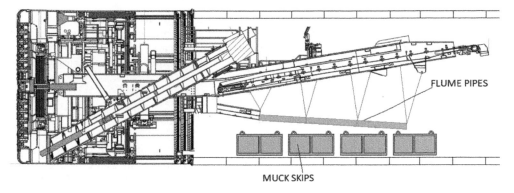

Figure 10. Overlapping flume pipes were used to funnel spillage into muck skips.

screw conveyor. This is because air losses and/or reduced face pressure allowed water in the surrounding geology to migrate into the chamber of the machine. When boring recommenced, this water mixed with excavated material and turned it into a slurry-like consistency, which resulted in significant amounts of spillage. As with the stretches containing insufficient fines, pumping polymer into the chamber and screw conveyor had no noticeable effect on the amount of spillage. Because the amount of spillage could not be reduced, the only remaining solution was to spend a set amount of time setting up a system to catch the spillage rather than spend substantially more time cleaning it up. As mentioned previously a catchment box and slurry pump were available at site but the amount of spillage greatly exceeded its capacity. At the start of boring a stroke the amount of spillage was as high as 200 L/minute consisting of approximately 90% ground water and 10% a mixture of silts, fines and small rock chips.

The solution applied was to fix a permanent chute below the transfer point of the screw conveyor/TBM belt conveyor. Before each boring stroke commenced four muck skips were placed on and behind the segment feeder. The segment crane and segment feeder were used to place the skips into the correct locations and remove them after the completion of the boring stroke. A series of overlapping flume pipes (half pipes) were used to funnel the spillage into the skips (see Figure 10). The excess water was allowed to flow over the upper edges of the skip and as each skip was filled with silt/fines/rock chips a flume pipe was removed to allow the next skip in line to be filled. The excess water was pumped into a settling tank before being pumped out of the tunnel.

Setting up the muck skips and flume pipes took approximately 30 minutes. Removing them at the end of the stroke and cleaning up the minor spillage took approximately 60 minutes. Another 30 minutes was lost loading a ring onto the segment feeder because the muck skips were located on the segment feeder while boring the stroke. A total delay of around two hours per stroke was a major improvement compared with up to eight hours that had previously been spent cleaning spillage after each stroke.

10 TAIL SKIN BRUSH LEAKAGE

Another problem that had to be overcome when Robbins took over the project was that the tail skin brushes had been damaged. Replacing damaged brushes in the geological and hydrological conditions present at the time of the takeover needed to be avoided if at all possible. It would have been an extremely difficult and time-consuming operation with a real risk of losing ground, resulting in an inundation of water and silts into the tunnel. To resolve this problem a back-to-basics approach was adopted. Rice chaff was sourced from a local farmer who had recently harvested and threshed his crop. Four bags (approximately 0.1 m³) of chaff was added to each 8 m³ batch of grout. Initially grout could be seen leaking through the tail skin brushes, before small quantities of chaff were also observed escaping through the brushes.

The amount of chaff escaping gradually reduced until both grout leakage and chaff leakage stopped completely. The use of chaff will continue until the machine enters geology that is suitable for replacement of the brushes. Preventing grout loss and silt ingress through the tail skin aided production rates significantly as cleaning of grout and silt from the ring build area and segment feeder had been a major issue.

11 CONCLUSIONS: IMPROVEMENT IN PRODUCTION

On the Sleemanabad project it had taken 6.5 years to complete just over 1.6 km of tunnel. Even after deducting almost two years of stoppages due to commercial issues the average production rates in the remaining 4.5 years equate to less than 30 m per month. The improvements described in this paper, along with mobilization of highly experienced personnel, improved maintenance regime, improved planning of production activities and improved utilization of downtime resulted in a huge improvement in production rates. During the first four months after the restart 400 m of boring was completed. Neither new technology nor additional equipment was deployed on the project. The principle reason for the improvement in production rates was the experience and skill set of the team that was mobilized. In an industry that is becoming increasingly driven by the introduction of the latest available technology it is important that we don't lose the skills that have been developed and passed down by generations of tunnellers. New technology often makes our task easier, but when operations don't go as planned, there really is no substitute for experience.

REFERENCES

Shanahan, A. 2010. Cutter instrumentation system for tunnel boring machines. Proceedings North American Tunneling 2010
Babendererde, T & Elsner, P. 2014. Keeping the face support in soft ground TBM tunnelling. Geotechnical Aspects of Underground Construction in Soft Ground – Yoo, Park, Kim & Ban (eds), Korean Geotechnical Society, Seoul, Korea

Tunnels and Underground Cities: Engineering and Innovation meet Archaeology,
Architecture and Art, Volume 7: Long and deep tunnels – Peila, Viggiani & Celestino (Eds)
© 2020 Taylor & Francis Group, London, ISBN 978-0-367-46872-9

Experimental determination and plausibility proof of the longitudinal displacement profile for deep tunnels for case studies at the Brenner Base Tunnel

T. Cordes, C. Reinhold & K. Bergmeister
Brenner Base Tunnel, BBT-SE, Innsbruck, Austria

B. Schneider-Muntau & I. Bathaeian
Unit of Geotechnical and Tunnel Engineering, University of Innsbruck, Austria

ABSTRACT: The loading of the tunnel support in deep tunnels is one crucial factor in tunnel design. For a realistic assumption of the tunnel loading the deformation response for deep tunnels due to unloading is of importance. One possible measure of this deformation is the radial displacement of the tunnel, both, the pre-deformations ahead of the face and the convergences in the excavated area. This measure is often predicted by theoretical longitudinal displacement profile approaches based on several simplifying assumptions and, therefore, should be treated with caution, especially in case of deep tunnels in anisotropic and inhomogeneous rock the accuracy has not been investigated so far. In this study, the measurements of longitudinal displacement profiles for two lithologies at the BBT are presented. The entire radial displacement distributions (ahead of the tunnel face to the decayed displacements in the excavated area) have been measured by 40 m long horizontal chain inclinometers installed just above the tunnel crown prior to tunnel excavation. The plausibility of the measurements of the longitudinal displacement profiles is investigated in this contribution and contributes to a better understanding of deep tunnel induced deformation.

1 INTRODUCTION

The Brenner Base Tunnel (BBT) is a flat alpine crossing rail tunnel from Innsbruck in Austria to Fortezza in Italy. This deep tunnel system has a total length of 230 km. Two single track tunnels and one exploration tunnel are excavated through four main lithology of: *Innsbrucker quarz phyllite, Bünder schist, Central gneiss* and *Brixner granites* (see Figure 1).

The above-mentioned rock structures are under stresses, caused by their overburden and their geological formation prior to the excavation. Along the axis of the Brenner Base Tunnel, the primary stress state reaches a magnitude of 30–50 MN/m² with an overburden of 1800 m. In comparison with the maximum bearing capacity of a concrete lining equal to 2,5 MN/m², the primary stress due to the overburden exceeds the bearing capacity by a factor of 12–20 times. The required unloading of the rock mass for the construction is achieved through excavation. This unloading process is for deep tunnels of important interest. Regardless of the tunneling method (TBM, ADECO or NATM), the expected deformations have great influence on the planning of the excavation process, the design of the support or on TBM specifications (e.g. shield length, overlapping space and drive force). One measure for the deformation response is the radial displacement of the walls of the excavated tunnel towards the cavity. Distribution of these radial displacements along the longitudinal axis forms the longitudinal displacement profile (LDP). LDP comprises the radial displacements before the tunnel face to the final point ahead of the face, where the convergence decays. This study describes the experimental determination of the LDP in two different lithology of the BBT by means of

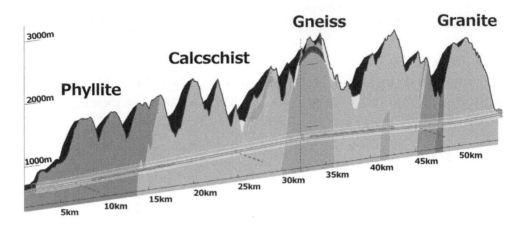

Figure 1. Lithology of the Brenner Base Tunnel.

40 m long horizontal chain inclinometers. The chain inclinometers were installed close to the crown prior to tunnel excavation. The measured radial displacements in dependence on the relative distance to the face are statistically evaluated. In this contribution approaches for a plausibility control of the measurements are presented. The measurements were conducted in a section of the tunnel in Austria built with NAT Method, by which face stabilization measures were not executed.

2   SET UP OF THE LPD MEASUREMENTS

In conventional excavation procedure of the tunnel, the installation of the convergence measurement points is conducted first after the blast and installation of the initial support systems. These measurement gauges can, therefore, only register the deformations that the rock-support system experiences due to excavations which are yet to happen. The determination of the deformations ahead of the face, the so-called pre-deformations, is done by use of the theoretical relations which are based either on simplified assumptions or on 3D finite element simulations. The determination of the pre-deformation is of great importance for deep tunnels, since either small displacements or large and uncontrolled displacements can have catastrophic consequences. Especially in case of machinery tunneling in deep tunnels, the determination of pre-deformation plays a more crucial role and it influences the decision on the type of tunneling machine (open or shield) and their determination can help avoid the danger jamming. In the special case of the Brenner Base Tunnel, by which about 60% of the tunnel is built by machinery methods, the determination of pre-deformation becomes more crucial. The determination of the pre-deformation is carried out in this study by horizontal inclinometers, which are installed prior to the excavation of the tunnel. This measurement method has already been used in shallow tunnels in soft rock (see Volkmann (2007), Oke (2016), Zhang (2016) or Lisjak (2015)). The installation of such inclinometers is not common in tunneling projects because of their complexity. Firstly, the boreholes should be drilled and in the second step the measurement elements must be placed in before the excavation can continue. This procedure leads to downtime periods of the excavation and consequently increases the costs of the project. The two sections of the Brenner Base Tunnel, Ahrental and Wolf had the advantageous situation that the installation of the measurement elements was possible without substantially delaying the excavation. The inclinometers could be installed parallel to the top of the prospective tunnel in transitional section from the larger profile to the smaller one. The larger sections of the cavern have a cross section of 230 m² in Ahrental and of 210 m² in Wolf and taper both to 60 m² for the investigation tunnel. In Ahrental section, the measurement line was chosen in the southern direction of the exploratory tunnel (see Figure 2) and in Wolf section, the transitional cavern between the

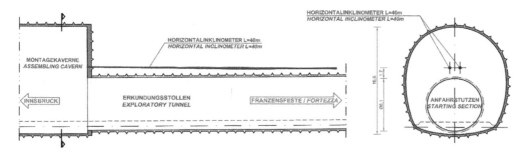

Figure 2. Measurement layout for Ahrental section.

exploratory and the entry tunnels of Wolf south was chosen (see Figure 3). The excavation sequences where top heading bench and invert and the investigation tunnel was supported with shotcrete and rock bolds. Two parallel horizontal inclinometers with a distance of 1.5 m to each other were installed starting from each cavern in order to make redundant measurements. All measurement sections have a length of 40 m with element length of 20 m.

2.1 Ahrental section

For the Ahrental section, the launching pipes of the **TBM-O** of the exploratory tunnel in southern direction were chosen for the arrangement of the measurement lines. The launching pipes are located right after the TBM assembly cavern, where the measurement elements can be located in the transitional area between the larger and the smaller profiles (see Figure 2). Furthermore, in order to measure the face deformations, which occur in axial direction, sliding micrometer and RH-extensometers have been installed, however, the results of these deformations are not considered in the scope of this study. The Ahrental section is located in the lower east alpine *Innsbrucker quartz phyllite* zone. The quartz phyllite intercalates with quartzite schists. The cleavage is mainly orientated flat to the NNE and flat against the driving direction. The faults and joints systems are dipping in both directions, east and west and are also dipping against the driving direction with no predominant direction. The interface – dominated rock mass behavior is characterized by jointed bodies which are falling or sliding out. The overburden is about 950 m. The measurements and the evaluation of those measurements in Ahrental section have been performed first and improvements for the second section could have been derived.

2.2 Wolf section

For the Wolf section, the transitional cavern between the exploratory and the Wolf southern entry tunnels were chosen for the arrangement of the measurement lines. The transitional cavern forms the cross section between the Wolf southern entry and both exploratory tunnel

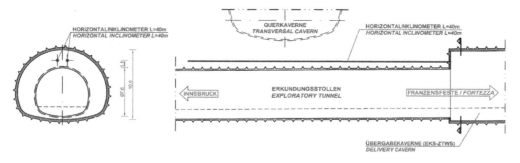

Figure 3. Measurement layout Wolf section.

sections of Wolf north and Wolf south. The measurement elements (the horizontal inclinometers) could be arranged in this section in the front wall appropriately (see Figure 3). The Wolf section is located in the lower Bündner Schist of the Glocknerdecke and the Tauern window and consisst of graphitic calcareous phyllites and calcareous schists with black phyllites. The cleavage is mainly steep orientated and dipping inclined into the driving direction. The fissures and joints are predominantly dipping medium steeply to the east and striking (sub) parallel to glancing to the tunnel axis (predominant orientation). The interface – dominated rock mass behavior is characterized by only small volume ruptures in form of surface spalling. The overburden is about 500 m.

3 ERROR SOURCES AND INPROVEMENTS

Based on the experiences made during the measurements in Ahrental section, the measurement for Wolf section was optimized. An extra convergence measurement point was fixed on the heads of the inclinometers in order to provide a redundant starting point (the conventional procedure for such inclinometers is to assume that the deepest point at the end of the borehole is the fixed point and the deformations are incrementally calculated relative to it. However, due to the extra geodetic measurement point in these measurements, it was possible to additionally determine the incremental deformations based on this point), which makes the calculation of measurement error possible. The displacement profile was anchored to the starting point. The error then follows out of the remaining displacement at the deepest point. The inclinometer measurement line was cased. After the completion of the measurements in Ahrental section, it was not possible to pull out the inclinometers and they had to be left in the boreholes. This happened due to the large local deformations, which could lead to either partial bending of the inclinometer or damages in the borehole. In order to avoid the reoccurrence of such damages and obtain reliable results, the boreholes were equipped with protective cases. The number of geodesic convergence profiles has been increased. The convergence profiles in Wolf section were configured in every 5 m on the walls of the tunnel. The shotcrete support has been executed with only 20 cm thickness, in comparison to 22 cm in Ahrental and with two deformation slots. With this flexible tunnel support on the one hand the work safety was ensured and on the other hand the influence on the tunnel deformation minimized for a most realistic deformation measurement of the rock.

4 RESULTS OF THE MEASUREMENTS

The chain inclinometers applied for both measurement lines had element lengths of two meter. Data registration was conducted automatically at 30-minute intervals; compare Huter (2018) and Unteregger (2015). After each blast, a radial displacement profile was produced. For each measurement line, an evaluation of the radial displacements relative to the distance from the face of the tunnel was conducted (Figure 4). In the Figures 4– 7 the tunnel face is marked with a black line and the unexcavated part of the tunnel is highlighted in grey. Displayed are the displacement profiles which fulfilled the accuracy criteria. Therefore, it was possible to offer one displacement profile for nearly each blast in the longitudinal direction. The changes in the radial displacements decrease by increasing distance in both directions from the face of the tunnel (with positive meaning to move away from the face towards the excavated area and negative meaning to move away from the face towards unexcavated rock). The maximum increase in the displacements occurs at the height of the face. By considering the aforementioned evaluated blast intervals, the validity of the measurements is statistically investigated. The average curves of both investigated sections and their corresponding deviation in terms of the confidence limits are shown in Figure 5. For the determination of the mean value the excluded displacement profiles were not taken into account.

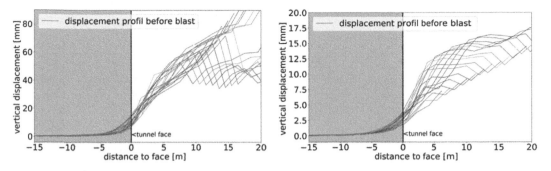

Figure 4. Results of all measured time steps, relative to the tunnel face (Ahrental left, Wolf right), and legend exemplary for all curves, slightly modified after Schneider-Muntau et al. (2018).

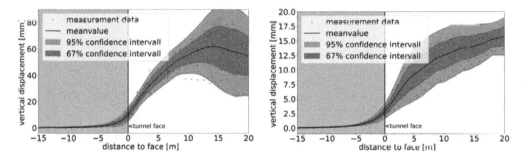

Figure 5. Mean value and standard variation of the LPDs (Ahrental left, Wolf right), slightly modified after Schneider-Muntau et al. (2018).

4.1 Implausible values for Ahrental section

The registered radial displacements in the Ahrental section demonstrate strong and insensible fluctuations in the excavated area (see Figure 4). According to Unteregger (2015), the inclinometers have been damaged during the excavation of the tunnel and, therefore, only the results right after the face are reliable. These fluctuations in the measurements occur after 20 m excavation of the tunnel, when the inclinometers have been exposed to several blasts and have become damaged. The evaluation of the LPDs has been discussed in detail in Schneider-Muntau et al. (2018). Furthermore, an axisymmetric FE simulation of the cavern and its unloading influence on the displacements shows that the influenced area ahead of the cavern is limited to almost 10 meters and after that the unloading influence caused by the cavern is negligible. Therefore, the data from a length of 12 m for Ahrental ahead of the face are not taken into account. The influence of the cavern is not only detectable through the FE simulation, but also the LDP evaluations confirm its influence on the displacements.

The comparison with the geodetic measurements in Figure 6 reveals theoretically not explainable results of the inclinometers. The inclinometer is situated 1.5 m above of the tunnel crown an records larger displacements than at the crown. This is not possible and must be due to the damage of the inclinometer.

Concluding it can be said for the Ahrental section that the results ahead of the tunnel face and at the tunnel face seem to be realistic even though an error control has not been possible due to the missing geodetic surveillance of the head point of the inclinometers. The results in the excavated part of the tunnel are not realistic due to damages on the inclinometer.

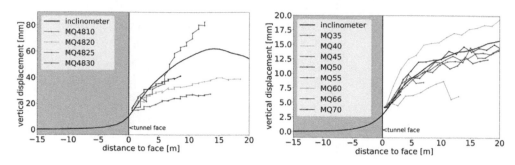

Figure 6. Mean values of the measured LPDs in comparison to the geodetic measurements (Ahrental left, Wolf right), modified after Schneider-Muntau et al. (2018).

4.2 Implausible values for Wolf section

In addition, for the measurement line in Wolf section, a geodesic observation of the starting point of the inclinometer was conducted. As a result, the error of the measurement can be determined since the data from the starting point and the deepest point of the inclinometer in the borehole are available, see Huter (2018). A distinct increase of the measurement error of the right inclinometer from TM 14 led to complete exclusion of this inclinometer from evaluation. The error of the left inclinometer lies below ±1 mm in the distance of 9.7 – 20.1 m of the excavated length of the tunnel and between ±2 mm in the distance of 20.1 – 30.5 m. The evaluation of the data is conducted for the distance of 9.7 – 30.5 m. The influence of the cavern is also apparent for the Wolf section; however, this influence is minute due to the smaller profile areas (see Figures 2 and 3).

The comparison with the geodetic measurements in Figure 6 shows are very good match of the measured displacements (inclinometer vs geodetic measurement). Also this inclinometer is situated 1.5 m above of the tunnel crown. One would expect a decrease of the deformation with increasing distance from the tunnel face. This has been shown in Finite Element calculation and also extensometer measurements placed perpendicular to the tunnel fortify this assumption. In this case we have to deal with blocky rock behavior, as has been described in section 2.2. Blocky rock behavior leads to an inhomogeneous distribution of the deformation over depth. When a block is moving in total, the displacements can be the same at the tunnel crown and at a certain distance from the tunnel crown.

Concluding it can be said that the improvements made for the construction of the Wolf section led to an improvement of the measurements results. The results are reliable even the already excavated part of the tunnel. The casing of the inclinometer avoided its damaging. A remaining open question is the influence of the stiffness of the tunnel support and of the inclinometer casing on the tunnel deformation.

5 SELECTION OF THEORETICAL APPROACHES FOR LPDS

There are several theoretical relations available to describe the displacement profile in longitudinal direction of a tunnel. Two of these relations Unlu and Gereck (2003) and Vlachopoulus and Diederich (2009) are selected for comparison in this study. The simplifying assumptions of both methods comprise, unsupported circular tunnel profile, no change in the primary stress state due to excavation, homogeneous and isotropic rock behavior. One of the variables of both relations is the maximum convergence, which is not known prior to the excavation, however, it can be approximated by means of theoretical relations (e.g. Sulem and Panet (1987), Salencon (1969)) or FE simulations. The relation suggested by Vlachopoulus and Diederich (2009) considers also the radius of the plasticified zone, which is unknown. Two radial displacement profiles were predicted by means of the two relations suggested by Sulem (1987)

3621

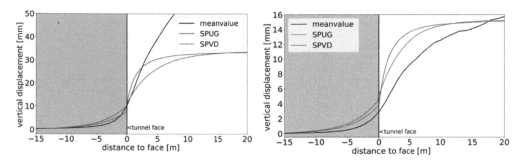

Figure 7. Comparison of the mean curve of the measured LPDs with the theoretical approaches (Ahrental left, Wolf right).

and Salencon (1969) in combination with (Panet (1995), Unlu (2003) and Vlachopoulus (2009)) for determination of maximum convergence and radius of the plasticified zone,

a) SPUG: Maximum convergence after Sulem and Panet (1987) and radial displacement profile after Unlu and Gereck (2003)
b) SPVD: Maximum convergence after Sulem and Panet (1987) and radial displacement profile after Vlachopoulus and Diederich (2009).

The results of the theoretical approaches have been discussed in detail in Schneider-Muntau et al. (2018). In Figure 7, the radial displacements obtained from the theoretical relations and from the measurements are compared (calculated for parameters in Table 1).

5.1 Comparisons for Ahrental section

It is very apparent, that the measurement results and the theoretical approaches fit well for the region ahead of the tunnel face and at the tunnel face, especially for Ahrental section. The displacement values at the face are in good agreement, see Figure 7. The large deviations in the excavated area of Ahrental section are due to the faulty measurements of the damaged inclinometers. A comparison of values displayed as relative values normalized with the maximum convergences (Table 2) has therefore to be considered with caution. The large range of the confidence interval (cf. Figure 5) leads to a not reliable maximum displacement measurement.

5.2 Comparisons for Wolf section

In the area ahead of the face and at the face, the theoretical relations deliver slightly overestimated values in comparison to the measurements, see Figure 7. In this section, although the displacements obtained by geodesic and inclinometer measurements differ from each other, since an error estimation of the measurements had been conducted; the resuLts are validated and are classified as reliable. The geodesic measurements in the excavated area should be

Table 1. Parameters for the calculation according to Schneider-Muntau et al. (2018).

	Ahrental	Wolf
Module of Elasticity	6.000 – 8.000 MPa	4.000 – 6.000 MPa
Poisson's ratio	0.18	0.18
Friction angle	34 – 38°	29 – 32°
Cohesion	1.5 – 2.5 MPa	1.3 – 2.1 MPa
Dilatancy angle	5°	5°
Primary stress state	27,000 MPa	12,500 MPa

Table 2. Comparison of ratio of displacement to the maximum displacement in percent at tunnel face and at the distance of one tunnel radius from the face at Ahrental section.

	$-r_0$	Face	$+r_0$
Measurements	3.10	16.43	56.08
SPUG	9.01	32.86	87.76
SPVD	11.10 – 11.40	33.19 – 33.30	68.64–75.53

Table 3. Comparison of ratio of displacement to the maximum displacement in percent at tunnel face and at the distance of one tunnel radius from the face at Wolf section.

	$-r_0$	Face	$+r_0$
Measurements	4.14	18.66	49.88
SPUG	9.01	32.86	87.76
SPVD	11.04 – 11.34	33.18 – 33.27	70.12–76.71

smaller than those obtained from the inclinometer in the rock. This could be attributed to the blocky behavior of the rock. As the maximum values in the excavated area coincide, the normalized values provide a good agreement, see Table 3. The experimental data shows, that the displacements have not reached their maximum after 20 m (as the theories predict). The maximum displacements are therefore slightly overestimated. Nevertheless, the narrow confidence interval in Figure 5 is sign for a good estimation of the mean value.

6 CONCLUSION

Determination of pre-deformations and rock unloading in the unexcavated part of the tunnel (ahead of the face) are of importance for deep tunnels. The available theoretical relations for determination of these deformations are based on simplifying assumptions, regarding the mechanical behavior of the rock (as e.g. homogeneity, isotropy), or are derived from FE simulations. On the other hand, validation of the theoretical relations is not simple, since the measurements must be carried out in the unexcavated part of the tunnel and near to the blasted face of the tunnel. In this study, the measured radial displacement profiles in longitudinal direction for two different lithological zones of the Brenner Base Tunnel (Ahrental and Wolf) are presented and compared with the results predicted by the theoretical relations. The displacement profiles show good qualitative agreement with the displacement profiles predicted by the theoretical approaches in the unexcavated part of the tunnel. Moreover, the displacements at the face of the tunnel could be well estimated by the theoretical relations. Since the two investigation areas have been performed one after the other valuable experiences could have been gained within the first experiment and the setup had been modified to obtain better results for the second investigation area. In the excavated part of the tunnel, the inclinometer measurements deviate from the geodetic measurement for the first section. After the improvements for the second section the results of the inclinometer corresponds to the results of the geodetic measurement. Comparing the measurements in this study with the theoretical approaches, the influence of the anisotropic and inhomogeneous rock may be responsible for the deviation from the predicted values by the theoretical relations, which are based on isotropic and homogeneous assumptions for rock behavior. These results could have become more comprehensive, if investigations in different lithology and for different overburdens were available. Our next step is towards determination of longitudinal displacement profiles by means of 3D-FE simulations.

This will also reveal the effects of actually unknown stiffness of the tunnel support (despite of the deformation slots). In the near future, a cross cavern will be excavated directly over the measurement line of the Wolf section and the inclinometer measurements will be continued for this section in order to investigate the influence and interaction of the new cavern on Wolf section. This information is of practical value regarding the interaction of crossing tunnels on each other.

REFERENCES

Hofstetter, G. & Neuner, M. & Schreter M. & Unteregger, D. 2016. Numerical Modeling of Deep Tunnels, final report on the interim research project, BBT, unpublished

Huter, M. 2018. Vergleich der experimentell ermittelten Verschiebung durch Gebirgsvorentspannung mit analytischen Ansätzen. Masterthesis. University of Innsbruck

Lisjak, A., Garitte, B., Grasselli, G., Müller, H.R., Vietor, T. 2015. The excavation of a circular tunnel in a bedded argillaceous rock (Opalinus Clay): Short-term rock mass response and FDEM numerical analysis, *Tunnelling and Underground Space Technology*. 45. page 227–248.

Oke, J., Vlachopoulos, N., Diederichs, M. 2016. Semi-analytical model for umbrella arch systems employed in squeezing ground conditions. *Tunnelling and Underground Space Technology*. 56, page 136–156.

Panet, M. 1995. Calcul des Tunnels par la Methode de Convergence-Confinement. Press. L'Ecole Natl. Des Ponts et Chaussee. page 178.

Salencon, J. 1969. Contraction Quasi-Statique d'une Cavite a Symetrie Spherique ou Cylindrique dans un Milieu Elasto-Plastique. Ann. Des Ponts et Chaussee 4, page 231–236.

Schneider-Muntau, B., Reinhold, C., Cordes, T., Bathaeian, I., Bergmeister, K. 2018. Validierung der Radialverschiebungen im Längsprofil durch Messungen beim Brenner Basistunnel. *Geomechanik und Tunnelbau*, in print

Sulem, J., Panet, M., Guenot, A. 1987. An analytical Solution for time dependent displacement in a circular tunnel. *Int. J. Rock Mech. Min. Sci. Geomech.* page 155–164.

Unlu, T., Gereck, H. 2003. Effect of Poisson's ratio on the normalized radial displacement occurring around the Face of a circular tunnel. *Tunn. Undergr. Sp. Tech.* 18, page 547–553.

Unteregger, D. 2015. In situ Messprogramm Anfahrtsstutzen EKS BBT, technical report, Innsbruck, unpublished

Vlachopoulus, N., Diederichs, M. S. 2009. Improved longitudinal displacement profiles for convergence confinement analysis of deep tunnels. *Rock Mech. Rock Eng.* 42, page 131–146.

Volkmann, G., Schubert, W. 2007. Geotechnical model for pipe roof supports in tunneling. *Underground Space – the 4th Dimension of Metropolises – Barták, Hrdina, Romancov & Zlámal (eds)*, Taylor & Francis Group, London

Zhang, Z.X., Liu, C., Huang, X., Kwok, F., Teng, L. 2016. Three-dimensional finite-element analysis on ground responses during twin-tunnel construction using the URUP method. *Tunnelling and Underground Space Technology*. 58. page 133–146.

Tunnels and Underground Cities: Engineering and Innovation meet Archaeology,
Architecture and Art, Volume 7: Long and deep tunnels – Peila, Viggiani & Celestino (Eds)
© 2020 Taylor & Francis Group, London, ISBN 978-0-367-46872-9

Performance of hard rock double shield TBM in tailrace tunnel of Uma Oya project, Sri Lanka

H. Darabi Kelareh, A. Rahbar Farshbar, A.H. Hosseini & F. Foroutan
Farab Co., Tehran, Iran

D. Dodangeh
Mahab Ghodss Co., Tehran, Iran

ABSTRACT: Uma Oya Multipurpose Development Project is under construction in Sri
Lanka. This project involves 2 Rolled Concrete Core (RCC) Dams connected through a
tunnel currently under construction by Drill and Blast method. The water is transferred from
first diversion dam to the 2nd regulatory dam and from which, it will be transferred through a
15.2 km headrace tunnel to the top of a 650 m deep drop shaft that feeds the high pressure
water to an underground powerhouse and turbine chamber for generation of 120 MW of electricity. Tailrace tunnel conveyance the water from the powerhouse to the outlet of the project.
The plan was to excavate headrace and tailrace tunnel by two 4.3m diameter double shield
TBMs. This paper will review the history of the Tailrace Tunnel along with the geology of the
site. The operation of TBM in Tailrace tunnel will be discussed in more detail, focusing on the
challenges of excavating through the hard rock.

1   INTRODUCTION

Uma Oya Multipurpose Development Project (UOMDP) is a hydro mechanical project in Sri
Lanka, targeted generating hydro-electric power, transferring water for irrigation purposes
and controlling seasonal devastating flood. This project lies in the south-eastern part of the
central highland region of Sri Lanka.

The project consists of two concrete roller dams. The first dam will be built on Uma Oya
River in Puhulpola area. Through a tunnel with 3.7km length (a transferring tunnel), the
water will be transferred from Uma Oya River to Dayaraba Dam Reservoir which is built on
Mahatota Oya River.

A long headrace tunnel with 15.2 km length and a vertical shaft with 618 m height will
transfer water from there to an underground power station. Through a tailrace tunnel with
3.7 km length, water in the power station will be directed towards Alikuta Oya River which is
a branch of Kirindi Oya River.

The excavation of the major tailrace tunnel is done by full mechanized method and by using
a double shield excavating machine with 4.30m diameter. This is the first use of a TBM in Sri
Lanka. Excavation by DS-TBM has become a standard method for long small to medium
sized pressure tunnels. Benefits comprise of the immediate support, the high flexibility of the
TBM and the lining which can be utilized as long-term support. The tailrace tunnel excavation
was commenced on 26 Oct 2013 and fulfilled on 18 Apr 2015.

The bedding of the segmental lining (Hexagonal type) is established by pea gravel filling of
the annulus gap between the segment and the rock at the side wall and crown segments immediately after ring building, while the invert segment is grouted by mortar.

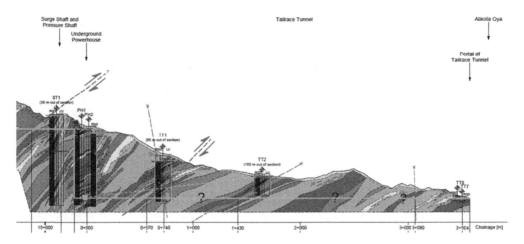

Figure 1. Tailrace tunnel geological section (question marks indicate heterogeneous nature of bedrocks).

2 GEOLOGY

The Highland Complex is composed of supracrustal and metaigneous rocks in Sri Lanka. Most of the rocks in the complex have ancient ages of 2.0–3.4 billion years. The supracrustal rocks of the Highland Complex comprise of shallow-water metasediments consisting of meta-quartzites derived from sandstones; quartzo-feldspathic gneisses derived from Arkose (type of sandstone); pelitic gneisses derived from greywacky and shale; marbles derived from limestone and dolomitic limestone; as well as a subordinate bimodal volcanic suite. The metaigneous rocks of the Highland Complex are now charnockites and Charnochitic rocks (granular rocks consisting mainly of quartz, feldspar, and orthopyroxene, plus other minerals). Figure 1 shows the geological profile of tailrace tunnel (JV, 2011a).

The overburden material is mainly formed by residual soils on the peneplains, a mixture of residual soils and colluvial material in the slopes of the valleys; alluvial material is subordinated and restricted to riverbeds and wider valleys. Bedrock units are as follows:

- Leucocratic garnet-quartz-feldspar gneisses
- Undifferentiated, dark gray Charnochitic gneisses
- Quartz-rich gneisses and pure quartzite
- Marbles and Calc-silicate gneisses/granulites

Due to the heterogeneity of the bedrock units and their rapid changes, it is not possible to predict the appearance of the different rock types along the tunnel alignment (JV, 2011b).

The most critical sections of the tunnel excavation has been encountered at the downstream part (portal zone) and in major joints and fault zones, characterized by dense fracturing and a high degree of weathering, including partial ingress of groundwater.

3 SCOPE OF THE WORK

According to the contract scope of work, mobilization included in 3 main parts as below:

- Portal preparation (digging, concreting & portal opening)
- Mobilization of the access road
- Assembly of the TBM

First 70m of the Tailrace tunnel excavated by D&B method to make a start chamber and gripping zone. Total length of the Tailrace tunnel is about 3718 which started from downstream of the tunnel and with 0.161% dip. 3401m out of 3718m was excavated by TBM and about 220m long by cut and cover and other parts by drill and blast methods

Excavation of the main part of the tunnel was based on mechanized excavation with a Herrenknecht Double Shield TBM in 4.3m diameter. It was equipped to 27 pcs of the cutter head and designed to excavation in Hard rock as the project geology.

Finally after finishing the excavation, plane was to made a cut & cover structure in 230m length and outlet structure and refilling the site area.

4 EXCAVATION

At chainage 0+150 (approximately CH 3+344 m) a high inflow of water occurred. Exact amounts were not measured, but the high inflow persisted. A probe drill was performed and revealed two weak zones (interpreted from drill probe penetration rates). After advancing with the machine a large ingress of water and sand occurred.

At chainage 0+164 (approximately CH 3+330 m) a large cavity was encountered (Figure 1). The cavity was located to the right side of the machine extending to 3.5 to 4 m above the top of the TBM. Visual inspection of the cavity by the technicians at the site revealed highly weathered surfaces of the cavern. It was not clear if the earlier water ingress and occurrence of the cavern are related.

The boreholes TT03 and TT04 show highly weathered and fractured rock masses and even residual soils up to depths below the alignment of the tunnel. Sketches made of the cavern could suggest a cave-in or wash-in of highly weathered material/residual soil possibly related to the earlier water and sand ingress. According to the technicians on the TBM this was not confirmed by observations made by technicians during excavation with the TBM up to the relevant chainage.

Excavation has progressed up to chainage 0+179.12 km (measured from first segment). The TBM is operating in single mode. The surrounding rock mass is too weak for the grippers to be used. As a consequence all thrust forces act directly and completely on the segments. The TBM passed a curve (radius 620 m) which causes unequal loading of the segments. The curve orientation was to the left relative to the boring direction.

Especially in zones of highly weathered or fractured rock mass where relatively large voids between the segments and the rock mass occur it was very important that backfilling was performed before advancing. When operating in single mode backfilling should be performed before advancing in any case. When it is possible, it should be checked at the tail shield if the back-fill is sufficient.

Excavated material at chainage 2882.55 consists of broken rock material in a sand/silt/clay matrix since approx. chainage 2895, confirming the expected fractured/sheared zone. So gripper could not be used due to insufficient wall support. TBM was operating in single mode and had dived approx. 23 mm overnight. Then at chainage 2878.35 excavated material consist of rock chips and some cobles and small boulders, indicating improving rock mass conditions. The rock face at the CH could be inspected only intact rock mass visible. Shear zone has been passed. Length approximately 15~17m.

5 EXCAVATION CHART

Excavation started on 4 March 2014 and breakthrough was on 8 of December 2014. Figure 2 shows the actual monthly boring progress in accordance with the actual plan. It has to be reminded that the less progress at the ends was because of some preparation in the TBM and disassembly chamber.

The best progress was 724m when the TBM excavated more than 2Km. It was based on the good learning curve and also the efficiency of the peoples with changing from 2 shifts to 3 shifts. In June 2014, excavation drop as the distance of the TBM from the portal increased. The variation of monthly advance rates is in Figure 3.

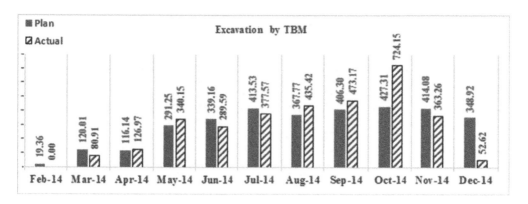

Figure 2. Monthly boring progress.

Table 1 shows the boring time analysis in different months. Also it is noted that with changing the working shifts from 2 to 3 in longer depth, efficiency and accordingly the excavation meters increased. Although the increased in July 2014, but with changing to 3 shifts, based on efficiency and skills of the personnel and good working curves, the excavation rate increased. Best records included in Table 2.

Table 3 summarized average time per day of all the activities in different months. Increasing the cutter changing time under the more efficiency of the personnel makes the more boring time and achieving more excavation rate in Tailrace Tunnel. Table 4 shows the monthly average Boring Time and Excavation rate of the TBM M-1684 in Tailrace tunnel boring of Tailrace tunnel of Uma Oya Project. Chart 1 illustrate the advance rate of the TBM based on the results of the data of Table 4. There is a correlation of the rock mass condition and the advance rate.

Figure 4 and Figure 5 show the utilization of the TBM in Tailrace tunnel based on 2 and 3 shifts, respectively. Also the results summarized in Figure 6.

Table 1. Boring time.

Description		Mar	Apr	May	Jun	Jul	Aug	Sep	Oct	Nov	Dec	Total
Working Days		28	30	31	30	31	31	30	31	30	8	280
Daily Shift (no)		1	2	2	2	3	3	3	3	3	3	-
Shift (Hrs.)		12	12	12	12	8	8	8	8	8	8	-
Excavation (m)		81	127	340	290	377	435	473	724	363	53	3264
Daily	Ave. (m)	2.9	4.2	11	9.7	12.2	14.1	15.8	23.4	12.1	6.6	11.7
	Max (m)	10.3	13.9	21.6	16.9	18.5	25.8	30	45.6	24.9	15.1	45.6
Shift	Ave. (m)	2.9	2.1	5.5	4.8	4.1	4.7	5.3	7.8	4.1	2.2	4.7
	Max (m)	10.3	10.8	14.4	10.8	8.4	12.1	12.7	16.8	12.6	12.7	16.8

Table 2. Excavation performance records.

Comparison of Excavation Performance Description	QTY (m)	Date
Average Daily Excavation	11.66	4/Mar/2014 ... 8/Dec/2014
Best Shift	16.81	14/Oct/2014 (Shift B)
Best Day	45.63	12/Oct/2014
Best Week (Monday To Sunday)	198.15	6/Oct/2014 ... 12/Oct/2014

Table 3. Daily average of the activity time for each activity based on each month.

Month	21~30 Apr	May	Jun	Jul	Aug	Sep	Oct	1~22 Nov	Total
Days	10	31	30	31	31	30	31	22	206
Excavation (m)	7.95	11.00	9.65	12.18	14.05	15.77	23.36	16.40	14.57
Available Time (Hr.)	20:24	23:14	21:36	22:27	22:27	22:24	22:43	22:11	22:27
Boring Time (Hr.)	4:32	4:31	4:53	6:26	6:59	7:45	7:40	8:11	6:34
Activities Time (Hr.)	3:05	1:57	1:46	2:16	2:13	2:45	3:38	3:30	2:33
Delay Time (Hr.)	11:29	14:05	10:11	7:20	9:04	7:12	6:36	5:03	8:39
Maintenance (Hr.)	0:16	0:37	2:24	2:54	2:53	3:03	3:11	1:46	2:26
Cutter Change (Hr.)	0:00	0:13	1:03	2:23	1:14	1:39	1:38	3:33	1:36
Stopped Time (Hr.)	3:36	0:46	2:24	1:33	1:33	1:36	1:17	1:49	1:33
Other Time (Hr.)	1:03	1:51	1:19	1:09	0:04	0:00	0:00	0:08	0:40
Total Time (Hr.)	24:00	24:00	24:00	24:00	24:00	24:00	24:00	24:00	24:00

Table 4. Monthly average boring time and excavation rate (Based on Table 3).

Month	Apr	May	Jun	Jul	Aug	Sep	Oct	Nov	Total
Excavation Total (m)	79.5	340.9	289.6	377.6	435.4	473.2	724.2	360.9	3,081.1
Boring Time Total (min)	2,715	8,400	8,785	11,955	13,000	13,960	14,255	10,810	83,880
Excavation Ave. (m)	7.9	11.0	9.7	12.2	14.0	15.8	23.4	16.4	14.3
Boring Time Ave. (min)	272	271	293	386	419	465	460	491	388

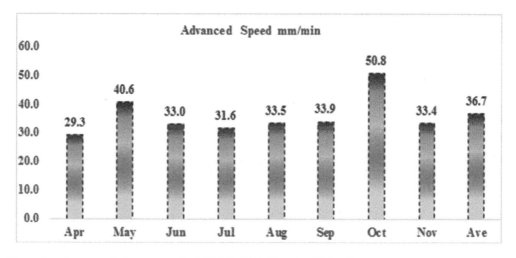

Figure 3. Average of advance rate by TBM M-1684 (Based on Table 3)

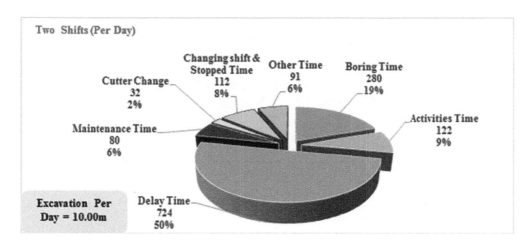

Figure 4. Utilization chart based on 2 shifts.

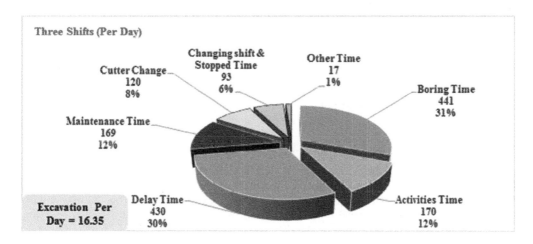

Figure 5. Utilization chart based on 3 shifts.

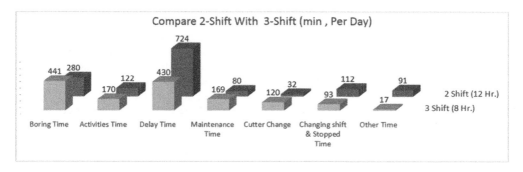

Figure 6. Comparison of TBM utilization based on 2 shifts and 3 shifts.

Table 5. Monthly Disc Cutter Norms.

Description		Mar 14	Apr 14	May 14	Jun 14	Jul 14	Aug 14	Sep 14	Oct 14	Nov 14	Dec 14	Total
Excavation (m)	Monthly	81	127	340	290	378	435	473	724	363	53	3264
	Cumul.	81	208	548	838	1216	1651	2124	2848	3211	3264	
Cutter Disc (No)	Monthly	0	0	8	22	40	27	37	27	44	11	216
	Cumul.	0	0	8	30	70	97	134	161	205	216	
Norm (meter)	Monthly	-	-	42.5	13.2	9.4	16.13	12.8	26.8	8.3	4. 8	15.1
	Cumul.	-	-	68.5	27.9	17.4	17.02	15.9	17.7	15.7	15.1	

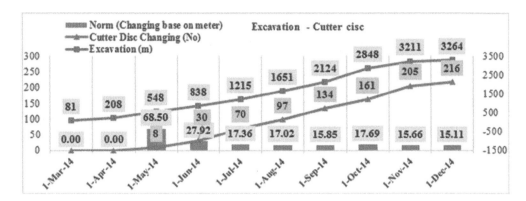

Figure 7. Diagram of Disc Cutter Norm.

6 DISC CUTTER

In this project all the standard 17 inch hard rock disc cutter used for the excavation. Table 5 shows the monthly norm. In October 2014, norm of the disc cutter was 27m, which the advance rate was 51mm/min (Chart 1) and best excavation rate was average 23 m/day which shows the good rock condition for the excavation. But in June 2014, although the advance rate was 33 mm/min and the norm of the disc cutter was 28m, delay times increased as it was referring to the efficiency also. Figure 7 shows the diagram of the disc cutters in Tailrace tunnel. Totally 216 Rings changed, which is 15 rings/day in 3263m excavation of hard rock.

7 OTHER MAIN ACTIVITY

One of the main other activity in this tunnel which had to be continuously done in parallel of the excavation, was the cleaning. Excavation of the hard rock material produce lots of the fine material which will settle in the tunnel. Especially when the TBM excavating upwards, this material will settle in whole the way and even make small dams, will not make more helps. The size of this material bakes a solution and they travel more than 3km. In this case all the way has to be clean because of the water level in the tunnel in small diameter and locomotive transport much easier.

Normally for this reason, more labors have to be involved. But in this tunnel with using the mechanized system by a small excavator, it was much easier and faster with less man power.

Design of the segment for this project was based on installation of 2 railway and always 2 line accessible. So cleaning was made a minimum delay in the excavation and in parallel of the maintenance or any stop time. It should be reminded that in both system, one of the line will be closed. To have more efficiency, it is suggested to clean fine material end of the backup before consolidation of this fine material. Figure 8 shows the cleaning system of fine material of the excavation material.

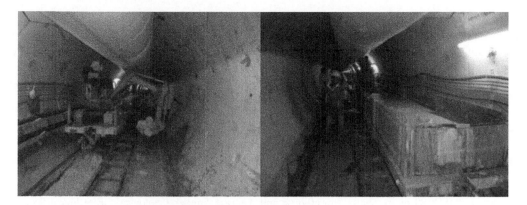

Figure 8.   Cleaning of the fine material with mechanized system.

 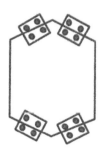

Figure 9.   Bracket Installation.

8   SEGMENT STEPPING

The other major difficulties in this tunnel was the stepping of the hexagonal segments. The most reasons were, namely:

– Excavation and segment installation in weak zone
  Shield vibration in hard rock boring
– Pea gravel quality

For the solution of this systematic matter, 2 solutions did in the tunnel. First solution was to install temporary bolt during the segment installation in the shield area After segment installation and injection of the pea gravel, this bolts removed.

Another solution was install of the steel bracket between the segment joints (Figure 9). This bracket bolted to the segments both sides the joints and after the grouting of the pea gravel behind the segments, the bolts and also the brackets removed. These system helps the segments to be fixed with the less deviation or steps in installation position and after caulking and contact grouting, they were removed.

9   CONCLUSIONS

This paper reviewed the Tailrace Tunnel of Uma Oya Project in Sri Lanka which was excavated by mechanized TBM as designed for Hard Rock. Geological condition and excavation parameters was reviewed. Also most difficulties and the site solutions described in this tunnel.

According to the experience of mechanized excavation method in this medium length tunnel, it is clear that to continue the efficiency of the tunnel, all the parameters should be controlled. Regarding to the personnel planning, now it is clear that with changing the shift from 2 shifts to 3 shifts, although the maintenance increased 100% and cutter change time more than 3 times, but delays decreased in 40% & boring time increased 60% accordingly. Also the other back up activity in parallel such as cleaning of the tunnel excavation discussed.

REFERENCES

Dietler, T. 2011a. Detail Design Report-Re-Design Tailrace Tunnel, Joint venture of Mahab-Ghodss consulting engineers and Poyry energy limited, DOC. NO: UMO-JV-II-800-10-DR-HD-002-0, Uma Oya Project Studies reports -Phase II.

Dietler, T. 2011b. Geology Shafts, Powerhouse and Tailrace Tunnel, Geological-Geotechnical Section, Joint venture of Mahab-Ghodss consulting engineers and Poyry energy limited, DOC. NO: UMO-JV-II-130-10-DG-HD-002-01, Uma Oya Project studies reports- Phase II.

Tunnels and Underground Cities: Engineering and Innovation meet Archaeology,
Architecture and Art, Volume 7: Long and deep tunnels – Peila, Viggiani & Celestino (Eds)
© 2020 Taylor & Francis Group, London, ISBN 978-0-367-46872-9

Lining stresses in a TBM-driven tunnel: A comparison between numerical results and monitoring data

V. De Gori
Geotechnical Design Group, Rome, Italy

A. De Lillis & S. Miliziano
Department of Structural and Geotechnical Engineering, Sapienza University of Rome, Italy

ABSTRACT: The paper deals with the evaluation of lining forces in TBM-driven tunnels in clayey soils. A 3D numerical model has been developed accounting for the main features of the excavation and construction, such as front pressure, cutterhead overcut, shield conicity, lining installation inside the shield, tail void grouting, grout hardening and injection distance from the shield. To accurately simulate the soil-shield and the soil-grout-lining interaction processes associated with the gap closure, the algorithm works in large-strain mode, thus updating the grid nodes position after each calculation step. A comparison between numerical results and monitoring data recorded during the construction of a hydraulic tunnel in fine-grained soils is presented. The collected data show the evolution of the lining forces due to the advancement of the excavation in undrained conditions and during the subsequent consolidation process. The results confirm the effectiveness of the proposed numerical approach as a design or analysis tool.

1 INTRODUCTION

Numerical modelling of tunnel excavation is a particularly complex issue. Two-dimensional modelling is currently the most common tool in tunnelling practice because of the low requirements in terms of computational effort and complexity of the models. Several methods have been developed to account for three-dimensional effects in simplified 2D analyses. A brief comparison of the main 2D numerical methods is given by Karakus (2007) and by Möller & Vermeer (2008). Still, plane-strain analyses cannot reproduce the three-dimensional complexity of tunnel construction procedures and at the same time 3D modelling is increasingly manageable thanks to the growth of computational power and to the diffusion of commercial codes with efficient computational algorithms.

3D numerical models can simulate the main excavation and construction features explicitly. With particular regard to mechanized tunnelling, the most influencing factors are: front pressure, TBM geometry, annular tail void grouting and grout hydro-mechanical properties, segmental nature of the lining. Several 3D models have been developed in the past years. Kasper & Meschke (2004) developed a 3D model for shield-driven tunnels accounting for the main components of the construction process and used it to investigate both the influence of TBM operation parameters (2006a) and that of grout properties and cover depth (2006b) on the ground surface settlements and on the internal forces in the lining. Advanced 3D models (e.g. Buselli et al. 2012) have also been used to study the interaction with existing buildings. The segmental lining ring has been modelled adopting various simulation strategies, usually involving spring elements to describe the interaction between segments (e.g. Arnau & Molins, 2012; Do et al. 2014). Kavvadas et al. (2017) compared the internal forces in the lining resulting from three different lining models (both including and neglecting the segmental nature of the lining ring) and showed that the presence of staggered joints has a limited effect on the lining forces and on the overall results.

This study focuses on the evolution of the soil stresses and of the lining forces due to the advancement of the TBM in undrained conditions and to the subsequent consolidation process. A 3D numerical model has been developed using the finite difference computer code FLAC3D (Itasca, 2012) and tested against monitoring data recorded during and after the mechanized excavation of a hydraulic tunnel in clayey soil. In the following, after a brief description of the case study's main features, the model is presented and the numerical results are discussed and compared with monitoring data, showing satisfactory agreement and providing direction for further improvements of the model predictive capability.

2 DESCRIPTION OF THE CASE STUDY

The investigated tunnel is a small diameter hydraulic tunnel recently excavated in southern Italy by a small TBM (diameter, D ≈ 4 m). The tunnel runs in a straight line for 4 km, entirely through stiff clays, under cover depths ranging from 10 to 140 m and piezometric heights between 3 m and 75 m. Figure 1 shows the geological profile and the section monitored and analyzed in this study. A detailed description of the case study is reported in de Lillis (2017).

An extensive geotechnical campaign, involving both in situ and laboratory tests, was carried out to investigate the soil affected by the excavation: a sub-appenninian Pliocenic clay, characterized by high stiffness and medium plasticity, extremely homogeneous along the path of the tunnel. The grain size distribution and the main physical properties are reported in Figure 2.

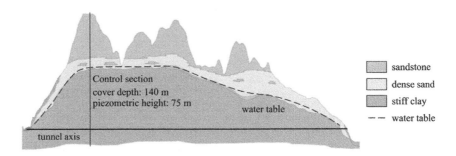

Figure 1.   Geological longitudinal profile.

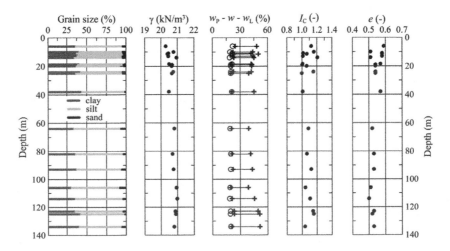

Figure 2.   Main soil physical properties.

The unit weight γ ranges between 20 and 21 kN/m³, increasing with depth; the void ratio e is between 0.5 and 0.6. The water content w is equal or slightly smaller than the plastic limit w_P, thus the consistency index I_c is stably above 1. In the control section, at tunnel depth (140 m), the soil is essentially normally consolidated. The geostatic stress state and the soil's cohesion c', friction angle φ', dilatancy angle ψ, undrained strength c_u, Young's modulus E' and Poisson's coefficient v' are listed in Table 1, together with the ranges of confidence adopted in the sensitivity analyses.

The TBM (Figure 3) operated in open mode. The machine's design and the construction procedures were aimed at minimizing the blockage risk, thus the cutterhead overcut and the conicity of the shield were significant (Table 2). This also increases the amount of stress relaxation around the tunnel, reducing the forces acting on the lining.

The lining ring is composed of 6 precast concrete segments, whose main geometrical properties are reported in Table 2. The lining's Young's modulus is 37.2 GPa. The tail void grouting was not automated. The injection was performed manually through the erector's pin holes, injecting one ring at a time (in average) and keeping a minimum distance of 3 lining rings from the tail of the machine. This design choice too, was aimed at maximizing the stress release in the surrounding soil, while still maintaining the benefits of the grout injection stabilizing effect.

2.1 Monitoring system and data

Eight instrumented lining rings were set up, each equipped with 12 strain gauges, 2 for each lining segment (installed 4 cm above and 4 cm below the center of the segment) and 6 optical prisms. In this paper, only the strains measured in the control section are taken into consideration. The measurements of the strain gauges have been used to calculate the forces in the lining (further details are reported in de Lillis, 2017), taking into account the temperature changes as indicated by several authors (e.g. Bilotta & Russo, 2012).

Table 1. Control section stress state and soil mechanical parameters.

σ_v (kPa)	σ_h (kPa)	u (kPa)	K_0 (-)	c' (kPa)	φ' (°)	c_u (kPa)	ψ (°)	E' (MPa)	v' (-)
2900	2050	750	0.6	0	21–25	600–750	0	170–250	0.3

Figure 3. a) TBM cutterhead; b) shield tail and hydraulic jacks.

Table 2. TBM and lining geometrical properties.

Cutterhead radius (m)	Shield front radius (m)	tail radius (m)	length (m)	Lining ring extrados radius (m)	width (m)	length (m)
2.06	2.045	2.02	6.40	1.95	0.25	1.20

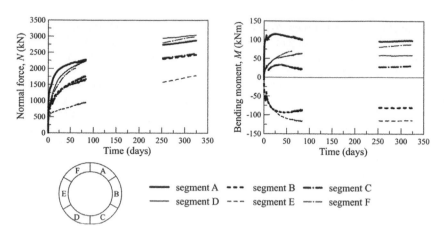

Figure 4. Monitoring data: normal force and bending moment in the lining.

Figure 4 shows the evolution of the internal forces in the lining due, at first, to the advancement of the excavation front in undrained conditions, and then, to the subsequent consolidation process. For about 5 months, the instruments didn't record any data due to technical difficulties.

Since the consolidation process is still evolving, monitoring data in drained conditions are not available; yet, the evolution gradient of the lining internal forces suggests that the remaining excess pore pressures are relatively small.

The monitoring data indicate that the consolidation process induces an increase in normal forces, which is almost independent of the position along the tunnel profile, while it does not have a significant effect on the bending moments. The maximum axial force, roughly 2900 kN/m, is registered near the crown and the invert (segments A, F and D). The bending moments are negative (internal fibers elongated) at the springline (segments B and E) and positive anywhere else; still, the absolute values are rather small.

3 NUMERICAL MODEL

The excavation was simulated using a three-dimensional model developed with the Finite Difference code FLAC3D. The model simulates the main features of the mechanized excavation, including front pressure, cutterhead overcut, shield conicity, annular tail void, installation of the lining inside the shield, tail void grouting and grout hardening over time. To simulate the interaction process associated with the soil closure onto the machine shield and onto the lining accurately, the algorithm works in large strain mode, thus updating the position of the nodes after each calculation step. This allows to detect the gap size automatically and, therefore, to identify *i)* when the soil and the shield start to interact; *ii)* the tail void volume to be grouted.

In a step-by-step approach, the starting calculation grid is usually designed with nodes coinciding with the excavation profile. As pre-convergences develop ahead of the excavation front, the nodes delimiting the portion of ground to be removed move inside the actual excavation profile, leading to the numerical misidentification of the correct excavation boundary. To tackle this issue, the modelling procedure described by de Lillis et al. (2018), was adopted. Briefly, the procedure consists in running a preliminary analysis to estimate the pre-convergences and then use them to modify the starting mesh in such a way that, after the pre-convergences, the profile of the excavation is accurately reproduced. This improves the accuracy of the solution, particularly in terms of lining stresses in deep tunnels (in which large deformations are induced by design).

The size and coarseness of the calculation grid have been optimized to minimize boundary effects performing a series of preliminary sensitivity analyses. The resulting transversal section is 80x80 m, while the longitudinal length is about 110 m (Figure 5a). The resulting grid size

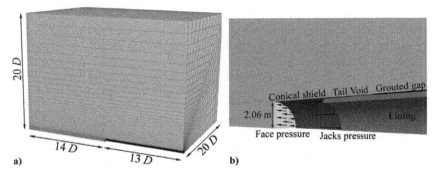

Figure 5. Numerical model: a) grid size; b) excavation scheme.

satisfies the criteria proposed by previous studies (e.g. Franzius & Potts, 2005; Zhao et al. 2012). The effects of gravity were neglected, thus only a quarter of the tunnel was modelled.

The soil is modelled as a simple linear elastic perfectly plastic medium with a Mohr-Coulomb failure criterion (the model parameters are reported in Table 1). The analyses are conducted in terms of effective stress.

The shield of the TBM is modelled using very stiff elastic continuum elements, reproducing its geometry explicitly, including the cutterhead overcut and the shield conicity (see Table 2). The cutterhead has not been modelled, in its stead a uniform pressure is applied. Since the TBM worked in open mode, the front pressure has been estimated as the difference between the total hydraulic jacks forces in service and the undrained adhesion between the shield and the surrounding soil; the former is a design value, monitored during the excavation, while the latter has been evaluated assuming that the undrained adhesion is equal to 0.25 c_u.

The lining is modelled using elastic continuum elements, equivalent to a cylindrical or conical shell (Augarde & Burd, 2001). The segmental nature of the lining is neglected, inducing limited errors on the analysis as reported by several studies (for instance Kavvadas et al. 2017). For the sake of simplicity, the forces of the hydraulic jacks on the lining are simulated as a longitudinal uniform pressure. The tail void is grouted at atmospheric pressure, injecting one ring at a time and keeping a distance of 3 rings from the tail of the shield. The grout mechanical characteristics are modelled following the hardening law proposed by Kasper & Meschke (2006a), assuming an advancement rate of the excavation of 10 m per day.

The excavation is simulated in detail (Figure 5b) adopting a step-by-step approach involving the following sub-steps: 1) the excavation advances one ring and the corresponding soil slice is removed; 2) the shield moves forward and the front pressure is applied to the new excavation face; 3) the last lining ring, activated during the previous step, is now outside the shield; a new lining ring is generated inside the shield and subjected to the hydraulic jacks pressure; 4) the tail void surrounding the third ring from the tail of the shield is injected with fresh grout at atmospheric pressure; previously injected grout is progressively hardened.

As the clayey soil around the tunnel has very low permeability, the excavation is carried out in undrained condition imposing the internal kinematic restraint of zero volumetric deformations, assuming that consolidation effects are negligible during the excavation. Once a stationary state of the interaction process is reached, the excavation is halted. Finally, long-term conditions are obtained restoring the initial pore pressure throughout the calculation domain and stepping for equilibrium, (the lining is impervious).

4 NUMERICAL RESULTS AND COMPARISON WITH MONITORING DATA

4.1 Numerical prediction

In this paragraph, some of the main results are described, focusing on the evolution of the stress state in the soil and in the lining due to the passage of the TBM in undrained conditions and on the long-term conditions.

In Figure 6 the longitudinal profile of the radial displacement of two soil points located at the crown and at the springline are reported. Since pre-convergences, starting from an aniso-tropic stress state, are non-symmetrical (higher at the crown and smaller at the springline), ahead of the excavation front the two points are located at different radial distances from the tunnel axis. After the development of pre-convergences, the two points are at 2.06 m from the axis, precisely on the excavation boundary. Following the excavation, the soil closes onto the machine shield. Behind the tail, the displacements are greater at the springline due to the development of plastic deformations. The injection of the grout at three rings from the tail, its subsequent hardening, and the advancement of the shield, allow to reach a stationary solution at roughly 3 diameters behind the excavation face.

Figure 7 shows the evolution of the stress state along two alignments (vertical, above the crown, and horizontal, at the right of the springline) in a reference transversal section crossed by the TBM.

The advancement of the excavation influences the stress state starting about 1 diameter ahead of the front (Figure 7, section A): radial and circumferential total stresses (σ_r, σ_θ) and pore pressure (u) increase about 10% near the tunnel. Once the shield is passed, the stresses decrease drastically. Section B is located 2 diameters behind the front, where the annular void has not been grouted yet. The radial stresses tend to zero near the crown and the springline, while the circumferential stress shows a significantly different behavior: at the crown, the decrease of σ_θ is limited to a very small area above the tunnel while at the springline its vari-ations propagate to about 3 diameters. In the area around 2 diameters from the springline, the mean pressure increases due to the arching effect, also inducing a rise of the pore pressure. Section C is placed $10D$ behind the front, where the solution is stationary. Comparing these results with those of section B, it can be noted that the injection of the grout (and the soil-grout-lining interaction) induces an increase of the stress state only in a small area around the tunnel.

The normal force N and the bending moment M in the lining in stationary undrained condi-tions are shown and compared with monitoring data in Figure 8. Since the asymmetry of the lining forces can only be caused by the assembly process (which is not included in the model), to facilitate the comparison the average of couples of monitoring data points symmetric with respect to the vertical axis is also shown.

The axial forces obtained numerically range between 1200 kN and 1500 kN; the minimum value being at the springline, while the maximum at the crown. The bending moment, quite small in absolute values, is negative (internal fibers elongated) at the springline and positive at the crown. The comparison with the internal forces obtained from the strain measurements is satisfactory in terms of bending moment, while the normal forces are overestimated by the model (20% more than the monitoring data in average).

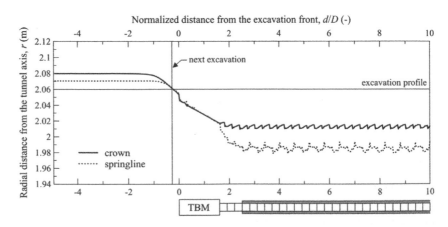

Figure 6.   Longitudinal displacement profile.

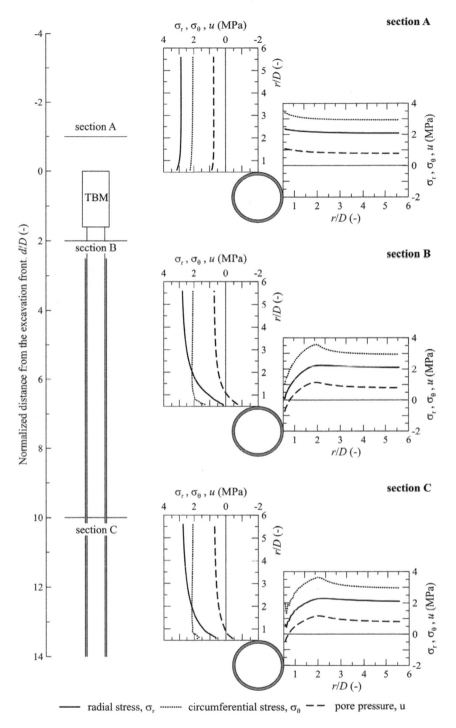

Figure 7. Vertical and horizontal stress profiles at different distances from the excavation front.

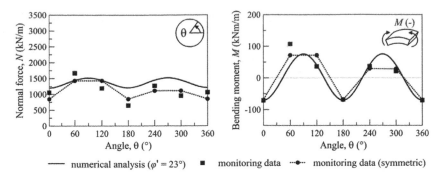

Figure 8. N and M in the lining in undrained conditions: numerical results and monitoring data.

4.2   *Sensitivity analyses: short-term conditions*

To study the influence of some soil mechanical parameters, which experimental determination has a range of confidence, several parametric analyses were performed varying the friction angle or the Young's modulus (Figure 9) from the base parameter set ($\varphi' = 23°$, $E' = 210$ MPa), both in short- and long-term conditions.

Varying the friction angle causes a change in the axial force of about 100 kN/m/° almost independently of θ (counterclockwise angle from the springline). The bending moment is less affected by the variations of φ'; its fluctuations along the tunnel wall are slightly reduced. The analysis performed adopting $\varphi' = 25°$ compares best with the average monitoring normal force, even though the numerical variations along θ are significantly smaller.

The influence of Young's modulus was investigated applying 40 MPa variations to the base set of parameters. Even though bigger percentage variations were chosen, the effect on the normal force in the lining is smaller than that of the friction angle.

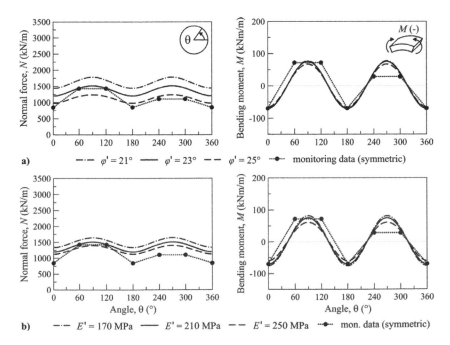

Figure 9. N and M in the lining in undrained conditions: a) influence of φ'; b) influence of E'.

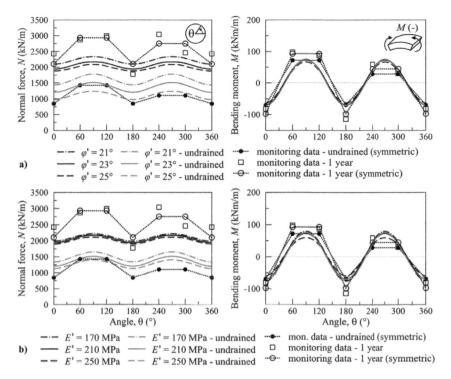

Figure 10. N and M in the lining in long-term conditions: a) influence of φ'; b) influence of E'.

4.3 Sensitivity analyses: long-term conditions

The long-term forces in the lining are shown in Figure 10 (for clarity, only one undrained M curve is shown), together with the monitoring data recorded 1 year after the excavation. The consolidation process induces a significant increase of N, while M is essentially unaffected; hence, the load eccentricity on the lining and the maximum compressive stress decrease appreciably, reducing the working rate of the lining.

The average ΔN associated with the consolidation process ranges between 600 kN/m and 870 kN/m, where bigger values correspond to higher friction angles. Furthermore, the consolidation process has a homogenizing effect, as confirmed by the reduction of the fluctuations of N along θ and by the negligible effect on M. The monitoring data instead show an average ΔN of about 1400, with significant variations along θ underestimated by the model. The numerical underestimation of ΔN is believed to be due to the overly simplistic evaluation of the excess pore pressure achievable using the Mohr-Coulomb soil constitutive model. The Young's modulus of the soil has a negligible effect on the long-term lining forces.

Both numerical results and monitoring data show higher values of N at the invert and at the crown than at the sides of the tunnel. This distribution of the axial forces (and also that of the bending moments), which is in contrast with what simpler models would predict, is due to the loss of stability associated with the advancement of the excavation front, which, after the grout injection, induces a slightly higher increase of radial stresses near the springline, where large plastic zones already developed (see also Figure 7, sections B and C).

5 CONCLUSIONS

A 3D numerical model was developed to simulate the mechanized excavation of deep tunnels. The model reproduces the main features of the excavation and works in large-strain mode to simulate the soil-shield and the soil-grout-lining interaction processes accurately.

In this study, the primary focus was on the evaluation of the forces in the lining. To this aim the model was tested against monitoring data recorded during the construction of a hydraulic tunnel in fine-grained soils, both in short- and long-term conditions. The influence of the friction angle and of the Young's modulus on the lining forces was also studied.

The overall response of the model was satisfactory, confirming the effectiveness of the proposed numerical approach. The model was able to reproduce the peculiar normal force distribution shown by the monitoring data (higher values at the crown and at the invert, lower values at the sides of the tunnel) which would have not been predicted by simpler models. On the other hand, the increase of normal force induced by the consolidation process was underestimated by the model, probably because of the underestimation of the negative excess pore pressures induced by the excavation in undrained condition. In this regard, more sophisticated soil constitutive models would certainly improve the prediction.

REFERENCES

Arnau, O. & Molins, C. 2012. Three dimensional structural response of segmental tunnel lining. *Engineering Structures* 44: 210–221.

Augarde, C.E. & Burd, H.J. 2001. Three-dimensional finite element analysis of lined tunnels. *International Journal of Numerical and Analytical Methods in Geomechanics* 25(3):243–262.

Bilotta, E. & Russo, G. 2012. Backcalculation of internal forces in the segmental lining of a tunnel: the experience of Line 1 in Naples. In G. Viggiani (ed.), *Geotechnical Aspects of Underground Construction in Soft Ground, Rome, 2011*. Rotterdam: Balkema.

Buselli, F., Logarzo, A., Miliziano, S., Formato, F., Simonacci, G. & Zechini, A. 2012. Prediction of the effects induced by the Metro C construction on an old masonry building. In G. Viggiani (ed.), *Geotechnical Aspects of Underground Construction in Soft Ground, Rome, 2011*. Rotterdam: Balkema.

de Lillis, A. 2017. Implementazione e utilizzo di un modello costitutivo avanzato per argille nella risoluzione di problemi al finito: studio del comportamento di gallerie realizzate mediante scavo meccanizzato. *PhD Thesis*. Sapienza University of Rome (in italian).

de Lillis, A., De Gori, V. & Miliziano, S. 2018. Numerical modelling strategy to accurately assess lining stresses in mechanized tunnelling. In A.S. Cardoso, J.L. Borges, P.A. Costa, A.T. Gomes, J.C. Marques & C.S. Vieira (eds.), *Proceedings of the 9th European Conference on Numerical Methods in Geotechnical Engineering, Porto, 2018*. London: CRC Press.

Do, N.A., Dias, D., Oreste, P. & Djeran-Maigre, I. 2014. Three-dimensional numerical simulation for mechanized tunneling in soft ground: the influence of the joint pattern. *Acta Geotechnica* 9(4): 673–694.

Franzius, J.N. & Potts, D.M. 2005. Influence of mesh geometry on three-dimensional finite-element analysis of tunnel excavation. *International Journal of Geomechanics* 5(3): 256–266.

Itasca Consulting Group, Inc. 2012. FLAC3D Version 5.0, Fast Lagrangian Analyses of Continua in Three-Dimensions, User's manual. Minneapolis.

Karakus, M. 2007. Appraising the methods accounting for 3D tunnelling effects in 2D plane strain FE analysis. *Tunnelling and Underground Space Technology* 22(1): 47–56.

Kasper, T. & Meschke, G. 2004. A 3D finite element simulation model for TBM tunneling in soft ground. *International Journal of Numerical and Analytical Methods in Geomechanics* 28(14): 1441–1460.

Kasper, T. & Meschke, G. 2006a. On the influence of face pressure, grouting pressure and TBM design in soft ground tunnelling. *Tunnelling and Underground Space Technology* 21(2): 161–171.

Kasper, T. & Meschke, G. 2006b. A numerical study of the effect of soil and grout material properties and cover depth in shield tunneling. *Computer and Geotechnics* 33(4): 234–247.

Kavvadas, M., Litsas, D., Vazaios, I. & Fortsakis, P. 2017. Development of a 3D finite element model for shield EPD tunnelling. *Tunnelling and Underground Space Technology* 65: 22–34.

Möller, S.C. & Vermeer, P.A. 2008. On numerical simulation of tunnel installation. *Tunnelling and Underground Space Technology* 23(4): 461–475.

Zhao, K., Janutolo, M. & Barla, G. 2012. A completely 3D model for the simulation of mechanized tunnel excavation. *Rock Mechanics and Rock Engineering* 45: 475–497.

*Tunnels and Underground Cities: Engineering and Innovation meet Archaeology,
Architecture and Art, Volume 7: Long and deep tunnels – Peila, Viggiani & Celestino (Eds)
© 2020 Taylor & Francis Group, London, ISBN 978-0-367-46872-9*

Numerical back-analysis of the Fréjus road tunnel and of its safety gallery

M. De la Fuente
*Ecole des Ponts ParisTech, Île-de-France, Champs-sur-Marne, France & Tractebel ENGIE,
Île-de-France, Gennevilliers, France*

J. Sulem
Ecole des Ponts ParisTech, Île-de-France, Champs-sur-Marne, France

R. Taherzadeh
Tractebel ENGIE, Île-de-France, Gennevilliers, France

D. Subrin
Centre d'Etudes des Tunnels (CETU), Auvergne-Rhône-Alpes, Bron, France

ABSTRACT: The present work aims at performing a numerical study of the response of the Fréjus road tunnel and of its safety gallery. This case study exhibits an interesting configuration of two parallel tunnels excavated in the same ground with different techniques: the Fréjus road tunnel was excavated by drill and blast methods in the seventies and its safety gallery was excavated with a single shield Tunneling Boring Machine (TBM) between 2009 and 2016.

Monitored convergences in the road tunnel are back-analyzed considering a visco-elasto-plastic constitutive behavior of the ground. This behavior is extrapolated to the neighboring zones of the safety gallery in order to predict the response in terms of stresses developed in the segmental lining. It is shown that the excavation method significantly affects the time-dependent parameters of the ground.

1 INTRODUCTION

The Fréjus road tunnel and its safety gallery are two examples of tunnels excavated in squeezing ground. The Fréjus road tunnel was excavated by conventional drill and blast methods in the seventies linking France and Italy under the Alps. Since the fire which took place in Mont-Blanc tunnel in 1999, a new safety regulation for tunnels was established. In order to be in accordance with it, a safety gallery running parallel to the road tunnel was excavated between 2009 and 2016 with a Tunneling Boring Machine (TBM). The average distance between the axes of both galleries is 50 m.

Squeezing ground behavior can be studied by analyzing convergence data retrieved during the excavation of the road tunnel. Convergences were monitored over a period of four months until the installation of the final lining. By processing the convergence data from the road tunnel, a good understanding of the face advance effects and of the time-dependent behavior of the ground can be obtained (De la Fuente et al., 2017). A visco-elasto-plastic numerical model is calibrated on the convergence data recorded in the road tunnel. This constitutive model is then applied to the neighboring zones of the safety gallery in order to numerically predict its response. The stress state monitored in the segmental lining installed in the safety gallery during its excavation can be accurately predicted with this procedure. The influence of the excavation method on the time-dependent tunnel response is explored and discussed.

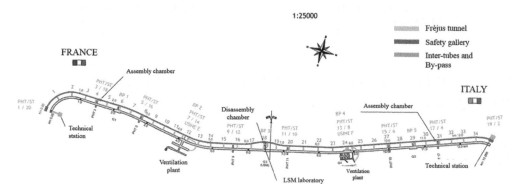

Figure 1.   Plan view scheme of Fréjus road tunnel and its safety gallery.

2   GENERAL CONTEXT

The Fréjus road tunnel links Modane (France) and Bardonnechia (Italy) under the ridge between the pic of Fréjus (3019 m) and the pic of Grand-Vallon in the Alps, following an average North-South direction. The tunnel is 12.87 km long and 11.6 m wide between the sidewalls with a two-lane classical horse shoe section. The overburden along most of the layout is over 1000 m (with a maximum of 1800 m). The Italian tunnel portal is at an altitude of 1297 m whereas the French tunnel portal is at an altitude of 1228 m. The tunnel slopes down 0.54% from Italy towards France (Sulem, 2013). The tunnel was supported with radial rockbolts before the final lining was installed at a distance of 600 m from the face of excavation. The average advancing rate was 5.6 m/day (Levy et al., 1981).

The safety gallery has a circular shape. Its diameter is 9.5 m and it has a length of about 13 km. The tunnel slope and the overburden is the same as for the road tunnel. The safety gallery is connected with the road tunnel by means of 34 inter-tubes spaced every 400 m. Furthermore, the existence of 5 by-pass allows the emergency team to access the road tunnel by vehicle from the safety gallery or *vice-versa*. Ten technical stations as well as two ventilation plants were also installed (Figure 1).

The first 650 meters of the gallery were excavated by conventional drill and blast methods. The rest of the tunnel was excavated with a single shielded TBM. The TBM was firstly used to excavate the 6.5 km of the French part of the tunnel. The average advancing rate was 12.9 m/day.

Both tunnels mainly cross a calcschists formation showing a schistosity which is oriented parallel to the tunnels axis. The schists present a dip angle of 45° towards the West. The calcschists result from a light metamorphism of marls and limy marls with the formation of phyllitous minerals (muscovite, chlorite) (Panet, 1996).

3   MONITORING DATA AND DATA PROCESSING

3.1   *Monitoring data and data processing of Fréjus road tunnel*

Convergences of 127 sections spaced 30 m apart were continuously monitored in the road tunnel. Measurements are carried out with invar tape extensometers until the installation of the final lining about 107 days after the excavation of the section. In average, the convergence rate at that moment is 0.2 mm/day.

The typical convergence curve is shown in Figure 2. The buckling of the schistosity planes causes the strongest convergence along the direction defined by targets 2 and 4 which is quasi perpendicular to the schistosity planes. Targets 1 and 4 define a direction which is parallel to the tunnel invert. In some areas of the tunnel, convergence along direction 1-3 has also been monitored. However, convergence data following this direction have been recorded over a shorter period of time.

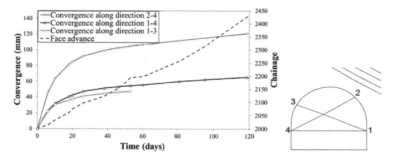

Figure 2.   Convergence curves and schematic distribution of the targets in section 13 (chainage 1998).

Very detailed data processing of the road tunnel has been carried out by De la Fuente et al., (2017). The semi-empirical law proposed by Sulem et al. (1987) (Equation 1) is used in the analysis of convergence data of the road tunnel by De la Fuente et al. (2017).

$$C(x,t) = C_{\infty x}\left[1 - \left(\frac{X}{x+X}\right)^2\right]\left\{1 + m\left[1 - \left(\frac{T}{t+T}\right)^n\right]\right\} \qquad (1)$$

where $C_{\infty x}$ represents the instantaneous convergence obtained in the case of an infinite rate of face advance (no time-dependent effect), X is a parameter related to the distance of influence of the tunnel face, T is a parameter related to time-dependent properties of the system (rock mass formation – support), m is a parameter which represents the relationship between the long term total convergence and the instantaneous convergence and n is a form-factor which is often taken equal to 0.3. Some "homogeneous" zones corresponding to similar values of the instantaneous convergence along the direction 2-4 have been identified, Figure 3.

3.2   Monitoring data and data processing of Fréjus safety gallery

During the excavation of the safety gallery, an important survey campaign was carried out: convergence data was retrieved at the inner face of the concrete ring, convergence data of the ground measured with hydraulic jacks through the TBM shield, cracks observation, monitoring data obtained from strain gauges embedded in the segmental lining of 49 sections which

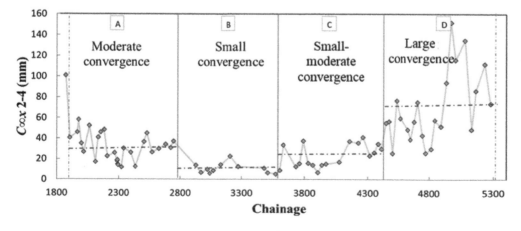

Figure 3.   Evolution of $C_{\infty x}$ along direction 2-4 along the road tunnel. The red dotted lines represent the average convergence value for each zone. After De la Fuente et al. (2017).

can provide information on the state of stress in the lining (Figure 4) and other information obtained during the excavation such as the thrust force and the torque exerted over the cutting head of the TBM. These measurements have three objectives: the collection of information to improve the excavation technique and/or the lining design during the excavation of the gallery, the prevention of risks that might be encountered during tunnel execution and the creation of a useful data base in order to back analyze the tunnel behavior.

Monitoring data from strain gauges represent the most reliable source of information. Six pairs of strain gauges were embedded in the segmental lining. Each pair in general represents the behavior of the extrados and intrados fibers of the segmental lining. It should be noted that many interruptions are observed in the retrieved strain data. The stress state in the lining can be obtained from strain gauges if concrete behavior is considered elastic. A mid-term Young's modulus of 20 GPa has been considered for the concrete.

Figure 5 shows some of the results from the data processing of the safety gallery (De la Fuente et al. 2017). The maximal compression stress in the lining is plotted and compared with the lateral friction exerted by the ground over the TBM and some approximate values of RQD

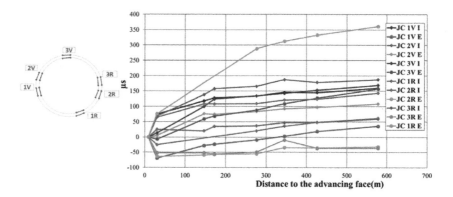

Figure 4.   Distribution of the strain gauges in the ring 1821, Chainage 3917 (raw data).

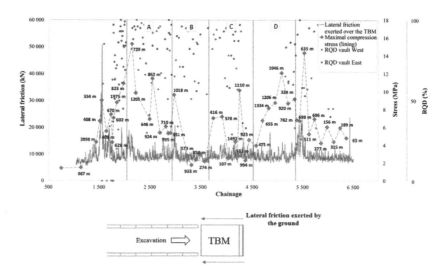

Figure 5.   Lateral friction exerted by the ground over the tail shield, maximal compression stress measured in the lining (the distance to the excavation face at which the stress has been retrieved can be found next to each point representing the stress state) and RQD values of the ground retrieved from the East or the West side of the vault during the excavation, as a function of chainage in the gallery after De la Fuente et al. (2017).

retrieved from small openings within the tail shield. Figure 5 also shows the previously identified "homogeneous" zones which are overlaid onto the safety gallery data. Monitoring data from both tunnels are in accordance. The areas of the road tunnel which exhibit large convergence correspond to the zones of the gallery where the stress state in the lining is higher.

We can observe that in the areas where lateral friction exerted by the ground over the TBM is higher, the values of the RQD are lower than the average of 70%. This can mainly be observed around chainage 1550 which corresponds to a very degraded rock. However, the RQD index is only representative of the degree of fracture and cannot describe the overall quality of the rock medium. Around chainage 1550 the highly fractured zone can also be identified with the increase of lateral friction over the TBM. Furthermore, the maximum friction which is observed around chainage 6430 is the result of the resumption of the excavation after a standstill of 126 days.

4 NUMERICAL BACK-ANALYSIS OF THE FRÉJUS ROAD TUNNEL

A 3D numerical simulation is carried out with FLAC^{3D} in order to simulate the behavior of the Fréjus road tunnel. The constitutive behavior for the ground is visco-elasto-plastic and anisotropic. The CVISC model is used to describe the behavior of the rock matrix. This model considers an elasto-plastic volumetric behavior and a visco-elasto-plastic deviatoric behavior.

The presence of weakness planes such as schistosity planes is taken into account by means of the so-called "ubiquitous joints model". It consists in a set of joints of a certain orientation which pass through any point in the rock mass. These joints are activated if a given yield criterion is reached. This constitutive model has already been successfully employed by Tran-Manh et al. (2015) in the simulation of the Saint-Martin-la-Porte access adit within the Lyon-Turin railway project.

Figure 6 and Figure 7 show the geometry of the model. The model is large enough in order to simulate the excavation and minimize boundary effects. Far field boundaries are placed at a distance of 28 radii (considering the vault radius) and the length of the model in the axial direction is 90 m. Mesh is discretized into small elements of 0.45 m (< 1/10 R). The in-situ stress state is initially imposed everywhere in the domain (average depth 1067 m and average specific weight of the ground 27 kN/m3). Gravity effects are disregarded. The step of excavation is 4.5 m and an advancing rate of 5.6 m/day is imposed in the computations in accordance with the average values observed during the excavation of the tunnel. The coordinates of the targets in the simulations are the average coordinates of all the targets along the tunnel (Figure 7).

The dip direction of the schistosity planes is parallel to the tunnel axis and its dip angle is fixed to 45°. The support composed of 20 rockbolts/m is simulated by introducing cable structural elements which are punctually anchored to the tunnel wall and to the ground. Each cable can yield in tension but cannot resist a bending moment. The length of the rockbolts is 4.65 m with a diameter of 20 mm and a strength limit of 450 MPa.

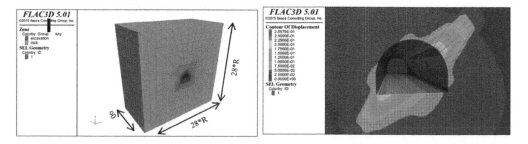

Figure 6.   Geometry of the model (left). Detail of the displacements around the tunnel during its excavation for section 12 (chainage 1976) (right). R is the radius of the vault of the excavated tunnel (5.8 m).

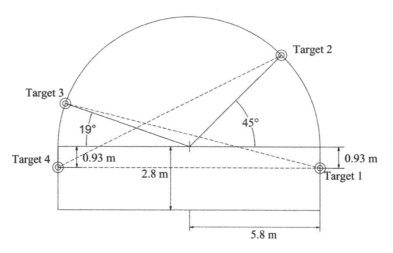

Figure 7.   Geometry of the tunnel and average position of the targets considered in the simulations.

Within the "homogeneous zone A", sections showing the largest and the smallest convergence have been identified by De la Fuente et al. (2018) (Figure 8).

These sections are back-analyzed by using the above constitutive model, Figure 9 and Figure 10. A horizontal pressure coefficient K_0 of 1.4 has been assumed. The values of the mechanical parameters of the joints are identical in both cases (c_j = 0,15 MPa, $_j$ = 20°, ψ_j = 5° and σ_{tj} =0.01 MPa). The values of some of the parameters of the rock matrix are assumed the same in both sections (E = 40 GPa, = 40°, ψ = 15°, σ_t = σ_c/10 and υ = 0.3). The four other parameters of the matrix differ from one section to another. The largest values (.)max of parameters c, η_K, G_M and η_M are assigned to the smallest convergence (section 29) and vice versa. The behavior observed in situ is accurately reproduced with the model.

Sections within zone A can be simulated by fitting the cohesion of the joints cj and a variability parameter α with values between 0 and 1, which can be seen as a variable describing the degree of damage of the ground, taking as reference values 0 for section 19 giving the smallest convergence and 1 for section 12 giving the largest one. This parameter permits to

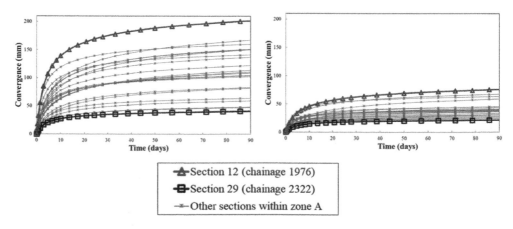

Figure 8.   Convergence curves in the "homogeneous zone A" along direction 2-4 (left) and along direction 1-4 (right)after De la Fuente et al. (2018).

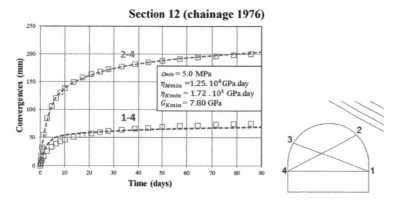

Figure 9. Back analysis of convergence data of section 12 (chainage 1976) (largest convergence) and schematic average position of the targets (right) in the section.

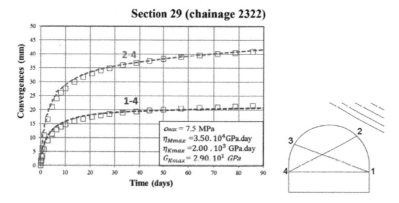

Figure 10. Back analysis of convergence data of section 29 (chainage 2322) (smallest convergence) and schematic average distribution of the targets (right) in the section.

simply evaluate the cohesion and the time-dependent parameters of the matrix and the matrix cohesion for all sections in zone A (Equation 2). The variability parameter α is evaluated for each section by fitting the convergence measured along direction 1-4. As this direction is sub-parallel to the weakness places, it is assumed that the convergence measurements along 1-4 are representative of the matrix behavior. Once parameter α is evaluated, cj is fitted from the convergence measurements along direction 2-4. The stronger the convergence along 2-4, the stronger the anisotropy of the section and the lower the value of cj. The other parameters remain the same for all the sections.

$$c = c_{\min}\alpha + (1 - \alpha)c_{\max}$$
$$G_k = G_{k_{\min}}\alpha + (1 - \alpha)G_{k_{\max}}$$
$$\eta_k = \eta_{k_{\min}}\alpha + (1 - \alpha)\eta_{k_{\max}}$$
$$\eta_M = \eta_{M_{\min}}\alpha + (1 - \alpha)\eta_{M_{\max}}$$

(2)

For instance, the behavior of section 43 in chainage 2682 can be simulated by attributing a value of 0,09 MPa to cj and taking $\alpha = 0.65$ (Figure 11).

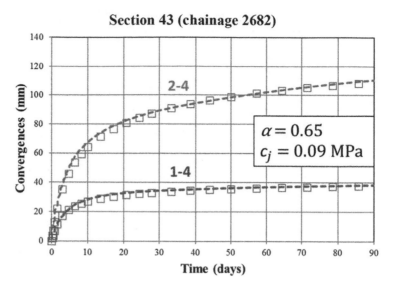

Figure 11. Back analysis of convergence data of section 43 (chainage 2682).

5 NUMERICAL PREDICTION OF THE RESPONSE OF THE FRÉJUS SAFETY GALLERY

The behavior of the Fréjus safety gallery has been simulated with a 2D model developed with FLAC^{3D}. An average value of 190 mm is assumed for the overcutting and an eccentricity of the lining of 0.095m with respect to the TBM cutting head is considered. The in-situ stress state is initially imposed everywhere in the domain (average depth 1067 m and average specific weight of the ground 27 kN/m3). Gravity effects are disregarded. The step of excavation is 1.8 m which corresponds to the transversal length of a segmental lining. An advancing rate of 12.9 m/day is considered in accordance with the average advancing rate observed during the excavation of the safety gallery.

The unsupported span can be considered to be 19.8 m as it was evidenced that the annular gap is completely filled up with the backfilling material only at that distance from the face, where it can be assumed that a deconfining rate of 100% has already taken place. A "sand-wich" type backfilling composed of gravel and mortar is considered in the simulations (Figure 12). The gravel and the mortar are assumed to have an elastic response with a

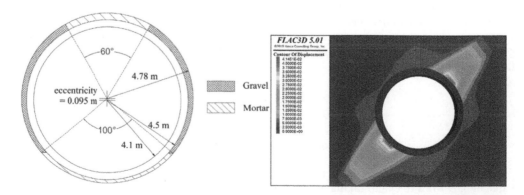

Figure 12. Scheme of the geometry of the lining and the backfilling in the safety gallery (left) and detail of the displacements around the gallery (section 1143 in chainage 2619) with the 2D model developed with FLAC^{3D} (right).

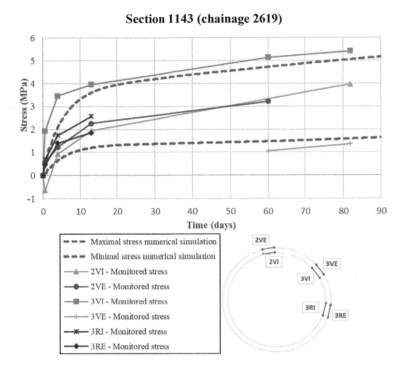

Figure 13.   Prediction of the behavior of section 1143 (chainage 2619) in the safety gallery.

Young's modulus of 100 and 500 MPa respectively. The installed elastic lining has a thickness of 40 cm and its Young's modulus is 20 GPa. The 2D numerical simulation is carried out in two steps: instantaneous excavation of the tunnel considering an instantaneous behavior of the ground (no-time dependent effects) and installation of the lining and the backfilling, then activation of the time-dependent behavior of the ground.

The ground behavior identified from the study of the road tunnel is extrapolated for the simulation of the safety gallery. From preliminary computations, it was obtained that assuming the same values for the constitutive parameters as those calibrated on the road tunnel leads to an overestimation of the stresses in the lining. As the lining is placed at a distance of more than two diameters to the tunnel face, its response is mainly controlled by the time-dependent behavior of the rock mass. Therefore, the instantaneous constitutive parameters are kept the same in both tunnels and the time-dependent parameters (η_M, η_k, G_k) are multiplied by a factor F. This is attributed to the fact that, when tunneling with a TBM the ground is less damaged than when tunneling by drilling and blasting and as a consequence time-dependent convergence is smaller.

The ground behavior obtained from the study of the road tunnel in chainage 2682 is extrapolated for the simulation of section 1143 in chainage 2619 (almost parallel) in the safety gallery (Figure 13).

A good agreement is observed between the numerical simulations and the monitored stresses in section 1143 (chainage 2619) in the safety gallery if we choose a factor F of 10. The order of magnitude of the monitored hoop stresses in the lining is well reproduced. Furthermore, the loading rate of the lining is also well simulated.

6   CONCLUSION

A numerical back-analysis of the convergence measurements of Fréjus road tunnel has been carried out in order to calibrate a constitutive model able to reproduce the instantaneous and the time-dependent behavior of the rock mass in one of the most complex areas of the tunnel.

The computed sections showing the smallest and the largest convergence successfully fit convergence data. The response of these extreme sections represents the envelope of convergences in the studied area. The obtained set of geotechnical parameters is realistic and is in accordance with the literature. The rest of the sections in the studied area are fitted by adjusting only two parameters: the joints cohesion which is related to the anisotropy of the section and a variability parameter which is representative of the magnitude of the convergences of the matrix which fall within the identified envelope of convergences.

Fréjus safety gallery response can be predicted by extrapolating the ground behavior identified from the study of Fréjus road tunnel. It is observed that although instantaneous parameters can be assumed the same in both tunnels, the time-dependent constitutive parameters of the rock mass to be considered in the numerical model depend upon the excavation process. Very good predictions of the stresses in the segmental lining are obtained when compared to the retrieved data.

Acknowledgements: The authors wish to thank the SFTRF for providing monitoring data on both tunnels and ITASCA for supporting this research through the Itasca Education Partnership Program.

REFERENCES

De la Fuente M., Sulem J., Taherzadeh R., Subrin D. (2017) Traitement et rétro-analyse des auscultations réalisées dans le tunnel routier du Fréjus et sa galerie de sécurité lors de leurs constructions respectives. *Congrés International AFTES 2017*, Paris 155 (in French)

De la Fuente M., Tahezadeh R., Sulem J., Subrin D. (2018) Analysis and comparison of the measurements of Fréjus road tunnel and of its safety gallery. *Eurock symposium 2018*, Saint Petersburg v2 1143–1148

Levy M., Courtecuisse G., Barral J. P.. (1981). Civil works of Frèjus road tunnel. *Annales de l'ITBTP*, 400 (TRAV PUBLICS-19) (in French)

Manh, H. T., Sulem, J., Subrin, D., & Billaux, D. (2015). Anisotropic time-dependent modeling of tunnel excavation in squeezing ground. *Rock Mechanics and Rock Engineering*, 48(6): 2301–2317

Panet M. (1996) Two case histories of tunnels through squeezing rocks. *Rock mechanics and rock engineering*, 29(3): 155–164

Sulem J., Panet M., Guenot A. (1987) Closure analysis in deep tunnels. *International Journal of Rock Mechanics and Mining Science Geomech Abstr* 24(3): 145–154

Sulem J. (2013) Tunnel du Fréjus: Mesures géotechniques et interprétation. *Manuel de Mécanique des Roches* Tome IV, chap. 7, Presse des Mines. (in French)

*Tunnels and Underground Cities: Engineering and Innovation meet Archaeology,
Architecture and Art, Volume 7: Long and deep tunnels – Peila, Viggiani & Celestino (Eds)*
© 2020 Taylor & Francis Group, London, ISBN 978-0-367-46872-9

Influence of the excavation rate on the mechanical response of deep tunnel fronts in cohesive soils

C. Di Prisco & L. Flessati
Politecnico di Milano, Milano, Italy

G. Cassani
Rocksoil S.p.A., Milano, Italy

R. Perlo
Officine Maccaferri S.p.A., Zola Predosa, Italy

ABSTRACT: In conventional tunnelling, the mechanical response of the tunnel front is a main concern and depends on both the geometry (tunnel diameter and cover) and soil mechanical properties. Moreover, in case of excavations in soils characterized by a low value of permeability, even the time factor plays an important role: in case of particularly problematic soils, the displacements of the front are expected to progressively increase with time and tunnel fronts stable under short term conditions potentially either develop unacceptable displacements or become unstable under long term conditions. In this paper, the mechanical response of deep tunnel fronts excavated in a homogeneous cohesive soil stratum are analysed by both performing experimental 1g small scale model tests and non-linear 3D FEM analyses. The numerical results are obtained by assuming the material to be isotropic, homogeneous and characterized by an elastic-perfectly plastic constitutive relationship. The results are presented in terms of the tunnel front characteristic curve, defined in analogy with the well-known characteristic curve for the tunnel cavity and by employing a suitable non-dimensional variable depending on the excavation time, the soil hydraulic/mechanical properties and the tunnel geometry. Finally, the authors introduce a rapid procedure allowing the front displacements estimation without performing ad hoc numerical analyses.

1 INTRODUCTION

When tunnels are excavated under particularly difficult ground conditions, the mechanical response of the front is a main concern.

In the past, numerous authors theoretically studied failure conditions for tunnel fronts. In particular, by employing the limit equilibrium method and by defining failure mechanisms suitable for shallow tunnels, Horn (1961), Anagnostou & Kovari (1996), Anagnostou (2012), Perazzelli et al. (2014) and Anagnostou & Perazzelli (2015) assessed the minimum pressure to be applied on the front to prevent its collapse. Failure conditions for the tunnel front were also studied by employing the limit analysis theory. As far as granular soils are concerned, lower bound solutions were introduced by Mühlhaus (1985), whereas upper bound solutions by Leca & Dormieux (1990), Wong & Subrin (2006) and Mollon et al. (2009). In contrast, cohesive materials were studied by Davis et al. (1980), Klar et al. (2007), Mollon et al. (2013) and Klar & Elkayam (2017), who proposed upper bound solutions. A different theoretical approach to study the mechanical response of the front was introduced by Leca & Panet (1988). These authors studied a spherical cavity excavated within an infinite and homogeneous soil domain subject to a uniform and isotropic state of stress and suggested to use the solution obtained in order to reproduce the front mechanical response.

From an experimental point of view, the approaches employed to study this problem can be subdivided in three categories: a) extrusion tests, b) centrifuge tests and c) 1g small scale model tests. The extrusion tests were first introduced by Broms & Bennermark (1967) and further developed by Attewell & Boden (1971) and Lunardi (1993). Experimental centrifuge tests on cohesive materials were performed by Mair (1979), Kimura & Mair (1981), whereas on frictional material by Chambon & Corté (1994), Nomoto et al. (1999) and Kamata & Mashimo (2003). 1g model tests for shallow tunnels in granular materials were performed by Sterpi & Cividini (2004), Kirsch (2009), Berthoz et al. (2012) and Chen et al. (2013), whereas, more recently, deep tunnels in cohesive soils were studied by di Prisco et al. (2018a).

The mechanical response of tunnel fronts was also numerically studied by performing FEM analyses. In particular, Vermeer et al. (2002), Yoo (2002), Sterpi & Cividini (2004) and Kirsch (2009) analysed the mechanical response of shallow tunnels in granular soils, whereas Ng & Lee (2002), Höfle et al. (2008) Schuerch & Anagnostou (2013), Callari (2015), Sitarenios & Kavvadas (2018), Flessati et al. (2017), di Prisco et al. (2018b) and di Prisco & Flessati (2018) analysed the mechanical response of tunnel fronts in cohesive soils.

As is well-known, the mechanical response of tunnel fronts excavated in cohesive soils is severely affected by the consolidation process taking place in the advance core. In particular, (i) for a nil stress applied on the front, displacements increase with time and (ii) the front mechanical response is severely affected by the excavation rate.

In this paper, only the influence of the excavation rate is investigated by both performing 1g small scale model tests (§2) and 3D FEM non-linear analyses (§3). In §4, on the basis of the numerical tests results, a procedure to assess the role of time in affecting the system mechanical response is suggested.

2 EXPERIMENTAL EVIDENCE

The experimental set-up conceived by the authors to reproduce the mechanical response of the front is illustrated in Figure 1. For the sake of brevity, only an essential description of the model and of tests procedure is hereafter reported. The details are illustrated in di Prisco et al. (2018a). The tests are performed by directly imposing, by means of the air cylinder of Figure 1, a horizontal force to a rigid brass plate modelling the front. This force is progressively nullified at a constant rate from an initial geostatic value. The time needed to completely unload the front is hereafter named (t_u). The force reduction induces front displacements (u_f), measured by employing a laser sensor (Figure 1).

The soil employed for the tests is a kaolin clay (plastic limit $w_p=34\%$ and liquid limit $w_l=73\%$) prepared from a slurry $(1.5w_l)$ and consolidated under its self-weight. To reduce the consolidation time, 19 vertical drains, removed before the tests, were introduced in the soil domain.

The tunnel cross section is not circular (Figure 1), its equivalent diameter (D) is equal to 19 cm and the cover diameter ratio (H/D) at the end of the consolidation process is equal to 5.

The authors performed many different tests by considering different front unloading times but, for the sake of brevity, hereafter only 4 tests (Table 1) are discussed.

The experimental results are illustrated in terms of the front characteristic curve, relating the mean value of the stress applied to the front σ_f (calculated by dividing the applied force by the area

Table 1. Test parameters.

	t_u (min)
A1	3.5
A2	9
A3	60
A4	180

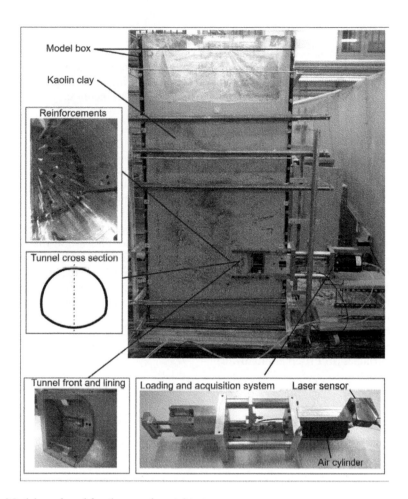

Figure 1.   Model employed for the experimental tests.

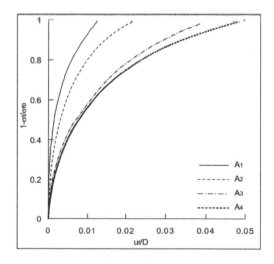

Figure 2.   Experimental results for different unloading times.

of the tunnel front) and the front displacement. The results are reported in the non-dimensional $1-\sigma_f/\sigma_{f0}-u_f/D$ plane (where σ_{f0} is the initial geostatic value for σ_f) in Figure 2.

The experimental results clearly highlight that the consolidation taking place, in the soil in the proximity of the front, governs the system response: by increasing t_u, the initial slope of the curve decreases and the residual displacements (i.e. for $\sigma_f=0$) increase.

3 NUMERICAL RESULTS

The mechanical response of deep tunnel fronts is also studied by performing 3D numerical analyses. The employed numerical model is reported in Figure 3. A homogeneous cohesive soil stratum characterized by a constant saturated soil unit weight $\gamma_{sat}=20\text{kN/m}^3$ is considered. The excavation process is modelled as a progressive reduction in the pressure initially applied to the front. The soil mechanical behaviour is modelled by means of an elastic-perfectly plastic constitutive relationship. The elastic properties (the Young modulus E and the Poisson's ratio v) are assumed to be constant along depth. The yield surface is defined according to the Mohr-Coulomb criterion (the internal friction angle is named ϕ', the cohesion is always assumed to be nil). A non-associated flow rule is considered and the dilatancy angle is assumed to be nil. The material is hydraulically isotropic and homogeneous. The water table is assumed to be coincident with the ground surface.

The numerical front characteristic curves, relating the average stress applied on the front (σ_f) and the average front displacement (u_f) are discussed by employing the following non-dimensional variables (di Prisco et al. 2018b):

$$Q_f = \left(1 - \frac{\sigma_f}{\sigma_{f0}}\right)\frac{\sigma_{f0}}{S_u^*} \tag{1}$$

$$q_f = \frac{u_f}{u_{fr,elu}}\frac{\sigma_{fo}}{S_u^*} \tag{2}$$

where $u_{fr,elu}$ is the residual elastic displacement under undrained conditions, whereas S_u^* is an "equivalent undrained strength for the system" (di Prisco & Flessati, 2018).

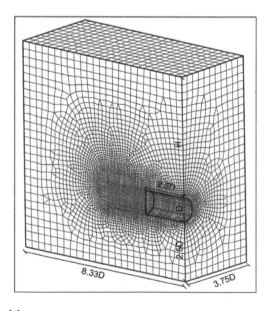

Figure 3. Numerical model.

The employment of these two non-dimensional variables, one for the stresses (Q_f) and one for the displacements (q_f) is particularly convenient since, under undrained conditions, in the Q_f-q_f plane the front characteristic curve does not depend on (i) system geometry, (ii) soil mechanical properties and (iii) initial stress conditions (di Prisco et al. 2018b).

Here in the following, the numerical analyses are performed by considering different unloading rates. In this case is convenient the introduction of a non-dimensional excavation rate (Flessati et al. 2017) defined as:

$$Y = \frac{\gamma_w D^2}{kKt_u},$$ (3)

where γ_w, K and k are the water unit weight, the elastic volumetric stiffness and the permeability, respectively.

The employment of this non-dimensional variable is particularly convenient, since if the plastic hydro/mechanical coupling is absent (i.e. the dilatancy at failure is nil), for a given Y value, a unique front characteristic curve is obtained (di Prisco & Flessati 2018).

The numerical results obtained by considering different t_u values are reported in Figure 4a (D=12m, H/D=5, ϕ'=25°, E=100MPa, v=0.3, k=10^{-8}m/s, γ_{sat}=20kN/m³, S_u^*=200kPa, k_0=0.58). In Figure 4b the results are also reported for Q_f<1 and q_f<1 to better appreciate the variation in the initial system response.

The numerical results of Figure 4 confirmed what was experimentally obtained: the system response is severely affected by the consolidation process. In fact, by increasing t_u, the residual displacements increase. This is not only due to the variation in the initial elastic slope (R) of the curves, but also to a variation in the curves shape. In fact, the (undrained) curve associated with Y=$1.7\cdot10^8$ is characterized by an indefinite hardening response due to the spatial propagation of the yielded soil domain (di Prisco et al. 2018b), whereas the (drained) curve associated with Y=$1.96\cdot10^{-7}$ is characterized by a horizontal asymptote, testifying an unstable response. It is worth mentioning that the value of the asymptote is practically coincident with Q_{L0} (the horizontal line of Figure 4a) defined as:

$$Q_{L0} - \left(1 - \frac{\bar{\sigma}' + u_0}{\sigma_{f0}}\right)\frac{\sigma_{f0}}{S_u^*},$$ (4)

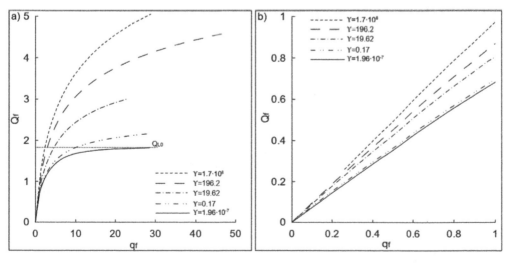

Figure 4. Numerical results for different unloading times (D=12m, H/D=5, ϕ'=25°, E=100MPa, v=0.3, k=10^{-8}m/s, γ_{sat}=20kN/m³, S_u^*=200kPa, k_0=0.58).

where u_0 is the initial steady-state average pore water pressure applied to the front, whereas

$$\bar{\sigma}' = \gamma' D \left(\frac{1}{9 \tan \phi'} - 0.05 \right) \qquad (5)$$

is the minimum effective pressure to be applied on the front to prevent its collapse (Vermeer et al. 2002).

4 PRACTICAL EMPLOYMENT OF THE NUMERICAL RESULTS

According to di Prisco & Flessati (2018), the numerical results can be interpolated by employing the following expression:

$$q_f = \begin{cases} Q_f / R(Y) & Q_f a_f(Y) \\ \frac{a_f(Y)}{R(Y)} e^{\frac{Q_f}{a_f(Y)} - 1} + \frac{Q_f - a_f(Y)}{Q_L(Y) - Q_f} & Q_f a_f(Y) \end{cases}, \qquad (6)$$

where function $R(Y)$ describes the dependence of the initial stiffness on Y, $a_f(Y)$ defines when the characteristic curves stop being linear and $Q_L(Y)$ represents a limit load for the system. These three functions are respectively defined as:

$$R(Y) = R_1 + (1 - R_1) \frac{R_2 Y^{R_3}}{R_2 Y^{R_3} + 1} \qquad (7a)$$

$$a_f(Y) = a_1 + \left(a_{fu} - a_1 \right) \frac{a_2 Y^{a_3}}{a_2 Y^{a_3} + 1} \qquad (7b)$$

$$Q_L(Y) = Q_{L0} + Q_1 Y \qquad (7c)$$

The values of the non-dimensional interpolating parameters R_i (i=1,3), a_i (i=1,3) and Q_1, calibrated on the numerical results, are reported in Table 2.

The comparison between the numerical results and the interpolating functions defined in Equations 6–7 is reported in Figure 5. As is evident, a very satisfactory agreement is obtained.

In Figure 6, Equation 6 is validated on the experimental results of Figure 2 (in Figure 6a and 6b the experimental results and the interpolating functions are reported, respectively). The curves corresponding to Equation 6 are obtained by performing a sort of blind prediction, that is the mechanical/hydraulic properties of the soil employed correspond to those experimentally obtained in the geotechnical characterization of the material (Flessati 2017). As is evident from Figure 6, the residual displacements estimated by employing Equation 6 are in good agreement with the experimental tests results. Nevertheless, initially the numerical curves are more rigid.

From a practical point of view, Equation 6 allows the front displacement estimation without performing any numerical analyses, once the system geometry, the soil mechanical properties, the initial stress condition and the excavation rate are assigned.

Equation 6 can also be represented in the Y-Q_f plane (Figure 7). The all iso-displacement curves of Figure 7 are characterized by two vertical asymptotes, one for Y $\rightarrow$ +∞ (undrained response), one for Y $\rightarrow$ 0 (drained response). In Figure 7, three zones can be individuated: for Y>10^4 the response is practically undrained (zone 1), for Y<10^{-1} the response is practically

Table 2. Values of the interpolating parameters.

R_1	R_2	R_3	a_1	a_2	a_3	Q_1
0.725	0.065	0.635	0.686	0.2	0.635	1.6

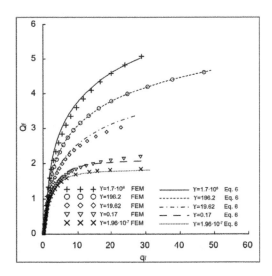

Figure 5. Comparison between the numerical results of Figure 4 and Equation 6.

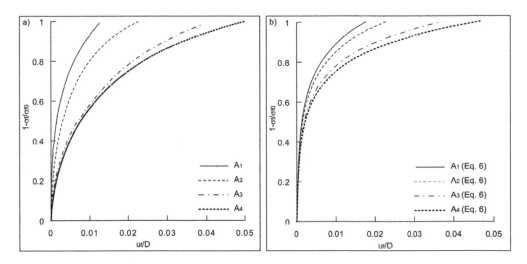

Figure 6. Comparison between the experimental results of Figure 2 (Figure 6a) and the curves corresponding to Equation 6 (Figure 6b).

drained (zone 3) and for $10^{-1} < Y < 10^4$ the response is "partially drained" (zone 2). In this last zone (in particular for $1 < Y < 10$) variations in the Y value (i.e. a variation in the excavation rate) significantly influence the front response.

To further clarify how Figure 7 can be employed from a design point of view, some practical cases are hereafter discussed.

In all the cases hereafter analysed:

- the water table is assumed to be coincident with the ground surface and an initial hydrostatic condition is considered;
- the at rest lateral earth pressure coefficient k_0 is assumed equal to $1-\sin 25°=0.58$;
- the excavation rate v, expressed in meter per day, is constant.

Under these hypotheses, once the parameters k, H/D, v and E (Table 3) are assigned, it is possible to practically employ the numerical results already illustrated in Figure 7.

As far as v is concerned, this must be related to t_u of Equation 3, that is the time need to excavate 1.5D.

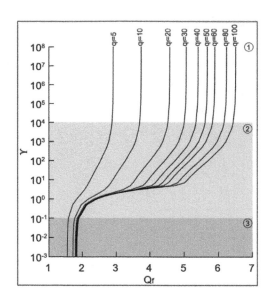

Figure 7.   Equation 6 in the Y-Q_f plane.

Table 3.   Values for the parametric study.

k [m/s]	H/D [-]	v [m/day]	E [MPa]
$10^{-7} \div 10^{-10}$	5	1	20

The dashed zone of Figure 8 is found by calculating the value of Y corresponding to $k=10^{-7}$ and 10^{-10} m/s. As is evident, for values of permeability smaller than 10^{-10} m/s the mechanical response of the front is practically undrained, whereas for $k>10^{-7}$ m/s the mechanical response is practically drained. In other words, very often for both geometries and excavation rate quite standard ($D=12$m $v=1$m/day) even in cohesive soils the excavation takes place under "partially drained" conditions.

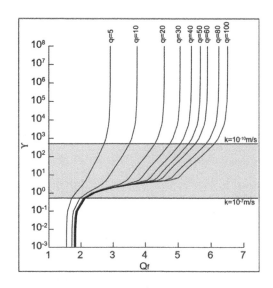

Figure 8.   influence of the permeability k ($D=12$m, $H/D=5$, $v=1$m/g, $E=20$MPa).

5 CONCLUSIONS

In this paper, the authors analysed the influence of the excavation rate on the mechanical response of deep tunnel fronts in cohesive soils, from both an experimental and a numerical point of view.

The experimental 1g small scale model test results clearly highlight that the system response is governed by the consolidation process taking place in the portion of soil in the advance core. In particular, by reducing the excavation rate, the initial slope of the front characteristic curve decreases and the residual displacements increase.

The experimental evidence was also confirmed by the numerical FEM results, which also permitted to introduce a simple expression allowing to estimate the front displacements without performing numerical analyses, once the geometry, the soil mechanical/hydraulic properties and the excavation rate are assigned.

Finally, it is shown that in most of the practical cases of deep tunnels in cohesive soils the excavation is under "partially drained" conditions and, therefore, the excavation rate may be employed as a design variable to limit the front displacements.

REFERENCES

Anagnostou, G. (2012). The contribution of horizontal arching to tunnel face stability. *Geotechnik*, 35(1): 34–44.

Anagnostou, G., & Kovari, K. (1996). Face stability conditions with earth-pressure-balanced shields. *Tunnelling and Underground Space Technology*, 11(2): 165–173.

Anagnostou, G., & Perazzelli, P. (2015). Analysis method and design charts for bolt reinforcement of the tunnel face in cohesive-frictional soils. *Tunnelling and Underground Space Technology*, 47: 162–181.

Attewell, P. B., & Boden, J. B. (1971). Development of stability ratios for tunnels driven in clay. *Tunnels & Tunnelling* 3: 195–198.

Berthoz, N., Branque, D., Subrin, D., Wong, H., & Humbert, E. (2012). Face failure in homogeneous and stratified soft ground: Theoretical and experimental approaches on 1g EPBS reduced scale model. *Tunnelling and Underground Space Technology*, 30: 25–37.

Broms, B. B., & Bennermark, H. (1967). Stability of clay at vertical openings. *Journal of Soil Mechanics & Foundations Division ASCE* 193: 71–94.

Callari, C. (2015). Numerical assessment of tunnel face stability below the water table. *In: Proc. 14th IACMAG*, 2007–2010, Kyoto.

Chambon, P., & Corté, J. F. (1994). Shallow tunnels in cohesionless soil: Stability of tunnel face. *Journal of Geotechnical Engineering*, 120(7): 1148–1165.

Chen, R. P., Li, J., Kong, L. G., & Tang, L. J. (2013). Experimental study on face instability of shield tunnel in sand. *Tunnelling and Underground Space Technology*, 33: 12–21.

Horn, N. 1961. Horizontaler erddruck auf senkrechte abschlussflächen von tunnelröhren. *Landeskonferenz der ungarischen tiefbauindustrie*: 7–16.

Davis, E. H., Gunn, M. J., Mair, R. J., & Seneviratine, H. N. (1980). The stability of shallow tunnels and underground openings in cohesive material. *Géotechnique*, 30(4): 397–416.

di Prisco, C., Flessati, L. (2018) A numerical tool for estimating deep tunnel front displacements: The role of the excavation rate in cohesive soils, *Acta Geotechnica*. under review

di Prisco, C., Flessati, L., Frigerio, G., Castellanza, R., Caruso, M., Galli, A., Lunardi, P. 2018a Experimental investigation of the time-dependent response of unreinforced and reinforced tunnel faces in cohesive soils *Acta Geotechnica*: 13(3) 651–670.

di Prisco C, Flessati L, Frigerio G, Lunardi P. 2018b A numerical exercise for the definition under undrained conditions of the deep tunnel front characteristic curve. *Acta Geotechnica*, 13 (3): 635–649.

Flessati, L. (2017). Mechanical response of deep tunnel fronts in cohesive soils: Experimental and numerical analyses. PhD thesis, Politecnico di Milano.

Flessati, L., di Prisco, C., Frigerio, G. and Lattanzi, A. (2017). Influence of time on the mechanical response of deep tunnel fronts excavated in cohesive materials. *In proc. EURO:TUN2017*, Innsbruck, Austria.

Höfle, R., Fillibeck, J., Vogt, N. (2008) Time dependent deformations during tunnelling and stability of tunnel faces in fine-grained soils under groundwater. *Acta Geotechnica* 3: 309–316.

Kamata, H., & Mashimo, H. (2003). Centrifuge model test of tunnel face reinforcement by bolting. *Tunnelling and Underground Space Technology*, 18(2): 205–212.

Kimura, T., & Mair, R. J. (1981). Centrifugal testing of model tunnels in soft clay. *In Proceedings of the 10th international conference on soil mechanics and foundation engineering* (pp. 319–322). ISSMFE: International Society for Soil Mechanics and Foundation Engineering.

Kirsch, A. (2009). On the face stability of shallow tunnels in sand, PhD thesis, Innsbruck University, Innsbruck.

Klar, A., & Elkayam, I., (2017). Tunnel face stability curves considering asymmetric yielding. In Euro: Tun 2017, *Proceedings of the 4th International Conference on Computational Methods in Tunnelling*, Innsbruck, Austria (pp. 75–82).

Klar, A., Osman, A. S., & Bolton, M. (2007). 2D and 3D upper bound solutions for tunnel excavation using 'elastic' flow fields. *International journal for numerical and analytical methods in geomechanics*, 31 (12): 1367–1374.

Leca, E., & Dormieux, L. (1990). Upper and lower bound solutions for the face stability of shallow circular tunnels in frictional material. *Géotechnique*, 40(4): 581–606.

Leca, E., & Panet, M. (1988). Application du calcul à la rupture à la stabilité du front de taille d'un tunnel. *Revue Française de Géotechnique*, 43: 5–19.

Lunardi, P. (1993). Fiber-glass tubes to stabilize the face of tunnels in difficult cohesive soils. In *Proceedings of the SAIE International Seminar on the Application of Fiber Reinforced Plastics (FRP) in Civil Engineering*, Bologna, Italy (pp. 22–23).

Mair, R.J. (1979). Centrifugal modelling of tunnel construction in soft clay. PhD thesis, Cambridge University.

Mollon, G., Dias, D., & Soubra, A. H. (2009). Face stability analysis of circular tunnels driven by a pressurized shield. *Journal of geotechnical and geoenvironmental engineering*, 136(1): 215–229.

Mollon, G., Dias, D., & Soubra, A. H. (2013). Continuous velocity fields for collapse and blowout of a pressurized tunnel face in purely cohesive soil. *International Journal for Numerical and Analytical Methods in Geomechanics*, 37(13): 2061–2083.

Mühlhaus, H. B. (1985). Lower bound solutions for circular tunnels in two and three dimensions. *Rock Mechanics and Rock Engineering*, 18(1): 37–52.

Ng, C.W.W., Lee, G.T.K. (2002). A three-dimensional parametric study of the use of soil nails for stabilizing tunnel faces. *Computers and Geotechnics* 29: 673–697.

Nomoto, T., Imamura, S., Hagiwara, T., Kusakabe, O., & Fuji, N. (1999). Shield tunnel construction in centrifuge. *Journal of geotechnical and geoenvironmental engineering*, 125(4): 289–300.

Perazzelli, P., Leone, T., & Anagnostou, G. (2014). Tunnel face stability under seepage flow conditions. *Tunnelling and Underground Space Technology*, 43: 459–469.

Schuerch, R., & Anagnostou, G. (2013). Analysis of the stand-up time of the tunnel face. *In World Tunnel Congress 2013*. London: Taylor & Francis Group.

Sitarenios, P. & Kavvadas, M. (2016). The interplay of face support pressure and soil permeability on face stability in EPB tunneling. *In World Tunnel Congress 2016*, San Francisco.

Sterpi, D., & Cividini, A. (2004). A physical and numerical investigation on the stability of shallow tunnels in strain softening media. *Rock Mechanics and Rock Engineering*, 37(4): 277–298.

Vermeer, P. A., Ruse, N., & Marcher, T. (2002). Tunnel heading stability in drained ground. *Felsbau*, 20 (6): 8–18.

Wong, H., & Subrin, D. (2006). Stabilité frontale d'un tunnel: Mécanisme 3D en forme de corne et influence de la profondeur. *Revue européenne de génie civil*, 10(4): 429–456.

Yoo, C. (2002). Finite-element analysis of tunnel face reinforced by longitudinal pipes. *Computers and Geotechnics*, 29(1): 73–94.

Tunnels and Underground Cities: Engineering and Innovation meet Archaeology,
Architecture and Art, Volume 7: Long and deep tunnels – Peila, Viggiani & Celestino (Eds)
© 2020 Taylor & Francis Group, London, ISBN 978-0-367-46872-9

Optimum tunnel system with regard to the entire lifecycle for long rail tunnels

H. Ehrbar & C. Tannò
ETH Zurich, Switzerland

H.-P. Vetsch
Vetsch Rail Consult, Bützberg, Switzerland

ABSTRACT: Since more than 30 years long tunnels with a total length of more than 50 kilometres exist. Many of them show a different tunnel system: double track tunnels with service tunnel, two single track tunnels and two single track tunnels with a service tunnel are the existing systems. The decision on the tunnel system of this long tunnels had to be taken at a time when only few information on operation and maintenance costs were available. Today more information on operation and maintenance should be available. The paper shows, how the decision-making process could be adapted today considering the criteria construction, operation and safety and life cycle. Recommendations on the selection of the tunnel system will be given, based on the available operation experience of the long tunnel railway tunnels.

1 MOTIVATION

For more than 100 years railway tunnels with lengths of 10 km and more have been built. To a large extent, these tunnels are still operating today (see Table 1). However, the demands posed on such tunnel systems have increased during the past years. For a long-time, the double track Tunnel without a service tunnel was the most popular system (variant 1A). Due to the higher safety standards such a system, even with an additional service tunnel, is no longer permissible nowadays unless drastic operating restrictions for mixed railway traffic apply (Ehrbar et al., 2016).

Today – similar to modern buildings –tunnel systems are highly developed technical systems with high demands. In order to thrive against competing transportation systems, the modern rail infrastructure must on the one hand, comply with all safety requirements and on the other hand, provide high availability and an economic operation. Thus, in the case of very long tunnels in particular the question arises, which tunnel system will be able to fulfil the large number of needs in an optimal fashion considering the entire lifecycle of the infrastructure.

2 TUNNELLING SYSTEMS

When choosing a system, there are theoretically no limits on the number of tubes and their configuration (see Figure 1). Systems with a pre-investment could also be made. Thus, a third tube could be created, which is not yet fully provided with railway equipment (as e.g. in the Lötschberg Base Tunnel on 40% of the total length the second single track tube is excavated, but no railway equipment has been placed in).

Railway tunnel systems with more than one tube can consist in a system of pure railway tunnels or in a mixed system of railway tubes and service tunnels.

The historic long railway tunnels such as the Mont Cenis Tunnel, the Gotthard Tunnel, the Arlberg Tunnel and the Lötschberg Tunnel were created as pure double track tunnels without

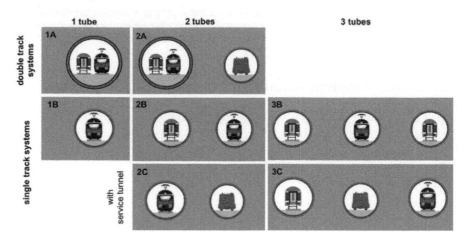

Figure 1.   Variants of railway tunnel systems.

a service tunnel. Only the 19.8 kilometres long Simplon Tunnel has a system with two separate traffic tubes. The decision on this system based on economic and logistical reasons (stages of the construction process, ventilation and cooling).

For the first time in history, in 1988 an over 50 km long railway tunnel was commissioned with the 53.8 km long Seikan Tunnel in Japan.

Parallel to this project in Japan the construction work on the 50.4 km long Channel Tunnel was started in 1987 crossing under the English Channel. The safety requirements for this tunnel exceeded all existing ones. The tunnel system was implemented with two single-track tunnels and one service and safety tunnel plus an extra complex ventilation system. The Channel Tunnel was commissioned in 1994.

In 2007 the Lötschberg Base Tunnel started the commercial operation with a mixture of the tunnel systems 2B (on 40% of the length without the installation of the railway installations) and 2C on 20% of the total length according to the definitions of Figure 1. The reasons for the selection of such a system were financial restrictions and political decisions.

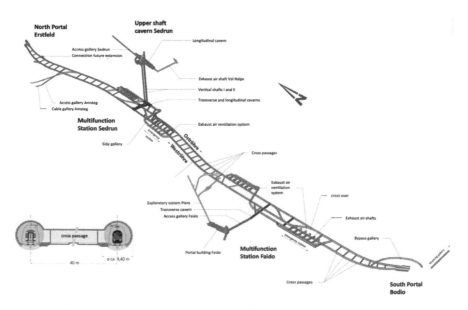

Figure 2.   Tunnel System Gotthard Base Tunnel (©Amberg Engineering, STS, 2016).

In 2016 the 57.1 km long Gotthard Base Tunnel (the longest railway tunnel of the world) followed. It was built following the principles of System 2B with two multifunction stations in the third pointes, dividing the tunnel in sections of 20 kilometres in the maximum (see Figure 2).

Other long tunnels following this construction principle are the TELT (Lyon Turin) and the Follow Line Tunnel in Norway (without multifunction station).

Table 1. Overview of operating long railway tunnels in Europe.

Project Name	Country	length (km)	Commissioning	Tunnel System
Mont Cenis Tunnel	France - Italy	12	1871	1A
Gotthard Tunnel	UK - France	14.9	1882	1A
Arlberg Tunnel	Austria	10.6	1884	1A
Lötschberg Tunnel	Switzerland	14.6	1913	1A
Simplon Tunnel	Switzerland – Italy	19.8	1906/1922	2B
Furka Base Tunnel[1]	Switzerland	15.4	1982	1B
Vereina Tunnel[1]	Switzerland	19.0	1999	1B

Table 2. Overview of operating very long railway tunnels (based on Tannò, 2018).

Project Name	Country	length (km)	Commissioning	Tunnel System
Gotthard Base Tunnel	Switzerland	57.0	2016	2B
Eurotunnel	UK – France	50.0	1994	3C
Lötschberg Base Tunnel	Switzerland	34.6	2007	2B 80%, 2C 20%
Guadarrama	Spain	28.4	2007	2B
Pajares	Spain	24.7	2011	2B
Seikan	Japan	54	1988	2A[2]
New Guanjiao-Tunnel	China	32.7	2014	2B
Qinling Tunnel	China	28.2	2016	2B
Taihang	China	27.8	2007	1A
Hakkoda	Japan	26.5	2010	1A
Iwae-Ichinohe	Japan	25.8	2002	1A
Lüliang-Tunnel South	China	23.4	2014	2B
Iyama	Japan	22.2	2015	1A
Dai-Shimizu-Tunnel	Japan	22.2	1982	1A
Wushaoling	China	22.1	2006	1A

Table 3. Overview of very long railway tunnels under construction in Europe (based on Tannò, 2018).

Project Name	Country	length (km)	Commissioning	Tunnel System
Brenner Base Tunnel	Austria – Italy	56	2026	3C
TELT Lyon – Turin	France - Italy	53	2026	2B
Koralm	Austria	32.8	2024	2B
Semmering Base Tunnel	Austria	27.3	2026	2B
Follo Line Tunnel	Norway	20.0	2021	2B

Table 4. Overview of very long railway tunnels for mixed traffic under design (based on Tannò, 2018).

Project Name	Country	length (km)	Location	Tunnel System
Finest-Link	Finland - Estland	100	subsea	3C
Gibraltar	Spain – Morocco	37.7	subsea	3C
Erzgebirgtunnel	Czech Rep. - Germany	24.7	mountain	2B
Bohai Tunnel	China	120	subsea	3C

The Brenner Base Tunnel follows the principles of System 3C, whereas the final use of the service tunnel, which is driven as exploratory and drainage gallery, is not yet fixed finally.

The Tables 1 to 4 show a trend from one tube systems to actually two tube systems and to tube systems for the future. What might be the reasons for this trend? Only the fact that most of them are subsea tunnels?

3 PROJECT REQUIEREMENTS AND STAKEHOLDERS INTERESTS

In order to explain the high variability of the tunnel systems of long railway tunnels one has to give a closer look on the project requirements of long railway tunnels.

The main goal of the implementation of a tunnelling project is to meet all project requirements within the agreed level of quality, design life and operational requirements (functionality) such as safety, operating (type of traffic, timetable, flexibility, costs) and maintenance etc. Other important requirements are the realization of the project within the fixed milestones and within the given cost budget, respecting the environmental aspects and the interest of the different stakeholders (see Figure 3). All these requirements are the boundary conditions for the definition of the tunnel system of a long railway tunnel.

Many stakeholders are involved in the processes for the realisation of a major tunnel project. Each stakeholder plays a different role and has his individual interests (see Figure 3).

3.1 *Financier*

An early stable financing of major tunnelling projects is crucial for a later successful realization. The railway operators usually are not able to create sufficient revenues with their transport services to payback the initial investment. The revenues should at least cover the operation costs. Therefore, almost all the projects get a public funding. Only the Eurotunnel was privately funded, with all the well-known financing problems 10 years after starting the commercial operation (Table 5).

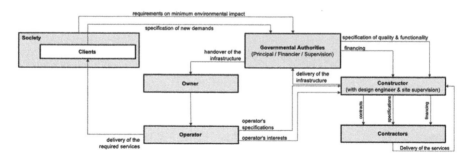

Figure 3.   Possible constellation of stakeholders for a public financed long rail tunnel project.

Table 5.   Overview on the total costs of selected very large tunnels (based on Tannò, 2018).

Project Name	Country	Total Costs [Bn EUR]	Prices from	Type of financing
Gotthard – Base Tunnel	Switzerland	7,0	1998	public
Lötschberg – Base Tunnel	Switzerland	4,3	1998	public
Seikan	Japan	4,7	1988	public
Eurotunnel	UK – France	4,7 Bn £	1994	private
Guadarrama	Spain	1,4	2007	public
Brenner– Base Tunnel	Austria – Italy	10	2018	public
Lyon – Turin	France – Italy	8,0	2018	public

3.2 Principal

The key interest of the principal is the implementation of his order within the required quality and functionality, on time and on budget (minim investment costs), considering the interests of the society. Often the principal is not the operator as he hands over infrastructure to a dedicated operator.

3.3 Constructor

The creation of a very long railway tunnel is often a project outside the field of action of the principal's organization. The long project duration allows to build up a specific, temporary organization for design, construction and commissioning. The constructor is the creator of the project, the overall project leader. The role of the creator is very demanding, as the existing worldwide knowledge is small. The creator has a pioneering role. A large number of processes have to be defined as usually structures and processes cannot be copied directly from other projects.

3.4 Operator

The operator takes over the responsibility for the operation of the infrastructure after completion of the construction work (commissioning process). Maintaining deadlines and the delivery of the mutually agreed quality are important to the operator. He has a high interest in a quick and smooth integration of the new infrastructure into the existing network. Finally, he interested in creating high profits. Therefore, the operator has a high interest on a high availability of the infrastructure, while minimizing the operation and maintenance costs. This requirement is usually in a direct contrast to the requirement of the minimization of investments.

3.5 Authorities

The authorities define the legal boundaries for the project by issuing the technical specifications and the approvals for construction and operation. The authorities check the compliance with the legal requirements. Since such centennial projects sometimes go beyond the current legal framework, it must be expanded or adapted.

3.6 Designers

The main interest of the designer is the utilisation of his resources, creating good references and financial profits. The designers are already involved in the very early project phases. They have a great influence on the project.

3.7 Contractors and Suppliers

The contractors' and suppliers' demand on the project are the utilisation of their resources or the delivery of their products (suppliers), to generate a reasonable profit and to receive a contribution for a good reputation. The constructability of the project should be confirmed already during the design phase in order to avoid time consuming and expensive changes during construction. Contractors knowhow should be used already in the design phase in such a way that conflicts on the procurement process can be avoided.

3.8 Experts

Experts bring an independent view on the project on special aspects, such as e.g. tunnelling, environmental and safety aspects. The expert's main interest is his good reputation.

3.9 Society

Large tunnel projects often affect also large regions and many people due to the environmental impact during construction and operation. Acceptance of the project by the public, politicians, industry and associations is important. A lack of acceptance can cause important delays in financing or in getting legal approvals.

3.10 *Customers*

The end customers are particularly interested in a high availability of the infrastructure and in cheap, comfortable and reliable transport services.

4 DECISION MAKING PROCESS

All the different interests of the various stakeholders create a complex situation for the decision-making process. Therefore, the decision-making process is highly depending on the complexity of the problem statement.

Several methods are available for taking decisions (see Figure 4). As long as costs are the only main target value, static (cost comparison studies, benefit comparison studies) and dynamic cost calculation methods (amortization studies, net present value studies) are helpful tools to create the information needed for the decision on different variants.

Often, the projects have to fulfil more than only one target value (usually costs), but also target values on construction, operation and safety (see Chap. 5) which cannot be measured only by cost elements. For such cases the value-benefit analysis is among other options an often-used powerful tool.

In the context of this paper, only the value-benefit analysis and the net present value method will be considered.

5 REVIEW OF THE DECISION-MAKING PROCESS FOR THE SWISS NRLA PROJECT

In the early seventies of the last century, when the Gotthard Road Tunnel was under construction, Swiss Federal Railway (SBB) elaborated the final design for a 46 km long Gotthard Base Tunnel with one double track tube and a service tunnel. The project was postponed by the political authorities but created the basis for the political decision on the New Railway Line through the Alps (NRLA) by a public vote in 1992. The preliminary design work started

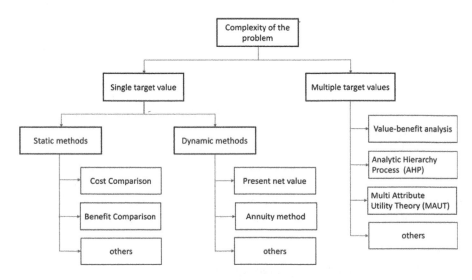

Figure 4. Overview of the various tools for decision-making (Tannò, 2018).

immediately after the positive decision by the swiss voters. Strategic decisions had to be taken on the alignment, the type and number of intermediate accesses and the tunnel system.

An expert committee of swiss and international experts was organised under the leadership of the Swiss Federal Office of Transport office (SFOT) to prepare the forthcoming decision on the basis of objective criteria.

Four alternatives were analysed in this process:

- **Variant 2A**: twin-track tunnel with service tunnel, derived from the 1975 project of Swiss Federal Railway (SBB) and most long railway tunnels built until then,
- **Variant 2B**: two single-track tunnels without a service tunnel but with two underground emergency stations at the third points.
- **Solution 3B:** three single-track tunnels in order to be able to keep two running tunnels open during maintenance,
- **Solution 3C:** tunnel system with two single-track tunnels and a service tunnel, similar to the Eurotunnel solution. The decision whether the service tunnel should be positioned in the middle or at the side, was not decided at this phase. For the cost calculation a lateral service tunnel was assumed.

A value-benefit analysis (point-scoring model) was used for decision making. This process has advantages when the target values are mostly difficult to be represented only by costs.

Table 6 shows, that construction costs were the objectives with the highest weight. This fact is determined by the political situation at that time, which was characterized by the fact that the cost budget of EUR 9 billion for both NRLA-axes should not be exceeded. Reductions in meeting the deadlines were accepted.

It is therefore not surprising that the system with the most favourable construction volume (tunnel system 2B) always achieved the highest score in the value-benefit analysis, also when

Table 6.   Objective system for the value-benefit analyses of the swiss NRLA base tunnels (Ehrbar et al., 2016).

Overall Target	Objective	Weighting	Detailed Objective	Weighting
Construction	Costs, Cost risks	0,70	Construction Costs	0,80
			Cost risks	0,20
	Project schedule	0,20	Construction time	0,80
			Time risks	0,20
	Environmental Impact	0,10	Management of spoil	0,80
			Impact on landscape at portal zones	0,10
			Material for embankments	0,10
Operation	Requirements of operation	0,30	Quality of production (timetable, travel time, comfort)	0,40
			Quantity of production (Capacity, complete blockings)	0,40
			Productivity (Energy, rolling stock, etc.)	0,20
	Maintenance & refurbishment	0,60	Operating impairment incidents	0,20
			Effort for maintenance	0,50
			Attractive workplaces	0,30
	Aero-/Thermodynamics	0,10	Effort for ventilation	0,80
Safety	Acceptance	0,30	Passengers	0,20
			Employees	0,80
	Risks	0,70	Train accident	0,20
			Fire	0,25
			Dangerous goods	0,30
			Accidents of persons	0,05
			Accidents at work	0,20

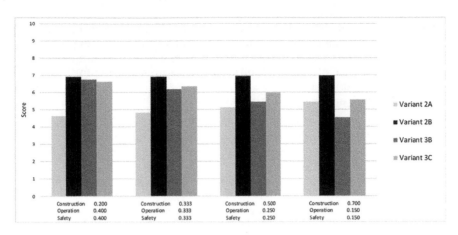

Figure 5.   Scores in value-benefit analysis for the Swiss NRLA-base-tunnels (based on EBP, 1993).

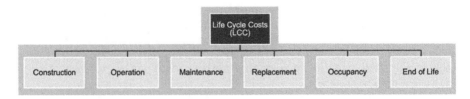

Figure 6.   Generic Structure for lifecycle benefit and costs.

the weighting among the overall goals construction, operation and safety was widely variated in the sensitivity analysis (see Figure 5).

The other cost elements of the lifecycle costs (operation, maintenance, replacement and dismantling) were at that time not considered as cost elements (see Figure 6).

6   ADDITIONAL STUDIES 2018

The question whether the changes in the boundary conditions within the last 20 years would have had an effect on the selection of the tunnel system arose the latest during the commissioning phase at the Gotthard Base Tunnel in 2015. The question was asked, not to question the system decision of 1993, which had a very stable basis, but to keep future owner's organisations of long tunnels away of copy paste approaches and to highlight the importance of a detailed study on the selection of the future tunnel system in the earliest design phases.

Various additional considerations were made during a master's thesis in 2018 at ETH Zurich (Tannó, 2018). In a first step the analysis of decision-making process for the projects Euro Tunnel, Gotthard Base Tunnel and Brenner Base Tunnel showed the high importance of this design step, showing also the changes of the selection of tunnel system due to an in-depth planning. At the Brenner Base Tunnel, the tunnel system of the Gotthard Base Tunnel (system 2B) was which copied first and later switched on a tree tube solution (system 3C) (see Figure 7).

In a second step a pure lifecycle cost analysis was tried to carry out, assuming that in the meantime since 1993 a lot information on operation costs should have been produced at the Lötschberg Base Tunnel and at the Eurotunnel. As maintenance budgets not compellingly correspond to the cost structure of the early design phase the usable information content of the provided data was lower than expected. It would be helpful to adapt the operators cost structure in a future digital world in order to create the information needed for such optimization studies.

But not only the lack of cost information limited the validity of such an analysis, but also the fact that the net present values to be determined in such an analysis depend to a very large

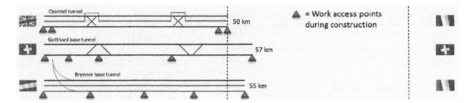

Figure 7. Systems layout of long tunnels in Europe with work access points during construction (FINEST LINK, 2018).

extent on the assumptions of interest rates and inflation rate (see Figure 8). Furthermore, a pure cost comparison study assumes the same benefit for all tunnel systems. This assumption is not correct as the availability of a system with three single track tubes is higher than with two single track tubes, creation also different earnings. Therefore, a pure cost analysis does not allow any compelling conclusions about the system selection and should not be used as a unique tool for the decision-making process.

Therefore, an adapted benefit-value-analysis was carried out with the following assumptions:

– Pure long double track tunnels are not approvable (see Technical Specifications for Interoperability, TSI, EU). Therefore, such tunnels were not part of the thesis anymore.
– The level of safety of the remaining systems was considered as more or less equal for all remaining systems.
– The overall target "safety" was therefore replaced by a new overall target "refurbishment" (see Table 7).
– A simplified scoring model was used with a maximum of 3 points instead of 10 points.

Similar to the benefit-value-analysis of 1993 the most recent studies showed also a clear favourite system, this time the system 3C instead of system 2B (see Figure 9).

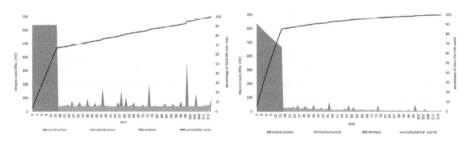

Figure 8. Life cycle cost (17 years construction time, 100 years of operation) without interests and inflation (left) and with (3% interests, 1% inflation (right) (Tannò, 2018).

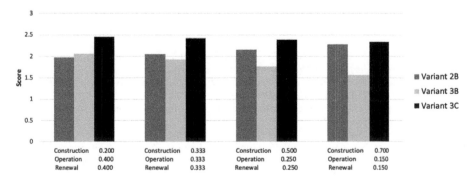

Figure 9. Scores in an updated value-benefit analysis for very long tunnels (Tannò, 2018).

Table 7. Objective system for the value-benefit analyses of the swiss NRLA base tunnels (Tannó, 2018).

Overall Target	Objective	Weighting	Detailed Objective	Weighting
Construction	Costs, Cost risks	0,70	Construction Costs	0,70
			Cost risks	0,30
	Project schedule	0,20	Construction time	0,70
			Time risks	0,30
	Environmental Impact	0,10		
Operation	Quantity of production	0,20		
	Maintenance	0,40	Maintenance effort	0,60
			Attractive workplaces	0,40
	Disruption of normal operation	0,30	Operating impairment	0,40
			Organization of the remedy	0,30
			Accessibility of the defect	0,30
	Effort for artificial ventilation	0,10		
Renewal	Renewal expenses	0,30	Major renovations	0,40
			Minor renovations	0,60
	Complexity of renewals	0,30	Major renovations	0,40
			Minor renovations	0,60
	Loss of capacity	0,30	Major renovations	0,40
			Minor renovations	0,60
	Working place conditions	0.1	Major renovations	0,50
			Minor renovations	0,50

This result is created by the fact that a third tube can contain parts of railway equipment which has to be placed in the driving tube for the solutions 2B and 3B. Solution 3C allows an independent access for tire vehicles which creates many benefits (fewer operating impairment, shorter intervention times).

Such a service tunnel should be placed below the railway tubes, in order to create a spatial separation of traffic and safety infrastructure and operation utilities. Therefore, it is very understandable, that the Brenner Base Tunnel switched to such a system which is from the authors point of view highly recommendable also for other future very long tunnels as long as the service tunnel is used also during operation.

If the boundary conditions do not allow the construction of an additional service tunnel, it should be considered to use also intermediate accesses as independent service accesses during operation. However, only the system with three single track tubes a complete separation of operation and maintenance for bigger renewal work.

NOTES

1 Narrow gauge railway (1'000 mm)
2 For the first time with an underground multifunction station, subsea tunnel

REFERENCES

Ehrbar, H. & Vetsch H. -P. & Zbinden, P. 2016. Long Railway Tunnel System Choice - Review for the future, *GeoResources Journal 2- 2016*. www.georesouces.net

Swiss Tunnelling Society (STS), 2016. *Tunnelling the Gotthard – The success story of the Gotthard Base Tunnel*. Bauverlag BV GmbH, Gütersloh, Germany

Tannó, C. 2018. Optimal choice of the tunnel system for long and very long railroad tunnels. *Master Thesis ETH Zurich*. Switzerland

Ernst Basler und Partner (EBP), 1993. *Tunnel system selection Gotthard Base Tunnel - Comparison of the tunnel system variants: benefit-value analysis*. Unpublished report. Zollikon, Switzerland

ATKINS, 2012. *Norway High Speed Rail Assessment Study: Phase III-Estimation and Assessment of Investment Costs*. London, UK

FINEST LINK, 2018, Helsinki-Tallinn Transport Link: *Feasibility Study-Final report*.

Mechanized tunnel excavation of lot KAT2 of the Koralm-Tunnel, Austria: Achievements and lessons learned

D. Fabbri & R. Crapp
Engineering Joint-Venture PG KAT/Lombardi Engineering Ltd., Switzerland

H. Hölzl
Engineering Joint-Venture PG KAT/Geoconsult Consulting Engineers, Austria

H. Wagner
ÖBB Infrastruktur AG, Austria

ABSTRACT: The 32.9 km long Koralmtunnel represents the key element of the new 130 km long railway line between Graz and Klagenfurt (Austria). The main lot KAT2, on Styrian side, includes a total of 4.5 km of main tunnel sections, 1 emergency-stop and 38 cross passages excavated by conventional methods, as well as 2x17 km of main tunnel excavated with two Double-Shield TBM (diameter approx. 10.0 m). The mechanized tunneling, in crystalline hard rock and with overburden up to 1'210 m, was successful and has overcome the pitfalls and difficulties of several fault zones of decametric extension. With the present contribution the authors intend to share the important goals achieved and the lessons learned.

1 INTRODUCTION

The Koralm railway as a part of the Baltic-Adriatic Axis is located in the south of Austria, linking the provinces of Carinthia and Styria. The tunnel with its length of 32.9 km crosses below the mountain range of the Koralpe, a crystalline massif in the Eastern Alps (see Figure 1).

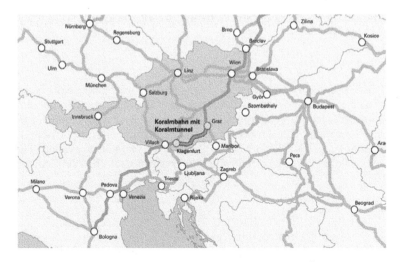

Figure 1. Koralm railway, in the core of the Baltic-Adriatic axis (source: ÖBB).

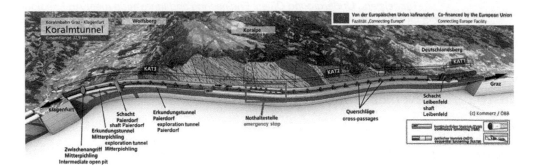

Figure 2. Project overview/schematic layout showing exploration tunnels and shafts, running tunnels, cross-passages, emergency stopping area as well as the sub-division of main contracts (source: ÖBB).

In October 2011, the European importance of the Koralm railway was acknowledged and the project has been added to the "Core-Network" of the European Union. In the suggestions of the European Commission the Baltic-Adriatic Axis with the Koralm railway has been ranked in first place in the top ten priority traffic projects. The route links northern Italy and the important Adriatic ports with the Vienna area, the Czech Republic, Slovakia and Poland and the Baltic States and thus the Adriatic with the Baltic Sea.

The tunnel is sectioned into three construction lots (see Figure 2): two on the Styrian side (KAT 1 and KAT 2) and one on the Carinthian side (KAT 3). The activities of the first smaller lot on Styrian side, KAT 1, were accomplished in 2013. The construction of both main lots (KAT 2 and KAT 3) is still under progress, the excavation activities of lot KAT 2 were finished with the first breakthrough at chainage 18'947 in the southern tube on August 14th, 2018. The second and breakthrough in the northern tube is expected early in 2020.

The tunnel system consists of two single-track tunnels, connected every 500 m by cross-passages. Additionally, an emergency stopping area is located app. in the middle of the tunnel.

2 THE LOTS

According to preliminary studies in the planning phase with respect to geological, topographic and logistic conditions, the Koralm Tunnel project was divided into three major construction lots i.e. KAT1, KAT2 and KAT3. The position and the sub-division of the entire project are shown in Figure 2. The railway equipment including slab track and feeder system is not part of the civil works contracts.

2.1 Lot KAT 1

The Lot KAT 1 is situated in the Western Styria area. Construction work started already in December 2008 with excavation works at the eastern portal. Subsequently, a cut and cover section of approx. 280 m length had to be completed within one year to meet requirements of the environmental impact assessment. The tunnel drives were launched in April 2010 and the first tube reached the neighboring Lot KAT2 in May 2012 after two years of excavation. The construction method was conventional tunneling following the principles of the New Austrian Tunneling Method (NATM).

2.2 Lot KAT 2

Lot KAT 2 is the largest construction contract of the Koralm tunnel. Works started in January 2011 with excavation from the existing exploratory shaft Leibenfeld. For logistic necessities the excavation of an additional twin construction shaft with a cross section of approx. 720 m² and a depth of 60 m was started (see Figure 3).

Figure 3. Construction shaft Leibenfeld (source: PG KAT).

From this shaft, which overlays both northern and southern tube, further conventional tunneling headings were launched. The driving was directed simultaneously towards the east (meeting KAT 1) and the west (advancing to the crystalline central part of the Koralm massif) containing 4.5 km of tunneling in Neogene sediments as well as in the transition zone to the crystalline rock mass.

The distance between the shaft and the break-through with Lot KAT1 is approximately 800 m. The westward headings had to advance for 1'082 m and 1'820 m in order to begin construction of the TBM assembly caverns in competent rock in the North and South tube, respectively. With a cross section of nearly 300 m² and a length of 40 m, the dimension of the two assembly caverns meet the necessities for the underground assembly of the two, about 180 m long TBM's. The underground assembly of the first TBM for the southern tube started in September 2012 and for the second TBM in November 2012.

The mechanized tunneling of the main tunnels to the west with 2 Double-Shield TBMs, outer diameter approx. 10 m, was originally planned with a length of 15.6 km in the North tube and 17.1 km in the South tube. Due to delays in excavation on both construction sites KAT 2 and KAT 3 it was decided by the owner to increase the length of mechanized tunneling in lot KAT 2 by an additional 1'500 m to 17.145 m in the North tube.

The construction shaft is the logistic center for the material management and supply of the two TBMs. The main construction site facilities are installed in the Leibenfeld area including the construction plant for segment fabrication. The selection of appropriate construction site facilities was undertaken in an early planning stage. Several aspects were considered such as the various impacts on residents, the vicinity of existing infrastructure like main roads and railway lines, the capacity of energy and water supply, topographic conditions as well as ecological criteria. The priority was to form a climate of understanding and tolerance of all parties, efficiency and capacity for constructional aspects with best-fitting safeguard and protection for residents in the surrounding cities and villages.

Additionally to the main tunnel drives, 42 cross-passages with a length of approx. 40 m have to be constructed. Furthermore, an emergency stop area with a length of app. 1 km has to be excavated in the middle of the tunnel.

2.3 Lot KAT 3

The third Lot KAT 3 is located entirely in the province of Carinthia and starts from the western portal in the Lavanttal valley. Almost 8 km of the southern tube already existed in form of the top heading which has been constructed as exploratory tunnels Paierdorf and Mitterpichling from 2004 to 2010. Bench and invert have been subsequently excavated conventionally within the Lot KAT 3. The remaining 2.7 km of the southern tube to the break-through point with Lot KAT 2 are excavated using drill and blast following the principles of NATM.

The northern tube with a length of originally app. 12 km (before variation of lot limits) is advanced with a Shield TBM, equipped with earth pressure components for the first five kilometers in the neogene sediments as well as the Lavanttal fault zone (Moritz et al. 2011). The EPB shield machine has been converted underground to a Single-Shield TBM for the remaining crystalline bedrock section with an original length of 7 km (before variation of the construction lot borders).

3 GEOLOGICAL INVESTIGATION AND MODEL

The geological model has been developed by a multi-stage process for each planning phase with the aim of achieving a three-dimensional understanding of ground conditions. This sequence is based on a stringent procedure to investigate with increasing detail and adapted to suit the required project phase and legal procedure. This enables an efficient and structured gaining of extensive knowledge of the ground conditions. In the course of the geological investigations, a total of 133 cores were drilled with a total length of 21.000 m and maximum depths of 1.200 m. Complementally, intense geophysical investigation was further applied for the specific conditions and the purpose of the investigations (Harer et al., 2009).

As an integral part of this stepwise investigation program, work started in 2003 with the excavation of a system of exploratory tunnels and shafts (see Figure 4). They served in particular to gain the necessary detailed geological and hydrogeological knowledge for the main tunnel construction.

The investigation of the Lavanttal fault system was in the main focus of exploratory works. For that purpose, the exploratory tunnels of Mitterpichling and Paierdorf with a length of almost 8 km and another 120 m deep exploratory shaft Paierdorf were excavated. The results of the investigations allowed a further calibration of geological predictions and represent a profound basis for excavation concepts and TBM specifications for the tendering of the main tunnel lots.

All required structures for the investigation are integrated in the further construction phases. The investigation tunnels were excavated as top headings in the southern tube. The two exploratory shafts located in Leibenfeld and Paierdorf are the main facilities for the definitive ventilation system for the tunnel operation.

4 GEOLOGICAL AND GEOTECHNICAL ASPECTS, DIMENSIONING

Along the alignment of lot KAT 2 the rock mass is mainly characterized from slightly fractured hard rock sections, Overburden up to 1'210 m, original (natural) rock temperature up to 35 °C, water pressure (undrained) up to 100 bar. Several fault zones were predicted: some a few decimeters long, others up to a few decameters long, in general consisting of cataclastic rock, embedded blocks and loose material, mainly sub-vertical, perpendicular or sub-parallel to the tunnel alignment (see Figure 4).

Concerning the mechanized tunneling in KAT 2, according to the developed geological model and to the predicted the geological and geotechnical conditions, "rock mass sectors" and the different expected fault zones have been submitted to rock mechanical analyses in order to permit a determination of the loads to be considered for the TBM-shields and for the segmental lining (see Figure 5).

For fault zones, favorable and unfavorable scenarios as well as fault zone length and orientation were considered. Critical fault zones were separately examined with numerical models (see Figure 6).

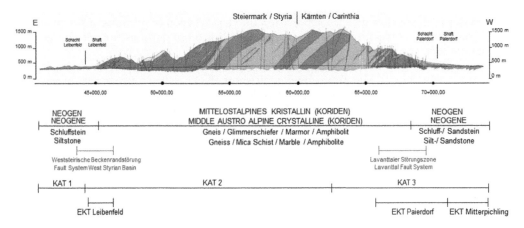

Figure 4. Geological longitudinal profile through the Koralm Tunnel (simplified) showing the main contract sections KAT 1, KAT 2 and KAT 3 and the section with the exploration tunnels at Leibenfeld, Paierdorf and Mitterpichling (source: ÖBB/PG KAT).

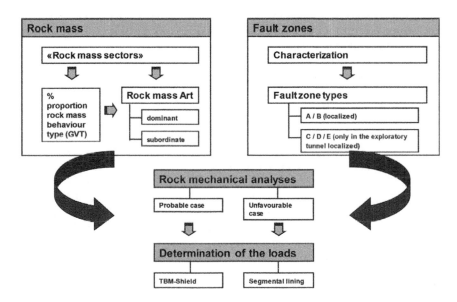

Figure 5. Rock mechanical analyses.

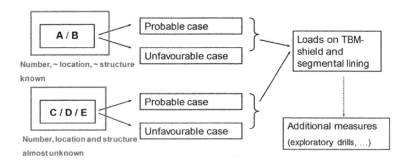

Figure 6. Fault zones analyses.

For major fault zones, a detailed subdivision into a succession of portions of fractured rock and of competent but weak rock has been assumed. For those portions the relevant geotechnical parameters have also been determined.

According to the geomechanical analysis, Critical fault zones were recognized. Fault zones from type A (the largest, significant ones with short thickness/high overburden respectively great thickness/moderate overburden) resulted decisive to establish the minimum requirement for the TBMs. For selected fault zones of type A the excavation by TBM were predicted as feasible but only with additional measures; otherwise an high risk of getting the TBMs embedded/blocked had to be accepted. Selected fault zones of type A and B resulted decisive in the dimensioning and the optimized use of the segments.

5 CONCEPTS AND TECHNICAL SPECIFICATIONS IN TENDER DOCUMENTS

In order to be ready to overcome expected and unexpected fault zones in lot KAT 2, some concepts und technical specifications were implemented in the tender documents and subsequently considered during the advancement of the TBMs:

- Overlapping exploratory drilling and drainage drilling ahead of the TBMs;
- Extra maintenance work on the TBM before entering fault zones in order to reduce the risk of unexpected maintenance stops (new cutters; control of conveyor belts; carry out repairs; etc.);
- Overcut 4–5 cm in radius with peripherical cutters;
- Lubricant for the reduction of shield friction;
- Conditioning of the cutter head against blockage;
- Silicate foam injection close to the cutter head in loose rock or for filling small over-profiles due to block fall;
- Injections and umbrella steel pipe roofing from the TBM area to improve the rock mass conditions.

6 OVERCOMING OF IMPORTANT FAULTS

Based on the calculations and assessments made in the context of the final/tender design, TBM stops were predicted in 3 fault zones and considered in the overall time schedule of the lot (assumed stop durations from 1 up to 3 months per single stop). Highest risk has been predicted in the last fault zone, near the construction lot border of KAT 2.

Actual stops occurred in 3 different fault zones, only the last one was approximately corresponding to the forecast.

The first TBM stop occurred in a fault zone at chainage 8'777 m (2014, tunnel south) and has to be attributed to a conveyor belt damaged by overwhelming excavation material, consequence of unstable front. The stop for repair allowed the convergences to raise and therefore led to a blockage of the rear shield of the TBM. The duration of the TBM stop was about 3 months. According to the experience acquired in the 17 km long TBM drive of both main tunnels, it can reasonably be assumed that without the conveyor belt damage, this fault zone would have been successfully overcome.

The second TBM stop occurred in fault zone at chainage 15'181 m (2016, tunnel south) and is to be attributed to a collapse of material on the front shield as consequence of unstable front (TBM excavation with negative balance, see Figure 7). Duration of the TBM stop was about 4 months (02.06.2016 - 06.09.2016).

The third and last, but multiple TBM stop occurred in a large fault zone at chainage 17 km (2017–2018, once in tunnel north and twice in tunnel south, see Figure 8). The cumulated duration of the stops altogether was about 9 months. In the present paper this third stop will be presented in a more extensive way.

To better understand what happened at the fault zone at km 17 it is appropriate to retrace the Chronology of advancement and events through the fault zone

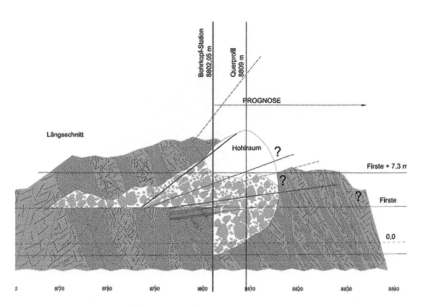

Figure 7. TBM stop at chainage 15'181 m; collapse of material as consequence of unstable front (source: PG KAT).

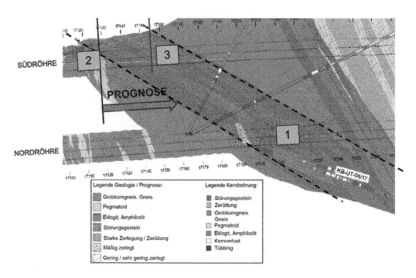

Figure 8. Multiples TBM stops at chainage 17 km; locations of the 3 stops (source: PG KAT).

- Event [1] May 2017: TBM North jammed in the middle of the fault zone;
- Event [2] May 2017: TBM South jammed at the beginning of the fault zone during execution of pipe umbrella;
- Freeing of both TBM by rescue tunnel above the TBM shields;
- Advancement of TBM stepwise with either anticipatory or pursuing tunnel above shields;
- October 2017: TBM North leaves fault zone and restarts regular advancement;
- Event [3] November 2017: TBM South advances without pursuing tunnel above shields and gets jammed again at the end of the fault zone;
- Freeing of TBM South by a by-pass tunnel from Tunnel North, combined with a rescue tunnel above the TBM shields;
- February 2018: TBM South leaves fault zone and the contingency is finally overcome.

Figure 9.   Event [1], damages to the segmental lining (source: PG KAT).

6.1   *Event [1] May 2017: TBM North jammed in the middle of the fault zone*

On May 22nd, 2017 the thrust forces began to rise over 100'000 kN together with a self-progressing excavation face. In the middle of the fault zone the thrust forces reached the maximum of 155'000 kN; damages to the segmental lining related to the uneven distribution of the thrust on the segmental ring were noticed (see Figure 9), so that the advancement was stopped. At that time the face was already advanced 4.5 m ahead of the cutter-head.

After securing the segmental lining (see Figure 9) an attempt to restart the excavation was done, but the shields didn't move again. Grippers and stabilizators were no longer able to be extended by the squeezing pressure. Several self-progressing instabilities were noticed leading to the formation of a collapse above the shields and the cutter-head. Attempts to correct the heading of the TBM during advancement led to a complete misalignment of shields (geometrical blockage).

Concerning the freeing of the TBM the following measures/procedures were adopted:

– Stabilization of the loose collapsed material above the TBM shields and ahead of the cutterhead (polymeric foam injections, cement grouting, pipe umbrella, GF-anchors, etc.);
– Realization of a rescue tunnel above the TBM for reduction of the lateral load on the shields (release of squeezing pressure), for cleaning of the telescopic shield and for release of geometrical constraints of tilted shield-structure (see Figure 12);
– Avoidance of repeated collapse of the face by an anticipatory tunnel above the TBM.

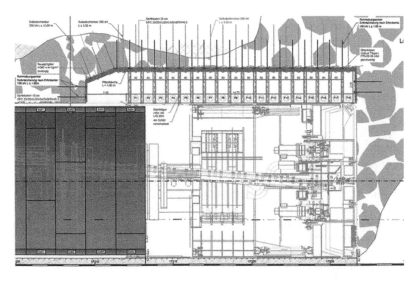

Figure 10.   Event [1], freeing of the TBM (source: PG KAT).

6.2 Event [2] May 2017: Tunnel South TBM jammed at the beginning of the fault zone during execution of pipe umbrella

On May 24th, 2017 the TBM stopped about 20–30 m before the expected fault zone (recognized with the northern tube) for the execution of additional measures such as a pipe umbrella and exploration drills. Restart of advancement after this stop was not possible due to excessive load on the shields. Thrust forces exceeded the installed capacity of 155'000 kN. Local instabilities at the excavation face propagated upwards.

Concerning the freeing of the TBM the following measures/procedures were adopted:

- Stabilization of the loose collapsed material above the TBM shields and ahead of the cutter-head (polymeric foam injections, cement grouting, pipe umbrella, GF-anchors, etc.);
- Rescue tunnel around the shields of the TBM for reduction of the lateral load on the shields (release of squeezing pressure), removal of collapsed material;
- Avoidance of repeated collapse of the face by an anticipatory tunnel above the TBM;
- In more stable condition pursuing tunnel to maintain rapid accessibility in case of collapsed face.

6.3 Event [3] November 2017: TBM South advances without pursuing tunnel above shields and gets jammed again at the end of the fault zone

Towards the end of September, less than 10 m from the end of the fault zone, it was decided to go ahead without the pursuing tunnel above the TBM since thrust forces were constantly below the specified limit of 60'000 kN.

At the beginning of October, however, thrust forces reached again a high level together with instabilities of the excavation face and above the TBM. Advancement had eventually to be stopped again when 155'000 kN of thrust were reached.

Self-progressing face (~4 m) and collapse (~8 m) occurred above the cutter-head.

Concerning the freeing of the TBM the following measures/procedures were adopted:

- Stabilization of the loose collapsed material above the TBM shields and ahead of the cutter-head (polymeric foam injections, cement grouting, pipe umbrella, GF-anchors, etc.);
- Rescue tunnel around the shields of the TBM for reduction of the lateral load on the shields (release of squeezing pressure), removal of collapsed material, stabilization of fault zone between TBM and end of fault, preparation of arrival of counter advancement, realignment of TBM drive (deviation of 10 cm to theoretical axis);
- Avoidance of repeated collapse of the face by a by-pass tunnel from the northern tube and a counter advancement towards the cutter-head.

Figure 11.   Event [2], rescue tunnel (source: PG KAT).

Figure 12.   Self-progressing face (source: PG KAT).

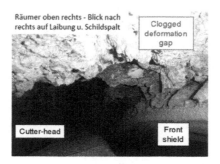

Figure 13.   Clogged deformation gap (source: PG KAT).

6.4   *Fault zone at km 17, Interpretation of rock mass behavior*

Jamming of the TBMs in the fault zone is referred to a combination of different mechanisms and not only to a single event! Two principal rock mass behaviors are usually involved when TBMs get stuck: gravity-induced behavior and stress-induced behavior.

In the presented case the following behaviors contributed to the jamming of the TBMs:

– Local instability of excavation face (so called "self-progressing" face, either gravity- or stress-induced)
   → Progressing local instability at face with negative balance of excavation material (actual volume exceeds theoretical volume);
– Global instability of the excavation face with progressive collapse above the cutter-head;
– Jamming of shields by squeezing rock-mass. In this encountered case:
   → contact between TBM shields and rock mass in squeezing conditions (see Figure 13).
   → critical conditions when entering and when leaving the fault zone.
   → core zone of the fault was nearly detensioned and gravitational effects prevailed.
   → Asymmetric squeezing led to deformation of shields and shield structure together with damaging of the lining.
   → Tilting between front and telescopic shield, consequently significant increase of thrust forces due to geometrical resistance.

7   LESSONS LEARNED AND CONCLUSIONS

Numerical analyses during Tender Design suggested that several fault zones would be difficult to pass with Double-Shield TBMs without additional construction measures. Several standstills had therefore to be expected. In the Tender Design construction schedule as well as in the contractual time schedule several months of advancement stop for such events have been considered. This has proven to have been a wise and right decision. For tunnels with high overburden in the alpine region standstills should always be considered in the contractual framework.

The geomechanical basis, in particular with regard to fault zones, has to be worked out in an early planning phase: Checking the feasibility of mechanical excavation with Shield-/ Double-Shield TBM. Consistent concepts for excavation, such as stepwise preliminary investigation and additional/special measures, must be integrated into the tender documents. Adjustments to these concepts during execution on the basis of the findings and experiences are to be provided and technical support by the designer is recommended.

After the first breakthrough at the Koralm Tunnel we can say: Yes, the use of a Double-Shield TBM in long deep tunnels is possible, even in the presence of several fault zones with high overburden. With a total excavation length of 17.127 m and 17.145 m of the both TBMs this outstanding performance can be considered as the world's longest single tunnel drives in history.

REFERENCES

UIC Kodex 779-11. 2005. Bemessung des Tunnelquerschnitts unter Berücksichtigung der aerodynamischen Effekte, Internationaler Eisenbahnverband, Februar 2005.

TSI Infrastruktur, Europäische Gemeinschaft Richtlinie 96/48/EG. 2008. Interoperabilität des transeuropäischen Hochgeschwindigkeitsbahnsystems, Entscheidung 2008/217/EG, Teilsystem "Infrastruktur", 19.03.2008.

Harer, G., Mussger, K., Hochgatterer, B., Bopp, R. 2008. Considerations for development of the typical cross section fir the Koralm tunnel. Geomechanics and Tunnelling, 4, 257–263.

Harer, G. 2009. Koralm Tunnel – Benefits of a structured investigation process for a large tunnel Project – the clients view. Proc. World Tunnel Congress 2009, May 23 – 28, Budapest.

Keiper, K., Wagner, H., Matter, J., Handke, D., Fabbri, D. 2009. Concepts to overcome squeezing conditions at the Koralm tunnel. Geomechanics and Tunnelling, 5, 601–611.

Keiper K., Crapp R., Amberg F. 2010: Assessment of the interaction of TBM and rock mass in rock tunnelling based on geomechanical calculations, 59th Geomechanics Colloquium, Salzburg, 07/08.10.2010, Austria

ÖGG. 2010. Guideline for the geotechnical design of underground structures with conventional excavation, 2nd revised edition, Austrian Society for Geomechanics.

Moritz, B., Handke, D., Wagner, H., Harer, G, Mussger, K. 2011. Criteria for the selection of tunnelling method through the example of the Koralm Tunnel, Geomechanics and Tunneling.

Pichler, W., Schöfer, H., Wagner, H. 2011. Eluatarmer Spritzbeton - Erkenntnisse aus zwei Jahren Baustelleneinsatz, Südbahntagung.

Vill, M., Schweighofer A., Pichler W., Wagner H., Huber H., Kollegger J. 2011. New development of a cracklimited invert slab, Geomechanics and Tunnelling

Keiper, K., Radoncic, N., Crapp, R., Hölzl, H., Moritz, B. 2015: TBM Tunnelling in tectonical faults – Design conclusions drawn from the observed system behavior Eurock 2015/64th Geomechanics Colloquium, Salzburg, 07.-10.10.2015

Fabbri D., Keiper K. 2016: Einsatz einer Doppelschildmaschine im druckhaften Gebirge am Beispiel des Koralmtunnels, Kolloquium Maschinelle Vortriebe, ETHZ, Zürich 19.05.2016

Fabbri D., Crapp R. 2018: Double-Shield TBM at the Koralm Tunnel (Austria): some experiences from excavation in a deep/35 km long tunnel, Master in TBM & Tunnelling, Politecnico di Torino, 20.03.2018

Tunnels and Underground Cities: Engineering and Innovation meet Archaeology,
Architecture and Art, Volume 7: Long and deep tunnels – Peila, Viggiani & Celestino (Eds)
© 2020 Taylor & Francis Group, London, ISBN 978-0-367-46872-9

The Susa-Bussoleno Interconnection Tunnel

A. Farinetti, P. Elia, M.E. Parisi & E. Gueli
TELT s.a.s.

ABSTRACT: The connection between the Lyon-Turin high-speed railway line and the historical railway is called Interconnection Tunnel, which foresees two tunnels, almost two-kilometer-long with excavation sections from 80 m2 up to 361m2. The interconnection tunnels underpass two existing tunnel infrastructure: the historical railway Tanze tunnel with an overburden of 70 m; two tubes of highway A32 Prapontin tunnels with an overburden of 25 m. The tunnels will be excavated using the D&B method with low vibration level allowed. The overburden between the Interconnection tunnels and highway A32 Prapontin Tunnels is around 25 m. Regarding geological and geotechnical information, the rock mass has a good/optimal quality with a compression uni-axial strength approximately about 70 MPa and GSI avg = 50. A 3D model for a back analysis process during the excavation phases will be necessary. To exclude any influence between the excavation of the interconnection tunnels and other existing underground structures, it is essential to monitor the deformation evolution, the rock mass-support system interaction, the stress analysis in the support structures and the plastic zone.

1 INTRODUCTION

The connection tunnel between the new Turin-Lyon high-speed line and the existing historic line is known as the "Interconnection Tunnel" (Figure 1).

The civil works consists in a twin-tubes rail link (even and odd tracks), between the new Turin-Lyon railway (NLTL) and the current Turin-Modane line, with the junction point corresponding to the future high-speed railway station in the Susa plain and the Bussoleno railway station.

The Interconnection Tunnel's layout has been defined starting from the connection with the NLTL and the historical railway, in such a way as to optimize the slopes and avoid interferences with the existing tunnels (the "Prapontin" motorway tunnel and the "Tanze" railway tunnel), and those of the NLTL that will be built at a later stage (the Orsiera Tunnel).

The result is two non-parallel tunnels built almost entirely on a curve, nearly 2 km in length from one portal to the other, and with variable slopes and radius of curvature. The T2 track tunnel is longer, as it is on the outside of the curve, and arches vertically to pass over the top of the future Orsiera Tunnel.

The Interconnection Tunnel project also includes branches linking the two tunnels, as well as TBM assembly chambers and tunnels predisposed to connect with the future Orsiera tunnel.

2 GEOLOGY

The Interconnection Tunnel crosses a geologically homogeneous sector, characterized entirely by the Mesozoic calcareous/micaceous schist layers (TCS) of the Dora Maira Massif, which constitute the Meana - Mt. Muretto Complex (Figure 2). This complex consists mainly of a strong sequence of metapelites with variable carbonate content, changing from calcareous schist to micaceous schist, to garnet ± chloritoid ± graphite.

The soil study revealed the presence of sporadic lenses (metric and pluri-metric dimensions) and boudinage basic and ultrabasic (TCS-G). Their presence was quantified in approximately

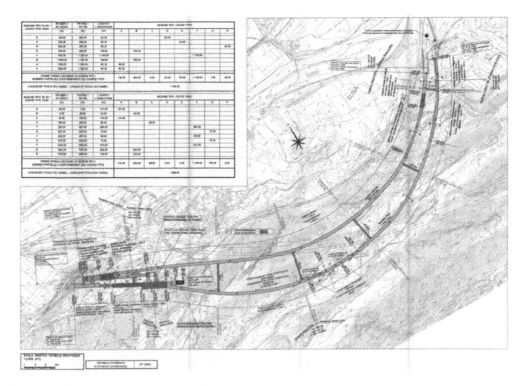

Figure 1.   Plan of the Interconnection Tunnel.

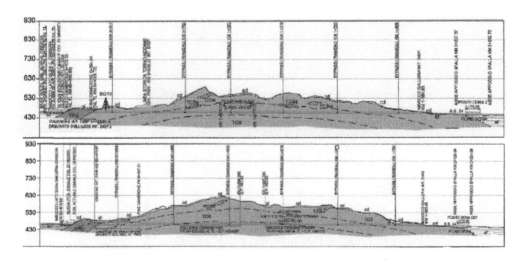

Figure 2.   Geological profile.

1% of the rock mass. The optical microscope analysis of the outcropping samples taken revealed an absence of amiantiferous minerals.

In the westernmost sector, corresponding to the Traduerivi portal, studies revealed the presence of paragneiss outcroppings (TPG), which are generally interspersed with calcareous/micaceous schist, especially towards the top of the sequence. These consist of leucocratic paragneiss associated with impure quartzite. The connection between the paragneiss and the calcareous/micaceous schist is visible between Borgata Sarette and Borgata Rodetti, SW of the Traduerivi portal sector. The contact is characterized by a strong zone with a few metres of thickness where carbonate layers alternate with layers of quartzites and/or paragneiss. The trend of this contact is orientated according to the main schistosity, which has a stratum inclination towards NNW, with an average slope of about 40°. According to the schistosity orientation, in relation to the tunnel's direction, it could happen located detachment events in roof and excavation face (risk of spalling). This critical aspect is linked to the particular hardness of the rock.

From a structural point of view, the area is characterized by a series of structures aligned in an ENE-WSW direction, the nature of which is not yet entirely clear.

The interconnection tunnel's sector of pertinence is located between two different structural domains; in fact, it falls within the transition area between the structural domains of Mompantero (at west) and Orsiera (at east). The geostructural measurements taken on the ground emphasise this particular location. It is clear how the sizes of the main schistosity's location vary from NW-NNW in the area of the Traduerivi entrance, and NNW-NNE in the area of the Bussoleno entrance.

The identified geological domains are:

• Mompantero structural domain;
• Orsiera structural domain;
• Sectors of entrance:
  ○ Traduerivi entrance;
  ○ Bussoleno entrance.

Concerning the structural layout, the sector's central fault system is orientated in an N - S direction with high angles. These structures have not been observed directly on the surface of the slope section affected by the excavation. However, significant structures belonging to this system have been detected to the east (the Rio Gerardo master joint). Furthermore, during the excavation of the Prapontin tunnel, specific layers linked to the cross of delicate fragile structures, which could belong to this main system of faults, were detected.

The mass is characterized by the presence of two leading families of joints (J1 and J2), with persistence and spacing ranging from 1 to 3 meters. The J1 system has an average inclination towards the east at a medium/high angle (average slope of 70°), with joined planes inclined to

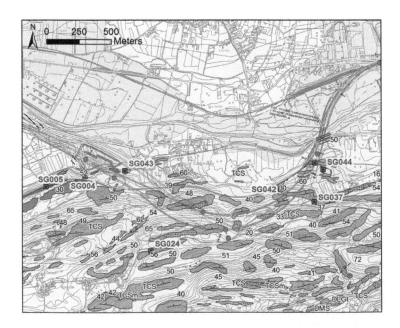

Figure 3. Excerpt from the geological map with the locations of the geomechanical surface stations (SG037).

W; J2 has an average inclination towards S - SW, with an average slope of 50°. The conditions of the rock mass are influenced by the degree of alteration, which, at the sequence of outcroppings, manifests itself as sandy particles or fractured rock along the joints, or the oxidation of the joints' walls alone. The rock mass ranges from unaltered to moderately altered.

2.1 Classification of the rock mass

The geomechanical parameters obtained from the data of previous studies contributed to the identification of 11 geomechanical units (GU). The paragneisses and calcareous/micaceous schists of the Dora Maira that will be affected by the interconnecting tunnels, are classified in the UG-D and are generally characterized by the parameters indicated in the Tables 1 and 2. The values have been defined according to the as-built data of the Tanze railway tunnel.

The parameters of the mass have also been determined according to the structural layout of the surface, which was evaluated using the data obtained from geostructural stations established at the most representative rock outcroppings (Figure 3).

The geomechanical characterization of the rock mass at the entrance was accomplished according to the parameters of the geomechanical stations. According to the measurements taken on the surfaces of the discontinuities, the UCS (uniaxial compressive strength) amounts vary between 40 and 200 MPa, with a dispersion that increases for the higher values up to ± 100 MPa. These amounts confirm what was observed during the excavation of the Tanze tunnel, for which the UCS range varied between approximately 45 and 120 MPa. (Tables 1 and 2).

Table 1. Values detected at the geomechanical stations established.

Geomechanical station	SG	code	SG4	SG5	SG24	SG37	SG41	SG42	SG43	SG44
Uniaxial compressive strength	UCS	Mpa	200	142	65	82	52	42	80	60
Number of discontinuity per cubic meter	Jv	no./m^3	8	25	6	08-ott	4	03-mag	5	3
Geological strength index (estimated)	GSI	-	80	50-55	70-80	75	70-90	4	70	60 70
Degree of mass alteration	-	-	weak	moderate	no	no	no	weak	no	weak
Mass moisture conditions	-	-	dry	dry	wet	dry		dry	dry	dry

Table 2. Reference and/or average values of the geomechanical parameters.

Geome-chanical unit	Lithology	UCS (MPa)			GSI			Cover (m)			RQD			RMR		
		avg.	min	MAX	avg.	min	MAX	avg.	min	MAX	avg.	min	MAX	avg.	min	MAX
UGD	TCS Calcareous/micaceous schists	72,4	42,8	117,5	61	52	69	82,4	0	165	-	-	-	60	52	67
	TPG Paragneiss	41,4	3,8	95,5	53	48	58	280	60	500	-	-	-	55	48	61

Geome-chanical unit	Lithology	G (kN/m^3)		E$_i$ (Gpa)		E$_{rm}$ (Gpa)		C (Mpa)		F (°)		Tensile strength (Mpa)		mi
		min	MAX	min	MAX	min	MAX	min	MAX	min	MAX	min	MAX	
UGD	TCS Calcareous/micaceous schists	27,6	28,5	10,7	29,4	3,7	21	0,4	3,5	61	50	2,5	16,2	7
	TPG Paragneiss	26,3	28,4	1,1	28,6	0,3	13,6	0,2	3	40	50	4	16,2	23

The characteristics of the rock mass are medium/good: the paragneisses fall within RMR class III, the calcareous/micaceous schists fall within an RMR class of Bieniawski III (prevalent), and class II (subordinate). There is a possibility of unstable rock wedges and detaching situations during the excavation phase (risk of spalling) based on the orientation of the schistosity according to the tunnel's axis and the elastic behaviour of the rock mass.

3 GEOMETRY

During the design of the Interconnection Tunnel first of all the overall geometries of the chambers and tunnels were determined based on the layout and shapes of the railway. Secondly to size and optimize them according to the limited knowledge and the interpretation of the geology and the mechanical parameters of the rock mass (to this date, no drilling surveys

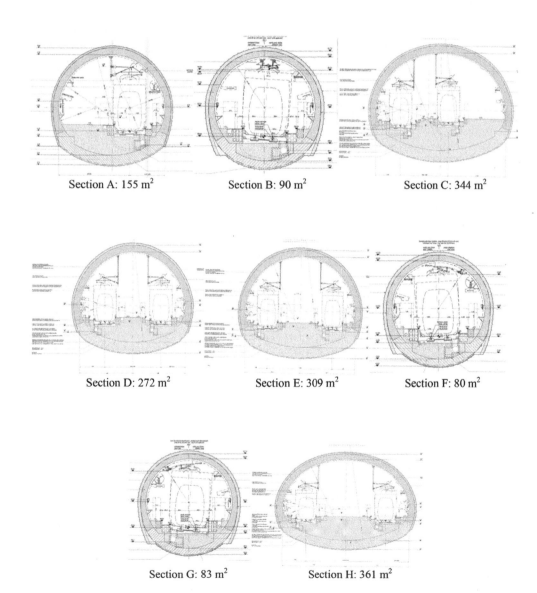

Figure 4.   Different excavation-sections typologies used in the Interconnection Tunnel.

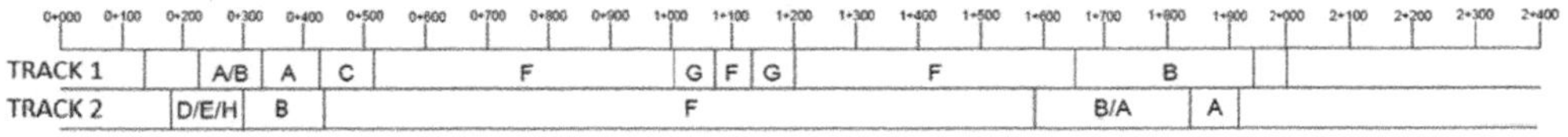

TRACK 1 | A/B | A | C | F | G | F | G | F | B

TRACK 2 | D/E/H | B | F | B/A | A

Figure 5. PK with the location of the different excavation-sections typologies for the tracks 1 and 2.

or in-situ tests have been done), as well as the information available regarding the construction of the Prapontin motorway tunnel and the Tanze railway tunnel.

The excavation-sections typologies (Figure 4) adopted for the chambers and tunnels of both Interconnection Tunnel's tracks are summarized below in the direction of the increasing PK (Figure 5), starting from the Susa side connection portal, up until the end of the traditional excavation on the Bussoleno side.

The project includes eight different excavation-sections typologies (labelled as A through H), with surfaces ranging from 80 m^2 to 361 m^2.

The two tubes will be built, frequently changing the excavation-sections typologies, to minimize the excavation volumes. Starting from the West portal, we have:

- The interconnection chambers, necessary to install the rail between Susa and the underground portal to the future Orsiera tunnel;
- Track with a constant single section of the same magnitude as the Base Tunnel (BT);
- Enlargement of the last part of the track 1 due to signal visibility issues.

4 EXCAVATION METHOD AND INTERACTION WITH EXISTING TUNNELS

The proposed excavation is traditional D&B method. Use of explosives is necessary for the following reasons:

- Jackhammer yields: not compatible with rock strength.
- The Hardness of the rock formation: according to the geological-geomechanical profile of the PD2, the GSI average expected is higher than 50, i.e., the one of a competent rock that is not suitable with jackhammer excavation.
- Previous experience: both the nearby Tanze railway tunnel and A32 Prapontin motorway tunnel were excavated entirely using explosives, except for the entrances, which were excavated with a jackhammer until conditions were suitable for excavation with explosives were encountered. Track geometry will be monitored in the area influenced by the excavation of the even track's large chamber where the Interconnection Tunnel will pass under the existing historical line.

4.1 Interference with the future Orsiera Tunnel

As the distance between the track 1 of the Interconnection Tunnel and the future Orsiera Tunnel will be limited (Figure 6), it will be necessary to reinforce the pillar between the Interconnection Tunnel along the portion affected by the future excavation of the Orsiera tunnel. This will reduce the risk of possible deformations and/or damages to the railway structure. For this purpose, two foundation beams were inserted along the contact between the impost and the invert, which will allow recovering the deformations induced in the transverse and longitudinal direction by the excavation of the future Orsiera's tunnels. Inclined micropile anchors were also inserted into the external extensions of the reinforcement beams.

4.2 Interference with the existing A32 highway Tunnel in Prapontin

The Interconnection Tunnel will pass under the existing Prapontin motorway tunnel of the A32 highway (Figure 7). In their closest point, the two galleries will be at a distance of about 25 meters (Figure 8).

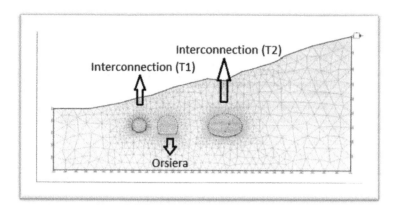

Figure 6. Section of the interference between the Interconnection Tunnel and the future Orsiera Tunnel.

During the Interconnection Tunnel design, as regards the two galleries' interaction, a numerical model was adopted describing the area where their distance was lowest. The first step was simulating the excavation of the A32 Highway tunnel, followed by the excavation of the Interconnection Tunnel. For the simulation, the same support used for the Interconnection Tunnel was considered for the A32.

Since the excavation of the Interconnection Tunnel will be drained, and that in long-term the area will be waterproofed, a hydrostatic charge was not considered in the model.

The model used the following sequence:

- Gravity. Initializing the strain in-situ and set the initial stress balance.
- Decrease of the A32 crown modulus of elasticity, to simulate the face effect.
- Excavation and support of the A32 crown.
- Reduction of the A32 bench modulus of elasticity, to simulate the face effect.
- Excavation and support of the A32 bench.
- Removal of bolts and supports. Covering activation. Rock mass strength resistance reduction (long-term).
- Movements reset.
- Decrease of the Interconnection Tunnel's modulus of elasticity, to simulate the face effect.
- Excavation and support of the Orsiera Tunnel.
- Removal of bolts and supports. Definitive covering activation.
- Rock mass strength resistance reduction.

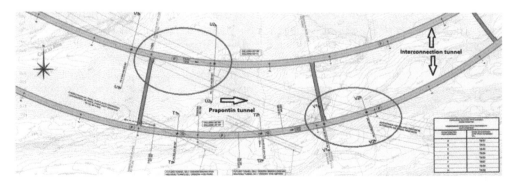

Figure 7. Plan view of the interference between the Interconnection Tunnel and the existing motorway Prapontin Tunnel.

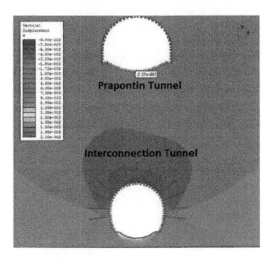

Figure 8. Interconnection and Prapontin Tunnel section.

The calculations have used 4-meters-long bolts with 1.5 x 2.5 m mesh. The support considered was composed of 15 cm of shotcrete with a 25 MPa resistance (C25/30), able to support the stresses with a security coefficient higher than 1.5, regardless of whether a conservative value for the stress relaxation coefficient was used.

Regarding the effects on the A32, the subsidence caused by the Interconnection Tunnel excavation is lower than 3 mm (as shown in Figure 8), so, it can be considered negligible. Concerning the increased stresses, the definitive covering barely suffers any changes, except for the inverted zone where an axial resistance increase, widely supported, is recorded.

4.3 Interference with the existing Tanze Tunnel of the historical railway

Near to the Interconnection Tunnel east entrance, we also find the Tanze Tunnel portal, belonging to the historical Turin-Modane railway (Figure 9). The two portals will be about 8 metres from each other and approximately at the same altitude (about 452 m above sea level). The closest point between the two galleries will be at the portals. Then the two tunnels will

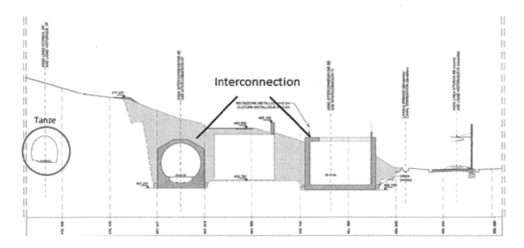

Figure 9. Tanze and Interconnection Tunnel section.

progressively move away from each other, as the Tanze Tunnel has a broader radius of curvature than the Interconnection Tunnel even track.

This geometry allows to reduce any kind of structural interactions between the two tunnels.

4.4 *Drainage water*

The drainage of the T1 track's water, will be collected and disposed by gravity, with collection and treatment occurring in front of the Bussoleno portal area. The water will be subsequently drained into the Dora river.

For the track 1, the water will be channeled from the high point of the arch over the future Orsiera tunnel (Pk 0 + 800) toward the portals. Towards Bussoleno, it will be collected by gravity, whereas towards Susa, at the low point of track 1, the water will be channeled towards the Orsiera TBM disassembly chamber. From there, water will be collected in an accumulation tank and then pumped towards the tunnel's entrance. Finally, it will be sent to a treatment facility for subsequent disposal.

5 EXCAVATION MONITORING

Monitoring is crucial for excavations using the traditional method. The monitoring of the behavior of the excavated and provisional support section allows verifying the assumptions and the criteria defined during the design process. If there are any differences between the behavior encountered and the one expected, the parameters and criteria used to determine the excavation behavior and the primary- lining must be reviewed. If convergences and first phase supports are greater than expected, a detailed investigation must be conducted to determine the reasons for the difference in behavior.

Furthermore, investigations must be carried out if necessary. If, on the other hand, the system's behavior is better than expected, analyses will still be helpful to understand the causes, and the results must be used to calibrate the model.

The tunnel's supplementary investigation plan has the following objectives:

- To conduct three-dimensional monitoring of the evolution of the cavity's deformation (face and profile of the excavation) in relation to both the time factor and the progressive distancing of the face, as well as the geostructural and geomechanical characteristics of the mass;
- To perform a 3D control of the development of the plastic area in advancement at the face of the excavation and surrounding it;
- To verify the mass-support system interaction;
- To analyse the state of stress within the support structures, with a particular focus on any stress anisotropy conditions.

6 CONCLUSIONS

This report summarizes the analysis conducted on different topics concerning the Interconnection Tunnel, its specific geology, its geomechanical parameters, how all these parameters have influenced the tunnel's geometry and the excavation method. The technical analysis of the interferences between the existing and future galleries and the Interconnection Tunnel is crucial. Finally, it looks at two critical additional elements: the drainage schemes and the excavation monitoring.

REFERENCE

LTF S.A.S. 2012. *Progetto definitivo PD2*. Torino, Italy.

*Tunnels and Underground Cities: Engineering and Innovation meet Archaeology,
Architecture and Art, Volume 7: Long and deep tunnels – Peila, Viggiani & Celestino (Eds)*
© 2020 Taylor & Francis Group, London, ISBN 978-0-367-46872-9

TBM steering difficulties. Innovative equipments: Strand jack and self retaining systems

A. Finamore & G. Bellizzi
Cooperativa Muratori e Cementisti, Ravenna, Italy

ABSTRACT: The project consists in the construction of 5 Km single-carriage way, including viaducts, embankments and tunnels, excavated with a single shield hard rock TBM, with 13.72m excavation diameter. The geological and geotechnical setting mainly consists in biotite-muscovite gneiss and amphibolites. The weak rock conditions did not allow the application of sufficient thrust force to steer the shield by means of the advance cylinders. Among the different countermeasures applied, the application of the *strand jack system* has solved the TBM steering difficulties and, in order to allow an easier installation of the equipments, a new technical solution has been designed, *self retaining system*, permitting the installation close to the shield and reducing the operative costs. During the tunnels excavation, both systems were together applied on the machine, with excellent results leading to the complete restoration of standard TBM steering conditions.

1 INTRODUCTION

1.1 *Location of the project*

The Project "SS1 Nuova Aurelia" is located in Liguria (Italy), in Savona district. The new route runs roughly parallel to the existing road "SS1 Aurelia" and intersects both the Genova-Ventimiglia railway and highway, with about 10m overburden. In particular, the Basci tunnel overlaps the existing railway tunnel, the San Paolo tunnel underpasses the highway existing tunnel with the same overburden and the Cappuccini tunnel underpasses the existing highway with about 35m overburden.

1.2 *Main features of the tunnels*

Main features of the tunnels are as follows:

- Basci Tunnel
  - Excavated length: about 490.00 m;
  - Excavation diameter = 13.72m;
  - Internal diameter = 12.60 m;
  - Concrete lining thickness = 35cm;
  - Horizontal alignment: horizontal curve radius = 380 m;
  - Longitudinal alignment: maximum slope of 4%
- San Paolo Tunnel
  - Excavated length: about 2000.00 m;
  - Excavation diameter = 13.72m;
  - Finished diameter = 12.60 m;
  - Concrete lining thickness = 35cm;
  - Horizontal alignment:

Figure 1.   project plan.

- maximum horizontal curve radius = 800 m
- minimum horizontal curve radius = 345 m
○ Longitudinal alignment: maximum slope of 5.5%

- Cappuccini Tunnel
  ○ Excavated length by TBM: about 850.00 m;
  ○ Excavation diameter = 13.72m;
  ○ Finished diameter = 12.60 m;
  ○ Concrete lining thickness = 35cm;
  ○ Horizontal alignment:
    - maximum horizontal curve radius = 370 m
    - minimum horizontal curve radius = 345 m
  ○ Longitudinal alignment: maximum slope of 6%

2   GEOMECHANICAL SETTING

According to baseline values of rock mass parameters and classification systems presented in the Final Design, the following Geomechanical units can be identified along the tunnels track:

- Geomechanical Unit 1 (GU1), about 40% of the tunnels: it corresponds to biotite-muscovite gneiss with very high intact rock strength (as underlined by high values of UCS) and high values of deformation modulus (around or above 10 GPa), the expected degree of jointing is rather low;
- Geomechanical Unit 2 (GU2), about the 25% of the tunnels: it corresponds to amphibolite with high values of intact rock strength and of deformation modulus;
- Geomechanical Unit 3 (GU3), about the 18% of the tunnels: it may include amphibolite and gneiss characterized by higher degree of jointing with respect to GU1 and GU2; reference values of UCS (100 – 175 MPa) coupled with values of RMR and GSI < 50 indicate a hard rock with rather high degree of jointing;
- Geomechanical Unit 4 (GU4), about the 18% of the tunnels: it corresponds to fracture and fault zones. The rocks within this unit are characterized by high values of intact rock strength but poor Geomechanical qualities as rock mass, as indicated by values of RMR., GSI and Q, corresponding to highly jointed and weathered rock masses with very poor geomechanical quality;

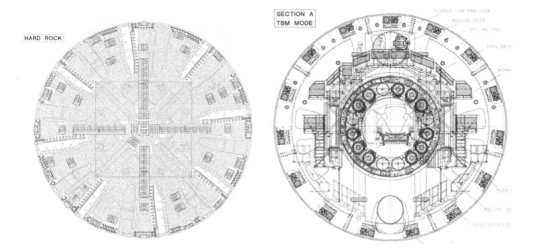

Figure 2. TBM cutterhead and main drive with motors.

3 TBM DESCRIPTION

The TBM involved in the project is an hard rock machine, named S-643, manufactured by Herrenchnekt AG, in year 2012. It is a single shield hard rock TBM, with five back up gantry, for a total length of about 133m. The configuration of the cutterhead is closed (10% opening ration), it is manufactured to bore in hard rock conditions. On the cutterhead n. 86 cutters, 17" diameter, and n. 8 buckets are installed. The maximum force per cutter is 250kN and the total thrust force on the cutterhead is 21.500kN. The nominal torque value is 21.000kNm while the breakthrough torque is 25.000 kNm. Eleven motors are installed on the main drive, corresponding to a total power of 3850KW. The length of the shiled is about 11m, while the shield diameter is 13.64m÷13.60m. On the TBM are installed n. 30 thrust cylinders, with a stroke of 1.70m. The maximum pressure for each cylinder is 350bar, leading to a total thrust force of 85000kN. The total TBM weight is about 1200 tons.

4 SHIELD STEERING DIFFICULTIES

4.1 *Early excavation phase – chronology of events*

The TBM launching at the Basci tunnel was in very weak rock conditions, comparable with soil. As from the start of the tunneling works up to advance n.33 (about 57m) the excavation process has been jeopardized by several events of face instability, that blocked the cutterhead. Generally, the cutterhead could be restarted just after having performed a proper cleaning, done manually. The anisotropic rock behavior and face instability encountered in the first tunnel stretch produced a downward deviation of the TBM axis respect to the project level. In particular, when the rear flap of the TBM shield was at the end of the concrete cradle, realized for the translation of the TBM to the tunnel face through the pre-tunnel, the vertical deviation was - 6.2cm, respect to the project level, considered as a normal deviation due to the tilting of the TBM coming out of the cradle. The excavation process up to advance n.22 (about 37m) was carried out using the following TBM parameters (given by the supplier):

Table 1. TBM parameters up to advance n.22.

Penetration (mm/rev)	10.0
Cutterhead revolutions (rev/min)	1.0
Thrust force (kN)	20.000
Cutterhead torque (MN/m)	7.85
Horizontal and vertical tendency (mm/m)	±3.0

Note: vertical tendency is the difference between project level and actual level in the section considered.

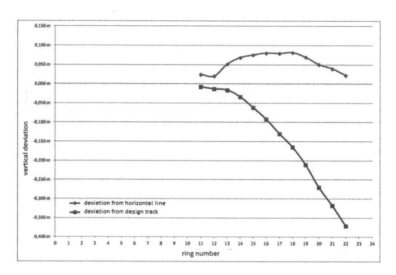

Figure 3. TBM vertical deviation up to advance 23.

Nevertheless the increasing of the thrust of the bottom cylinders with respect to the upper ones, at the end of advance n.22 the vertical deviation from the project level reached the value of – 37cm. During the advance n. 23 an important collapse occurred at the tunnel face and the cutterhead was blocked (maximum torque exceeded) because big blocks has crushed in front of it. The vertical tendency was -3cm/m.

Several actions have been applied, such as manually cleaning of the crushed materials, filling and consolidation at the tunnel face by means of foams and cement and reduction of the cutterhead openings, to unlock the cutterhead and then the excavation activities restarted. After two strokes, at the end of the advance n. 25, the vertical deviation was -56.6cm and the vertical tendency has been increased to -3.6cm/m. During advance n. 26 a new collapse at the tunnel face occurred, but it was possible to unlock the cutterhead and to continue the excavation process up to advance n. 34, using only the bottom thrust cylinders. Unfortunately, the cutterhead contact force at the tunnel face was very low and did not allow to steer the TBM up, hereby, the vertical deviation, at the end of the advance n. 34, was about -135cm and the tendency was increased to -5.7cm/m. Despite of elevated values of applied total thrust, contact force at the tunnel face remained rather low, not allowing the steering of the TBM and it was decided to suspend the excavation activities until a solution was found.

4.2 Balance of forces and weights

The following Figure 4 shows the forces and the moments acting on the TBM during the excavation process. F_G is the weight, applied at the center of gravity, $F_{Cylinder}$ is the resultant of the thrust cylinders force,F_{cw} is the ground reaction, consisting in two components: the contact force at the tunnel face and the friction force along the cylindrical shield surface. Considering the total weight of the TBM, about 1250 tons, the actual dimensions and assuming

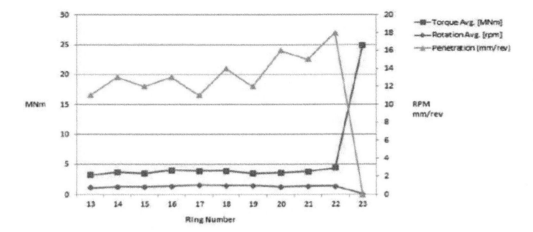

Figure 4.   Cutterhead parameters up to advance 23.

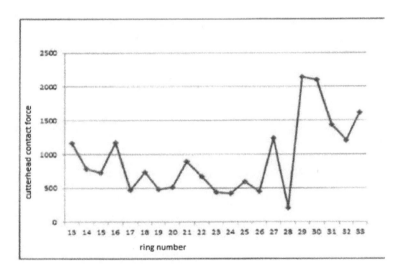

Figure 5.   Cutterhead contact force at the tunnel face during the excavation.

positive rotations counterclockwise by convention, the balance of the moments (tilting and stabilizing) is as follows.

$$M_A = m * F_G * arm = 34.300kNm \qquad (1)$$

is the moment related to the friction force, in which "m" is the friction coefficient equal to 0.4 and F_G is the TBM weight equal to 1250 tons

$$M_B = F_{Cylinder} * arm = -60.300kNm \qquad (2)$$

is the moment related to the thrust force, in which $F_{Cylinder}$ is the actual value of the thrust force equal to 18.000kN

$$M_C = F_{CW} * arm = 1.900kNm \qquad (3)$$

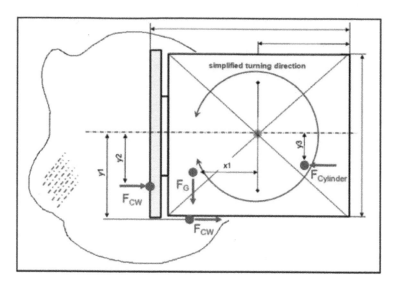

Figure 6. forces and moments on TBM.

is the moment related to the contact force at the tunnel face, in which F_{CW} is the actual average value of the contact force equal to 3.800kN

$$M_D = F_G * arm = 63.750 kNm \qquad (4)$$

is the moment related to the weight of the TBM, in which F_G is equal to 1.250kN. The resultant moment is:

$$M_A + M_B + M_C + M_D = F_G * arm = 39.650 kNm \qquad (5)$$

It is a positive value, that means for the TBM a counterclockwise rotation trend, hereby, a tendency to overturn downward. The balance of the moments on the TBM leads to the following considerations:

- both the friction force along the shield and the contact force at the tunnel face are relevant for the steering of the TBM;
- the resultant application point of the friction force is in the front bottom part of the shield, producing a "jibbing" effect on the TBM, which tends rotate downwards during the advance;
- weak conditions of the rock at the tunnel face are related with low value of the resultant contact force and moves its point of application in the bottom part of the cutterhead;
- in this conditions, the moment related with the contact force is not high enough to balance the tilting moments due to weight and friction force, but, even, could have a downward tilting effect, depending on the point of application of the resultant force. Anyhow, the effect on the TBM is a rotation downwards leading to a steering difficulties, if using only the thrust cylinders.

5 STRAND JACK SYSTEM

In the Basci tunnel, the *strand jack system* was installed on the TBM to compensate the lack of contact force at the tunnel face. The force applied by the *strand jack system* generates a negative moment, which is opposed to the moments related to the friction force and weigth of the TBM, hereby, it helps to steer the TBM up. The system consists in four hydraulic cylinders installed on the steel thrust structure at the Basci tunnel inlet, appropriately reinforced, steel

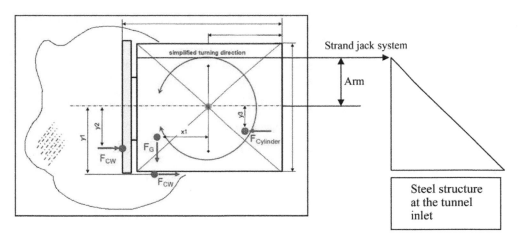

Figure 7.   force applied by the strand jack system.

ropes and four hydraulic cylinders installed on the TBM bulkhead, just behind the cutterhead. In addition, to giving a contribution in terms of upwards rotation of the TBM, it also provides a reduction of the advancing speed, while keeping the same force on thrust cylinders.

5.1   *Strand jack system: how it is done and how it works*

Referring to the Figure 8, the strand jack system is composed by:

1. Anchorage of the steel ropes at the thrust structure, by means of four hydraulic cylinders (bracket number 1 in Figure 8);
2. steel ropes running along the tunnel, supported by means of steel structures anchored on the tunnel crown (bracket number 2 in Figure 8);
3. anchorage of the steel ropes at the TBM bulkhead, by means of four hydraulic cylinders (bracket number 3 in Figure 8).

In the steel thrust structure, located at the tunnel Basci inlet, four hydraulic cylinders are installed, each one with 1200 kN pulling capacity (total pulling capacity installed equal to 4800kN), in order to pull the steel ropes during the advance of the TBM. Each rope is composed by twelve steel strands. An hydraulic power unit, with control panel, was installed to manage the stroke of the pistons in the cylinders. At about 50 mm from the pistons end of stroke, the steel strands are blocked on the cylinders, by means of brakes and disconnected

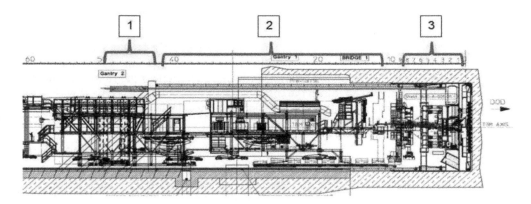

Figure 8.   strand jack system distribution along the TBM.

from the pistons, in order to close the pistons to the initial position. Then, the steel strands are reconnected to the pistons and the brakes released, in order to start a new stroke. The mentioned "re-gripping" of the pistons takes about two minutes, in which the stretching of the steel ropes is guaranteed by the four hydraulic cylinders installed on the TBM bulkhead. At the end of the TBM excavation stroke, in order to allow the installation of the ring that makes up the tunnel concrete lining without interference, the steel ropes running along the tunnel are loosened by opening the pistons at the bulkhead. At this stage, the same are fixed by means of brakes positioned on the TBM back up, so as don't lose completely the tensioning of the steel ropes. Once the ring installation is completed, the steel ropes are reconnected to the pistons on the bulkhead, in order to start a new TBM stroke.

6 EFFECTIVENESS OF STRAND JACK SYSTEM

Several countermeasures have been evaluated to manage the shield steering difficulties:

- injection of inert mortar to push up the TBM;
- reduction of the overcut of the cutterhead;
- combination of segments on erector and full retracted cutterhead in order to increase the upward tilting moment;
- increased pressure of bottom groups of thrust cylinders;
- consolidation of instable and weak rock conditions at the tunnel face;
- hydraulic cylinders and steel ropes pulling system (*strand jack system*).

The *strand jack system* was selected as preferred solution because it has an immediate effect and it is most liable and independent from the TBM operative procedure. Using the *strand jack system* it was possible to improve the TBM performance through tunnel stretch with unstable face condition and to improve the steering of the TBM, so as to recover the design level in the tunnel Basci. Regarding the first of the two previous issues, the installation of the strand jack system allowed to reduce the TBM advancing speed and, therefore, the disturb on the rock mass deep in front of the cutterhead, while maintaining high values of thrust force, necessary to steer the TBM up. The result was a reduction of the collapses at the tunnel face and, consequently, the improving of TBM advancing performances.

Regarding the second issue, since the negative values of TBM vertical tendency reached at the advance n. 34 in Basci tunnel, the maximum vertical deviation, between actual and design level in tunnel Basci, reached after *strand jack system* installation was -3.45m. After that, the design level was recovered in about 200m TBM advancement and the actual tunnel track realized, after *strand jack system* installation, has been checked, with success, by the designers, in order to verify the respect of gabarit and road parameters in the regulation ranges. At the moment of installation, the vertical tendency was negative with minimum value about -55 mm/m. After installation the vertical tendency began to rise and, once the design level has been reached, it has stabilized around value 0.00 mm/m.

It's worth remembering that the effectiveness of the *strand jack system*, because of the load losses distributed along the steel ropes track, decreases with the increasing of the distance between the TBM and the anchoring point of the rear pulling cylinders and, hereby, it is inversely proportional to the steel ropes length. It was found experimentally, during Basci tunnel excavation, that the maximum ropes length which guarantees the full functionality of the system is about 450m÷500m. Since the total length of Basci tunnel is 490m, it was possible to excavate the whole tunnel with the rear cylinders anchored at the inlet, without compromising the functionality of the *strand jack system*.

7 SELF RETAINING SYSTEM

At the completion of tunnel Basci excavation, performed with the help of the *strand jack system* anchored at the tunnel inlet, in order to reduce the total length of the steel ropes and,

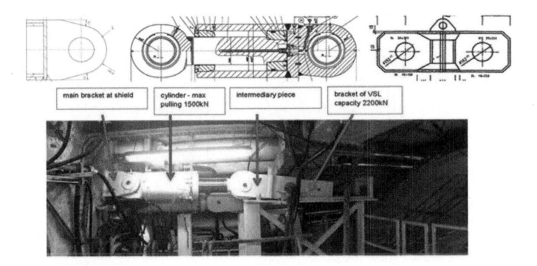

Figure 9. self retaining system – anchorage on TBM bulkhead.

consequently, the loss of pull along the tunnel, that limit its use to a maximum length of about 400m, a new alternative technology, named *self retaining system*, has been designed, for the managing of steering problems during the excavation of San Paolo and Cappuccini tunnels. The new technical solution was based on the following issues:

- more easy installation of the console;
- installation close to the shield;
- activation suitable also for short use;
- backup gantry n. 1 used as counterweight;
- quick delivery on site;
- low operative costs (if compared with strand jack system).

The *self retaining system* is "united" with the TBM and, therefore, do not depends from the steel ropes length, while, in case of *strand jack system* use, the ropes length increase with the excavation advancement. The *self retaining system* is composed by two hydraulic cylinders, anchored at the TBM bulkhead in the upper part, just behind the cutterhead, each with a pulling capacity of 1500 kN (total pulling capacity installed equal to 3000kN) and steel ropes fixed on the backup gantry number 1, properly reinforced. The operation is very similar to that described for the *strand jack system*, but since the *self retaining system* is "joined" with the TBM, it provides a contribution to the upward tilting moment, but does not act on the TBM advancing speed.

8 SIMULTANEOUS APPLICATION ON THE TBM OF THE TWO PULLING SYSTEMS.

During the early excavation phase of San Paolo tunnel (about 160m) only the *self retaining system* was installed on the TBM and it was enough, because the good rock conditions encountered in this first stretch of the tunnel (high GSI and UCS values, stability at the face, etc.) allowed the advancing of the TBM without problems for the steering of the machine, neither for the cutterhead blockage. But after the first 110m excavation of San Paolo tunnel the rock mass condition at the tunnel face have worsened again, with very low value of GSI, blocks instability and collapse and very low value of the contact force. In these conditions the only *self retaining system*, with at most 3000kN pulling capacity, helped not enough to steer the TBM up and a new loss in the tunnel vertical alignment occurred. In order to recover the design level and proceed with the excavation of San Paolo tunnel avoiding further steering problems, a new configuration of the *strand jack system* was installed, coupled with the *self*

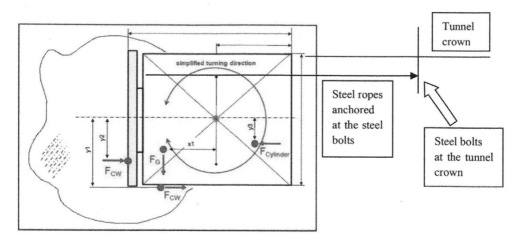

Figure 10.   self retaining system scheme.

retaining system, already acting on the TBM. Two hydraulic cylinders, each with pulling capacity of 1200KN, has been added on the TBM bulkhead, next to the *self retaining system* cylinders, while, the rear cylinders of the *strand jack system* has been anchored on the tunnel crown, by means of steel structures properly assembled, which, if necessary can be disinstalled and re-assembled in a new position, closer to the TBM. Using the described configuration of the *strand jack system* it was possible, notwithstanding San Paolo tunnel length of 2000m, to avoid the loss of pulling capacity, which, as said, increases with the advancing of the TBM and cuts down the effectiveness of the system up to about 500m distance between the rear cylinders position and the tunnel face. In fact, whenever the aforementioned limit distance was reached, the TBM advancing was stopped and the anchorage position of the rear cylinders was moved just behind the TBM back up, in order to restart the excavation activities with the full pulling capacity available. In order to reduce the stop time of the TBM, needed for the translation of the anchoring point in tunnel crown, two couples of steel anchoring structures has been constructed: one installed on the tunnel crown and the auxiliary one on site, ready to be installed. Immediately after the reaching of the limit ropes length, the TBM stops and the onsite auxiliary anchoring structure is installed on the tunnel crown, just behind the TBM back up. The rear hydraulic cylinders and the steel ropes are disconnected from the rear anchoring structure and installed on the one just assembled on tunnel crown behind the TBM. The excavation process restarts with the strand jack system anchored on the steel structures just behind the back up, while the rear anchoring steel structure is disassembled, waiting for to the next alternation. Using both the systems, coupled installed on the TBM, the whole San Paolo and Cappuccini tunnels were excavated without relevant steering problems.

9   CONCLUSIONS

Respect to TBM steering, some difficulties have been encountered since the first meters of TBM excavation. Among the different countermeasures applied in order to reduce steering problems, *strand jack pulling system* was selected due to the immediate effect on the TBM operative procedure. The *strand jack system* and the *self retaining system*, that found their first application in Aurelia Bis Savona tunnels, have proven their effectiveness in helping excavation of high diameter hard rock TBM through weak soil and unstable face conditions, especially regarding the contribution in solving steering problems, mainly related to the low values of contact force at the tunnel face. In the opinion of the authors, the systems could be improved and, in the near future, when the geotechnical conditions require it, could be applied on the TBMs since their manufacturing at workshop.

Tunnels and Underground Cities: Engineering and Innovation meet Archaeology,
Architecture and Art, Volume 7: Long and deep tunnels – Peila, Viggiani & Celestino (Eds)
© 2020 Taylor & Francis Group, London, ISBN 978-0-367-46872-9

Risk assessment in the unlocking of the double shield TBM within the Los Condores project, Chile

A. Focaracci & M. Salcuni
Prometeoengineering.it Srl, Rome, Italy

ABSTRACT: Los Cóndores will be a hydroelectric plant in Chile with an installed capacity of 150 MW. As part of the project a headrace tunnel (length 11 km) is excavated by means of a double shield TBM with an excavation diameter of 4.56m. After 1300 m in which a daily production up to 25 rings/day was reached the TBM blocked due to a fault zone with particularly poor geomechanical characteristics and significant water flow rates. Two different solutions have been proposed and developed for the unlocking of the TBM both aimed at the improvement of the mechanical characteristics of the soil at the front of the TBM. The paper shows how the analysis of all the potential technical risks has allowed to provide useful indications for optimizing the design and the operating procedures and minimizing the identified risks.

1 INTRODUCTION

In this paper, the risks associated with the unlocking of the double shield TBM, by which an headrace tunnel is being built within the Los Condores project, are analyzed and evaluated.

Los Cóndores will be a hydroelectric plant using the resources from Laguna del Maule with an approximate installed capacity of 150 MW.

As part of the project a headrace tunnel (length 11 km) is under excavation by means of a double shield TBM. The tunnel excavation was halted since the TBM blocked at pk. 6+683 due to geological/geotechnical problems. Until that moment the TBM advanced for about 1300m with daily production up to 25 rings/day.

The TBM has an excavation diameter of 4.56 m, and the tunnel lining is made by precast reinforced concrete segment with thickness of 25 cm and length 1.2 m, in a 4+1 configuration.

In the area where the excavations interrupted, the tunnel has a cover of about 326 m and the water table is located about 120m above the shell.

During the works fault areas have been met and overcame, with flow rate up to 3500 l/min. However, at the point abovementioned, the TBM has probably encountered an additional fault zone with particularly poor geomechanical characteristics, consisting of altered and degraded inconsistent material for a distance of about 8–10m from the cutting edge. From March 2017 the water flow rates had been measured between 1600 and 2000 l/min.

Two different solutions have been proposed and developed for the unlocking of the TBM and the recovery of the excavation operations:

- Solution 1: soil consolidation injections at the front of the TBM by expansive resins, with operations carried out from the inside of the cutter shield. Scheduled timing for completing the operations is about 2 weeks;
- Solution 2: excavation of a 30 m long tunnel starting from the TBM and soil consolidation injections at the front of the TBM performed by a chamber located at a distance of about 4 m from the cutter head. Scheduled timing for completing operations about 4 months.

Both the solutions aimed at the improvement of the mechanical characteristics of the soil at the front of the TBM.

The purpose of the study was therefore the analysis of all the potential risks associated with both the design solutions, their frequency of occurrence, and to assess the possible consequences of unfavourable events in order to provide useful indications for optimizing the design and the operating procedures and minimizing the identified risks.

To assess the structural and geotechnical construction risks and the specific risks at which workers may be subject to during their work, an overview of the geological, hydrogeological and geomechanical model of the site where the TBM is located will be provided to evaluate the foreseeable response of the rock mass and of the structures to the interventions that will be realized for the unlocking of the TBM.

Subsequently, individual risks to which workers may be subjected during work will be analyzed and the protection and prevention measures to be implemented to reduce or eliminate the risks will be identified.

2 DESCRIPTION OF THE TWO DESIGN SOLUTIONS

2.1 Solution 1: intervention from the TBM

The project consists of injections (minimum 8) to be realized directly from the TBM cutter head prior execution of drainage drilling at the front. At the end of consolidation injections, n.3 further injections had to be made to evaluate the validity of the treatment.

To allow operations inside the TBM and to facilitate the escape under emergency conditions some parts of the TBM had to be dismantled (eg. strip transportation) to allow for the installation of walkways to reach the drainage and injecting position points.

As an additional security measure, a further lighting lamp and a CCTV camera was installed inside the cutting shield, a telephone equipped with a dedicated emergency line had to be used and a manually activated alarm button was installed with an acoustic and light system visible in the tunnel. An Emergency Plan had also been prepared to transport any injured.

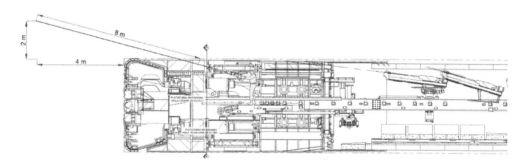

Figure 1.   drainage drilling.

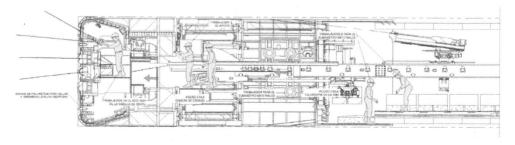

Figure 2.   drillings for injection tubes.

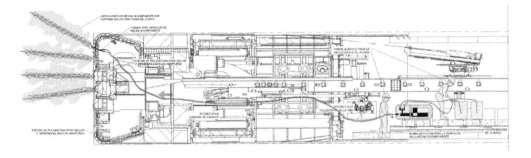

Figure 3.   ground consolidation.

2.2   *Solution 2: excavation of a rescue tunnel*

The design consists in the construction of a service tunnel of 2.30x2.50m dimensions about 30 m long up to the cutter head, at a lateral distance of about 4 m from it. From the terminal area of the tunnel, the injections for the rock mass consolidation had then to be performed.

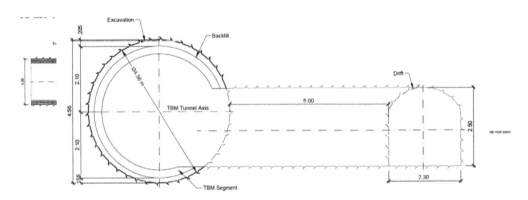

Figure 4.   plan and profile of the lateral by-pass.

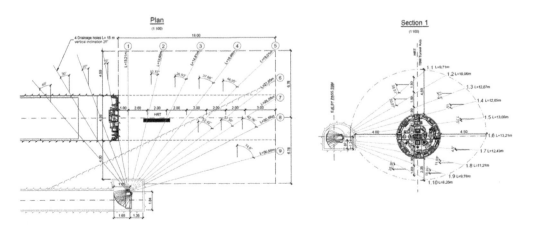

Figure 5.   geometry of injections.

3 GEOLOGICAL/GEOTECHNICAL OVERVIEW

3.1 *General project geology*

In the area where the TBM was blocked the main units are the volcanic rocks of the Trapa - Trapa (Mtt), the Campanario (Mpc) and the Cola de Zorro geological complexes, together with various units from the Pleistocene – Oligocene. The Campanario Formation and the Trapa – Trapa Formation show intruders made of bodies with diorite and rhyolite composition.

Through the quantitative observation of the outcrops in the area and from the geological surface study it has been possible to recognize 4 faults along the tunnel path. Among these structures, the Falla Lo Aguirre and Falla Quebrada formations are recognized in the bibliography while the other two have been deduced from the morphological analysis of the area. The studies conducted during the project did not show any relevant tectonic structures in the area where the TBM was blocked. Nevertheless specific and deep surveys in the area showed that for about 10 m from the front of the TBM the rock mass was fractured and degraded, presumably due to the presence of a geological or tectonic structure. Beyond this area the rock mass still appeared locally fractured but with stone consistency.

4 HAZARD, RISK AND SAFETY IN GEOTECHNICAL PRACTICE

Safety means the complex of conditioning actions that population's behavior, the structural solutions, the technological systems and control and management procedures apply to risk.

The two concepts are interrelated according to the formula as follows:

Risk = Hazard x (Safety) – 1

The above formula allows to comprehend that nil risk is impossible to attain.

Risk is not a physical parameter, therefore not quantifiable; however, it is possible to mathematically define risk using the group theory. According to this theory, risk is defined as an application of the group of hazardous events and the consequences' group. Both of these groups are probabilistic. The consequences' group defines the potential damage that can be related to a system of possibly hazardous events.

In analytical terms, the Risk is always defined as function of the probability of occurrence and of the magnitude (severity of the damage) relative to the single danger through the expression: $R = f(D,P)$. The severity of damage D can be expressed as a function of the number of subjects involved in a certain type of danger and of the level of damage that can caused to them.

Safety engineering must provide the correct scientific elements and mental processes to evaluate the factors on which to intervene.

From literature we can find countless different definitions of risk: definitions of NS5814 and ISO 31000 states: "Risk expresses a combination of the likelihood and consequences of an incident" (NS5814); "Risk is effect of uncertainty on objectives" (ISO 31000).

Risk management involves thinking more broadly about risk, not just spotting work-related hazards. Think about the root cause of any harmful event, the likelihood it will occur, the consequences if it does and the steps to take to eliminate or minimize the risk.

The various risk assessment methods differ for purpose, completeness and use, but all methods have the same sequence of logical. The following points are important in the risk management process in accordance with ISO 31000:

- Step 1: Risk Identification: Potential threats and hazards related to the activity under investigation and the associated risk factors are systematically identify. What can happen?
- Step 2: Risk analysis: Probabilities, potential consequences and risks are combined. How likely is it and if an unwanted event happens, what are the consequences?
- Step 3: Risk Assessment: The risk is assessed in relation to the criteria for acceptable or tolerable risk. Are the risks acceptable?
- Step 4: Risk reduction measures: danger are sorted by decreasing risk values and for all cases where the level of risk is unacceptable, appropriate corrective actions are identified. What can be done to get the risk down to an acceptable level?

The following figure shows the risk management process, as defined in ISO 31000.

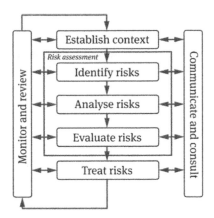

Figure 6.   The ISO 31000:2009 risk management process.

4.1   *Hazard identification*

The first step in the risk management process is to identify hazards which could injure or harm anyone. A good hazard identification process is the key to risk management.

During an excavation, to manage the risk it is critical to consider all relevant matters including the:

- Nature of the excavation;
- Means of entry into and exit from the excavation;
- Working in an excavation being exposed to contaminants that take oxygen from the environment.

4.2   *Risk Assesment*

Risks to health and safety arise from people being exposed to hazards (sources of harm).

A risk assessment must be carried out when:

- it is uncertain a hazard may cause injury or illness;
- a work activity involves different hazards, and the workers involved do not know how those hazards interact to produce new or greater risks;
- workplace changes may impact on the effectiveness of controls;
- new or different risks are associated with a change in work systems or work location.

Risks should be eliminated so far as is reasonably practicable. If a risk cannot be eliminated, it must be minimized so far as is reasonably practicable.

To decide what is "reasonably practicable" all relevant matters must be weigh up. Those matters include, but are not limited to:

- How likely the hazard or risk is to happen;
- What degree of harm the hazard or the risk might cause;
- How much is known about the hazard or risk and how to eliminate it;
- What ways are available to eliminate or minimize the risk;
- What ways are suitable to eliminate or minimize the risk.

A risk assessment will help to:

- Determine what sources and processes are causing that risk;
- Determine the severity of the risk;

- Identify if and what kind of controls should be implemented;
- Check the effectiveness of existing controls.

4.3 *Risk reduction*

Risks can be minimized through prevention and control method.

Some 'general principles of prevention' are reported in the EU 'OSH Framework' Directive 89/391:

- avoiding risks;
- evaluating the risks which cannot be avoided:
- combating the risks at source;
- adapting the work to the individual, especially as regards the design of work places, the choice of work equipment and the choice of working and production methods, with a view, in particular, to alleviating monotonous work and work at a predetermined work-rate and to reducing their effect on health.
- adapting to technical progress;
- replacing the dangerous by the non-dangerous or the less dangerous;
- developing a coherent overall prevention policy which covers technology, organization of work, working conditions, social relationships and the influence of factors related to the working environment;
- giving collective protective measures priority over individual protective measures;
- giving appropriate instructions to the workers.

Some risk reduction measures are more effective than others.

Different hierarchies of prevention and control measures have been developed by different institutions. The BS OHSAS 18001 hierarchy can be seen as a typical system and the risk can be minimized by one or a combination of the following:

- elimination;
- substitution;
- engineering controls;
- signage/warnings and/or administrative controls;
- personal protective equipment(PPE).

5 RISK ANALYSIS AND RISK ASSESSMENT FOR THE INTENDED ACTIVITIES

5.1 *Assessment criteria*

The risk factors assessment is aimed at eliminating or minimizing risk through the adoption of prevention and protection measures. The adequacy and reliability of these measures will reduce the probability of occurrence and the magnitude of a particular unfavourable event. These measures are defined with a priority deriving from:

- the severity of the damage (D);
- the probability of occurrence (P);
- the number of workers exposed;
- the complexity of protection and prevention measures to be taken.

For the quantitative evaluation of each risk an operating method based on a "Criticality Matrix" was used, combining the probabilities and severities in an indexed way (criticality index). The method allows to determine the severity of the risks identified and consequently the priority of the prevention measures to make the risk acceptable.

The matrix contains the probability of occurrence (P) in the columns and the expected impact (level of damage, D) in the lines and it is therefore possible to determine the level of risk (R) of each work activity by the following formula:

$$R = P \times D$$

Through the criticality matrix it is also possible to establish a threshold of attention that defines the level of risk considered acceptable. Each risky event beyond the threshold of attention will require, in the course of work, a more careful attention from the manager of the work.

Since the level of probability of damage is directly related to the identified deficiencies and to the damages estimated, while the magnitude of the damage depends on the effects of the incident and/or exposure of workers, for the rating of probability (P) and level of damage (D), quantitative scales are used, both for incident analysis and type of damage caused.

The probability of the damage occurrence (P) is evaluated according to a quantitative scale that concerns the existence of a correlation more or less direct between the identified deficiencies and the damage.

Also the level of damage (D) is evaluated with a severity scale, which refers to the reversibility of the damage, distinguishing between injury and acute or chronic exposure.

The probability of the damage occurrence (P) and the level of damage (D) can then be combined in the "Criticality Matrix" for the risk assessment.

Table 1. Probability of damage occurrence values.

Value	Level	Definition
4	Highly probable	Direct correlation between the failure and the occurrence of the estimated damage. The damage has occurred in other circumstances when the unfavourable event happened.
3	Probable	The detected failure can cause damage, even if not directly
2	Low probable	The detected failure event can cause damage only if other unfavourable events occur
1	Improbable	The detected failure can only cause damage for an unlikely coincidence of multiple independent events

Table 2. Probability of damage occurrence values.

Value	Level	Definition	Cost and time increase
4	Very serious	Chronic exposure with lethal and/or totally disabling effects. Injury or episode of acute exposure with lethal effect or total disability	8–10 %
3	Serious	Chronic exposure with irreversible and/or partially disabling effects Injury or episode of acute exposure with partial invalidity effects.	4–8 %
2	Medium	Chronic exposure with reversible effects. Injury or episode of acute exposure with reversible disability effects.	1–4 %
1	Light	Chronic exposure with rapidly reversible effects. Accident or episode of acute exposure with rapidly reversible disability effects	1 %

P					
	4	4	8	12	16
	3	3	6	9	12
	2	2	4	6	8
	1	1	2	3	4
		1	2	3	4
					D

5.2 Mitigation measure

In order to eliminate or minimize the risks associated with the various processes, mitigation measures must be implemented; this measures can be prevention measures (reducing the probability of occurrence of unfavourable events) and protection measures (reducing the severity of the damage associated with the unfavourable event).

In the risk assessment the mitigation measures are represented by the correction factors: the training factor (F_f) and the organizational factor (F_o) quantify the effect of prevention measures while the correction factor (F_{mp}) quantifies the effect of protection measures.

5.2.1 Preventive measures

Preventive measures include:

- Information, education and training;
- Internal organization (procedures, programming instructions, etc.)

The correction factor on workers training is characterized by the parameters shown in the following table:

Table 3. Formation factors.

Formation Factor	F_f
Information, education and training procedures repeated periodically through the organization of courses and meetings with learning verification	0,50
Information procedures repeated periodically through the organization of courses with learning verification	0,40
Informed/trained personnel through the participation in a course with learning verification	0,30
Information/personnel training through courses with programmed learning verification, but not yet realized	0,15
Staff informed through an information brochure	0,00

The correction factor for organizational measures is characterized by the parameters shown in the following table.

The correction factors identified by the preventive measures adopted result in a reduction in the probability that a dangerous event may occur and therefore a reduction in the risk index.

The residual probability factor is given by:

$$P_r = P /(1 + F_f + F_o)$$

5.2.2 Protective measures

The protective measures include:

- Active protection measures (education and training of first-aid emergency crews, emergency procedures, etc.)
- Passive protection measures (collective protection devices, provision of PPE, protection and safety distances to be met during the exercise of the activities and presence of structural protection devices such as rails, non-slip strips, safety ropes, etc.)

Table 4. operational factors.

Operational Factor	F_o
The organization has a Safety Management System (Emergency Plan)	0,50
Adequate organizational measures	0,40
Organizational measures expected to be appropriate, but in the process of completion	0,30
Insufficient organizational measures	0,15
Absent organizational measures	0,00

Table 5. protection measures.

Active protection measures	Passive protection measures			
	Appropriate	Sufficient	Not sufficient	Absent
Appropriate	1,00	0,75	0,50	0,35
Sufficient	0,75	0,50	0,35	0,15
Not sufficient	0,50	0,35	0,15	0,00
Absent	0,35	0,15	0,00	0,00

Passive protection measures include the provision of PPE which are a necessary condition for applying the F_{mp} factor. Therefore, if the staff does not have PPE, the F_{mp} factor calculation is not applicable.

The correction factors identified by the protection measures adopted result in a reduction in the severity of the damage relative to its theoretical value, and therefore in a reduction in the risk index.

The residual damage is given by:

Dr = D / (1 + FMP)

Through the values obtained from the F_{mp} factor, the value of Dr may be between 0.5D and D since in any case, while using correct and complete protection systems, the severity of residual damage cannot be considered less than half of the theoretical damage. In order to reduce further the severity of the damage, it would then be necessary to redesign the activity.

5.2.3 Operational mitigation measures

These measures are both preventive and protective measures aimed at risk mitigation. In the following paragraphs a series of operational measures is associated to each identified unfavourable event that could occur.

Moreover attached to the report there are some observations about the designed solutions and some aspects that should be examined in depth.

5.2.4 Residual risk index

The residual risk rate is given by:

Rr = Dr Pr

If the value of Rr is higher than the acceptable risk threshold, it will be necessary to review preventive and protective measures and/or implement new ones in order to further reduce the associated risk.

5.3 Analyses results

All the risks associated with the both the proposed solutions have been analysed and evaluated to highlight possible problems, propose solutions and assess the safety of the procedures.

Both the proposal are considered attainable even if the Solution 2 is characterized by a lower level of risk.

According to the critically matrix abovementioned, indeed, for the Solution 1 the sum of the values for the identified risks without any preventive or protection measure was determined equal to 232, with n.12 elements whose consequence can be classified as very serious (red part of the matrix) and n.16 serious elements whose consequence can be classified as serious (orange part of the matrix). Introducing the safety preventive and protective measures the global sum of values associated to the different risks is reduced to 121, no element determines very serious while n.15 element can determine serious consequences. For the Solution 2, instead, the sum of the values for the identified risks without any preventive or protection measure was determined equal to 186, with n.8 elements whose consequence can be classified as very serious (red part of the matrix) and n.18 serious elements whose consequence can be classified as serious (orange part of the matrix). Introducing the safety preventive and protective measures the global sum of values associated to the different risks is reduced to 95, no element determines very serious while n.10 element can determine serious consequences.

The realization of a service tunnel that allows to reach the front of TBM to requalify the rock mass ahead of it is considered a definitive intervention. Nevertheless, even if the Solution 1 was characterized by activities in narrow spaces manually executed, the intervention could be carried out if proper mitigation measures were implemented to reduce the risk at an acceptable level.

Moreover, within the proposal aimed to reduce the risks for workers some comments and suggestions were provided to further reduce the risks:

- availability of preventers at site during the injections at the front, if needed during the drilling to control possible flow of water or material;
- repetition of shotcrete after each advancement after the cut of the lining during the advancement of the rescue tunnel and, if needed, repetition of the front consolidation at short distance;
- modification of the geometry of consolidation during the excavation by adding a second layer of anchor;
- installation of a monitoring system (such as optical targets) in order to evaluate soil response to the excavation analysing deformation and pressure to adjust the number of GFR anchors to further guarantee safety for workers.

6 CONCLUSIONS

Los Cóndores will be a hydroelectric plant using the resources from Laguna del Maule with an approximate installed capacity of 150 MW.

As part of the project a headrace tunnel (length 11 km) is under excavation by means of a TBM with an excavation diameter of 4.56 m.

The tunnel excavation blocked after more than 1000 m of excavation due to a faulty zone with significant water flow rates and two different solutions have been proposed and developed for the unlocking of the TBM and the recovery of the excavation operations.

Due to the complexity of the work and to the associated risks for the workers for both the solutions a decision making process was conducted through a specific methodology in order to analyse all the possible risks and the associated consequences and to identify and define all the preventive and protective measures aimed to elevate the level of safety as much as possible giving to the Contracting Authority all the elements (safety, economic, temporal) to determine the best possible solutions by means of an cost-benefit analysis.

Tunnels and Underground Cities: Engineering and Innovation meet Archaeology, Architecture and Art, Volume 7: Long and deep tunnels – Peila, Viggiani & Celestino (Eds)
© 2020 Taylor & Francis Group, London, ISBN 978-0-367-46872-9

Back-calculation model for instantaneous TBM cutter wear

H.I. Frostad, A. Bruland & P.D. Jakobsen
Norwegian University of Science and Technology (NTNU), Trondheim, Norway

F.J. Macias
JMConsulting-Rock Engineering AS, Oslo, Norway

ABSTRACT: This paper describes the calculation model for instantaneous cutter wear designed as a back-calculation and database tool for cutter consumption in TBM tunneling. The software is designed to process larger amounts of data from e.g. TBM cutter logging. The NTNU instantaneous cutter consumption model calculates the instantaneous cutter consumption along the tunnel and between cutter changes. The model expresses the instantaneous cutter ring life in terms of tunnel meter per cutter ring, hours per cutter ring and solid cubic meter per cutter ring. Based on the NTNU instantaneous cutter consumption model, computes the software calculation model these parameters as well as cutter ring per tunnel meter, actual penetration rate, cutter wear based theoretical penetration rate and the wear process indicator (theoretical/actual penetration rate). The paper will focus on the algorithms of the back-calculation model, and compare software results from different cases with the NTNU instantaneous cutter consumption model.

1 INTRODUCTION

In hard rock TBM tunnel projects, one of the challenges is to estimate the individual cutter ring consumption on the cutter head. A more accurate model for cutter ring consumption estimates will have major influence on the planning and risk management of TBM excavation projects.

Several different prediction models for estimation of cutter wear and performance is developed e.g. Q_{TBM} (Barton, 2000), The NTNU Model (Bruland, 1998, Macias, 2016), Colorado School of Mines (CSM) (Rostami and Ozdemir, 1993, Rostami, 1997), Gehring (Gehring, 1995), (Hassanpour et al., 2016), (Farrokh et al., 2012), RME (Bieniawski et al., 2006), Frenzel (Frenzel, 2010, Frenzel, 2011, Frenzel, 2012), Alpine Model (Wilfing, 2016).

The NTNU instantaneous cutter consumption model calculates the instantaneous cutter consumption along the tunnel and between cutter changes (Bruland, 2000, Macias, 2016). This paper describes the calculation model for instantaneous cutter wear designed as a back-calculation and database tool for cutter consumption in TBM tunneling. The software is designed to process larger amounts of data from e.g. TBM cutter logging (Frostad, 2013). The paper will focus on the algorithms of the back-calculation models, and compare software results from two different cases with the NTNU instantaneous cutter consumption model.

The first case is from Meraaker hydropower project in Norway, d_{tbm}=3.5 meter and N_{tbm}=25 cutters. The second case is based on project data from an another hard rock TBM tunnel boring project, however, the project information is confidential.

2 THE NTNU INSTANTANEOUS CUTTER CONSUMPTION MODEL

The NTNU instantaneous cutter consumption model calculates the instantaneous cutter consumption along the tunnel and between cutter changes (Bruland, 2000, Macias, 2016). The

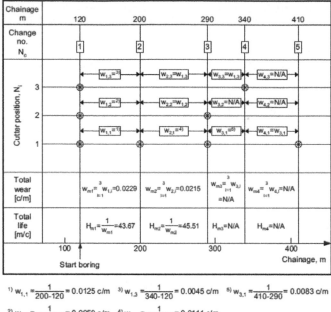

$^{1)}$ $w_{1,1} = \frac{1}{200-120} = 0.0125$ c/m $^{3)}$ $w_{1,3} = \frac{1}{340-120} = 0.0045$ c/m $^{5)}$ $w_{3,1} = \frac{1}{410-290} = 0.0083$ c/m

$^{2)}$ $w_{1,2} = \frac{1}{290-120} = 0.0059$ c/m $^{4)}$ $w_{2,1} = \frac{1}{290-200} = 0.0111$ c/m

$\otimes$ = cutter position replaced

Figure 1. Cutter consumption in cutter/meter for a theoretical cutter-head with three cutters (Bruland, 2000).

calculation is between cutter change, tunnel inspection and cutter repair log at the cutter workshop (Macias, 2016). The model expresses the instantaneous cutter ring life in terms of tunnel meter per cutter ring, hours per cutter ring and solid cubic meter per cutter ring. These parameters refer only to the cutter consumption of cutter rings. Based on a theoretical cutter-head with three cutters, calculates the instantaneous cutter consumption model the cutter consumption (Bruland, 2000). The model requires as a minimum input data for each cutter change: Cutter position(s) replaced, chainage in meters and machine hours. By substituting chainage (meters) with corresponding machine hours for each cutter change calculates the same model cutter per hour (Bruland, 2000).

Figure 1 describes the basic calculation model for cutter consumption in cutter/meter for a theoretical cutterhead with three cutters (Bruland, 2000).

2.1 Algorithms of the back-calculation model

The NTNU instantaneous cutter consumption model calculates the instantaneous cutter consumption along the tunnel and between cutter changes (Bruland, 2000, Macias, 2016).

Cutter consumption, equation (1):

$$W = \frac{1}{m} \tag{1}$$

W = Cutter consumption in cutter/meter.

Total cutter consumption, equation (2):

$$W_{m1} = \sum_{i=1}^{n} W_{1,i} \tag{2}$$

W_{m1} = Total cutter consumption in cutter/meter.

Total cutter ring life, equation (3):

$$H_{m1} = \frac{1}{W_{m1}} \tag{3}$$

H_{m1} = Total cutter ring life in meter/cutter.

Based on the TBM cutter change log in this case from the Meraaker hydropower project in Norway with d_{tbm} = 3.5 meter and N_{tbm} = 25 cutters, calculates the cutter consumption, as shown in Table 1 (Bruland, 2000). The data file in Table 1 shows the minimum input data necessary for calculation of cutter consumption. Notice that Cutter Change No. 1 defines the boring start and the term cutter change refers to replacing one or more worn out or damage cutters at a given chainage. At the actual cutter means "1" that the position has been replaced and "0" that no cutter position replaced (Bruland, 2000).

The project data in Table 1 is from the Meraaker hydropower project in Norway with d_{tbm} = 3.5 meter and N_{tbm} = 25 cutters. Project data in this case is from a hard rock TBM tunneling project with extremely hard strong rock, medium abrasivity, high cutter change frequency and very low cutter life as shown in Table 1, Table 2 and Figure 2 (Bruland, 2000).

The results from calculation of instantaneous cutter consumption in Table 2 and Figure 2 is based on the data in Table 1. Figure 2 shows that the representative cutter life is approximately 1.2 – 1.3 hours/cutter for the tunnel section until chainage 300 (Bruland, 2000).

2.2 The wear process indicator I_c/I_n (theoretical/actual penetration rate)

The quality ratio of calculated cutter ring life is evaluated by I_c/I_n. The cutter wear conditions are defined as normal when I_c/I_n is close to 1.0. If I_c/I_n is less or larger than 1.0 is this an indication of abnormal cutter wear, e.g. blocked cutter(s). It can also be an indication of irregular cutter replacement, e.g. one or more cutter ring replacements before it is worn out. In evaluation of representative sites for rock sampling to compare cutter life with geological parameters, should the ratio I_c/I_n be used (Bruland, 2000).

2.3 Software description

This section provides a general description of the calculation module of the software (Frostad, 2013), designed as a back-calculation and database tool for cutter consumption in TBM

Table 1. Calculation of cutter consumption from data file for a TBM with 25 cutters (Bruland, 2000).

Cutter Change No.	Chainage (m)	Machine hours (h)	Positions 1 - 25
1	122.3	2186.5	1111111111111111111111111
2	164.4	2208.4	0000000000100000000000000
3	179.5	2214.4	1111100000011111111111100
4	181.7	2215.0	1010000000000000000000000
5	195.6	2223.1	0000000000011111111100000
6	219.0	2238.0	0000000000000011111100000
7	228.0	2242.8	0000000000001001000000000
8	232.6	2244.6	0000000000000111111111100
9	244.4	2252.2	0000000000000000010110000
10	265.0	2264.1	0000000000000111111110000
11	288.0	2277.1	0000000000000000011110000
12	297.8	2283.9	1111111111111111111111111
13	305.1	2290.4	0000000000000011111000000
14	311.1	2293.0	0000000000001000000100000
15	316.5	2296.8	0000000000000011100000000
16	325.0	2300.5	0000000000000001111111000
...	...	...	...

Table 2. Calculation of cutter consumption along the tunnel, data file from a TBM with 25 cutters (Bruland, 2000).

From Chainage (m)	To Chainage (m)	H_f (m³/c)	H_h (h/c)	H_m (m/c)	W_m (c/m)	Interval Length (m)	Cutters Changed	% Changed	I_n (m/h)	I_c (m/h)	I_c/I_n
122.3	164.4	26.7	1.38	2.77	0.361	42.1	1	4	1.92	2.01	1.05
164.4	179.5	27.9	1.44	2.90	0.345	15.1	17	68	2.52	2.02	0.80
179.5	181.7	6.2	0.22	.65	1.544	2.2	2	8	3.67	2.91	0.79
181.7	195.6	14.8	.0.84	1.53	0.652	13.9	8	32	1.72	1.83	1.06
195.6	219.0	20.5	1.30	2.13	0.469	23.4	6	24	1.57	1.64	1.05
219.0	228.0	13.9	0.75	1.45	0.692	9.0	2	8	1.87	1.93	1.03
228.0	232.6	12.3	0.60	1.28	0.781	4.6	10	40	2.56	2.12	0.83
232.6	244.4	17.4	1.11	1.81	0.552	11.8	3	12	1.55	1.63	1.05
244.4	265.0	21.7	1.32	2.26	0.443	20.6	8	32	1.73	1.70	0.98
265.0	288.0	21.9	1.32	2.28	0.439	23.0	4	16	1.77	1.72	0.97
288.0	297.8	14.3	0.96	1.48	0.674	9.8	25	100	1.44	1.54	1.07
297.8	305.1	9.4	0.75	.98	1.022	7.3	5	20	1.12	1.31	1.17
305.1	311.1	13.7	0.81	1.43	0.700	6.0	2	8	2.31	1.76	0.76
311.1	316.5	14.9	0.84	1.55	0.646	5.4	3	12	1.42	1.85	1.30
316.5	325.0	14.9	0.76	1.55	0.646	8.5	7	28	2.30	2.04	0.89

H_f = Solid cubic meter per cutter (m /c).
H_h = Hours per cutter ring (h/c).
H_m = Meter per cutter ring (m/c).
W_m = Cutter ring per meter (c/m).
I_n = Cutter ring per meter (c/m).
I_c = Theoretical penetration rate (m/h).
I_c / I_n = Theoretical penetration rate (m/h).

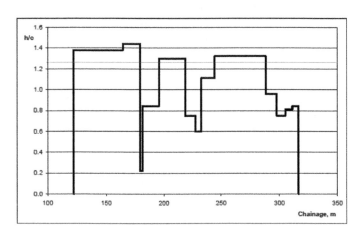

Figure 2. Instantaneous cutter consumption along the tunnel, based on data from Table 1 (Bruland, 2000).

tunneling. The program is designed to process larger amounts of data from e.g. TBM cutter logging. Based on the NTNU instantaneous cutter consumption model in Figure 1, computes the software's calculation module the parameters described in Table 2 (Bruland, 2000). The software exists in a version for Microsoft Excel and a version for Microsoft Visual studio (Frostad, 2013).

3 RESULTS

In this section will results from two different hard rock TBM tunneling cases be compared with the NTNU instantaneous cutter consumption model. The first case is from Meraaker hydropower project in Norway with d_{tbm} = 3.5 meter and N_{tbm} = 25 cutters, based on Table 1, Table 2 and Figure 2 (Bruland, 2000). The NTNU instantaneous cutter consumption model assumes that all cutter rings are new when boring starts at Cutter Change No. 1, as shown in Table 1 (Bruland, 2000). In this specific case has all the cutters been replaced at Cutter Change No. 12. Table 1 shows a section of the tunnel and indicates that there are several cutter changes after Cutter Change No. 16 (Bruland, 2000).

In this case is back-calculating from Cutter Change No. 47 (chainage 1076.3 where 24 out of 25 cutter is changed) (Bruland, 2018) (Frostad, 2013) necessary to accomplice the corresponding values in Table 2 (Bruland, 2000). Figure 3 shows that the representative cutter life is approximately 2.5 – 2.6 hours/cutter for the tunnel section from chainage 400 to chainage 1250 (Frostad, 2013). Figure 3 (Frostad, 2013) shows that the representative cutter life is approximately 1.2 – 1.3 hours/cutter for the tunnel section until chainage 300 as shown in Figure 2 (Bruland, 2000).

Special focus in selection of tunnel sections used in evaluation of cutter life have to be taken to avoid incorrect cutter life results, from short and preliminary defined sections (Macias, 2016). Figure 3 show in addition to the project data in Figure 2 (Bruland, 2000) an extended tunnel section overview until approximately chainage 1250 (Frostad, 2013). Notice in Figure 3 the particularly low hours per cutter value at e.g. Cutter Change No. 4 (Figure 4) and at e.g. Cutter Change No. 35 (Figure 4). The value of hours per cutter e.g. for Cutter Change No. 4 is 0.22 and for e.g. Cutter Change No. 35 is the value 0.26 (Frostad, 2013).

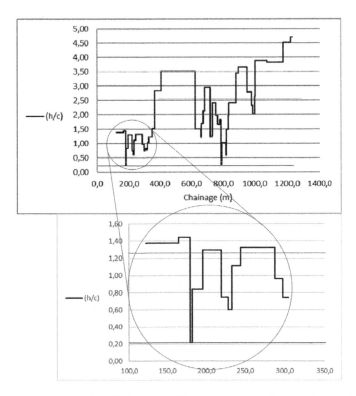

Figure 3. Instantaneous cutter ring life in terms of hours per cutter ring, section results from software (Frostad, 2013).

Figure 4 show software results of instantaneous cutter ring life in terms of solid cubic meter per cutter ring (Frostad, 2013). Cutter Change No.1 – 16 corresponds with Cutter Change No. 122.3 – 325.0 in Table 2. The results of solid cubic meter per cutter ring for Cutter Change No. 1 – 50 refer to the corresponding extended tunnel section overview in Figure 3 (Frostad, 2013).

Notice in Figure 4 the particularly low value of solid cubic meter at e.g. Cutter Change No. 4 and at e.g. Cutter Change No. 35. The solid cubic meter e.g. for Cutter Change No. 4 is 6.2 m^3/c and corresponds in Figure 3 to an hours per cutter ring value of 0.22 (Frostad, 2013).

Figure 5 show software results of instantaneous cutter ring life in terms of ratio I_c/I_n (theoretical/actual penetration rate). Notice in Figure 5 the particularly low value of I_c/I_n at e.g. Cutter Change No. 4 and at e.g. Cutter Change No. 32. The ratio I_c/I_n e.g. for Cutter Change No. 32 is 0.66 and corresponds in Figure 3 to an hours per cutter ring value of 1.97. Notice also in Figure 5 the particularly high value of I_c/I_n at e.g. Cutter Change No. 23 and at e.g. Cutter Change No. 26. The ratio I_c/I_n e.g. for Cutter Change No. 26 is 1.57 and corresponds in Figure 3 to an hours per cutter ring value of 2.13 (Frostad, 2013).

The second case is based on project data from an another hard rock TBM tunnel boring project, however, the project information is confidential. Figure 6 shows that the representative cutter ring life is approximately 1.4 –1.5 hours/cutter for the tunnel section from chainage "2000"

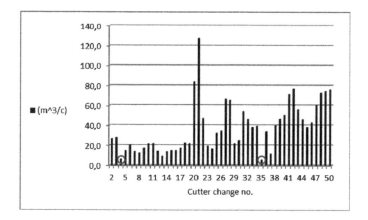

Figure 4.   Instantaneous cutter ring life in terms of solid cubic meter per cutter ring, results from software (Frostad, 2013).

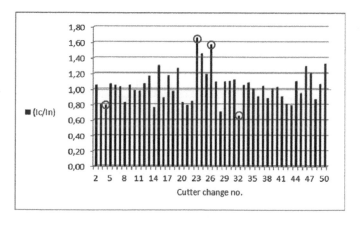

Figure 5.   Instantaneous cutter ring life in terms of Ic/In (theoretical/actual penetration rate), results from software (Frostad, 2013).

to chainage "2200" (Frostad, 2013). Figure 6 shows that the representative cutter ring life is approximately 1.3 – 1.4 hours/cutter for the tunnel section until chainage "2100" (Frostad, 2013).

Special focus in selection of tunnel sections used in evaluation of cutter life have to be taken to avoid incorrect cutter life results, from short and preliminary defined sections (Macias, 2016). Notice the particularly low value of hours per cutter ring in Figure 6 with an approximately value of 0.42 (Frostad, 2013).

Figure 7 show distribution (in number of cutter change) related to reasons of cutter replacement from project data of instantaneous cutter consumption, based on Figure 6.

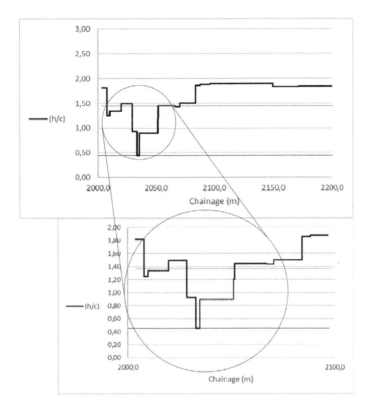

Figure 6.   Instantaneous cutter consumption in terms of hours per cutter ring, section results from software (Frostad, 2013).

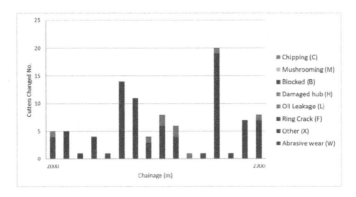

Figure 7.   Distribution of cutter replacement (in number of cutter change) from project data of cutter consumption, based on Figure 6.

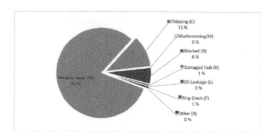

Figure 8. Distribution of cutter replacement distribution (in percent) from project data of cutter consumption, based on Figure 6.

Figure 8 show distribution (in number of cutter change) related to reasons of cutter replacement from project data of instantaneous cutter consumption, based on Figure 6.

4 ANALYSIS AND DISCUSSION OF RESULTS

This section analyses and compare software results (Frostad, 2013) from two different cases with the NTNU instantaneous cutter consumption model (Bruland, 2000). The algorithms of the back-calculation software model (Frostad, 2013) is based on the NTNU instantaneous cutter consumption model (Bruland, 2000).

4.1 *Case 1*

The first case is from Meraaker hydropower project in Norway with $d_{tbm} = 3.5$ meter and $N_{tbm} = 25$ cutters. Project data in this case is from a hard rock TBM tunneling project described in chapter 2.1 with extremely hard strong rock, medium abrasivity, high cutter change frequency and very low cutter life (Bruland, 2000). Based on Table 1, Table 2 and Figure 2 presents Figure 3 section results from software (Frostad, 2013) of the NTNU instantaneous cutter consumption model (Bruland, 2000). Figure 3 (Frostad, 2013) shows that the representative cutter life is approximately 1.2 – 1.3 hours/cutter for the tunnel section until chainage 300 as shown in Figure 2 (Bruland, 2000).

Special focus in selection of tunnel sections used in evaluation of cutter life have to be taken to avoid incorrect cutter life results, from short and preliminary defined sections (Macias, 2016). The software shows in Figure 3 an extended tunnel section overview, in this case from chainage 122.3 – 1250. Notice that the representative cutter life is approximately 2.5 – 2.6 hours/cutter for the tunnel section from chainage 400 to chainage 1250 (Frostad, 2013).

Notice in Figure 3 (Frostad, 2013) the particularly low hours per cutter ring value at e.g. Cutter Change No. 4 (Figure 4) and at e.g. Cutter Change No. 35 (Figure 4). The value of hours per cutter ring e.g. for Cutter Change No. 4 is 0.22 and for e.g. Cutter Change No. 35 is the value 0.26 (Frostad, 2013). When comparing the parameters for cutter change in Table 1, Table 2 and Figure 2 (Bruland, 2000), this particular low cutter life at e.g. Cutter Change No. 4 relates to relatively limited cutter change interval for these cutter(s). When comparing software results with the TBM log, also the particularly low cutter life at e.g. Cutter Change No. 35 relates to relatively limited cutter change interval for these cutter(s).

Figure 3 (Frostad, 2013) shows higher value of hours per cutter ring at e.g. chainage 400 – 600. Data from software results compare with the TBM log show no cutter change in this section. There are several options for this e.g. cutter change policies, abrasive wear, geology and TBM operation practice. Figure 3 shows very varied values of hours per cutter at e.g. chainage 600 –1000. When comparing data from software results with the TBM log show this a systematically cutter replacement of the same previously cutters on the cutterhead, however the number of replaced cutters are relatively low. In this case the geology is of special interest in addition to e.g. cutter change policies, abrasive wear and TBM operation practice.

Figure 4 (Frostad, 2013) show software results of instantaneous cutter ring life in terms of solid cubic meter per cutter ring. Notice in Figure 4 the particularly low value of solid cubic meter per cutter ring at e.g. Cutter Change No. 4 and at e.g. Cutter Change No. 35. The solid cubic meter per cutter ring e.g. for Cutter Change No. 4 is 6.2 m^3/c and corresponds in Figure 3 to an hours per cutter value of 0.22.

Figure 5 (Frostad, 2013) show software results of instantaneous cutter ring life in terms of ratio I_c/I_n (theoretical/actual penetration rate). Figure 5 shows a particularly low value of I_c/I_n at e.g. Cutter Change No. 4 and at e.g. Cutter Change No. 32. The ratio I_c/I_n e.g. for Cutter Change No. 32 is 0.66 and corresponds to hours per cutter value of 1.97. In Figure 5 is there a particularly high value of I_c/I_n at e.g. Cutter Change No. 23 and Cutter Change No. 26. The ratio I_c/I_n e.g. for Cutter Change No. 26 is 1.57 and corresponds to hours per cutter value of 2.13.

In this case are e.g. Cutter Change No. 4 and Cutter Change No. 32 less than 1.0, and larger than 1.0 at e.g. Cutter Change No. 23 and Cutter Change No. 26. From theory is there if I_c/I_n is less or larger than 1.0 an indication of abnormal cutter wear, e.g. blocked cutter(s) (Bruland, 2000). It can also be an indication of irregular cutter replacement, e.g. one or more cutter ring replacements before it is worn out. This gives an indication to evaluate the cutter change policies in relation to projects time and cost estimates.

4.1 *Case 2*

The second case is based on project data from an another hard rock TBM tunnel boring project, however, the project information is confidential.

Based on project data presents Figure 6 (Frostad, 2013) results of hours per cutter ring from software of the NTNU instantaneous cutter consumption model (Bruland, 2000). The previous case 1 forms the basis for assessing the results in case 2. Figure 6 shows that the representative cutter life is approximately 1.4 –1.5 hours/cutter for the tunnel section from chainage "2000" to chainage "2200". Figure 6 shows that the representative cutter life is approximately 1.3 – 1.4 hours/cutter for the tunnel section until chainage "2100". The presented results correspond well with the NTNU instantaneous cutter consumption model (Bruland, 2000).

The same effect as shown in case 1 can also be found here in case 2. Notice the particularly low value of hours per cutter ring in Figure 6 (Frostad, 2013) from software where the approximately value is 0.42. When comparing data from software results with the TBM log, relates this particular low cutter life to a relatively limited cutter change interval for this cutter(s).

Figure 7 show distribution of cutter replacement (in number of cutter change) and Figure 8 (in percent) related to reasons for cutter changes from project data of cutter consumption based on Figure 6 (Frostad, 2013). In this case, abrasive wear is the largest part of cutter consumption, however notice the part related to e.g. blocked cutters, which occurs in rock types transitions and high cutter thrust (Macias, 2016).

5 CONCLUSIONS

In hard rock TBM tunnel projects, one of the challenges is to estimate the individual cutter ring consumption on the cutter head. A more accurate model for cutter ring consumption estimates will have major influence on the planning and risk management of TBM excavation projects.

The focus in this paper have been on the algorithms of the back-calculation models, and compare results from different cases with the NTNU instantaneous consumption model. The calculation tool has been a software designed as a back-calculation and database for cutter consumption in hard rock TBM tunneling.

Project case 1 where required basic data is available show the accuracy in the calculate results from the software compare to the NTNU instantaneous consumption model. This case will work as a reference case when evaluating the second case. The calculated software results in case 1 and 2 compared to the NTNU instantaneous consumption model have the same accuracy.

The calculated results from the back-calculating model is dependent of the accuracy in the available project data. It is also important to show caution when selecting tunnel sections that is short or preliminary defined, to avoid incorrect cutter life results.

The results of the back-calculation model may be used for e.g.

- Evaluation of cutter change policies
- TBM operation practice
- Indication of rock mass conditions
- Disc cutter design and quality
- And more...

REFERENCES

Barton, N. 2000. TBM tunnelling in jointed and faulted rock. A.A. Balkema, Rotterdam (2000).

Bieniawski, Z. T., Celada, B., Galera, J. M. & Alvarez, M. 2006. Rock mass excavability indicator: new way to selecting the optimum tunnel construction method. *Tunnelling and Underground Space Technology Vol. 21, no.3–4 (2006), pp* 237.

Bruland, A. 1998. *Hard rock tunnel boring.* 1998:81, Vol.1., Vol. 2, 1A–98., Vol. 3, 1B–98., Vol. 4, 1C–98., Vol. 5, 1D–98., Vol. 6, 1E–98., Vol. 7, 1F–98., Vol. 8, 13A–98., Vol. 9, 13B –98., Vol. 10, 13B –98., Norwegian University of Science and Technology, Department of Building and Construction Engineering.

Bruland, A. 2000. *Hard rock tunnel boring: Vol. 6: Performance data and back-mapping,* Trondheim, Norwegian University of Science and Technology, Department of Building and Construction Engineering.

Bruland, A. 2018. Personal communication.

Farrokh, E., Rostami, J. and Laughton, C. 2012. Study of various models for estimation of penetration rate of hard rock TBMs. *Tunnelling and Underground Space Technology Vol. 30 (2012), pp* 110–123.

Frenzel, C. 2010. Verschleißkostenprognose für Schneidrollen bei maschinellen Tunnelvortrieben in Festgesteinen.

Frenzel, C. 2011. Disc Cutter Wear Phenomenology and their Implications on Disc Cutter Consumption for TBM. Colorado School of Mines, Golden, CO, USA. ARMA 11-211. Copyright 2011 ARMA, American Rock Mechanics Association. This paper was prepared for presentation at the 45th US Rock Mechanics/Geomechanics Symposium held in San Francisco, CA, June 26–29, 2011.

Frenzel, C. 2012. Modeling uncertainty in cutter wear prediction for tunnel boring machines. Department of Mining Engineering, Colorado School of Mines, 1600 Illinois St, Golden, CO, 80401. GeoCongress 2012 © ASCE 2012.

Frostad, H.-I. 2013. *TBM Kutterslitasje database. MSc Thesis.* Norwegian University of Science and Technology (NTNU), Trondheim, Norway (2013), (in Norwegian).

Gehring, K. 1995. Leistungs- und Verschleißprognosen im maschinellen Tunnelbau. Felsbau 13 (1995), No. 6, pp. 439–448.

Hassanpour, J., Vanani, A.A.G., Rostami, J. and Cheshomi, A. 2016. Evaluation of common TBM performance prediction models based on field data from the second lot of Zagros water conveyance tunnel (ZWCT2). *Tunnelling and Underground Space Technology Vol. 52 (2016), pp* 147–156.

Macias, F. J. 2016. *Hard rock tunnel boring: performance predictions and cutter life assessment.* 2016:350, Norwegian University of Science and Technology, Faculty of Engineering Science and Technology, Department of Civil and Transport Engineering.

Rostami, J. 1997. Development of a force estimation model for rock fragmentation with disc cutters through theoretical modelling and physical measurement of crushed zone pressure, PhD Thesis. Colorado School of Mines, Golden, Colorado, USA (1997).

Rostami, J. & Ozdemir, L. 1993. A new model for performance prediction of hard rock TBMs. Proceedings, rapid Excavation and Tunnelling Conference (RETC 1993), Boston, Massachusetts, USA (1993), pp 793–809.

Wilfing, L. 2016. *The Influence of Geotechnical Parameters on penetration Prediction in TBM Tunneling in Hard Rock.* PhD Thesis. Technical University of Munich, Munich, Germany (2016).

Tunnels and Underground Cities: Engineering and Innovation meet Archaeology,
Architecture and Art, Volume 7: Long and deep tunnels – Peila, Viggiani & Celestino (Eds)
© 2020 Taylor & Francis Group, London, ISBN 978-0-367-46872-9

Brenner Base Tunnel, Italian Side: Construction methods for a railway 22 km long tunnel

S. Fuoco & R. Zurlo
BBT SE, Bolzano, Italy

M. Moja & E.M. Pizzarotti
Pro Iter, Milan, Italy

ABSTRACT: The Brenner Base Tunnel is mainly composed by two single track Railway Tunnels and an Exploratory Tunnel; Lots Mules 1 and 2–3 and previous Preparatory Lot concern a 22 km long stretch on the Italian side. They cross the South part of mountainous dorsal between Austria and Italy, under over-burdens up to 1850 m, consisting of rocks both of the Southalpine and Australpine domains, separated by the major Periadriatic Fault. More than 30 km of tunnels have to be carried out with dimensions ranging from 7 m (Exploratory Tunnel) to 20 m (Logistics Caverns). This results in an extremely hard logistic, geological and geomechanical challenge. During design, different construction options were evaluated, from all the technical, economic and schedule point of view, duly considering the construction risks. The paper explains the design choice of a mix of Mechanized (TBM) and Mining Methods excavation, identified as the most convenient solution.

1 INTRODUCTION

The Brenner Base Tunnel (BBT) system consists of two 55 km long parallel Railway Tunnels (at a distance from 40 to 70 m), and an Exploratory Tunnel, which during construction will also play an important role in construction logistics, while during railway operation it will host the water disposal system and allow an access to other work.s The Lot Mules 1 was the continuation of the preliminary works for the construction on the Italian side; its key features were the creation of large cavities, the presence of several intersections among the cavities, and above all the Exploratory Tunnel crossing the Periadriatic Tectonic Lineament. The Lot Mules 2–3 on Italian side includes 22 km of Railway Tunnels, from km 32 (Austrian border) to km 54 (close to the South access), and 17 km of Exploratory Tunnel, from km 10.5 to km 27 (Figure 1a).

For most part of their length Railway Tunnels only host one track each, but close to the adjacent lot 'Underground Crossing of the Isarco river' they become two-track and three-track.

The Exploratory Tunnel is located between the two main ones, 12 m lower than the main axes (Figure 1b); only close to the South entrance (from km 51+600 Eastern pipe) the Exploratory Tunnel deviates from its central position and is no longer aligned with the Railway Tunnels up to its Aica South entrance.

Approximately at the Eastern tunnel chainage km 44+700 there will be a 530 m long Emergency Stop with connection tunnels for passenger evacuation and ventilation tunnels, which shall be used to manage emergency situations in case of fire. The Stop is connected to a ventilation and access tunnel that runs parallel to the line tunnel for approximately 4.5 km and subsequently inserts onto the 1720 m long Mules access tunnel, which connects the system to the outside. The tunnel system is completed by 69 cross passages (one approximately every 333 m) that provide an escape between the pipes and host technical rooms and firefighting water tanks. For plant connections and hydraulic purposes, approximately every 2000 m there

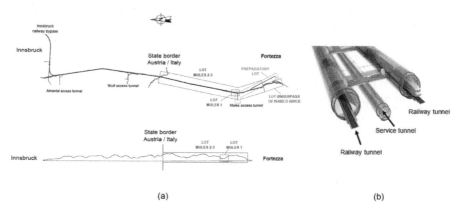

Figure 1.   (a) BBT key plan with identification of Lots. (b) General system configuration.

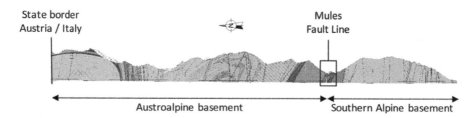

Figure 2.   Geological section along tunnel.

will be vertical shafts connecting the cross tunnel with the underlying Exploratory Tunnel. Close to the Mules access tunnel there is an underground ventilation station, directly connected to the outside by means of a 60 m long vertical shaft. The intersection between Mules access tunnel and Exploratory Tunnel/Railway Tunnels hosts a system of tunnel and chambers which allow to transfer underground most of construction logistics.

From a geological point of view, the BBT develops across the main tectonic units forming the Alpine chain, with a maximum overburden of 1850 m. These units, which consist of several overlapping layers, are what remains of the collision between the European plate and the Adriatic (African) one; in the design area they form a dome, at the center of which it is possible to identify the Pennidic and Subpennidic units of the Tauern window, i.e. the deepest tectonic units that form the core of the Alps. Southward, the BBT crosses the fault zone that forms the Periadriatic Lineament, which separates the Austroalpine basement from the Southern Alpine one (Figure 2).

Several typical sections were designed, which vary based on both tunnel type (Railway Tunnels, Exploratory Tunnel, Cross Tunnels, Emergency Stop and other logistic works), excavation method (in part mining methods, but mainly with TBM) and geological and geomechanical conditions.

2   INTRODUCTION

The main lot of the BBT, Italian side, (Lot Mules 2–3) involves the construction of a section of about 22 km of Railway Tunnels, in addition to the Exploratory Tunnel and numerous other works. The realization of such a complex work required some important preparatory works, with logistic functions at the service of the following construction phases; the two lots in which these preparatory works have been carried out are briefly described below.

2.1 Preparatory Lot

The first step for completing the BBT connection on the Italian side was the Preparatory Lot, realized in the period 2008–2011 and consisting of:

- Exploratory Tunnel from the Aica adit to the logistic hub of Mules (10.5 km).
- Mules Access Tunnel (1.7 km).
- Unterplattner Tunnel (0.4 km).

The main function of this lot was the preparation of the logistic works necessary for the succeeding construction of the Railway Tunnels. In particular, this section of Exploratory Tunnel connects the underground logistics hub of Mules with the Unterplattner construction site and, through the Unterplattner Tunnel, with the main spoil repository and logistic area of Hinterrigger where the excavation material will be conveyed. The Mules Access Tunnel, on the other hand, connects the same logistics hub with the Mules external construction site, the main access route to the construction sites on the Italian side.

As for the construction methods, the length of the Exploratory Tunnel, combined with the excellent characteristics of the rock mass crossed (Bressanone Granite) led to the choice of an excavation with a double shield TBM. The other two works of this lot have been made via mining methods because of the reduced lengths and, for the Mules Access Tunnel, also due to the considerable longitudinal slope.

2.2 Lot Mules 1

The Lot Mules 1, completed in 2011–2015, was the second lot completed on the Italian side. The two main purposes of this lot were:

- Completion of the logistic works necessary for the construction of the main section on the Italian side (Figure 3).
- Crossing, with the Exploratory Tunnel and partly with the Railway Tunnels, of the Mules Fault Line, considered one of the most demanding geological sections of the BBT on the Italian side.

In this phase the underground logistics hub of Mules was completed with the construction of three logistic caverns connected with the Mules Access Tunnel and of one Connection Tunnel with logistical functions between the Exploratory Tunnel and the Western Railway Tunnel.

On the other hand, a chamber for the final phase ventilation was realized along the Mules Access Tunnel, having also logistic functions during the construction phase.

In this lot, moreover, both the Exploratory Tunnel and the two Railway Tunnels entered the Mules Fault Line; although without completely crossing the tectonized area, these excavations provided fundamental indications for the succeeding Lot Mules 2–3.

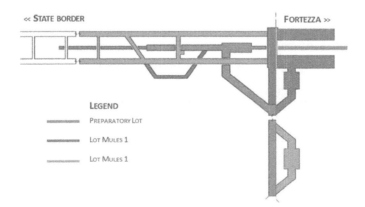

Figure 3.   Mules underground logistic hub.

(a)   (b)

Figure 4. (a) Logistic cavern. (b) Front face in Mules Fault Line. (Images by PAC SpA).

All the works in this lot were made with mining methods, and this for three reasons: the geometric variability of the works, with dimension ranging from 7 m to 20 m (Figure 4a); the reduced lengths; the uncertainty about the geological and geomechanical characteristics of the rock masses constituting the Mules Fault Line (Figure 4b).

3 CURRENT ACTIVITIES

3.1 *Lot Mules 2–3*

The Lot Mules 2–3 includes excavation and/or final lining of: 22.0 km of Railway Tunnels; 17.0 km of Exploratory Tunnel; a 530 m long Emergency Stop with tunnels for ventilation and passenger evacuation; 4.5 km of ventilation and access tunnel; 69 cross tunnels. The final lining of the works created in the Preparatory Lot and Lot Mules 1 is also planned. The work for Lot Mules 2–3 started in 2016; the end of the works is scheduled for 2023.

It must be noted that South of Lot Mules 2–3, the Lot "Underground Crossing of the Isarco river" is presently being excavated; this lot links the new BBT to the existing Brenner line in Fortezza railway station and is scheduled to be completed in 2022. For this reason, the only accesses for the construction of Lot Mules 2–3 are the Exploratory Tunnel from the South adit in Aica (which lies far from Railway Tunnels South adit) and the Mules Access Tunnel, that is not barycentric to the lot (Figure 5), is 1.7 km long and has a 9 % longitudinal slope.

3.1.1 *Design phase – Excavation methods*
The excavation methods foreseen in the Detailed Design (DD) phase are mining methods and TBM, according to the subdivision summarized in Table 1 and Figure 5.

In general, the excavation of the Exploratory Tunnel and Railway Tunnels northwards was planned with TMB due to the length of the sections exceeding 10.0 km. The choice of the type of machine to be used derives from a cost - benefit evaluation that has considered several

Table 1. Lot Mules 2–3: excavation methods 3.

Excavation method	Work	Stretch
Single shield TBM	Railway Tunnels (North)	From 150 m North of the Emergency Stop to the state border (12.0 km each tunnel).
Single shield TBM	Exploratory Tunnel	From the end of the Mules Fault Line to the state border (13.9 km).
Open TBM	Railway Tunnels (South)	From the Mules underground logistics hub to the southern limit of the lot (3.5 km each tunnel).
Mining methods	All other works	-

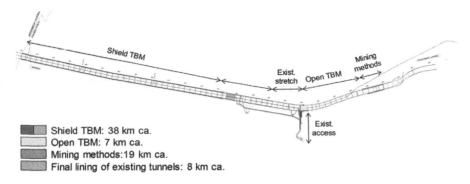

Shield TBM: 38 km ca.
Open TBM: 7 km ca.
Mining methods:19 km ca.
Final lining of existing tunnels: 8 km ca.

Figure 5.   Lot Mules 2–3: excavation methods.

parameters, including the methods for crossing difficult geomechanical sections, the possibility of performing ground injection and other ahead rock mass improvements, the machine costs and the complexity of usage; at the end of the analysis, three single-shield TBM were considered the best option.

The excavation of the first section of the Exploratory Tunnel northwards was planned with mining methods because, as previously mentioned, in Lot Mules 1 it was not possible to identify the end of the Mules Fault Line. It was therefore decided to excavate a further 830 m long section with mining methods (precautionally estimated length basing on the data available during the design phase), in order to approach the excavation with TBM in a rock mass having better geomechanical qualities.

As a result of the three design choices described below, the beginning of the TBM excavation of the Railway Tunnels northwards was not placed at the end of Lot Mules 1, but approx. 3.0 km further north:

• Carry out the Emergency Stop section with mining methods, due to both the greater width of the escape way dock and the numerous openings necessary for the construction of the cross tunnels and ventilation tunnels.
• Given the choice to carry out the initial section of the Exploratory Tunnel with mining methods, the anticipation of the departure of the TBM of the Railway Tunnels would have implied the need to stop the Exploratory Tunnel excavations to respect the design constraint providing that the Exploratory Tunnel must always precede the Railway Tunnels.
• The area intended for the Emergency Stop is characterized by possible geological and geomechanical problems (tectonic transition between the paraschists and the amphibolites of the crystalline basement and the paraschists of the Glockner Stratum).

Upon completion of the excavations northwards, for the Access Tunnel to the Emergency Stop the excavation was planned with mining methods due to the reduced length of the section (4.5 km), the presence of a high slope stretch at the crossing point of the Western Railway Tunnel and the variability of the sections along the works progress.

As for the Railway Tunnels to the South, a mechanized excavation was chosen because, although these works were not on the critical path as far as the respect of times was concerned, the request to eventually anticipate the setting up (arriving from the South) favoured a solution that would guarantee shorter construction times. This choice made also possible to have always a maximum of five simultaneously active advancements fronts, thus improving the efficiency of the ventilation system. Considering the quality of the rock mass crossed (Granite of Bressanone), an open TBM (to be used for both tunnels) was chosen.

On the other hand, the excavation with mining methods was maintained in the final Railway Tunnels section, on the border with the Lot "Underground Crossing of the Isarco river" (1.2 km each tunnel), because in that section the tunnels house two/three tracks, with consequent modifications of the excavation size.

3.1.2 *Design phase – Logistic*

Construction site logistics for long-distance tunnels is strongly influenced by the large amounts of excavated material and the need to supply the building materials necessary for the construction works. During the Lot Mules 2–3 project, a study of the logistics was developed with the aim of verifying that the construction phase could be organized efficiently and consistently with the construction needs. In particular, the organization of the construction sites and of the materials' transport system were carefully studied, in order to verify the effective feasibility of the construction works under the logistics profile.

The first objective of the logistics study is to organize the available building sites in the most possible rational and operative way, so as to allow the management of the complex activities planned.

The second objective is to reduce the impact on the territory. For this reason, it was first of all decided to re-use the material of good quality as much as possible, while distributing the material no longer usable on the shortest route (this objective was also a prescription of the authorities authorizing the works). It was also decided to mechanize as much as possible the transport system of spoils and supply materials reducing the road transport and therefore the impact on the existing road system.

The areas identified for the construction logistics are three (Figure 6):

- Mules area, including the Mules construction site, the Sachsenklemme base camp and the Genauen 2 temporary repository.
- Fortezza Area, consisting of the homonymous base camp
- Aica area, including the Unterplattner construction site and the Hinterrigger site/repository (temporary and final).

For the excavation materials management, it was envisaged the use of conveyor belts from the excavation fronts to the switches (exchange points) in correspondence with the Mules underground logistics hub, which allow to convey the spoils in two directions (Figure 7):

- The materials necessary for the concrete production go back to the Mules Access Tunnel up to the Mules construction site, from which they reach the temporary repositorye site of Genauen 2 through another belt. This belt must be reversible in order to allow the materials stored at Genauen 2 to be conveyed again to the aggregates treatment and manufacture plant at the Mules construction site, from which the aggregates are then transported to the concrete mixing plants located in the underground logistics hubs via a second belt in the Mules Access Tunnel.
- Materials that are not suitable for use as concrete aggregates are conveyed on the conveyor belts of the Exploratory Tunnel, cross the Unterplattner construction site and the Unterplattner Tunnel until the final repository site of Hinterrigger.

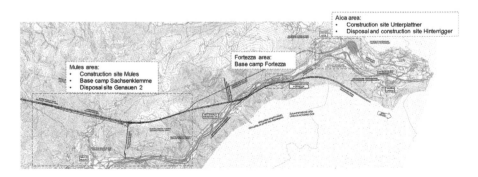

Figure 6.   Lot Mules 2–3: logistic areas.

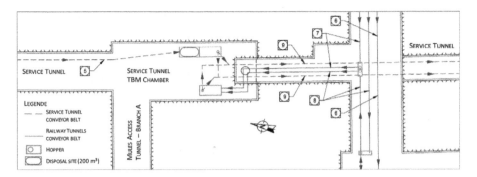

Figure 7. Detail of Mules underground logistic hub. 5 = Conveyor belt from Exploratory Tunnel TBM. 6 = Conveyor belt from logistic chamber to Mules Access Tunnel. 7 = Conveyor belt from logistic chamber to Exploratory Tunnel. 8 = Reversible Conveyor belt from Exploratory Tunnel to Mules Access Tunnel. 9 = Conveyor belt from Exploratory Tunnel to Hinterrigger repository site.

However, the system requires a great deal of flexibility, as the transfer of materials depends not only on the quality of the material itself, but also on the need of supply of the construction sites and on the availability of deposits in the various phases.

The ideal size of the spoils for belt transport is 150–200 mm; the system is therefore optimal for TBM excavations, while for advancements with mining methods primary mobile crusher-mills will be provided close to the front in order to reduce the size. In general, a railway system with switch lay-bys and shuttle trains is provided to supply the fronts with all the necessary material (in particular the prefabricated segments for the advancements with shield TBM and the concrete, both sprayed and cast, for the other fronts):

- In the Exploratory Tunnel, the system is a single track in the section between the Mules underground logistics hub and the state border (Figure 8b), while it is a double track in the section between the Mules underground logistics hub and the Aica adit (Figure 8a).
- In the Railway Tunnels northwards the system is double.track.

Please note that for the Railway Tunnels to the south, road transport is envisaged instead of railway system because of the short length of this stretch.

The main shuttle train station is located in Unterplattner, just after the bridge over the Isarco river. The connection between the railway system of the Exploratory Tunnel and that of the Railway Tunnels takes place in two ways: through the Connection Tunnel excavated within the Lot Mules 1 (Figure 8c) or using a specific shaft equipped with a travelling crane in correspondence with the by-pass 48/3, which transfers the materials from the Exploratory

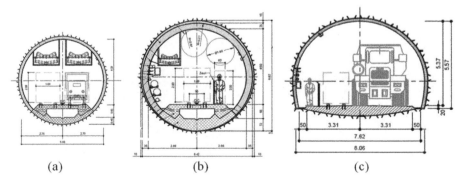

(a)   (b)   (c)

Figure 8. (a) Exploratory Tunnel between Mules underground logistics hub and Aica portal. (b) Exploratory Tunnel between Mules underground logistics hub and state border. (c) Connection Tunnel.

Tunnel trains to those of the Railway Tunnels. This ensures the redundancy necessary for the expected large circulation of materials (especially prefabricated segments).

In order to ensure maximum redundancy and efficiency of the railway system, an independent train circuit is envisaged in the TBM assembling chambers of the Railway Tunnels northwards; this allows to accumulate in the same chambers a certain number of trains thus satisfying the needs of supplies towards the final part of the tunnels and avoiding the long routes that the trains would otherwise have to pass through.

3.1.3 *Construction phase*

During the construction phase, the Contractor committed to realize what had been defined in the DD, however with the possibility of proposing upgrading alternatives that could reduce construction times and/or costs. This contractual availability to propose improvements, allowed the Client to obtain advantages both in programmatic and economic terms. The main alternatives proposed by the Contractor are summarized below.

In the section north of the Mules Access Tunnel there are three main changes:

- The use of double shield TBM instead of single shield (Figure 9).
- The anticipation of the TBM excavation both for the Exploratory Tunnel and for the Railway Tunnels.
- The casting of the final lining of the section excavated with mining methods while advancing with the excavation, and not at the end of the excavation as provided for in the DD.

The first change was proposed by the contractor following in-depth evaluations carried out with the TBM producers, which led to the formulation of an innovative proposal combining the operational flexibility of a double shield TBM with adequate reliability in difficult geomechanical situations. In particular, some features have been defined (thrust system with operating capacity equivalent to that of the single shield TBM, reduction of shield length) such as to allow the machine, when forced to operate in single shield mode by the geology crossed, to really guarantee the performances of a single shield TBM.

Further improvements have been made to the machine, aimed at increasing the efficiency and reliability of the production process, such as: reducing the overall length of the TBM, increasing the conical shape of the shields, increasing the number of positions from which to carry out ground improvements and investigation.

The motivations for the second change (displacement of the TBM drive starts - Figure 10) are different for the three works excavated northwards:

- For the Exploratory Tunnel, it is a simple optimization of the project, as the crossing of the Mules Fault has terminated earlier than expected. It should be noted that, as explained above, the length of the section in mining methods before the start of the TBM had been cautiously estimated on the basis of the data available at the design stage; in the construction phase, once the tectonized section was passed, it was agreed to build the TBM

(a)   (b)

Figure 9.   Double shield TBM for the excavation of Exploratory Tunnel (a) and Railway Tunnels (b).

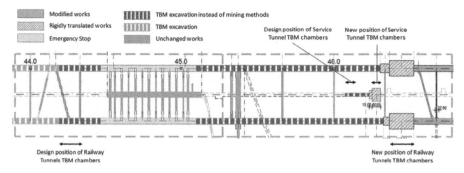

Figure 10. Scheme with the TBM start chambers displacement.

assembly chamber 200 m further south than expected, allowing an earlier departure of the excavation with TBM.

• For the two Railway Tunnels, the change derives instead from an in-depth analysis on the execution methods of the Emergency Stop carried out by the Contractor to verify the possible execution with TBM. The proposal formulated essentially consists in introducing a slight planoaltrimetric misalignment between the railway axis and the excavation tracking axis; in this way it is possible to optimize the position of the railway lay-out, allowing the housing of the internal clearance of the Emergency Stop, which has the escape way dock wider than the standard section. In this case, the Client accepted the internal concrete cast lining to be entrusted just to guaranteeing the fire resistance requested in Emergency Stop, with a nominal thickness reduced to only 23.5 cm; a connection between final lining and prefabricated segments was provided by means of connectors (studs).

The third proposed change (casting of the final lining along with the advancement of the mining methods sections) derives from the optimization of the execution phases suggested by the Contractor considering its organizational skills. On the one hand, this change has allowed the construction times to be reduced, and on the other to use immediately part of the reusable spoils to produce the aggregates already stored in the Genauen 2 deposit after the excavations of the previous lots.

Regarding the section south of the Mules Access Tunnel, the main change refers to the decision to provide the Railway Tunnels with mining methods instead of open TBM. In this case, the change essentially derives from an increase in efficiency of the ventilation system, which has allowed to provide six simultaneous advancements. It was therefore possible to excavate the

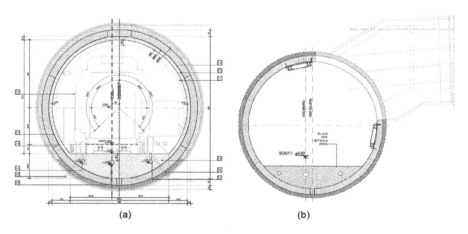

Figure 11. (a) Plano-altimetric misalignment between the railway axis and the excavation axis. (b) Example of excavation of a ventilation tunnel.

Figure 12. Segments prefabrication plant at Hinterrigger.

two Railway Tunnels at the same time, with considerable savings in costs compared to the excavation with TBM, but ensuring compliance with the planned construction times, thanks also to the possibility of proceeding with the casting of the final lining along with the excavation.

From a logistical point of view, the study provided during the design phase had the purpose of verifying the feasibility of the works from a logistical point of view. In any case, the Contractor had the possibility to set up its own logistics according its own equipment and organization, considering three constraints:

- The areas available for construction sites and deposits cannot be changed compared to those identified in the project.
- It is necessary to ensure compliance with the requirements, especially the environmental ones, imposed by institutions at the various stages of the design.
- The amendments cannot entail aggravations for the Client in terms of construction times and/or costs.

The main change proposed by the Contractor is the construction of a prefabrication plant for the TBM segments at the Hinterrigger construction site (Figure 12). This alternative, defined by the Contractor on the basis of its operating organization, has allowed the Client to achieve a great saving in time and costs, since it eliminates the need to transfer the aggregates to an external prefabrication plant from which the segments are then supplied; at the same time, the impact on the territory is also reduced, as the handling of the aggregates and segments takes place entirely within the job site.

4   CONCLUSIONS

The construction of the BBT railway tunnel, of a length of 22 km, operating form just one point of attack, represented by the Mules access tunnel, is characterized by considerable lengths and both geological-geomechanical and logistical complexity. The definition of the construction methods of these works must therefore take into consideration a great number of factors, such as the advancement safety and the respect of times and costs of construction.

To this goal, only the synergy between Client, Designer and Contractor can guarantee the optimal achievement from a performance and implementation point of view.

REFERENCES

Lombardi A., Franze F., Fuoco, S., Pasquali, F., Rausa, L., Pizzarotti, E.M. 2008. Il cunicolo di Aica: il primo passo verso Monaco. *Strade e autostrade.*

Tunnels and Underground Cities: Engineering and Innovation meet Archaeology, Architecture and Art, Volume 7: Long and deep tunnels – Peila, Viggiani & Celestino (Eds)
© 2020 Taylor & Francis Group, London, ISBN 978-0-367-46872-9

Presentation of the successful crossing by the "Federica" TBM of a geological accident in Saint-Martin-la-Porte construction site

F. Gamba
Alpina S.p.A./Egis

E. Hugot, P. Gilli & C. Salot
TELT s.a.s.

G. Giacomin
Ghella S.p.A.

ABSTRACT: The 57.5 km long Mont Cenis Base Tunnel will link up Saint-Jean-de-Maurienne in France to the Susa plain in Italy. In 2015 additional exploratory work was launched from the Saint-Martin-la-Porte decline access tunnel. This included TBM excavation of a 9 km stretch of the southern tube of the Base Tunnel linking the existing access tunnels of Saint-Martin-la-Porte and La Praz. The purpose of this work was in part to check assumptions related to the geology and to acquire the experience required to excavate the Base Tunnel by means of a TBM through the Briançonnais Houiller zone. After excavating close to 300 metres, the TBM encountered a significant geological accident which disrupted progress and required adaptation work to be carried out. The geological context is presented along with the characteristics of the TBM. The major stages which enabled the geological accident to be successfully overcome will then be detailed along with the significant milestones involved in crossing the fault and the means used to strengthen supporting structures and terrains. The technical changes made to the cutting wheel which enabled boring to be resumed will also be detailed. The additional geological exploratory work by surveys and measurements of the deformations observed are presented.

1   PRESENTATION OF THE CROSS-BORDER SECTION

1.1   *The Mont Cenis Base Tunnel*

The planned Lyon-Turin link has a common French-Italian section between Saint-Jean-de-Maurienne and Susa-Bussoleno, which is the first functional phase of the entire project. This cross-border section includes the 57.5 km Mont Cenis Base Tunnel comprising two one-track tubes and also involves the excavation of three decline access tunnels, one adit and two air ventilation shafts (Figure 1).

Studies carried out during access tunnel work helped improving knowledge of the terrain, but these access tunnels will also be used to excavate the Base Tunnel on several fronts and once the tunnel has been commissioned they will be used for ventilation, maintenance and emergency service access, if needed.

The findings from access tunnel excavation work showed the specific and complex nature of a particular geological formation encountered when boring the Saint-Martin-la-Porte access tunnel. Initial work at the Saint-Martin-La-Porte site in Savoy took place between 2003 and 2010 with the construction of a 2,380-metre-long decline access tunnel. Boring work related to this access tunnel encountered quite exceptional difficulties when crossing the so-called "Productive" Houiller terrain. These difficulties come in the form of very high amplitude

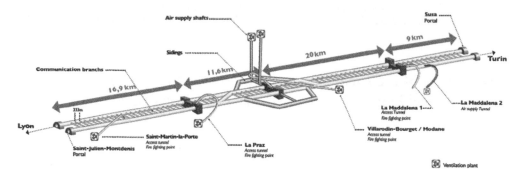

Figure 1. The Base Tunnel.

convergence phenomena in a sector in which the thickness of the overburden terrain was moderate, less than 300 metres (Figure 2).

In 2015, a second additional exploratory phase was therefore launched. As project owner, TELT entrusted the project management to Egis/Alpina and the works to a consortium of six public works companies - three French and three Italian: Spie Batignolles TPCI (agent for the consortium), Eiffage Génie Civil, Ghella SpA, CMC di Ravenna, Cogeis SpA and Sotrabas.

The goals in pursuing exploratory work in this sector are numerous:

- verify assumptions related to the geology,
- specify the geotechnical data concerning the Productive Houiller at the level of the Base Tunnel,
- acquire the experience required to excavate the Base Tunnel using a TBM,
- determine the characteristics and adaptations required for the TBMs,
- adapt the excavation section, geometry and mechanical characteristics of the lining segments;
- analyse the behaviour of the terrain crossed and the force to be supported at the Houiller face and other areas,
- consolidate the working methodology based on the conditions encountered,

Figure 2. Convergence problems encountered in the Saint-Martin-la-Porte access tunnel.

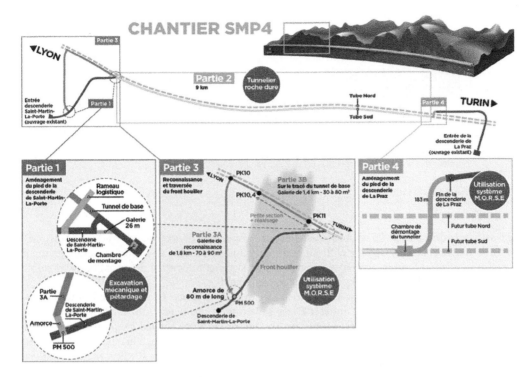

Figure 3. Presentation of the SMP4 site.

- explore any karst areas and underground water movements in the vicinity of the Base Tunnel,
- test different sealing and/or terrain consolidation products in the delicate areas.

1.2 *The Saint-Martin-la-Porte 4 site (SMP4)*

SMP4 exploratory construction site can be divided into four parts (Figure 3).

The main work is focused on parts 3b and 2. Part 3b concerns the excavation of a 1.4 km adit using the so-called "conventional" methods. Part 2 consists in using a TBM to bore a 9 km exploratory adit along the axis and to the diameter of the future South tube of the Base Tunnel. The TBM, baptised "Federica", was given the task of excavating this adit and was designed to cope with the special geological constraints in this area. Built in France, in the NFM Technologies du Creusot plant (Saône-et-Loire), it has a 11.25 metre diameter cutterhead and 70 disc cutters, with a power output of 5 megawatts.

2 GEOLOGICAL CONTEXT OF THE SMP4 SITE

Exploratory work on this site aims at improving geological, hydrogeological, geomechanical and geotechnical knowledge of the sector. Figure 4 shows the projected geological context of SMP4 structures.

The Saint-Martin-la-Porte access tunnel sector is characterised by the Briançonnais Houiller zone overlapping the sub-briançonnais zone by means of the Houiller Face, a tectonic accident made up of hectometric anhydrite layers at the tunnel level.

The excavation of part 3b, using conventional methods, crosses the last carbonated formations of the sub-brianconnaise unit (mostly limestone, calcareous shale and dolomites) then the anhydrites of the Houiller Face before entering the so-called Productive Houiller unit (Encombres hE Unit). The latter is characterised by a predominance of schists and carbonaceous facies (60%) as

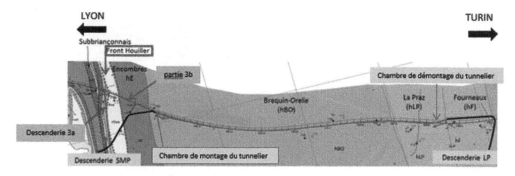

Figure 4. Geological context.

well as sandstone facies (25%) and a significant proportion of tectonised crushed levels (15%). These formations usually appear without continuity and in a highly disrupted structure giving rise to major convergence phenomena requiring specific systems to control deformations and stabilise the structure as well as protect the resources implemented.

As regards TBM excavation, the lithostratigraphic sequence to be crossed is well known and comprises metasandstone, more or less sandstone-type black schists with carbonaceous levels in the Brequin-Orelle unit. However, the sector is characterised by a large topographic overburden (between 700 and 1200 m) hence the difficulty in extrapolating from the surface geological and structural observations. On the basis of the terrain studies and on analysing the boreholes, a progressive increase in the sandstone-type fraction was discovered within the Briançonnais Houiller zone going from west to east, i.e. from the Brequin-Orelle unit to the La Praz unit. A similar progressive reduction in the carbonaceous levels in terms of quantity and thickness was also detected.

3 CROSSING THE GEOLOGICAL ACCIDENT

3.1 The context

The TBM excavation went normally for the first 300 m, crossing rocks with good geomechanical characteristics comprising schist-bearing sandstone masses with some rare levels of black schists.

From PK 12+085 (ring 193) the excavation came upon alternating destructured, even crushed, rocks comprising black schists and carbonaceous schists. This degradation in geomechanical conditions is associated to a sudden change in stratification, observed in the tunnel face surveys: on a large scale, the layers are subhorizontal before PK 12 +085, approximately, and become subvertical in correspondence with the geological accident.

Initially excavation work continued up to ring 200 with several difficulties:

- significant over-excavations from PK 12+099 (ring posed 196/ring excavated 202);
- frequent stoppages of the conveyor belt due to excess material, difficulties in driving the belt, spillage of materials, blocking of hoppers;
- an increase in friction and the torque up to the blockage of the cutting wheel while in rotation;
- the presence of water and slurry.

Face surveys were carried out on the lower part of the TBM's cutting wheel: the face can be partially observed through the peripheral openings used to evacuate the spoil. The predominant lithologies observed before PK 12+085 comprise sandstones and, to a lesser extent, black schists, black carbonaceous schists and coal (Figure 5).

Figure 5.   Materials encountered.

Beyond PK 12+085, the formations turn up (orientation of the dip vector N 225/235, dip varying between 70 and 80. During face surveys, sudden changes of direction and dips were observed as well as developments of subsidence at the vault with water seepage, spontaneous collapses and the rupture of the face with collapses over the entire face.

This geological framework is coherent with the data available from continuous core drillings performed before tunnel boring work began and the destructive soil investigations performed during excavation as the boring progressed.

One of this survey, called SMP2 in the previous figure, stopped due to the borehole instability at PK 12+100 in carboniferous layers associated to a sub-vertical foliation, different from the previous zones (quite sub-horizontal). Anyhow, those important elements available in advance from the borehole didn't allow to deeply understand the geological context around the tunnel such as thickness, orientation and inclination of those layers and the geometry of a probably open fold. In particular, the fault zone met during the TBM excavation wasn't identified by this survey: this fact, associated to carboniferous layers and sub-vertical foliation, led to the encountered difficulties, further worsened by the unexpected presence of water at significant pressure in the carboniferous layers.

In face surveys, the rock was classified according to the RMR and GSI classifications. The values attributed to the rock are shown in the table below. We can note the rapid change in the quality of the cluster as a result of lithological and structural variations (Table 1).

Table 1.   RMR and GSI changes.

PM	12002	12017	12034	12045	12064	12086	12092	12098	12101	12103
RING	137	147	158	166	178	193	197	201	203	204
RMR	65	62	59	41	51	24		36	23	27
GSI	65–70	60–65	55–65	55–60	50–60	35–40	30–40	20–30	25–35	30–35

Tests to close openings on the cutting wheel were carried out with an expanding polyurethane foam in order to reduce the volume of incoming materials (12 of the 16 arms were closed).

After several attempts to resume boring to excavate ring n. 200, millimetric advances were recorded associated to off-limit parameters (torque, friction and extraction rate), the appearance of numerous cracks on the lining segments and a non-negligible oval shaping of the rings. All these conditions, associated to significant withdrawals (some 6,600 tonnes excavated compared to the 2,000 tonnes in theory), meant that the boring was stopped to consolidate the terrain and technically improve the TBM before resuming.

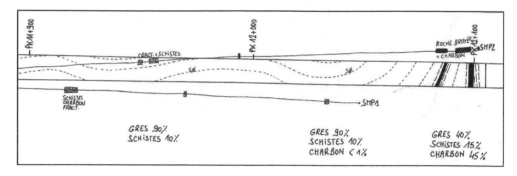

Figure 6. Attitude of the stratification (long profile).

The main stages in crossing the fault are shown in diagram form in Figure 7 then described in the successive chapters.

Figure 7. Geological accident: the main stages in crossing the fault.

3.2 Strengthening the support

Several phenomena were observed on the support (mismatching between rings, millimetric cracks, concrete fragments from the lining segments, deformations and oval-shaping of lining segments) at the rear of the TBM up to PK 12+075, i.e. 30 m before the cutting wheel stopped in the fault at PK 12+ 105 (Figure 8). The large volumes withdrawn actually caused a broad destructuring of the zone and the force was carried over to the lining segments to the rear of the TBM, in the terrain already excavated.

The support installed up to ring 197 (concrete segments C45/55, ~100kg/m^3 of steel) had to be reinforced by installing HEB 180 or 240 ring beams, then more resistant rings (concrete C80/95 with 160 to 190 kg/m^3 of steel) were used (Figure 9).

This work required the implementation of systems designed with the TBM and available on the site as well as adaptations to the TBM to enable anticipated installation behind the shield tail of invert segments to support the ring beams.

Indeed, Federica is designed to power the lining segment erector via a conveyor table located behind the tail: the installation of invert segments is planned between 25 and 30 m from the face and a ring beam erector is available between 35 and 40 m from the face. Following the deformations and cracks, the support required rapid reinforcements behind the tail: for this reason the segment feeder was dismounted and the invert segments were installed back to the rear of the tail using a specific hoist. The lining segment erector was modified and adapted to install ring beams just behind the shield tail.

These operations were accompanied by radial bolting of the vault rings with reinforcement and drainage self-drilling bolts 3 to 4 m long as well as a control of the ring grouting with re-grouting if required.

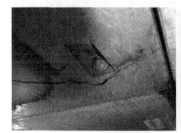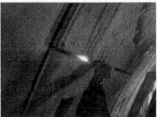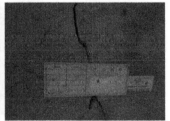

Figure 8.   Phenomena observed on the support.

Figure 9.   Strengthening the support by laying ring beams.

3.3 Non-destructive soil investigation campaign

Simultaneously with the work to reinforce the support and reinforce the face, non-destructive soil investigations were performed to understand the nature and geometry of the geological accident. One of the TBM's two core drilling rigs used for destructive soil investigations was adapted to enable non-destructive soil investigations to be performed at the front of the TBM, from the reentrants planned at the level of the shield, in order to explore the extension and nature of the geological accident.

An additional rig installed on a gateway at the rear of the shield enabled areolar surveys to define more precisely the geometry of the fault and check the possible presence of empty spaces (Figure 10).

Overall, 310 m of non-destructive soil investigations were performed in one month enabling the geological context of the accident area to be defined with a good degree of reliability: a plurimetric passage of carbonaceous and graphitous black schists and coal, folded and upright in this area, cut into by a major sub-vertical accident (around 75°) and therefore subparallel to the cutting wheel (Figure 11).

Figure 10. Non-destructive soil investigation workshop and example of core materials extracted.

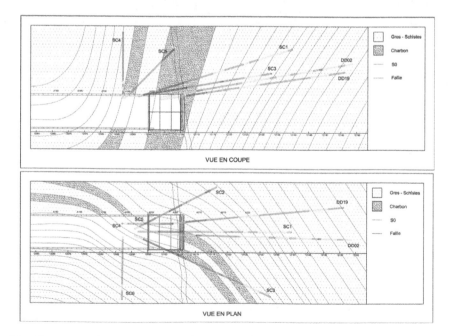

Figure 11. Summary of non-destructive soil investigations performed and assumption related to the geological context.

The characteristics of the fault show a fragile type of displacement, with the formation of gouge and clay breaches at the core of the accident. The thickness of this accident is limited: the plastic carbonaceous levels concentrated most of the deformation and limited the extension of damaged zones. Despite the volume of over-excavations recorded, the non-destructive soil investigations did not identify empty zones above the TBM, since collapses most likely filled the withdrawal areas.

3.4 *Grouting*

Given the presence of empty spaces around the cutting wheel and the nature of the materials extracted, several grouting operations were performed in the following order:

– expanding polyurethane foam grouting just in front of the cutting wheel to fill in empty spaces and seal the cutting wheel;
– grouting using a dual-component cement-based resin that is very fluid and with low viscosity for the part in front of the zone filled with the foam in order to penetrate the materials and consolidate them;
– grouting with pressurised water reactive resin through the vault reentrants in order to finalise the treatment in front of the cutting wheel.

These grouting operations were performed from reentrants crossing the upper part of the shield that were used for surveys and through self-drilling fibreglass bolts performed to match the disc cutters. They helped improve the characteristics of the terrain to be bored and limit, in association with the closing of the cutting wheel, withdrawals when resuming boring. This objective was in fact essential in order to control the extraction flow rate, conveyor management and therefore the regularity of the boring.

3.5 *Modifications to the cutting wheel*

After the first withdrawals leading to blockages and the cutting wheel being immobilised, it was necessary to reduce materials flowing into the excavation chamber. The disc cutter housings were therefore blocked with blocks of polystyrene and expanding polyurethane foam was injected into the booms. These operations were carried out on 1 boom in 3 but, upon restarting, they turned out to be insufficient because the foam was compressing rapidly on contact with the terrain at the level of the buckets, the material penetrated again and the torque force of the wheel exceeded the limits for the machine once more. So it was decided to reduce the opening of the cutting wheel from 8 to 4.5% in several stages:

– preparatory work (inspection, manual draining, cleaning, cutting out old scraper supports and steel plates);
– preparing new supports and steel plates for reducing the openings of the buckets;
– welding of the new elements (scraper supports, plates, brackets, grizzly bars) involving adding for each of the 16 booms 2 additional steel closing plates compared to the initial configuration.

3.6 *Monitoring deformations*

As anticipated in § 3.2 several phenomena and deformations were observed on the support right from the first surveys. For these reasons, simultaneously with the lining segment reinforcement work by installing ring beams, specific monitoring was carried out to see how these deformations evolved and to estimate the strain levels by installing:

– Saugnac gauges on the cracks;
– optical prisms on segments and ring beams;
– strain bars on ring beams and on the shield;
– strain bars sunk into the concrete and pressure cells at the extrados of the segments;

- flat cylinders in the segments;
- probes to measure pore pressure from incoming water.

These instruments were used to continually monitor how the degradations changed over time while the other operations were being performed then during the unblocking and excavation resumption phase. The deformations of the rings were also continually monitored by a so-called RCMS system, comprising biaxial inclinometers installed on each segment of the same ring, enabling the oval-shaping of the rings to be assessed (Figure 12).

The concrete lining segments underwent up to 100mm in cord variation, became oval shaped under the pressure of more significant horizontal strains associated to management of the vault ring grouting. Indeed, this grouting, initially planned between the tenth ring behind the tail, was brought up to the second last ring laid to reduce oval-shaping and improve the distribution of forces in the ring. The maximum strain measured through the bars sunk in the segment concrete was around 15 MPa of compression, confirmed by the flat cylinder tests. The maximum pressure levels measured at the extrados of the segments (in contact with the mass) are around 1000–1500 kPa. The maximum strain levels measured for the reinforcement ring beams placed on the intrados of the rings were around 20 MPa.

Simultaneously with the monitoring in the tunnel, inspections were carried out on the surface to monitor settling and piezometric levels, which did not record any notable variations in relation to the volumes withdrawn.

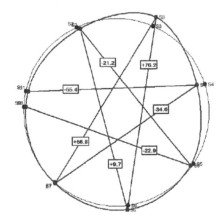

Figure 12.   Chart of convergences and visualisation of the deformation (in mm).

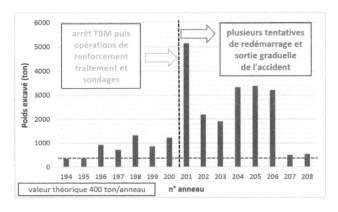

Figure 13.   Chart of the excavated weight per ring in the accident area.

3743

3.7 Resuming boring

The boring work resumed slowly with careful management of the volumes of materials withdrawn and TBM parameters through the accident area. Boring conditions went back to normal from ring 207 approximately (Figure 13).

For the accident area, which concerned some 15 rings, the excavated weight was some 20,000 tonnes more than in theory, including a part comprising drained water which was assessed as being 20% of the total.

4 CONCLUSIONS

The difficulties encountered by the TBM as it approached the 300 m mark required boring to be stopped for several months, a range of different operations to be performed and stages to be crossed before boring could be resumed. The successful crossing of this accident enable to reach several goals in the project, in particular the exploration and crossing of the Houiller with an 11.25 m diameter hard rock single shield TBM, the ability to cross fault zones and carbonaceous passages with over 700 m of overburden, the possibilities of adapting the machine and support reinforcement.

Above all, the lessons learned from this operation were applied when a new fault around PK 15+120 (approximately ring 2209) similar to the previous one was encountered, with the necessary operations and adaptations enabling this new accident to be crossed in less than one month.

5 ACKNOWLEDGMENTS

Authors would like to acknowledge the dedicated work of all the people involved in the design and construction process of the SMP4 works, in particular the contracting authority (LTF first and later TELT), the construction contractor (joint venture Spie Batignolles TPCI, Eiffage TP, Sotrabas, Ghella, CMC, Cogeis) and the work management companies (Egis Structures & Environnement and Alpina S.p.A.).

REFERENCES

Barla, G. 2009. Innovative tunnelling construction method to cope with squeezing at Saint Martin La Porte access adit (Lyon-Turin Base Tunnel). *In* Vrkljan (ed.): *Eurock 2009 – Rock Engineering in Difficult Ground* Conditions – *Soft Rocks and Karst, pp. 15–24*. CRC Press/Balkema.

Barla, G., Bonini, M. & Debernardi, D. 2010. Time dependent deformations in squeezing tunnels. *International Journal of Geoengineering Case Histories 2, No. 1, pp. 40–65. DOI: 10.4417/IJGCH-02-01-03*.

Brino, L., Monin, N., Fournier, C. & Bufalini, M. 2013. Nuova Linea Torino-Lione - Ritorni d'esperienza, Congresso Internazionale Società Italiana Gallerie, *Gallerie e spazio sotterraneo nello sviluppo dell'Europa, 17-19 ottobre 2013, pp. 595–604*, Pàtron Editore, Bologna.

Descoeudres, F., Giani, G.P. & Brino, L. 2015. Il tunnel di base del Moncenisio per la nuova linea ferroviaria Torino-Lione: aspetti geomeccanici e confronto con i grandi trafori svizzeri, *Gallerie e Grandi Opere Sotterranee, settembre 2015, n. 115, pp. 21–31*, Pàtron Editore.

Gamba, F., Brino, L., Triclot J., Hugot, E., Barla, G. & Martinotti G. 2017. A TBM assembly cavern in the French Alps, *Geomechanics and Tunnelling, Base Tunnels, Vol.10, n.3, giugno 2017, pp. 256–264*, Ernst & Sohn Berlino.

Monin, N., Brino, L. & Chabert, A. 2014. Le tunnel de base de la nouvelle liaison ferroviaire Lyon-Turin: retour d'expérience des ouvrages de reconnaissance, *Congrès International AFTES 2014, 13-15 ottobre 2014*, Lyon.

Rettighieri, M., Triclot, J., Mathieu E., Barla G. & Panet M. 2008. Difficultés liées aux fortes convergences rencontrées lors du creusement de la descenderie de Saint Martin La Porte. *Congrès International AFTES 2008*. Monaco.

Subrin, D., Vu, T.M., Sulem, J., Robert, A., Monin N. & Brino L. 2009. Geometrical treatment of convergence and levelling data for the description of the anisotropic behaviour of carboniferous coal schists met in the St-Martin-La-Porte access gallery, *AITES-ITA 2009 World Tunnel Congress, 23-28 maggio 2009*.Budapest.

Tunnels and Underground Cities: Engineering and Innovation meet Archaeology,
Architecture and Art, Volume 7: Long and deep tunnels – Peila, Viggiani & Celestino (Eds)
© 2020 Taylor & Francis Group, London, ISBN 978-0-367-46872-9

Use of expanded clay as annular gap filling. Design and application at the Brenner Base Tunnel

F. Gasbarrone, A. Oss & L. Ziller
SWS Engineering SpA, Trento, Italy

D. Buttafoco & J. Debenedetti
B.T.C. Brennero Tunnel Construction, Rome, Italy

ABSTRACT: When crossing squeezing rock masses the risk of overstressing the lining is always relevant. In case of mechanized excavation with shield TMBs one of the possible counter-measure is the use of compressible filling for the annular gap. This paper presents the application of the expanded clay as filling material for the construction of the main tunnels in the sector of the Trens emergency stop of the Brenner Base Tunnel which will be constructed with mechanized excavation using a double shield TBM. The emergency stop is located in a particularly complex geomechanical context, characterized by the presence of fault zones and squeezing rock masses.

1 INTRODUCTION

When the excavation of a tunnel involves squeezing rock masses, the risk of overstressing the lining is always relevant. This condition is especially challenging for shield tunnel boring machine (TBMs) excavation because even small convergences may lead to difficulties for the machine's advancement.

The paper describes the design of squeezing countermeasures provided for the mechanized excavation of the Trens emergency stop at the Brenner Base Tunnel.

The Brenner Base Tunnel, which is currently under construction, is an Alpine base railway tunnel and it will link Fortezza (Italy) with Innsbruck (Austria). Total length is approx. 64 km. The Trens emergency stop is one of the three emergency stops foreseen for safety reasons along the whole tunnel (Innsbruck and St. Jodok are the others) and is the only one located in the Italian side. The emergency stop is a very complex work because is composed by the main tunnels and by several kind of transverse cross passage, which linked the main tunnels with a safety central tunnel. The main tunnels of the emergency stop will be excavated using a double shield TBM in a particularly complex geomechanical context with squeezing rock masses and the presence of fault zones. Multiple near tunnels are also present and the mutual interaction conditions between tunnels overstresses the linings.

After a description of the project, the paper presents the use of the expanded clay as annular gap filling for the construction of the main tunnels built by mechanized excavation.

The results of the laboratory tests and the numerical analyses performed during the design phase will be described in detail in order to show the key aspects of the expanded clay application.

2 DESCRIPTION OF THE TRENS EMERGENCY STOP

The Trens Emergency Stop (FdE) purpose is to allow the evacuation of passengers in emergency or critical situations, through the lateral platform inside the enlarged section of the main tunnels and then through the escape tunnels.

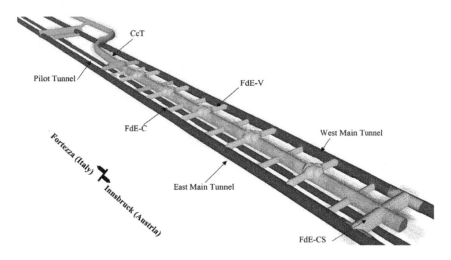

Figure 1.   3D model Trens emergency stop.

The main tunnels and the pilot tunnel, which is placed about 12 m lower, along the route of the underground infrastructure are excavated by mechanized method using double shield TBMs while other tunnels are excavated as conventional method.

The total development of the Trens emergency stop is about 470 m. It consists of:

- 2 emergency stop areas where the main tunnels have a larger cross section compared to the standard one in order to have the space for a lateral platform of about 2.69 m;
- Central tunnel over the pilot tunnel with the purpose of safety area between the two main tunnels (Trens Central Tunnel - CcT). This tunnel leads the passengers outside to the Mauls area;
- 6 cross passages with a spacing of 90m (FdE-C) that connect the emergency stop area to the safe area inside the CcT for people evacuation;
- 6 ventilation tunnels (FdE-V) in order to evacuate the smokes generated during fire events;
- Access tunnel: the circulation compartment is used for the intake of fresh air in the FdE, while the ceiling space is separated and used for the passage of polluted air;
- Ventilation outflow tunnel (FdE-CS) located at the end of the emergency stop that connects the two main tunnels to guarantee the air supply in the central tunnel and in the waiting area. Its aim is to decrease overpressure as well.

A 3D view of the Trens emergency stop have shown in Figure 1.

3   GEOLOGY CONDITIONS

The Brenner Base Tunnel crosses the central part of the Eastern Alps between Fortezza (Italy) and Innsbruck (Austria).

The tunnel will drive through all the geological units belonging to the eastern Alpine Area. Most of these are metamorphic rocks, but plutonic rocks are also present. The geological profile of whole tunnel with the expected rock sequences during the excavation has shown in Figure 2.

The contact between plutonic and metamorphic rocks is the so called "Periadriatic Fault" which represents the tectonic border between African and European plates. The area is intensely fractured so the rock masses have very low geomechanical properties.

The geology of the area crossed by Trens emergency stop is just as complex and characterized by faults of the "Avenes System": subvertical strike-slip faults striking to NNE-SSW and subvertical normal faults dipping to WNW-ESE. The main expected lithologies are calcischists

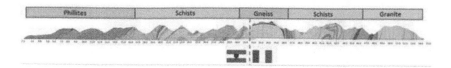

Figure 2.   Geological profile of the Brenner Base Tunnel with the main lithologies (Zurlo et al. 2018).

(calcareous and poor calcareous), Quartzite, Phyllites. A little percentage of Anhydrites and incoherent fault rocks as Carniole, Siltstones and Claygouge is expected.

The sequence of rocks units encountered along the Trens emergency stop is from South to North:

- GA-BST-KS-8f: composed by alternance of calcischists, quartzite and phyllites. This is the main rock formation.
- GA-BST-KPH-8f: composed by calcischists poor calcareous and phyllites.
- GA-T-A-8f: composed by schists with anhydrites.
- GA-T-R-8f: composed by carniole, siltstones, claygouge.

The rock units named GA-T-A-8f and GA-T-R-8f are expected to be excavated for very short lengths. The main properties of the rock units encountered along the Trens emergency stop have shown in Table 1.

Table 1.   Main properties of rock units along the Trens emergency stop (average values).

Unit	Expected (%)	Unit weight (kN/m^3)	UCS (MPa)	Elastic modulus (GPa)	RMR (-)
GA-BST-KS-8f	84	26.6	41	43	60
GA-BST-KPH-8f	14	27.3	54	39	45
GA-T-A-8f	1	28.2	48	46	60
GA-T-R-8f	1	25.0	15	5	30

* UCS (Unconfined compressive strength) and Elastic modulus are related to the intact rock conditions.

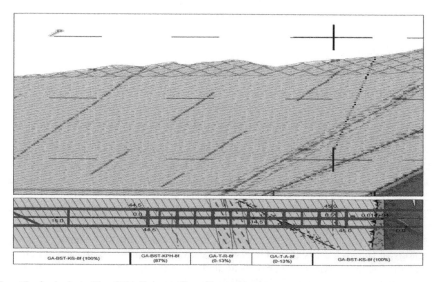

Figure 3.   Geological profile of the Brenner Base Tunnel in the emergency stop sector.

The most part of the formations tend to exhibit squeezing behavior due to medium-low strength properties associated with a high overburden (roughly constant overburden equal to 1110 m along the emergency stop).

The drained water along Trens emergency stop are expected in the order of 0.16 liter per sec/10m with a maximum of 2 liter per sec/10m for calcareous calcischists. Crossing the fault zones drained water of 2–10 liter per sec/10m are expected.

4 TBM EXCAVATION THROUGH SQUEEZING CONDITIONS: PROBLEMS AND POSSIBLE SOLUTIONS

Due to specific geological and geomechanical conditions and the high overburden, preliminary design analyses have shown that squeezing conditions can be expected for the excavation of Trens emergency stop. "Squeezing" refers to the phenomenon of large deformations of the bored profile due to the overstressing of the ground surrounding the tunnel (Ramoni and Anagnostou 2011). Deformation may terminate during construction or continue over a long-term (Barla 1995). The conditions for squeezing occurring are weak rocks with high deformability and low strength in combination with high overburden (Barla 2001; Kovari 1998).

Squeezing conditions can be a challenging problem for the excavation with TBM machine compared to conventional tunneling due to following reasons:

- Large convergences may lead to difficulties for the machine's excavation (i.e. sticking of the cutterhead, jamming the shield, inadmissible convergences of the bored profile).
- Damage of the tunnel support (TBM installs directly the final lining composed by precast concrete segments).

Further problem of the Trens emergency stop is the excavation of multiple tunnels with a low spacing and the interaction effects related to the excavation development of big convergences and damage of the tunnel linings become important problems to solve in conjunction with squeezing conditions.

The excavation with double shield TBM under squeezing conditions can be done, essentially, improving the TBM design and developing special tunnel lining.

Some key aspects that have been taken into account for the proper TBM design under squeezing are:

- Length of the shield. The TBM was designed so that, compatibly with the choice of a double-shield machine, the length of the shield was as short as possible, in order to reduce the surface of the TBM exposed to rock mass pressure.
- Overboring. The TBM cutterhead is provided by extendable cutter, in order to increase the boring diameter and create a space for expected ground deformations with the aim to reduce the risk of TBM's blockage.
- Trust force and torque. Installation of high trust forces and torques allow to excavate when bored profile closes to shield.
- Injection of lubricating agents (as bentonite, foams) through the shield in order to reduce considerably the friction between shield and ground profile.

The afore-mentioned aspects influenced the type of TBM chosen (in general TBM's are gripper, single shield or double shield machines) and the TBM equipping.

About the tunnel lining in squeezing conditions, two ways of lining are possible: "resistance supports" and "yielding supports". In the first case the lining is a very rigid support which shall be strong enough to withstand the ground pressure acting. In the second case the development of excessive ground pressure shall be avoid allowing a certain amount of ground deformation.

Resistance supports are thick linings with high stiffness and strength properties. Segmental linings for TBM usually are classified resistance supports as for their geometry (circular and closed ring's form) as for the thickness (typical range is 30–50 cm) and very high concrete

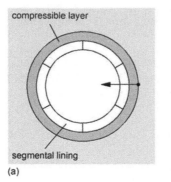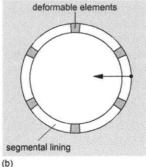

Figure 4. Concepts for achieving deformable linings: (a) Compressible layer between ground profile and lining; (b) Special deformable elements in the joints between the segments (taken from Ramoni and Anagnostou (2010).

class (compressive strengths up to 50 MPa) due to prefabrication process. Usually this kind of linings supports even high ground pressures.

Yielding supports become a convenient solution when ground (and water) pressures are higher than segments strength or may necessitate very thick segments which are difficult to install and require to increase the boring diameter of machine. A deformable segmental lining can be do according two ways: (a) arranging a compressible layer between ground profile and lining or (b) arranging special deformable elements in the joints between the segments.

A compressible annulus layer (case a) must have a high deformability to allow big convergences but also usual properties of gap grouting materials: easy pumpability and high stability of the material in order to embed the lining. Different kind of materials are suggested in literature to fill the annular gap: polyurethane foam (Lombardi 1981), expanded clay (Cucino et al. 2012), light weight concrete (Strohhäusl 1996), compressible mortar with expanded polystyrene patented as "Compex" (Schneider et al. 2005) and finally a cement-based pumpable mortar with expanded pearls and foam patented as "DeCo Grout" (Billig et al. 2007a). An important requirement of the material can be permeability if a drained tunnel is required. Expanded clay has this requirement.

Special deformable elements (case b) can be of different types. Among others, recently highly deformable concrete elements composed by a high-strength concrete matrix with porous aggregates have proposed (Kovari 2005).

5 APPLICATION OF EXPANDED CLAY AS COUNTERMEASURE IN SQUEEZING CONDITIONS

The excavation in the area of the Trens emergency stop mainly interests the complex of the calcareous calcischists and potentially crosses two fault zones at the chainage 45 + 000 and 44 + 800.

The rock masses belonging to the calcischists complex mainly present a relevant squeezing behavior, with important plastic zones associated and instability of the face. Furthermore, the presence of anhydrite within the same rock mass could lead to long-term swelling problems.

The numerical analyses performed (see chapter 6) show how the geomechanical conditions are not particularly favorable and that there is a relevant mutual interference between the different excavations (main tunnels, pilot tunnel, Trens Central Tunnel, complex of cross passages and ventilation tunnels of the FdE) involving high deformations and high loads on the linings.

In order to limit the stress acting on the lining the annular gap of the TBM tunnels will be filled with a highly deformable material, such as expanded clay.

The expanded clay is a granular material consisting in an external "shell", more resistant and less porous, and an internal matrix, characterized by a high void index and a high deformability.

It follows that the expanded clay grains exhibit a low unit weight and a rather low modulus of elasticity.

The main characteristic of expanded clay is that, due to its high deformability, it allows the rock mass to deform and to release its stress implying a lower loads on linings. Furthermore, the permeability guaranteed by the material is high and comparable with pea-gravel material.

In civil engineering applications, this material has proven to have good mechanical properties and the ability to be combined, even at different grain sizes, with other construction materials to achieve the required behavior.

From the results of literature oedometric tests performed on samples of expanded clay of different grain size distributions, and on samples of clay mixed with other materials (foamed glass, polystyrene, calcarenite, crushed clay – (Cucino et al. 2012)), the oedometric modulus was estimated equal to about E_{oed} = 5.4MPa. The corresponding Young's modulus is equal to E = 4.0 MPa.

In order to compare the properties of the material between design stage and construction stage, specifically about the deformability, BTC has commissioned a series of laboratory tests: grain size distribution analysis, oedometric tests and permeability tests.

The laboratory tests were performed on a sample marked as "Leca Laterlite 8–20", having a 10–20 particle size distribution. An oedometric modulus equal to 5.3 MPa has been measured according to the test results for this material. The related Young's modulus estimated starting from this value is equal to 4.0 MPa.

The permeability measured both before (K=2.21E-03 m/s) and after (K=8.42E-04 m/s) the oedomertic tests is compatible with that of a granular material, and therefore the drained conditions on the section are guaranteed.

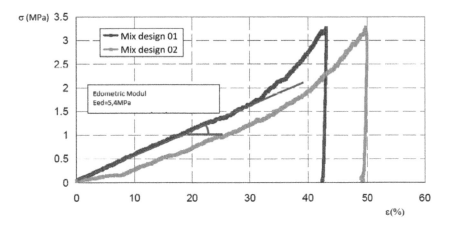

Figure 5.   Estimation of the oedometric modulus from the results of the oedometric tests.

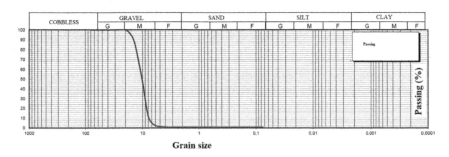

Figure 6.   Grain size distribution.

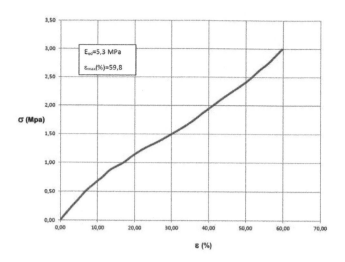

Figure 7. Results of the oedometric test performed by BTC.

The characteristics of the grains in terms of size and density, are compatible with the pumping system used for the pea-gravel and are however irrelevant for the purpose of the function required for the use of the material.

6 DESIGN OF TBM EXPANDED CLAY FOR TRENS EMERGENY STOP

In order to investigate in detail the behavior of the excavation, a series of numerical analyses has been performed during the design stage. The specific aims of the analyses are:

*   Evaluate the effect of the expanded clay as annular gap filler;
*   Evaluate the mutual interaction effects between multiple tunnels.

The numerical models had to take into account all the construction phases and the exact sequence of realization of the different tunnels. Due to the reduced distance between the tunnels that compose the emergency stop, each excavation has a relevant impact on the others.

Analyses were performed with Phase[2] by Rocscience, a structural/geotechnical 2D-finite elements software which allowed a reliable simulation of the evolution of rock mass convergence and efforts on linings for each excavation phase. Figure 8 shows the typical model layout in the excavations area while Figure 9 shows a detail of the mechanized excavation; it clears the relationship between the rock mass, the annular gap filling and lining. The rock mass and the annular gap filling have been modelled with soil elements while the lining has been modelled with beam elements.

The model has a width of 300 m and a height of 200 m. The in situ vertical stress has been evaluated as the product of the unit weight and the overburden while the ratio of the horizontal stress to the vertical stress is expressed by the coefficient at rest K_0 equal to 0.75. Since the overburden is equal to approximately 1100 m in the emergency stop sector, the in situ vertical stress is equal to about 30 MPa. Under these conditions, also with fair rock mass quality, there is a high squeezing potential because the state of stress is often very high when compared to the rock mass strength. Three different sections, corresponding to three different rock mass qualities, have been considered in order to assess the behavior of the expanded clay in the different conditions that characterized the sector of the emergency stop.

The rock mass has been discretized with 6-nodes elements, very fine refined close to the tunnels in order to have the maximum dimension of a side equal to 50 cm. An elastic-plastic

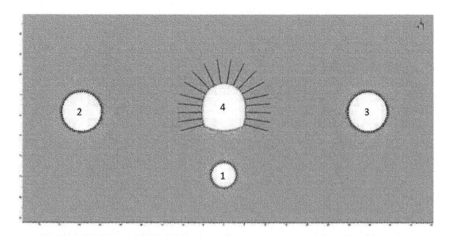

Figure 8. Brenner Base Tunnel – Emergency stop FEM model: General layout (numbers in the different tunnels indicate the construction sequence).

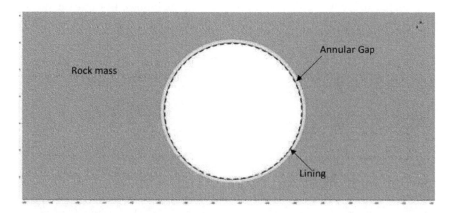

Figure 9. Brenner Base Tunnel – Emergency stop FEM model: Detail of mechanized excavation modelling.

behavior with the Mohr-Coulomb failure criteria has been used to model the rock mass. The lining of the main tunnels has a thickness of 45 cm while the lining of the pilot tunnel is 30 cm thick. The lining has been modelled by an elastic behavior. The flexural stiffness of the segmental linings has been reduced in accordance with the Muir-Wood formulation (Muir-Wood 1975).

The parameters of the rock mass for the different conditions have shown in Table 2.

The stress release to simulate the 3D behavior of the excavation has been determined according to Panet's stress release approach (Panet 1995): the stress release has been evaluated using a longitudinal displacements curve by axisymmetric analyses combined with a ground reaction curve by plain strain analyses. The lining is installed immediately at the end of the

Table 2. Rock mass parameters used in the numerical simulations.

Unit	Unit weight (kN/m^3)	Elastic modulus (MPa)	Friction angle (°)	Cohesion (MPa)
GA-BST-KS-8f	26.6	19822	34	3.00
GA-BST-KPH-8f	26.6	11504	28	2.19
GA-T-R-8f	25	209	13	0.69

shield that has a total length of approximately 12 m. It is interesting to highlight that in this condition there is almost a complete release of the rock mass prior to the lining installation. Indeed, most part of the load that acts on the lining in the final phase of the analysis, is caused by the mutual interaction between the tunnels that can be reduced significantly by the application of a "damping" material around the lining.

The annular gap filling composed by expanded clay has been modelled by an elastic material with a Young's modulus of 4 MPa and a Poisson's ratio of 0.3. This leads to an oedometric stiffness modulus of about 5.3 MPa.

Some relevant results of the analyses have shown below: Figure 10 shows the plastic zones while Figure 11 shows the total displacements. Both are referred to the rock mass GA-BST-KPH-8f after the excavation of all the tunnels. The stated results show a relevant squeezing behavior of the rock mass and a high mutual interaction between the different tunnels.

In each analysis the results were examined to check if the following items were respected:

- The tunnel deformation prior to the lining installation should be lower than the shield "conicity". This to have enough space to install a filling material around the lining;
- The volumetric strain of the expanded clay shall be lower than 30% in order to be compliant with the lab test results which shows a linear behavior up to this deformation value.

The expanded clay has proven to be very effective in limiting the mutual interaction between the tunnels reducing significantly the loads on the linings. In this case, the load on the linings is equal to about 20% of the load compared if the annular gap was filled by pea-gravel.

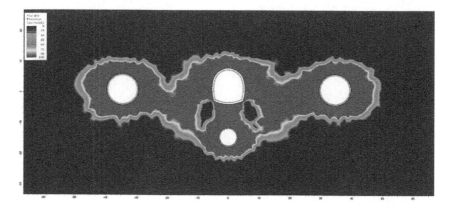

Figure 10.   Brenner Base Tunnel – Emergency stop FEM model: Plastic zones (final phase).

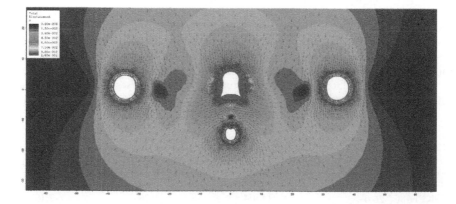

Figure 11.   Brenner Base Tunnel – Emergency stop FEM model: Total displacements (final phase).

For example, for the rock mass GA-BST-KPH-8f, the expected maximum load has been reduced from 1200 kPa to 250 kPa.

7 CONCLUSIONS

The paper has described the use of expanded clay as filling material for the excavation of the Trens emergency stop apart of the Brenner Base Tunnel. The main tunnels of the emergency stop will be excavated using a double shield TBM in a particularly geomechanical context with squeezing rock masses and the presence of fault zones. Multiple near tunnels are also present and the mutual interaction conditions between tunnels overstress the linings.

As part of the development of the detail project, the use of expanded clay was introduced to fill the annular gap for the precast segments of the main tunnels and the pilot tunnel, both excavated by mechanized method, with the aim of containing the pressures on linings.

This design solution was studied by the execution of laboratory tests and numerical fem analyses. Laboratory tests have allowed to evaluate both mechanical and permeability properties of the material. Permeability properties are very important because the lining was designed as drained section. Numerical analyses have performed on the most critical sections in order to capture the behavior of the expanded clay about the development of convergences and loads acting on linings.

The results of the analyses have shown that expanded clay is very effective to fill the annular gap of the mechanized excavation for different reasons:

- The grain size distribution is compatible with ordinary pumping systems used for pea-gravel injection;
- The permeability as a granular material allow to guarantee a drained section;
- The mechanical properties allows to limit the deformations on linings due to squeezing conditions and due to effects of the mutual interaction between the different tunnels.

REFERENCES

Barla, G. 1995. Squeezing rocks in tunnels. *ISRM News Journal* vol. 2 n. 3&4: 44–49.

Barla, G. 2001. Tunnelling under squeezing rock conditions. In: *Eurosummer-School in Tunnel Mechanics, Innsbruck. Logos Verlag*: 169–268. Berlin

Billig, B. Ebsen, B. Gipperich, C. Schaab, A. Wulff, M. 2007a. DeCo Grout – Innovative grout to cope with rock deformations in TBM tunnelling. *Proc. ITA World Tunnel Congress 2007*. Prague.

Cucino, P. Eccher, G. Castellanza, R. Parpajola, A. Di Prisco, C. 2012. Expanded clay in deep mechanised tunnel boring. *Proc. ITA World Tunnel Congress 2012*. Bangkok.

Lombardi, G. 1981. Bau von Tunnel bei grossen Verformungen des Gebirges. In: *Tunnel 81, Internationaler Kongress*. Düsseldorf.

Kovari, K. 1998. Tunnelling in squeezing rock. *Tunnel* 5(98): 12–31.

Kovari, K. 2005. Method and device for stabilizing cavity excavated in underground construction. *US Patent 20050191138*.

Muir-Wood, A.M. 1975. The circular tunnel in elastic ground. *Geotecnique* 25(1): 115–127.

Panet, M. 1995. Le calcul des tunnels par la méthode convergence-confinement. *ENPC, Paris*.

Ramoni, M. & Anagnostou, G. 2010. Tunnel boring machine under squeezing conditions. *Tunnelling and Underground Space Technology* 25(2010): 139–157.

Ramoni, M. & Anagnostou, G. 2011. The Interaction Between Shield, Ground and Tunnel Support in TBM Tunnelling Through Squeezing Ground. *Rock Mechanics and Rock Engineering* 44(2011): 37–61.

Schneider, E. Rotter, K. Saxer, A. Röck, R. 2005. Complex support system. *Felsbau* 23(5): 95–101.

Strohhäusl, S. 1996. TBM tunneling under high overburden with yielding segmental linings; Eureka Project EU 1979. Tunnel boring machines – *Trends in design & construction of mechanized tunneling, International lecture series TBM tunnelling trends*. Rotterdam: Balkema.

Zurlo, R. Marottoli, A. Rughetti, E. Fuoco, S. 2018. Impact of the Rock Mass Quality on the cost of the Conventional Excavation: The experience acquired during the execution of the Brenner Base Tunnel. *Proc. ITA World Tunnel Congress 2018*. Dubai.

Tunnels and Underground Cities: Engineering and Innovation meet Archaeology,
Architecture and Art, Volume 7: Long and deep tunnels – Peila, Viggiani & Celestino (Eds)
© 2020 Taylor & Francis Group, London, ISBN 978-0-367-46872-9

The Brenner Base Tunnel: Back-analyses of excavation data and TBM parameters in the Aicha-Mules exploratory tunnel

G.M. Gaspari
Arup, Toronto, Canada

K. Bergmeister
BBT SE, Innsbruck, Austria
University of Natural Science, Vienna, Austria

ABSTRACT: The data collected from the Aicha-Mules pilot tunnel were back-analyzed to assess the performances of the mechanized tunneling excavation and to evaluate the suitability of the chosen double-shield TBM adopted in the first Italian 10km of exploratory tunnel of the Brenner Base Tunnel (BBT) project. An extension of Rock Mass Excavability (RME) parameter to predict TBM advance is presented in the article in correlation with the "Average Rate of Advance (ARA)". RME, in combination with ARA, was calibrated on the data obtained from the exploratory tunnel and showed good correlation with the rock quality daily recorded at the tunnel face in terms of RMR (Rock Mass Rating). A study was developed adapting the RME parameter to correlate rock quality data and advance rate, to understand if TBM performances were appropriate to the specific rock conditions and support decision making at next construction stages. A novel proposal, not based on RMR or Q, is finally presented on the basis of back-analyses.

1 THE BRENNER BASE TUNNEL PROJECT

1.1 *Description of the Tunnel System*

The BBT is spread over a length of about 64 km, the tunnel portals are located in Tulfes and near the stations in Innsbruck (Austria) as well as in Fortezza (Italy), respectively South and North of their existing rail yards. The system consists of two tunnels connecting a single track, with the existing ring tunnel of Innsbruck, on the South side of the city of Innsbruck.

The average slope of the twin tunnels will correspond to about 6,7 ‰ in Austria and 3,9 ‰ in Italy, according to technical specifications developed by Onwer's designers. The configuration of the tunnel has two main single-track tunnels placed at intervals between 40 m and 70 m and linked together by cross-passages every 333 m of connection necessary for safety in tunnels (see Figure 1, Bergmeister, 2011).

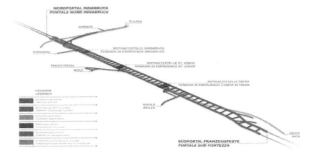

Figure 1. The system of tunnels with the exploratory tunnel (yellow line).

1.2 The Aicha-Mules Pilot Tunnel

Figure 2 shows a pilot tunnel with a diameter of about 6 m located just underneath the two rail tunnels: it has been excavated from both portals, on the Italian and on the Austrian side.

The 6m diameter exploratory tunnel of the BBT, in the underlying position of the twin rail tunnels, is composed of a total of 6 routes with access excavation including Aicha and Mules in Italian territory, for total 30.3 km, Wolf, Ahrental and Innsbruck in Austria, for total 28,7 km

With the goal of performing an extensive geological prospection for the acquisition of additional geotechnical and hydro-geological information along the chosen alignment, the exploratory tunnel was excavated with a selected methodology that allowed both the safety of the crew and the collection of data providing inputs for selecting excavation methods and support systems based on a high-level technical/economical comparison, summarized in this paper.

The tunnel, in addition to fulfilling the unique role of soil tests to full-scale axis in the path of the main galleries, serves more important functions during construction and operation of the Brenner base tunnel, including:

1) underground transportation means for excavated material and supply of building materials as an alternative to roadways,
2) pre-consolidation of the rock mass of the tunnels through injections, precuts or bolting, allowing a more stable face for the main tunnels excavation,
3) selective drainage of water.
4) maintenance
5) other services

The resulting environmental impact has been decisive for the acceptability of the project, becoming the mandatory requirements under the EIA procedure.

The BBT Italian side first lot of excavation works started in August 2007 and essentially consisted in setting up and running the exploratory tunnel Aicha - Mules for an approximate length of 10.5 km. The same lot included the construction of the access tunnel of Mules (about 1.8km).

In Austria the excavation of the exploratory tunnel started in December 2009 after getting the EIA approval. The chosen excavation approaches for the exploratory tunnel Aicha - Mules were SEM (Sequential Excavation Method – drill and blast) conventional tunneling for the Mules access tunnel and double shield TBM (Tunnel Boring Machine) mechanized technique for the Aicha pilot tunnel, with a conventionally excavated pre-tunnel. The combination of high uniaxial compressive strength and limited fractures in the rock brought reduced the average excavation rate to 15 m/day for the TBM tunnel, lower than achievable with this method.

The pilot tunnel is currently being used during the construction phase of the two rail tunnels for transporting the spoil material and logistics as well as for drainage water.

1.3 The geological geotechnical reference model (GGRM) and the RMR classes

The Aicha-Mules tunnel has been excavated through the junction between the European tectonic plate and the Adriatic (African) one. The tunnel cover reached about 1800m and the main formation encountered was the *granite of Bressanone*, featuring granite with high percentages of quartz and granodiorite with biotite and amphibole, affected by weak veins, pegmatitic aplitic and a high angle faults that cross it generally from NE to SW (refer to Figure 3).

Figure 2.   View of the transversal section of the tunnels with the pilot tunnel under the running tunnels.

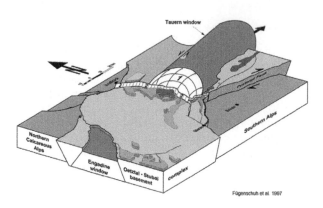

Fügenschuh et al. 1997

Figure 3.   General geological view of the Brenner Region.

The average uniaxial compressive strength of the rock mass was often exceeding 200MPa, confirming a very tenacious material, with negligible level of fracturing.

In terms of geotechnical properties of rock-mass along the project, it is possible to identify a classification ranging between the V and the II class of Bieniawski Rock Mass Rating (RMR), in different proportions. The RMR classification is often adopted to assess the conditions of face stability in the tunnel and the type of supports to be adopted. This classification consists in calculating an index value ranging between 0 and 100, derived as the sum of six partial indices that are estimated based on uniaxial compressive strength, quality index (RQD), spacing of discontinuities, condition of discontinuities (roughness, alteration, filling, persistence, openness), hydraulic conditions and finally location of the discontinuities.

In accordance to the contractual technical specifications, every day both the Client and the Contractor with engineers and geologists have been inspecting the tunnel face to determine these parameters, thus correlating the results with the parameters of the TBM excavation and with the pre-cast liner (PCL) support to be adopted accordingly with expected bending moments development. For values between 100 and 81 the class I (good quality) was defined, between 80 and 61 the class II (good quality), between 60 and 41 the class III (moderate quality), between 40 and 21 the class IV (poor quality), for values less than 20 the class V (very poor quality).

2   THE SELECTION OF THE TUNNELING METHOD

2.1   *The definition of a metric to select the excavation method in Alpine Tunnels*

In the evaluation of the best tunneling method, it is important to consider several different aspects at the same time and to develop a proper risk analysis for each suitable method. The recent and invaluable experience gained from a series of accidents in tunneling worldwide has made everyone aware that the TBM is simply not a fully mechanized tool integrating the various operations of the conventional excavation method for excavating more rapidly, and overcoming all (or almost all) the well-known problems and uncertainties. Instead, the TBM and the tunnel to be excavated constitute a delicate and sensible, unitary system, which should be managed with a new approach, rationally organized and scientifically sustained, in a unified context of research and design of the tunnel, the machine, and the environment. For the selection of the most appropriate excavation method, only few publications are known in the literature (Flora, 2013). It is commonly assumed that mechanized method is the most performing both in terms of advance rates and costs, thanks to the reduced supports to be installed to guarantee the tunnel stability, but the performance is very depending on the rock behavior.

Designing Alpine Tunnels of high capital value is a critical activity that requires comparing production rates and cost/efficiency ratio of the different excavation methods due to the massive length and very differentiated geological/hydrogeological problems that

3757

could make the project very difficult to be completed in acceptable time and with sustainable costs.

When thinking about costs, furthermore, it is not possible to limit their identification only to the monetary ones: social costs are one of the most important to be considered, particularly because of the local population behavior. As the construction may last for decades, population must be properly informed, advised and educated in order to get a proper understanding of the risks they can be affected by during the construction and the impact the new tunnels will have on their everyday life. Furthermore, populations leaving nearby the new tunnels for centuries should be trained to learn how their temporary disadvantage can impact large-scale economy of whole Countries and should be informed about countermeasures and trade-offs they could get from tunnel operativity.

On the Brenner Base Tunnel proper action startet already many years ago; only 2007 more than 150 interactions with the local population were organized (Bergmeister, 2018).

In fact, when analyzing advantages and disadvantages of construction methods, impact in terms of social and environment hazards shall be considered: the mechanized method requires much more space than the conventional one. In tunneling, differently than any other civil work, the job site is not visible, but there is still the need of space outside of the tunnel and it is much lower in the case of conventional excavations, as the job sites of Aicha and Mules respectively show.

Another important aspect to be considered when comparing mechanized and conventional excavation methods is the flexibility and adaptation to geological uncertainties and sudden unforeseeable hydro-geological or geotechnical problems (Bergmeister, 2015). It is the main motivation which made the BBT designers choosing a drilling and blasting technique of excavation for the "Mules Window" and the "Periadriatc Alignment" in order to be ready to deal with a conventional excavation method while approaching the faulted and squeezing area at the border between the African and Euro-Asiatic tectonic plates, few km North of the Fortezza portal.

2.2 *Adapting the Risk Matrix approach to long and deep tunnels*

The specific risks associated with long and deep tunnels can be identified either through a qualitative or a quantitative method. The qualitative method is based on the definition of Risk Matrixes that can visually describe the product between impact and probability to fall into a pre-defined risk tolerance. Quantitative analyses can fit into the definition of the impacts.

Different categories of hazards are commonly identified, including: a) general; b) environmental; c) construction; d) health & safety; e) lifecycle.

Three main evaluation criteria are commonly adopted when approaching long and deep tunnels: 1) Engineering judgment; 2) Probabilistic analyses, 3) Deterministic analyses.

Frequency of occurrence is assessed following standardized intervals, referred to the anticipated number of times the hazard would be encountered during the construction of the project. ITA's Guidelines for Tunneling Risk Management (2004) suggest that consequences be classified into five factors which address a breadth of impacts from encountering these hazards, thus leading to 5 different risk matrixes with different rationales defining the acceptability limits (Figure 4):

- Economic loss: economic losses to third parties and/or to the Owner.
- Delays: delays in construction and in operational efficiency of the project.
- Health & Safety: personal injuries and loss of life as well as permanent health conditions.
- Environment: pollution, tailing/mucking/disposal, damage to flora and wildlife.
- Social: residents' complaints and political influence on the project schedule and budget.

A rating criterion can be applied to each factor to quantify/qualify the intensity of the impacts and the perceived weight of the risks in different scenarios accordingly to specific project context.

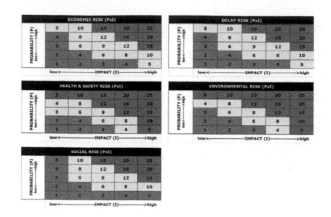

Figure 4.   Typical Risk Matrixes as proposed for the innovative approach based on 5 scenarios

3   ROCK MASS CLASSIFICATION SYSTEMS AND TUNNEL ADVANCE RATE

3.1   *Efficienciesand limitations of different indexes and models to predict tunneling advance*

In order to support the selection of the excavation method most suitable for the BBT, a number of indexes and relevant studies in literature have been analyzed and compared. While RMR, RMi and Q system are mostly indicated for conventional tunneling and several world-wide case histories can correlate their values to actual average advance rates, the rock TBM performances have not been aggregated into a unique universally recognized method.

In the case of mechanized tunneling, methods for TBM performance prediction are based on one or more of the following main principles: 1) Field mapping and/or -testing; 2) Small scale laboratory testing ("index testing"); 3) Large scale laboratory testing; 4) Empirical methods; and 5) Theoretical models. Many researchers have independently worked on their own indices and tests to be able to predict the performance and economic factors associated with boring rock tunnels. Therefore, a wide variety of performance prediction methods and principles are used in different countries and by the various research institutes and TBM manufacturers. Some of the methods are based mainly on one or two rock parameters (for instance uniaxial compressive strength and a rock abrasion value), while others are based on a combination of comprehensive laboratory, field- and machine-data. The effect on the boring rate from jointing is a factor which has been adopted for many years. It has always been recognized that the presence of joints improves the boring rate. However, "in the interest of conservatism in most analyses, the improvement in boring rate due to jointing has been neglected by testing unfractured specimens of solid rock and basing predictions on the strength characteristics of intact rock" (Robbins, 1980). This has probably caused some of the problems in comparing the various models, as reported in several papers.

Of the many models existing, the NTH one (Norwegian Institute of Technology, 1994) is the closest relation to RMi system and its parameters, thus offering a reliable way to compare conventional and mechanized tunneling methods. The main advantages of the NTH model for TBM performance prediction are the generally very comprehensive empirical data-base, where the important influence of rock jointing can be easily included (Nilsen and Ozdemir, 1993). The model is based primarily on empirical correlations between geological/rock mechanical parameters and actual tunneling performance. Time and cost curves for the various tunneling operations have been established by collecting and analyzing a large amount of data on tunneling performance and rock mass properties from tunneling in Europe. The model has been continuously revised and improved as new tunneling data and TBM modifications become available. However, it is very difficult to characterize and apply the great variations in rock mass properties in a simple model or method. Often, rapid changes occur in the rock properties as well as in the jointing features. Hence, simplifications of the real conditions using average values may introduce errors. In addition, there may be errors connected to the way descriptions and characterizations are performed and how they are quantified. The use of block volume as the measure of the degree of jointing causes a problem where the joints do not delimit defined blocks. This happens when only one or two joint sets occur, for instance in schistose rocks (such as mica schist and mica gneiss) without other discontinuities than foliation partings. Errors may also be introduced in the laboratory tests. As testing of small samples introduces significant scale effects (Stavrou, Murphy, 2018).

3.2 The Rock Mass Excavability and the Average Rate of Advance

In the years, other systems were developed to facilitate prediction of TBM advance rates in rock, including Q_{TBM} from prof. Barton and RME from Prof. Bieniawski.

The Q_{TBM} method is based on an expanded Q-system and average cutter force in relations to the appropriate rock mass strength. Orientation of fabric or joint structure is accounted for, together with the compressive or point load strength of the rock. The abrasive or non-abrasive nature of the rock is incorporated via the cutter life index (CLI). Rock stress level is also considered. The new parameter QTBM, can, according to Barton (1999), be estimated during feasibility studies, and can also be back calculated from TBM performance during tunneling. This method received most attention but was also severely criticized (Blindheim, 2005). In this research, it was also tested but without success mostly due to non-uniquely definition of the *parameter rock mass strength* σ_{MASS}, based on "inversion of σ_c to a rock mass strength, with correction for density".

Bieniawski (2004) introduced the concept of Rock Mass Excavability (RME) based on the RMR as adjusted for TBMs. To utilize the RME as a tool for predicting TBM advance, it is commonly employed the concept of Average Advance in a tunnel section, otherwise known as "Average Rate of Advance (ARA)". The ARA is calculated by dividing the length of a characteristic tunnel section over the time of completion of the excavation, expressed in meters per day, m/d.

RME was then presented at ITA WTC (World Tunnel Congress) in Seoul (2006), while afterwards some modifications were offered at RETC Toronto (2007). Finally, in Agra WTC (2008) some specific correlations developed to predict the Average Rate of Advance (ARA) for open TBMs, double-shields and single-shields, using the RME index were presented. In the following years continued investigations allowed to improve by introducing an objective criterion to attribute the RME 10 points concerning the homogeneity of the excavation front, by estimating of the advances of tunnel-boring machines in faults zones and cutters' consumption.

However, there is convincing evidence that complex equations combining rock mass quality RMR or Q, with additional parameters related to TBM characteristics, are not an effective approach. In other words, it is doubtful that one formula can include all the factors pertinent to rock mass quality as well as those influencing TBM choice and performance. In fact, expert opinion (Grandori and Mendaña, 2005) holds that the RMR and Q systems are most effective as they are commonly used, consistent with the purposes for which they were developed.

Hence, adjusting these systems for TBM-sensitive parameters, such as rock abrasivity and cutter thrust, may be counter-productive and may only create confusion, as indicated above concerning the Q_{TBM}. Moreover, while abrasivity is an important parameter for TBM performance and cutter wear estimates, it is not a deciding consideration for the selection of TBM type. Consequently, a specific indicator, not based on RMR or Q, is preferable and it is the purpose of this paper to present one.

3.3 The proposal for a modified RME in hard rock masses

RME input parameters and their ratings were derived from extensive investigations involving statistical analyses of data from 387 tunnel case records (geo-structural regions), with a cumulative length of 22.9 km, mainly excavated by double-shield TBM operating in two modes: double-shield and single shield. RME is usually reported to provide a tool which enables tunnels designers and constructors to decide the most effective tunnel construction method estimating TBM performance. Its purpose is to evaluate Rock Mass Excavability in terms of TBM performance and to serve as a tool for choosing the type of TBM most appropriate for tunnel construction in given rock mass conditions. In the BBT exploratory tunnel through the *Granite of Bressanone* formation, it was chosen to record the rock quality through a daily study of the face of the tunnel in terms of RMR. This method has limitations in the description of TBM performances:

1) in fact, the RMR was specifically created to evaluate the rock quality in case of conventional tunneling, while in this case it was applied for a mechanized excavation;
2) the rating given to water presence can easily unbalance the calculated values of RMR and result in RMR classes which do not correspond to the effective risk of TBM tunneling;
3) quality of face investigation might be limited by ease of accessibility, use of foams and additives to condition the excavated material and interface with maintenance operations;
4) for very high values of RMR, as driven by high uniaxial compressive strength, correlations of data from double-shield TBMs is less effective than open machines.

Nevertheless, the back-analysis developed on the data supplied relevant feedback on the TBM performances in very hard little fractured rock, leading to a proposed modified RME index, which would include two more factors as the correlations adopted in this paper:

- the type of TBM: in the analyzed scenario, for example, due to the behavior of the rock mass, the machine worked with high trust and very little use of the back cylinders as the regripping was almost everywhere allowed and a single shield way of excavating was usually adopted;
- the condition of the TBM: in the specific case, the machine never recovered completely from the accident related to a sub-parallel fault that in August 2009 at chainage 6150 damaged both the segmental lining and the shield to the point that, independently from the geomechanical condition and of the classes of rock mass, the advance rates were sensibly lower than before (Figure 5).

As shown in Figure 7, the proposed correlation between RME and ARA can be applied to different rock classes: Figure 6 shows their distribution along the alignment. The graph describes an increase of ARA as RME decreases, up to an "optimum" defined by a minimized RME value, when ARA is equal to 20–25 m/day per each rock mass class. After this point, an increase in TBM production is only possible with an increase of RME. This result proves that optimization of efficiency in a double-shield TBM in rock has limitations due to the combined inverse effect of the drillability index and the fractured rock mass, which facilitates the excavability only up to a limit imposed by the face stability and consequent thrust and torque to be applied (refer to next chapter).

A novel correlation between RME and ARA considering rock mass strength and type of TBM have been elaborated and is shown in Figure 8.

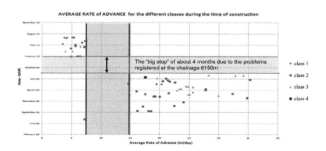

Figure 5.   Correlation with time of average rates of advance and RME for different RMR classes.

Figure 6. Rock mass classes identified along the alignment of the Aicha-Mules exploratory tunnel.

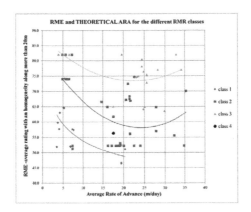

Figure 7. Data regression correlating ARA and RME for different rock classes in a double-shield TBM.

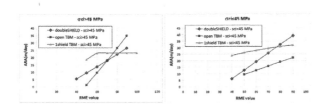

Figure 8. Proposed correlation between RME and ARA considering rock mass strength and type of TBM.

4 THE DOUBLE SHIELD TBM AND THE RESULTS OF THE EXPLORATION

4.1 *The TBM characteristics and its performances in the mostly granitic geology*

The machine used for the Aicha-Mules tunnel was a double shielded TBM Seli-Wirth model TB-630 E/TS with an excavation diameter of 6.3 meters. The TBM back-up had a length of 134 meters and a total weight of 550 tons. The cutter head was equipped with 45 cutters, removable from the inside. The TBM had 7 motors with variable frequency, with a power of 280 kW each. The maximum speed of rotation of the head was 7.3 rpm. The thrust force was provided by 14 cylinders that allow a theoretical 25,330 kN thrust. The maximum force, however, did not exceed 10,000 kN in order not to overload the cutter head, producing exponentially higher wear to the cutting tools. The regripping or the repositioning of the second shield close to the first one was allowed by 19 auxiliary cylinders.

The muck-out was achieved through a system of conveyor belts, which were divided into five distinct bands, with a load capacity of 700 m³/h and a speed of 2.5 m/s, segmented in accordance to different wear-rates into the TBM, along the tunnel, at the river overpass outside the portal, and at connection to the temporary storage area of Unterplattner.

The transportation in the tunnel of materials, equipment and pre-cast segmental lining was done by train. The high level of traffic and massive length of the tunnel imposed the installation of two California switches to optimize train circulations maintaining constant supply to the tunnel front.

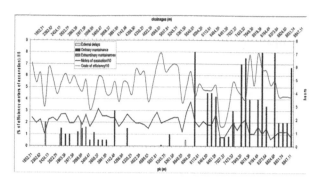

Figure 9. Data correlations for the Aica-Mules exploratory tunnel in terms of excavation efficiency.

During the excavation of the tunnel, a series of monitoring systems were set in place to allow a better understanding of the rock mass properties, the risks for the main tunnels excavation and the safety of the workers. The systems included the following: 1. Gas monitoring and Radioactivity probing in relation to possible leakages of methane gas and hydrogen sulfide (H_2S) or presence of radon; 2. Beam, to ascertain the rock quality ahead of the tunnel face, investigating a length up to three times the diameter of the cutter-head; 3. Seismic surveys and probe drilling ahead of the face every 50 meters of TBM advance for lengths of up to 100–150 m.

4.2 *The TBM efficiency back-analysis and the application of the results to the next phases*

Results in terms of production rate from the TBM excavating the exploratory tunnel where back-analyzed by means of statistical regression and literature correlations with machine's excavation parameters such as torque and trust, showing (Figure 9) a solid effect of the higher maintenance stops required after the accident at chainage 6150.

Machine performance data were cleaned up by engineering judgement: in fact, despite the wide use that is made of RME method, some limits need to be overcome. First, the impact of rock stresses was introduced which in future can be extended to describe overstressing effects (rock bursting, squeezing). Another important limitation discovered is the impact of faults and weakness zones as no specific parameter for such features is considered. In the BBT Project, it was chosen to evaluate the rock quality through a daily study of the face of the tunnel in terms of RMR, which is deemed more appropriate for conventional than mechanized tunneling. A study was developed to monitor geological and TBM data and recalculate RME to correlate advance rates and rock quality. The results brought to a desktop study that helped understanding the performances of different machines to select appropriate methods for next phases specific conditions.

5 CONCLUSION

In recent years tunnel designers and consultants have been feeling the need for an aid to assist their decision making when facing with a choice between mechanized (TBM) and conventional (SEM) construction methods. Today, tunnel boring in rock masses is successfully carried out even with uniaxial compressive strengths exceeding 300 MPa, and with tunnel diameters larger than 10 m. Technically, it is two decades that TBMs can be said to have reached a stage of development where a tunnel can practically be bored in any rock and ground. Still, however, performance prediction is an important part of any TBM project. This is due to the general need for cost- and schedule-evaluations at the various planning stages of a tunnel project, as well as to develop the information necessary for a reliable comparison between alternative tunnel construction methods to achieve supply chain optimizations and economy of scale.

This issue became even more important when in 2005 the Aicha-Mules exploratory tunnel was recognized by BBT-SE as a fundamental and urgent mean to achieve a proper knowledge of the rock masses behavior along the longest tunnel ever designed in the World, the Brenner Base Tunnel. As part of a master thesis in 2008, whose results were un-disclosable till date, the first data collected from this tunnel were back-analyzed by the Authors to assess the performances of the mechanized tunneling excavation and evaluate the suitability of the chosen method of double-shield TBM adopted in the first Italian 10 km long exploratory tunnel between Aicha and Mules.

The purpose of the present work is to generalize the results to set up a conventional metric that can be adopted for investigating the main difficulties that can be encountered in long and deep tunnels under the big orographic chains, like the Alps, and achieve efficiencies through proper selection of construction methodology.

REFERENCES

Alber, M., 2000. *Advance rates for hard rock TBMs and their effects on project economics.* Tunnelling & Underground Space Technology, Vol. 15 (1), pp. 55–60.

Barton, N., 1999. *TBM performance estimation in rock using QTBM.* Tunnels & Tunnelling, September 1999, pp. 30–34.

Barton, N., 2005. *Comments on 'A critique of Q_{TBM}'.* Tunnels & Tunnelling Int., 2005, pp.16–19.

Bergmeister, K. 2011. *The Brenner Base Tunnel.* Published Tappeinerverlag, Lana

Bergmeister, K. 2015; Brenner Base Tunnel - Life Cycle Design and innovative Construction Technology. Swisstunnel Congress

Bergmeister, K. 2018. *Best practice of communication for a complex transnational project – the Brenner Base Tunnel.* WTC 2019

Blindheim O.T., 2005. A critique of QTBM. Tunnels & Tunnelling International, June 2005, pp. 32–35.

Bruland A., 1998. *Hard Rock Tunnel Boring,* Dr. ing. thesis, 10 Volumes of Project reports, Dept. of Building and Construction Engineering, NTNU, Trondheim.

Bieniawski, Z.T., Celada, B., Galera, J.M. and Álvarez, M. 2006. *Rock Mass Excavability (RME) Index,* ITA World Tunnelling Congress, Seoul, Korea.

Bieniawski, Z.T., Celada, B. and Galera, J.M. 2007. *TBM Excavability: Prediction and Machine Rock Interaction.* Proc. RETC, Toronto.

Bieniawski Z.T. and Grandori R. 2007. *Predicting TBM Excavability in Ethiopia.* Tunnel and Tunnelling International, December, p.15.

Bieniawski, Z.T., Celada, B., Galera, J.M. and Tardáguila, I. 2008. *New applications of the excavability index for selection of TBM types and predicting their performance.* ITA-AITES WTC, Agra, India.

Fink, M., 2013. *Dynamisches Entscheidungsmodell zur Auswahl der Tunnelvortriebsmethode mit Validierung am Erkundungsstollen Ahrntal des Brenner Basistunnels.* PhD-thesis University of Innsbruck

Gaspari, G., 2010. *Technical and economical comparison between conventional and mechanized tunneling. The case of the Brenner Pilot Tunnel Aica - Mules,* MSc. thesis, Tunneling Master, Polytechnic of Turin.

Gaspari, G. & Lavagno, A. 2018, Small diameter TBM tunneling. Risk Management approach to face geological uncertainties. Proc. North American Tunneling Conference, NAT, Washington D.C.

Grandori, R, 2007. *TBM Performances and RME Classification System.* Jornada sobre 'Experiencias recientes con tuneladoras'. CEDEX, Madrid.

ITA (International Tunneling Association) Working Group 2, Research & Development, eds. 2004, *Guidelines for Tunneling Risk Management.*

Mendaña, F., 2004. *Double shield TBMs in the construction of the Guadarrama tunnels.* Proc. Int. Congr. On Mechanized Tunnelling, Torino, pp. 207–224.

Norwegian Rock Mechanics Group, 2000: *Engineering geology and rock engineering.* Handbook. Editors: Palmstrom A. and Nilsen B. Norwegian Rock and Soil Engineering Association. 250 p.

Palmstrom A., Blindheim O.T. and Broch E., 2002. *The Q-system - possibilities and limitations.* Norwegian National Conference on Tunnelling, 2002, pp. 41.1 – 41.43. Norwegian Tunnelling Association.

Russo, G. & Grasso, P. 2007. *On the classification of rock mass excavation behavior in tunneling.* Proceedings of the 11th Congress of International Society of Rock Mechanics ISRM, Lisbon.

Stavrou, A.; Murphy, W., 2018. *Quantifying the effect of scale and heterogeneity on the confined strength of micro-defected rocks.* International Journal of Rock Mechanics and Mining Services, Vol. 102, 02/2018, pp. 131–143

*Tunnels and Underground Cities: Engineering and Innovation meet Archaeology,
Architecture and Art, Volume 7: Long and deep tunnels – Peila, Viggiani & Celestino (Eds)*
© 2020 Taylor & Francis Group, London, ISBN 978-0-367-46872-9

Constrains, design and construction methods of the Cigéo five deep shafts

C. Gaudry, H. Ouffroukh & L. Moscone
ARCADIS ESG, Le Plessis Robinson, France

R. Taherzadeh
Tractebel Engineering, Gennevilliers, France

ABSTRACT: Within the framework of CIGEO project, 5 shafts with depth over 500 m and effective diameter between 6 and 8.50 m are to be constructed, to meet the needs for the construction works and the operation phase. The last experiences in deep shaft construction in France are 15 years old, there is currently no project including such structures. Moreover, the evolution of the regulations and the specific requirements of the CIGEO project are increasing the complexity of these structures. The issues in terms of security of the works, durability, water tightness, ventilation, and works extension are essential aspects of the design. Even if the shafts are sunk independently from the other structures construction, they must be efficiently integrated in the global project. This article will focus on the sinking methods and the design chosen to fit the best to the project, and how they were adapted as the constraints and the requirements evolved.

1   INTRODUCTION

The project of industrial center for geological radioactive waste disposal (Cigéo) is a nuclear underground installation and is meant to be a long-term solution for this specific type of waste. After a primary construction phase that will last for 8 years (1st Phase), the installation will be operated and extended over 130 years. The total amount of tunnels planned at the end of the extension will be around 260 km.

1.1   *The shafts in the project: construction and operation sequence (1ˢᵗ Phase and Operating phases)*

Cigéo project has been thought out as a two-level architecture. A part of the installations is located on the surface, while the storage part is located around 500 m below. To create a connection between these 2 different levels, it is necessary to build several SUL (Surface-Underground Link) in order to ensure the different flow of personnel, equipment/material and ventilation, as well as the transport of the conditioned radioactive waste underground. A total of 7 SUL (5 shafts and 2 declines) are planned in the architecture of the installation.

The project is sequenced in two major phases. The 1st Phase is the construction of the main installations and tunnels that are required to start operating the industrial process. The following phases (Operating phases) will result in the development of the underground infrastructure, simultaneously with the storage of the conditioned waste. The expansion of the installation over time is following the requirements for storage. A physical barrier between the operating area (waste disposal) and the construction area (tunnel excavation and lining for the expansion of the installation). The 5 shafts are sunk at the beginning of the 1st Phase and are on the construction planning critical path.

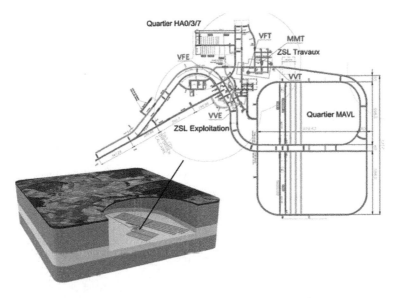

Figure 1. Location of the shafts in the underground architecture of the 1st Phase.

1.2 Geometry and equipment of the shafts

The five shafts to be constructed are divided in 2 different categories:

- The shafts used for the underground works (extension of the installation) connecting the surface to the Construction Logistic Area (CLA) underground:
  - VFT-Shaft ($\varnothing_{eff}$ = 8.00 m): fresh air ventilation shaft and transport of the personnel, equipped with a main lift and an emergency lift;
  - VVT-Shaft ($\varnothing_{eff}$ = 6.00 m): stale air ventilation shaft, no equipment;
  - MMT-Shaft ($\varnothing_{eff}$ = 8.50 m): transport of the material (excavated material, concrete. . .) and equipment, equipped with a main lift and 2 skips;

- The shafts used to operate the waste disposal process connecting the surface to the Operating Logistic Area (OLA) underground:
  - VFE-Shaft ($\varnothing_{eff}$ = 8.00 m): fresh air ventilation shaft and transport of the personnel, equipped with a main lift and an emergency lift;
  - VVE-Shaft ($\varnothing_{eff}$ = 6.50 m): stale air nuclear ventilation shaft, no equipment;

The effective dimensions of the shafts are determined by the requirements for the ventilation and/or the size of the equipment.

1.3 Most recent deep shafts sunk in France

The Cigéo shaft are sensitive structures because of their impact on the construction planning. On top of that, there is a really few feedbacks of recent shaft sinking in France (see Table 1), making this kind of structures remarkable.

Shaft sinking was historically related to the extraction of material in the great mining basin. Since the beginning of the 21st century and the closing of the main coal mines (Provence in 2003 and Lorraine in 2004), the industry has collapsed in France and the shaft sinking has followed. In terms of depth and sinking methods, the only recent shafts are Andra's Laboratory Shafts from 2000 to 2004. In terms of diameter, the last references are from the years 1980. The loss of expertise in shaft sinking in France, and especially for the drill and blast method, is relevant through these feedbacks. Mechanized construction methods such as raise drill are often chosen in the most recent projects.

Table 1. Main characteristics of the most recent shafts build in France.

Name	Depth [m]	Diameter Exc - Eff [m]	Year	Sinking Method	Shaft function
Livet Gavet					
- *Cheminée d'éq.*	180	5.5 – 5.0			
- *Puits Blindé*	163	4.5 – 3.3	2017 - 2018	Raise drill	Hydroelectric plant
Passy Shaft	216	3.5 - 2.4	2016 - 2017	Raise drill	Penstock
Andra Laboratory Shafts					- Personnel and material transport, ventilation
- *Main Shaft*	510	6.0 - 5.0			
- *Auxiliary Shaft*	510	5.0 - 4.0	2001 - 2004	Drill and Blast	- Emergency exit
Aven d'Orgnac Shaft	125	5.5	1999	Drill and Blast	Lift Shaft
Lavera Shaft	172	6.5	1996 - 1997	Drill and Blast	Propane storage
La Houve Western Shaft	522	8.5 - 6.5	1987 - 1989	Freezing & Drill and Blast	Ventilation (mine)
Z-Shaft Gardanne	879	7.5 - 6.5	1982 - 1984	Drill and Blast	Extraction, material transport (mine)
Y-Shaft Gardanne	1109	11.0 – 10.0	1981 - 1983	Drill and Blast	Personnel and material transport (mine)

2 CONSTRAINS OF THE PROJECT AND EVOLUTION

2.1 Regulatory framework

One of the main consequences of the absence of deep shaft projects in France is the ack of regulatory framework for this kind of specific works, requiring staff working deep down the shaft. Whereas the mine shafts were regulated by the mining Code (Extraction Industries General Regulation, last update in 1980), Cigéo project is not ruled by this regulation, even if it can be used as an interesting reference source.

For the shaft sinking, the design of the extraction and hoisting system must be in conformity with the regulation for suspended loads lifting equipment (EN 13001-1 from July 2015), as well as the machines regulation (2006–42), even if these texts do not concern shaft sinking equipment. Safety of work aspects are controlled by professional organizations such as DIRECCTE and CARSAT, like it was done for Andra's Laboratory shafts construction.

The Contractor in charge of the works will have to fix the lack of regulatory framework with a detailed risk analysis for the choice of the construction methods and the design of the sinking equipment. This task will be achieved in collaboration with the prevention and work safety organizations.

2.2 Geotechnical and hydrogeological context

A shaft can be compared with a large-scale deep opening as they cross all the geological strata from the surface to the underground. The main formations are:

- The Tithonian (or Barrois limestone) is a limestone layer with a thickness between 50 and 70 m on the location of the shafts. A karstic network is supplied by two different aquifers more or less connected.
- The Kimmeridgian consisting into the marlstone part with a thickness of approximately 100 m and the limestone part with a thickness of 20 m, both layers being aquiferous.
- The Oxfordian, a 260 m thick limestone layer, bearing a non-productive aquifer with water production associated with the porous strata.
- The Callovo-Oxfordian (COx), an argillite layer with a thickness of 160 m and a soft rock behavior. This stratum has no aquifer and very low permeability. It is the layer where the nuclear waste will be disposed.

2.3 Main requirements

The design of the shafts needs to fulfill the requirements and constrains of the project. These requirements can be either delivered by the client Andra or be prescribed by other expertise fields such as Health and Safety, Environment, Radiologic Protection… Most of the requirements concern the water seepage, vertically from one aquifer to another along the shaft lining, or horizontally from the ground to the inside of the shaft through the lining. The main concerns of these requirements are the protection of the groundwater table supplying the local needs for water, and long-term maintenance and operating costs. These requirements are driving two main measures:

- The shafts must be watertight in order to reduce the maintenance issues caused by radial drains. This is a lesson from the Andra Laboratory Shafts that are designed with drains subjected to quick calcification and requiring a lot of maintenance to keep them operating.
- The design of the shafts must include two vertical watertight cores, one between the Tithonian and the Kimmeridgian strata, and the other one between the Oxfordian and the Callovo-Oxfordian strata. They allow to reduce the communication of the aquifers, and the water seepage in the COx which is very sensitive to water.

Finally, another requirement is to prevent the use of blasting method to sink the shafts in the COx in order to reduce the risk of damaging the ground (cracks in the rock are increasing the permeability and can be an issue for the radiologic barrier role of this ground layer). This requires changing the sinking method in the argillites.

2.4 Construction planning of the 1st Phase

Having in mind these requirements, the construction planning resulting from the basic preliminary design ended up being extended because of restrictive operations such as the casting of a watertight lining or the modification of the drilling tools in the COx in order to fulfil the requirements. Therefore, the shafts are reaching their final depth late, with an important impact on the global schedule. In fact, the declines are not sufficient to supply the material flow required for a lot of simultaneous tunnel excavation faces (the architecture underground is close to a grid, requiring many excavations faces to allow for a faster construction). As a solution, it was decided to use the shafts to initiate excavation of the two Logistic Areas, allowing a faster construction of the underground structures and achieve all the works in the given time. Consequently, all the shafts (except the VVT-Shaft) are used right after the end of their construction to transport the material required for the excavation and the lining of the galleries. The sinking speed of the shafts has also been increased in order to start the excavation of the galleries earlier.

2.5 Main design choices et evolution of the requirements

After a detailed analysis of the project context and all the constrains, most of the design choices regarding the construction methods have been made to go further in the design. A specific study of the sinking method was achieved and is discussed in this paragraph.

Amongst all the shaft sinking methods, the ones using a pilot opening (raise boring, downward reaming) have been removed because of the construction phasing and the planning that provides no early access at the final depth level of the shaft. Moreover, these methods require a good quality of the unsupported rock unlike the COx which is a weak rock. Finally, these methods do not fulfil the requirement on the water communication between aquifers and water seepage in the COx, as there is no possibility to control the water. Therefore, the methods without pilot opening (blind sinking) have been studied.

There are several different blind sinking methods. The large diameter drilling method called "Big Opening" cannot achieve the required large diameters (> 6.00 m) and high depths (> 500 m), it's been rejected as well. The shaft boring machine (SBR developed by Herrenknecht) has been studied, but because the feedbacks on this machine are not sufficient (only 2 shafts have been sunk with this machine so far), it could not be the solution to keep in the first place. Finally, the conventional Drill and Blast method has been chosen for the project. It was

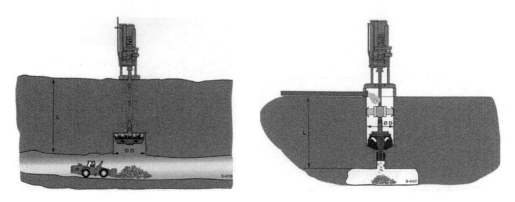

Figure 2.   Raise boring and downward reaming construction methods.

decided to modify the method in the COx argillites according to the requirement and use a hydraulic hammer instead of blasting to excavate the last 100 m of the shafts. For the water-tight lining, it was decided to use a thin non-structural steel liner placed between the ground and the concrete cast lining. It is actually the only type of watertight liner with a feedback of more than 150 years which is the lifespan of the project structures.

In order to fulfil the planning requirements, it was finally decided, in collaboration with the client Andra, to reasonably relax some of the constrains. The Drill and Blast method has been extended to the COx to avoid a modification of the sinking equipment and a slower excava-tion rate. In return, the blasting sequence will be studied and tested to optimize the efficiency of the blasting energy and reduce the damaged zone in the argillites. This optimization is quite important due to the absence of modification of the sinking platform and tools during the works, and due to the high excavation rate of the blasting method.

The second modification is related to the watertightness of the shafts, initially planned on the 500 m of the shaft. The installation of the steel liner is slow as many welds are required to assemble the steel rings. It was decided to keep the steel liner on the first 70 m of the shafts, where the permeability of the limestone is high due to karstic network and the water table must be protected, but to remove it on the rest of the shaft. The watertightness is then only guaranteed by the concrete lining itself. The construction joints are hence equipped with a gutter to collect the water that may infiltrate at this weak point of the concrete lining. The only exception made is the VVE-Shaft where the steel lining is maintained on the wopening shaft, due to the difficult access for maintenance in a nuclear ventilation shaft.

3   SUPPORT AND LINING DESIGN OF THE SHAFTS

3.1   *In the "Cover soils"*

In the Cover soils (i.e. Tithonian, Kimmeridgian and Oxfordian), the dimensioning of the tem-porary support was achieved with the RMR method and based on feedback. The temporary support consists in rock bolts, wire mesh to prevent the fall of small rock blocks, plus shot-crete only for the pre-shaft of the shaft (from 0 to -60 m, corresponding to the Tithonian stra-tum). The final lining thickness is calculated to hold an hypothetical water pressure (up to 37 bars) due to the watertightness of the lining. Hence the calculated thickness is 0.30 m of C30/37 concrete for the pre-shaft, and 0.70 m of C60/75 for the rest of the shaft.

3.2   *In the COx argillites*

The behavior of the COx was studied, and it appeared that it is a squeezing soft rock, with an unsupported short-term radial convergence around 0.30 m (in diameter). The argillites also have a fragile behavior with a fast spalling effect if the ground stays unconfined. The long-term

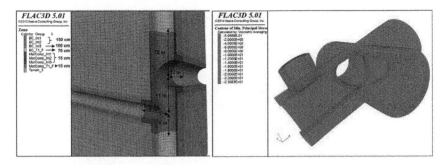

Figure 3.   Geometrical configuration and principal stress in the lining at the -500 m level (MMT-Shaft).

behavior is favorable if the strains are not blocked by a lining. Otherwise, if a rigid support is installed, high concrete thickness is required (more than 1.00 m) to ensure the long-term stress doesn't exceed the concrete resistance over 150 years. The behavior of the COx is modelled with a Hoek & Brown criterion and the creep behavior is described with Lemaitre law. This behavior is imposed fir the design of all the structures in the argillites.

The solution considered to control the creep of the rock and reduce the concrete thickness consists in using a compressible material between the temporary support and the final con-crete lining, in order to control the long-term strains with maintaining a pressure between 1.5 and 3 MPa, corresponding to the compression plastic yield of the material.

The common cross-section of the shafts was calculated with a simple 2D finite elements model, in order to determine the support and the final lining thickness. The results are giving a thickness of 0.70 m of concrete associated with a 0.15 m layer of compressible material. The intersection areas with the CLA and OLA galleries required a more complex model and were modelled with FLAC3D in order to take into account the complexity of this specific geometry. The main phenomenon observed is an important stress transfer, leading to locally excessive stress in the concrete. To ensure the concrete resistance is not exceeded, a 1.50 m thick lining concrete associated with 0.30 m of compressible material is required.

4   DESCRIPTION OF THE SINKING METHODS AND EQUIPMENTS

4.1   Construction of the shaft pre-shaft (from 0 to -60 m)

The upper part of the shaft corresponds to a geological and regulatory limit: the depth allows to reach the base of the Tithonian limestone layer, but it is also the maximum height allowed for lifting loads without a guiding system. In addition, this depth is required to install the gal-loway (sinking platform). As a consequence, the construction methods for this specific part of the shaft are different than the rest of the shaft. The handling movements will be operated from the headframe which will be adapted to operate as a gantry. The excavation rate will be reduced by the number of handling movements to lower and raise the excavation equipment in the shaft because of the absence of sinking platform.

The Drill and Blast method is used for the shaft pre-shaft, with 2.0 m high rounds. The drill-ing is achieved by a vertical drilling rig lowered on the shaft excavation face. The mucking can be done from the surface (up to 30 m depth) or by a power shovel and a skip lowered in the shaft at each round. The support composed of rock bolts, wired mesh and shotcrete is installed with specific machines. Once the sinking of the shaft pre-shaft is achieved, the vertical watertight core associated with a concrete foundation is cast at the top of the Kimmeridgian stratum. The steel lining can then be installed: the steel rings are welded in a collar excavated in the upper part of the shaft. The rings are lowered one by one, then embedded with a mortar cast between the ring and the ground. Finally, the concrete lining is cast at the intrados of the shaft.

The upper part of the shafts presents an interface with the final cover structures (headframe and its foundations, air duct) so attention must be given to the design of this specific part. The

3770

Figure 4.   Excavation of the pre-shaft with minimal equipment.

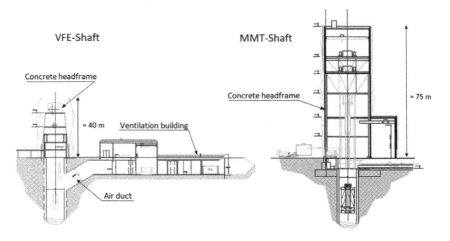

Figure 5.   Final structures at the surface for VFE-Shaft and MMT-Shaft.

needs for the shaft construction and the needs for the operating phase are considered in the design, for instance the collar used to weld the steel rings is also designed to integrate the connection between the vertical shaft and the horizontal air duct. For the MMT-Shaft, the final 75 m high concrete headframe stands on massive foundations. The collar is used as a basement for the building and integrates a part of the equipment required for the operation of the shaft lift. These structures (headframes and ventilation buildings) are constructed after the needs to transport the material for the excavation and the lining of the galleries underground is over.

4.2   *Sinking and lining cycle in the Kimmeridgian and Oxfordian (from -60 m to -440 m)*

As a consequence of the modification regarding the watertightness of the shafts, the sinking method has been optimized. The lining now only composed of concrete, is cast downwards (from top to bottom) close to the excavation face. Every 3 excavation rounds, the formwork is lowered, and the lining is cast, allowing to increase the global sinking speed. The formwork is yet to be reinforced to avoid the damages from the blasting.

4.3   *Sinking and lining cycle in the Callovo-Oxfordian (COx) (from -440 m to -540 m)*

The sinking sequence in the COx is similar to the pre-shaft. The shaft is excavated with the Drill and Blast method until the final depth. Every round is 2.00 m high, and the temporary

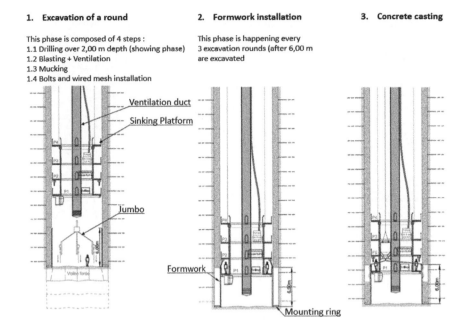

1. Excavation of a round

This phase is composed of 4 steps :
1.1 Drilling over 2,00 m depth (showing phase)
1.2 Blasting + Ventilation
1.3 Mucking
1.4 Bolts and wired mesh installation

2. Formwork installation

This phase is happening every
3 excavation rounds (after 6,00 m
are excavated)

3. Concrete casting

Ventilation duct
Sinking Platform
Jumbo
Formwork
Mounting ring

Figure 6.   Sinking and lining cycle in the Kimmeridgian and the Oxfordian

support made of bolts, shotcrete and compressible blocks, is installed after the blasting and mucking. The particularity of the COx argillites is the behavior that leads to short-term radial convergences around 0.20 m, and a continuous creep for 150 years. As a consequence, the lining of the last part of the shaft is done after a 2-months break, allowing most of the convergences to happen and reducing the stress in the concrete. During this time, the sinking platform is adapted for the lining phase.

The lining of the shaft, as described previously, is made of a compressible layer at the extrados of the concrete lining. The compressible layer will absorb the convergences of the rock and limit the stress in the concrete. This lining is installed from the bottom to the top until the watertight core. The compressible layer is installed as a ring made of many individual elements bolted to the ground. The concrete is then cast between the formwork and the compressible layer on a height of 4 m for each round (see Figure 7). Prior to the lining installation, the cross section modified by the convergences will be corrected to reach a circular shape and to allow for a smoother installation of the compressible rings.

4.4   Sinking equipment

4.4.1   Temporary headframe and winches
The equipment required to sink the shafts is specific and must fulfill the requirement of the works in a vertical configuration at great depth. The lifting operations between the ground level and the excavation face are requiring a great number of winches fixed at the surface all around the shaft. The cables connecting the winches to the loads are held by a steel structure headframe located over the shaft. The design of this headframe and the associated winches must be in coherent with the sinking methods and equipment chosen. The main equipment for hoisting is:

- The sinking platform winches, allowing for the movement of this platform;
- The extraction hoists with double-cables, operating the movements of the material, the mucking buckets and the transport of the workers thanks to a dedicated lift;
- The emergency winch with double-cables and a dedicated lift, powered by a specific generator;
- Diverse winches: ventilation duct, compressed air, dewatering, electricity, blasting line…).

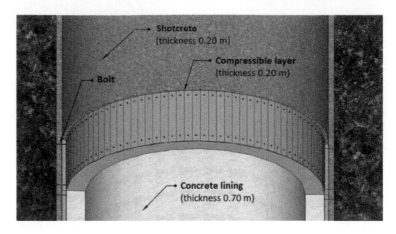

Figure 7.   Installation of the compressible ring in the COx

Beyond 50 m depth (height of the pre-shaft), every load moving in the shaft must be guided. The buckets and the personnel lift are guided from the surface to the sinking platform thanks to cursors sliding along the platform cables. The emergency lift is guided by a weighted cable hanging all the way down to excavation face. From a regulatory point of view, the design of these specific equipment must be in compliance with the "machines regulation" (2006–42).

4.4.2 *Shaft cover deck and collar*

The shaft cover deck is a steel structure covering the top of the shaft at the ground level, in order to avoid any fall of objects or people. Many openings are provided into the cover deck, in particular the automatic hatches opening to allow for the buckets to be taken out of the shaft, the openings for the diverse cables, pipes, wires and air duct. A specific opening for the emergency lift is also provided. The deck also holds a rail system used to place the mucking buckets on a lorry and unload them on a specific area distant from the shaft. Another opening is provided to lower the steel ring segments in the collar for the watertight lining. The arrangement of the openings must be compatible with the deck structural beams, but also with the openings provided in the sinking platform. The shaft collar is covered by the steel deck and is mostly used for the welding of the steel rings. The diameter of the collar is around 15 m (wider than the shaft itself) and its depth around 6.00 m allows to assemble steel rings up to 4.00 m high. The collar is required in order to lower the rings without removing the ropes, air duct and pipes hanging in the middle part of the shaft.

4.4.3 *Sinking platform (Galloway)*

The excavation of the shaft requires a working platform suspended close to the excavation face thanks to a set of cables. This platform, also called Galloway, is made of 4 different decks, each deck having a specific function depending on the sinking stage (see Table 2). Some openings are provided for the mucking buckets, the ventilation duct, the emergency lift and the diverse pipes. Decks n°1 and n°2 must provide an access to the shaft wall due to their function. Therefore, the diameter of these decks will be enlarged by 2 or 3 m. This can be achieved by fixing retractable platforms (see Figure 8).

In any case, the Galloway is an individual equipment that will be designed by the Contractor. The design will have to be done according to the excavating and lining equipment chosen by the Contractor to sink the shafts, and the different construction phases. Attention must be paid to this design as it the main sinking equipment that will lead the advancement rates.

Table 2. Main functions for each deck of the sinking platform.

Deck n°	Excavation (↓)	Excavation + Lining (↓)	Lining (↑)	
			Steel + concrete	Compressible + concrete
4	Reception of the equipment for the galleries expansion			
3	Dewatering pump	Dewatering pump Concrete receipt	Dewatering pump Concrete receipt	Dewatering pump Concrete receipt
2	Compressors Hydraulic power Shotcrete mixing	Compressors Hydraulic power	Handling/welding of the steel rings	Handling and installation of the compressible elements
1	Sinking machines storage	Machines storage Formwork move-ment Concrete pouring	Mortar pouring Formwork move-ment Concrete pouring	Formwork movement Concrete pouring

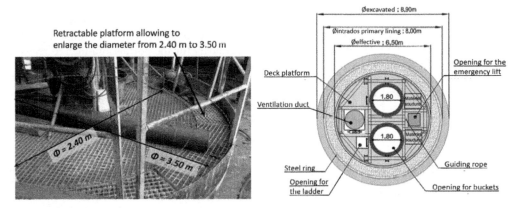

Figure 8. Retractable platform for the Galloway enlargement (Passy shaft) and deck arrangement

5 CONCLUSION

The construction of the five shafts of the project is a challenge in many aspects: these struc-tures are remarkable and complex, there is no recent feedback for such structures, and many constrains must be considered. The design of the sinking equipment will require a great expert-ise regarding the construction methods, but also a deep understanding of the regulations. Finally, the construction planning is short and must be secured by detailed studies because of its impact on the whole underground extension works.

REFERENCES

AFTES. 2008. *Recommandations sur les puits profonds et galeries inclinées.* Tunnels & Ouvrages Souter-rains, mars/avril, pp. 94–126
AFTES. 1996., *Sennecey-Le-Grand Histoire d'un puits.* Tunnels & Ouvrages Souterrains, juillet/ aout, pp. 217–222
Prevotat, M. 1982. *Le fonçage des puits.* INTRAFOR-COFOR

Special intersection formworks for cast in situ with cross vault shape

M. Granata
CIFA S.p.A., Senago, Italy

ABSTRACT: Since 2012 Italy has been involved in a new high-speed line project called "Terzo Valico", which develops the railway connections between the port of Genoa and the main railway lines of Northern Italy and Europe. The Contractor Consorzio COCIV awarded CIFA the design and construction of the formworks for the casting of the intersection areas between the main lining tunnels and the perpendicular cavern tunnels. The two intersection formworks (bolted to the cavern formworks) stand out for their geometric and functional complexity. They are firstly used as a template for the installation of the rebars for the portals and then to realise a monolithic cast in situ with cross vault shape. The casting of these portals allows the subsequent digging of the main tunnel in safety conditions. The technical solution has been considered so fit to be chosen for both Polcevera and Vallemme, Cravasco and Castagnola cavern tunnels.

1   THE PROJECT

Since 2012 Italy has been at the centre of one of the biggest European projects, to upgrade the high-speed and high-capacity railway. This project, called "Terzo Valico", has the target of transferring a substantial share of traffic of goods from the road to the railway.

The project concerns the construction of railway connections between the port of Genoa and the port of Rotterdam, joining the major cities of Northern Italy with Northern Europe. Thanks to this infrastructural work, Italy will be part of the so-called Rhine-Alps corridor (TEN-T core network), which connects the most populated and industrialized European countries.

The project mainly involves underground works for a total length of approximately 36 km and it is developed with two single-track tunnels side by side, connected each other at regular intervals, by perpendicular by-passes, so that each tunnel will also deal as safety way out for the other.

For this layout, as much as 64% will be realized with traditional excavation and casting in situ. The main works of the project "Terzo Valico" concern 3 natural tunnels: the "Galleria di Valico", to be realized with both traditional and mechanized excavation and about 27 km long; the "Galleria Serravalle" to be built on the Piemontese territory and about 7 km long and the "Galleria Campasso", about 716 m long.

The General Contractor of these works is COCIV (Consorzio Collegamenti Integrati Veloci), a Joint Venture constituted by different Italian Companies.

COCIV has awarded to CIFA SpA (the Italian Company of the Zoomlion Group, leader in the design and construction of tunnel formworks) the construction of more than 23 formworks for the lining of the single-track tunnels, the "adits", the widening tunnels, as well as the intersection with the main lines.

Among the main equipment provided along the layout of the "Galleria di Valico", it is worth mentioning in particular: the formworks for the caverns type 2 and 3, equipped with hydraulic stop-ends; the formwork for the interconnection tunnel; the formwork for the Polcevera single-track tunnel, specially designed for the passage of the ventilation pipe; the formworks for the adits, able to overcome longitudinal slopes up to 12.5%, through the use of special retractable slides.

Figure 1.   Cavern formwork type 3.

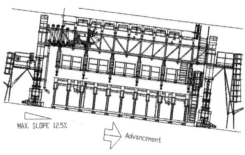

Figure 2.   Formwork for adit.

Figure 3.   Polcevera single-track tunnel form.

Figure 4.   Truck mixer and truck-mounted pump.

Having a full line product range for tunnel works, CIFA has also supplied truck mixers, 4 truck-mounted pumps, 8 Spritz Systems CSS3 and 8 electric truck-mounted concrete pumps.

2   ADITS AND TRACKS FOR THE INTERSECTION WITH THE MAIN TUNNEL

Along the "Galleria di Valico" the project foresees the construction of secondary tunnels, so called "adits" which connect the single-track tunnels to the outside. These adits have a dual operative and safety purpose. In fact, during the construction works, they have to guarantee access to the excavation fronts of the "Galleria di Valico". When in operation, they must deal as access to the railway line with function of service, emergency and safety (e.g. to guarantee an emergency escape route in the event of accidents, as well as the ventilation and the flow of people). All in order to ensure the highest safety standards according to the current Italian and European regulations.

The project concerns the construction of 4 adits at a distance of about 3-5 km from each other: the adit of Polcevera (up-to-now completed), and the adits of Vallemme, Cravasco and Castagnola.

Excavations and coatings have started from the adit tunnels. The latter, in the area of intersection with the line tunnels, widen, becoming circular chambers with a radius of 6.31 m, to be realized in blocks of variable length up to 20 m, according to the job-site specifications.

The enlargement of the tunnel section from adit to cavern is due to technological needs, among which the construction of a "false ceiling" after completion of the work, to position an elevated walkway that allows the staff to bypass the tracks in safety conditions. This enlargement also satisfies the logistical needs of the job-site during the construction tunnel phases, as the circulation of the job-site vehicles in the intersection area becomes easier. For similar logistical needs, even the main line tunnel is transformed into a cavern in the area of intersection with the adit. The crossroads of the two tunnels becomes an enlargement area with the same section on all four arms.

In this area the project involves the construction of reinforced portals made with monolithic cast in situ having a cross vault shape. These portals will allow the yard to safely perform the excavation of the single-track line tunnels, even in presence of loose soil or with little coverage.

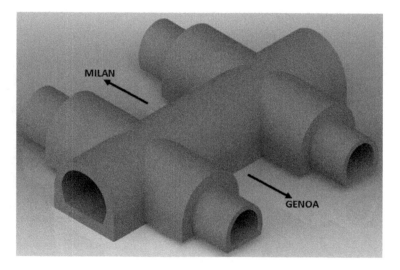

Figure 5.   3D view of the intersection between the adit and the main tunnels.

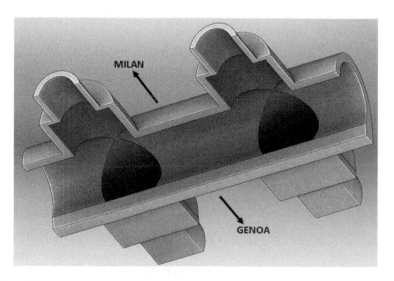

Figure 6.   Cap with cross vault shape in the intersection area - 3D Section.

3 ANALYSIS OF THE EQUIPMENT

The particular type of crossed tunnels has represented a design challenge for CIFA, which has devised and built a complex equipment made of a self-supporting cavern formwork and two intersection formworks, which work together to cast the intersection area having a cross vault shape.

The cavern formwork has a radius of 6.31 m, a height of about 10 m from the sliding surface and it is dimensioned in order to support a maximum concrete thickness of 1.8 m on the cap. The shutter has been realized with modular and flangeable rings, so that it can be used for blocks of longitudinal length equal to 4, 9, 10, 11, 19 and 20 m according to the job-site layout.

The formwork is made up of four vault forms and two piers (each one composed of two bolted elements), moved by means of hydraulic cylinders, in order to facilitate the setting-up and dismantling of the equipment.

For the transport and handling of this formwork, CIFA has devised a special wheeled carrier with diesel engine and turret. With this system it is possible to move a maximum of three shutter rings at a time (total longitudinal length equal to 4.5m). The advantage of such a system is that the turret allows both a transverse and longitudinal adjustment of the formworks, as well as a 90° rotation around its axis, thanks to the revolving turntable of the turret. This means that the forms can be rotated perpendicularly to the direction of tunnel advancement, thus allowing the transfer of the formwork through reduced width sections. This solution has been essential to allow the perpendicular transfer of the shutter from the adit to the main line tunnel.

The equipment is also supplied with a dual system of hydraulic lifting cylinders: four cylinders are positioned on the turret and further four cylinders on the wheeled carrier. This dual lifting system has made it possible to move the shutter without interference with the excavation and to carry out the telescopic movement phases under the set-up formworks.

CIFA has further designed and built two self-supporting steel intersection formworks also necessary for the casting of the reinforced portals. Each of these forms has a radius of 6.31 m, a height of 8.33 m and a total weight of about 10 tonnes. The intersection formworks are composed of two vaults and two piers, each one divided into three bolted elements, to facilitate the transport and the assembly of the equipment. The two piers are connected to each other by means of a horizontal strut. The shutter is supported by four wedge-boxes, which allow the vertical load due to the weight of concrete and to the rebars to be transferred on the ground.

Every intersection formwork is equipped with a steel stop-end, inclined as to the rib, to follow the excavation profile of the single-track tunnel and to guarantee the casting of the portals with cross vault shape, according to the customer's requests. On the vault this

Figure 7. Cavern Formwork during set-up phase. Figure 8. Cavern formwork rings rotated of 90°.

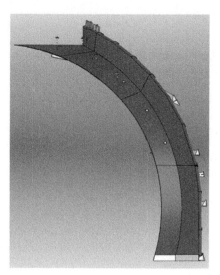

Figures 9 and 10.   Intersection formwork.

Figure 11.   Intersection formwork bolted to the cavern formwork.

stop-end is of tipping type, moved by means of a jack and it is perpendicular to the mantle. On the contrary, on the piers the stop-end is welded to the formwork, reaching a maximum length of 1.4 m at the base.

The complex geometry of the intersection formworks has also required a considerable commitment for the static calculation, performed with an analysis by means of three-dimensional FEM models.

Each intersection formwork has been dimensioned considering the different casting phases both in symmetrical and asymmetric loading conditions, with a concrete pressure equal to 38.5 kN/m^2 on the piers and 80 kN/m^2 on the cap. Particularly, the asymmetrical loads, associated with the concrete pressure acting on the stop-end, showed an anomalous behavior of the constraints compared to a traditional formwork. The anomaly is not due to the reaction values, but to the fact that on changing the concreting level, every constraint would have assumed reactions substantially different in terms of direction, changing from compression to tension and vice-versa. Having had only compression-resistant one-sided constraints, this effect would have meant that in every load case only some of the constraints would have been really effective for the structure. In order to limit the effects of the concrete pressure, to reduce the number of constraints and to guarantee the stability of the structure, one horizontal steel strut was introduced between the two piers. In addition, some props have been located at the base, as well as six wooden struts, which have been constrained to the excavation, in order to transfer the concrete pressure of the stop-end.

4 ASSEMBLY OF THE EQUIPMENT AND OF THE REINFORCED REBARS OF THE PORTALS

In order to meet the design requirements of the construction site, CIFA evaluated and agreed with the customer a procedure to perform the operations of setting-up and casting in a functional and safe way. For this reason, the equipment supplied was firstly used as a template for the installation of the reinforced rebars of the portals. Once the setting-up of the rebars was carried out, the equipment could be used to realise the monolithic cast of the portals with cross vault shape.

The assembly of the equipment was carried out by CIFA in the casting area inside the tunnel. During the design phase, it was necessary to evaluate the problems related to the dimensions of the various formwork elements, since they had to pass through the already cast adit section.

The first challenge that CIFA had to face concerned the assembly times that had to be minimized, to start the excavation and the casting of the single-track sections in the shortest possible time. Once the vaults were assembled, the revolving turntable of the turret could rotate the structure of 90°, thus allowing the assembly of the piers without space limits.

A further aspect that CIFA focused was to guarantee the stability of the intersection formworks during the assembly phase. Because of their irregular geometry, in fact, the center of gravity is located outside the supporting base. For such a reason, it was planned to bolt the intersection formworks to the cavern formworks. Precisely, the intersection vaults were bolted to the two central rings of the cavern shutter (3 m long) which were rotated of 90° as to the advancement of the tunnel. This allowed the operators to perform the assembly without space limitations.

The assembled equipment was then rotated by 90°, in order to place it in casting position and the piers of the intersection formworks were bolted to the above mentioned vaults.

Figures 12 and 13. Set-up phases of the intersection formwork on the cavern formwork.

Since the first operation to be carried out was the setting-up of the reinforcement rebars of the portals, CIFA devised an assembly solution for the rebars to satisfy the operative requirements of the construction site.

Once the 3 m long cavern formwork with the two intersection formworks had been set up, at the two ends of the 20 m long block, one formwork ring on each side was positioned (rings A and B of Figure 14). The reinforcement rebars of the portals could be installed on the equipment made up as described. The great advantage of such a solution is that the operators and the construction vehicles could move easily in the free space between the central rings and the side rings A and B.

After positioning the rebars of the portals, it was then possible to complete the assembly of the cavern formwork with the missing rings, in order to proceed with the subsequent casting phase.

The dismantling phase also required careful planning by CIFA. In this case, the cavern formwork (20 m long) was completely dismantled, moving it to an area far from the casting one, leaving the two intersection formworks set up. Since the center of gravity of the intersection formworks is located outside the supporting base, CIFA identified a technical solution allowing the equipment to remain stable after casting without the cavern shutter. This was achieved by placing anchors on the vaults let into concrete, thus allowing to keep these intersection formworks stable until completely disassembled.

Figure 14.   Assembly solution for the set-up of the rebars of the portals.

Figure 15.   Setting-up of the rebars.

Figure 16.   Intersection area cast with cross vault shape.

Figures 17 and 18.   Dismantling of the intersection formworks.

5   CONCLUSIONS

The equipment designed by CIFA was used for the first time in May 2016 for Polcevera cavern tunnel. The customer considered this technical solution so fit to choose it also for Vallemme cavern tunnel. To date, the equipment supplied by CIFA will be reused also for Cravasco and Castagnola intersection caverns.

This project demonstrates how the tunnelling field with casting in situ is offering still today new design challenges, requiring the highest standards of efficiency and safety, as well as innovative solutions CIFA has achieved with this prototype equipment for the monolithic cast with cross vault shape.

ACKNOWLEDGEMENTS

Thanks to the staff of Polcevera and Vallemme construction sites, with whom there has been a close collaboration during the development of the project and the commissioning of the equipment.

REFERENCES

www.terzovalico.it (last time visited 04/05/2018)
www.stradeeautostrade.it/cementi-e-calcestruzzi/pianeta-cassero (last time visited 04/05/2018)

*Tunnels and Underground Cities: Engineering and Innovation meet Archaeology,
Architecture and Art, Volume 7: Long and deep tunnels – Peila, Viggiani & Celestino (Eds)*
© 2020 Taylor & Francis Group, London, ISBN 978-0-367-46872-9

Extreme ingress: Managing high water inflows in hard rock TBM tunneling

B. Grothen
The Robbins Company, USA

D. Kough
Kiewit, USA

ABSTRACT: Managing water inflows is not new to TBM tunneling, but today there are an increasing number of methods and best practices to handle potentially high water inflows efficiently and safely. High volumes of water can be safely contained or managed in hard rock TBM tunneling, but this requires the proper foreknowledge and planning. This paper will outline how machines can be designed ahead of time for expected high water, and how risk can be mitigated during tunneling. We will also cover the importance of pre-planning and include a look into the future of water control methods. Case studies of hard rock tunneling with heavy water inflows will be examined, with a focus on New York, USA's Delaware Aqueduct Repair. The 3.8 km long bypass tunnel below the Hudson River requires excavation through limestone rock at water pressures of up to 20 bar. A unique 6.5 m diameter Single Shield TBM, sealable for high pressure excavation, is boring and lining the tunnel.

1 INTRODUCTION

Probe drilling and pre-grouting is an essential part of drill and blast tunneling and the most important defense against weak rock mass and water leakages in D&B tunneling. The first use of pre-grouting in a TBM application was on the Oslo sewer tunnels from 1977–1981 in Norway. Due to the risk of ground settlement the project owner required a system on the TBM capable of extensive pre-grouting in order to qualify any bids. The solutions chosen by the contractors and machine manufacturers were mainly to place a drill jumbo behind the TBM, guide the drill string through the TBM and bore through the cutterhead. This proved to be a bad solution and eventually the chosen contractor decided to force the drilling rod into the tunnel wall behind the grippers. The project was successful at limiting the water ingress, but the extensive pre-grouting limited the advance rate.

Since this time many machines and projects have faced similar circumstances, where the systems or procedures that were in place were less than optimal for advancing in the difficult conditions. In some particularly difficult projects heavy modifications or changes needed to be put into place to help the projects advance. During the process a better understanding of what was needed and the tools to complete the job were developed that would make for more effective mechanized tunneling. Some projects of particular note and in no special order are:

- Arrowhead Project, California USA
- Kargi HEPP, Turkey
- Hallandsas, Sweden
- Gerede, Turkey

A machine feature proven particularly valuable by these projects is the ability to seal the machine from water flows, both at the heading and behind the machine using sealed segments.

Drilling and grouting is made more effective by monitoring the drilling parameters, monitoring while drilling (MWD), probing longer distances, and supporting an increased grout pattern. These methods, along with new kinds of grout and improved machine systems like auxiliary thrust, variable speed cutterhead torque, and dewatering systems are integrated together so that machines can more effectively manage the potential for high water inflows.

An ongoing project in the US, the Delaware Aqueduct Repair, is a prime example of the use of a machine that has brought these features together, and some new ones, to address the geology.

2 CASE STUDY: DELAWARE AQUEDUCT REPAIR

In 1990 a utility worker at a power plant along the Hudson River found water shooting up from the ground in an area where a large water line ran. At that time, the New York City Department of Environmental Protection (NYCDEP) was treating the water in the Delaware Aqueduct with copper sulfate, which was used to prevent algal growth at some of the city's reservoirs. The water samples tested positive for copper sulfate, and as the NYCDEP were the only ones using copper sulfate in the area there was good reason to believe it was coming from the Delaware Aqueduct.

At 137 km long, the Delaware Aqueduct is cited in the Guinness Book of World Records as the world's longest continuous tunnel. Since 1944 the Delaware Aqueduct has supplied water to New York City and today it accounts for more than 50% of the water supplied. The Delaware Aqueduct is a gravity-fed water supply line of 4.1 m diameter that conveys water from several reservoirs in the Delaware System and across the Hudson River. The tunnel was constructed using drill and blast methods and most of the rock was competent ground made up of schist and shale, but as the teams approached the Hudson River, the ground gave way to faulted limestone. Work crews documented groundwater inflows of 7.5 to 15 million liters day into the tunnel, and they invented some creative solutions to deal with it. The ground was lined with concrete, and then a steel interliner was placed in the tunnel before a final concrete lining was set on top, effectively creating a sandwich of concrete with steel inside. While the method was groundbreaking for the time, the steel liner ultimately did not extend far enough through the troublesome rock formation to prevent leaks.

For more the 25 years leakage has been monitored in the section that connects the West Branch Reservoir to the West Branch Tunnel. It was determined that approximately 132 million liters per day was lost with approximately 95% coming from two sections of this tunnel, the Roseton and Wawarsing areas, which are just outside of the lined section. Compared to the overall flow this was not a critical amount, but the location and nature of the leaks was a cause of concern. The repair would be anything but simple. To address the issue it was decided to bypass the Roseton and Wawarsing areas with a newly constructed tunnel, the Rondout West Branch Bypass Tunnel (RWBT).

The project was separated into two distinct phases and two separate contracts were tendered, BT-1 and BT-2. The first phase, contract BT-1, consists of completing the shaft sinking to a depth of approximately 275 m at shaft 5B, near the town of Newburgh and 213 m at shaft 6B near the town of Wappinger. The excavation connecting the shafts is approximately 3,810 m of tunnel at a diameter of 6.58 m. This phase also includes the installation of 2,800 m of 4.87 m diameter steel interliner pipe through the new bypass tunnel with cast-in-place concrete liner for a finished diameter of 4.27 m. Access chambers at the top of shafts 5B and 6B will be constructed for access and housing of the mechanical and electrical equipment for supporting pumps and valves. Figure 1 shows the location and layout of the project.

The second phase, BT-2, includes an additional 30 m of excavation from shafts 5B and 6B to the existing tunnel and will be completed during a scheduled shutdown of the main tunnel. Additionally, construction of the permanent plugs within the RWBT will be undertaken. Figure 2 shows the layout of the work. Access to the work is through the deep shafts.

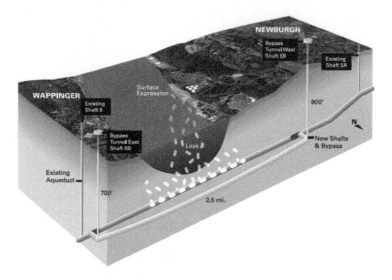

Figure 1. Project location and layout.

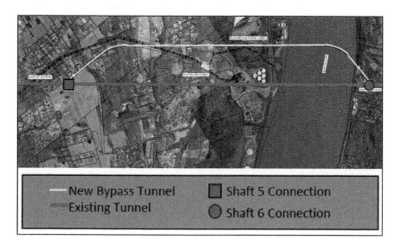

Figure 2. Shaft and connecting tunnel locations.

3 SIGNIFICANT PROJECT ASPECTS

The geologic formations encountered on the project mainly consist of the Wappinger group (dolomite, dolomitic limestone and limestone). At the beginning and the end the tunnel excavation will be in the Normanskill and Mount Merino Formations, respectively, which are made of slatey shale, argillite and sandstone. The GBR states that there will be some areas of extremely hard rock with UCS values as high as 372 MPa, but averages in the Wappinger formation are a more reasonable 241 MPa with low to medium boreability ratings.

From Figures 1 and 2 it can clearly be seen that the tunnel alignment passes under the Hudson River at a depth of 183 m, having a width of 1,052 m at that location. The groundwater head along the rest of the tunnel ranges from 267m on the west to 213 m on the east side. The most challenging tunneling conditions are anticipated to be the high conductivity of the Wappinger formation with the fault zones in the Normanskill Formation. Per the owner's tender documents, the selected contractor had the option to excavate the tunnel between the shafts with either a TBM or drill and blast (D&B). The protential for high water inflows combined with high head conditions, due to the depth and large source, made water control a driving factor. To handle these potential inflows the owner prescribed probing ahead and intensive pre-excavation grouting. These methods were introduced as requirements and were intended to lower the groundwater inflow potential of the rock mass, not necessarily to increase its strength or improve stability. The difficulties of water control are compounded by the fact that the tunnel is being bored downgrade and ground water will need to be pumped to shaft 5B and on up to the surface more than 267 m above.

While still significant the TBM method of excavation had lower dewatering requirements of 9500 l/min. The contractor, a Joint Venture between Kiewit and J.F. Shea Construction, determined that mechanized excavation would be the most economical and safest solution to control ground water when compared to D&B multi-stage excavation methodology.

The gasketed segmental lining installed behind the TBM would help to minimize the water inflow, however these geological challenges at the face in concert with high water pressures (over 20 bar), required the TBM to be designed with many special features for drilling, pre-excavation grouting and dewatering.

4 TBM DESIGN

To ensure the best chance of project success the NYCDEP and Kiewit/Shea were heavily involved in the specification and design elements of the TBM. The basic Robbins Single Shield TBM specifications and features are listed below with the TBM general layout shown in Figure 3.

- Bore Diameter 6,583 mm
- Cutterhead Speed 0–8.8 rpm
- 9 x 330 kW Drive Motors
- Pressure compensated 19-inch disc cutters
- Sealable TBM design – 30 bar
- Large dewatering system capacity
- Mucking via muck train
- Forward and aft drill positions
- Drill-in ports in cutterhead and forward shield
- Dedicated dewatering ports to dewater from the heading
- Water handling at conveyor discharge

4.1 Special TBM Features

The owner's specifications concerning water drove much of the TBM design in water pressure and flow as well as degree of grouting that needed to be planned for on the project. The

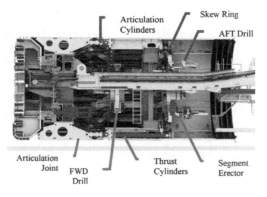

Figure 3.   General TBM arrangement.

priority was placed on sealing the TBM from the high pressure/high flow water at the face, but the presence of water was also taken into account for the design of the dewatering system. This was because efficiency in managing the muck-laden water inflows from the heading was also important. KSJV worked closely with Robbins to develop and install these systems and optimize the planned operations as well as others on the machine not closely related to the expected water. Some examples of this include the cutterhead, which was designed to prevent clogging in shale, and the design around segment handling that included a decked back-up and rapid segment unloading system.

4.2 Design Considerations for High Water Pressure

The owner also required that the TBM be capable of withstanding 30 bar of hydrostatic pressure (20 bar with a 1.5 safety factor). Due to the 20 bar static water pressure expected on the project, a new main bearing sealing system was engineered for the project and is comprised of multiple rows of traditional lip type seals and emergency inflatable seals. The inflatable seals are not in running contact with moving parts of the sealing system during boring but can be activated like a "parking brake" when needed for additional protection of the TBM's main bearing. The seals are flushed and lubricated with grease for protection when exposed to the fines-laden water expected on the project. As such, the use of pressure compensated disc cutters became a necessity. Their unique design incorporates a pressure equalization system to keep water out and protect the bearings when the pressure is high.

4.3 Design Considerations for High Water Inflow

The TBM was designed to be quickly sealed to protect the TBM and personnel from sudden inrushes of water. The steps listed below are required to seal the TBM during high inflows and are shown in Figure 4:

1. Close knife gates over muck chute
2. Retract conveyor frame
3. Retract belting out of cutting chamber
4. Retract bulkhead sealing plate
5. Close stabilizer doors

Due to the water inflows measured during the geotechnical investigation, the contract specifies a very high dewatering capability. To achieve this, dewatering is controlled by two separate systems that can work independently or in concert:

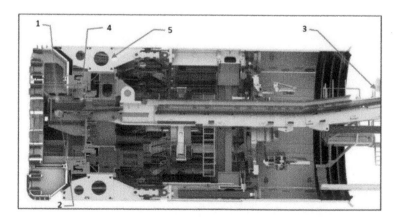

Figure 4. TBM sealing sequence.

- Primary Dewatering – 0–3,000 l/min during boring without impact on boring or ring building
- Emergency Dewatering – 9,500 l/min emergency capacity for water from heading and construction water

The primary dewatering system can collect water from the cutterhead chamber, shields/ring build area and the transfer point between machine conveyor and transfer conveyors. The system is designed to transfer fines up to 7 mm in diameter through the piping, tanks and on to the tunnel dewatering system. The TBM is equipped with 20 m³ of dewatering storage capacity through the combination of two, 10 m³ tanks. Each tank is fitted with mixing pumps to prevent settlement of fines and reduce shutdowns for maintenance and cleaning. Since the inflow of water is variable and the required size of the pump is quite large the primary dewatering pump is controlled with a VFD to effectively manage lower water flow events.

The emergency dewatering system bypasses the dewatering tanks and transfers water directly to the tunnel dewatering system. A telescoping pipe extender on the back-up allows the TBM to advance 6m before adding another 305 mm tunnel line. Figure 5 shows a general system schematic of the dewatering system.

4.4 Drilling and Grouting Methodology

The project specification requires a mandatory probe drilling program for the entire tunnel alignment that includes water inflow measurements at the probe hole locations. The TBM is required to drill four probe holes every 60 m to measure water inflows. When water inflows exceed contract-allowable values, grouting will be required to reduce water inflows to acceptable levels. The TBM can then advance inside the grouted area of the alignment. The TBM is equipped with two types of grouting systems. The pre-excavation grouting (PEG) system is a mono-component grout system used to grout ahead of the TBM. The two-component (A+B) grout system is used to backfill the annular gap between the segmental lining and the bored tunnel. A detailed drill scheme to allow drilling under pressure was developed during the TBM design phase and has been further refined during boring operations.

The first step in the probe cycle is to rotate the cutterhead into either the A or the B positions to align the eight associated ports in the forward shield and locations in the cutterhead. A 127 mm diameter casing is then installed through the shield up to the face. The casing is in 1.22 m long sections. A blow-out preventer (BOP) installed on the forward shield port is inflated to lock the casing in place.

The drill—a down-the-hole hammer, the W70 manufactured by Wassara, with a bit diameter of 104 mm—is then used to collar the hole and drill 2.74 m deep. The 127 mm casing is then removed and a drill-through inflatable packer is installed. The packer is made up of five sections, which are screwed together. Once installed the packer is inflated with water to 20 bars. The BOPs are also inflated again around the packer. At the end of the packer there is

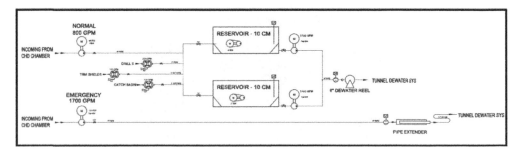

Figure 5. Dewatering schematic.

a 3-inch ball valve, a T with a hose for cuttings, and then an additional BOP, which is installed after the hammer and centralizers. Once the hammer is to the face drilling can commence. Drilling out with the probe drill requires 130 sections of 1.5 m long drill steel, hoses to bring cuttings out of the shield and filter bags for drill cuttings. Drilling out ahead of the machine is done using Wassara W50 DTH hammers. The contractor has been drilling around 91 m with the cuttings coming back and being diverted through the T in front of the BOP and back to filter bags in the bridge area. Each probe hole produces around 0.75 m^3 of cuttings and requires 5 to 6 hours to complete (see Figure 6).

4.5 *Drilling Systems*

The TBM is equipped with two drill systems for probing and grouting operations. The forward drill system is used for drilling operations through the tunnel heading at angles of 0 degrees and up to 5 degrees measured relative to the tunnel alignment. The drill system consists of two independent drill positioners mounted on a fixed ring that can position each drill 360 degrees radially to drill and grout through 16 cutterhead drill ports. This system is the primary drilling and grouting system used on the TBM and will be used to probe and grout along the tunnel alignment.

The aft drill system is a single drill permanently mounted to the segment erector. The segment erector is used to position the drill for drilling and grouting operations through 14 peripheral shield ports. The ports are at seven degrees to the tunnel alignment and are used for umbrella drilling and grouting to form a grout curtain, thus cutting off water inflows surrounding the tunnel alignment.

The down-the-hole water hammer is ideally suited to this project for the following reasons:

- Down-the-hole drilling reduces size of drill equipment inside TBM
- Water used to power drill also provides flushing of cuttings
- Drills longer, straighter holes compared to top hammer type drills
- Water used to power the drill will not erode borehole
- Core drill units are relatively short and compact, making them better suited to fit inside TBM
- API drill rods are used in drill string to power drill with high pressure water

Drill testing completed with water hammer drills near the jobsite area verified the drill performance and suitability for the project.

The bridge area of the TBM was also designed to allow for radial drilling with a portable drilling platform in order to verify backfill grouting. Forward shield drilling/grouting and dewatering ports are shown in Figure 7.

Figure 6.   Drilling with pressure.

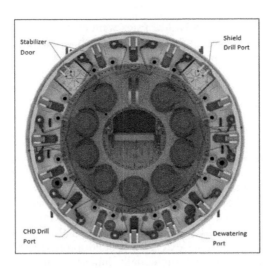

Figure 9. Forward shield drilling and dewatering ports.

4.6 Grouting Systems

The TBM is equipped with two PEG mixing and grouting plants for grouting ahead of the TBM. In addition, the PEG system can be used for proof grouting the segmental lining to counteract high water pressure and to mix and inject bentonite around the TBM shields to reduce friction in squeezing ground conditions.

The A+B grout system is supplied from a batch mixing plant at the surface and pumped directly to the TBM to help simplify the already-complicated shaft logistics. A+B storage tanks on the TBM back-up are equipped with level sensors that start and stop pumps at the contractor-supplied surface grout batch plant. This ensures grout is available when required for grouting operations on the TBM.

Due to high static water pressure the contractor required the TBM to have the ability to backfill grout through the TBM tail skin, which is more common on machines that excavate in soft ground to prevent subsidence. Although not an issue on this project this was done due to the concerns with the grout penetration plugs becoming dislodged at high velocities, which could endanger worker safety.

5 TUNNEL EXCAVATION

The TBM is currently in 1,219 m with a best month of production being 228 m. Reach 1 is composed of Normanskill Group shale and is around 762 m in length; the GBR in this section indicated max heading inflows of 946 l/min. During excavation two 91 m long probe holes with a 6 m overlap were drilled and no water was encountered.

Reach 2 is composed of a dolomitic limestone and is part of the Wappinger Group and is 1189 m long with anticipated steady inflows of 6.21 l/min/meter and up to 3,785 l/min during probing. This reach requires four probe holes per cycle and with half of the reach completed no water has been encountered.

There have been challenges on the project, including the extensive probe drilling and the number of required cutter changes. As expected, mucking out the TBM from the bottom of a 274 m deep shaft while delivering segments and supplies slows down the mining process. Each TBM stroke is around 100 m^3 of muck, which fills seven muck boxes. Currently the project is running two trains with a service loci to assist in loading segments and material. The probe drilling has been a challenge due to the setup required to handle potential high water/pressure inflows. The tight space also allows only 1.5 m drill steels to be used, which means 60 steels per hole. Mining advance rates went as expected during the softer shale of Reach 1 with limited cutter wear/changes. During reach 2 there has been much greater than expected cutter wear and advance rates have been cut in half.

6 CONCLUSION

All tunneling projects come with challenges that need to be overcome with water inflows being one of the primary concerns that needs to be considered. Three key areas in managing this risk are isolating the tunnel from the water with the use of grouting, preventing the water from entering the tunnel with the use of a sealed machine and lining, and finally managing the water once it has entered the tunnel with a dewatering scheme. How these different elements are brought together with the rest of the machine features can make a successful project even in difficult conditions. The Delaware Aqueduct Repair is no exception, requiring several innovative solutions with water driving much of the design. This includes drilling and grouting systems that are able to support an extensive ground improvement scheme. The machine's ability to handle high-water pressures is equally important and requires the TBM to have many features typically associated with pressurized face TBMs. Muck removal is, however, still done using a belt conveyor to minimize wear and maintenance due to the hard rock. Finally, the planning and integration of a dewatering system helps manage the water if it does make it to the tunnel heading. Few projects have all the same constrains, but these features along with others reviewed in the paper are valuable tools that with the right collaboration and planning can be applied to make for more successful future projects.

REFERENCES

Willis, D. 2017. *The Big Repair*. Tunneling Journal North America, April/May 2017: 26–30.
Terbovic, T., Scialpi, M., 2017. *Rondout West Branch Bypass Tunnel*. Rapid Excavation and Tunneling Conference 2017 Proceedings.
Brennan, D., Brennan D., 2018. *High-Capacity Hoisting at Rondout West Branch Tunnel Project*. North American Tunneling Conference 2018 Proceedings.
Dowey, T., Kharivala, B., and O'Connor, P., 2014. *Rondout West Branch Bypass Tunnel Construction and Wawarsing Repairs, Delaware Aqueduct Bid Set*. New York City Department of Environmental Protection.

Tunnels and Underground Cities: Engineering and Innovation meet Archaeology,
Architecture and Art, Volume 7: Long and deep tunnels – Peila, Viggiani & Celestino (Eds)
© 2020 Taylor & Francis Group, London, ISBN 978-0-367-46872-9

Numerical investigations on the system behaviour of a ductile shotcrete lining with yielding elements

A.-L. Hammer & M. Thewes
Institute for Tunnelling and Construction Management, Ruhr University Bochum, Germany

ABSTRACT: The choice of a suitable lining concept is elementary in conventional tunnelling under squeezing ground conditions. The use of yielding elements to create a targeted flexibility of the shotcrete lining is state of the art. The defined flexibility results in an overall different kinematic load-bearing capacity of the shotcrete lining. The functional relationship between the time-dependent material behaviour of the shotcrete and the load-dependent deformation behaviour of the yielding elements is decisive for the optimization of the lining concept. The aim of the work is to analyse the relationships between the yielding elements and the system behaviour of the shotcrete lining. With a three-dimensional numerical model, the kinematic behaviour of the overall system is analysed, with a focus on the interaction between ground, shotcrete lining and yielding elements. The influence of different ground and system parameters on the load acting on the ductile shotcrete lining is investigated.

1 INTRODUCTION

In case of squeezing ground the conventional support method with shotcrete, rock bolts and steel arches cannot absorb the rock pressure. This can result in spalling or even destruction of the shotcrete lining. The shotcrete lining in combination with yielding elements is internationally state of the art and proved to be an economical and safe alternative to a rigid lining with a monolithic shotcrete lining (Schubert 1996, Budil et al. 2004, Barla et al. 2008). Yielding elements installed in deformation slots of the shotcrete lining allow controlled rock deformation and reduce the load acting on the young shotcrete. The defined yieldingness leads to a divergent kinematic overall bearing behaviour of the shotcrete lining, which has to be investigated in more detail. In addition, a yielding support concept requires a revision of existing calculation models, taking into account the latest material developments, and a uniform analysis of the load-deformation behaviour of yielding elements. (Radoncic & Schubert 2011, Hammer 2018)

The analysis of the system behaviour and in particular the displacement prognosis of the rock mass is the basis for the design of the tunnel lining. A realistic estimation of the rock mass behaviour in interaction with the planned lining is therefore required for a technically reasonable and economic dimensioning. In the case of a shotcrete lining with yielding elements, there is a functional relationship between the time-dependent material behaviour of the shotcrete and the load-deformation behaviour of the yielding elements. This relationship should be taken into account when investigating the overall kinematic system; no further details are given in the norms and recommendations. There are approaches in the literature for the implementation of the shotcrete lining with yielding elements, but the functional relationship between the time-dependent material behaviour of the shotcrete and the load-deformation behaviour of the yielding elements has so far only been considered in isolated cases. However, the interaction between the shotcrete and the yielding elements is decisive for the support concept. (Panet et al. 2001, Wittke et al. 2012)

Therefore, this publication analyses the effects of the squeezing ground behaviour on the load-bearing system. First, the influence of the geotechnical boundary conditions in squeezing ground on the shotcrete lining with yielding elements is investigated and then the most important influencing factors are identified by carrying out a variation study with a plain strain numerical model. Based on these results, the behaviour of the kinematic load-bearing system is analysed on the basis of the identified influencing factors. In addition, the stress and deformation history at the interaction surface between rock mass, shotcrete and yielding element will be investigated. The numerical models are described in detail by Hammer & Thewes (2018) and the analysis by Hammer (2018).

2 TUNNELING IN SQUEEZING GROUND

2.1 Squeezing ground

Squeezing behaviour is a "time-dependent large deformation occurring around the cavity" as defined by the International Society of Rock Mechanics (ISRM). This ground behaviour can be observed in rock mass types with creep behaviour, low strength, and high ductility. The ground convergence due to squeezing behaviour may occur during excavation or continue for a long period. The geotechnical factors influence the squeezing behaviour of rocks including rock mass properties, groundwater table, pore water pressures and in-situ ground stresses. The convergence's magnitude and rate, as well as the size of the plastic zones around the tunnel excavated through squeezing ground, depend on the influencing geotechnical properties. (Barla 2016)

The squeezing ground behaviour requires that yielding elements must be designed for a corresponding load-deformation behaviour. In order to be able to determine and evaluate this, a precise knowledge of the geotechnical factors influencing the squeezing ground behaviour is required. (Hammer & Thewes 2018)

2.2 Support concept of a shotcrete lining with yielding elements

In addition to the slotted shotcrete lining, in which yielding elements are inserted, the lining concept consists of radial rock bolting and steel arches or lattice girders. The stiffness of the yielding elements enables the mobilization of the lining resistance by utilizing the load-bearing capacity of the young shotcrete while initial deformations occur. The initial force level of the yielding elements can be selected variably and the force level increases in the process of deformation.

The requirement for this is that the load on the shotcrete lining must be below the time-dependent shotcrete strength at all stages. In particular, the initial resistance of the yielding elements must therefore not be too high, as the young shotcrete has a low initial strength. In the further process, the load-deformation behaviour of the compression elements should be based on the strength development of the shotcrete in order to make the best possible use of the mobilized support capacity. (Wiese 2011, Radoncic 2011, Hammer 2018)

In practice, different forms of the load-deformation behaviour of the lining with yielding elements were observed, which are compared in Figure 1. In case A, the yielding elements at the interaction surface (highlighted in grey) cause concentrations of the tangential displacements and high relative displacements between rock mass and shotcrete (slip). This has a decisive influence on the distribution of the internal forces and the general mobilization of the lining resistance. The rock bolts (highlighted in blue) are subjected to high shear stress due to the convergences that occur (Pöttler 1997). The radial strains can lead to a rapid loading of the rock bolts beyond the elastic limit (John & Mattle 2007). The large tangential displacements in the lining also cause considerable settlement of the top heading feet (highlighted in green), which results in them cutting into the underlying rock mass and the lining being practically punched into the rock mass (Radoncic & Schubert 2011).

In case B, the displacements occur uniformly over the edge of the excavation. The rock mass including the lining deforms into the cavity, whereby the yielding elements allow a radial shortening of the lining. The reduction of the load acting on the lining by the yielding elements is lower in this system behaviour. In addition, the general load on the lining is higher. Here also comes to the concentrations of tangential displacements at the yielding elements (highlighted in grey), but these are lower due to the uniform rock mass deformation. The normal force load of the rock bolts (highlighted in blue) is higher in this case and can lead to a full

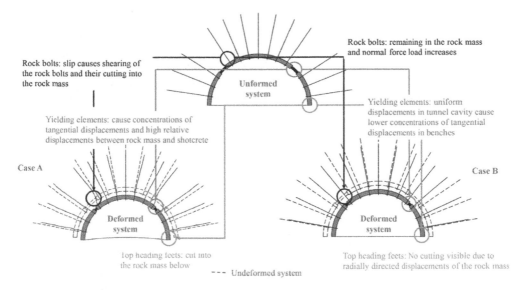

Rock bolts: remaining in the rock mass and normal force load increases

Rock bolts: slip causes shearing of the rock bolts and their cutting into the rock mass

Unformed system

Yielding elements: uniform displacements in tunnel cavity cause lower concentrations of tangential displacements in benches

Yielding elements: cause concentrations of tangential displacements and high relative displacements between rock mass and shotcrete

Case A

Case B

Deformed system

Deformed system

Top heading feets: cut into the rock mass below

Top heading feets: No cutting visible due to radially directed displacements of the rock mass

--- Undeformed system

Figure 1. Variants for the load-bearing behaviour of the overall system based on Radoncic & Schubert (2011) (Hammer 2018).

utilization of the load capacity. Due to the radially directed rock mass displacement, no cutting of the top heading feet (highlighted in green) could be detected (Hammer 2018).

The observed deformation figures seem to result from different loading situations. In case A high tangential displacements occur, in case B the displacements are radial. The deformation figure could be correspondingly dependent on the existing stress conditions in the rock mass, case A would occur with small horizontal stresses and case B with uniform loading around the cavity. Which case occurs under which boundary conditions is therefore investigated in the present work.

2.3 *Numerical calculation of a shotcrete lining with yielding elements*

The support concept of the shotcrete lining with yielding elements is determined by the interaction of these two elements. When implementing the elements in the calculation models, it is a central task to accurately map the initial stiffness of the yielding elements and at the same time to describe the time-dependent strength and stiffness development of the shotcrete as realistically as possible. If the initial stiffness of the yielding elements is set too high, this leads to an early overload of the shotcrete lining.

John et al. (2004) present numerical calculations in the planning of the Strenger Tunnel, in which the yielding elements are determined by limiting the normal force and the bending moment in individual bars. Brandtner & Lenz (2017) implement the yielding elements using beam elements with a nonlinear force-displacement relationship. Both Radoncic (2011) and Barla et al. (2011) use a perfectly plastic Mohr-Coulomb material for the modelling of the yielding elements by determining stiffness and strength values from the load-deformation behaviour of the elements. Further numerical calculations with implementation of yielding elements are shown by John & Poscher (2004), Cantieni (2011) and Likar et al. (2013), however the modelling of the elements is not explicitly described here.

3 INFLUENCE OF GEOTECHNICAL BOUNDARY CONDITIONS IN SQUEEZING GROUND ON THE SHOTCRETE LINING WITH YIELDING ELEMENTS

3.1 *Description of the selected two-dimensional calculation model*

The parameter study is carried out on a finite element model in the plain strain state, which was created with the DIANA software. The implementation of the essential components of the ductile support in the calculation model is described in depth by Hammer & Thewes (2018) and Hammer (2018). For modelling of rock mass, the Hoek-Brown failure criteria was used. Creep effects of the ground were neglected in this model. The shotcrete was modelled linear-elastic, whereby the stiffness was adapted in different time steps. After excavating the top heading, the shotcrete lining with its initial stiffness (1 d) was installed. The construction sequence was modelled considering a phase analysis based on nine phases and the depending phase factors were adjusted to reproduce the field measurements.

In detail, the modelling of the geotechnical boundary conditions and the tunnel geometry as well as model dimensions, mesh geometry and the modelling of the support are dealt with. Furthermore, the simulated construction stages and the individual calculation steps are explained. Figure 2 shows the numerical 2D model with which the parameter study was carried out.

3.2 Parameter study on the influence of the geotechnical boundary conditions on the system behaviour of the ductile shotcrete lining

Squeezing ground behaviour can occur in a differentiated way. On the one hand, this depends on the chosen advance and support concept and on the other hand on the geotechnical boundary conditions. A parameter study is used to numerically determine the influence of the geotechnical parameters on the load on the ductile shotcrete lining. For this purpose, the validated numerical model is used, whereby the input values are changed and roughly adapted to the mean values of the variation parameters (Figure 3, values highlighted in green).

The parameters and the selected boundaries are shown in Figure 3, where only one parameter is varied at a time. Other factors, such as the influence of tectonic forces, could be taken into account by means of a discrete simulation of the joints or faults. However, since a homogeneous model is chosen for the rock mass and a discrete installation of the interfaces or possible disturbances is not possible due to a missing information, this factor is not investigated. The influence of the unit weight of the rock mass is estimated to be less effective due to similar sizes or the low bandwidth of individual hard rocks.

3.2.1 Influence of geological factors
Below are the influences of geological factors:

- Modulus of elasticity of the rock mass E,
- Lateral pressure coefficient k_0,
- Geological Strength Index GSI,
- Hoek-Brown strength parameters m_i,
- Uniaxial rock strength σ_i

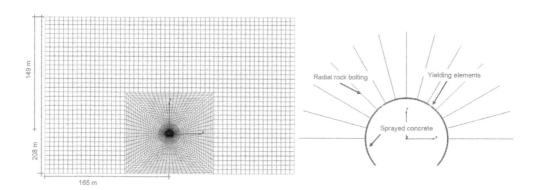

Figure 2. Discretization of the 2D-model (left) and support elements in the 2D-model (right) (Hammer & Thewes 2018).

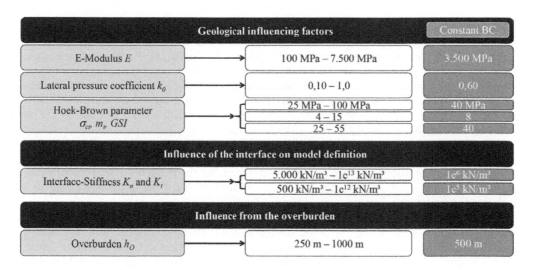

Figure 3. Input parameters for the numerical parameter study, variation range and constant output conditions according to Hammer (2018).

on the stress distribution in the rock mass, the internal forces in the shotcrete lining and the displacement development in chosen reference points. Figure 4 shows the displacements in reference points as a function of the modulus of elasticity. In addition, three qualitative deformation curves around the cavity are shown as examples. In rock mass with a low modulus of elasticity, large displacements occur which decrease with increasing modulus of elasticity. The loosened zone around the tunnel becomes increasingly smaller with a higher modulus of elasticity, the displacements only occur around the tunnel, as can be seen in the comparison of the deformation curves for E = 100 MPa and E = 7,500 MPa. A better load transfer in the rock mass is achieved by larger stress redistributions with a higher modulus of elasticity. The magnitude of stress around the tunnel is higher with a low modulus of elasticity and the redistribution area is smaller than with a high modulus of elasticity.

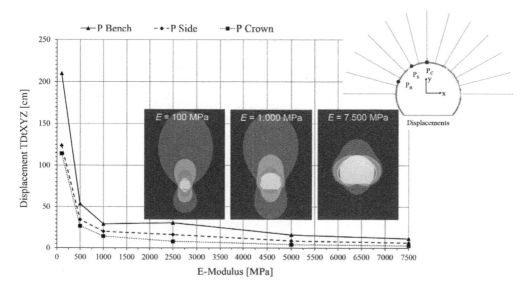

Figure 4. Displacement development in the reference points as a function of the modulus of elasticity E (Hammer 2018).

Figure 5 shows the stress distribution σ_y in the rock mass in the crown, the bench and in the area of the yielding elements as a function of the distance d. Therefore, the vertical stress is plotted both in the contour line plot and in the charts. In the bench, the stresses are highest in comparison. In the case of larger E-modules, the redistribution takes place at greater distance from the tunnel, while the initial stresses are smaller. The redistribution area of the stresses in the crown is more significant; after nine meters, there is no uniform stress level for the different moduli of elasticity. In the area of the yielding elements, the redistribution takes place with higher moduli of elasticity further away from the cavity.

In summary, it can be stated that the modulus of elasticity of the rock mass has a significant influence on the squeezing ground behaviour. With a low modulus of elasticity, high displacements and internal forces occur in the shotcrete lining. As the modulus of elasticity increases, the stress redistributions take place at a greater distance from the cavity, resulting in less impact on the shotcrete lining.

Figure 6 shows the displacement changes as a function of the lateral pressure coefficient k_0. The total displacements extend with increasing lateral pressure coefficient in crown, side and bench. The deformation figures and the analysis of the displacements for the limit values of the parameter study show that the convergences occurring increase with a larger lateral pressure coefficient. With smaller lateral pressure $k_0 = 0.1$, the vertical displacement in the yielding element has a share of approx. 85 % of the total displacement, with a large lateral pressure

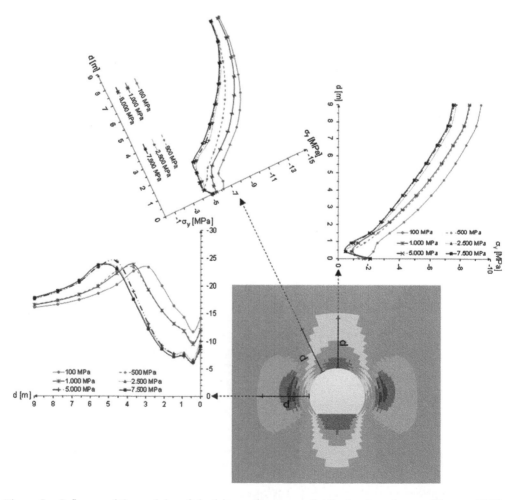

Figure 5.   Influence of the modulus of elasticity on the stress redistribution in rock mass (Hammer 2018).

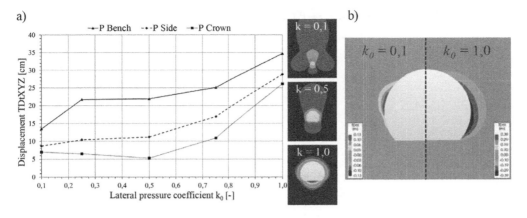

Figure 6. (a) Influence of the lateral pressure coefficient k_0 on the displacement development and (b) horizontal displacement development in comparison at $k_0 = 0.1$ and $k_0 = 1.0$ (Hammer 2018).

coefficient of $k_0 = 1.0$, the share is approx. 60 %. The displacements occurring in the yielding element are accordingly more evenly distributed and radially directed at high lateral pressures. For smaller values, there is a vertically directed displacement.

In summary, it can be stated that the magnitude of the lateral pressure coefficient has a significant influence on the squeezing ground behaviour. On the one hand, loads and the resulting displacements become higher as a result of increases; on the other hand, the ratio between horizontal and vertical displacement changes. In the case of the ductile support, this can result in the described variants for the load-bearing behaviour of the ductile overall system resulting from different lateral pressure coefficients.

The Hoek-Brown failure criterion m_i strength parameter, the Geological Strength Index GSI and the uniaxial rock strength σ_{ci} are varied successively and analysed in one step. The uniaxial rock strength σ_{ci} influences the displacement development at small values of $\sigma_{ci} = 25$ MPa especially in the bench, with an increase of σ_{ci} the displacements decrease and have a similar order of magnitude at all measuring points. The variation of GSI leads to an approximately linear decrease of the displacement in the bench and the side, the displacement in the crown hardly changes with the increase of GSI. With low values for m_i, large displacements occur which stagnate with an increase from m_i to 7 or 10 and decrease with $m_i = 12$ or 15 and remain constant.

3.2.2 *Influence of the selected interface between rock mass and shotcrete*

The interface elements define the connection between the rock mass and the shotcrete. A high interface stiffness describes a good bond between the two elements, a low stiffness allows displacements in normal and tangential direction. The relationship between the normal stiffness K_n and the tangential stiffness K_t is defined as 10:1 in the parameter study and remains unchanged (DIANA 2014).

Figure 7 shows the displacement development as a function of the interface stiffness. It is noticeable that with a small stiffness the displacements in the bench are largest and decrease with increasing stiffness. The displacements in the side, at the installation site of the yielding elements, increase with increasing stiffness, since the convergences are mainly absorbed by the yielding element. A relative movement between the rock mass and the shotcrete lining is no longer possible with an assumed solid bond. The evaluations show that the choice of the interface stiffness has a large influence on the internal forces and the deformation figure of the shotcrete lining. The effects on the load-bearing behaviour are examined more closely with the spatial model.

3.2.3 *Influence of rock mass overburden*

The influence of the overburden is only estimated to be low in combination with largely intact rock and rock mass with high stiffness (Hammer 2018). A change in the primary stress state by increasing the superposition has such an effect on the displacements in the shotcrete lining

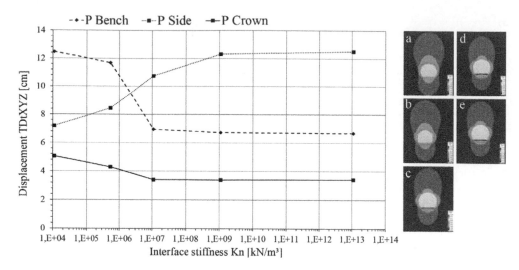

Figure 7. Dependence of the displacement development on the interface stiffness K_n (Hammer 2018).

that these increase almost linearly at all measuring points up to a superposition of 750 m and then the gradient increases (Figure 8a). The situation is similar with the normal forces in the shotcrete lining, which are small with a small superposition and increase approximately linearly up to 750 m and strongly increase from a superposition of 750 m (Figure 8b). Due to the increasing load, the plasticized area in the rock mass increases and causes the increase in displacements and normal forces.

3.2.4 Discussion

Squeezing ground behaviour can be differently pronounced and depend on different factors. The parameter study shows the influence of the investigated geotechnical parameters on an example tunnel with otherwise constant boundary conditions. In summary, it can be stated that all parameters influence the pressure behaviour of the rock mass and the resulting load on the ductile shotcrete lining, but in different ways and intensities.

The influence of the modulus of elasticity on the intensity of squeezing ground behaviour is particularly large; with a low modulus of elasticity, the deformability of the rock mass is high and the convergences and loads on the shotcrete lining increase sharply. The variation of the lateral pressure coefficient shows that as the value increases, the effects on the tunnel structure increase and the load changes, leading to a different load-bearing behaviour of the overall system. The influence of the lateral pressure coefficient depends on the position of the tunnel and the tectonic processes in the rock mass.

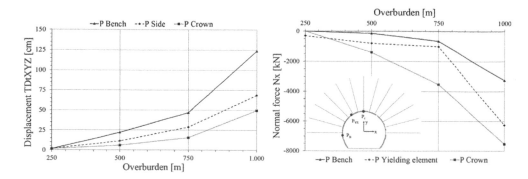

Figure 8. Influence of the overburden height on (a) the displacements and (b) the normal forces in the shotcrete lining (Hammer 2018).

4 THREE-DIMENSIONAL NUMERICAL ANALYSIS OF THE SYSTEM BEHAVIOUR OF A DUCTILE EXTENSION WITH YIELDING ELEMENTS

4.1 *Description of the selected three-dimensional calculation model*

The investigations on the system behaviour of the kinematic overall system consisting of shot-crete lining, yielding elements, rock bolts and lattice girders are carried out on a three-dimensional finite element model in order to consider the effects of the spatial stress redistributions. This model, like the model in the plain strain state, is created with the DIANA software and is shown in Figure 9. (Hammer & Thewes 2018, Hammer et al. 2018, Hammer 2018)

4.2 *Analysis of the overall kinematic system*

The evaluation of the variation study with the two-dimensional numerical model has shown that the variation of the lateral pressure coefficient and the associated changed load has a decisive influence on the deformation figure and on the kinematic overall load-bearing behaviour. In this section, the kinematic system behaviour is analysed using the reference example. In addition, the lateral pressure coefficients are varied in order to determine the influence on the load-bearing behaviour with the three-dimensional model.

Shortly before the bench is driven up, the behaviour of the top heading feet at the measuring point is analysed. Settlements of a maximum of 3.5 cm occur at the feet of the shotcrete lining. The top heading sole rises in the middle by 7.9 cm, but this is partly due to the choice of the elasto-plastic Hoek-Brown material model. Consequently, the top heading feet settle, but an incision as observed by John & Poscher (2004) and Radoncic (2011) cannot be shown for the example considered.

The concentration of the displacement at the yielding elements inevitably leads to shear movements of the rock bolts in the vicinity of the element, which leads to exceeding of the plasticity limit of the bolt bar and to local failure of the rock bolt. However, a general failure of the rock bolts embedded in the mortar does not occur. The relative displacements between the shotcrete lining and the rock mass are small, but this is also due to the design of the interface elements. In summary, the system behaviour of this calculation example does not resemble either case A or case B. A combination of both deformation figures results.

Based on the results of the variation study with the two-dimensional model, the lateral pressure coefficient is increased to k_0 =1.0 and reduced to k_0 =0.25 to analyse the influence on the overall system behaviour. Figure 10 shows the results of the two calculations. The evaluation of the two models shows that the deformation figure changes as a function of the lateral pressure coefficient. The equal load at k_0 =1.0 leads to a uniform displacement into the tunnel cavity. With a lower lateral pressure coefficient, smaller displacements occur in the bench, but

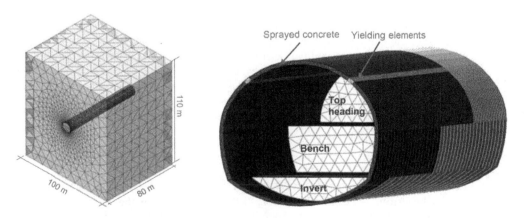

Figure 9.   Discretization of 3D-model with different excavation steps (Hammer & Thewes 2018).

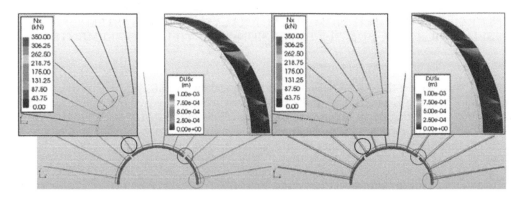

Figure 10. Deformation figure of the top heading shotcrete lining and the system rock bolting in the final state for lateral pressure coefficients (a) $k_0 = 1,0$ und (b) $k_0 = 0,25$ (Hammer 2018).

higher settlements at the ridges, which leads to an uneven displacement of the shotcrete lining. The breaking load of the rock bolt bars is 350 kN, the normal forces in the rock bolts are shown. The red areas show an exceeding of the rock bolt core-breaking load. In both cases the breaking load of the rock bolts adjacent to the yielding element is exceeded, whereby the forces in the rock bolts are greater at a higher lateral pressure coefficient.

4.3 Discussion

In this publication the behaviour in the interaction area between shotcrete, rock mass and yielding element is numerically investigated. The relative movements between the shotcrete lining and the rock mass are marginal, at the rock bolts shear displacements lead to local failure. This speaks in favour of a combination of case A and B.

5 SUMMARY AND OUTLOOK

5.1 Summary

Especially in squeezing ground with large convergences, neglecting the three-dimensional effects can lead to inadequate results. The calculations shall analyse the system behaviour of the shotcrete lining with yielding elements and provide quantitative evidence of the bearing capacity of the ductile lining as well as of the serviceability with regard to deformations to be maintained. In the final detailed design, three-dimensional numerical models are recommended in the main calculation for squeezing ground conditions in order to be able to consider the construction sequence and the stress redistributions. In this work a model is developed with which the displacement development can be well represented for that reference project.

This model is used to analyse the influence of rock mass and system parameters on the load of the ductile shotcrete lining. Important influencing factors are the modulus of elasticity, the lateral pressures and the description of the interaction between rock mass and lining by interface elements. Based on these results, a differentiated consideration of the kinematic overall system of a shotcrete lining with yielding elements is carried out by means of a three-dimensional model.

5.2 Outlook

Squeezing ground is characterized by heterogeneity and a conspicuous microstructure as well as by the rock types, which tend to deformation. In this thesis a homo-genic soil model with elastoplastic material behaviour is chosen, since only little information on the rock and rock mass properties was available. In general, with a detailed state of knowledge, the further information on microstructure, pore water pressures, etc. should be taken into account in the modelling.

Higher-quality constitutive models, which take the creep potential of the rock mass into account, can only be used for a realistic and comparable description of the mechanical rock mass behaviour in the implementation planning if sufficient information on the geological and geotechnical properties of the rock mass is available. Therefore, the measuring programs must be adapted to the planned calculations in order to investigate the required input parameters. Variation studies for the evaluation of the influence of individual parameters have to be carried out in an iterative process with advance exploration.

REFERENCES

Barla, G. 2016. Challenges in the Understanding of TBM Excavation in Squeezing Conditions. 16th ISRM Online Lecture. International Society for Rock Mechanics. http://www.isrm.net/gca/?id=1276.

Barla, G. 2011.Contributions to the understanding of time dependent behaviour in deep tunnels. In *Geomechanics and Tunnelling 4 (3)*: 255–265.

Barla G., Di Torino, P., Rettighieri, M., Fournier, C., Fava, A., Triclot, J. 2008. Saint Martin squeeze. In *Tunnels & Tunnelling International* (May): 15–19.

Brandtner, M. & Lenz, G. 2017. Checking the system behavior using a numerical model. In *Geomechanics and Tunneling 10 (4)*: 353–365.

Budil, A., Höllrigl, M., Brötz, K. 2004. Strenger Tunnel-Gebirgsdruck und Ausbau. In *Felsbau 22 (1)*: 39–43.

Cantieni, L. 2011. Spatial effects in tunneling through squeezing ground. Ph.D. Thesis. Switzerland.

TNO DIANA (ed.) 2014. Diana - Finite Element Analysis. User´s Manual. Release 9.6. Material Library.

Hammer, A.-L. 2018. Untersuchungen zum Einsatz von Stauchelementen in einer nachgiebigen Spritzbetonschale bei druckhaften Gebirgsverhältnissen. Ph.D. Thesis. Ruhr University Bochum, Germany.

Hammer, A.-L., Hasanpour, R., Hoffmann, C., Thewes, M. 2018. Numerical analysis of interaction behavior of yielding supports in squeezing ground. In: *9th European Conference on Numerical Methods in Geotechnical Engineering. Porto, Portugal, 25–27. June*: 1319–1327.

Hammer, A.-L. & Thewes, M. 2018. Integration of yielding elements in various computational methods for calculations in different planning and construction phases. In *ITA-AITES World Tunnel Congress. Dubai, United Arab Emirates, 21–26. April*.

Hasanpour, R., Hammer, A.-L., Thewes, M. 2018. Analysis of multilateral interaction between shotcrete, yielding support and squeezing ground by means of two different numerical methods. In *ITA-AITES World Tunnel Congress. Dubai, United Arab Emirates, 21–26. April*.

John, M. & Mattle, B. 2007. Auswirkungen stark druckhafter Gebirgsverhältnisse auf den TBM-Vortrieb. In *Felsbaumagazin* 25 (6): 14–21.

John, M., Spöndlin, D., Mattle, B. 2004. Lösung schwieriger Planungsaufgaben für den Strenger Tunnel. In *Felsbau 22 (1)*: 18–24.

Likar, J., Marolt, T., Likar, A. 2013. Adequacy of the yielding elements selection for underground construction in high squeezing grounds. In *Proc. of the 12th international conference underground construction, Prague, Czech Republic*.

Panet, M. et al. 2001. Recommendations on the convergence-confinement method. AFTES Report, V 1.

Pöttler, R. 1997. Über die Wirkungsweise einer geschlitzten Spritzbetonschale. In *Felsbau 15 (6)*: 422–429.

Radoncic, N. & Schubert, W. 2011. Novel method for ductile lining pre-design. In *Geomechanics and Tunnelling 4 (3)*: 195–210.

Radoncic, N. 2011. Tunnel design and prediction of system behaviour in weak ground. Ph.D. Thesis. University of Technology, Graz, Austria.

Schubert, W. 1996. Dealing with squeezing conditions in Alpine tunnels. In *Rock Mechanics and Rock Engineering 29 (3)*: 145–153.

Wiese, A.-L. 2011. Vergleichende Untersuchungen von Stauchelementen für den Einsatz in druckhaftem Gebirge. In STUVA (ed.), *Forschung + Praxis, Vol. 44. Berlin, Germany*: 148–156.

Wittke, W., Schmitt, D., Wittke-Schmitt, B., Wittke, M. 2012. Tragverhalten eines nachgiebigen Ausbaus im druckhaften Gebirge. In Deutsche Gesellschaft für Geotechnik e.V. (DGG) (ed.), *Taschenbuch für den Tunnelbau. Essen: VGE Verlag GmbH*: 95–129.

*Tunnels and Underground Cities: Engineering and Innovation meet Archaeology,
Architecture and Art, Volume 7: Long and deep tunnels – Peila, Viggiani & Celestino (Eds)
© 2020 Taylor & Francis Group, London, ISBN 978-0-367-46872-9*

Optimizing TBM cutterhead design for application in very strong and abrasive rocks, case study of Kerman Water Tunnel

J. Hassanpour
School of Geology, College of Science, University of Tehran, Tehran, Iran

J. Rostami
Earth Mech. Inst. (EMI), Department of Mining Engineering, Colorado School of Mines, Golden, CO, USA

Y. Firouzei
School of Geology, College of Science, University of Tehran, Tehran, Iran

H.R. Tavakoli
SOI Consulting Engineers, Tehran, Iran

ABSTRACT: Kerman water conveyance tunnel (KrWCT) is located in central Iran. The tunnel is under construction both from northern and southern portals using two refurbished TBMs. In order to excavate northern section, a double-shield machine was employed, which was used in Alborz Mountains in northern Iran. Geological conditions of KrWCT is quite different from previous tunnel, and a long section of tunnel will pass through massive, very strong and abrasive rocks. Preliminary investigations revealed the necessity of improving the old machine particularly its cutterhead design and disc cutter arrangement on the head. In this paper, using empirical performance prediction models, the effect of cutterhead design and various operational parameters on penetration and advance rates of the machine is evaluated; Furthermore, optimum cutterhead design parameters are suggested to cope with KrWCT tunnel geological setting. Eventually, the effect of modifications on boring operation is discussed by comparing performance of new cutterhead and the original head in two short sections.

1 INTRODUCTION

Kerman water conveyance tunnel is the main part of a water transfer system to supply drinking water for Kerman city which is located in an arid area in the center of Iran. This very long tunnel (total length=38 km) is under construction in two sections (each 19 km in length) using two refurbished machines overhauled in northern and southern portals. During design phase of project, it was predicted that a long part of this tunnel will pass through very strong and very abrasive rocks which belong to intrusive rocks of Sahand-Bazman zone in central parts of Iran.

It is clear that due to the specific geological conditions of these two sections of the KrWCT tunnel, it was necessary to employ suitable machines with sufficient capacity to excavate the rock, especially in the hard and abrasive igneous formations. Therefore, before starting the tunnel excavation, a question had to be answered. Are these two TBMs suitable for new geological conditions? To answer this key question, the suitability of used machines for overcoming adverse geological conditions of the new project was checked using a methodology that was based on our previous experiences and a local, empirical TBM performance prediction model.

In this paper, just the specifications of machine employed in northern section and its suitability for the new geological condition of KrWCT-N will be described and evaluated.

To achieve optimum advance rates and shortest time schedule for excavation of the tunnel, the effect of variations in some of the key features of the machine (including cutter load, RPM, number and diameter of the disc cutters) on TBM performance will be evaluated and the minimum values that are required for completing this tunnel on time, will be determined.

Furthermore, a new methodology for making a relationship between four sets of parameters including TBM performance, operating, cutterhead design, and geological parameters, which is based on available empirical TBM performance prediction model will be described. The proposed model was employed for assessing suitability of machine and optimizing its cutterhead design for the northern section of KrWCT tunnel. In this paper, the prediction model used to determine the performance parameters will be described, followed by geological characteristics of the tunnel route, and finally, the influence of different machine design and operating parameters on construction schedule will be evaluated.

2 PROJECT DESCRIPTION

Kerman water conveyance tunnel is a very long tunnel with total length of about 38 km and is the main part of a water transfer system which has been designed to transfer about 23 m³/s of drinking water from Safa Dam, constructed near Rabor city in the South central part of Iran, to Golzar area and then to Kerman city. The tunnel is being constructed in two sections with almost equal lengths using two refurbished double shield machines with different diameters. Both sections are being excavated under positive slope conditions. An inclined access tunnel (wilt length of 900m) is also being constructed (using drill and blast methods) to pull out two machines after completing two sections. As of September 2018, about 10000 m of the Northern section and 5000 m of Southern section have been excavated and lined. Both machines install universal, 25cm-thick pre-cast concrete segments, in their tail shields as they advance.

3 TBM SPECIFICATIONS

The machine that was used to complete Karaj Water Conveyance Tunnel (KWCT, a 30-km tunnel in pyroclastic and igneous rocks of Karaj Formation, in Alborz zone), was employed again for excavating Northern section of Kerman tunnel. This machine is a double shield TBM (model S-323) manufactured by Herrenknecht, which had been designed for hard rock

Table 1. Specifications of employed machine in northern section.

Parameter	Value
Machine diameter	4.65 m
Cutters diameter	432 mm
Number of disc cutters	31
Disc nominal spacing	90 mm
Max. operating cutterhead thrust	16913 kN
Cutterhead power	5 * 250 = 1250 kW
Cutterhead speed	0 to 11 rpm
Cutterhead torque (nominal)	1029 kN.m (11 rpm)
Thrust cylinder stroke	1400 mm
Conveyor capacity (approx.)	200 m3/h
TBM weight (approx.)	170 tons

Figure 1. TBM assembling in Northern portal.

conditions (Figure 1a). The original cutterhead, as shown in Figure 1(b), featured 31 disc cutters, each 432 mm (17 inches) in diameter, with average spacing of 90mm. Other main technical specifications for the TBM are summarized in Table 1. As will be explained later, due to presence of very hard rocks in extended reaches of the tunnel alignment, a modification of the machine was necessary to improve its performance. The most important modification was rearranging of disc cutters with larger diameters on the cutterhead.

4 GEOLOGICAL SETTINGS

KrWCT project is located in Sahand-Bazman magmatic belt (SBZ), also known as Urumieh-Dokhtar zone (UDZ), which is a ~2000 km long and 50-200 km wide zone of Tertiary volcanic rocks, which extends from Uromieh in NW to Baluchistan in SE of Iran. Magmatic events in this zone started in Late Paleocene and peaked during Late Eocene. This zone lithologicaly composed of different volcanic, pyroclastic and plutonic rocks. So, it was predicted that the tunnel will pass through the kinds of rocks shown in Figure 2. High strength is the typical characteristic of granitic rocks which form almost 30% of tunnel length. Geotechnical investigations and some lab tests showed that strength of some rocks can reach up to 350 MPa. Values of CERCHAR Ablativity Test varies between 3 and 4, the highest values are up to 4.5 which are related to granites and values less than 2 was observed in test. These CERCHAR values are Considered to show intermediate to very high abrasivity.

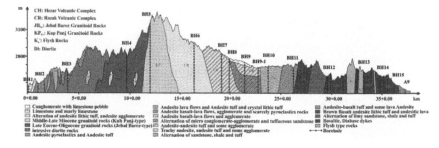

Figure 2. Geological cross section along the tunnel (SOI, 2016).

5 PREDICTION MODELS AND THEIR APPLICATION

The characteristics of a machine and its proper operation in different geological conditions (which is the subject of this study) is critical to successful operation of the TBMs. Hence, it was necessary to examine the relationship between operational parameters and machine performance in the given formations. As shown in Figure 3, the performance of a tunnel boring machine depends on the three main set of parameters: 1) design parameters of the cutterhead; 2) operating parameters and 3) geological and geotechnical parameters of the rock mass along the tunnel alignment.

First set of parameters are determined during design of the machine based on geological conditions and rock mass properties and other project requirements and are usually fixed throughout the project. Of course, depending on the case and if there are major problems in designing the machine, these parameters can be redesigned and machine can be improved by expending a high cost and stopping the project for a long time. The machine operating parameters (2nd set), during tunnel boring, are controlled by the operator of the machine to achieve optimum penetration rates and achieve safe conditions for the stability of tunnel walls. In other words, an experienced operator during tunneling, in accordance with the geological conditions of the tunnel, sets the parameters of the machine in such a way that the best performance of the machines are achieved. The most important of these parameters, as mentioned above, are the cutterhead rotation speed (RPM), thrust force, torque and power of the cutterhead. The third group of parameters which influence on TBM performance are geological parameters. Intact rock strength and fracturing degree are two important parameters in this group.

Considering the three sets of controlling parameters plus TBM performance, there will be four sets of parameters, and a prediction models are the connection or a link between these four sets of parameters. By definition, a prediction model is an equation, a chart or table or a combination of these tools that uses input parameters from rock mass and machine for estimating penetration in rock to estimate TBM performance, which is the key for preparing a time schedule for excavating the tunnel (Figure 4a). When TBM design parameters and geological parameters are known, by assuming reasonable ranges for TBM operating parameters, performance parameters of machine

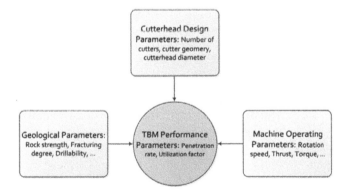

Figure 3. The relationship between the three sets of parameters and TBM performance.

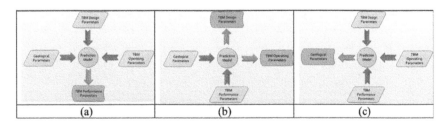

Figure 4. different applications of TBM performance prediction models.

can be estimated using an appropriate prediction model. On the other hand, if a reasonable TBM Performance in a known geological condition is needed, TBM operating or TBM design parameters can be determined using the selected prediction model (Figure 4b).

If the actual variations of operating parameters and machine performance are known, it is possible to estimate general rock mass condition and its boreability and answer to some important geological questions using these prediction models (Figure 4c). This is based on back analysis of machine performance to estimate rock mass conditions.

6 RELATIONSHIPS BETWEEN MACHINE PARAMETERS AND GEOLOGICAL PARAMETERS

Further details of the relationship between the above-mentioned parameters are shown in the flowchart presented in Figure 5. This chart in current form explains how to calculate minimum operating and cutterhead design parameters to reach a desired performance. As shown, by specifying the input parameters (yellow boxes) and taking into account the limits of the machine (red boxes), the operating parameters and the design of the cutterhead (green box) can be calculated using the selected prediction model (blue boxes). The white boxes in flowchart contain an equation for calculating the target parameters. This process can also be reversed and, by fixing the operating parameters of the machine, the performance parameters are predicted and a construction schedule for tunnel can be prepared.

7 APPLIED METHODOLOGY

The methodology for checking the main operating and design parameters of a hard rock TBM to reach a given penetration rate was explained above (Figure 5). As stated, this methodology can be used for preparing construction schedule of tunnel in reverse direction. On the other hand, the main object of this study was to achieve the optimum advance rates and shortest

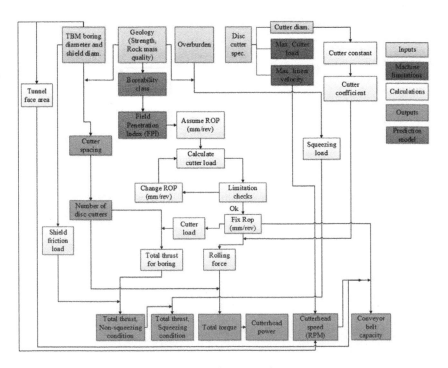

Figure 5. Flowchart of the relationships between four sets of parameters.

tunnel construction schedule, by changing operating and cutterhead design parameters. To reach this goal, several scenarios had to be examined. As will be explained later, by assumption of different number of cutters, different diameters and load capacities for disc cutters and also different cutterhead speeds, the time schedule for each scenario must be prepared and compared to select best one. To prepare a construction schedule following steps must be done:

7.1 Providing Input Data

The first step was to provide required input data. As can be seen in upper part of flowchart, the main data for calculating machine design parameters and estimating machine performance are geomechanical properties of identified engineering geological units, tunnel diameter and overburden. These kinds of data are usually available in design phase of any mechanized tunneling project. As described before, a detailed geological and engineering geological studies and geotechnical investigations were performed in this project. According to results of these studies, several different units with a wide range of engineering properties have been identified along the tunnel.

7.2 Selecting TBM performance prediction model

If required input data are available, the performance of the machine can be estimated in each engineering geological unit by using TBM performance prediction models. Many prediction models have been proposed by researchers and research institutes in the world. Some of them are very simple equations and are based on one or two input parameters, and some are more complicated. In this study the model presented by Hassanpour e al. (2011, 2015) was selected for predicting TBM performance in different units. This model has been developed based on the data collected from main tunneling projects in Iran, including projects that have been excavated in similar conditions to KrWCT project, and thus can be very useful for predicting machine performance in this special project. The main chart of this model is presented in Figure 6. As can be seen, this model is based on two main rock mass properties, including Uniaxial compressive strength of intact rock (UCS) and rock quality designation (RQD), to predict machine performance. The estimated rate of penetration is specifically through the use of field penetration index (FPI) for each rock mass. FPI is a composite parameter and calculated as follows:

$$FPI = \frac{F_n}{ROP(mm/rev)} \tag{1}$$

Where ROP is the penetration rate (mm/rev), Fn is cutter load (KN) and FPI is the field penetration index (kN/cutter/mm/rev).

FPI ranges can be estimated from Figure 6, and is an important and reliable parameter in evaluating performance of rock TBMs. Using this parameter, the penetration rate, thrust force and rotational speed of cutterhead are accounted for in the analyses. Therefore, estimating the penetration rate will be possible by identifying geomechanical properties of engineering geological units (UCS and RQD) and taking into account the machine specification and operating parameters. Based on this model (Figure 6), rock units can be categorized into 6 classes from B-0 as toughest rocks for boring to B-V as easiest rock for boring. As shown, the geological units identified along KrWCT-N project have a wide range of FPI and fall into four classes of B-0 to B-IV. As shown in this chart, part of the alignment is through very strong and massive rocks with very high values of FPI (B-0 class, more than 70 kN/c/mm/rev). Consequently, in these sections, penetration is not easy and high cutter loads are needed for efficient chipping process.

7.3 Checking machine limitations

To assure proper handling of the machine's and prevent unexpected damages to different parts during its operation, it is necessary to monitor machine limitations. One of the most

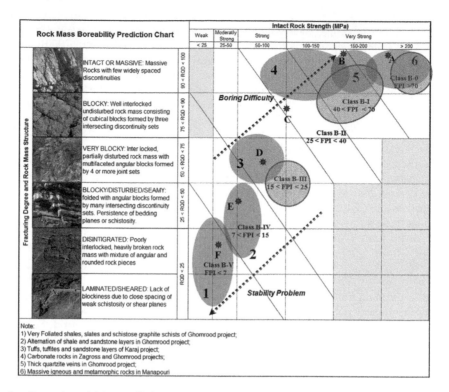

Figure 6. General model for predicting FPI and boreability class of rock masses.

important limitations is the maximum penetration rate in softer rock, which is referred to as pump limit and usually must be less than 15 mm/rev. Another limitation is the maximum cutter load which is usually less than 250 kN for 17- inch cutters. The next limitation is cutterhead torque, which always must be less than maximum installed torque of the machine. High values of torque are usually caused by high thrust force and deep penetration of cutters into the rock and subsequently increased rolling forces. Also, the volume of the muck produced at higher penetration rates should be monitored to avoid exceeding the capacity of the conveyor belts. To meet these limits of a machine, one approach is to look for maximum penetration rate within the maximum capacities, assuming that the machine always works at full capacity. The more appropriate approach is to use part of the machine's capacity during tunnel boring and not always push the machine to its limits. Obviously, in this approach, the operational limits are set and operators directed to observe these limits, which are often slightly lower than the nominal capacity of the machine, to achieve proper operation.

7.4 Preparing construction schedule

To prepare construction schedule it is necessary to know TBM advance rate in each unit. Advance rate or AR (m/day) can be calculated using following equation:

$$AR = \frac{PR \times U \times n_s \times h_s}{100} \qquad (2)$$

Where PR is penetration rate (m/h), U is utilization factor (%), ns is number of shifts per day and hs is number of hours per shift. Number of days that machine will be excavating through each geological unit (N_{day}) can be estimated using Eq. 3. N_{day} is the most important parameter for preparing tunnel construction schedule. Knowing this parameter, it is possible to calculate TBM advance during construction period.

$$N_{day} = \frac{L_{sec}}{AR} \quad \text{Completion time} = \sum N_{day} \qquad (3)$$

Where Lsec is length of given geological unit along the tunnel (m).

8 RESULTS OF STUDY

8.1 *Preparing construction schedule for different scenarios*

The above mentioned methodology was used in KrWCT-N project to check suitability of old cutterhead and possible improvement of machine performance by a new cutterhead. To achieve this goal, several scenarios were checked. As shown in Table 2, in each scenario, different values for operating and cutterhead design parameters (such as the rotation speed (RPM), cutter load capacity, diameter and number of cutters) were assumed and the penetration rate in each geological unit along KrWCT-N project was estimated based on the engineering properties of that unit. A suitable utilization factor was assumed for estimating advance rate and tunnel construction schedule for each section based on the recent experiences in the area. In addition, the max trust, torque, capacity of conveyor, construction period, average monthly advance, max monthly advance and net boring time for each section have been estimated and listed in Table 3.

It must be noted that some of the assumptions made for comparing different scenarios are unlikely choices. For example, due to limited space in cutterhead structure, it's not possible to install 31 of 19-inch cutters on the cutterhead. Table 3 and the graphs presented in Figure 7 show the effect of these changes on the construction schedule of north section of tunnel. As shown, the construction schedule is very sensitive to changes in these parameters, and by modifying the design of the cutterhead, a considerable reduction in construction schedule occurs.

Among different scenarios, the options with best performance of machine or shortest construction period could be selected as a basis for making a decision for using old cutterhead or designing a new one. By considering some design requirements and limitations of cutterhead structure, the option no 4 which recommends installing 28 disc cutters with 19 inch diameter was selected.

Table 2. Different scenarios for calculating machine performance and preparing construction schedule.

No	Cutterhead diam. (m)	Number of cutters	Cutter diam. (inch)	Cutter load (kN)	RPM
1	4.67	31	17	200	5.5
2	4.67	31	17	200	6.5
3	4.67	31	19	250	6.5
4	4.67	28	19	250	6.5
5	4.67	26	19	250	6.5
6	4.67	34	17	200	6.5

Table 3. Different scenarios for calculating machine performance and preparing construct ion schedule.

No	Max.thrust (kN)	Max.torque (kN.m)	Max.load of conv.(m3/h)	Const.period (month)	Av.monthly advance (m)	Net boring time (h)
1	6800	1176	119	54	352	6501
2	6800	1176	140	46	413	5501
3	8350	1470	140	34	557	4539
4	7600	1323	128	37	500	4966
5	7100	1115	140	40	476	5271
6	7400	1419	140	43	442	5063

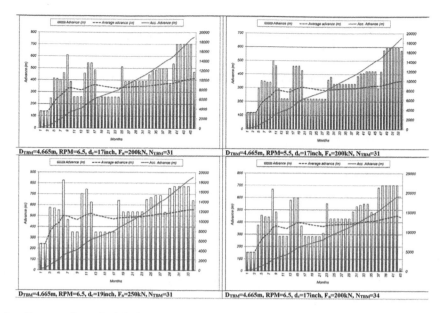

Figure 7. Construction schedule for some of the selected scenarios.

Therefore the cutterhead had to be replaced by a new one. The new cutterhead was equipped with 19 inch disc cutters which have higher load capacity and it is possible to apply higher thrusts to reach more penetration in rock in each revolution of machine and due to sufficient penetration, chipping process can occur more effectively and machine performance improve accordingly.

8.2 *Evaluating the performance of new cutterhead*

Based on results obtained from above mentioned analyses, a new cutterhead was designed and ordered to manufacturer. As time schedule for completing the project was very tight, before completion of design and manufacturing of cutterhead, the contractor decided to start the project using old cutterhead for saving the time. The machine started to excavate the tunnel, and it was

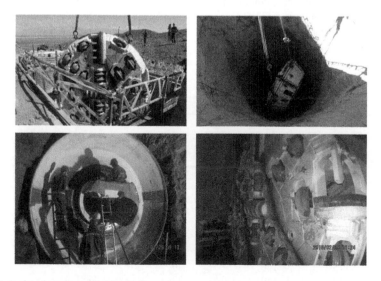

Figure 8. Cutterhead replacement in shaft.

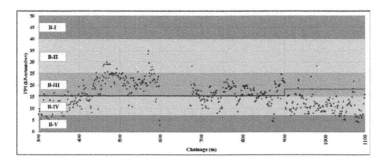

Figure 9.   Variations of FPI in tunnel for each stroke before and after replacing of cutterhead.

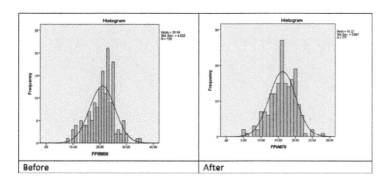

Figure 10.   Frequency distributions of FPI values in 200m before and after of cutterhead replacement.

planned to replace new cutterhead by old one in a 30m shaft which constructed in a distance of 600m from portal where overburden is not very high. The shaft was constructed in advance, the machine arrived at the shaft and operation of cutterhead replacement was conducted in the shaft (Figure 8).

Based on performed geological investigations it was observed that the geology in the first 900m of tunnel of tunnel has not changed and the completed section had passed through basaltic rocks with more or less similar engineering geological properties. So, it can be assumed that any change in machine performance was related to changes in cutterhead not ground.

Figure 9 shows variations of FPI in tunnel for each stroke before and after replacing cutterhead. These FPI values are calculated using formula No. 1, in which actual mean value of F_n (cutter load) is used which can be obtained by dividing total thrust force (After deducting shield friction) by number of discs. Also, two histograms presented in Figure 10, show frequency distributions of FPI values in 200m before and after of cutterhead replacement. As shown, the average of FPI values has reduced after cutterhead replacement and about 25% reduction can be observed in average values. It means that the new cutterhead offered better performance as predicted.

9   CONCLUSION

The results of this study showed that the change in the adjustable parameters such as the machine's rotation speed or the number and diameter of cutters can influence construction schedule. These parameters directly affect the penetration rate of the machine and thus change the advance rates. In this study, the changes were made in such a way that machine limitations such as penetration rate, conveyor capacity, cutter load capacity are considered. Also, to protect the mechanical integrity of the TBM, the maximum capacities of the machine, such as the maximum thrust capacity, maximum torque, and max RPM was not used and the values of these parameters during the boring operation was within a range that was below the maximum installed capacity on the machine. Another important point is to look beyond the effective parameters on the penetration rate, and incorporate the parameters that affect the utilization factor (delay times), which also affect the daily advance rate and thus, tunnel construction schedule. In this study, the change in cutterhead and using more suitable cutterhead for the anticipated geology has shown strong gains in rate of penetration and daily advance rates.

REFERENCES

Hassanpour, J. Rostami, J. & Zhao, J. 2011. A new hard rock TBM performance prediction model for project planning. *Tunneling and Underground Space Technology* (26): 595–603.
Hassanpour, J. Rostami, J. Zhao, J. & Tarigh Azali, S. 2015. TBM performance and disc cutter wear prediction based on ten years experience of TBM tunneling in Iran. *Geomechanics and Tunnelling* (8): 239–247.
SOI Company. 2016. Engineering geological report of Kerman water conveyance tunnel. unpublished report.

*Tunnels and Underground Cities: Engineering and Innovation meet Archaeology,
Architecture and Art, Volume 7: Long and deep tunnels – Peila, Viggiani & Celestino (Eds)*
© 2020 Taylor & Francis Group, London, ISBN 978-0-367-46872-9

Key construction technology of Long Metro Subsea Tunnel in complex site conditions

W. He, C. Song & B. Du
China Railway Liuyuan Group Co.ltd, Tianjin, China

ABSTRACT: Subsea tunnel is generally large, having complex site conditions and lacking of engineering experience, the construction method is directly related to the success or failure of the project. At present, there is no metro subsea tunnel built in China. On the background of XiamenLine3subsea tunnel, construction method selection, ventilationand evacuation, complex geological response in the sea area, durability design are analyzed by geological analysis, engineering analogy and comprehensive comparison.A combination of the shield and mining methods is proposed to combine the geological conditions of different sections. The complex geology of the sea area is used in the targeted design, including a deep weathering slot, a water-rich sand layer, a hard rock and uneven stratum, and the development of solitary rocks. The durability design of the tunnel structure and the limit of the bearing capacity are treated with equal emphasis to consider reserving complementary spaces. The study method and conclusions can provide references to support tunnel project decision and similar project.

1 INTRODUCTION

In recent years, with the development of rail transit, subsea tunnel is more and more used in China when rail transit crosses the sea area. At present, the main construction methods of the cross-harbor interval tunnels are the mining, shield, and immersed pipe methods. As the construction of such a tunnel is a large-scale project with a large investment and a high risk, the demonstration of key technology in its design stage is related to the success or failure of the project[1].

As early as the mid-30s, developed countries began to construct harbor tunnels, such as the shield tunnel of Japan Kan-Mon Tunnel and the Netherlands Rotterdam immersed tunnel. In 1988, Japan completed the longest tunnel in the world, Seikan Tunnel, with a total length of 54 km[2], by using the borehole-blasting method on the Tsugaru strait. The representative harbor tunnels constructed using the shield method include the 49.2-km-long Anglo-French Channel Tunnel and the 7.9-km-long Danish Strait project, all of which are railway tunnels with a double pipe diameter of 8.5 m in the circular section[3,4].In China, Liuyang River Tunnel of Wuguang High-speed Railways is the first underwater high-speed tunnel, with a span of 14.8 m and excavated section of 170 m^2, developed using mining construction[5]. Further, Shiziyang Tunnel of Guang-zhou-Shenzhen-Hong Kong High-speed Railways is the longest underwater high-speed railway tunnel in China with a length of 10.8 km[6]. The technology of shield tunneling in the opposite direction, docking in the ground, and disintegrating in the tunnel was adopted for the first time for the construction of this tunnel. Five subway cross-harbor tunnels built in Hong Kong were constructed by the immersed tube method using round steel pipe and rectangular concrete pipe sections[7]. Qingdao Metro Line 1 and Xiamen Rail Transit Line 2 and Line 3 need to cross the sea area and are currently under construction. Therefore, the construction of large cross-harbor tunnels needs to be demonstrated in detail. The related studies focus on the feasibility analysis, construction, bridge and tunnel selection, and route selection[8,9,10,11]. Ref. [12] introduces the characteristics and difficulties of the construction of Xiamen Xiangan Tunnel, the first undersea tunnel in the mainland. Refs. [13,14,15] introduce the concrete technology of a harbor tunnel, such as grouting, calculation of water pressure, and prefabrication of an immersed tube tunnel. There are few articles on the key technology of cross-harbor subway tunnel design. In this paper, we introduce the detailed key technology of the design of a very long cross-harbor subway tunnel in a complex site environment in combination with the specific project of Xiamen Rail Transit Line 3, which provides a reference for similar projects.

2 PROJECT PROFILE

2.1 Interval design profile

Wuyuan Bay ~ LiuwudianStation tunnel of Xiamen Rail Transit Line 3 crosses Xiamen eastern waters, connects the island and Xiang'an, which is located on the northwest side of the Xiang'an harbor tunnel; the length of the interval is 4.95 km, and the length of the intersection with the sea is approximately 3.6 km. The number of main tracks in the interval is two lines, the maximum design speed is 80 km/h, and six cars are grouped in type B vehicles. One inclined shaft, one ventilating shaft, ten service channels, and three waste water pumping houses and substations are set in the interval.

2.2 Geology and environment

1. The cross-sea section belongs to the coastal accumulation area, and the terrain of the inland and the outer continental segment is flat; the depth of the sea area is 20 m, and the deepest point is approximately 25 m deep.
2. The geological conditions of the interval crossing strata are complex and changeable, the islands of the Xiamen segment and the Xiang'an bedrock surface are low, the bedrock scar of the most moderately and slightly weathered sea regions fluctuates considerably, and there are rocks, reef plates, and weathered deep grooves.
3. The tunnel sea area passes through four groups of deep grooves of fault weathering. The weathering groove is mainly composed of completely weathered granite and scattered and cracked strongly weathered granite, which contains dense sand and gravel mixed clay.
4. The groundwater is a mainly cataclysmic strong weathering zone and includes the fissure water of the bedrock below. The permeability of other strata is poor, except for the sand bed and the possible water-rich bedrock fracture zone in the sea area.
5. The sea area passes through the breezy granite layer and the weathered deep trough. The strength of the breezy bedrock is between 63.5 MPa and 204.2 MPa, the average strength is 110 MPa, and the rock strength is high. The Xiangan side tunnel mainly passes through a sand layer, round gravel, and fully weathered stratum, with strong permeability and is connected with the sea water.
6. Marine chloride corrodes the environment, and the environmental action grade is III-E.

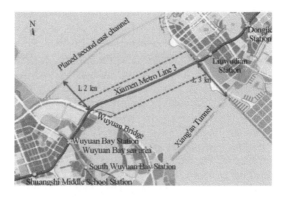

Figure 1.   Wuliu interval tunnel plane location diagram.

7. The crossing sea area is the core protection area of Chinese White Dolphin. The surface pipeline is dense and complex in the island, and Xiangan is formed by reclamation, which requires high environmental protection.

2.3 Design features and standards

1. The length of the tunnel crossing the sea area is 3.6 km with a high construction risk and a high requirement of operational disaster prevention and evacuation.
2. The tunnel is constructed to be deep and large and bears high water pressure. The maximum depth of the cross-sea is 25 m, and the maximum depth of the tunnel is 69 m.
3. The design working life of the main structure is 100 years.
4. The structure's waterproofing grade is grade II.
5. The seismic intensity of the interval structure is 7°, and the seismic structural measures are adopted according to 8°.

3 OPTIONAL DESIGN FOR INTERVAL TUNNEL CONSTRUCTION

The selection of the construction method is a key problem related to the success or failure of the project. The commonly used construction methods of an underwater tunnel are mining method, shield method, and immersed pipe method. The selection of various methods depends to a great extent on the geological conditions of the site, the condition of the buildings on both the coasts, tunnel function and engineering costs, construction period and risk, etc. The immersed tub method is excluded considering the influence of the buried depth, protected area, main waterways, and the underwater reef explosion at both ends of the station. Either the mining method or the shield method is mainly selected as the tunnel construction method.

The interval crossing geology is complex and variable; it is generally divided into three sections. The overburden thickness in the continental segment of the island is 11.5 m to 21.5 m, and the underlying bedrock is either completely weathered or highly weathered granite. The

Table 1. Suggested values of the rock–soil parameters.

Rock–soil name	Natural density (g/cm³)	Cohesion (kPa)	Inner friction angle(°)	Permeability coefficient K (m/d)	Poisson's ratio	Lateral pressure coefficient	Compressive strength (MPa)
Sand filling	1.97	0	26	10	0.30	0.43	–
Medium coarse and gravelly sand	2.00	1	34	15	0.30	0.40	–
Circular gravel	2.05	0	35	30	0.25	0.33	–
Completely weathered granite	1.92	22	21	0.1	0.28	0.40	–
Scattered strongly weathered granite	2.03	30	26	0.15	0.25	0.33	–
Cataclastic strongly weathered granite	2.21	50	28	1.0	0.2	–	–
Moderately weathered granite	2.58	70	30	0.1	–	–	50
Breezy granite	2.68	120	35	0.08	–	–	110

middle part of the sea area mainly passes through medium weathered and breezy bedrock and seven weathered deep troughs. The thickness of the Xiang'an section is approximately 15–30 m, and the lower part is completely weathered and highly weathered granite.

Approximately 1.1 km of the surrounding rock of the underwater and shoal sections of the Xiang'an tunnel is a permeable sand layer and fully to heavily weathered ground, and the construction difficulty and risks are considerable in the case of the mining method. The interval crossing hard rock segment is 1703 m long, and the average strength of the rock is more than 110 MPa; therefore, it is difficult to construct a shield with a low efficiency (long construction period, serious abrasion of the cutter and the other tools, shield autorotation, etc.) by using the shield method. The tunnel water pressure is more than 0.4 MPa for a total length of 2435 m. The risk coefficient of the cutter with a pressure bin is high, and risk control is difficult.

A comprehensive analysis is performed from eight aspects, including the influence of the project on the environment, construction risk, implementability, lining quality, waterproofness, influence on the stations at both ends, duration, and cost, and a combination of the shield and mining methods considering the geological conditions of the different sections is proposed. The earth pressure shield construction is used in the section of the land area of the interval, which is approximately 870 m long. On one side of Xiang'an, the water-rich sand bed is crossed by using slurry balance shield construction (approximately 1450 m long). The middle section of the interval is constructed using the mining method and is approximately 2600 m long. The tunnel crossing stratum is shown in Tables 2 and 3, The intermediate sea area mainly passes through the breeze granitic stratum and constructed using the mining method; the two ends of the shield tunneling have a total length of 148 m, and only one of them is in the hard rock section. The other is in a sand bed and fully to heavily weathered ground with strong shield tunneling. The longitudinal section of the tunnel is shown in Figure 2. The construction of the mining section is organized using inclined well and air shafts, and the shield is organized using the two ends of the station. The shield receiver is located in the mining tunnel with the construction in the sea area completed. The construction plan is shown in Figure 3.

Table 2. Distribution of tunnel mining section crossing strata.

Lithology	Sand seam	Fully strong weathering layer	Intermediate weathering layer	Breeze layer	Upper-soft ground	lower-hard
Length/m	0	235	167	2017	174	

Table 3. Stratigraphic distribution of right line crossing in shield tunneling section (Left line equivalent).

Lithology	Sand seam	Fully strong weathering layer	Intermediate weathering layer	Breeze layer	Upper-soft ground	lower-hard
Length/m	482	785	20	128	36	

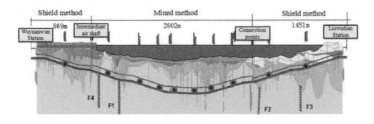

Figure 2. Vertical section diagram of tunnel with shield mining combination construction method.

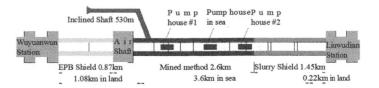

Figure 3. Construction organization plane diagram of tunnel with shield mining combination construction method.

4 DISASTER PREVENTION VENTILATION AND EVACUATION IN THE ULTRA-LONG SUBSEA TUNNEL

The sea-crossing section in the Wuhui section is 4.95 km long. It accommodates three trains simultaneously during peak hours. Ventilation and disaster prevention requirements are that only one train travels in each ventilation section at one time. According to the vertical ventilation mode in conventional segmentation, at least two air shafts are needed in the middle of tunnel; however, since the tunnel crosses the sea, conditions are poor for air shafts and excavation is highly difficult.

Solutions: 1) This tunnel covers 3.9 km of subsea area, where it is not feasible to construct a middle air shaft. Instead, a special civil construction air shaft is set above the construction tunnel segment using the mine method. Segmented vertical ventilation and the smoke discharge mode is adopted to solve disaster prevention and ventilation problems in the long-term sea-crossing section. 2) Encryption setting of service channels and increasing vertical evacuation platform width are considered from the perspective of disaster prevention and evacuation, due to increased evacuation capacity. A total of 11 service channels are set in the Wuhui segment. The average cross aisle space is about 420 m. The evacuation platform width in the mine method tunnel is 2000 mm (Figure 4) and the evacuation platform width in the shield method tunnel is 950 m (Figure 5). The inclined shaft is used as the evacuation exit and its width is 4 m. 3) Many safety regions and emergency exits are set, including Wuyuan Bay Station, Exhibition Center Station, non-disaster fire tunnel, interval air shaft and inclined shaft. During a fire disaster in the sea-crossing interval, if the train still has power it must drive to Wuyuan Bay Station or Exhibition Center Station for evacuation and rescue. If the train loses power and cannot reach a station, it must stop immediately for passengers to disembark via the side or rear train doors onto the evacuation platform or track bed, for evacuation to a non-fire tunnel through the transverse service channel.

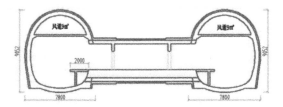

Figure 4. Cross-section of tunnel using mine method (unit: mm).

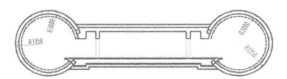

Figure 5. Cross-section of tunnel using shield method (unit: mm).

5 COMPLEX GEOLOGICAL RESPONSES IN SEA AREA

5.1 Key technology of weathering deep groove construction in mining tunnels

Over the six weathered deep grooves in the mining section of the over sea area, the main weathered trough material is completely weathered granite and strongly weathered granite; further, the rock mass is broken, and the storage of pore water is abundant.

Based on the design principle of "safety first, prevention first," the advanced geological forecast is regarded as the first working procedure before the construction. The forecast is based on a geological analysis, a combination of long-distance macro forecast and short-distance accurate prediction, a combination of advanced hole exploration and geophysical exploration, a mutually complementary verification of various geophysical exploration methods, and a comprehensive advanced prediction scheme based on a combination of qualitative and quantitative methods. Further, a variety of traditional and advanced detection methods such as geological surveys, TSP, high-resolution DC electrical method, infrared water detection, geological radar, and advanced hole detection are integrated and applied to identify the location range of the rich water zone, fault fracture section, and other poor geological bodies as accurately as possible.

The weathering trough section adopts the advanced full section grouting to reinforce the stratum after it. Ordinary cement single slurry, superfine cement single liquid slurry, special sulfoaluminate cement single liquid slurry, ordinary cement, and water glass double liquid slurry are selected as the grouting materials. The ordinary cement single slurry is used in the small amount of water flowing through the hole section. The superfine cement single slurry is mainly used for the surrounding rock with poor intense weathering and permeability; it can also be used for pre-grouting simultaneously to break through the cracks in the dense rock mass and then inject the ordinary cement slurry. The special sulfoaluminate cement single liquid slurry is mainly used in the high-pressure surrounding rock and the sea water connecting section of the water exploration hole, and the superfine cement slurry is used to realize the controllable region grouting. The common cement–water glass double liquid slurry is mainly used for sealing the palm face, anchoring the orifice tube, and the water grouting of the probe hole, and simultaneously, the cement–sodium silicate double liquid slurry is mainly used for sealing the face, anchorage hole pipe and probing hole, and top water grouting. In the meantime, the cement–sodium silicate double liquid slurry with good injectability and high early strength of stone can be used in some drilling holes in front of the face of the palm.

5.2 Key technology of shield tunneling crossing undersea unfavorable stratum construction

There are some poor physical conditions such as water-rich sand layer, hard rock and uneven soft and hard strata, and solitary stone strata in the construction site of Xiamen Line 3. The safety risk of the shield construction is high.

The clay content of the sand layer and the gravel layer is not high, and the slurry shield is selected from the stratum properties. From the analysis of the permeability coefficients of the strata, the permeability coefficients of sand and circulargravel are inferred to be approximately $(1.2–5.8) \times 10^{-2}$ cm/s, and the percolation coefficient of granular strong weathering is $(1.2–5.8) \times 10^{-3}$ cm/s, which is close to or larger than the maximum allowable range of the earth pressure balance shield. The maximum water pressure is approximately 0.53 MPa. The tunnel passes through the sand layer and the circulargravel stratum with strong water permeability, and the groundwater is closely related to the sea water; therefore, it is suitable to select a mud shield. Considering the complex site environment and the geological conditions of shield tunneling, it is recommended to choose a compound mud water balance shield.

For the passing sand layer of the shield tunnel, the technical strategies are mainly adopted considering the aspects of the tunneling support pressure, tunneling parameters and tapping slag control, mud quality control, grouting quality control, shield sealing management, and so on. The balance chamber pressure of the mud shield is adjusted dynamically according to the change in the groundwater level in the intertidal zone. The shield tunneling machine maintains a constant driving speed, and the mud discharge and the mud delivery in the mud circulation system maintain the relative balance. The grouting behind the segment adopts special dispersion-resistant, rapid condensation grouting and timely supplementary grouting. The shield seal brush is regularly checked and replaced, and the lubricating grease is replenished between two seal brushes in time to maintain the pressure.

For the hard rock and the uneven soft and hard strata of the shield, technical measures are mainly taken considering the aspects of advanced detection, configuration of cutter head, tunneling parameters, control of slag quantity, and pretreatment of bedrock. Owing to the installation of advanced geophysical prospecting equipment in the shield cutter head, the scope of the hard rock in front of the tunnel can be checked in advance with construction prospecting. The hobbing tool with the rock breaking ability is arranged for the section of hard rock. Changes in the shield tunneling speed and the slag output are closely monitored to prevent the collapse of the upper soft soil. Meanwhile, for areas with high strength and difficult areas or isolated rock areas in which a shield machine is driven directly, the method of simultaneous grouting preprocessing after deephole blasting at sea is used as the preliminary method, and if necessary, the method is adopted.

5.3 Connections of different sections in the sea

The slurry shield starts from the Xiang'an side. The shield section and mine section are connected in the sea. To ensure safety and waterproof connections, the connecting point and procedures are considered comprehensively.

Solutions: Strata with complete bedrock and hard overlying rock are chosen as the connection point, with the exact location flexible depending on the actual schedule of the shield and mine methods. 1) If the mine method comes first, the expansion and reception section is reinforced. 2) If the shield method comes first, the rock wall thickness of the connection point is verified and the surrounding curtain grouting is performed thoroughly. Additionally, the shield structure is equipped with a detachable cutter head for the convenience of disassembly in the hole.

5.4 Passive cutter replacing technology in abnormal condition

During shield tunneling, the cutter replacing is decided according to the tool wear detection device and the tool inspection at intervals of a certain distance.The cutter replacing technologyunder hyperbaric condition is adopted to ensure the normal driving of the shield machine.In particular, when the sea shield is trapped due to cutter wearing, the sea surface vertical freezing can be used to reinforce the surrounding strata of the shield machine, and then repair the cutter and cutter

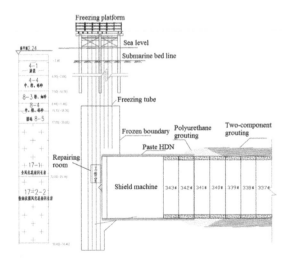

Figure 6.   Scheme of repairing cutter for vertical freezing reinforcement on sea surface.

head using chamber-entering technology under the normal pressure.. At the same time, the leading hole in the tunnel is frozen horizontally, and the repairing room is excavated and repaired under normal pressure.In addition, the adjacent tunnels can be used to freeze the surrounding strata horizontally and then repaired under normal pressure. These methods have been successfully applied in this project (Figure 6).

6 DURABILITY DESIGN

6.1 Design philosophy

The durability and the ultimate bearing capacity design of the tunnel structure are treated with equal emphasis. Meanwhile, the repairable requirements of structures and components are also met, and the reinforcement space is reserved for the subsea tunnels to increase their service life. The key joints of the reinforced concrete structure and the drainage system, the segment connecting member, and the segment sealing material constitute a durable system, in which the failure of a certain link not only directly affects the normal use of the project but may also accelerate the deterioration of other parts; therefore, the service life of the tunnel is limited by the weakness of the durable system.

6.2 Durability design of mining tunnel supporting structure

The mining tunnel adopts a compound lining structure, in which the secondary lining is designed for the bearing capacity and durability according to the full load. The durability of the initial support lacks a quantitative execution standard, only as an auxiliary measure to further improve the durability and the structural safety of the project.

6.2.1 Design measures for durability of initial support

In order to reduce the soil permeability and the water pressure behind the lining, advanced curtain grouting is used to seal the groundwater in the strongly permeable stratum such as the mine tunneling crossing weathered slot. It is recommended that slurry materials with good durability of ordinary and sulfoaluminate cement slurry be used to control the dosage of the cement–sodium silicate double slurry.

Initial support of the weak surrounding rock section adopts the type-8 grid steel frame, the anti-seepage grade of shotcrete is not lower than P6, and the wet spraying technology is used. The net protective layer thickness of the inner and outer sides of the steel frame is not less than 40 mm.

6.2.2 Measures for durability design of secondary lining

According to the environmental activity grade, the concrete strength and the construction measures are reasonably selected. The second liner is made of C50 and P12 reinforced concrete with a crack width limiting value of 0.15 mm and a net protective layer thickness of 60 mm.

In order to reduce the hydration heat of the molded concrete, ordinary Portland cement with low hydration heat is used in the project. The ratio of water to the binder is controlled within 0.36, and the amount of binding material is controlled at 360–480kg/m^3.In the meantime, a large number of high quality fly ash is mixed with 20%–30% of the binding material. The curing time of concrete is not less than 14 days, humidification and curing up to 28 days of the standard strength of 50%, and not less than 7 days.

In order to further improve the anti-crack and impermeability of concrete, polypropylene fiber is added at a dosage of 0.9kg/m^3.

6.2.3 Durability design of waterproofing and drainage system

Here, 2-mm-thick ECB waterproofing roll and waterproofing tape are considered to be resistant to sea water. It needs to be soaked in a 10% NaCl solution at 23°C for 168h. The tensile strength retention rate is not less than 80%, and the elongation at break is not less than 90%.

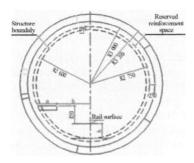

Figure 7.   Cross-sea shield section of Xiamen Rail Transit Line 3.

6.3   *Durability design of shield tunnel structure*

The durability design of the shield tunnel is mainly aimed at a segmented structure with a joint sealing material and a connecting bolt. In the past projects, the corrosion resistance of the latter two was ignored. The first submarine shield tunnel, the Shenzhen Qianwan cable tunnel, completed in 2005 is taken as an example. The carbonation and strength test results of the segment concrete were good on February 2014 and did not require any repairs. The corrosion of the connecting bolts between the shield rings was serious in the tunnel. The results show that the average corrosion rate of the bolt ends reached 80%, the average corrosion rate of the nuts reached 92.3%, the average corrosion rate of the bolt gasket reached 99.5%, and some severely corroded bolts were loose and leaked.

6.3.1   *Design measures for durability of segments*
The segment is made of C55 and P12 reinforced concrete, the width limit of crack developing is 0.15 mm, and the thickness of the net protective layer is 50 mm.

The durability indexes such as the water–binder ratio of concrete, amount of cementitious material, and chloride ion diffusion coefficient are strictly controlled. Additives such as polycarboxylic acid, a high-performance water reducer, and steel bar rust inhibitor, and admixtures such as fly ash and polypropylene fiber are reasonably selected. The required steam and tank curing temperature and time are ensured.

Anticorrosive and impermeable coatings with good alkali resistance, adhesion, and corrosion resistance, such as high-permeability modified epoxy and polyurethane waterproof coating, are applied on the outer side of the concrete segments.

6.3.2   *Design measures for durability of bolt and joint sealing materials*
At present, the shield bolts are usually coated with an anticorrosion material, which is limited by the thickness of the coating and years of protection, and does not meet the requirements. The project adopts 316L stainless steel bolts, which are resistant to chloride salt corrosion, and involves passivation treatment on their surface.

Micro-expansive cement mortar is used to block the bolt hand hole.

Materials of the sealant gasket and the sealing rod at the joint are considered to be resistant to sea water, and the inside of the segment is seamed.

6.3.3   *Reinforcement and anticorrosion measures*
Next, a 250-mm-thick reinforcement space is reserved for the clearance in the tunnel and for later maintenance reinforcement and extended structural service life (Figure 7).

7   CONCLUSIONS

The Wuliu section of Xiamen Rail Transit Line 3 is the first batch of cross-sea tunnel projects under construction in China, and the key technical problems are relatively new. On the basis of a geological analysis, an engineering analogy, and a comprehensive comparison, the key technical problems of the cross-sea section design are analyzed in this paper:

1. Combined with the geological conditions of different sections, the combination of the shield and miningmethods is used, and the single hole and single line type is adopted in the section layout. The mining section is used to increase the section, set the flue, and meet the needs of the second lining.
2. The drainage prevention and drainage in the mining section take into account the advanced grouting to control the displacement, and the drainage system is maintainable.
3. Targeted design measures are taken for the complex geology of the sea area including a weathered deep trough, a water-rich sand layer, hard rock and uneven soft and hard strata, and development of solitary rocks.
4. The durability design of the tunnel structure and the limit condition of the bearing capacity are treated with equal emphasis to consider the reservation of complementary spaces.

The construction of the undersea tunnel is large in scale, the train of thought of construction shall be high standard, strict requirement, technical innovation, mechanized matching construction, and promote the development of tunnel construction technology in China. At present, the construction of the auxiliary inclined well of the tunnel is in progress. The construction of undersea tunnels still has some problems in China, such as lack of experience and considerable safety risk. The technical design problems discussed in this paper need to be accumulated, categorized, and summarized in a future study.

REFERENCES

[1] Wang Mengshu, et al. 2010. Construction technology of tunnel and underground engineering in China. Beijing: People's Transportation Press.
[2] Zhang Jiuchang, SHI Junling, QU Yunteng. 2017. Technical specifications and operation of Seikan submarine tunnel. Chinese Railways. (5):91-96. (in Chinese)
[3] LI Yan, MU Huansen. 2009. The channel tunnel: construction process and innovation experiences. Journal of Engineering Studies. 1(1):90-96. (in Chinese)
[4] Wang Mengshu. 2014. Tunneling by TBM/shield in China: state-of-art, problems and proposals. Tunnel Construction. 34(3):179-188. (in Chinese)
[5] Huang Kan, Ding Guohua, Peng Limin, et al. 2011. Safety accessment of Liuyanghe tunnel construction through weatheredslot. Journal of Central South University. 42(3):803-809. (in Chinese)
[6] LI Guangyao. 2010. Technology for slurry shield to passing through up-soft-down-hard stratum in construction of Shiziyangtunnel. Railway Standard Design. (11):89-94. (in Chinese)
[7] Yang Wenwu. 2009. Development of immersed tube tunneling technology. Tunnel Construction. 29(4):397-404. (in Chinese)
[8] Huang Jun, LIU Hongzhou. 2006. Feasibility research on shield scheme of sumergedtunnel. Highway. 6(6):190-194. (in Chinese)
[9] Wang Mengshu. 2008. Current developments and technical issues of underwater traffic tunnel-discussion on construction scheme of Taiwan strait undersea railway tunnel. Chinese Journal of Rock Mechanics and Engineering. 27(11):2161-2172. (in Chinese)
[10] Tian Yuanjin, Guo Xiaohong, Liao Chaohua, et al. 2008. Proposal selection of Dalian bay subsea tunnel. Journal of China and Foreign Highway. 28(3):124-128. (in Chinese)
[11] Zhang Yueling. 2013. Proposal selection of planning programs for Bohai strait tunnel. Railway Economics Research. (1):12-16. (in Chinese)
[12] QU Shouxin. 2009. Preliminary conclusions of construction technology in Xiamen Xiangan subsea tunnel. Engineering Science. 11(7):24-30. (in Chinese)
[13] Wang Qian, QU Liqing, Guo Hongyu, et al. 2011. Grouting reinforcement technique of Qingdao Jiaozhou bay subsea tunnel. Chinese Journal of Rock Mechanics and Engineering. 30(4):790-802. (in Chinese)
[14] Song Chaoye, Zhou Shuming,Tan Zhiwen. 2008. Discussion on calculation of hydraulic load upon subsea tunnel lining. Modern Tunnelling Technology. S: 134-138. (in Chinese)
[15] Chen Junsheng, MO Haihong, Liu Shuzhuo, et al. 2012. Studies on the site selection of a segment prefabrication factory for the immersed tunnel of the Hong Kong-Zhuhai-Macao bridge project. Modern Tunnelling Technology. 49(6):122-127. (in Chinese)

Tunnels and Underground Cities: Engineering and Innovation meet Archaeology,
Architecture and Art, Volume 7: Long and deep tunnels – Peila, Viggiani & Celestino (Eds)
© 2020 Taylor & Francis Group, London, ISBN 978-0-367-46872-9

Excavations of the Povazsky Chlmec highway tunnel in Slovakia

M. Hilar

Czech Technical University in Prague and 3G Consulting Engineers s.r.o., Prague, Czech Republic

M. Srb

3G Consulting Engineers s.r.o., Prague, Czech Republic

ABSTRACT: The paper is focused on the bid preparation and the construction of the Povazsky Chlmec tunnel situated on the highway D3 in Slovakia. The tunnel has two double-lane tubes of length 2 x 2.2 km, the tunnel was excavated conventionally. The tunnel construction started in May 2014. Due to very short time schedule the excavation was allowed from 8 faces (4 faces from the central pit). The tunnel was opened for operation in December 2017. The tunnel was tendered using Yellow FIDIC Book (e.g. Design and Build contract), therefore contractor was responsible for detail technical solutions. The paper is focused on interpretation of site investigation and optimisation of excavation and support classes prior the bid submission. Consequently optimisation of conventional excavations during tunnel construction is described. In this project, contractual conditions allowed flexible and efficient tunnel construction with benefits to all project partners.

1 INTRODUCTION

The tunnel Povazsky Chlmec is situated on the section Zilina (Strazov) – Zilina (Brodno) of the highway D3. This section of the highway with the length 4 250 m goes over the bridge (1500 m long) across the water reservoir Hricov on the river Vah. Consequently the highway goes to the west portal of the tunnel Považský Chlmec. The tunnel includes two curves with opposite directions. From the east tunnel portal the highway continues via 400 m long bridge across the river Kysuce.

The Povazsky Chlmec contains two tubes with opposite direction of traffic, two lanes are in each tube, the tubes are connected with cross-passages (Figure 1). The tunnel is excavated conventionally with tunnel face sequencing to top heading and bench. The tunnel construction started in May 2014 and tunnel opening for operation was in December 2017. The excavation was allowed from 8 faces due to extremely short time schedule (two faces from the west portal, two faces from the east portal and 4 faces from the central pit). Cut and cover sections are in area of portals and also in the central pit. The South tunnel tube (STT) is 2186.5 m long including 2120.5 m long mined section. The North tunnel tube (NTT) is 2249 m long including 2200 m long mined section. The tunnel face area varies from 83.1 m^2 (the excavation class 4.1 without invert for the best ground) to 105.2 m^2 (the excavation class 6.3 with invert for the worst ground).

The maximum tunnel overburden is about 125 m. Surface above the tunnel is covered by meadows, forests, cottages, various ways, and water supply pipes. The tunnel level lies in altitude 341 to 352 m above sea level. Cut and cover sections were partly realised using the turtle method (tunnel excavation under ceiling generated from monolithic reinforced concrete) – Marik 2015. The turtle construction was used on the west portal (37.5 m on the STT and 50.0 m on the NTT). The turtle construction was also used in the central pit on the NTT in direction to east in length 34.5 m.

Figure 1. Primary lining of the tunnel Povazsky Chlmec.

The client of the new D3 highway section is NDS (Slovakian Highway Agency), the contractor is JV of companies Eurovia SK, Hochtief CZ and Stavby mostov Slovakia. The tunnel Povazsky Chlmec is constructed by Hochtief CZ, which executed 4 excavations from the central pit. Excavations from the west portal and the south portal were executed by TuCon (subcontractor of the Hochtief CZ). The detail documentation was prepared by IKP Consulting Engineers and Hochtief CZ. Geotechnical supervision of the tunnel excavation is done by 3G Consulting Engineers which are together with Hochtief CZ responsible for decision about application of designed excavation classes. Geotechnical monitoring and tunnel face mapping is done by Arcadis. Surveying works (tunnel alignment control, monitoring of the tunnel lining deformations, monitoring of overbreaks, etc.) is done by Angermeier Engineers. Construction supervision for client is done by JV of EUTECH&ESP&MULLER&API-D3.

2 GEOLOGY

The Povazsky Chlmec tunnel is situated in the Pieniny Klippen Belt. The northern and southern part of the tunnel route is formed of Pieniny formations with flysch beds of calciferous claystones and sandstones. In their upper part (along the corridor axis), there are spots of exotic conglomerates (age: Coniacian – Lower Santonian). The topmost part of the Pieniny beds is lined by exotic conglomerates in the south. The southern side of the Pieniny beds is lined by beds of varied marls (age: Upper Santonian – Maastrichtian). The maximum thickness of the flysch bed with inserts of varied marls and with exotic conglomerates (Figure 2) is around the village of Divinka (1,500 m). Its thickness is decreasing eastward, ca. 400 m. In the Kysuca unit of the Pieniny Klippen Belt, there are sandstones on top – 10 to 16 cm. These are grouped into smaller megacycles consisting of 3 to 9 beds. In the zone of thicker beds of up to 200 mm, the sandstones are solid in the whole thickness. Solid beds are frequent in complexes where sandstones prevail over claystones (in a ratio of 5:1). Laminated layers present in the marlstones are of siltstones. It is characteristic of the Pieniny Klippen Belt that the thickness of the individual beds is invariable. The main component of the sandstones is quartz, fragments of carbonaceous rocks, granitic and metamorphic rocks and fragments of volcanic rocks. The cement is carbonic, the share of clay matrix is from 0 to 13%. According to distribution of the main components, the sandstones are typical calciferous lithic sandstones.

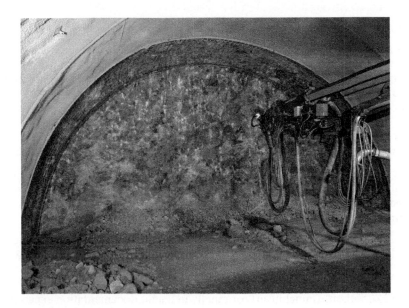

Figure 2.   Tunnel face with exotic conglomerates.

Conglomerate flysch of the Kysuca unit is very coarse. It consists of clusters and grains of large size, 100 – 400 m, and volume of 5 to 8 cubic kilometres, often with blocks and boulders. The individual rock bodies are 2 to 12 m thick, they are often gradually layered or with an inversion of the gradation with rounded boulder-sized fragments (up to 2.5 m) to fine powder. Flysch conglomerates are immature, both in terms of their petrographic composition and structure. Fragments of labile rock complexes mixed with lower quantities of plutonic rocks and metamorphic rocks are predominating in them. In the conglomerates, carbonaceous fragments prevail over volcanic rocks, clastic rocks, and intrusions and metamorphic rocks. Quartzite fragments and quartz conglomerates of significant sizes are present as well and locally also vein quartz. The water table was identified only locally at the depth ranging from 4m to 15m, on the base of terrace sediments, or in the underlying layers of the Mesozoic complex. When the hydrogeological survey was being carried out, not a single sample of water exhibited aggression to concrete structures.

3   TENDER

The contract for the construction of the D3 motorway section between Zilina (Strazov) and Zilina (Brodno) went out to tender according to the FIDIC Yellow Book. The tunnelled section forms 50% of the alignment and its cost forms a significant part of the overall tender cost. Therefore the company Hochtief CZ a. s. utilised external partners for the bid preparation. The company 3G Consulting Engineers s.r.o. interpreted the results of the site investigation and optimised the excavation classes. The most important result of 3G Consulting Engineers s.r.o. was expected utilisation of tunnel invert which has a very important impact on the final tunnel cost. The company IKP Consulting Engineers, s.r.o. optimised the scope and method of the support of construction pits, the scope of cut-and-cover tunnel sections, the tunnel cross-section and the final lining dimensions, to forecast the application of the unreinforced final lining, to optimise the block diagram and safety elements (the number and location of cross passages, emergency stopping lay-bys, fire hydrant niches, SOS boxes, drainage cleaning recesses etc.). A prognosis of tunnel sections with various final lining reinforcement contents or an unreinforced final lining was developed on the basis of the prognosis of the division of the mined tunnel part into excavation support classes. The Germany-based Hochtief head office also participated in the bid preparation.

The company 3G Consulting Engineers realised namely the following tasks:

- Analysis and interpretation of geological conditions with regard to the tunnel excavation
- Evaluation of the swelling risk, presence of clay minerals
- Optimisation of designed excavation classes, volumes and prices of proposed support measures
- Evaluation of geotechnical conditions and optimisation of excavation classes along the STT and NTT
- Evaluation of the further muck utilisations
- Numerical modelling of the primary lining in 2D (5 models in various cross-sections)

The most important impact on the proposed tunnel cost had a length of section with and without tunnel invert. The distribution of excavation classes with and without invert proposed by 3G Consulting Engineers was the following:

South tunnel tube (STT):

- 1631 m (79%) without tunnel invert (originally in tender documentation expected 47%)
- 438 m (21%) with tunnel invert (originally in tender documentation expected 53%)

North tunnel tube (NTT):

- 1735 m (82%) without tunnel invert (originally in tender documentation expected 43%)
- 475 m (18%) with tunnel invert (originally in tender documentation expected 57%)

The proposal was discussed, during a bid preparation was adopted more conservative proposal (with higher utilisation of the tunnel invert). Realised conventional excavations of the Povazsky Chlmec tunnel showed even better geology in comparison to 3G proposal, finally tunnel invert almost has not been utilised.

4 REALISED CONVENTIONAL EXCAVATIONS

The tunnel construction started in May 2014 (Petko & Pastrnak 2015). The conventional excavations of the tunnel started after constructions of the central pit and the west portal including turtle structures. The realised schedule of conventional excavations of the tunnel tubes was the following:

Excavations from the central pit:

- STT towards west started 16/2/2015, top heading excavation was finished 20/8/2015, excavated 567 m (including one emergency bay) – average progress about 95 m per month
- NTT towards west started 7/4/2015, top heading excavation was finished 26/9/2015, excavated 566 m (including one emergency bay) – average progress about 99 m per month
- STT towards east started 12/3/2015, top heading excavation was finished 6/1/2016, excavated 719 m (including one emergency bay) – average progress about 80 m per month
- NTT towards east started 27/4/2015, top heading excavation continues, k 31/1/2016 excavated 691 m (including one emergency bay) – average progress about 77 m per month

Excavations from the west portal:

- STT started 21/3/2015, top heading excavation was finished 18/8/2015, excavated 515 m – average progress about 103 m per month
- NTT started 29/3/2015, top heading excavation was finished 3/9/2015, excavated 590 m – average progress about 118 m per month

Excavations from the east portal:

- STT started 17/10/2015, top heading excavation was finished 15/1/2016, excavated 303 m – average progress about 101 m per month
- NTT was not excavated from the east portal

Table 1. Comparison of proposed and realised distribution of excavation classes in STT.

	Tender documentation - 2006 (m)	Tender documentation - 2006 (%)	3G 2013 (m)	3G 2013 (%)	Reality 2015/16 (m)	Reality 2015/16 (%)
Cut and Cover section	27.0	1.3	73.0	3.4	66.0	3.0
Turtle section	0,0	0.0	0.0	0.0	37.5	1.7
MP 1 (with invert)	83.0	4.0	0.0	0.0	12.1	0.6
6/3 (with invert)	210.0	10.0	0.0	0.0	17.1	0.8
6/2 (with invert)	396.9	18.9	438.0	20.4	31.2	1.4
6/1 (with invert)	162.0	7.7	0.0	0.0	44.7	2.0
5/2 (with invert in tender, without invert later)	240.0	11.4	495.0	23,1	143.1	6.6
5/1 (without invert)	573.1	27.3	481.0	22,5	586.7	26.9
4/2 (without invert)	327.4	15.6	425.0	19.8	599.9	27.5
4/1 (without invert)	80.0	3.8	230.0	10.7	644.0	29.5
TOTAL	2 099.3	100.0	2 142.0	100.0	2 182.3	100.0
With invert	1 092	53	438	21	143	7
Without invert	1 007	47	1 704	79	2 040	93

All excavations were completed in 2016. The average progresses of conventional excavations were affected by available number of miners and machines, usually only 3 excavations were ongoing from the central pit in one time (one face was usually stopped), also all emergency bays were excavated from the central pit. Comparison of expected and realised distribution of excavation classes in the STT is presented in Table 1.

A very significant part of excavations was realised in the excavation class 4.1 with the length of advances in top heading up to 3.5 m. The excavation class 4.1 has one layer of meshes without lattice girders, 10 cm of sprayed concrete and 5/6 rockbolts with length 3 m. The accuracy of drilling works for explosives was extremely important due to higher length of advances (to minimise overbreaks). Boomers Atlas Copco were used with semiautomatic setting of drilling schemes. The software in boomers allowed comparison of theoretical and realised drilling schemes. All excavated profiles were recorded by surveyors. Comparison of results of surveys with drilling records allowed identifying reasons of overbreaks (geological or technological) and optimisation of excavations.

Table 1 indicate that distribution of excavation classes in tender documentation was quite conservative. Tender documentation proposed utilisation of invert on 53% of the STT, in reality only 7% of STT was excavated in classes with proposed invert. But finally no invert was applied at the STT, as all lower parts of the tunnel profile are in the reasonably strong rock and convergences stabilised on low values. Evaluation prepared by 3G Consulting Engineers prior start of excavations expected 21% of the STT with invert which was significantly closer to reality in comparison to tender documentation.

5 CONCLUSION

The tunnel Povazsky Chlmec was opened for operation in December 2017. Experience from its conventional excavations showed that participation of competent external subjects can be beneficial for contractors in stages of the bid preparation and tunnel construction. Especially in case of the contract according the Yellow FIDIC book (e.g. Design and Build contract) the effectiveness of the tunnel excavation is extremely important for the final economical result.

Cost-effective conventional excavations (optimisation of advance length and applied support) tends the contractor to accept higher risks (overbreaks, collapses). This factor can be compensated by involvement of external geotechnical experts, who are not directly financially involved in economic aspects of excavation (speed, support), thus can be more objective in comparison to the contractor regarding acceptable risk. This approach had an impact on the completed tunnel excavations; optimisation was done according to behaviour of the open tunnel face together with monitoring results. Majority of conventional excavations could be realised with minimum support in excavation classes 4.1 and 4.2 with no significant over-breaks or collapses.

The Povazsky Chlmec tunnel excavations were realised according original time schedule which was from the beginning very tight. The tunnelling problems on the Povazsky Chlmec tunnel were very small in comparison to other neighbouring tunnelling projects near Zilina on the highways D1 and D3. This fact was partly caused by favourable geology, but also by detailed preparation, optimal design and good construction management including competent external subjects.

ACKNOWLEDGEMENTS

This paper was prepared with the support of the research grant TACR TE01020168.

REFERENCES

Marik L. 2015 Povazsky Chlmec tunnel on Zilina (Strazov) – Zilina (Brodno) section of the D3 motor-way – from design to realisation. *Tunel 24 (3): 88–103*.

Petko, A. & Pastrnak, V. 2015 Technology, procedure and specifics of the Povazsky Chlmec tunnel exca-vations. *Tunel 24 (3): 81–87*.

Tunnels and Underground Cities: Engineering and Innovation meet Archaeology,
Architecture and Art, Volume 7: Long and deep tunnels – Peila, Viggiani & Celestino (Eds)
© 2020 Taylor & Francis Group, London, ISBN 978-0-367-46872-9

Requirements and methods for geometric 3D modelling of tunnels

M. Hofmann, R. Glatzl & K. Bergmeister
Brenner Base Tunnel BBT SE, Innsbruck, Austria

Š. Markič & A. Borrmann
Chair of computational modelling and simulation, Technische Universität München, Germany

ABSTRACT: A BIM model contains information relevant to planning, construction and operation. 3d geometry is one of the most important information without many BIM applications would not be possible. An important trend in the building industry is parametric modelling in order to adapt models quickly to changing constraints. Especially in tunnelling, changes in geometry are possible at any time, for example adjustments of the alignment caused by measurement inaccuracies. In case of tunnelling, the arc length of the horizontal alignment (stationing) is the most important parameter. The paper highlights requirements and describes different aspects of geometric modelling for tunnels. A focus is on geometric modelling approaches of the alignment as the basis of 3d models. Furthermore, the current status of open standards such as IFC alignment for infrastructure buildings is presented.

1 INTRODUCTION

The Brenner Base Tunnel (BBT) is a major European infrastructure project of the Helsinki (Finland) - La Valletta (Malta) North-South TEN corridor. The corridor connects the economic centres and ports in Italy with those in Germany and Scandinavia. The BBT's two 64 km long, parallel, single-track railway tunnels enable freight and passenger trains to cross the Alps between Austria and Italy without overcoming the 1371 m high Brenner Pass (Bergmeister 2011). As shown in Figure 1, the exploration tunnel for geological and hydrogeological exploration is excavated in the middle, 12 m below, between the single-track railway tunnels. Approximately 50% of the 230 km long tunnel system of the BBT will be excavated by tunnel boring machines and approximately 50% by drilling and blasting. By October 2018, approximately 40% of the entire length of the tunnel system had already been excavated.

It is planned to apply BIM methods for planning and construction in the future. A Building Information Model contains information relevant for planning, construction, operation and maintenance. Three-dimensional geometry is one of the most important information, without many BIM applications would not be possible (Borrmann et al. 2015). Important advantages of the availability of a 3d model are for example:

- 2d plans (views, sections) can be derived consistently
- simpler and more precise quantity determination and plausibility checks
- basis for further analyses and calculations (e.g. geological investigations, excavation data, static investigations, ...)
- basis for visualizations and reporting tools.

Possible BIM applications include:

- as-built documentation
- building inspections
- public relations/visualisations

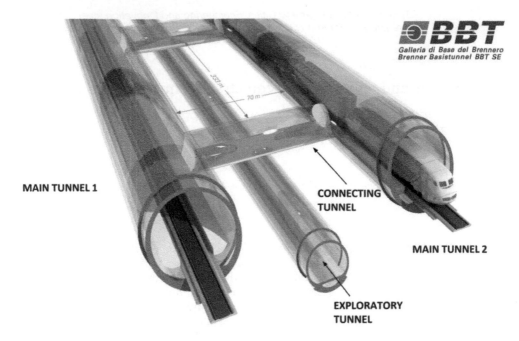

Figure 1. Schematic cross-section of the Brenner Base Tunnel.

- documentation of excavation process
- administration of tunnel scan data

The contribution is structured as follows: In the next section the geometric elements needed for modelling a tunnel are reviewed. In Section 3 modelling approaches to represent the geometric elements in the context of CAD or BIM models are given. Section 4 examines the applicability of the IFC format to infrastructure buildings.

2   GEOMETRIC ELEMENTS OF TUNNELS

2.1   *Alignment*

The alignment is the backbone of every infrastructure object as it describes the object's base curve. It is its top most abstraction and serves as a positioning reference. It can be compared to the structural axes in the building sector. A 3d model of a tunnel system can be constructed by extrusion of cross sections along the alignment.

2.1.1 *Geometric elements*

An alignment can be accurately represented as a 3d curve in the engineering Cartesian coordinate system (x,y,z). It is usually a superposition of two planar curves: the horizontal alignment (HA) and the vertical alignment (VA). HA is the projection of the aforementioned 3d curve onto the horizontal (x,y) plane. As usual in practice, we define a new coordinate axis s along the HA called the stationing axis. VA is the projection of the 3d curve on the curvilinear (s,z) plane, i.e. a ruled curve.

Each alignment consists of an ordered array of elements of three types: straights, curves, and transition curves. There are many types of curves and transition curves used in practice. The used types depend on the type and category of the infrastructure asset and are presented in Table 1. Knowing the order, the types, and the parameters of each individual element in both HA and VA the resulting alignment is unambiguously defined.

Table 1. Different types of elements for rail alignments with their parameters.

Alignment	Plane	Curve	Transition curve
Horizontal	(x,y)	circular arc	clothoid, Bloss curve, Vienna curve*, sinusoid, Cosinusoid, cubic parabola, biquadratic parabola
Vertical	(s,z)	circular, parabolic arc	-

* The Vienna curve ("Wiener Bogen") is recent development in Austria and used by ÖBB and Wiener Linien

2.1.2 *Alignments and axes*
In the context of tunnels, several alignments and axes are of interest which may or may not coincide. The axes commonly used in tunnelling are: (i) track axis, (ii) tunnel axis and (iii) profile axis respectively. The alignment typically refers to the track axis and each track has its own alignment. In general, there may be several track axes and thus alignments in a tunnel. The alignments are in principal independent. Therefore their gradients may differ within a cross section as indicated in Figure 2.

2.2 *Cross section*

Along the alignment different cross sections may be defined. Cross sections include information of the geometric elements of tunnel support system like shotcrete and inner lining. Cross sections are referenced to intervals of the alignments and its stationing respectively. Between different cross section a transition zone may be defined.

2.2.1 *Geometric elements*
Cross sections are typically built up of circular arcs and straight lines. The arcs are usually tangential as shown in Figure 3.

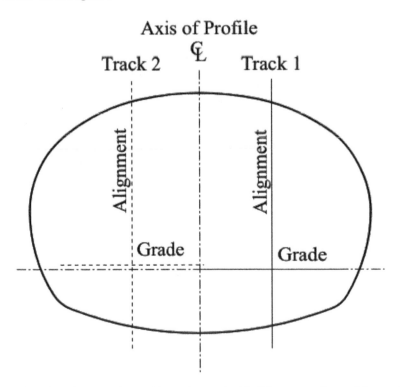

Figure 2. Track and profile axes of a double track tunnel with different gradients of the tracks.

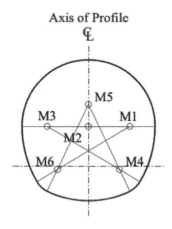

Figure 3.   Geometric elements of a cross section built up with circular arcs.

2.3   *Cant/transverse slope*

As a rule, railway tracks have so-called cants, which serve to minimise transverse acceleration forces when bends are made. Transverse slope is defined as the twisting of the tunnel profile relative to the vertical. This is usually used in road construction for roadway drainage.

2.3.1   *Geometric elements*

Cant of the rail influences the vertical position of a track. The transverse slope corresponds to a twisting of the profile. Both are generally related to the curvature of the alignment. However, there are exceptions, e.g. in the case of large radii, the rail may have no cant. Therefore, the cant and transverse slope should be modelled independently of the curvature of the alignment, depending on stationing.

2.4   *3d geometry*

Given an alignment, cross sections, cant and transverse slope are defined for intervals of the stationing. All together define implicitly the 3d geometric elements of a tunnel. Whole tunnel systems may be generated by boolean combinations of tunnels, e.g. by intersection of two tunnels.

2.5   *Coordinate systems*

Alignments are commonly defined within a project coordinate system (Markič et al., 2019). Cross sections are referenced to intervals of the stationing of an alignment. Local stationing may be

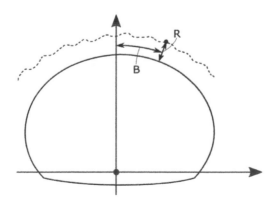

Figure 4.   Sketch of relating scan data to reference surfaces using a local coordinate system (B,R).

additionally defined for constructions work (*Tunnelmeter*). Cross sections have their own local cartesian coordinate system. The amount of coordinate systems indicates the special importance of dealing with and transforming geometry and informations between these coordinate systems.

2.6 *Further geometric elements*

In the sections above the most basic geometric elements of a tunnel system are given. Depending on the application on hand further geometric elements may exist. For instance tunnel scans with terrestrial laser scanning deliver point cloud data. It is advantageous to relate these data to reference surfaces as indicated in Figure 4 to efficiently detect over- and underprofiles. Due to the massive amount of data using scanning techniques data management is of special importance.

3 GEOMETRIC MODELLING OF TUNNELS IN THE CONTEXT OF BIM

In architecture, engineering and construction industry, a solid modelling approach is usually chosen.

3.1 *Implicit an explicit geometric modelling*

The geometry of solids can be modelled either by explicit or implicit methods. The so called Boundary Representations (BRep) is a popular choice for explicit methods. Implicit schemes are based on the construction sequence of a building and are also called procedural methods. Both approaches are used in BIM systems and associated data exchange formats like the IFC-standard. Explicit and implicit methods have their specific pros and cons. Thus a hybrid approach is often used in BIM systems. The geometry and its parameters are defined in an implicit way whereas the geometric representation is internally converted to an explicit representation like BRep (Borrmann et al. 2015).

3.2 *Parametric modelling*

An important achievement in the area of building industry is parametric modelling. Geometric models are designed with dependencies and constraints in such a way that a flexible model is created that can be quickly and easily adapted to changing constraints. In building construction, parameters include, for example, the height, width, length and position of a column. For civil engineering constructions, the most important parameter is the stationing.

3.3 *Geometric modelling of an alignment*

3.3.1 *Implicit modelling of an alignment*

The horizontal and vertical alignment can be unambiguously represented either by segments or point of intersection (PI). The first option considers each alignment's elements to be individual segments with their own parameters. The second option represents the HA and VA with an ordered array of points – points of horizontal intersection (PHIs) and points of vertical intersection (PVIs), respectively. These can be obtained by extending all sections of zero curvature to obtain a polygon. With these polygons together with "fillet" parameters an alignment can be defined.

Both options are exemplary shown on Figure 5 for a simple alignment. The HA includes elements of type straight, circular arc, and clothoid transition curve, while the VA includes elements of type straight and parabolic curve. The used notation is explained below.

- A_i^b and A_i^a or A_i are the transition curve parameters before (b) and after (a) the curve at i-th PI or at j-th segment. Here exemplary the clothoid parameter $A \geq 0$ in [m], however other parameters may be needed for other transition curves.

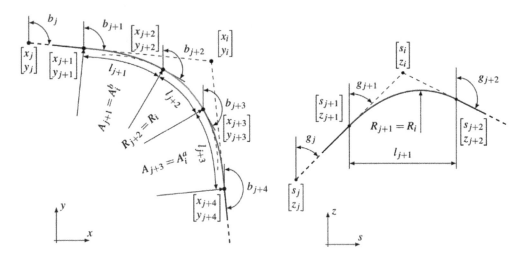

Figure 5. The types of segments for HA (left) and VA (right) with their parameters marked. Each segment has a starting point, a direction and defined length. Curved segments also include information about the underlying curve, for example radius for circular arc. The PIs are marked with dashed lines.

- b_j is the direction of the tangent at the beginning of the j-th segment in HA, i.e., the azimuth angle from the North direction.
- g_j is the grade of the tangent at the beginning of the j-th segment in VA, i.e., the slope in the direction of s-axis.
- i is the index of a PI.
- j is the index of a segment.
- l_j is the length of the j-th segment. The length is always measured along the s-axis.
- R_j or R_i is the radius of the curve at the i-th PI or at j-th segment (in [m]). In case of a parabolic curve, the value denotes the radius at its vertex.
- $[x_i, y_i]^T$ and $[s, z_i]^T$ are the coordinates of the i-th PI in HA and VA, respectively.
- x_j, y_j^T and s_j, z_j^T are the coordinates of the beginning of the j-th segment in HA and VA, respectively.

3.3.2 *Explicit modelling of an alignment*
For further geometric operations in CAD or BIM systems like extrusion of cross sections or intersection of two tunnels it is beneficial to model the 3d alignment geometrically explicit. Then these operations can be accomplished by means of standard CAD methods.

3.3.2.1 APPROXIMATION OF A 3D ALIGNMENT WITH POLYGONS

The simplest method to approximate a 3d alignment is a polyline (i.e. element-wise linear curve), consisting of straight lines between sufficiently precise calculated points on the axis. From a mathematical point of view, this polygonal approximation is C0 continuous, i.e. the function values are continuous, tangents and higher derivatives and thus the curvature not. Although the position of the axis can be described sufficiently precise with this approach, the error in transverse direction becomes larger, since the tangent direction is approximated more coarsely.

Another disadvantage is, that a relatively large number of pre-calculated vertices are required compared to the approximation method presented in the next section. Due to the size of the BBT, this can lead to considerably greater memory requirements and thus longer processing times for a 3d model based on it.

3.3.2.2 HIGHER ORDER APPROXIMATION OF A 3D ALIGNMENT

Due to the aforementioned disadvantages of the polygonal approximation, a geometric approximation of the axis with higher continuity requirements is advantageous. Some alignment elements

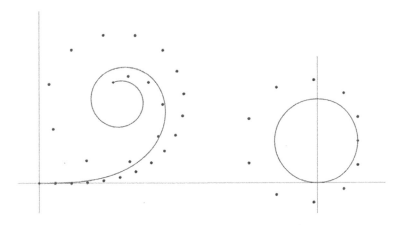

Figure 6. Modelling a clothoid and a circle with a Bézier curve of order 25 and 11 respectively.

(straight lines, parabolas) are polynomial. Thus a polynomial modelling approach with stationing as a parameter is promising. In addition, power series expansion for circles and clothoids converge very quickly, so that only a few terms are required. This approach is usually used in surveying.

In CAD systems, curved geometries are also referred to as freeform curves and surfaces. The most common types of freeform curves are Bézier curves, B splines and NURBS (Non-Uniform Rational BSplines). All three types are defined by a series of control points, with the first and last control point lying on the curve and the remaining control points defining the path between them. These free-form curves are mathematically described as parametric curves. Bézier and BSpline curves and are element-wise polynomials. They are therefore ideally suited for modelling the alignment of infrastructures. It is important that the used CAD system supports higher order (>10) curves, in order to achieve sufficient accuracy for strong curved elements.

In Figure 6 the approximation of a clothoid with a Bezier Curve of order 25 and order 11 respectively is shown. The dots are control points of the Bezier curves.

The Figure suggests that Bézier curves are suitable for modelling alignment elements. With this approach the alignment of the BBT system was modelled. Figure 7 shows a view of the 3d alignment of the hole BBT system and Figure 8 a detail of main, connecting, access and exploratory tunnels in the area of Ahrental. Dots indicate the start and end points of curve segments.

A property of Bézier curves is that they are affine invariant. This means that affine transformations such as translation and rotation can be applied by applying the respective transform on its control points. Transformations between coordinate reference systems are a common task in tunnelling (Markič et al. 2019). Helmert transformation is often used for this task. It is a simplified affine transformation. Thus the presented explicit geometric modelling approach is well suited for such transformations.

3.4 Geometric modelling of cross sections

Cross sections consist usually of circular arcs and straight lines, i.e. a subset of geometric elements of an alignment. Thus geometric modelling of cross sections is similar to modelling an alignment and will not be discussed further here.

3.5 Geometric modelling of a 3d model

With a suitable geometric model of an alignment and corresponding cross sections a 3d model of a tunnel can be obtained with standard CAD methods like extrusion, sweep and lofting operation. 3d models of multiple tunnel segments can be jointed with boolean operations to obtain a model of a tunnel system. Figure 9 shows a 3d model of an area of the BBT tunnel system (main and connecting tunnels at Ahrental) obtained with these standard CAD methods.

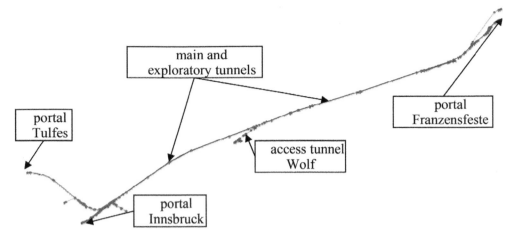

Figure 7.   Alignment of BBT tunnel system modelled with Bézier curves.

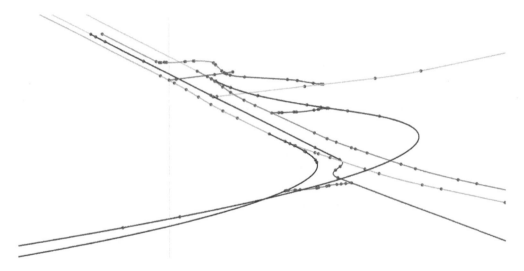

Figure 8.   Detail of the alignment of a main, connecting, access and exploratory tunnels at Ahrental.

4   OPEN BIM – OPEN FORMAT – IFC ALIGNMENT

A 3Dd polyline represents the alignment in a simple yet sufficiently precise manner depending on the density of points. However, changes to such representation are very demanding as each individual point needs to be adjusted. Therefore, a parametric alignment model that retains as many of the design parameters intact as possible is preferred for data exchanges.

There are many formats available which include a model of an alignment, like industry foundation classes (IFC), LandXML and Objekt Katalog Straße (OKSTRA) (AMANN ET AL., 2014). The chosen format we had a critical look at, was the recently developed IFC Alignment (IFC schema version 4x1.0), as it is becoming internationally very well accepted. The IFC standard extension opted to model the alignment entities according to the segment definitions schema explained in Section 3 (Liebich et al., 2017).

In the IFC schema documentation standard, both these representations of an alignment are foreseen to be supported by software vendors:
– IfcAlignmentCurve as a 3D (HA and VA) or a 2D (only HA) curve,
– IfcOffsetCurveByDistances as a curve relative to some IfcAlignmentCurve, and

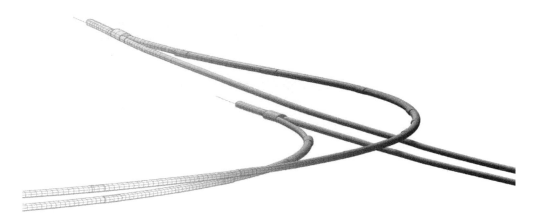

Figure 9.   3d model of main and connecting tunnels at Ahrental.

– IfcPolyline as a 3D or a 2D curve.
– the superposition of two planar curves as described in Section 2 and as a polyline

The schema itself allows for any curve to be used (all entities deriving from IfcCurve), though, and a side note in the documentation allows for the list above to expand in the future. We would like to encourage adding the IfcBSplineCurve to this list and thus supporting the representation presented in Section 3.3.2.

With the basis curve well defined, further volumetric objects like IfcSweptAreaSolid or IfcSectionedSolid can be produced and exchanged.

5   CONCLUSIONS

In this contribution the required elements for geometric modelling of a tunnel are described. For large tunnel systems like the Brenner Base Tunnel the accurate and efficient geometric modelling as basis for a BIM model is of special importance. The essential geometric constituents such as alignments and cross sections to obtain a 3d model are presented. A central topic of the contribution is geometric implicit and explicit modelling approaches for the alignment.

It is planned to model the whole tunnel system of the Brenner Base tunnel in the next future as a 3d model. Based on this, information will be added to the model to obtain "real" BIM models and use them operationally.

REFERENCES

Amann, J., Flurl, M., Jubierre, J. R. & Borrmann, A. 2014. An alignment meta-model for the comparison of alignment product models. *In: Proc. of the 10th European Conference on Product & Process Modelling*, Vienna, Austria.

Amvrazis, S, Bergmeister, K., Glatzl, R. W. 2019. *Optimizing the Excavation Geometry using Digital Mapping*. ITA-AITES World Tunnel Congress, Naples, Italy

Bergmeister K. 2011. Brenner Basistunnel - Brenner Base Tunnel - Galleria di Base del Brennero. *Tappeiner Verlag*

Borrmann A., König M., Koch C. & Beetz, J. 2015. Building Information Modeling: Technologische Grundlagen und industrielle Praxis. [Building Information Modeling - Technological foundations and industrial practice]. *VDI-Buch, Springer Fachmedien*, Wiesbaden.

Markič, Š., Borrmann, A., Windischer, G., Glatzl, R. W., Hofmann, M., Bergmeister, K., 2019. *Requirements for geo-locating long transnational infrastructure BIM models*. ITA-AITES World Tunnel Congress, Naples, Italy

Liebich, T., Amann, J., Borrmann, A., Chipman, T., Hyvärinen, J., S. Muhič, Mol, L., Plume, J. & Scarponcini, P. 2017. IFC Alignment 1.1 Project, IFC Schema Extension Proposal.

Tunnels and Underground Cities: Engineering and Innovation meet Archaeology,
Architecture and Art, Volume 7: Long and deep tunnels – Peila, Viggiani & Celestino (Eds)
© 2020 Taylor & Francis Group, London, ISBN 978-0-367-46872-9

Brenner Base Tunnel – Key aspects of the guide design

R. Insam, W. Eckbauer & T. Gangkofner
BBT SE, Innsbruck, Austria

K. Matt
ILF Consulting Engineers, Rum, Austria

ABSTRACT: The 64 km long Brenner Base Tunnel is the heart of the SCAN-MED Trans-European Corridor. To the north, the BBT tunnel system is linked not only with Innsbruck central station but also to the access routes in the Lower Inn Valley and, to the south, with the planned new railway stretches running toward Verona. Changes were made to the project during the authorization procedures. In addition to this, BBT SE had already been considering optimizations based on experience obtained during other long tunnel projects and the current technological state of the art. There was also the need for greater in-depth uniformity of specifications and construction details as a basis for the tender procedures for the main tubes in both Austria and Italy. This contribution is meant to give an overview of the experience and results from transnational guide design planning, divided into various working groups.

1 DEFINITION OF TASKS

The preliminary project for the construction of the Brenner Base Tunnel, which includes key route definitions, potential construction methods and an initial cost calculation, was created in the years between 1999 and 2002 (Phase I) (Bergmeister. 2011).

The submission plans for the Brenner Base Tunnel on both the Austrian and Italian sides of the project were drafted between 2005 and 2008 and submitted to the various authorities. In 2009, the permits were granted by the Austrian Federal Ministry for Transport, Innovation and Technology (BMVIT) and the Italian Interministerial Committee for Economic Planning (CIPE) (Phase II).

In 2008, construction of the first exploratory lot "E91 Aicha-Mauls" began in Italy. In Austria, construction of the first explanatory lot "E41 Erkundungslos Ahrental" began in December 2009 (Phase IIa).

Phase III, which involved the construction of the main tunnels, commenced in March 2011. From this point on, the construction tender specifications were also prepared for the main tunnels.

In the course of and after the approval procedures, the tunnel system was optimised and greatly simplified on both sides in a collaboration between BBT SE, the Austrian Federal Railways (ÖBB) and the Italian Railway Network (RFI) based on previous experience from other long tunnel construction projects.

In the course of the 2013 cross-border guide design procedure, the changes resulting from the approval procedures and the project optimisation measures taken up to this point were incorporated into the project. This resulted in a new basis for cross-border planning. This later formed an essential basis for the main lots which were to be tendered on both the Austrian and Italian sides of the project.

Up to the end of August 2018, 88 km out of a total of 230 km have been excavated. 85% of the overall civil engineering works have been awarded.

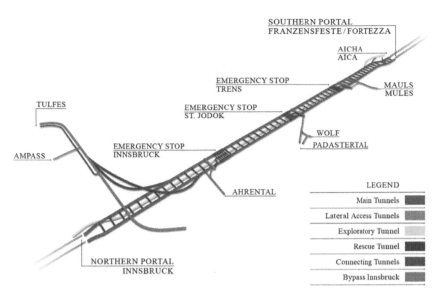

Figure 1.   Brenner Base Tunnel system.

2   2013 CROSS-BORDER GUIDE DESIGN PROCEDURE

The 2013 cross-border guide design procedure commenced in mid-2011 and was completed at the end of 2013. Other optional services, such as the review of various tender documentation for the main lots, were provided by the planning group until 2017.

Since different laws, standards and regulations apply in both Austria and Italy and different rules for railway tunnel projects also apply, one of BBT SE's goals was to establish a project-specific design guideline which would harmonise existing regulations in Austria and Italy where possible and, where this was not possible, adopt national and international legislation (Matt et al. 2014). On the basis of a life-cycle of 200 years, a specific safety concept has been worked out on the basis of EC 1 (Bergmeister. 2017).

These framework conditions and specifications resulted in the following main objectives for the guide design procedure:

- the incorporation of the changes and measures from the approval procedures.
- the incorporation of the project optimisation measures coordinated by ÖBB, RFI and BBT SE.
- the drafting of additional project optimisation measures.
- the drafting of additional cross-project basic technical specifications which were pertinent to the construction contract for the future tender and execution documentation for the main lots

The following project optimisation measures were incorporated and developed in the course of the cross-border guide design procedure and therefore in the 2013 project stage:

- the simplification of the multifunctional stations (decoupling of the emergency stop/cross-over) (Figure 2).
- the reduction of cross-overs in the main tunnels (from three to one).
- the optimisation of the Innsbruck connecting tunnels: the creation of a standard cross-section with a continuous wall to separate the travel and escape routes resulting in the avoidance of very long and complicated connecting side tunnels (Figure 3).
- the simplification of the mountain and track drainage system in terms of a reduction of future maintenance work.

Approved project 2009

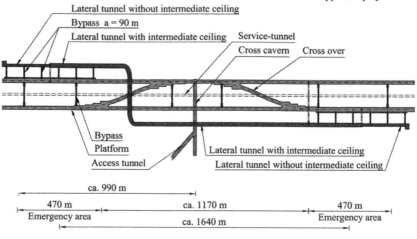

ca. 990 m

470 m
Emergency area

ca. 1170 m

ca. 1640 m

470 m
Emergency area

Project status 2013

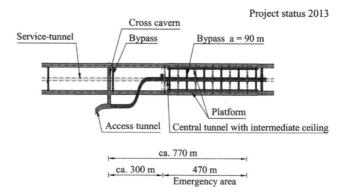

ca. 770 m

ca. 300 m | 470 m
Emergency area

Figure 2.   Comparison of multifunction station/emergency station.

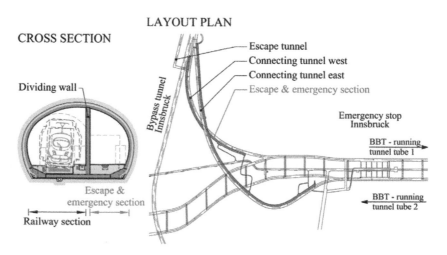

Figure 3.   Connecting Tunnel Innsbruck with railway section and escape & emergency section.

- the optimisation of cable paths and ducts in the main tunnels and at the junctions with the connecting side tunnels.
- the simplification and standardisation of the integration of the connecting side tunnels into the main tunnels.

3 ORGANISATION

The cross-border guide design procedure was divided into seven working groups (WGs) with different objectives. They consisted of various representatives of BBT SE and the planning consortium. In addition, additional experts were consulted for specific tasks.

The scheduled drafting of the guide design procedure in the various working groups, in which BBT SE, the planning group and experts actively took part in a discussion process on the scope of services and the results of such, was found tob be a correct decision (Matt et al. 2014). On the one hand, this work encouraged discussions in the following areas:

- Technical aspects,
- National characteristics and specifications and
- Experience from other projects (Gotthard Base Tunnel, Lötschberg Base Tunnel, Koralm Base Tunnel, Semmering Base Tunnel).

On the other hand, it offered the opportunity to incorporate BBT SE's knowledge and experience from other completed projects.

4 OVERVIEW AND DESCRIPTION OF THE DIFFERENT WORKING GROUPS (WG)

4.1 *Working group WG 1 – Project Management:*

The objective of this working group was to coordinate all other working groups. In addition, this working group had to identify the various conditions and measures from the approval procedures and take them into account in the various documents which had to be prepared.

4.2 *Working group WG 2 – Alignment*

The objective of this working group was the planning of the route for the entire tunnel system in the new BBT coordinate system, taking into account project changes and project optimisation measures.

4.3 *Working group WG 3 – Geomechanics*

The objective of this working group was:

- The development of geomechanical guidelines which were to form the basis for the geomechanical planning in both countries.
- The creation of a fault zone management system as an aid for the optimised management of geological/hydrogeological/geotechnical fault zones.
- The on-site evaluation of geological, hydrogeological and geotechnical documentation for a standardised process for determining parameters.

4.4 *Working group WG 4 – Design*

The objective of this working group was:

- Development of sizing principles for load-bearing ability, serviceability and durability for a life cycle of 200 years.
- Updating project and usage requirements.
- The development of system and type plans, such as standard profiles based on driving methods (conventional and machine driving with TBM-O, TBM-S/DS) for the main tunnels, Innsbruck connecting tunnels, connecting side tunnels and emergency stops.
- The creation of detailed plans, such as cleaning shafts, cable shafts, joint and sealing details, etc.
- The creation of schematics, such as drainage systems, lateral walkway heights, cable ducts, etc.
- The creation of specifications for the dimensioning and structural design of the outer and inner shell (shotcrete, in-situ concrete, single and double-shell segmental lining).
- The harmonisation of quality standards for concrete, sealing and drainage, cable conduits and shaft coverings including the test conditions and frequency.
- The specification of project-specific tolerance standards taking into account measurement tolerances depending on the driving method (conventional and machine driving with TBM-O, TBM-S/DS) for all standard profiles based on the current construction program.
- The preparation of specifications for profile measuring with a tunnel scanner.

4.5 Working group WG 5 – Technical contract terms and service book

The objective of this working group was different for Austria and Italy and involved the preparation and creation of:

- Quality standards for construction products for the lining and supports including test conditions, provided these were not handled by working group WG 4.
- Project-specific standard specification books based on national standard specification books.
- Technical terms of contract and billing terms.

4.6 Working group WG 6 – Tunnel safety and ventilation

The objective of this working group was the preparation and creation of:

- Variant studies on safety-related system components such as location and types of emergency stop, location and number of cross-overs.
- Aerodynamic and thermodynamic simulations in the main tunnels depending on the operating program, maintenance and climatic conditions in the portal areas.
- Determination of project-specific pressure and suction loads in the main tunnels, connecting side tunnels and access tunnels depending on the operating program.
- Updating the ventilation concepts for the construction and equipment phase taking into account the current construction program as well as for the operational phase.
- Updating the tunnel safety concept for the operational phase based on the implemented project changes and project optimisation measures.

4.7 Working group WG 7 – Equipment

The objective of this working group was:

- The planning of all system engineering, railway operational and railway technical schemes.
- The preparation of all railway technology specifications relating to construction (equipment).

Table 1.   Overview of the Working Groups (WG 1 – WG 4) and their activities.

Working Group		responsible for
WG 1	Project Management	Project management, Coordination and Authority provisions
WG 2	Alignment	Routing and coordinate transformation of the existing documents to the system of BBT SE
WG 3	Geomechanics	Geomechanical guideline, fault zone management - to help optimise the geological / hydrogeological / geotechnical management of fault zones. Evaluation of geological, hydrogeological and geotechnical documentation on site for a consistent method for defining the characteristic values.
WG 4	Design	Compilation of the basic requirements of the different tunnel structures Typical drawings Project optimisations, such as • Drainage system (carriageway and groundwater) • Water supply system (fire water) • Cross passages incl. connection to exploratory tunnel Design of the emergency stops

Table 2.   Overview of the Working Groups (WG 5 – WG 7) and their activities.

Working Group		responsible for
WG 5	Technical contract terms and service book	Quality requirements for construction products for lining and support incl. testing procedures provided that they are not dealt with by WG4 • Project-specific bills of quantities based on national standard bills of quantities • Technical contractual and invoicing provisions
WG 6	Tunnel safety and ventilation	Study of alternatives for safety related system components Review of aerodynamic requirements of the main structures Aero- and thermodynamic simulations Updating of the ventilation concepts for the construction, equipment and operation phase Updating of the safety concept
WG 7	Equipment	Electrical Equipment like traction power and 50 Hz main power supply, signalling and telecomm-unications equipment • Planning and design of all railway operational and systems schemata • Compilation of all railway system requirements (equipment) relevant for the construction • Development of new concepts • for the equipment in the cross passages • for the earthing • for the cable routing

5   RESULTS AND EXPERIENCES FROM THE 2013 GUIDE DESIGN PROCEDURE

The guide design procedure resulted in an uniform cross-border project basis being created which included the changes and measures from the approval procedures, the various project optimisation measures and the results from the individual working groups. This in turn formed the basis for the various tender and execution documentation for the main lots in Austria and Italy.

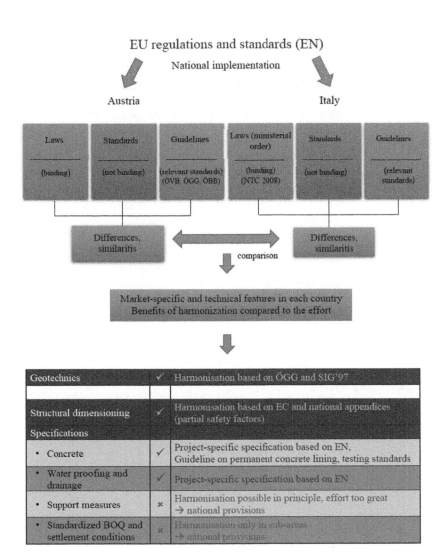

Figure 4. Overview of the harmonisations.

Despite European regulations and standards, there are sometimes considerable differences in implementation and handling on a national level. As a result, there are differences not only in the requirements for materials but also differences in testing and the corresponding documentation which, for concrete in particular, with the corresponding test standards for the starting materials, resulted in extensive processing.

In the processing stage the existing planning concepts and applicable laws, standards and regulations in Austria and Italy are compared with each other and evaluated with regard to their potential for harmonisation. In addition to the fundamental possibility of a harmonisation based on the applicable statutory provisions, the following criteria also apply to the evaluation:

– Market-specific and technical characteristics in the individual countries.
– Comparison of the benefits of a harmonisation versus the expenses of creating a harmonised, project-specific planning basis.

In the course of the guide design a harmonised basis was created as a result of these comparisons for geotechnical planning, design concepts and the specification of the standard components, such as concrete parts, sealing and drainage systems, cable conduits, shaft covers, etc. The other specifications were created separately for Austria and Italy according to national provisions.

The same apples to the drafting of the contractual bases (standard specifications book and billing terms). Here, too, separate documents were drawn up for Austria and Italy.

6 RESULTS AND EXPERIENCES FROM WORKING GROUP WG 4

Working group WG 4 (Design) has been able to develop a wide range of constructional details, design principles and technical specifications.

6.1 *Optimisation of the standard profiles of the main tunnels*

During the development of the new standard profiles for the main tunnels using conventional (blasting) and machine (TBM-O and TBM-S/DS) driving methods, experience from the construction of other long tunnels, such as the Gotthard Base Tunnel, Ceneri Tunnel, Koral Base Tunnel and the Semmering Base Tunnel was also taken into account, without having to change the structure gauges for the travel area as well as the escape and service route from the submission plans. The following changes were made to the standard profiles from the submission plans from a construction standpoint and with regard to future maintenance:

- Elimination of the central mountain water collecting pipe and the discharge of mountain water exclusively via the two tunnel side wall drainage systems across the connecting side tunnels and then down to the exploratory tunnel. Thanks to the elimination of the mountain water collecting pipe, the new standard profile in the cyclically driven areas can be formed for the most part with a flat bottom slab instead of the earlier proposed bottom invert.
- Central, instead of lateral, placement of the railway drains resulting in the creation of larger and more maintenance-friendly cleaning shafts between the platforms.
- Laying of cable conduits in the side walkways resulting in the elimination of the originally proposed cable ducts and the corresponding covers.
- Creation of uniform, maintenance-friendly cleaning shafts for lateral side tunnel wall drainage systems at regular intervals of 111 m.

During the creation of the various standard profiles, the tolerances, among other things, from the surveyors' measurements, machine travel, geometry and manufacturing were examined in detail and considered for each tunnelling section.

In addition, a comparison of the advantages and disadvantages of the various standard profiles in the BBT depending on the tunnelling method and machine type was produced. From this comparison, it emerged that in the sections which are machine driven, the goal should be to have a single-shell sealed tubbing ring lining over an extensive area, taking into account project-specific framework conditions, such as future maintenance, predicted geology and hydrogeology as well as engineering and construction logistics. For the single-shell sealed segmental lining which usually serves as the sole consolidation measure for the driven cavity and which must meet the criteria for fitness for use for the entire service life, minimum requirements were subsequently defined for concrete compressive strengths, permissible crack widths, concrete covering, manufacturing tolerances, installation conditions of the lining segment (joint offset, ovalisation), joint sealing systems, etc.

6.2 *Design principles for shotcrete, in-situ concrete and segment shells*

Also the processing of guidelines were the principles for the dimensioning and the construction design including load cases are specified for:

- the primary support measures, like shotcrete lining,
- the final cast in place lining concrete and
- the segmental lining.

was a task of the WG 4. Due to the special features of a roughly 64 km long tunnel system, detailed studies are carried out to specify the load cases during operation; such as seasonal and operational distribution of air temperature in the tunnel, or the dynamic loads on escape doors due to the variations of the pressure in the tunnel.

6.3 *Harmonisation of the technical specifications for concrete and sealing and drainage systems*

One of the biggest challenges of the guide design was the development and preparation of an harmonized project concrete specification. Between Italy and Austria there are significant discrepancies as regards the regulation from

– the ambient conditions (limit values for determining the exposure class for chemical attack) which require specific exposure classes - which are expected in the BBT over long distances and
– "Verification" that the concrete meets the respective exposure class requirements and -determination of suitable concrete mixes.

For example in Austria they verify that the hardened concrete fulfils the specified requirements (XA) with a defined water penetration depth on solid concrete samples whereas in Italy we verify the concrete composition requirements.

With the allocation of the exposure class as per UNI 206 and UNI EN 11104 the following is defined as well

– the minimum compressive strength = concrete class.
– the minimum cement content and the water/cement ratio.

With regard to the long transport routes and the planned use of tunnel excavation material the standard specification for concrete were judged to be inadequate. Due to the national discrepancies in the concrete specification and the relevant Italian standard, it was decided to create a project-related specification including testing procedure based only on the EN standards.

For the verification of the exposure classes and the durability of the concrete, the so called "performance related design method", which means testing on hardened concrete according to EN 206, point 5.3.3 and Annex J shall be used. The application for this alternative method is in line with the requirements according EN 206, Annex J.

This has the effect that regarding testing procedures no references to national (Italian) standards were possible. Therefor all necessary specification resulting from the implementation of the "performance-related design methods" are determined with the documents of the guideline design.

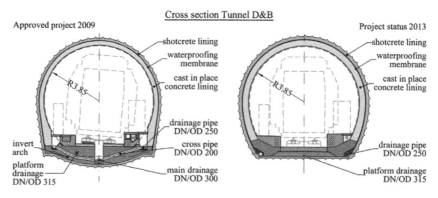

Figure 5. Comparisation of the standard profiles for the main tunnel according to the approved project 2009 and the state of the design in 2013 (conventional driving method).

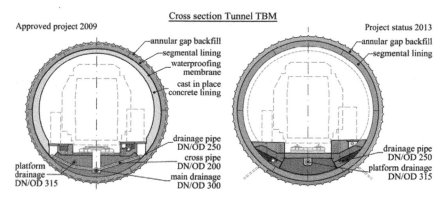

Figure 6. Comparisation of the standard profiles for the main tunnel according to the approved project 2009 and the state of the design in 2013 (machine driving method).

For example the following specifications have been made with the concrete specification of the guideline design:

- restriction of permitted cement types.
- selection of concrete class depending on exposure classes.
- definition of minimum requirements for concrete composition, like
- cement content,
- Water/cement ratio, Water/(cement + k x addition) ratio,
- k-value – that specifies the quantities of additions, like fly ash to be used in concrete
- requirements for production, processing and installation.
- monitoring and controlling of quality.

7 CONCLUSIONS

The 2013 guide design procedure was developed within a period of approximately 2.5 years. They therefore represented a new stage in the project and simultaneously formed an essential basis for the ensuing tender and execution plans on both the Austrian and Italian sides. Different planning consortia where therefore able to produce these tender and execution plans in very short periods and in an objective-oriented manner.

In the course of the cross-border guide design procedure, the project participants were able to gain an insight into the different approaches of the individual countries, which also required them to accept that other approaches could also lead to the achievement of the objectives. At the same time, the partially binding national guidelines as well as the project framework conditions showed the limits, from a technical point of view, of a meaningful harmonisation (Matt et al. 2014).

From the point of view of the project participants, the technical and economic project optimisation measures, as well as an extensive harmonisation of the planning principles could be achieved within the framework of the available possibilities.

REFERENCES

Bergmeister, K. 2011. Der Tunnel kommt. *Brenner Basistunnel*. Lana: Tappeiner Verlag.
Bergmeister, K. 2017. Optimized design of the Brenner Base Tunnel through numerical modelling. Innsbruck: EURO:TUN 2017. Computational Methods in Tunneling and Surface Engineering.
Matt, K. & Corsi, A. & Rudin, Ch. & Eckbauer, W. & Mattle, B. 2014. Erfahrungen aus der Regelplanung. Innsbruck: Brenner Congress 2014.

Tunnels and Underground Cities: Engineering and Innovation meet Archaeology,
Architecture and Art, Volume 7: Long and deep tunnels – Peila, Viggiani & Celestino (Eds)
© 2020 Taylor & Francis Group, London, ISBN 978-0-367-46872-9

Brenner Base Tunnel – an example of planning and implementation of project optimizations

R. Insam & W. Eckbauer
BBT SE, Innsbruck, Austria

D. Zierl & M. Ebner
OEBB, Vienna, Austria

ABSTRACT: The 64 km long Brenner Base Tunnel is the heart of the SCAN-MED Trans-European Corridor. The actual planning has been underway since 2005, whereas construction began in 2007. The paper is a general overview of the cross-border project optimizations carried out so far and also developed in cooperation with the Austrian (ÖBB) and Italian state railway companies (RFI). The main focus of the project optimizations is the simplification of the tunnel system and the railway equipment and outfitting and a reduction of future maintenance costs, considering the general and project-specific framework conditions which have changed due to the lengthy project development. The Brenner Base Tunnel system, its construction and its railway equipment and outfitting have a significant impact on future maintenance requirements, which in their turn impact operations and availability.

1 INTRODUCTION AND TASKS

In large infrastructure projects, the costs of construction are a tiny fraction of the total costs to be sustained over the service life of the structure. Therefore, the implementation of big transport projects has to be focussed on sustainability. Transport infrastructures are sustainable when operations are safe, if the structure is highly available, with easy maintenance and low operational costs. (Eckbauer et al. 2014).

Based on the example of the cross-border project "Brenner Base Tunnel" (BBT), this paper shows different ways a project can be optimised in terms of structural engineering, railway technology and operations, carried out in the last ten years by the Austrian railway company Österreichischen Bundesbahnen (ÖBB), the Italian railway company Rete Ferroviaria Italiana (RFI) and the Brenner Base Tunnel company (BBT SE).

Project optimisation is also built upon an enhanced technical state of the art that has been updated over the years and on the experience gained in other long tunnels.

2 PROJECT OVERVIEW

The Scan-Med Corridor (Scandinavian-Mediterranean Corridor) is the longest north-south connection through Europe, ranging from Helsinki to La Valletta. This Trans-European connection links the urban centres of Finland, Sweden, Denmark, Germany, Austria and Italy with the sea ports in Scandinavia and the Mediterranean Sea. The Brenner Base Tunnel from Innsbruck to Fortezza with a length of 64 km is the core of the SCAN- MED Corridor and overcomes a natural barrier: the Alps. Therefore, the Brenner Base Tunnel is one of the most important infrastructural projects and of high priority within the EU.

To the north, the BBT is linked not only to Innsbruck Central Station but also to the access routes in the Lower Inn Valley and, to the south, with the planned new railway stretches running toward Verona and with the railway station in Fortezza, on the existing railway line.

The preliminary project for the construction of the Brenner Base Tunnel, which includes key route definitions, potential construction methods and an initial cost calculation, was created in the years between 1999 and 2002 (Phase I) (Bergmeister. 2011). The submission plans for the Brenner Base Tunnel on both the Austrian and Italian sides of the project were drafted between 2005 and 2008 and submitted to the various authorities. In 2009, the permits were granted by the Austrian Federal Ministry for Transport, Innovation and Technology (BMVIT) and the Italian Interministerial Committee for Economic Planning (CIPE) (Phase II). The construction works began in 2007.

The BBT is composed of two single-track tubes, lying 70 m apart and connected with each other every 333 m by cross passages and in contrast to other long tunnel projects - a central exploratory and service tunnel running between 10 and 18 m beneath (Figure 1). The four northern and southern connecting tunnels that link the tunnel with the existing lines are part of the tunnel system, too. Furthermore, the Innsbruck bypass tunnel, which was built just over 25 years ago has been upgraded with safety measures and will be incorporated into the tunnel system.

Three emergency stops are located every 20km: Innsbruck, St. Jodok and Trens. Each emergency stop has a so-called cross-chamber that houses many of the technical operations and maintenance systems. Each emergency stop is accessible through the cross-chambers and the access tunnels with road vehicles. The access tunnels have a maximum slope of about 10% and they run directly to the underlying exploratory tunnel. In the Brenner Base Tunnel, a crossover point is planned south of the emergency stop St. Jodok as a strategical reserve option.

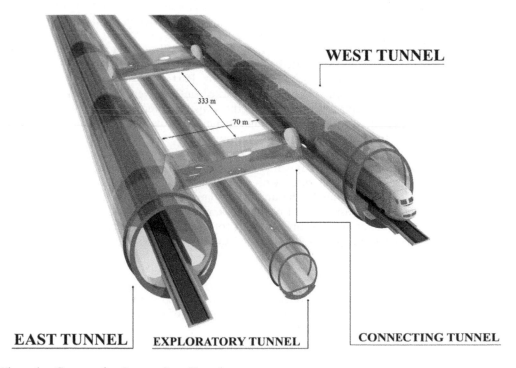

Figure 1.   Cross section Brenner Base Tunnel.

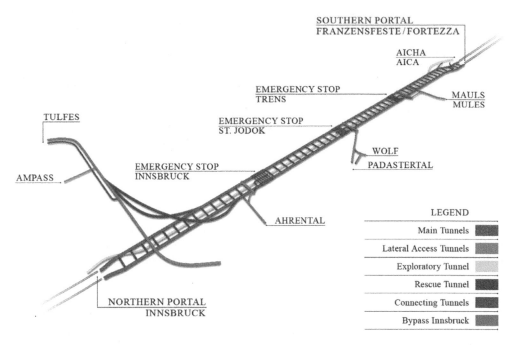

Figure 2. Brenner Base Tunnel system (state of project design 2013).

3 PROJECT OPTIMISATIONS

The **BBT** is a very complex tunnel system and has a sensitive interface between the areas of responsibility of the future tunnel operator and the neighbouring railway networks regarding signalling and command and control and energy systems, maintenance and emergency management (Eckbauer et. al. 2014). Furthermore, there are the different legal frameworks and the different ways of approaching the tasks in the two states.

The application planning for the Brenner Base Tunnel in both Austria and Italy was drawn up between 2005 and 2008 and submitted to the different pertinent authorities for authorization. It comprised a very detailed construction planning. Part of the electrical and mechanical and the other railway equipment systems were worked out in a very detailed way. In many technical sectors, the planning approaches and the usage requirements were taken from the Gotthard Base Tunnel, at that time under construction: e.g. the configuration of the multifunction stations and the number of the crossover points.

In 2009, authorisations were issued in Austria and Italy. However, already during authorisation procedures, ÖBB, RFI and BBT SE had begun the optimisation and simplification of the tunnel system on both project sides, with several planning consultants and experts concerning the specific topics required. The applied optimisations are based on technical, constructional and operational aspects and on the experience gained in other long tunnels that have become operational in the meantime, such as for example the Gotthard Base Tunnel, the Lötschberg Base Tunnel, the Lainzer Tunnel, the Eurotunnel etc.

3.1 *Project optimisation from 2008 to 2009*

Before the submission of the project to the authorities for their approval, ÖBB and RFI had been able to identify the first project optimisations in terms of routing and construction, so as to integrate these into the final planning:

– Reduction of the main slope of the main tunnels from 7.4 ‰ to 6,7 ‰ (in Austria) and from 5,0 ‰ to 4,0 ‰ (in Italy).

- Re-location of the multifunction station from Steinach to St. Jodok (about 4 km southwards) due to new geological-hydrogeological discoveries.
- Assignment of the new Wolf South access tunnel to the relocated St. Jodok multifunction station.
- Maintaining the Wolf North access tunnel to the eliminated Steinach multifunction station as an intermediate heading for the excavation of the main tunnels.

The elimination of the Vizze access tunnel (about 3.7 km in length) in compliance with authorization constraints led to a re-location of the multifunction station from Vizze to Trens (about 2.5 km southwards) and to the accessing the Trens multifunction station through the Trens access tunnel (about 3.6 km in length), via the Mules access tunnel.

The approved project (2009) is shown in Figure 3.

3.2 Project optimisation from 2009 to 2013

Between 2009 and 2013, ÖBB, RFI and BBT SE made further project optimisations to simplify and reduce structural complexity and to enhance tunnel safety, such as for example:

- New routing in the following areas: "Innsbruck main station junction" and "Innsbruck bypass junction". This will lead to less structural complexity and to a simplification of operations within the Innsbruck node.
- Optimisation of the multifunction stations and the emergency stops at Innsbruck, St. Jodok and Trens.
- Reduction from three crossover points to one in the tunnel near St. Jodok as a strategical reserve option.
- Elimination of the passing loops in the tunnel near Steinach.
- Optimisation of the standard profile for the main tunnel (Figure 4), connecting tunnel Innsbruck, emergency exits.

The elimination of the crossover points and the passing loops made it possible to reduce the number of tunnel doors from 15 to 4 and the number of the switches from 26 to 6, as well as

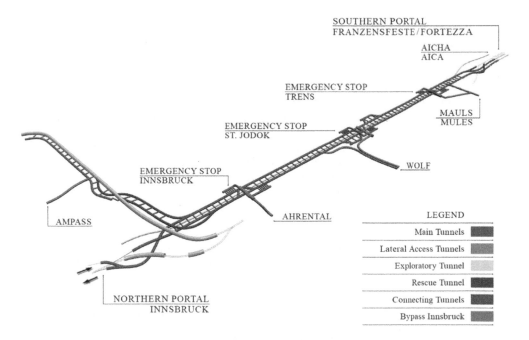

Figure 3. Brenner Base Tunnel system (approved project design 2009).

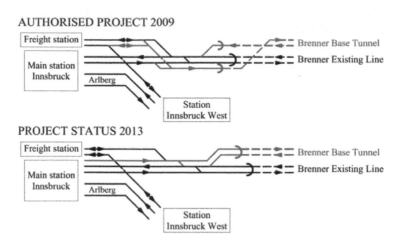

AUTHORISED PROJECT 2009

Freight station

Main station Innsbruck

Arlberg

Station Innsbruck West

Brenner Base Tunnel

Brenner Existing Line

PROJECT STATUS 2013

Freight station

Main station Innsbruck

Arlberg

Station Innsbruck West

Brenner Base Tunnel

Brenner Existing Line

Figure 4. Track layout for the link to Innsbruck Main Station from the project design approved in 2009 and the state of the project design in 2013 (Eckbauer et al. 2014).

Figure 5. Optimised standard profile for the main tunnel (figure: planning consortium Brenner Basistunnel Nord – PG BBTN, c/o Amberg Engineering AG).

simplify the technical systems for ventilation during railway operations. This substantially reduces the future maintenance cost (Eckbauer et. al. 2014).

Between 2011 and 2013, the cross-border design planning was carried out, too. This means that the project optimisations that had been implemented up to then were implemented into a new project status in 2013. This new project status and the cross-border elaborations from the design planning, such as for example harmonised sizing principles (Bergmeister, 2013), technical specifications etc. were an important basis for the main construction lots in Austria and Italy that had yet to be tendered at that time.

3.3 Project optimisation from 2013 to today

The maintenance plan was adapted on the basis of the new project status of 2013 - starting from a study conducted by the Technical University of Graz, together with ÖBB (Berghold. 2013), including an estimation of the maintenance costs. The following knowledge was gained (Eckbauer et. al. 2014):

– Requirement of a time-concentrated, massive personnel deployment in the weekly maintenance windows, at the weekends.
– Very high maintenance costs for the drainage system in the main tunnels.

From 2013 on, more project optimisations and planning principles focussing on railway outfitting, operations and emergency management have been identified and detailed in several working groups that consisted of alternating members of ÖBB, RFI and BBT.

The following results have been achieved:

– Definition of minimum requirements and elaboration of technical specifications for ballast-less track as basis for the future tender.
– Comparison of the advantages and the disadvantages of the classical contact line and the feeder rail. As a result, the feeder rail will be used for the BBT.

There have been several other activities so far, but no definitive decision has been made with regard to the implementation of the solution proposals. Among them, for example:

– Detailed studies regarding the relocation of technical systems from the main tunnel system to the exploratory tunnel located below. The special situation of the BBT, with a exploratory tunnel and respectively service tunnel lying underneath the main tunnel tubes provides a high optimisation potential, as compared to other long railway tunnels, that can be put to good use for an efficient and economical maintenance of the technical equip-ment and the drainage system. The objective is the planning of the maintenance works of the technical systems from the service tunnel if possible (Figure 6) without having to close the main tunnel tubes to rail traffic. The maintenance of technical equipment through the service tunnel is possible at any time, on all days of the week and through several accesses. In the main tunnel tubes and the cross-passages, on the other hand, maintenance works can be executed only in certain, limited time periods, for example at the weekend, and thus always lead to a tunnel closure. This reduces the availability of the tunnel system. Moreover, the limited time frames require a high number of technical personal and instru-ments. Maintenance works are executed on rail-based vehicles that travel from the main portals.
– Optimisation of the drainage system in the main tunnel tubes and in the Innsbruck connect-ing tunnels. On the basis of the TU-Graz study carried out in cooperation with ÖBB (Berg-hold. 2013), several solution approaches have been studied within the last years for an optimisation of the drainage systems from a constructional and maintenance related point of view and in consideration of the project-specific framework conditions, and have partly been integrated into the planning process (Insam et al. 2018). At the moment an attempt is ongoing to optimize the drainage system in the Innsbruck connecting tunnels, which are currently under construction, on the basis of the experience gained over the past years with the maintenance of long tunnels. The Innsbruck connecting tunnels - unlike the main tunnel tubes - do have a standard profile which is divided by a partition wall into a tube for rail traffic and a safety corridor. The project optimisation includes the flushing of the two pier wall drainage channels of the Innsbruck connecting tunnels from the safety corridor

Figure 6.   Description of a technical cross passage with a standard interval of 2 km with a vertical con-nection shaft to the underlying exploratory tunnel (figure: planning consortium Brenner Basistunnel Nord – PG BBTN, c/o Amberg Engineering AG).

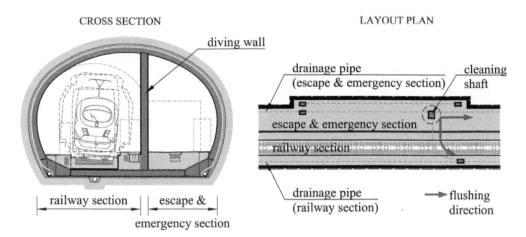

Figure 7.   Pier wall flushing scheme in the Innsbruck connecting tunnel.

without affecting the traffic in the connecting tunnels (see Figure 7). Similar solutions that allow for the flushing of drainage pipes outside the traffic tunnels were implemented in the last years. Innovative flushing systems allow flushing lengths of about 500 m (long-distance flushing systems). Further developments up to 1.000 m and over are predicted.
– Minimization of the operational interruption times for the maintenance of the drainage pipes. The maintenance of the drainage pipes can be carried out in the safety corridor, unaffected by railway operations.
– Maintenance in long, interrelated stretches and therefore more efficient.
– More safety because the works are carried out outside the rail area.
– More flexibility because of the use of road-based flushing vehicles outside the rail area and therefore elimination of rail-based flushing vehicles.
– Study of the installation logistics for ballastless track and the different railway outfitting systems as a function of the tunnel system and the available construction site areas. Due to the lack of space in the portal area of the main tunnels, most of the installation logistics is carried out through the access tunnels and the exploratory tunnel. The decision to install the ballast-free track using concrete slabs makes it possible to bring several slab elements of the ballast-free track at the same time into the tunnel systems, through several accesses, and subsequently install them in the main tunnel tubes. In that way, the continuously welded tracks in the main tunnel tubes are available at a very early stage through the main tunnel portals for the following railway outfitting works.
– The Innsbruck by-pass can be integrated in the logistic concept. Construction materials or e.g. long tracks can be brought by rail through the existing Innsbruck by-pass to the double-track Aldrans junction, re-loaded there and brought directly into the BBT tunnel system.

Further optimisations, especially optimisations of maintenance measures will be detailed further in the future planning and execution phases, together with the Austrian and Italian railway companies, ÖBB and RFI.

4   RESULTS AND EXPERIENCES, FINAL NOTES

The objective of BBT SE and the future infrastructure operator is the creation of a sustainable, maintenance-friendly and highly operational railway system.

In the last 10 years, many mainly structural optimisations have been implemented in tender documents and executive planning. Up to the end of August 2018, 88 km out of a total of 230 km have been excavated. 85% of the overall civil engineering works have been awarded.

The project optimisations carried out so far have considerably simplified the tunnel system, enhanced tunnel safety, reduced the need for future maintenance works and increased the availability of the tunnel infrastructure for rail traffic.

Compared to other long tunnels, the **BBT** has the advantage that, besides the two main tunnel tubes for rail traffic there is a third tube that will be integrated in the overall tunnel system. This third tube is accessible from the portals and from three access tunnels. Its first function is as an exploratory and logistic tunnel. Later on, it will be used as a service tunnel, where different maintenance activities can be carried out regularly or flexibly over several days in a week, without interrupting rail operations.

The project optimisations have been identified and drawn up in an interdisciplinary planning process between ÖBB, RFI and BBT SE, with several planning consultants and experts concerning the specific topics required. Because of the high technical complexity, technical experts for railway equipment and outfitting were brought in already in the conceptual planning phase, so at a very early stage.

Further optimisations are currently being prepared. They refer mainly to the parts of the railway outfitting systems. At the moment, ÖBB and RFI are revising and optimising the exact interfaces for operations management and the associated railway outfitting system and for emergency management.

REFERENCES

Berghold, A. 2013. Zwischenergebnisse Sperrpausen im BBT. Graz: TU Graz (unveröffentlichter Bericht).

Bergmeister, K. 2011. Der Tunnel kommt. *Brenner Basistunnel.* Lana: Tappeiner Verlag.

Bergmeister, K. 2013. Life-cycle design for the world's longest tunnel project. *Life-Cycle and Sustainability of Civil Infrastructure Systems.* London: Taylor &Francis Group.

Eckbauer, W. & Insam, R. & Zierl, D. 2014. Planungsoptimierungen beim Brenner-Basistunnel aus Sicht der Instandhaltung und Nachhaltigkeit. *Geomechanics and Tunneling* Volume 7: 601–609.

Insam, R. & Carrera, E. & Crapp, R. 2018. The Brenner Base Tunnel Drainage System. Luzern: Swiss Tunnel Congress 2018.

Tunnels and Underground Cities: Engineering and Innovation meet Archaeology,
Architecture and Art, Volume 7: Long and deep tunnels – Peila, Viggiani & Celestino (Eds)
© 2020 Taylor & Francis Group, London, ISBN 978-0-367-46872-9

Brenner Base Tunnel – interaction between underground structures, complex challenges and strategies

R. Insam
BBT SE, Innsbruck, Austria

R. Wahlen & G. Wieland
Amberg Engineering AG, Regensdorf-Watt, Switzerland

ABSTRACT: The 64 km long Brenner Base Tunnel is the core of the SCAN-MED Trans-European Corridor. This railway tunnel consists of two single track tubes, cross passages every 333 m, an exploratory tunnel – which lies between 12 to 18 m below the main tunnels, and three emergency stops. Especially in the area of the emergency stops, a variety of underground construction activities are carried out in the immediate vicinity, partly in very difficult geological conditions. The complexity of the surrounding conditions, with distances between cavities of a few meters only, intersecting structures, excavation sections up to 300 m², over-burden up to 900 meters, in partly squeezing rock conditions lead to a significant interaction between the structures. To cope with these complexities is a major challenge for all involved. The paper describes and shows in several examples, the problems and the measures applied to the planning and construction process.

1   INTRODUCTION

The Brenner Base Tunnel is a complex underground tunnel system consisting of various structures with excavation cross-sections between approx. 25 m² and 300 m². The complexity of the surrounding conditions, with distances between cavities of a few metres only, many structure intersections, overburden up to 1700 metres and, in part, squeezing rock conditions lead to a significant structure interaction. Managing these complexities is a great challenge for everyone involved.

During the excavation of the Innsbruck emergency stop, the first structural interactions with the underlying exploratory tunnel were detected, which led to local damage on the rock support – coupled with subsequent refurbishment measures. This article outlines and explains the interaction between the structures using the example of the node of Wolf which will be excavated by drill and blast as part of the H51 "Pfons-Brenner" construction lot. All the parallel running excavation works have to be considered in the context of the overall system and not as a traditional singular tunnel excavation.

2   PROJECT OVERVIEW

2.1   *Brenner Base Tunnel System*

The Brenner Base Tunnel is the core of the Pan-European Transport Corridor Helsinki – Valletta. To the north, the BBT tunnel system is not only linked to the Innsbruck Central Station but also to the access routes in the Lower Inn Valley and, to the south, to the planned new railway stretches running toward Verona and connecting the railway station in Fortezza, on the existing railway line.

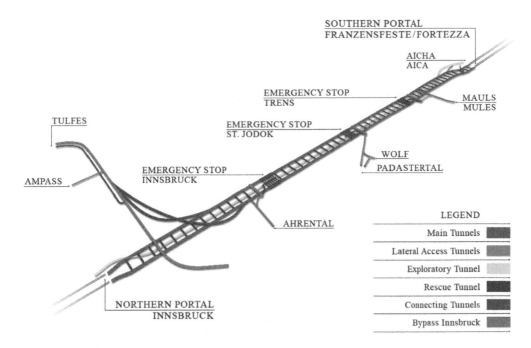

Figure 1. Brenner Base Tunnel system.

The tunnel system (Figure 1) is divided into two 55 km single-tube main tunnels, four connecting tunnels in the north and south, which join the existing lines and a deeper exploratory tunnel, which is located between the main tunnel tubes and runs underneath them. In addition, the twin-track Inntal Tunnel, which was completed in 1994, will be upgraded with structural safety measures and integrated into the tunnel system, resulting in a total tunnel length of 64 km between the portals in Tulfes (Austria) and Fortezza (Italy) (Bergmeister 2011).

The main tunnel tubes run 70 m apart from one another and are linked every 333 m by cross passages. The exploratory tunnel runs between the main tunnel tubes at a depth of approx. 12 to 18 m below them.

Three emergency stops are located at maximum intervals of 20 km: south of Innsbruck, St. Jodok, and Trens. A cross cavern is located at each emergency stop, which houses most of the technical operation and maintenance systems. Each emergency stop is accessible from the cross cavern through an access tunnel, and can be reached with road vehicles. The access tunnels have a maximum slope of about 10% and run directly to the underlying exploratory tunnel. In the Brenner Base Tunnel, a tunnel cross over is planned south of the St. Jodok emergency stop.

2.2 H51 "Pfons-Brenner" Construction lot

The H51 "Pfons-Brenner" construction lot (Figure 2) which stretches from km 13.5 to the state border at km 32.1 is the largest construction lot on the Austrian side of the project and was awarded to an Austrian-Italian joint venture in spring 2018 for just under one billion Euro. Construction will begin in Autumn 2018 and is expected to take 6 years.

The following works will be carried out on the construction lot:

- Drill and blast excavation of the exploratory tunnel towards the south (L = approx. 5.5 km) and towards the north (L = approx. 3.2 km);
- Drill and blast excavation of the node of Wolf which includes the St. Jodok emergency stop and tunnel cross-over with all the corresponding structures;

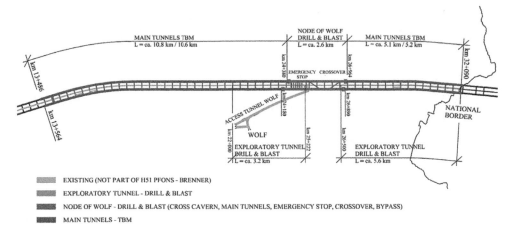

Figure 2.   Overview of the H51 "Pfons-Brenner" construction lot.

- TBM excavation of the main tunnel tubes using shielded tunnel boring machines (TBM-S) and segmental lining towards the south (L = approx. 5.2 km) and towards the north (L = 10.8 km);
- Drill and blast excavation of 55 connecting cross passages;
- Internal lining for all tunnel structures in the construction lot, such as the main tunnels, cross passages, St. Jodok emergency stop and tunnel cross over, exploratory tunnel as well as the lining of the Wolf access tunnel, including the lining of the ventilation and junction chambers and other auxiliary structures which were excavated in a preliminary lot.

In addition, approximately 500 m have to be excavated in the Hochstegen area, which must first be sealed from the exploratory tunnel by means of rock grouting works. An approx. 30 m thick fault zone consisting of rauhwacke is predicted at the northern range of the Hochstegen area, which requires consolidation and sealing grouting works.

The technical cross passages, which are generally 2.0 km apart, are also directly connected to the exploratory tunnel by vertical shafts which have an internal diameter of at least 3.90 m.

The construction works will be carried out with a maximum of seven parellel running drill and blast excavations which will lead to a great effort in terms of logistics. All the parallel running excavation works must be carried out within the context of the overall system and taking into account different framework conditions from the construction process.

2.2.1 *Node of Wolf*
The "Node of Wolf" includes-drill and blast excavations for the following structures (Figure 3):

- Main tunnels:
    - East tunnel (from km 24.365 to km 26.979), length: 2614 m, excavation area: approx. 90 m²
    - West tunnel (from km 24.165 to km 26.905), length: 2740 m, excavation area: approx. 90 m²
- St. Jodok cross cavern (at km 25.243) length: 100 m, excavation area: approx. 230 m²
- St. Jodok emergency stop:
    - Extended main tunnel in the emergency area, length: 470 m, excavation area: approx. 100 m²
    - 1 central tunnel, length: 900 m, excavation area: approx. 116 m²
    - 1 pressure relief tunnel, length: 60 m, excavation area: approx. 116 m²
    - 6 exhaust cross tunnels, length: 60 m, excavation area: approx. 41 m²
    - 6 connecting tunnels, length: 60 m, excavation area: approx. 38 m²
    - 1 turning niche
    - 2 niches for railway tunnel doors

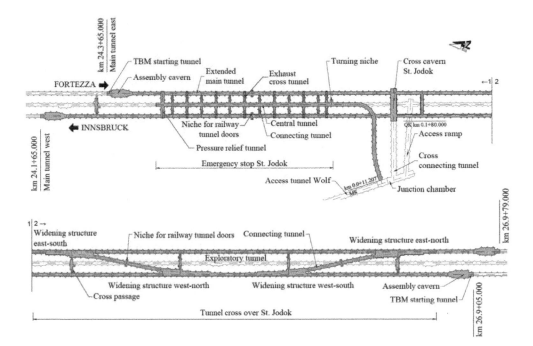

Figure 3.   Overview of the Node of Wolf.

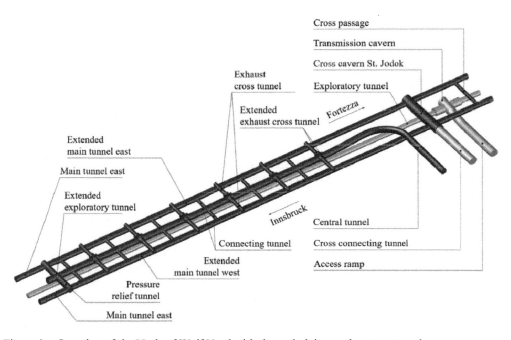

Figure 4.   Overview of the Node of Wolf Nord with the underlying exploratory tunnel.

– St. Jodok tunnel cross over, total length: 1200 m:
  – 4 widening structures with excavation areas of up to approx. 300 m²
  – 2 connecting tunnels, excavation area: 90 m²
  – 2 niches for the railway tunnel doors, each one in the middle of the connecting tunnel
– 4 assembly caverns and TBM starting tunnels

The exploratory tunnel is located directly below the central tunnel of the emergency stop.

3 GEOLOGICAL, HYDROGEOLOGICAL AND GEOTECHNICAL BOUNDARY CONDITIONS

3.1 *General overview of the geological and hydrogeological conditions of the H51 "Pfons-Brenner" construction lot*

The project area for the construction lot ranges from the Pfons area in the north to the main chain of the Alps in the south. Figure 5 shows a schematic geological longitudinal section of the Austrian project area. Therefore, the H51 "Pfons-Brenner" construction lot cuts through (from north to south):

- the East Alpine quartz phyllite zone from km 12.8 to km 14.0. This largely consists of metamorphic sediments from the Palaeozoic Era;
- the Pennine Upper Schieferhülle from km 14.0 to km 28.5. This mainly consists of the Bündnerschiefer Group and Triassic formations which represent the basal detachment horizon;
- the Sub-Pennine Lower Schieferhülle from km 28.5 to km. 30.3. This can be subdivided along the line into the lying Hochstegen zone and the hanging Flatschspitze nappe with interjacent Triassic units;
- the Sub-Pennine Basement in the central Tux gneiss from km 30.3 to km 32.1.

From a hydrogeological point of view, it can be assumed for the H51 "Pfons – Brenner" construction lot that:

- dry to dripping and slightly trickling rock mass conditions prevail over long sections (class 0–0.2 l/s/10 m);
- unsteady inflows can arise from the individual discrete zones which, at a maximum, can be assigned to the class 5 l/s to 10 l/s;
- the largest ingress volumes are to be expected (10 to 50 l/s) in the section of the Triassic nappe boundary and Hochstegen marble without precautionary rock mass sealing and improvement measures to reduce hydraulic permeability.

In summary, it can be stated that the exploratory tunnel and the main tunnel tubes of the H51 "Pfons-Brenner" construction lot are located in dry to moist rock mass conditions for up to approx. 95% of the stretch which has to be excavated.

The largest overburden is found in the area around the border in the Sub-Pennine Basement in the central Tux gneiss and lies between 800 m and 1700 m.

4 GEOTECHNICAL DESIGN AND INTERACTION OF STRUCTURES

The geotechnical design was carried out according to the Geotechnical Design of Underground Structures with Conventional Excavation of the Austrian Society for Geomechanics. The basis for the design of the tunnel structures is the geological and hydrogeological forecast which includes, among other things, the following information:

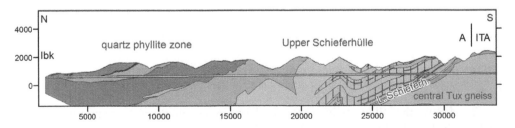

Figure 5.   Schematic geological longitudinal section on the Austrian side of the project.

- Tectonic unit/subunit
- Lithological description
- Information on the discontinuities
- Information on the fault zones
- Water ingress and permeability
- Forecast of the rock mass types (rock mass with similar characteristics)

The following parameters have been derived for all types of rock mass based on the geological exploration campaigns and extensive laboratory tests:

- Mechanical properties of the rock
- Orientation, distance and mechanical properties of the discontinuities
- Mechanical properties of the rock mass

The Node of Wolf is situated in the area of the Upper Schieferhülle, where predominantly Bünder schist and Triassic rocks are predicted. The overburden fluctuates depending on the location of the mountain valley between approx. 400 m and 900 m. Predominantly calcareous (GA SH-KS) and non-calcareous (GA SH-KPh) Bünder schist and subsidiary black phyllite (GA SH-SP), chlorite schist (GA SH-SC), rauhwacke (GA SH-RW), calcareous marbles (GA SH-M) and fault zones are forecasted for this area.

Based on the distribution and the properties of the rock mass types (mechanical properties of the rock mass, orientation and properties of the discontinuities), the hydrogeological conditions and the primary state of stress, the ground behaviour has been determined according to the Guideline for the Geotechnical Design of Underground Structures with Conventional Excavation of the Austrian Society for Geomechanics and has been divided into behaviour types (BT) using suitable demarcation criteria. No lining or support measures have been considered in the analysis of the ground behaviour. The GSI (Geological Strength Index) value, rock mass loading, depth of the yielding zone and radial displacements were used as the main demarcation criteria. The main instrument of the analysis was the analytical ground reaction curve. The following behaviour types were determined according to the Geotechnical Design of Underground Structures with Conventional Excavation of the Austrian Society for Geomechanics:

Predominant – for the rock mass types SH-KS, SH-KPh, SH-SP, SH-CS, SH-M:

- BT 2: Potential of discontinuity controlled block fall – Voluminous discontinuity controlled, gravity induced falling and sliding of blocks, occasional local shear failure on discontinuities;
- BT 3: Shallow failure – Shallow stress induced failure in combination with discontinuity and gravity controlled failure.

Secondary – for the rock mass types SH-RW and fault zones:

- BT 4: Voluminous stress induced failure – Stress induced failure involving large ground volumes and large deformations;
- BT 9: Flowing Ground – Flow of intensely fractures, poorly interloocked rocks or soil with high water pressure.

Thus, in the area of the Node of Wolf, can be predominantly expected discontinuities dominated ground behaviour and only secondarily stress dominated ground behaviour in the area of fault zones and poor BT3. The interaction of the structures was not considered in the analysis of the ground behaviour (analytical ground reaction curve). The construction concept (lining, subdivision of excavation profile, round length, etc.) was subsequently designed based on the identified ground behavior types in accordance with the Guideline for the Geotechnical Design of Underground Structures with Conventional Excavation of the Austrian Society for Geomechanics. As discontinuities dominated ground behaviour was predominantly expected, a rigid lining concept consisting of a reinforced shotcrete lining, steel lattice girders and system bolting was designed for a large part of the structures. Consequently, the system behaviour (behaviour of the rock mass in combination with the rock mass support) was investigated for the selected construction concept using wedge analysis and numerical 2D and 3D

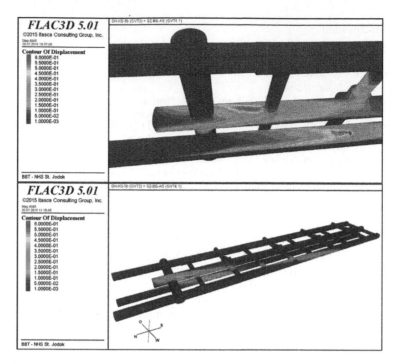

Figure 6.   Numerical 3D FD-Model from the emergency station.

calculation models. The whole tunnel system was modeled in the numerical calculations and thus the interactions of the structures could be fully captured.

The numerical calculations have shown that significant mutual interactions of the structures have to be expected based on the partly short distances between the structures combined with the size of the excavation profiles. The rigid lining in previously excavated structures will be so highly loaded, as a result of the stress redistribution caused by subsequently excavated neighbouring structures, that the shotcrete shell will be overloaded. Even increasing the thickness of the shotcrete lining to a technically and economically useful level, it was still not possible to proof the structural safety. In addition, it became apparent that the construction process has a decisive influence on the loading of rock support. The construction concept has been revised with regard to these findings.

1. A ductile rock support with open slots has been planned for the structures which are situated in relatively competent rock mass conditions with a ground behaviour which is prevalently dominated by discontinuities (BT2 and good BT3) and where an interaction of structures is expected. As the expected additional displacements caused by the interaction of neighbouring structures are relatively small (a few cm), no yielding elements have been planned for smaller profiles. For structures where no mutual interactions are expected, a rigid lining concept has been chosen.
2. A reinforced ductile rock support with yielding elements has been planned for structures in relatively poor rock mass conditions with ground behaviour which is prevalently stress dominated (poor BT3 and BT4). As the expected displacements caused by the excavation of the structures (up to 40 cm) as well as the mutual interactions of the structures are relatively large (up to 10 cm), a controlled guidance of the shotcrete lining using yielding elements will be required.

The numerical analyses have shown that the exploratory tunnel directly below the central tunnel is a particularly critical area. The central tunnel will be previously excavated to the exploratory tunnel. It runs approximately 600 m directly above the exploratory tunnel with a remaining rock bar thickness between the exploratory tunnel and the central tunnel of about 3 m.

While the tender planning of the H51 construction lot was taking place, the H33 Tulfes-Pfons construction lot was already under construction and the Innsbruck emergency stop, which has essentially a similar layout to the St. Jodok emergency stop, was being excavated. Even if the geological situation is different, the findings on the interaction of the structures from the H51 tender planning were essentially confirmed. Especially in the previously excavated exploratory tunnel, where a rigid rock support has been installed without deformation slots or yielding elements, the shotcrete lining became overstressed over wide ranges at the crown area due to the excavation work in the overlying central tunnel. For occupational health and safety reasons, the damaged area had to be refurbished by completely replacing the shotcrete lining in the crown area.

The exploratory tunnel is a central element in the logistics concept for the construction lots to the north of the Brenner pass. Due to the high relevance of the structure, the computational investigations and the excavation of the Innsbruck emergency stop in the neighbouring H33 Tulfes-Pfons construction lot (Lussu et. al. 2019), appropriate measures were evaluated during the H51 tender design to minimize the risk of damage in the exploratory tunnel. The following framework conditions had to be considered:

– The risk of damage in the exploratory tunnel must be minimized;
– Occupational health and safety in the exploratory tunnel must be ensured at all times;
– A change in horizontal alignment was not possible;
– A change in vertical alignment was possible in principle, whereby fixed points in the previously excavated access structures and maximum gradients (construction operation) had to be considered.

Different variants were analyzed with regard to the location of the structures and the lining concept. The evaluation showed that the following measures could minimize the risk of damage in the exploratory tunnel taking into account the aforementioned framework conditions:

– The vertical alignment of the exploratory tunnel was adjusted to maximize the distance in the area of influence of the central tunnel. It was possible to increase the rock bar between the exploratory tunnel and the central tunnel approximately from 3 m to 6 m (Figure 7).
– A ductile rock support concept was planned, consisting of a steel fibre reinforced shotcrete lining with an additional layer steel mesh on the cavity side, mortar rock bolts and 4 rows of deformation slots with or without yielding elements depending on the expected system behaviour. After the completion of the tunnelling works in the exploratory tunnel in the area of influence of the central tunnel, a rockfall mesh will also be installed on the interior edge of the shotcrete lining as an occupational health and safety measure (Figure 8).

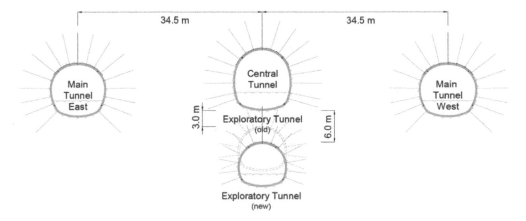

Figure 7. Cross-section through the St. Jodok emergency stop – Lowering of the exploratory tunnel.

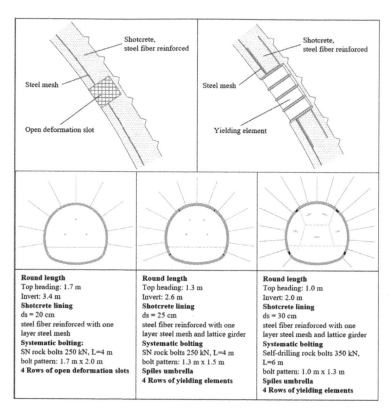

Figure 8. Ductile lining in the exploratory tunnel in the area of influence of the overlying central tunnel.

5 CONCLUSIONS

During the tender design for the H51 "Pfons-Brenner" construction lot, very detailed investigations were carried out for the Node of Wolf, which consists of several complex structures, such as a cross cavern, emergency stop and tunnel cross over, taking into account the overall tunnel system, the geological conditions and the construction processes. It emerged that the excavation of different structures can partly lead to significant interactions with adjacent previously excavated structures and the underlying exploratory tunnel. Based on these investigations, various measures for excavation and rock support were subsequently identified and established. The selected measures are aimed primarily to minimizing the risk of damage on the rock support and at ensuring occupational health and safety in the various working areas. The structures which have been excavated in the meantime in the neighboring construction lots have shown that the influence between adjacent structures can be significant and confirm the results of the conducted investigations as well as the selected construction measures for the H51 "Pfons-Brenner" construction lot.

REFERENCES

Austrian Society for Geomechanics 2010. Guideline for the Geotechnical Design of Underground Structures with Conventional Excavation.
Bergmeister, K. 2011. Der Tunnel kommt. *Brenner Basistunnel*. Lana: Tappeiner Verlag.
Lussu, A. & Kaiser, C. & Gruehlich, S. & Fontana, A. Innovative TBM Transport Logistic in the constructive lot H33 – Brenner Base Tunnel. Neaples: WTC 2019.

Tunnels and Underground Cities: Engineering and Innovation meet Archaeology,
Architecture and Art, Volume 7: Long and deep tunnels – Peila, Viggiani & Celestino (Eds)
© 2020 Taylor & Francis Group, London, ISBN 978-0-367-46872-9

Logistic management in the longest drives of the Mont Cenis Base Tunnel

M. Janutolo Barlet, G. Seingre, P. Bourdon & M. Zampieri
ALLTI GEIE, Aix-les-Bains, France

E. Humbert & C. Pline
TELT SAS, Le Bourget-du-Lac, France

ABSTRACT: Long tunnels make great demands upon the logistics, for which tunnel muck-ing-out and supply of materials and equipment (concrete, shotcrete, bolts, steel rings, etc.) have to be carefully planned. For the 57 km long Mont Cenis Base Tunnel and especially the drives from the Modane underground security area in France to the Clarea underground security area in Italy, the construction logistics represents one of the biggest challenge. In this section, the drives of the two hard rock TBMs are planned to reach more than 18 km. Cast of final concrete lining and cross passages construction will be concurrent with TBMs' excavation. The sole link with the surface will be the already excavated 4 km long Villarodin-Bourget/Modane access gal-lery, which will serve also a network of galleries and caverns forming the Modane security area. The purpose of the paper is to present the analysis of the logistics at the design stage.

1 INTRODUCTION

For the construction of long and deep tunnels as the Alpine Base Tunnels, the overall logistics plays a great role and can affect the advance rate if not carefully planned. The complexity and magnitude of the site involves an extremely high diversity of activities and worksites, spread across the underground network, operating in parallel for a long period.

Furthermore, to meet construction's schedule expectations, in the case of the Mont Cenis Base Tunnel, similarly to the Gotthard Base Tunnel (Gruber at al., 2011), installation of the tunnel lining has to run concurrently with the excavation with a lag of about one kilometer. In addition, D&B excavation of cross passages is conducted between the TBM face and lining installation worksite. This simultaneity presents special challenges to the site logistics.

As for an industrial facility, site installations (underground and outside) have thus to be correctly dimensioned and construction processes perfectly defined to guarantee the expected performance.

For such project, logistics, often underestimated, is therefore of similar importance than geome-chanics for the success of the project. But hopefully, the design of the Mont Cenis Base Tunnel can take full advantage of the experience of already completed important projects in Switzerland as the Nant de Drance power station and the Loetschberg, Gotthard and Ceneri Base Tunnels.

2 THE SPECIFIC SECTION OF THE BASE TUNNEL WITH THE LONGEST DRIVES

ALLTI GEIE (Alliance Lyon-Turin Ingénierie European Economic Interest Group), com-posed a dedicated team of Arcadis, Amberg, BG, Lombardi and Tecnimont engineers, has been appointed by TELT SAS (Tunnel Euralpin Lyon-Turin), for the overall management of

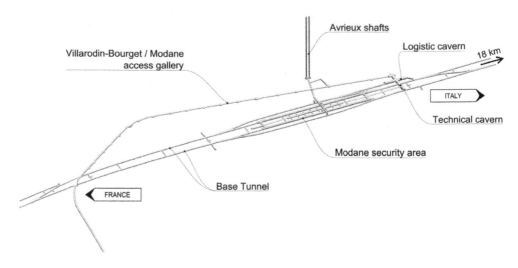

Figure 1. Layout of the underground works in the Modane sector.

the construction of the operational site CO 5 (CO: "Chantier Operationnel") of the Mont Cenis Base Tunnels on the new rail link Lyon-Torino.

Located mainly in the French section of the tunnel and starting from the foot of the already excavated Villarodin-Bourget/Modane (VBM) access gallery, the works consist mainly of:

– Hard rock 10.4 m diameter TBM tunnelling (most likely of gripper type) of the main tubes from the Modane security area to the Clarea security area in Italy (near the border), including the cross passages every 333 m. The drives' length is about 18 km, representing the longest drives for this Base Tunnel.
– At the other site, drill & blast tunneling activities for 3km of twin tunnels and the construction of the Modane security area, comparable to a deep underground station, made of a complex network of caverns and galleries.

Preliminary works will be carried out through a separate contract (overall management by Egis and Alpina engineers) to prepare for the excavation of the main works. These early works consist in excavation of main tunnels and central gallery of the Modane security site to allow for site installation and launching the TBMs, including also the Avrieux ventilation shafts in the sector (see Figure 1). The main works, as described in the previous chapter, will then been carried out from these points towards the completion of the final lining in all areas. These will represent the main civil works activities; all minor civil works for equipment, the equipment themselves and the rail systems are foreseen to part of a specific single contract (named CO 12).

Figure 1 shows the underground works in the Modane sector, while Figure 2 shows a simplified schedule for the Base Tunnel including the specific section described in this paper.

The civil works concept design studies ("études de projet – PRO") and the preparation of the construction contract tender documents ("Dossier de Consultation des Entreprises – DCE") are carried out by ALLTI GEIE.

3 TUNNEL LINING ACTIVITIES

The tunnel lining design is composed of:

– a temporary support system by means mainly of rock bolting, shotcrete, from time to time completed by heavy duty steel arches;
– and a permanent cast-in situ lining, covered by a waterproofing layer so as to properly drain the water inflow towards the lateral drains.

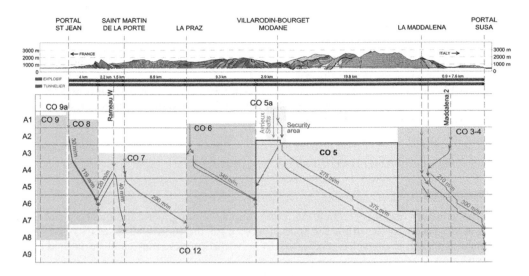

Figure 2. Simplified time schedule for the Mont Cenis Base Tunnel works.

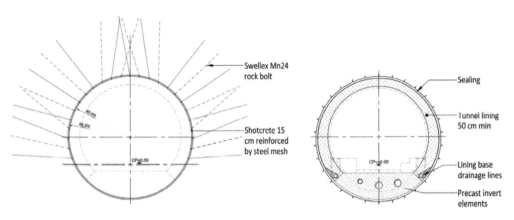

Figure 3. Example of temporary support system (in the left) and permanent cast-in situ lining (in the right): section named TS3 for TBM excavation.

Permanent lining activities are carried out in a very similar way to the Bodio and Faido sub-sectors of the Gotthard Base tunnel (Gruber et al., 2011). Thus at the back of the **TBM**, 100 m away from the cutterhead, a specific construction unit follows to allow the invert installation and all activities related to the cast-in-situ lining.

The actual chain of activities is:

– precast invert elements installation. A precast solution has been preferred compared to a cast-in-situ solution for logistical reasons, among which a simplified supply of material, early traffic capacity etc.
– Profile inspection and checking of the sealing mountings, including any corrections necessary (local reprofiling, etc. . .);
– Waterproofing installation and checking;
– Installation of lining base drainage lines;
– Installation of tunnel lining reinforcement, where necessary;
– Casting and concreting of the lining kickers;

- Casting and concreting of the tunnel lining;
- Curing of concrete.

The installation units are designed in such a way that the TBM supply lines and ventilation ducts remain continuously in operation.

As described in Gruber et al. (2011), due to its specific mode of movement, as well as the considerable length, in the Gotthard Base Tunnel the lining installation unit earned the nickname "worms" ("Wurm" in German).

4 MUCKING-OUT AND HAULING OF MATERIALS

4.1 *Introduction*

The different means necessary to haul materials, supply of equipment and mucking-out of the spoil have obviously to fit within the space foreseen for the tunnels and security site. To be economically viable, areas solely dedicated to construction activities shall be limited to the strict minimum.

In this case, the construction sites can be split into three key zones:

- TBM drives (including cross passages): 1.1 % slope max, up to 18.8 km of distance from the foot of the VBM access tunnel;
- Modane security area: galleries with slopes ranging from 0 to 12 %, up to 3 km of distance from the foot of the VBM access tunnel;
- VBM access tunnel: 12 % slope, 4 km long.

4.2 *TBM drives*

Where possible, belt conveyor systems are generally used for such long drives for their efficiency. However, in this case, the lining construction unit running concurrently to TBM tunneling prevents such concept to be implemented. Such solution would have required regular belt transfers through the lining construction units, or through cross-passages from one tube to the other, putting both tunnels on the critical paths and linking unnecessarily their progression.

An alternative solution was then to be found, and only two solutions were left: rail or wheels. The later has been preferred for its flexibility.

The use of multi-service vehicles (MSV) has considerably increased in the last few years. MSVs are currently used on a portion of the Mont Cenis Base Tunnel with ongoing excavation (from Saint-Martin-La-Porte to La Praz, see Figure 4; Martin & Lescaillier, 2016), for segmental lining's, invert's and steel arches' supply and for workers transportation.

Whereas MSVs prevent track laying operation and maintenance, they nevertheless require centralized traffic regulation, as for rail transport. Their configuration can be adjusted to site needs (people transportation, concrete supply etc.) and they can be custom made to fit within the TBM back units and the lining construction units as shown on Figures 5 and 6.

4.3 *D&B drives (Modane security area)*

For the main tunnels and caverns of the security area, a more elementary but robust solution is foreseen using loaders, dumpers and crushers. The cinematic for the D&B sections, on the west side of the security are, is thus more standard, using firstly loaders, then dumpers into a crusher feeding to a conveyor belt.

4.4 *VBM access tunnel*

For the transport along the access gallery, the logistic is made of a combination of:

Figure 4.   MSV for segmental lining supply in the current works of the Mont Cenis Base Tunnel.

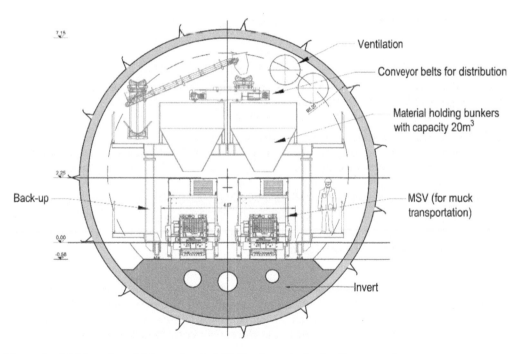

Figure 5.   Multi-service vehicles under the TBM back-up during loading operations

– Conveyor belts for spoil removal and aggregate supply towards the underground concrete plant;
– And trucks for supply of materials and equipment.

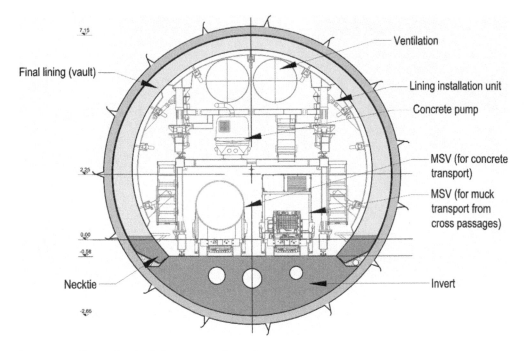

Figure 6.   MSVs under the lining construction unit.

4.5 *Logistic assumptions*

Because of the concurrence of TBM, D&B and lining activities within the same time frame, various scenarios were to be considered.

Whereas the key assumptions were easy to identify:

– 24h/day excavation operation, in three shifts of 8 hours;
– 6 working days for the excavation, the remaining for TBM ordinary maintenance and potential complementary investigation;
– 5 working days for lining;
– A delay of 1 year between the two TBM's launching.

All logistic configurations were to be covered, from material supply, spoil removal to people transportation etc. using an optimized number of MSVs, considering the travelling distances but always meeting the peaks' demands.

4.5.1 *TBM drives*

When sizing the complete logistic chain necessary for TBM operations one has to first assess the possible advance rates, which is itself highly dependent of the expected cutters' penetration rate. The assessment of the latter has to consider the uncertainties around geology, which, for such deep tunnels, can be rather significant, leading to a penetration rate which can vary of a factor of 3. But in the case of the Mont Cenis TBM drives, where the rock mass is mainly within a RMR of class III, an average peak advance rate of 36 m/day and 4 m/h has been ascertained.

To prevent heavy duty maintenance on TBM's, regular maintenance activities in the TBM were considered. According to the experience on similar projects, these operations lead to the production interruption during 4 to 6 hours per day.

Lining operations can take place within a 6 hours shift, using two pumps of 24 m³/h and 10m length pour, and can therefore fit the TBM maintenance period, allowing transfer of dry and wet TBM supply pipes, as for many other activities which breakdown is provided in table 1.

Table 1. Breakdown of activities for TBM drives (with cross passages).

Time interval	During 18 h	During the remaining 6 h
Activities in the TBM	Excavation	Maintenance
Activities at the back of the TBM	Support installation	
	Invert installation	
	Prefabricated drainage pits installation	
	Collecting pipes installation	
Activities in the cross passages	Reinforcement (if needed)	Excavation
		Support installation
		Concreting (previously excavated cross passage)
Activities in the lining installation unit ("Wurm")	Waterproofing	Vault concreting
	Lateral drains' installation	
	Reinforcement	
	(if needed)	

To be on the safe side, the concept also considers concurrent activities of the two tubes, but in reality, using an efficient production system, the contractor will most likely alternate the cycles, reducing accordingly the peak's demands.

To secure the key parameters influencing the logistic performance, MSV suppliers were also consulted in order to define the main performance of these vehicles leading to a maximum payload of around 80t and a maximum speed (full load) of 15 km/h in flat ground.

Finally to complete the assessment, cross passages excavation and lining were also taken into account.

This detailed analysis has confirmed the highest demand of MSVs, being the spoil management phase when the travelling distance reaches 18 km (at the end of the drive). At this stage, the maximum frequency of MSVs in the main tubes is around 5 minutes leading to the need of 12 MSVs per tube.

Concrete supply for temporary and permanent lining operation requires at peak 4 MSVs per tube every 30 minutes.

The other traffic demand for prefabricated inverts, material supply (steel sets, bolts, etc..), prefabricated drainage pits, waterproofing, pipes, reinforcements, etc. have been proven below spoil management demand, as such supply can convey in many instances within the same MSV trip.

4.5.2 *D&B drives*
For D&B excavations, the maximum quantity per hour of muck that can be hauled after each blast is mainly dependent of the loader capacity. In this case, as these sections are far less demanding than the TBM drives, even if advance rates are lower and distances are shorter, the concept design was mainly directed towards sizing the conveyor belt in the access gallery and the capacity of the concrete underground plant.

4.5.3 *Access tunnel*
The bottleneck of the muck management being the sole access gallery, a detailed analysis of construction program has been carried out to determine the required capacity of the final conveyor belt systems. It leads to a maximum number of concurrent excavations of four: two on the TBM drives and two on D&B drives in the security area.

This configuration of activities requires around 2500 t of muck to be evacuated per hour. The capacity of the two conveyor belts in the access gallery has thus been fixed to 1350 t/h.

5   UNDERGROUND INSTALLATIONS

The particular layout of the tunnel and especially its depth and length of access gallery requires site installation to be located within the Modane security area footprint to guarantee a proper supply and underground logistic.

Various caverns are used to host the main underground site installations (see Figure 7 for their location). These are:

- The technical cavern
- The logistic cavern
- The intersection caverns.

The VBM access tunnel will also host several elements, as described in the following.

5.1   Technical cavern

The technical cavern, thanks to its center position is the crucial point of the site.

This is the place where the wagons of the MSVs coming from TBM drives will transfer the muck into hoppers, using laterally self-tippers. A chain of conveyor belts will transfer them to the conveyor belts of the access gallery.

The crusher for materials coming from the D&B excavations is located in the so-called intertube gallery (see Figure 7). At the exit of the crusher, the muck is then transferred into the conveyor belts of the technical cavern.

A buffer stock pile area is located in the so-called "niche de pied" (see Figure 7).

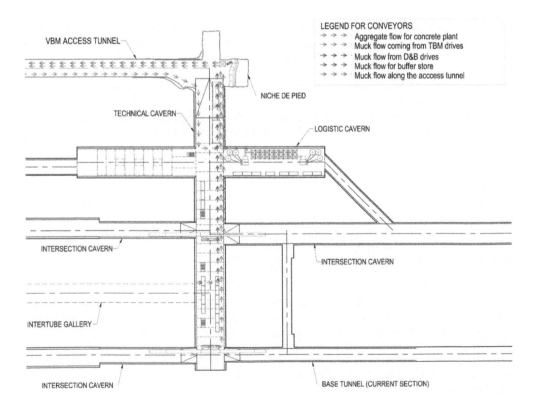

Figure 7.   Flows of the excavated materials within the technical cavern.

To take full advantage of the cavern size, an upper floor will be created (by a metallic structure), in which the different control centers will be located inside containers. These centers will control: cooling/ventilation, conveyors, crushers, buffer stores, safety/communication, MSVs and other transportations.

5.2 Logistic cavern

Within its East wing, the logistic cavern will host the concrete plant. The assessment of the concrete needs in all the tunnels lead to a required capacity of 2200 m³/day. It is to be noted that the underground concrete plants of the Nant de Drance and Ceneri (Grosso, 2012) sites had similar daily capacity. And as in the Nant de Drance project (Bertholet, 2012), the plant will be equipped of two mixers.

The west wing of the cavern will host the area for transfer and storage of different materials, e.g. bolts, steel sets, pipes, etc… A metallic structure allowing storage at different levels will be created. A lifting gantry will ensure the transfer of materials and supply from the trucks coming from the surface to the relevant storage area, with the help of loaders with forks.

Materials and supply will be then be transferred to the MSVs using the same means.

5.3 Intersection caverns

The intersection caverns will host the MSVs' area, allowing 9 MSVs to be parked and five to be maintained.

5.4 Current section of the access gallery

The access gallery will host:

– Two conveyor belts of capacity 1350 t/h for muck transport;
– A ventilation duct, separated by a false ceiling;
– Conveyor belt to supply the concrete plant with aggregates (capacity: 300 t/h);
– Pedestrian passage L=0.8m x H=2 m;
– Pipes for evacuating water (mainly mountain water);
– Pipes to supply water for some activities (in the first months, the mountain water is not sufficient);
– Dry networks;
– Lighting;
– Compressed air.

and will allow for a clear width of L=6m x H=4 m for truck movement.

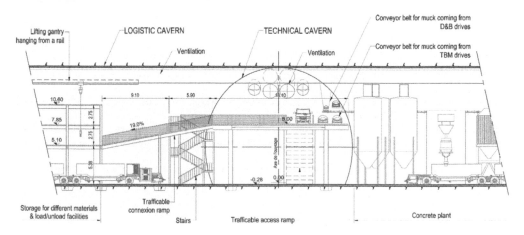

Figure 8. Longitudinal section of the logistic cavern.

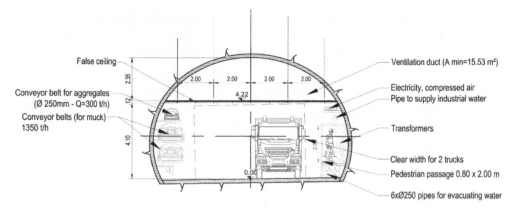

Figure 9. Cross section of the Villarodin-Bourget/Modane access gallery.

6 CONCLUSIONS

During the design development of the section of the Mont Cenis Base Tunnel starting from the foot of the already excavated Villarodin-Bourget/Modane (VBM) access gallery, a comprehensive logistic study was conducted to define and size the key equipment necessary for construction, from the supply of materials and equipment (concrete, shotcrete, bolts, steel rings, etc.) to the muck removal. A detailed analysis of the underground site installation was also carried out to guarantee the performance of the overall logistic, which can be easily compared to an industrial facility. A summary of this study is provided in this paper.

The study also allowed to verify the adequacy of the different tunnel sections to the construction needs. This concerned especially the access gallery, already excavated and all underground facilities within the security site. The study also encompassed cooling, ventilation, water management, excavation material management and surface installations, not presented in detailed in this paper.

Finally, it also allows to confirm the key construction rates, securing accordingly the project program.

ACKNOWLEDGMENTS

Special thanks go to Etienne Garin for his supervision and advisor role and to Roberto Serra for the quality of the drawings.

REFERENCES

Bertholet F. 2012. Technical challenges in constructing the new pumped-storage power station Nant de Drance. In: *Swiss Tunnel Congress 2012 Proc.*: 34-47.
Grosso N. 2012. Aspects of geological risks in advancing with high overburden – Ceneri Base Tunnel. In: *Swiss Tunnel Congress 2012 Proc.*: 104-112.
Gruber L., Böckli O., Spörri D. 2011. Lining installation as excavation continues – Technical and logistical challenges at the Bodio and Faido Sub-sectors of the Gotthard Base Tunnel. In: *Swiss Tunnel Congress 2011 Proc.*: 38-48.
Martin F. & Lescaillier J. 2016. Véhicules sur pneus de nouvelle génération dans la construction du Lyon-Turin, *Tunnels et espace souterrain* n° 258: 393-395.

Tunnels and Underground Cities: Engineering and Innovation meet Archaeology,
Architecture and Art, Volume 7: Long and deep tunnels – Peila, Viggiani & Celestino (Eds)
© 2020 Taylor & Francis Group, London, ISBN 978-0-367-46872-9

Varies challenges for a new railway connection between Divača and Koper, Slovenia

P. Jemec
Elea iC d.o.o., Ljubljana, Slovenia

D. Dvanajščak
2TDK, Družba za razvoj projekta, d.o.o., Ljubljana, Slovenia

E. Škerbec
SŽ - Projektivno podjetje d.d., Ljubljana, Slovenia

ABSTRACT: The Koper-Divača link is a bottleneck on the core TEN-T network, impeding cargo transport from emerging port of Koper to their customers in the middle European countries. Karst edge with world famous natural phenomena presents the biggest obstacle for implementing a new railway line. Climbing more than 400 m in a short distance means that most of the track is pushed underground. The ground water table in karstic features can oscillate up to 120 m above tunnel level and due to environmental reasons the ground flow should not be changed. Therefore an undrained concept is foreseen. Many different challenges were addressed during the investigation and design phase and will be reflected during the construction, especially karstic ground with high water table and/or filled with soft material and others like squeezing in over thrusted zones, excavation with low cover below highway and other infrastructure, underground excavation along steep slopes, etc.

1 INTRODUCTION

Activities for increasing the capacity of the railway line Divača - Kopar were started in 1996 with the Feasibility Study on "Increasing the capacity of the single-track railway Divača - Koper". After a number of studies and analyzes, it was decided that only the construction of a new railway line was good solution and that modernization of the existing railway line was not sufficient to increase capacity. During this time, all projects

Table 1. Technical construction characteristics.

Line length	27.200 m
Number of overpasses	1
Number of underpasses	2
Number of bridges	3
Number of tunnels / length of all tunnels	8 / 20.472 m
Longest tunnel	6.714 m
Tunnel ratio	75.28 %
Number and length of all viaducts	2 / 1.100 m
Longest viaduct	647 m
Ration of viaducts and bridges	4.81 %
Length of all tunnel tubes	37.375 m (service and exit tubes included)
Length of new required roads	20.013 m

Table 2. Technical traffic characteristics.

Max. train speed	160 km/h
Max. track slope	17 ‰
Max. axle load	22.5 t (D4)
Structure clearance	GC
Number of stations and passing points	1
Journey time	17–18 min (passenger trains)
	25–34 min (freight trains)

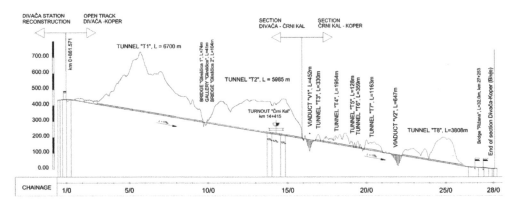

Figure 1. Longitudinal profile.

have been made and confirmed from concept to main, so that today's railway is practically in front of the start of construction. The length of the track with a maximum slope of 17 prom and for a speed of up to 160 (100) km/h with category D4 and a GC profile is approximately 28 km, 20 km of which as 8 tunnels with parallel service tubes or emergency exists, all together 37 km of tunnels. Two valleys are crossed with viaducts. In all structures a slab track is foreseen.

The alignment passes in the upper part through one of the most karstic regions. Carbonate rocks are over thrusted on flysch rocks which continue on the lower part. At all stages, it was necessary to apply TSIs for tunnel safety, as well as other European regulations.

2 GEOLOGICAL AND HYDROGEOLOGICAL CONDITIONS

2.1 General

Railway line starts west of Divača in carbonate rocks as an open track. At most upper part a laminated limestone will be dominating. Thickness of layers increases going south towards tunnel T1.

From the northern portal at about km 3.0 up to km 5.0 tunnel T1 runs through the liburnian layer with bituminous limestone and limestone breccia; both of them are strongly cavernous. Larger pockets of plastic clay and clayish gravel, which extend from the surface, can be expected. Further 2.0 km tunnel runs through nummulitic limestone. Probability for karst phenomena is lower compared with previous section. South to the flysch rocks at km 9.0 again more karstified limestone is predicted. Southern portal of tunnel T1, bridge with covered gallery over Glinščica valley and northern part of tunnel T2 will be driven mainly through layers of sandstone and marls with several subvertical faults, especially at transition from/to carbonates. Interface between impermeable flysch and permeable karstified rock, ie. contact karst is mainly characterized by underground forms, such as caves, channels, abysses and shafts up to the surface.

About 1.0 km from the northern portal of tunnel T2 contact karst again presents the transition to nummulitic limestone, which is expected almost to the south portal. Degree of karstification along the contact and trough the limestone is very high, with prediction of 5-10 phenomena with diameter up to 10 m along 1 km of alignment. This is the area of some already known cave systems. At km about 14.0 over trusted fault zone consisting of thin laminated marl cuts the limestone section. Due to sub horizontal orientation of about 30 m thick layer 240 m of excavation of double track profile will be challenged by high squeezing conditions. Last 0.5 km at south portal will be again excavated in over trusted zone of tectonically disturb marl and slate limestone. Alignment with bridge V1 passes Osapska valley which divides karst and flysch Slovenian Istria.

Lower part of the section is in terms of geological formations more monotonic. As afore mentioned, 5 tunnels, namely from T3 to T8, will be driven through flysch rocks. Surface is covered with 1 to 8 m thick soil like material, deeper several meters of weathered rock mass is found. Alignment is parallel to the steep surface and passes numerous smaller stream erosion valleys. Therefore several short slope tunnels with low overburden and high precuts with

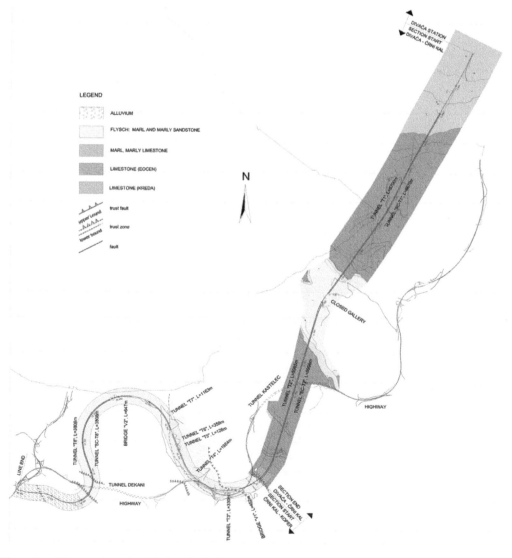

Figure 2.   Alignment on simplified geological map.

unstable natural slopes are foreseen. Tunnels T3 to T6 are within undertrusting zone with a thickness of about 250 m which is a result of the undertrusting Istria under the Dinaric Mountains. Consequently strongly tectonically disturbed rock, where marl is dominating over sandstone, is expected. In tunnels T7 and T8 relationship between marl and sandstones changes and sandstone prevails. Due to intense tectonic dips of layers are quickly changing.

2.2 *Karst*

Almost entire area of future railway track section between Divača and Črni Kal consists of water permeable carbonate rocks where numerous karst caves in which water flow is predominant can be found. In general trough entire alignment in carbonate rocks karst phenomena of different sizes and shapes can be expected. In the interpretation several section with higher probability are defined where caves of size over 10 m are expected.

Also surface water from a surrounding low permeable flysch areas sinks at the contact with karst. Underground water then flows over a larger area of Slovenian southwest and supplies

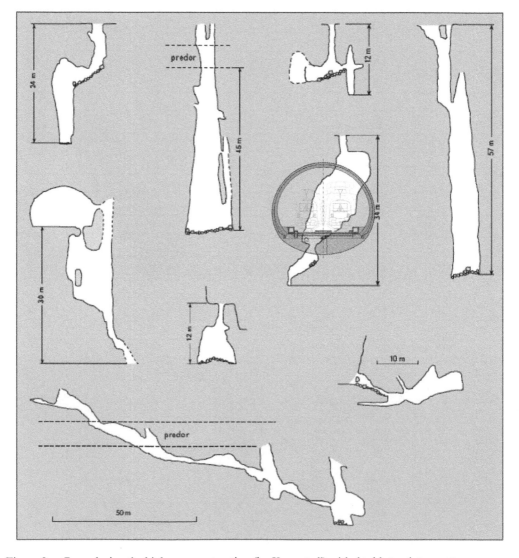

Figure 3.   Caves during the highway construction (by Knez et.all) with double track turnout.

some, in terms of water supply very important karst streams and at the same time feeds common underground karstic aquifer which represents an important source of drinking water for this part of the country. Based on past experience and investigations, openings, empty or filled with water and/or alluvium are certainly expected.

Wider considered area has 177 registered caves while along the alignment three areas have higher caves density. One of those was found during the construction of the highway tunnel Kastelec with a cave longer than 500 m. Near by, in one of the investigation borehole, a bigger opening at depth of 195 m was found as well. At this point should be mentioned that foreseen alignment underpass tunnel Kastelec with tunnel T2. For the risk assessment needs to be pointed out that when 70 km of highway was built in these area 350 caves was found (see some examples on Figure 2).

2.3 Underground water

Determined hydrogeological characteristics of the area of proposed alignment and the follow-up experiments results showed a good permeability of the karst aquifer through which the groundwater is rapidly transmitted via vadose and phreatic zones towards karst springs. Spring of Rižana river should be highlighted as a water supply for all coastal region. All underground water south from Beka-Ocizelj cave system flows towards Rižana. This section of the alignment is therefore part of a wider water protection area.

In terms of ground water inflow into the tunnel karstic limestone is the most problematic. Underground water can appear very locally and at high levels, suddenly and in large quantities. Water ingress can occur also in fault zones, in over-trusted zones of limestone and flysch, from siphon channels and water accumulations in unsaturated zones.

During the initial and complementary investigation campaigns piezometers were installed and a continuous monitoring of underground water levels is carried out. Tunnels are subject to a sudden and high oscillation of water level, which in case of borehole T2-90 can be 70 m above or 20 m below tunnel level (Figure 4 shows results for tunnel T2). On the other hand one can see other sections where water table is constantly above tunnel level, with exceptions of boreholes T2-10 and T2-11. Additionally to the high water pressures an increase in time is important as well. Water table increase rate is in low karstified rock from 0.5 m/h to 1.7 m/h, in medium system with local dewatering up to 9.1 m/h and in good connected underground systems with wider contribution area up to 42 m/h, even up to 69 m/h.

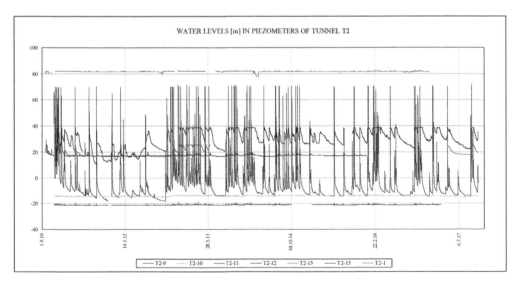

Figure 4.   Water table fluctuation in piezometers of tunnel T2.

During the excavation water accumulations can be drained and will occur again during/ after rainy period, on the other hand constant water inflows have to be considered. Latest can occur when tunnel excavation would pass the caves system which drains water from wider area, not only local.

3 TUNNELS CONCEPT

3.1 *General and excavation*

According to geological prognosis majority of tunnels length will be excavated conventionally using drill and blast technique and by means of New Austrian Tunneling Method (NATM). Additionally for portal areas, some low overburden sections and faulted section backhoe and hammer will be used. Excavation profiles are generally divided into top heading, bench and invert. Support types are defined based on Austrian norms and guidelines.

In general tunnels are designed as drained, ie. side drainages are installed at bottom side-walls. Due to environmental reasons, relatively high amount of water in karstic features above tunnel level, an undrained concept is foreseen. The design principles and other applied measures are described in the following chapters.

3.2 *Cross sections*

Basic regular cross section was initially defined for the lower line section, mainly in flysch rocks. There, as usually, only "inner boundary" conditions had to considered, such as traffic clearance profile, additional space for emergency walkways and electromechanical installations, with respect to static shape as well. This lead to a relatively elliptic profile, as shown left on Figure 5. Profile is symmetrical with respect to the vertical axis, regardless the super elevation.

On the upper, karstic part of the line, high water pressures had to be considered. In order that the ground water flow will not be changed an undrained almost circular cross section was chosen (middle profile on Figure 5). Thicker reinforced inner lining is designed to withstand water pressures up to 120 m above tunnel level. Side drainages are omitted, only main collector is driven trough. Providing the uniform hydraulic pressure around the tunnel perimeter a stubbed membrane is foreseen. This concept is limited to karstic phenomena with water, rest of the tunnel will be built more economically as drained, however inner shape of the profile remains the same (last profile on Figure 5).

Should undrained profile be used can be known already during the pre-investigations in front of the face, however most likely finally confirmed after passing through or near the karstic phenomena, based on additional geological, hydrogeological and speleological analyses. Thus part of the tunnel will be excavated in smaller profile and will be later re-profiled to accommodate thicker inner lining. Undrained profile which goes through karst feature or very cracked zone

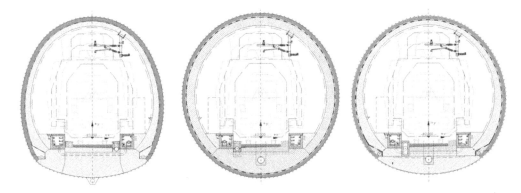

Figure 5. Development of tunnel cross section from left to right.

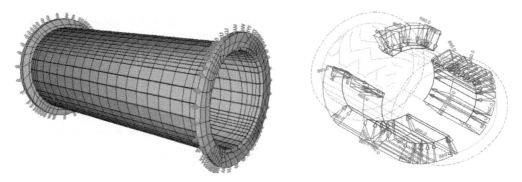

Figure 6.   Secondary lining model in SOFiSTiK (left) and Asymmetrical water pressure load cases (right).

should be extended in compact rock for at least half of the tunnel diameter. Inner lining will be cast without construction or structural joints. Both ends of the undrained section are structurally strengthened with a circumferential ring serving as a water barrier as well. In order to provide a proper seal, ground around the ring will be additionally consolidated with grout.

Apart from hydrostatical pressure on the secondary lining at several heights, additionally some asymmetrical water pressures load cases were analyzed (Figure 6 right). Finally a 50 cm thick reinforced secondary lining for single track and 60 cm thick for double track was proposed.

3.3   Other measures

In compliance with EU directive distance between save areas in parallel tubes should not exceed 500 m. Since the connection of cross passage with main tubes is not possible as an undrained solution, would mean very high costs, a shorter distance is proposed. Distances were reduced down to 450-470 m allowing some freedom to move if karstic feature is meet.

Long tunnels in karst will be generally excavated from both sides, meaning excavation upwards and downwards. The highest risk of sudden water inrush is on the tunnel face or when the excavation passed some channel where water table can rise latter. High water amounts prediction is certain threat to the safety of workers and equipment and consequently causing longer downtime. In this potential scenario fast retreat on the safe area is required. Since this distance can be quite big a special concave barrier is proposed. Thick concrete wall with stiff steel doors should in normal advance allow communication with heading. In case of water inrush or when water rises in karst channel, workers and possibly equipment is pulled back, behind the barrier. With closing the doors also spill over bigger length of the tunnel is prevented.

Due to lower construction costs water will be drained and collected in the central tube wherever possible. As afore mentioned, karst region represents an aquifer for coastal region water supply. Therefore drained water should not be discharged to the surface, but it needs to be recharge back to the underground. Tunnel alignment will cross karst phenomena which forms karstic underground aquifer. Keeping underground water flow unchained as much as possible, drained water from the main collector will be discharged in one of them. If the water table in that feature can arise at the tunnel level, a non-return valve preventing back flow is proposed. Adequate locations of this system will be determined by the specialist at the time of construction.

4   CONCLUSIONS

In this paper we focused mainly on karst, though there are many other interesting challenges on the whole railway line between Divača and Koper. For the first time in Slovenia an undrained concept was introduced, considering high water pressures, which for this type of tunnels is not that common elsewhere as well. For particular cases some special measures are foreseen, mainly related to high and water tables and its fluctuation.

Obviously one of the very important tasks will be geotechnical monitoring in terms of investigations infront of the excavation face. This will be done by predrilling and additionally by geophysical investigation technologies. To insure long term serviceability and stability ground below tunnel invert should be checked as well. For this purposes microgravity can be used.

REFERENCES

Knez, M. & Slabe, T. 2006. Karstological research during the construction of motorways crossing the Slovene karst. *Annales, Ser. Hist. Nat.*, 16: 259–266.

Prestor, J., Ratej, J., Knez, M., Vukadin, V., Celarc, B. 2018. Preliminarno sintezno poročilo (s področja hidrogeologije, krasoslovja, strukturne geologije in geomehanike). *Geološki zavod slovenije, Irgo consulting d.o.o., Gradbeni inštitut zrmk d.o.o., Znanstveno raziskovalni center SAZU, Ljubljana*

Petkovšek, B. & Žibert, M. 2012. Predor T1, geološko-geotehnični elaborat. *Zavod za gradbeništvo Slovenije & Elea iC d.o.o., Ljubljana*

Žibert, M. & Vukadin, V. 2012. Predor T2, geološko-geotehnični elaborat. *IRGO Consulting d.o.o. & Elea iC d.o.o., Ljubljana*

Kočevar, M. 2010. Geotehnični elaborat o geološki zgradbi prostora in geotehničnih pogojih projektiranja ter izvedbe predora T3, *Geoinženiring d.o.o., Ljubljana*

Žibert, M. 2010. Geološko-geotehnični elaborat predora T4, *Elea iC d.o.o., Ljubljana*

Žibert, M. 2010. Geološko-geotehnični elaborat predora T7, *Elea iC d.o.o., Ljubljana*

Vukadin, V. 2010. Geološko-geotehnični elaborat za predor T7 in IPC 7, *Irgo consulting d.o.o., Ljubljana*

Prestor, J., Ratej, J., Mavc, M., Juvan, G., Janža, M., Rot, M., Meglič, P. 2011. Analiza tveganja za onesnaženje podzemne vode in vodnega zajetja rižana zaradi gradnje 2. tira železniške proge Divača – Koper. *Geološki zavod Slovenije, Inštitut za rudarstvo, geotehnologijo in okolje, Ljubljana*

Knez, M. 2010. Sintezno poročilo s področja krasoslovja; Drugi tir železniške proge Divača – Koper, odsek Divača – Črni Kal. *Inštitut za raziskovanje krasa ZRC SAZU, Ljubljana*

Žigon, A. & Žibert, M. 2011. Načrt predora T2 s servisno cevjo. *Elea iC d.o.o., Ljubljana*

Žigon, A. 2010. Načrt predora T4 z izstopnima cevema IPC-T4a in IPC-T4b. *Elea iC d.o.o., Ljubljana*

Tunnels and Underground Cities: Engineering and Innovation meet Archaeology,
Architecture and Art, Volume 7: Long and deep tunnels – Peila, Viggiani & Celestino (Eds)
© 2020 Taylor & Francis Group, London, ISBN 978-0-367-46872-9

The feeder 9, River Humber, replacement pipeline project, United Kingdom

S. Jukes
PORR UK Ltd., London, UK

ABSTRACT: The Feeder 9, River Humber Pipeline project is required to replace the existing Feeder 9 gas pipeline with a new 1,050 mm high pressure gas pipeline under the estuary of the River Humber. To avoid any impact on the local environment which includes a designated RAMSAR 8 and SSSI site the pipeline will be installed inside a precast concrete lined tunnel. The tunnel is being excavated utilising a 4.4 m diameter Slurry Pressure Balance Machine (SPBM) to provide a tunnel of 4.9 km in length with an internal diameter of 3.65 m. Following completion of the tunnelling works the project team will insert the pipeline in a continuous 4,992 m string into the water filled tunnel before connecting at each end to the existing Feeder 9 pipeline. The project commenced in May 2016 and is programmed to be complete by November 2020.

1 PROJECT OVERVIEW

A strategic component of the United Kingdom Gas National Transmission System (NTS) is the Feeder 9 pipeline that crosses the Humber estuary near Kingston Upon Hull, Figure 1: Project location overview. In 2009, underwater surveys highlighted an unprecedented amount of erosion near Feeder 9 which had exposed sections of the pipeline in the navigation channel. Following the discovery of this erosion, National Grid implemented a temporary remedial solution in 2010, which required further work in 2012. The pipeline although currently stable, is subject to two main risks, further erosion and third-party impact, both of which could result in catastrophic failure of the pipeline. The impact of a catastrophic failure would be significant with consequences potentially greater than any previous incident within the gas industry, both in terms of disruption to gas supplies and the physical consequences of rupture.

Based on the above, it was necessary to find a long-term solution for Feeder 9, whilst monitoring the existing pipeline and having all measures in place to undertake further remedial works and isolate the pipeline if necessary. Strategic optioneering started in 2011 to find the best possible option to replace the pipeline. After a rigorous analysis process, a tunnelled solution was determined to be the most; economical, environmental and safe way to proceed. In 2015, the process of obtaining a Development Consent Order (DCO) was started, which concluded in August 2016, with the DCO being awarded by the Secretary of State.

In parallel with the DCO, several procurement and design activities were undertaken to mitigate the timeline risks associated with the existing pipeline. This resulted in the NEC 3, Option C, Contract for the main works being awarded in May 2016 to a Joint Venture comprising PORR Deutschland GmbH, Skanska and A. Hak. Skanska are the major civils contractor, A. Hak brings expertise in pipeline testing and insertion and PORR brings significant experience and expertise in tunnelling activities. The project delivery phase for the Feeder 9 pipeline replacement is currently in progress with several key milestones already achieved, most notable of which is the commencement of tunnelling in April 2018.

The Feeder 9 pipeline project, upon completion in 2021, will be the longest pipeline in a tunnel in the world and will transport up to 20% of the UK gas supply. The new pipeline will

Figure 1. Project location overview.

not be subject to the uncertain conditions of the Humber Estuary and will thus ensure the reliable and safe transportation of gas for the foreseeable future. This regulatory submission made under Special Condition 5E of the NTS Transmission License is for the design and build of the Feeder 9 replacement pipeline. The submission excludes DCO spend and any prior RIIO-T1 spend, the total funding request is for £139.9m in 09/10 prices.

As well as being economically significant, the Humber Estuary and the intertidal mudflats surrounding the area are of significant ecological importance for many species including birds, mammals (seals and otters), and fish. As such it is afforded some of the highest levels of environmental protection available through International, European and National legislation. The Humber Estuary is an internationally designated RAMSAR8 site, a European designated Special Area of Conservation (SAC), a Special Protected Area (SPA), a nationally designated Site of Special Scientific Interest (SSSI) and an Important Bird Area (IBA).The replacement scheme will comprise of the following:

– Construction of a concrete lined tunnel up to 30m deep below the Humber for nearly 5 km
– Installation of a 1,050 mm diameter concrete weight coated pipeline with a maximum operating pressure of 70bar
– Connection of the new pipeline to the existing connections approximately 120 m onshore at Goxhill and 400 m at Paull
– Decommissioning of the existing Feeder 9 pipeline
– Cathodic protection facilities for the new pipeline
– Two construction compounds, one each side of the river at Goxhill and Paull, adjacent to the existing AGIs
– Significant environmental works to mitigate the impact on the existing protected environment
– Associated works for permanent and temporary accesses, highway works, drainage works, temporary spoil storage, temporary lay-down areas and ancillary works

For the offshore pipeline under the Humber Estuary, design standards require that the tunnel will be located at such a depth as to give a depth of cover of not less than 7m from the true bed of the watercourse, after the removal of any silting, to the top of the tunnel. The pipeline, when installed onshore, will be typically laid to contour at a depth of cover not less than 1.2 m from the original surface to the top of the pipe. The trench will be excavated so that pipes are evenly bedded throughout their length.

Minor variations in contour will be excavated in order to minimise field bending. The minimum depth of cover for the onshore section of the pipeline in relation to a public highway is 2 m. The depths of highway crossings are also influenced by third party services and adjacent existing ditches. The pipeline is designed to have a minimum operation life of 40 years and the tunnel a minimum design life of 120 years.

2 TUNNELING ACTIVITIES

The tunnel between the launch pit in Goxhill and the reception pit in Paull has an overall length of 4,862 m and follows a nearly straight horizontal alignment. The planned vertical alignment follows a decline of the gradient of approximately 4% for a length of 450 m. Beneath the estuary the tunnel drive is virtually horizontal and will revert to an inclined gradient of approximately 4% on the north bank of the estuary for the last 600 m. Excavation is mainly within the Burnham and Flamborough with a minimal overburden to the seabed of approximately 10 m. However, within the launch pit and reception shaft the tunnel passes through layers of the Alluvium and Glacial Deposits, Figure 2: Geological section of the tunnel drive.

The TBM is commences its drive from a ramped launch pit at Goxhill, this pit provides Multi Service Vehicle (MSV) access via a graded ramp from the surface to a depth of approximately 13 m. The launch pit has been constructed utilising a combination of secant and steel sheet piling and is completely water tight. At the reception chamber on the North of the river at Paull we have constructed a secant piled shaft 15 m diameter and approximately 13 m deep. The tunnel will be excavated utilising a Slurry-TBM with an excavation diameter of 4.38 m and an internal diameter of 4.38 m. Four rhomboidal and two trapezoidal precast concrete segments will be assembled in the tail skin of the shielded machine to build a complete tunnel ring of 1.20 m in length. The precast concrete segments are 225 mm thick, and installed utilising the vacuum lifter within the TBM shield. Tunnelling commenced in April 2018 and is planned to be completed by May 2019. Following the completion of the detailed design and an assessment of the Geotechnical Baseline Report (GBR), the Joint Venture opted for a new slurry TBM, fabricated specifically for the project. As part of the Early Contractor Involvement, the Joint Venture developed the TBM delivery plan. These proactive measures assisted in the timely design, manufacture and Factory Acceptance Testing (FAT) of the TBM prior to shipping to the project.

The selection of a Slurry Pressure Balance Machine (SPBM) was made as a result of analysis of the GBR and on the basis that the alluvial and glacial deposits at either end of the tunnel would require the Joint Venture to drive through particularly soft and highly permeable material. Another consideration that influenced the decision was the high hydrostatic head (water pressure) from groundwater particularly whilst driving beneath the Humber. In conclusion, although an Earth Pressure Balance Machine (EPBM) typically has a lower capital cost than an SPBM, it is more susceptible to damage from flint and interventions, therefore not necessarily the lowest whole life cost. Furthermore, it was determined that an EPBM would not have been as efficient in the expected ground conditions identified within the GBR. Figure 3: Slurry-TBM 4.38 m during factory acceptance.

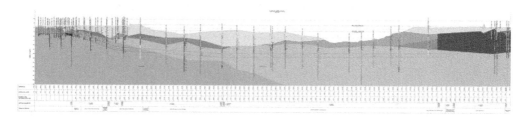

Figure 2.   Geological section of the tunnel drive.

Figure 3. Slurry-TBM 4.38 m during factory acceptance.

3 CHALLENGES DURING TUNNELING

The main challenge of the planned tunnel drive is the small tunnel diameter compared to its tunnel length of nearly 5 km without intermediate structures such as shafts or emergency exits beneath the estuary of the River Humber with a tidal impact of up to 6.4 m. During the planning phase of the project which also involved the Health and Safety Executive (HSE) it was risk assessed that the tunnel had to have an inner diameter of at least 3.65 m and following a detailed design phase the diameter was confirmed as sufficient for the tunnelling activities.

Decision making was mainly influenced by installations in the tunnel which had to be considered during the TBM tunnelling activities as well as possible emergency scenarios and PORR brought to bear their experience on the Emscher Sewer Project BA 40 in Germany with similar conditions in regards to length and diameter of the tunnel.

As a result of the tunnel being excavated with a SPBM and the depth of the tunnel, several booster pump stations for the slurry feed and discharge line had to be incorporated into the design. These installations inevitably led to a reduction in the free cross section of the tunnel and as such required a detailed review of how to feed the TBM during production to ensure logistics did not limit the output of the machine. Figure 4: Tunnel cross-section in booster pump location.

A further impact on the tunnel cross section is the requirement for a ventilation duct to provide primary ventilation which is installed during the drive in the crown of the tunnel. The duct provides a supply of fresh air to the working chamber of the TBM and is designed based on the number of people working within the TBM (16) and the diesel kw power in the tunnel.

The logistic supply of segments, materials and consumables for the tunnel excavation works was also considered and resulted in the decision between utilising conventional rail bound logistics or rubber tyred vehicles (so called MSV or Multi Service Vehicles).

The Joint Venture concluded that an MSV would provide the most cost effective and reliable solution to tunnel logistics and has become the first project in the UK to utilise this technology.

4 TUNNEL LOGISTICS

As described above the JV chose to utilise MSVs for the supply of the tunnelling activities, an approach PORR has successfully utilised on the Doha Green Line Project in Qatar. The main benefits of the MSVs on this project were evaluated as follows:

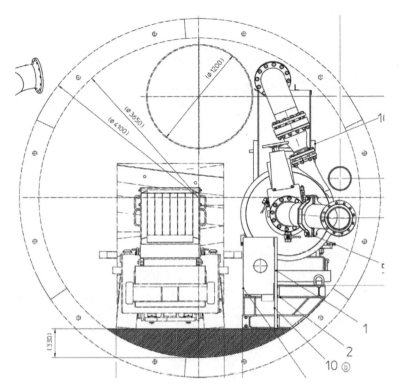

Figure 4.   Tunnel cross-section in booster pump location.

- Improved breaking performance even on wet or muddy surfaces
- Smaller diesel engine required for the MSV resulting in a smaller ventilation duct within the tunnel
- Clear visibility in the direction of travel as a result of operators cabin at either end of the vehicle
- Personnel, obstacle and backup detection with automatic slow-down of the vehicle to crawling speed
- Avoidance of more than 10,000 vertical lifts into a shaft by utilising a ramp at the launch site

The use of Multi Service Vehicles in the tunnelling industry has been increasing over the last few years but mainly in larger diameter tunnels. On this project the MSVs have a length of 17,150 mm a width of 1,260 mm and a maximum height (including load) of 1,900 mm. The vehicle has a tare weight of 12 tons with a maximum permissible payload of 20 tons, design speed when loaded is 15 km/h.

The MSV comprises of an operator's cabin at each end of the vehicle where the rear cabin is fixed and the front cabin is foldable allowing the segments to be lifted over the operators seat. Adjacent to the fixed cabin is a fixed personnel cabin for up to four operatives. The main loading area of the MSV has the ability to transport two stacks of 3 segments and during shift change the segment stack holder is replaced with a personnel cabin for up to 12 operatives, Figure 5: MSV with personnel cabin. The MSV is also utilised to transport slurry and water pipes in a specifically designed support frame as the TBM progresses. Smaller items such as grease barrels, IBC containers for the B-component of the annulus grout and tools are transported on an additional platform behind the operators cabin.

The operator of the MSV sits facing the direction of drive (DoD) and has therefore direct sight on the situation in front of the vehicle. Nevertheless, for additional safety reasons the vehicle is equipped with cameras and a personnel and obstacle detection system which automatically slows down the MSV to crawling speed when activated by an operative or obstacle in front of the vehicle.

Figure 5. MSV with personnel cabin.

Figure 6. Multi-Service Vehicle loaded with segments and accelerator.

The vehicle is also equipped with a two-stage tilt-monitoring-system which in the first stage of tilt provides an audible warning and in the second stage shuts off the engine. During travel through the TBM gantries the MSV is guided by means of a guidance system which maintains an equal distance between the MSV and gantry structures and both the front and rear of the vehicle. Additional guide rollers are installed on the MSV as mechanical guidance support in case of any contact between the vehicle and gantries. Figure 6: Multi-Service Vehicle loaded with segments and accelerator.

5 HEALTH & SAFETY ASPECTS

A detailed Health and Safety Management Plan has to be developed to enable effective management of the severe conditions encountered in the confined space of the tunnel environment. In order to ensure that the management plan will deliver the appropriate responses in the event of an emergency the site team carry our regular emergency response drills in both the mock up and within the tunnel itself (Figure 7: Emergency Response Drill in the tunnel mockup). In addition to health surveillance, training and inductions the following health & safety equipment and installations have been considered:

5.1 *Emergency Control Centre:*

An Emergency Control Centre (ECC) has been installed on the surface where all relevant data from the TBM and tunnel such as fire detection systems, air quality and gas monitoring are collected and displayed on screens. The ECC can control the smoke curtains in the tunnel or the primary ventilation fan on surface to facilitate the management of an emergency. The

Figure 7.   Emergency Response Drill in the tunnel mockup.

ECC also provides the interface for radio communication between TBM, MSV operators and any other persons working in the tunnel.

In an emergency case the ECC will become the dedicated control room for the rescue team to organise and guide any rescue intervention.

5.2   Self-Rescue Equipment

In case of a smoke, fire or gas incident in the tunnel the first target is the safe evacuation of personnel from the tunnel. For this reason, self-contained oxygen breathing apparatus are provided for all personnel who enter the tunnel. The oxygen self-rescue sets are issued at the beginning of the shift in the Tally Hut before entering the tunnel and returned to the Tally Hut at the end of the shift where they are checked and properly stored for the next shift. The rescue sets provide for a 60 minute duration in case of an emergency and as such when considering an evacuation speed of 40 meters per minute through the length of the tunnel additional rescue stations will be provided where replacement self-rescuers are stored.

Additional Oxygen self-rescuers are located in the fixed personnel cabin of each MSV, in the MSV operator cabin and in the Refuge Chamber on the TBM.

5.3   Refuge Chamber on the TBM

In case of an incident in the tunnel where the evacuation from the TBM is not possible the operatives are required to stay on the TBM awaiting help from external rescue teams. For this purpose, a refuge chamber has been provided at the working end of the TBM. Due to the small diameter of the tunnel the location for the refuge chamber was chosen on gantry 2 on a platform located above the segment feeder as this provided the only available space. All the other gantries behind gantry 2 have a portal structure to allow the logistical supply of the TBM and therefore do not provide sufficient space for a refuge chamber.

The design of the refuge chamber took into consideration that an incident could occur during shift change with a maximum of two times eight people on the TBM therefore a maximum of 16 people. The refuge chamber has two access doors one at each end to allow easy access from both the forward and rear working areas.

The refuge chamber can be operated with 16 people on board in stand-alone mode for up to 28 hours. Stationary gas monitoring devices (inside and outside) assist in the operation of the chamber in the event of an emergency.

5.4   Fire Detection and Suppression system

Fire detection and fire suppression systems are installed on the TBM in vulnerable areas where there is a high risk of fire. These areas include the gear box motors in the shield, hydraulic units in the shield and on gantry 1, the main and sub-distribution on gantry 6, 7, 8 and 12, the transformer on gantry 7 and the emergency power generator on gantry 13.

Tunnel fire detection and fire suppression systems are also installed in the booster pump locations, the electrical cabinets and on the transformer.

The fire detection system consists of a heat detection thermo-element which in the event of the design temperature being exceeded releases an aerosol preventing the chemical reaction required for combustion. The aerosol works by preventing chemical combustion reaction at molecular level without depleting the oxygen content. The benefit of this system is that it is harmless to the personnel involved in the operation.

On the Multi Service Vehicles two independent fire detection and suppression systems are installed. One system is installed to protect the hydraulics of the plant whereas the second system is installed over the rubber tyres as a fire involving the combustion of rubber would cause toxic fumes to be created in the confined space of the tunnel environment.

5.5   Water/Smoke Curtain, Sprinkler System

The TBM is equipped at the rear end with a water curtain which can be manually activated in case of an emergency when leaving the TBM. In addition to this there is a sprinkler system over the entire length of the walkway on the TBM to allow the operatives to evacuate safely from the face.

Additional smoke curtains will be located within the tunnel in the downstream area of each of the four booster pump stations dividing the tunnel into five smoke-free sections of approximately 800 m – 1000 m in length. The smoke curtains can be manually activated by passing tunnel operatives or by the Emergency Control Centre (ECC). The smoke curtain consists of water nozzles around the intrados of the segments (similar to the water curtain on the rear of the TBM) and is connected to the cooling water circuit inlet pipe.

5.6   Gas Monitoring System

A stationary multi gas monitoring system is installed on the TBM. The station is configured for environmental monitoring of oxygen, methane, carbon monoxide, carbon dioxide, nitrogen dioxide, sulfur dioxide, hydrogen sulfide, nitric oxide and consists of a climate sensing module. The location of the sensors has been selected following consideration of the areas likely to have a high risk of gas collection.

Additional stationary gas monitoring systems will be provided in the tunnel downstream of each of the 4 booster pump stations. Following a detailed evaluation of the risks, sensors for oxygen, carbon monoxide, carbon dioxide, nitrogen oxide and nitrogen dioxide are to be provided.

The data of the TBM and tunnel gas monitoring stations is displayed and monitored in the Emergency Control Centre.

To increase the safety of personnel working in the tunnel in locations between the stationary systems additional mobile gas monitoring devices are provided for each work crew.

5.7   Communication System

The communication system has been specifically designed to meet the project Works Information requirements. The system is based on a high bandwidth fibre optic cable in combination with a high-speed WI-FI radio server to provide voice and data communication on site. Wi-Fi tunnel stations will be located every 250m inside the tunnel connected with fibre optic cables to the surface.

The tunnel stations can provide a Wi-Fi bridge to each other in case of cable failure or cable damage and they are equipped with a battery backup system to maintain the radio and telephone communication even in the case of a power loss. These tunnel stations can also be monitored from the Emergency Control Centre.

Voice communication will be realised by Voice over IP telephones and multichannel radios (RoIP) on site and on the TBM. Surface communication will be a radio based system (RoIP) which includes eight wireless surface stations to provide a radio communications over an IP network on site and additionally there are VoIP telephones in the ECC, workshop, stores and other installed plant on site.

6 PIPE INSERTION

Following completion of the tunnelling drive and strip out of the tunnelling services preparations will begin for the pipeline insertion. No pipe welding activities are possible within the tunnel and therefore the pipes (each 12 m long) will be welded into strings of between 612 m and 624 m on the surface whilst tunnelling activities continue.

The pipes have been delivered to site by truck and are made up of two different types; concrete weight coated (CWC) for installation inside the tunnel with a pipe weight of 16 tons, and fusion bonded epoxy (FBE) coated pipe for installations outside the tunnel with a pipe weight of 5.9 tons.

Unloading and stockpiling in the pipe lay-down area was carried out utilising a crane with a vacuum lifter for the FBE pipes and traditional lifting equipment for the CWC pipes.

In the pipe stringing yard eight strings will be constructed providing two lines of 624 m in length and six lines 612 m in length, Figure 8: Installation of the pipe strings in the pipe stringing yard. The pipes will be semi-automatically welded allowing for a tolerance or mis-alignment of 0.5°. All welds will be inspected and tested using automatic ultrasonic testing.

A total of 10 winches or strand jacks will be used to move the pipe strings laterally on bogies across the foundations in the yard. The bogies run on a rail system which will be laid perpendicular to the line of insertion.

Prior to the insertion of the pipes into the tunnel the following activities must be completed:

– Strip out/Removal of all utilities from the tunnelling works
– Installation of cathodic protection (CP) equipment (monitoring cables, anodes with power cables, reference cells)
– Installation of a gravel bed

Figure 8.   Installation of the pipe strings in the pipe stringing yard.

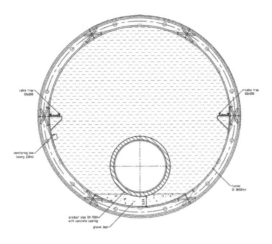

Figure 9.   Gas pipe installed within the tunnel.

- Installation of an internal ramp at Goxhill
- Partial construction of a tunnel bulkhead at the launch shaft in Goxhill
- Filling the tunnel with water

The pipe string will then be winched from the stringing yard to the pipe thrusters in the launch ramp. Prior to installation of the pipe the tunnel is filled with water and a bulkhead installed at each end. The total amount of water required is approx. 51,000 m³ and will be supplied from either the local potable water supply or boreholes.

Two pipe thrusters with a capacity of 750 tons and 500 tons will be installed in the launch ramp to push the pipe gently into the tunnel and as the first section is pushed into the tunnel a tie-in weld to the second string will be completed. In total seven tie-in welds are required before the pipeline is completely inserted into the water filled tunnel, Figure 9: Gas pipe installed within the tunnel. Note that the pipeline will rest on the invert of the tunnel as there is no direct connection between the tunnel and the pipeline. The pipeline will remain in the water filled tunnel for the duration of its lifetime and is monitored through the cathodic protection system at the surface.

The inserted pipe is sealed to the tunnel by means of a 1,336 mm pipe sleeve cast into the bulkhead wall and provides a 50 mm gap around the gas pipe. Following intensive inspection and testing of the pipeline and cathodic protection system, the team will complete the installation of the tunnel bulkheads and connect the pipeline to the existing Above Ground Installations (AGI) in Paull and Goxhill. The installation on completion will be completely buried with the area being returned to arable farm land and as such will be maintenance free and apart from the AGI's not visible at the surface.

7   CONCLUSION

The project delivery phase for the Feeder 9 pipeline replacement, which will be the longest pipeline in a tunnel in the world, is well underway and several key milestones have been reached:

- The detailed design is approaching completion, with uncertainty of the permanent design resolved on key aspects including the TBM, Slurry Treatment Plant, pipeline installation and cathodic protection. The final steps including CAT 3 checks and design approval and appraisal have also been substantially completed.
- The TBM passed the Factory Acceptance Testing and was successfully delivered to site on time, in preparation for Site Acceptance Testing and launch.

- The launch and reception pits have been successfully constructed, in particular the dewatering system has been completed and is operational in "passive" mode as per the DCO condition and to the acceptance of the EA.
- The STP has been designed and constructed with supplementary leachate and is working as expected within the design.
- The risk range of the STP arisings has been refined and is resulting in the arisings being treated as inert or active waste
- The pipeline has been delivered, welded into the 8 strings and testing successfully completed.
- Tunnelling is now nearing the half-way point and the next phases of the project will be progressed in the coming months, with a planned project completion date of November 2020.

Despite the huge technical challenges involved in this project the site team in collaboration with National Grid have developed a solid technical solution to the installation of a circa 5 km pipeline into a flooded tunnel, an achievement that has not been equaled anywhere in the world. Not only will this project set the bar in respect to overcoming technical challenges never before encountered it will provide a long-term sustainable solution to the transmission of gas across the United Kingdom.

Tunnels and Underground Cities: Engineering and Innovation meet Archaeology, Architecture and Art, Volume 7: Long and deep tunnels – Peila, Viggiani & Celestino (Eds)
© 2020 Taylor & Francis Group, London, ISBN 978-0-367-46872-9

Simulation of TBM operation to assess the impact of geology on the muck transportation

A. Khetwal
P.hD. student, Colorado School of Mines, Golden, USA

J. Rostami
Associate Professor, Colorado School of Mines, Golden, USA

O. Frough
Research Associate, Colorado School of Mines, Golden, USA

ABSTRACT: Tunnel Boring Machine (TBM) is one of the most efficient methods for underground excavation. The estimation of the performance of TBM requires accounting for a large number of factors like geology, machine design and configuration, and operations. Simulation of tunnel activities considering TBM to be a tunneling factory can be a reliable method for estimation of TBM utilization factor. The subsurface geotechnical conditions that control machine penetration rate and muck transportation system are the two most important technical factors affecting the overall performance of the machine in a routine operation. This paper focuses on establishing the impact of geology on the muck transportation system to observe the interdependence between TBM penetration rate and transportation delays. The muck haulage system should be designed to incorporate the changes in the excavation rate in different geology to minimize related delays. The simulations were performed using discrete event simulation software, ARENA, where various tunnel activities were modeled in their relevant sequence with varying geological conditions. The simulation accounts for the number of trains and addition of California switch to examine the impact of the setting on transportation-related delays in various formations.

1 INTRODUCTION

Tunnel boring machines (TBMs) have dominated the tunneling industry since they were first introduced in 1950s and recent technological advancement have made them more efficient to meet increasingly challenging conditions and operational demands. Justifying the use of TBMs as the tunneling method and selection of machine type, specs and back-up system requires accurate assessment of anticipated productivity and performance of machine in given geological conditions. There are several models for TBM performance prediction that allow for estimation of rate of penetration. The most commonly used models for this purpose include the Colorado School of Mines or CSM model, Field Penetration Index (FPI) model, and Norwegian University of Science and Technology or NTNU model (Xue et al., 2018). These models estimate rate of penetration (ROP). Different models are needed to estimate TBM utilization considering the machine downtimes and delays (Rostami, 2016). However, the available models for estimation of TBM utilization are insensitive to project specific parameters such as geology, machine characteristics, and site set up and do not account for all the individual tunneling activities and downtimes. This means that accuracy of TBM utilization factor estimated by available models is not very high. The details of shortcomings of these models are discussed by Farrokh (2012). Hence, there is a need to establish a model that is capable of estimating utilization factor considering site specific geological and machine parameters.

The rock mass condition has a significant impact on machine downtime and utilization factor (Frough et al., 2018). The muck transportation system and whole back-up system is designed according to the ground condition and project requirement. The table provided by Rostami (2016) in his paper shows general guidelines of estimation of TBM utilization for uniform and complex ground conditions and different muck haulage systems. However, this table is general in nature and does not take into account the intricacies of the machine, site set up and detailed geology.

The muck generated by the hard rock TBM excavation is generally well-graded, and contains large elongated chips with low percentage of fines. If the excavation cycle of a given stroke is longer (i.e in harder rocks), muck haulage system will have sufficient time for trip to the portal (or shaft), disposal, and return to the heading. As such, the operation does not incur any delays, whereas if the formation is softer and time for advancing a stroke is short, then trip time to the disposal site may register as downtime and lower utilization. This simple example shows how the ground condition could make transportation system the bottleneck in overall TBM performance. As pointed out by Faddick (1978), the transportation system is a potential limiting constraint in the TBM operation and therefore studying its impact under different geological conditions is important. As such, the focus of this study is to evaluate the effect of different ground conditions on muck transportation system using the recently developed discrete event simulation for modeling TBM performance in a given geological setting.

2 TRANSPORTATION TYPES

Muck transportation system plays a vital role in estimation of TBM utilization factor. The different types of muck transportation system includes track bound logistics systems, trackless logistics system and conveyor belt. While haulage system from TBM face to the backup area is usually by a conveyor belt (TBM belt), the muck is carried through the tunnel to the dumping area by different means of transportation (tunnel haulage). Among the available options, the most commonly used tunnel haulage system is track bound logistics systems or simply, trains. The objective of studying the transportation system is to minimize the delay caused by tunnel haulage. This can theoretically be achieved by having the cycle time of one train less than or equal to the cycle time of one TBM stroke. (Rostami et al., 2014). Table provided by Rostami (2016) shows that overall utilization rate is higher in case of conveyor belts when compared with train haulage system. The table also shows that utilization factor reduces with the complexity in the ground condition. However, the relation between the transportation system and ground condition suggested in the proposed table was not well defined and is general in nature.

3 EFFECT OF GEOLOGY ON TRANSPORTATION

To represent the impact of ground conditions on TBM operation, it is assumed that with lower RMR value, the penetration rate will increase but considering the presence of joints and instability, the TBMs are driven at reduced thrust and RPM (Farrokh, 2012). Frough et al. (2018) established that highest utilization and lowest downtime can be achieved with RMR between 40 and 60 for the cases that were studied, where a double shield TBM was utilized for tunneling. In the case of open TBMs, the time required for ground control at lower RMR reduces the utilization factor and thereby advance rate. This in turn, gives more time for the transportation system to reach from portal to heading and vice versa.

4 DISCRETE EVENT SIMULATION APPROACH

The analysis was done by formulating a simulation model using ARENA which is discrete event simulation (DES) software. The simulation approach is proposed to minimize the uncertainties related to project logistics, resources, cost, time, and overall project management. DES provides flexibility and reduces the complexity of any system (Sharma, 2015) thereby proving a promising approach for simulating TBM operations. The model accounts for various tunneling activities that were set as events and inter-connected in series or parallel similar to real case scenarios. The simulation of transportation activity for a hard rock TBM was introduced by Rostami et al. (2014 and 2016) using Arena. The proposed approach allows for optimization of the number of trains and California switches. Comprehensive simulation model was prepared by Rahm et al. (2016) depicting the activities of a mechanized tunneling operation. The models however did not incorporate different ground conditions, maintenance delays and other operational parameters and hence it was unable to predict the effect of geology on the transportation system.

In this study, a model is prepared for a double shield machine used in Karaj water conveyance tunnel project in Iran (see Figure 1). The project had two sections with 16 km and 14 km long tunnels that was excavated using double shield TBM. The excavation and final diameter of the tunnel were 4.66 m and 3.90 m, respectively with 250 mm of lining thickness. The adverse geological conditions prevailing in the area was raveling, block failure, and some groundwater inflow (Frough et al., 2013). Different geological units with their average RMR values are listed in Table 1. Details of Lot 1 (16 km) were taken for the analysis.

Time distributions related to each tunnel activity were prepared using Karaj water tunnel project – Lot 1 data. Tunnel activities include boring time and downtimes related to different geological conditions, transportation, segment installation, re-gripping, unexpected breakdown, maintenance, surveying, utility extension and muck disposal that are inter-linked on the basis of field data. Data obtained from project site were analyzed to formulate time distributions of each tunnel activity. The best fitting curve with least square error was considered in the analysis. Example of time distributions for key tunneling activities are shown in Figure 2. Table 1 summarizes the time distributions of geology related delay for each geotechnical unit (Hassanpour et al., 2009).

Figure 1. Double shield machine and site set-up for Karaj water conveyance tunnel project (Frough et al., 2014, 2018).

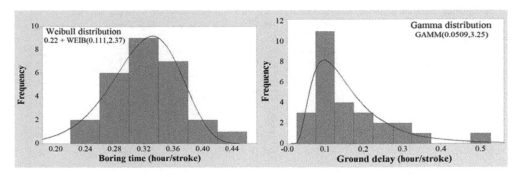

Figure 2.   Example of time distributions prepared to be set as input for the simulation model.

Table 1.   Engineering geological units encountered along the Karaj project alignment (Hassanpour et al., 2009).

Rock mass unit	Av. RMR	Remarks on rock properties and descriptions	Time distribution of ground related delays
Gta1-1	35	Weak to moderately strong rock, thin to moderately bedded, fractured	-0.001 + WEIB(0.596,0.245)
Gta1-2	50	Moderately strong rock, thin to moderately bedded, fractured	-0.001 + WEIB(0.0284,0.359)
Gta2	49	Moderately strong, thick to moderately bedded, moderately fractured, stable	-0.001 + EXPO(0.258)
Gta3	64	Moderately strong, thick to moderately bedded, moderately fractured	-0.001 + LOGN(0.298, 0.866)
Gta4	75	Very strong thick to moderately bedded	-0.001 + 10 * BETA(0.17, 4.16)
Sts1	57	Weak to moderately strong rock, thin to moderately bedded, fractured	-0.001 + EXPO(0.386)
Sts2	72	Very strong, thick bedded	-0.001 + WEIB(0.284, 0.736)
Mdg	63	Moderately strong, thick to moderately bedded, moderately fractured	GAMM(0.0509, 3.25)
Tsh	46	Weak to moderately strong rock, thin -moderately bedded, foliated, fractured	LOGN(0.242, 0.197)
Cz	21	Very weak strength, unstable rock	-0.001+143*BETA(0.0719,0.303)

5   SIMULATION MODEL

The main idea behind the simulation of tunneling operation with emphasis on muck transportation was to investigate the impact of train travel time. The cycle time of train less than or equal to that of one TBM stroke was not considered to cause delays. The simulation model was prepared in ARENA as shown in Figure 3. In total 19 models were prepared based on the increasing distance and varying geological units and number of trains to make the model compatible with the site conditions. The models showed good correlation with the recorded field data. The analysis showed that downtimes due to transportation and geology was 13% as shown in Figure 4. Five scenarios were considered to assess the interaction of geological and transportation related delay and the resulting overall utilization factor. The cases are:

Case I: 1 train with 5 different formations and related geology related delays
Case II: 2 train with 5 different formations and related geology related delay
Case III: 2 trains with 1 California Switch and 5 different formations
Case IV: 5 geological units arranged in different sequences with 2 trains (Figure 5)

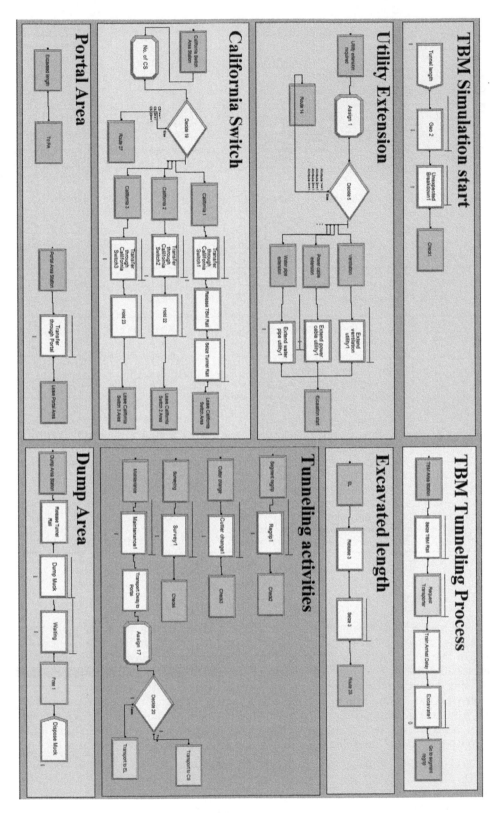

Figure 3.   Karaj project simulation model in Arena.

3899

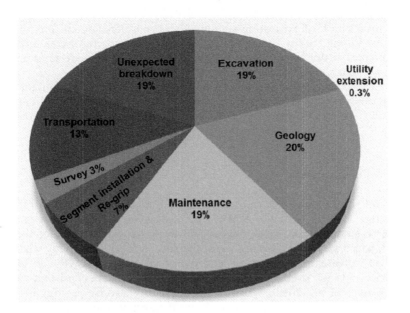

Figure 4. Karaj project overall downtime analysis.

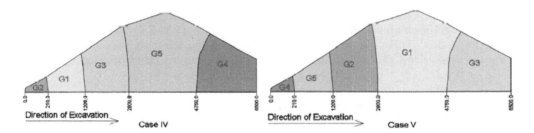

Figure 5. Different sequence of geological units along the tunnel alignment (Cases IV versus Case V).

Case V: 5 geological units arranged in different sequences with 2 trains (Figure 5)

For Cases I to III, the geological units selected were units Gta1-2, Sts-1 and Gta4 and RMR values were used as the base for the selection of these formations. The distance from the portal was kept same to compare the three cases. The top speed of the train for modeling was 14 miles/hour, which is typical of many muck haulage trains. The position of switch was kept the same i.e. behind TBM backup area. The total tunnel length was 6.5 km for all the cases.

For Case IV and V, the effect of distance from the portal was analyzed with variable sequence of geological units. Two (2) scenarios were modeled and compared as shown in Figure 5. Two trains without California switch was considered for this analysis. Five geological units with variable length along the tunnel alignment was considered as given in

Table 2. Input parameters of geological units for Case V.

Geological units	RMR	Time Distribution for geological delay	Time Distribution for Boring
G1	21	-0.001 + 143 * BETA(0.0719,0.303)	LOGN(1.13,0.943)
G2	35	-0.001 + WEIB(0.596,0.245)	LOGN(1.51,1.2)
G3	50	-0.001 + WEIB(0.0284,0.359)	0.1 + LOGN(0.303,0.0971)
G4	65	-0.001 + LOGN(0.298, 0.866)	0.22 + LOGN(0.123, 0.0497)
G5	75	-0.001 + 10 * BETA(0.17, 4.16)	0.04 + ERLA(0.0316, 11)

Table 2. The total length of excavation simulated was considered equal to 6.5 km. The time distributions for boring, utility extension, maintenance, unexpected breakdown, surveying, regular maintenance, cutter change, and transportation delay were modeled as per field observations in Karaj project.

A total of 10 replications for each case was executed to obtain the resultant distribution. Time distributions were taken on the basis of tunnel length and geological conditions. The simulation model shown in Figure 3 was considered as the base scenario for the analysis for all the cases.

6 RESULT & DISCUSSION

In Case I, II and III, the delay due to transportation is inversely proportional to the geology related delays (see Table 3). At lower RMR, significant delays are caused by ground support issues in adverse ground where face collapse and increased maintenance of the cutter discs were expected, which reduces utilization rates. Increase in RMR, especially when RMR goes higher than ~50 increases the utilization, and it plateaus around RMR ~70 (depending on other issues, specially rock strength and abrasion). These results are consistent with the trends discussed by Frough et al. (2018). When delays caused by adverse geology is the same, the difference in overall utilization rate as a function of transportation can be assessed and the impacts observed easily (Table 3). The average completion time reduces with addition of number of trains and further addition of California switch. The boring time remains the same but the overall percentage in time allocation increases as the transportation delay reduces. Significant reduction in the transportation downtime can be seen when California switch is added showing its significant applicability in the optimization of the TBM operation.

Case IV and V: This case was focused on assessing the effect of sequence of variable geological units along the tunnel alignment considering similar transportation system. As

Table 3. Sensitivity of geology on transportation systems.

Case No.	Transportation system	RMR 50	RMR 65	RMR 75
I	1 train	33%	32%	30%
II	2 trains	23%	22%	21%
III	2 trains + 1 California Switch	17%	16%	15%
Increased Utilization factor with change in transportation system		5%	3%	4%

Notes: The transportation delay is reduced to almost half but utilization is not increased accordingly due to other delays being more prevalent.

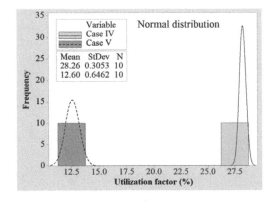

Figure 6. Utilization factor variability with change in sequence of geological units (Case IV versus Case V).

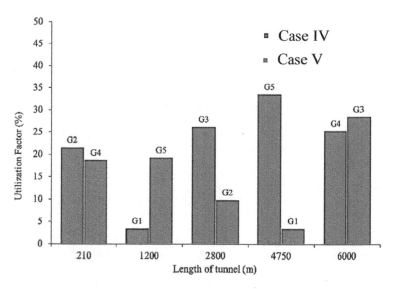

Figure 7. Case V: Utilization factor variability with change in sequence of geological units.

can be seen from Figure 6 and 8 that changing the sequence of units can significantly change the overall utilization factor of a TBM operation. Considering one geological setting, it was observed that the utilization factor varied as location of the unit changed along the alignment. Units with RMR 65 and 75 resulted in lower values of utilization factor when they were near the portal. This is due to the fact that the excavation cycle was faster and TBM had to stop for the train to arrive at the backup area thereby reducing the utilization. It must be noted that during initial excavation, the TBM operation is still under learning curve and many other activities lead to reduction in overall utilization factor.

7 CONCLUSION

The current study was conducted to assess the interaction between the geology and transportation system in TBM operation. Adverse geology represented by lower RMR values can have different effect on time for the train to transport the muck to the shaft or portal, depending on the site settings. In weak grounds, delays and lower utilization factor is mostly due to ground control issues. In the case that was modeled in this study, given the machine type and geological settings, as RMR value increased to ~70, the transportation system became the main bottleneck thereby contributing to increased delays in this sub-system. With further increase in RMR value, the rock mass becomes stronger and perhaps with high abrasivity, that causes longer boring cycles and while utilization factor might increase, the daily advance rate might be lower. The time allocated for the cutter inspection and frequency of cutter change increases in this scenario that could be the main component of delay i.e. geology. With addition of number of trains and California Switch, the delay due to transportation is reduced as the enhancement in the transportation subsystem will address the bottleneck and thereby increases the overall utilization of the machine.

The study shows that optimizing the transportation system by increasing number of trains and adding California Switch, increases the TBM utilization and hence notable reduction in the completion time can be realized. The sequence of geological formations along tunnel profile also impacts utilization factor, which is often very difficult to quantify using conventional performance prediction approaches. As such, the simulation approach allows for optimization of the transportation system, or perhaps other subsystems, for the geological settings and site arrangements.

REFERENCES

Faddick, R.R., Martin, J.W. 1978. The transportation of tunnel muck by pipeline. Transportation research board, National technical information service, DOT-TSC-UMTA-78-7

Farrokh, E. 2012. Study of Utilization factor and advance rate of hard rock TBMs. Dissertation in Energy and mineral engineering, Pennsylvania State University

Frough, O., Torabi, S.R. 2013. An application of rock engineering systems for estimating TBM downtimes. Engineering Geology 157, 112–123

Frough, O., Rostami, J. 2014. Analysis of TBM performance in two long mechanized tunnels, case history of Karaj water conveyance tunnel project lots 1 and 2 (Iran). World Tunnel Congress, Foz do Ignacu, Brazil, 177

Frough, O., Rostami, J. 2018. Study of the correlation between RMR and TBM downtimes, North American Tunneling Conference, Washington D.C, 2018, 58–65

Hassanpour, J., Rostami, J., Khamechian, M., Tavakoli, H.R., 2009. TBM performance analysis in pyroclastic rocks: case history of Karaj water conveyance tunnel. Rock Mechanics and Rock Engineering Journal 43 (4), 427–445

Rahm, T., Scheffer, M., König, M., Thewes, M., Duhme, R., 2016. Evaluation of disturbances in mechanized tunneling using process simulation. Computer-Aided Civil and Infrastructure Engineering 31, 176–192

Rostami, J. 2016. Performance prediction of hard rock Tunnel Boring Machines (TBMs) in difficult ground. Tunnelling and Underground Space Technology 57, 173–182

Rostami, J., Farrokh, E., Laughton, C., Eslambolchi, S. S. 2014. Advance rate simulation for hard rock TBMs. KSCE Journal of civil engineering 18 (3), 837–852

Sharma, P., 2015. Discrete event simulation. International Journal of Scientific and Technology Research 4 (4), 136–140

Xue, Y., Zhao, F., Zhao, H., Li, X., Diao, Z. 2018. A new method for selecting hard rock TBM tunneling parameters using optimum energy: A case study. Tunnelling and Underground Space Technology 78, 64–75

Tunnels and Underground Cities: Engineering and Innovation meet Archaeology, Architecture and Art, Volume 7: Long and deep tunnels – Peila, Viggiani & Celestino (Eds)
© 2020 Taylor & Francis Group, London, ISBN 978-0-367-46872-9

Analysis of micro slurry TBM excavation parameters in subsea tunnel

K.Y. Kim, H.H. Ryu, D.S. Bae & S.A. Jo
Korea Electric Power Corporation Research Institute, Daejeon, Republic of Korea

ABSTRACT: There are some differences between inland and subsea tunnel under construction using mechanical excavation method. High water pressure due to sea water besides underground water become a trouble factor and it is difficult to plan countermeasures in the subsea tunnel construction. So, it is important to prevent unexpected trouble in advance. To prevent construction trouble, analysis of geological and mechanical data of TBM under construction are necessary. This paper introduces the first time slurry TBM method for drilling micro subsea tunnel under expected 6 bar in the Yellow sea whose location is the west of Korea peninsula. Analysis of TBM excavation parameters including net penetration rate, penetration depth, thrust force, torque, RPM and face pressure were performed based on in-situ geotechnical condition and mechanical data. The surface layer of seabed ground in the Yellow sea of Korea peninsula was composed of clay. Since this layer formed an impermeable zone, face water pressure up to expected 6 bar did not occur. The acquired subsea tunnel database from in situ will be used as a lessons learned for the future subsea tunnel project.

1 INTRODUCTION

Mechanical excavation method like Tunnel Boring Machine (TBM) has been used in the whole countries. About three decades have passed since TBM was used to excavate the tunnel in KOREA. New planned tunnel by TBM will increase gradually (Kim et al. 2013). Subsea tunnel is increasing owing to development of cities. There are some differences between inland and subsea tunnel under construction using mechanical excavation method. High water pressure due to sea water besides underground water and its countermeasures are difficult in the subsea tunnel construction. So, it is important to prevent unexpected trouble in advance. To prevent construction trouble, analysis of geological data and mechanical data of TBM under construction are necessary. In recent, the subsea tunnel subject to about 11 bar of face pressure was completed in Turkey (Burger & Arioglu 2015). This paper introduces the micro slurry TBM method for drilling micro subsea tunnel under expected 6 bar in the Yellow sea whose location is the west of Korea peninsula. Analysis of TBM excavation parameters including net penetration rate, penetration depth, thrust force, torque, face water pressure and RPM are performed based on in-situ geotechnical condition and mechanical data. The acquired subsea tunnel database from in situ will be used as a lessons learned for the future subsea tunnel project.

2 SUBSEA TUNNEL PROJECT

This tunnel is constructed by the micro slurry TBM method of which the outside diameter of 3.59 m. Excavation depth is about 60 m and tunnel length is 1,831 m. The analysis data was collected from an approximate length of 1,190 m, which corresponded to 65 % of the total tunnel length. The depths of the shafts for carrying the TBM machine in and out were 63.7 m and 57 m, respectively. The slope of excavation tunnel is 0.3 % upward excavation, the RMR ranged from II to IV.

2.1 Slurry Tunnel Boring Machine

For drilling micro subsea tunnel under expected 6 bar in the Yellow sea whose location is the west of Korea peninsula, the slurry type TBM was applied for the excavation. The slurry TBM have lots of merits that working pressure is systematic, less power at cutter head, less torque required, easier to operate and less frequent interventions (ITA 2001, Townsend & Jenkins 2009). The micro slurry TBM with an outer diameter of 3.53 m is refurbished by RASA heavy industries in 2016. Rotation per minute (RPM) at cutter head can be operated up to max 4.6. The rated torque of cutter head is 1,410 kN-m, the electric motor is 4 x 440 kW. Total thrust force 12,000 kN is generated by 12 jacks. The allowable normal load of installed disc cutter is above 240 kN considering soft and hard rock. The disc cutter is applied to 27 pieces of 15.5 inch. A schematic view of Slurry TBM is given in Figure 1 and Table 1, which summarizes the TBM machine specifications.

2.2 Geology of project

11 Boreholes in the depth ranging 29 m to 63 m was bored to investigate geological and geotechnical properties along the tunnel route. Ground in the studied area consists of the multiple layers with different thickness such as deposited clay, weathered soil, weathered rock, soft rock and hard rock. In particular, the thickness of deposited clay was about 3 m. The geotechnical characteristics for each ground layer were shown in Table 2.

The rock stratum of this site was formed in the Precambrian, most of bedrock was composed of gneiss and schist. Rock Mass Rating (RMR) was identified as 11 to 77 (II~IV class).

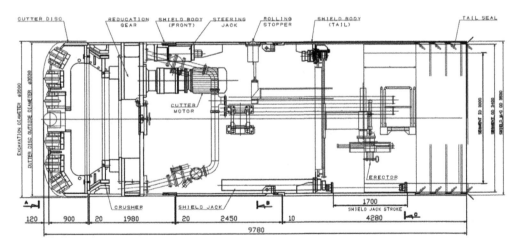

Figure 1.   Slurry TBM schematic.

Table 1.   Tunnel project and TBM profile.

Project	○○ micro tunnel	Tunnel Length	1,831 m
TBM type	Slurry TBM	Manufacturer	RASA(2016)
Out diameter	3,530 mm	Raidus	200 m
Max RPM	4.6 rev/min	Max stroke	1,700 mm
Max Torque	1,410 (Norm. 915) kN·m	Thrust per jack	1,000 kN
Cutter type	CCS (15.5 inch)	Total thrust	12,000 kN
Cutter No.	27EA	Power	440 kW

Table 2. Geotechnical characteristics for the studied area.

Ground type	Unit weight (kN/m³)	Cohesive strength (kPa)	Friction angle (°)	Uniaxial compressive strength (MPa)	Thickness (m)
Deposited clay	18.0	20	15	-	1.5~3.0
Weathered soil	19.0	20	29	-	1.2~7.0
Weathered rock	21.0	30	30	-	3.5~19.0
Soft rock	23.0	210	35	25.9	2.1~21.4
Hard rock	26.0	540	41	43.9	2.1~9.0

Table 3. Distribution ratio of RMR class along the tunnel line.

RMR	Tunnel length(m)	Composition ratio(%)
I	-	-
II	70	4
III	541	29
IV	1,090	60
V	130	7
Total	1,831	100

The distribution ratio of RMR class along the tunnel route was 4 % of II class, 29 % of III class, 60 % of IV and 7 % of V class, as shown in Table 3. In other words, the III to IV classes of RMR occupied 89 % in the tunnel route. Maximum tide level is about 9.6 m. From Figure. 2, anomaly zones were identified at 2 places in the tunnel site. Limestone cavities and fault fractures were found in Anomaly-1 and also the fracture zone was found at around 1,500 m from the shaft # 1 (Anomaly-2).

3 ANALYSIS OF EXCAVATION PARAMETERS

The database was compiled using the data measured at the drilling borehole and the recorded machine data which was collected between December 2016 and August 2017. Analysis of the correlation between some factors such as net penetration rate, penetration depth, RPM, torque, thrust force and face pressure was performed. The values required for predicting the TBM performance were calculated using the operation parameters (e.g., thrust, torque, RPM, and power) recorded in the TBM black box.

$$PR \ (mm/min) \ = \ L \ / \ t \tag{1}$$

$$Pe \ (mm/rev) \ = \ PR/RPM \tag{2}$$

Where, PR is net penetration rate and Pe is penetration depth per revolution, L is segment length (mm), t(min) is penetration time per segment, RPM is rotational speed (rev/min).

From the results of the statistics analysis PR, Pe, RPM, Torque(Tq) and Thrust(Th) values were in the 14.1±3.2 mm/min, 5.3±1.7 mm, 2.7±0.4 RPM, 96.4±2.4 KN-m and 5,378±1,010 kN range, respectively. Whereas thrust per cutter showed values of 199.153.9 ± 37.5 KN and maximum face pressure occurred up to 4.2 bar.

3.1 *Net penetration rate(PR) and penetration depth(Pe)*

The graphs in Figure 2 represent the PR depending on the tunnel length. As you can see in the Figure 2 the average PR was displayed above 15 mm/min line until the first fracture zone at around 800 m. After that, the average PR was displayed under 15 mm/min. After passing the first anomaly zone, the average PR came back up to above 15 mm/min. When passing the second anomaly zone, the average PR was displayed under 15 mm/min again. While the construction problem when passing through the anomaly zone had been predicted from the borehole investigation, slurry TBM was operated carefully and any construction troubles did not occur. The graphs in Figure 3 represent the Pe depending on ground condition. As you can see in the Figure 3, the Pe was displayed lower values relatively at around the anomaly zone and RMR II region.

3.2 *RPM and torque*

The graphs in Figure 4 & 5 represent the RPM & Tq depending on the tunnel length. As you can see RPM values were ranged 2 to 3 value, whereas Tq were ranged 90 to 110 kN-m value. When passing the first anomaly zone, the RPM was operated at about 2 RPM to ensure safety excavation. When passing the second anomaly zone, the RPM was operated around 3 RPM without difficulty.

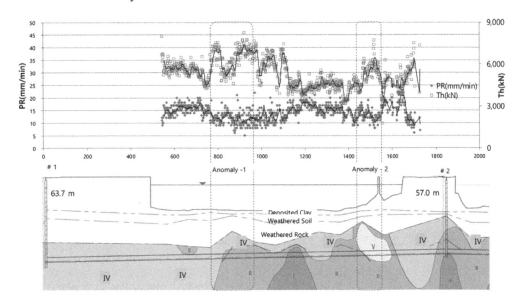

Figure 2.   Geology, net penetration rate (PR) and Thrust force (Th) (KEPCO, 2014).

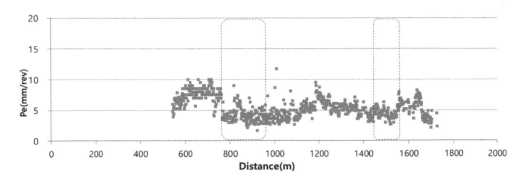

Figure 3.   Penetration depth depending on the tunnel length.

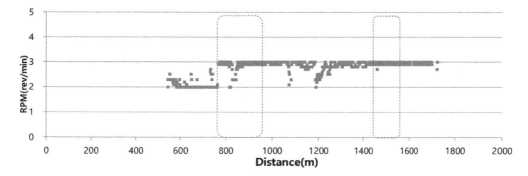

Figure 4. Cutterhead rotation per minute depending on the tunnel length.

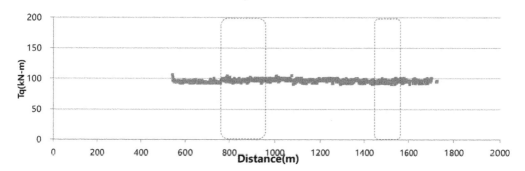

Figure 5. Torque depending on the tunnel length.

Looking at the relation between PR and thrust force when passing through the rock, the PR decreased with the increase of thrust rather (Figure 6). The reason why is that the rock strengths of tunnel route were not homogeneous. But, Rostami (2016) reported that rate of penetration will increase with thrust/cutter load in a given homogeneous rock and rock mass. The Torque did not affect the PR values in this site. The PR increased partially with the increase of RPM value as Figure 7.

3.3 Face pressure

The expected maximum face pressure at this site was 6.0 bar, but actually it was 4.2 bar under construction of subsea tunnel. Fortunately, the maximum face pressure passing through the

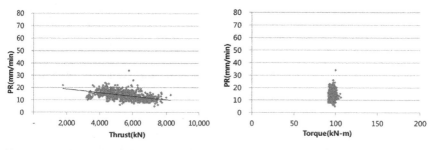

(a) Net penetration rate depending on Thrust force   (b) Net penetration rate depending on Torque

Figure 6. PR (Net penetration rate) depending on Thrust force and Torque.

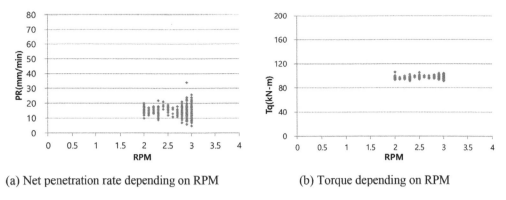

(a) Net penetration rate depending on RPM   (b) Torque depending on RPM

Figure 7.   PR (Net penetration rate) and Tq (torque) based on RPM.

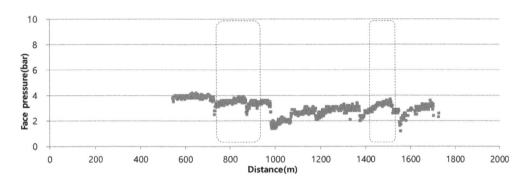

Figure 8.   Face pressure depending on the tunnel length.

anomaly zone was under about 4 bar (Figure 8). The reason why the surface layer of seabed ground in the Yellow sea of Korea peninsula is composed of clay. Since this layer forms an impermeable zone, face water pressure up to 6 bar did not occur and only hydrostatic pressure would have worked. On the other hand, condition of the anomaly zone might be better than expected status.

4   CONCLUSION

This paper introduces the first time micro slurry TBM method for drilling subsea tunnel under expected 6 bar in the Yellow sea whose location is the west of Korea peninsula. Analysis of TBM excavation parameters including net penetration rate, penetration depth, thrust force, torque, face water pressure and RPM was performed based on in-situ geotechnical condition and mechanical data of slurry TBM. The acquired subsea tunnel database from this study will be used as a lessons learned for the future subsea tunnel project. The summary results are as below.

Looking at the composition ratio of RMR depending on the route, 4 % of II grade, 29 % of III grade, 60 % of IV and 7 % of V grade was identified. The III to IV grade of RMR occupied 89 % in the tunnel route.

From the results of the statistics analysis, PR, Pe, RPM, Torque (Tq) and Thrust (Th) values were 14.1±3.2 mm/min, 5.3±1.7 mm, 2.7±0.4 RPM, 96.4±2.4 kN-m and 5,378±1,010 kN respectively. Whereas thrust force per cutter showed values of 199.153.9 ± 37.5 KN and maximum face pressure occurred up to 4.2 bar. The expected maximum face pressure at this site was 6.0 bar, but actually it was 4.2 bar under construction subsea tunnel. The reason why the surface layer of seabed ground in the Yellow sea of Korea peninsula is composed of clay. Since this layer forms an impermeable zone, face water pressure up to 6 bar did not occur and only hydrostatic pressure would have worked.

To overcome the construction problem which was predicted from the borehole investigation when passing through the anomaly zone, Slurry TBM was driven carefully and any construction troubles did not occur.

5 ACKNOWLEDGEMENT

This research was supported by a grant (17SCIP-B105148-03) from the Construction Technology Research Program, funded by the Ministry of Land, Infrastructure, and Transport of the Korean government.

REFERENCES

Burger, W.E. Arioglu 2015. Istanbul Strait Road Tunnel Crossing Project: Challenges and TBM Solutions, *AFTES Journees Techniques*, Lyon, 1–12.

ITA Working Group No.14. 2001. *Recommendations and Guidelines for Tunnel Boring Machines*, Bron: ITA-AITES.

KEPCO. 2014. *Cable Tunnel Geotechnical Investigation Report*, Daejeon: KEPCO.

Kim, K.Y., Lee, D.S., Cho, J., Jeong, S.S. & Lee, S. 2013. The effect of arching pressure on a vertical circular shaft, *Tunnelling and Underground Space Technology* 37: 10–21.

RASA. 2016. *DHL-3500 SLURRY SHIELD MACHINE INSPECTION REPORT*, Tokyo: RASA.

Rostami, J. 2016. Performance prediction of hard rock Tunnel Boring Machines (TBMs) in difficult ground, Tunnelling and Underground Space Technology 57, 173–182

Townsend, B.F. & Jenkins, P.E. 2009. *TBM Selection and Specification*, retrieved from http://www.hkieged.org/download/workgroup/TBM%20selection%20and%20specification.pdf.

Tunnels and Underground Cities: Engineering and Innovation meet Archaeology,
Architecture and Art, Volume 7: Long and deep tunnels – Peila, Viggiani & Celestino (Eds)
© 2020 Taylor & Francis Group, London, ISBN 978-0-367-46872-9

Countermeasure of jammed TBM in rock tunnel excavation: Feedback from two cases study

C.H. Lee
United Geotech, Inc., Taipei, Taiwan.

Y.C. Chiu & T.T. Wang
National Taiwan University, Taipei, Taiwan.

ABSTRACT: Tunnel Boring Machine (TBM) has been widely used in long tunnel excavations in many countries for its advantages in speed, price, and safety. However, there are some difficulties, risks, and challenges in TBM excavation. When encountering grounds of fault, fracture zone, water ingresses, or ground with high variation in rock strength and lithology, collapses may trap TBM frequently. Such incidents decelerate TBM advancing and raise project expenses in most of the case. This manuscript introduces experiences from two tunneling cases, Hsuehshan tunnel, the longest road tunnel in Taiwan with a 12.9 km length, and Tsengwen tunnel, the hydraulic tunnel of the highest overburden 1300 m in Taiwan. Releasing methods carried out in the cases, including top heading tunnel, bypass tunnel and others, and measures to prevent further TBM jam, such as forepoling and ground improvement are presented here for future TBM excavations in long and deep tunnels.

1 INTRODUCTION

Tunnel boring machine (TBM) is a machine which has revolutionised the tunneling industry both making tunnelling a safer, more economic solution for creating underground space and opening the possibility of creating tunnels where it was not feasible before (Spencer *et al.*, 2009). TBM method has recently become the mainstream approach for the construction of long tunnels in the international tunnel engineering community. Compared with the New Austrian Tunneling Method commonly used in tunnel construction, the main advantages of TBM method include: high excavation speed under uniform geological conditions, small over-excavation and disturbance to rock mass, less impact on the environment, better working environment and safer construction. However, the TBM method also has its construction bottleneck. If the geological variability along the tunnel is large, or when it encounters a fault or shear zone, a fracture zone or high pressure water ingress, the construction speed may greatly reduce and even cause TBM jam.

This manuscript collected dozens of foreign and domestic tunnelling cases which encountered TBM jam. The causes of jam and corresponding countermeasures were sorted out, and the more effective countermeasures were summarized into four main categories, forepoling method, ground improvement method, bypass tunnel work method and top heading tunnel method. This article will introduce the countermeasures for TBM jam according to jam causes, and discuss the effectiveness of each extrication method through actual cases.

2 CASE COLLECTION AND ANALYSIS

2.1 Case collection

Despite significant progresses in the development of TBM with shield, the use of these machines through weak grounds and adverse geological conditions is still risky. The presence of the shield limits accesses to the tunnel walls, and blocks observations on geological conditions and ground behaviors. Meanwhile, the excessive convergence of weak ground under high in-*situ* stresses can impose high levels of load on the shield, which makes the machine susceptible to entrapment in weak rocks, especially under large overburden. It results in machine jamming and imposes high economic costs on tunneling companies (Hasanpour et al., 2017).

This manuscript first collects relevant cases of TBM jammed in rock tunnels worldwide, organizing them according to the project name, country, TBM type, length and diameter, overburden, geology condition, and description of the jammed conditions, as shown in Table 1.

2.2 Case analysis

According to the results of different TBM jammed cases, the influencing factors of TBM are often not singular, and the TBM type, geological structure, lithology, geostress, groundwater and other factors encountered are related to each other. A proper design of the TBM and the geological exploration during the construction are also very important prevent methods. Furthermore, different engineering geological conditions result in different jammed situations and different types of possible damage encountered by TBM. Shang *et al.* (2007) pointed out that the adverse engineering geological conditions that may be encountered during TBM excavation can be divided into geological structure, lithology, geostress and groundwater factors, and the factors can be divided into various categories and forms. When the composition of factors is likely to cause the TBM to be jammed, the corresponding method forestalling jammed should be established for the influencing factors. The methods of extrication in case of being jammed is an important issue before planning the TBM method for construction.

3 COUNTERMEASURE OF JAMMED TBM IN ROCK TUNNEL

Figure 1 is the prevention methods and countermeasures of TBM jammed in rock tunnel in this study. The main contents include forepoling, ground improvement, bypass tunnel and top heading tunnel. The first two auxiliary methodologies are preventive methods to forestall TBM jammed, and the latter two methodologies are for extrication after the TBM is jammed. The various methodologies are described below.

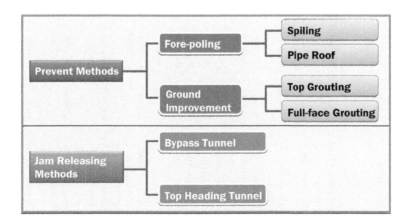

Figure 1.   The prevention methods and countermeasures of TBM jammed in rock tunnel.

Table 1. Representative cases studied of TBM jammed in rock tunnels.

No.	Project name	Country	TBM type	Tunnel length (km)/Diameter (m)	Overburden (m)	Geology	Jammed description	References
1	Hida Tunnel	Japan	Single shielded	10.7/4.5	1000	Four types of rock: Shira-kawa granite, Nohi rhyolite, intrusions of granite porphyry, and Hida gneiss.	Twenty collapses and nine TBM jamming events occurred.	Terada et al., 2008
2	Yacambú – Quibor Tunnel	Venezuela	Gripper	24.3/4.8	400	Graphitic phyllite	Along 250 m, bored profile practically closed within 30 days.	Ramoni, 2010
3	Stillwater Tunnel	USA	Double shielded	12.9/2.91	800	sandstone, siltstone, blocky clayey schist, fault zones	Convergences of 4 %; convergence rate of 0.4 %/d after 1 day and of 0.09 %/d after 10 days.	Ramoni, 2010
4	Evinos – Mornos Tunnel	Greece	Double shielded	29.4/4.04	950	Flysch, 650 m long overthrust zone	Convergences of 15 cm within less than 1 hour at a distance of 1–2 m from the working face.	Ramoni, 2010
5	Guadiaro – Majaceite Tunnel	Spain	Double shielded	12.2/4.88	400	Sandy and clayey flysch, claystone	Gap between shield and ground closed at a distance of 1 m from the working face; radial displacements of 5–6 cm of the damaged segmental lining.	Ramoni, 2010
6	Verena Tunnel (Nord Section)	Switzerland	Gripper	11.6/7.64	1250	Heavily fractured crystalline rock	Radial displacements of 20 cm at the cutter head during standstill due to raveling or squeezing ground.	Ramoni, 2010
7	Umiray – Angat Tunnel	Philippines	Double shielded	13.2/4.88	1200	Basalt combined with other metamorphic volcanic rocks	Convergence rate of 20 cm/h; in two cases reduction of the bored profile of 12 cm at a distance of 1 m from the working face, in a third case convergences of 37 cm within less then 2 hours in the machine area.	Ramoni, 2010
8	Fujikawa Transport and Pilot Tunnels	Japan	Double shielded	4.5/3.5	300	100 m long fault zone with clayey material	Reduction of the bored profile by 60 cm with a convergence rate of 29 cm/d; the gap between ground and tail shield became closed within 1 day.	Ramoni, 2010
9	Nuovo Canale Val Viola Tunnel	Italy	Double shielded	18.8/3.6	800	Pelitic and phyllitic rock	The gap between shield and ground (8 cm in diameter) closed at a distance of 9 m from the working face (at the end of the shield) within 6 to 10 hours.	Ramoni, 2010
10	Shanggongshan Tunnel	China	Double shielded	13.8/3.65	250	Alternating stratification of sandstones and clayey schists with cataclastic shear zones	Gap between shield and ground (5–10 cm) closed practically instantaneously.	Ramoni, 2010
11	Ghomroud Tunnel (Sections 3 and 4)	Iran	Double shielded	16.5/4.5	650	Graphitic schists and sandstones	The TBM was already blocked for eight weeks 300 m before in the same geology due to instability of the working face and of the tunnel walls.	Ramoni, 2010
12	Gilgel Gibe II Tunnel	Ethiopia	Double shielded	25.8/6.98	670	weathered, brecciated and decomposed basalt	Extrusion rate of the core of 4–6 cm/h; TBM pushed back more than 60 cm and displaced laterally more than 40 cm.	Ramoni, 2010

(Continued)

Table 1. (*Continued*)

No.	Project name	Country	TBM type	Tunnel length (km)/Diameter (m)	Overburden (m)	Geology	Jammed description	References
13	Uluabat Tunnel	Turkey	Single shielded	11.47/5.05	300	The majority parts of tunnel alignment are situated at Karakaya formation of Triassic-aged metadetritic rocks like fine-grained meta-claystone, meta-siltstone, meta-sandstone, and graphitic schists.	The TBM became jammed 18 times at different tunnel locations.	Hasanpour et al., 2017
14	Lake Mead	USA	Single shielded	4.7/7 2	150	Metamorphic rocks over the first and over the central part of the alignment, and sedimentary rocks in the rest. Two major fault zones of completely sheared rock in the metamorphic rocks at the begin of the TBM drive.	Jamming of the shield. The onset of jamming was observed in correspondence of the less competent zones of the metamorphic.	Seingre et al., 2017
15	Kargi tunnel	Turkey	Double shielded	7.8/9.84	610	The tunnel route lies very close to the East-West trending main branch of the world-wide known, active North Anatolian Fault Zone and within highly tectonized area where complex ophiolitic/metamorphic rock formations are dominating around the region covered by younger volcanic and volcano sedimentary unit.	All squeezing and CH blockage phenomenas occurred at extensively tectonize ophiolitic rocks. Note that this phenomena occurred sometimes in low overburder, thus most probably horizontal stress is greater than vertical stress. The TBM rescued with recovery galleries.	Seingre et al., 2017

3.1 Fore-poling method

When the ground ahead of the TBM is weak or when there is a large rock wedge above the tunnel, which may cause the machine to be jammed, it is necessary to install steel bars or steel pipe roofing as forepoling of the tunnel, and the steel bar, steel tube or drilled hole may be grouted with cement mortar or silica resin to form a reinforced umbrella-shaped protection arch in advance. When the required improvement range is short, use the forepoling steel bar metho (as shown in Figure 2(a)). When the required improvement range is long, it is more suitable to use the forepoling steel pipe roofing meth for support (as shown in Fig. 2(b)).

3.2 Ground improvement method

If the ground ahead of the TBM is weak and fractured, or the plastic zone of the rock mass is too large, the grouting hole can be set forward from the side of the TBM body to carry out the ground improvement work on the front and surrounding rock masses to increase the self-stand up time of the rock mass excavation, and quickly pass through. The improvement range of the ground can be in the upper half section (Fig. 3(a)) or full-section (Fig. 3(b)). For better results, the grouting material can be mix of cement and bentonite, or ultra-fine cement.

3.3 Bypass tunnel method

While carrying out tunnel excavation operations, the TBM may encounter disaster ahead of the cutter head such as a fractured zone or large-scale water ingress resulting in massive collapse of debris ahead or debris accompanied by a substantial water ingress which could possibly trap and jam the machine. In order to get the TBM out, a bypass tunnel can be excavated from the side of the TBM or the side of the rear ring to the front of the TBM, and remove the fractured rock mass in front of the cutter head to relieve the TBM. Generally speaking, the entrance of the pilot tunnel should be as close as possible to the machine head. And sometimes the execution of the bypass tunnel is in bad rock conditions or in presence of water, it should be consider to combine with prevent measures. Refer to Figure 4 for the schematic diagram of the bypass tunnel method.

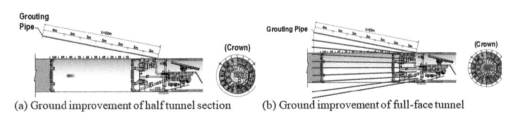

(a) Ground improvement of half tunnel section   (b) Ground improvement of full-face tunnel

Figure 2. Schematic diagram of steel pipes method and fore-poling method.

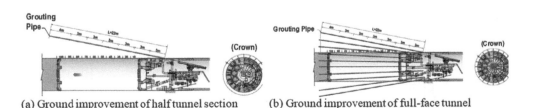

(a) Ground improvement of half tunnel section   (b) Ground improvement of full-face tunnel

Figure 3. Schematic diagram of ground improvement.

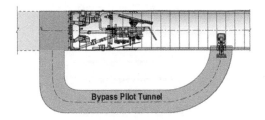

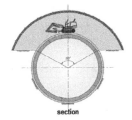

Figure 4. Schematic diagram of bypass tunnel method.

Figure 5. Schematic diagram of top heading method.

3.4 Top heading tunnel method

When the TBM is jammed due to excessive forward force because of fracture of the surrounding rock on the top arch, the top heading tunnel method is a more suitable methodology for extrication. The method is to remove and support the fragile ground above the TBM. If necessary, it is also possible to carry out consolidation grouting to strengthen the overlying rock on the top arch. Then, after the excavated top heading tunnel is backfilled and consolidated, the TBM can be removed and smoothly advance forward. The schematic diagram of the top heading tunnel is shown in Figure 5.

4 CASE STUDY OF JAMMED TBM IN ROCK TUNNEL

4.1 Example of bypass tunnel

Before the excavation of a tunnel in the northeastern part of Taiwan passes through many geological structures with abundant groundwater, a pilot tunnel was first excavated by a double shielded TBM to prevent water ingress. However, the TBM was jammed 13 times during the construction. The longest rescue action for the TBM was 290 days. The causes of jam can be summarized as follows: (1) when the TBM passed through shear or fault zones, massive debris landed on the cutter head of the machine and made it unable to rotate; (2) when the TBM passed through the hard rock zone, where the rock mass was fractured and held abundant groundwater, high pressure and massive water ingress and debris rushed into the TBM, flooding the motor and equipment, and jammed the TBM (Lin & Yu, 2005).

In this case, the longest duration of TBM jam was due to fractured rock and massive water ingress (150 l/sec), which caused the TBM to be jammed from the nose to the tail, and the arched top ring to fall. The entire TBM machine was jammed. In order to get the TBM out, the bypass tunnel method was immediately adopted. Since the TBM was jammed in an extremely bad geological section with abundant pressurized groundwater, collapsing occurred even after excavation of bypass tunnel. Henceforth, a number of bypass pits were excavated to solve the problem (Shen et al., 2005). A total of 6 bypass tunnels were excavated to extricate the jammed TBM with a total length of over 100 m. Refer to Figure 6 for relevant locations of the bypass tunnels. Refer to Figure 7 for the principle of formulating the bypass tunnel method in accordance with the excavation experience of constructing multiple bypass tunnels for this case.

4.2 Example of top heading tunnel

A single-shield slurry-supported TBM with a diameter of 5.46 m was used in a tunnel in southwestern Taiwan for excavation. The expected length of the excavation was 4–6 km. The lithology that TBM passed was mainly sandstone and shale with the geological structure of one syncline and one anticline. The maximum overburden is 1,300 m. It was estimated that the

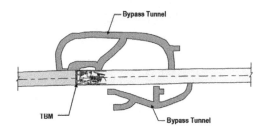

Figure 6. Schematic diagram of case tunnel jammed and using bypass tunnel method.

Figure 7. Flow chart of treatment of bypass tunnel method.

TBM had the possibility of encountering extruded ground and high earth temperature (Lee & Wang, 2015).

When the TBM excavated up to 1,624 m, it was jammed. Figure 8 shows the collapse and accumulation of the rock debris surround the advancing face when jammed. The main caving section concentrated on the upper left side from the advancing direction of TBM. It was estimated that the cause of the jam was mainly due to the accumulation of clay-containing caving material in the gap between the front and rear of the TBM body and the rock mass. The water ingress was around 10 l/min near the right shoulder in front of the excavation face. The rock mass was fractured above the excavation section, and the loose rock mass produced an earth pressure after excavation, resulting in excessive friction at the front and rear of the TBM body. At the initial stage of the jam, relevant countermeasures were taken, mainly: (1) lubricant injection to reduce the friction between the rock mass and the TBM; (2) moving operation of the forepoling and each jack, by a combination of different thrusts to overcome the pressure of the rock mass extrusion; (3) high pressure water cleaning between the shield body and the rock mass to remove the fractured rock between the shield body and the rock mass; (4) additional jacks to increase the thrust of the TBM. However, the above-mentioned response measures still could not overcome the extrusion of the rock mass that the shield shell was subjected to.

In order to formulate a proper TBM extrication plan, the geological characteristics of the jammed section must be understood. Therefore, in this case, the condition of the rock stratum, range and extent of looseness of the jammed section were explored through drilling investigation, elastic wave velocity detection, and borehole deformation test were performed in the borehole. According to the investigation, the looseness range of the overall rock stratum was evaluated. In this case, the top heading tunnel was used to extricate the TBM. The main process is as following.

(1) Pre-preparation operations: including the construction of the working platform, the laying of steel plates, etc., and the subsequent operations of removing the ring and expanding excavation of the upper part. At this stage, the excavation drilling machine and the automatic measuring system above the TBM body were first removed, and the steel plate was laid at the removal position to protect the hydraulic system.

(2) Rock mass stabilization countermeasures in the ring removal section: Necessary measures to stabilize the rock mass above the ring. Figure 9 shows the range of the forepoling method of this case. A total of 9 self-drilling rock bolts (L=2.0 m) with a diameter of 60 mm were arranged in the 40° range from the middle of the ring, and a grouting operation from the top of the rock bolt was carried out with silica resin.

(3) Ring Washing and Demolition: The ring was split to reduce the weight of the ring and possible hazard it may cause.

(4) Excavation and erection of support in the gradational section: The gradational section refers to the oblique section of the original TBM travel route and the top heading tunnel route. Figure 10 is a schematic diagram of this part of the excavation. Since the gradational section was still inaccessible to large-scale construction machinery, therefore, the hand-held breaker was used for manual operation. During the construction process, the

surface was sprayed with concrete according to the geological conditions to prevent shredding and collapsing of the stone. Upon completion, a thickness of 10 cm of shotcrete was sprayed to ensure the stability of the rock mass. If necessary, the thickness of the shotcrete was further increased and the fore-poling methodology applied.

(5) Standard section excavation and erection support: Figure 11 shows the process of the top heading tunnel (standard section). The hand-breaker and the small excavator were simultaneously used to carry out excavation, and the gravel was transported by trolley or simple conveyor belt. During the excavation process, erection support, shotcrete and fore-poling method was used depending on the geological conditions. After the excavation was completed, the mirror surface grouting operation of the excavation surface commenced.

(6) Blocking and backfilling operation of the excavation section: through the application of the above-mentioned top heading tunnel, the surrounding rock crushing section above the top arch was smoothly removed to minimize extrusion pressure, and the excavation area was blocked. In this case round gravel was used for backfilling, and LW grouting material (cement + water glass) was injected to fill the gap, as shown in Figure 12; on the other hand, the removed ring above the TBM was restored.

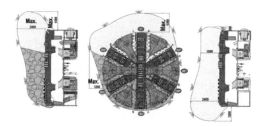

Figure 8. Rock avalanche and accumulation condition of TBM jammed.

Figure 9. Steady countermeasures of rock mass in segment removal section.

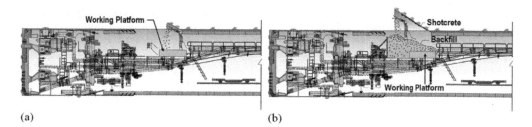

(a)　　　　　　　　　　　　　　　　　　(b)

Figure 10. Schematic diagram of excavation at top heading transition section.

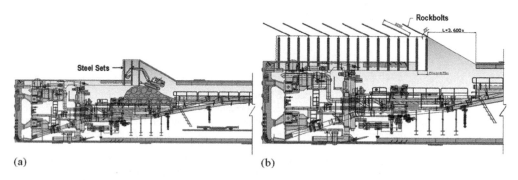

(a)　　　　　　　　　　　　　　　　　　(b)

Figure 11. Schematic diagram of excavation at top heading standard section.

3918

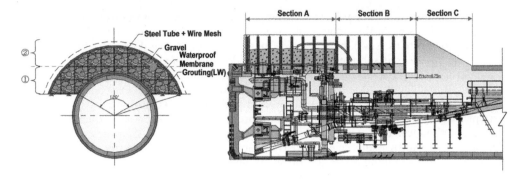

Figure 12.   Schematic diagram of block up and backfill work of top heading method.

In this case when the TBM was jammed, the top heading tunnel was used to smoothly over-come the problem. The duration of the interruption is 90 days. This successful case may be used as a reference when the TBM encounters fracture rock ahead, the lithology with high muddy content and extruded rock mass potential during excavation of rock tunnel.

5   CONCLUSION AND SUGGESTION

The TBM method can improve the efficiency of rock tunnel excavation, reduce the disturbance to the rock mass, and greatly improve the construction quality and safety of the rock tunnel. However, during the initial period of tunnel excavation, when assessing whether TBM method should be adopted or not, it is imperative to draft strategy for possible geological bottlenecks and for extrication from jammed during the planning and design stage of rock tunnel.

This study collects TBM jammed cases in rock tunnels to understand that TBM jam occurs under different combinations of engineering geological conditions, and there are also different jam conditions and damage types. Wherever the geological variability is difficult to grasp, pre-ventive measures (forepoling method, ground improvement method) are required in the early period while extrication programs are required for the later period (such as bypass tunnel method, top heading tunnel method). This has been extensively applied to construction of rock tunnels.

Through two actual cases study of the TBM excavation in rock tunnel, it is confirmed that the bypass tunnel method and the top heading tunnel method can be effectively applied to TBM jammed in construction of rock tunnels. However, in the future, when the evaluation is selected, it is necessary to consider the factors and mechanisms of the TBM jammed, the char-acteristics of surrounding rock, and the presence or absence of water ingress.

In addition, with the concept of pre-grouting being gradually accepted by tunneling projects in various countries. By applying the geological exploration results in the early stage geo-logical survey and construction, using the auxiliary methods in advance as suggested in this paper can drastically reduce the risk of jam during TBM construction.

REFERENCES

Hasanpour, R., Schmitt, J., Ozcelik, Y. & Rostami, J. 2017. Examining the effect of adverse geological conditions on jamming of a single shielded TBM in Uluabat tunnel using numerical modeling. *Journal of Rock Mechanics and Geotechnical Engineering* 9: 1112–1122.

Lee, C.H. & Wang, T.T. 2015. Application of drilling survey system for D&B and TBM tunneling in Taiwan. *ISRM Congress 2015 Proceedings - Int'l Symposium on Rock Mechanics, Quebec, Canada.* 206.

Lin, C.C. & Yu, C.W. 2005. Discussion and solution of the TBM trapped in the westbound Hsuehshan tunnel. *2005 International Symposium on Design, Construction and Operation of Long Tunnels, Taipei, Taiwan.* 383–393.

Ramoni, M. 2010. On the feasibility of TBM drives in squeezing ground and the risk of shield jamming. *PhD thesis of ETH Zurich.*

Seingre. G., Schivre. M., Classen, J., Grothen, B., Lecomte, B., Pigorini, A. & Schuerch, R. 2017. TBM excavation of long and deep tunnels under difficult rock conditions. *ITA Working Group n°17 Long Tunnels at Great Depth.*

Shang, Y., Yang, Z., Zeng, Q., Sun, Y., Shi, Y. & Yuan, G. 2007. Retrospective analysis of TBM accidents form its poor flexibility to complicated geological conditions. *Chinese Journal of Rock Mechanics and Engineering.* 26(12): 2404–2411.

Shen, T.Y., Lee, P.S. & Yu, C.W. 2005. Failure modes of TBM and remedial measures used in the Hsuehshan tunnel. *2005 International Symposium on Design, Construction and Operation of Long Tunnels, Taipei, Taiwan.* 373–381.

Spencer, M., Stolfa, A., Bentz, E., Cross, S., Blueckert, C., Forder, J., Wannick, H., Guggisberg, B. & Gallagher, R. 2009. Tunnel Boring Machines, *IMIA Conference Istanbul.*

Terada, M., Matsubara, T., Moriyama, M. & Nakata, M. 2008. Excavation of the long evacuation tunnel in the underground with high pressure ground water by TBM of 4.5 m in diameter. *World Tunnel Congress 2008 - Underground Facilities for Better Environment and Safety - India.* 1107–1118.

Tunnels and Underground Cities: Engineering and Innovation meet Archaeology,
Architecture and Art, Volume 7: Long and deep tunnels – Peila, Viggiani & Celestino (Eds)
© 2020 Taylor & Francis Group, London, ISBN 978-0-367-46872-9

Site observations and physical model tests of the friction coefficient between concrete-type element and foundation layer for immersed tunnel

M. Lin, W. Lin & X. Wang

CCCC HZMB Island & Tunnel project, 1699 Tangqilu, Xiangzhou district, Zhuhai 519015, China

ABSTRACT: There is a growing number of immersed tunnel that used concrete-type struc-ture and gravel bed. However, the friction coefficient between tunnel element and gravel bed, which is an important parameter for the appropriate design and preparation of construction, was missing. In this paper, the author reported extensively on the data of friction coefficient between concrete slab and gravel layer. The data were observed from site work of installing the thirty-three immersed tunnel elements on Hong Kong-Zhuhai-Macao Bridge Project, i.e. pulling connection and re-alignment of tunnel element, and from small-scale physical model tests. In the tests, the influential factors of the friction coefficient were studied such as contact pressure, geometrical character of foundation layer, sediments, and nominal diameter of gravel. Further, an observation of a special case during the work of closure joint were dis-cussed to reflect the overall friction coefficient of tunnel elements after backfilling.

1   INTRODUCTION

There is a growing number of the immersed tunnel that used a concrete-type structure and gravel` bed. However, the friction coefficient between the tunnel element and gravel bed, which is an important parameter for the appropriate design and preparation of construction, was missing.

On the Hong Kong-Zhuhai-Macao Bridge project, friction was tested before the design of the pulling system and re-alignment system (Shang & Wang 2015). Further, the pulling force was recorded, based on which the friction coefficient can be calculated. Moreover, after the installation of the closure joint (Lin et al. 2018a), unexpectedly a slightly opening of immer-sion joint was observed. In this paper, we report our observations on the friction coefficient.

2   TEST

From 13 June to 23 July 2012, 12 batches of 127 times of tests were carried out.

2.1   *Test approach*

2.1.1   *Model*
Test set-up is shown in Figure 1. The small-scale model was used. The element was repre-sented by a concrete slab with an outline of 8.55×1.803×0.2 m, on top of which a steel ballast tank with an outline of 8.51×1.5×0.4 m was set to simulate various vertical contact pressure (Figure 1a). In order to test the friction underwater, a tank was prepared (Figure 1b).

2.1.2   *Way of movement*
Generally speaking, the way of movement of the model can be divided by plane translation (Figure 2a-b) and rotation (Figure 2c-h)

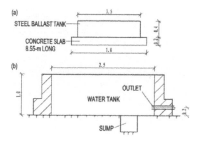

Figure 1.   Test model set-up.

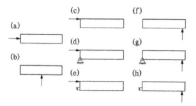

Figure 2.   Test scenarios (a) longitudinal translation; (b) transverse translation; (c) end rotation, free; (d) end rotation, hinged; (e) end rotation, monotonic constraint in longitudinal displacement; (f) side rotation, free; (g) side rotation, hinged; (h) side rotation, monotonic constraint in longitudinal displacement.

2.1.3 *Types of foundation bed*

The foundation bed were elaborated in Figure 3 and below.

Six types were used: ①Gravel size 2~6cm, fully covered; ②Gravel size 2~6cm, with grooves and berms, to which model was pushed in parallel and perpendicular; ③Gravel size 1~3cm, with grooves and berms; ④Gravel size 1~3cm, fully covered; ⑤Gravel size 2~6cm, with mud (representing sediments); ⑥Sand layer.

The thickness of the foundation bed, for the test, was 35cm. Its underneath was hard base made of concrete. The underneath condition is similar to the actual condition of the immersed tunnel on the HZMB project. Because the solution of the composition foundation bed was used (Zhu et al. in press, Lin et al. 2018b, Lin & Lin 2017, Lin et al. 2017). Figure 4 shows the two types of foundation bed as previously mentioned. The size of the groove and berms of the tested gravel bed are the same as that of the gravel bed adopted in the project.

Figure 3.   Preparation of foundation layer of the test (a) fully placed; (b) with grooves; (c) with sediments; (d) sand layer with grooves; (e) lifting element model into the tank; (f) lifting lug that can measure the weight of element model.

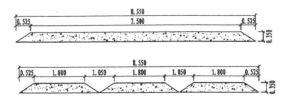

Figure 4.   Section of two types of foundation layer of the test.

Table 1.   Friction test results summary, translational movement.

Case	Diameter (cm)	Shape 1st time	Shape 2nd time	Material	Flatness (cm)	Pushing direction	Vertical load (kN)	Horizontal load (kN) 1st time	Horizontal load (kN) 2nd time	Friction coefficient 1st time	Friction coefficient 2nd time
T1	2~6	Full cover (FC)	Full cover	gravel	±1	Longitudinal (L)	12.56	4.94	4.61	0. 39	0. 37
T2	2~6	FC	FC *	gravel	±4	L	18.77	7.02	7.84	0. 37	0.42
T3	2~6	FC	FC *	gravel	±4	L	25.07	9.95	10.71	0.4	0.43
T4	2~6	FC	FC *	gravel	±4	L	31.27	12.97	14.17	0.41	0.45
T5	2~6	Berm and Groove (BG)	-	gravel	±1	L	12.86	5.05	-	0.39	-
T6	2~6	BG *	B&G	gravel	±4	L	13.54	5.81	5.29	0.43	0. 39
T7	2~6	BG	B&G	gravel	±1	L	14.54	5.40	5.81	0.4	0.43
T8	2~6	BG	B&G	gravel	±1	L	18.01	7.15	7.92	0.4	0.43
T9	2~6	BG	B&G*	gravel	±1	L	22.43	9.42	9.74	0.42	0.43
T10	2~6	BG	B&G*	gravel	±1	L	27.08	11.44	11.73	0.42	0.43
T11	2~6	BG	B&G*	gravel	±4	L	37.06	15.94	17.71	0.43	0.46
T12	2~6	BG	B&G*	gravel	±4	Transverse (T)	13.51	5.27	5.80	0.39	0.43
T13	2~6	BG	B&G*	gravel	±4	T	18.05	7.10	6.97	0.39	0. 39
T14	1~3	BG	B&G*	gravel	±1	L	13.53	5.55	6.41	0.41	0.47
T15	1~3	BG	B&G*	gravel	±4	L	18.04	7.38	8.13	0.41	0.45
T16	1~3	BG	B&G*	gravel	±4	L	22.56	9.64	10.56	0.43	0.47
T17	1~3	BG	B&G*	gravel	±4	L	27.06	10.98	12.59	0.41	0.47
T18	1~3	FC	FC*	gravel	±4	L	18.80	7.57	8.42	0.4	0.45
T19	1~3	FC	FC*	gravel	±4	L	25.05	10.30	11.45	0.41	0.46
T20	1~3	FC	FC*	gravel	±4	L	31.32	12.77	15.10	0.41	0.48
T21	1~3	FC	FC*	gravel	±4	L	37.58	15.76	17.73	0.42	0.47
T22	2~6	BG	BG*	Mud	-	L	13.53	3.85	5.52	0.28	0.41
T23	2~6	BG*	BG*	Mud	-	L	18.04	5.52	6.18	0.31	0. 34
T24	2~6	BG*	BG*	Mud	-	L	22.55	5.56	7.57	0.25	0. 34
T25	2~6	BG* BG*(3rd time)	BG*	Mud	-	L 18.04 L	3.67 5.57 (3rd time)	4.60	0.2 0.31 (3rd time)	0.25
T26	-	BG	BG*	Sand	±1	L	13.53	5.91	6.17	0.44	0.46
T27	-	BG	BG*	Sand	±1	L	18.04	8.37	8.41	0.46	0.47
T28	-	BG	BG*	Sand	±1	L	22.55	10.60	10.65	0.47	0.47

Note: ①The contract face of the model is 8.550×1.803 m, based on which the bottom pressure can be calculated; ②The * means that the foundation layer was not replaced from the previous test; the foundation layer without the * means it was replaced before the test.

2.1.4 *Bottom pressure*

Considering the possible negative buoyancy of the tunnel element in installation, we simulated various degree of bottom pressure, with a range of 0.9~2.4 kPa. The pressure was defined as the vertical weight divided by the gross area where the tunnel element bottom meets the gravel bed. The change of the area due to the deformation of a gravel bed being compressed was neglected.

2.1.5 *Test process*

In the test, the model was lifted and placed on the pre-placed gravel bed in the water tank. Then, the jack and pressure meter was fixed in the bracket. The pressure meter was set against one end of the model of tunnel element. On the other end of the model, the displacement sensor was set. The load was incrementally added while the pressure and displacement were recorded.

2.2 *Test results and discussions*

2.2.1 *Translational movement*

The results were summarized in Figure 1. It can be seen that the increase of bottom pressure generally increased the friction coefficient (Figure 5). For other factors as described in Section 2.1.3, the change of them was not sensitive to the change of friction coefficient. Furthermore, it was found that the friction still slowly increases after it reached the turning point in Figure 6. One possible explanation is that the gravels under the model were rolling, in the beginning, the movement of the model causes both the rolling and friction of gravels, with the increased pushing distance, the component of rolling decreases while the component of friction increases. Also, the sediments are likely to reduce the friction while the sand layer has a relatively higher friction coefficient.

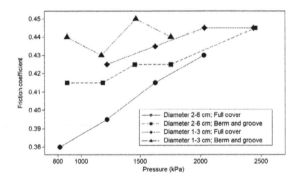

Figure 5.   Friction coefficient versus bottom pressure.

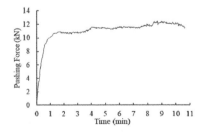

Figure 6.   Typical test force-time curve (Corresponds to Table 1-T3).

Table 2. Friction test results summary, rotational movement.

Cases	Gravel diameter (cm)	Foundation bed	Flatness (cm)	Vertical load (kN)	Constraint	Load position offset	Load (kN) 1	2	3	4	5
R1	2~6	Berms and grooves (BG)*	±4	13.51	None	Secondary end (S) 3.6m	2.46				
R2	2~6	BG*	±4	18.08	None	S 3.6m	3.96				
R3	2~6	BG*	±4	22.55	None	S 3.6m	5.01	4.96			
R4	2~6	BG	±4	13.51	Longitudinal (L)	S 3.6m	2.68				
R5	2~6	BG	±4	18.08	L	S 3.6m	4.05				
R6	2~6	BG*	±4	22.55	L	S 3.6m	4.45	5.90	6.17		
R7	2~6	BG**	±4	13.59	Rotation (R)	S 3.7m	2.54	3.01			
R8	2~6	BG*	±4	13.56	R	S 4.2m	2.57	2.79	3.05	3.19	3.28
R9	2~6	BG**	±4	18.05	R	S 3.8m	3.70	3.98			
R10	2~6	BG*	±4	18.04	R	S 4.2m	3.78	3.83	4.05	4.71	
R11	2~6	BG**	±4	22.55	R	S 3.7m	4.56	5.27			
R12	2~6	BG*	±4	22.56	R	S 4.2m	4.65	5.50	4.94	5.46	
R13	2~6	BG**	±4	13.54	R	Primary end (P) 0.9m	12.62	14.00			
R14	2~6	BG*	±4	13.53	R	P 0.9m	14.37	12.59	14.79	14.71	13.82
R15	2~6	BG**	±4	18.03	R	P 0.9m	17.48	16.97			
R16	2~6	B&G*	±4	18.04	R	P 0.9m	19.57	20.54	19.73	19.91	
R17	2~6	B&G**	±4	22.55	R	P 0.9m	24.68	25.70			
R18	2~6	BG*	±4	22.56	R	P 0.9m	24.67	24.64	24.77	26.32	
R19	2~6	BG*	±4	13.53	None	P 0.8m	5.66	5.73	6.03	6.09	
R20	2~6	BG*	±4	18.04	None	P 0.8m	8.61	9.05	8.98	8.73	
R21	2~6	BG*	±4	22.55	None	P 0.8m	10.51	10.50	10.85	10.29	
R22	2~6	BG*	±4	13.53	L	P 0.8m	4.62	5.34	5.42	5.06	
R23	2~6	BG*	±4	18.04	L	P 0.8m	7.38	7.81	7.84	8.12	
R24	2~6	BG*	±4	22.55	L	P 0.8m	9.70	10.05	9.82	10.00	

Note: The contract face of the model is 8.550×1.803m, based on which the bottom pressure can be calculated; the * means that the foundation layer was not repaved from the previous test; the foundation layer without the * means it was repaved before the test.

2.2.2 End rotation

The results of the rotational movement of the model were summarised in Table 2. It was observed that the movement of the model was a combination of translational movement and rotational movement. The data are to some extent scattering. The reason might be that the force required by the realignment depends on the moment which influenced by the frictional force. The frictional force was governed by the moving direction of the model. Consequently, the contributions of translation and rotation changed with the movement.

3 OBSERVATION OF FRICTION DURING PULLING OPERATION OF ELEMENT

3.1 Description of the pulling operation

During immersion, by ballasting water, the tunnel element was lowered down and submerged, with a negative buoyancy of around 8000~10000 kN. Before it touched the ground, the negative buoyancy was balanced by the lifting force of the vertical lines. After the element touched down the ground, where the contact area at element bottom is around 4000~6700 m^2, part of the lifting force remained. A pair of pulling jacks on new element stretched out to reach the bases on top of the existing element. Subsequently, the pulling process started. The process

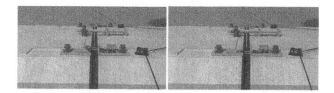

Figure 7. Pulling jacks operation illustrative drawing.

can be divided into the following stages: the pulling jacks extended out and connected with the bases keeping a pretension force; the new tunnel element was pulled towards the existing element until its Gina gasket reach the end frame of the existing element. In this process, the absolute value of the force of the pulling jacks was basically equal to that of the frictional force (i.e. a coupled force) and was the concern of our study. After this operation, the involvement of Gina gasket reaction caused a further increase in the pulling force of jacks, which are, not interesting to our study.

3.2 Typical pulling process description

As shown in Figure 8, in stage I, the pulling jacks connected with their bases; in stage II, the element starts to move by the pulling force, until the nose of Gina gasket touched on the opposite side; in stage III, to perform the underwater check, the pulling force was unchanged with the unchanged stroke value; in stage IV, start pulling again until a complete compression of the nose of Gina gasket; in stage V another underwater check; in stage VI a hydraulic connection was made, followed by the subsequent works in stage VII.

3.3 Summary of observed data of jack-pulling thirty-three tunnel elements

Table 3 summarised the concerned data regarding friction of jack-pulling the thirty-three tunnel elements. Some data were missing. For the rest data, calculations were made to estimate the frictional force and friction coefficient. The calculation of frictional force accounts for the force component of element submerged weight along the inclined longitudinal slope. The calculation of friction coefficient accounts for the lifting force that should be subtracted from the negative buoyancy. The results show that the frictional coefficients are in the range of 0.4~0.8, most of them around 0.5.

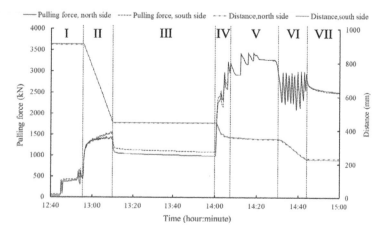

Figure 8. Tunnel element E25 pulling/connection process: jack force and end clearance versus time, 31st March 2016.

Table 3. Calculation of frictional force and friction coefficient based on records HZMB tunnel thirty-three elements pulling operations.

Tunnel element	Negative buoyancy (kN)	Longitudinal gradient (‰)	Vertical force① (kN)	longitudinal force② (kN)	lifting lines force-1③ (kN)	lifting lines force-2④ (kN)	Normal force⑤= ①-③-④ (kN)	Contact area (m²)	Pressure on bottom* (kPa)	Average⑥ (kN)	maximum⑦ (kN)	Averaged ⑧=②+⑥ (kN)	Maximum⑨ =②+⑦ (kN)	Averaged= ⑧/⑤ (kN)	Maximum= ⑨/⑤ (kN)
							Foundation layer pressure			**Pulling jacks force**		**Calculated friction**		**Friction coefficient**	
E1	7848	2.996%	7844.5	-235.0	·	·	·	4070.1	·	·	·	·	·	·	·
E2	7848	2.465%	7845.6	-193.4	·	·	·	4070.1	·	·	·	·	·	·	·
E3	9810	1.614%	9808.7	-158.3	·	·	·	4070.1	·	·	·	·	·	·	·
E4	7848	1.447%	7847.2	-113.5	·	·	·	4070.1	·	·	·	·	·	·	·
E5	7848	1.500%	7847.1	-117.7	·	·	·	4070.1	·	·	·	·	·	·	·
E6	7848	1.517%	7847.1	-119.0	·	·	·	4070.1	·	·	·	·	·	·	·
E7	7848	1.708%	7846.9	-134.0	·	·	·	4070.1	·	·	·	·	·	·	·
E8	8338.5	1.848%	8337.1	-154.1	·	·	·	4070.1	·	·	·	·	·	·	·
E9	8338.5	1.908%	8337.0	-159.1	·	·	·	4070.1	·	Error/Abnormal		·	·	·	·
E10	8436.6	1.443%	8435.7	-121.7	·	·	·	4070.1	·	1764.1	2079.7	1642.4	1958.0	·	·
E11	9810	0.773%	9809.7	-75.8	·	·	·	4070.1	·	1596.5	2403.5	1520.7	2327.6	·	·
E12	9810	0.287%	9810.0	-28.2	·	·	·	4070.1	·	Error/Abnormal		·	·	·	·
E13	9810	-0.098%	9810.0	9.6	·	·	·	4070.1	·	2477.6	3237.3	2487.2	3246.9	·	·
E14	9810	-0.325%	9809.9	31.9	·	·	·	4070.1	·	Error/Abnormal		·	·	·	·
E15	9810	-0.300%	9810.0	29.4	·	·	·	4070.1	·	2885.9	3874.8	2915.4	3904.2	·	·
E16	9810	-0.300%	9810.0	29.4	·	·	·	4070.1	·	1555.6	1893.3	1585.1	1922.8	·	·
E17	9810	-0.277%	9810.0	27.2	·	·	·	4070.1	·	2647.1	2770.9	2674.3	2798.1	·	·
E18	9810	0.000%	9810.0	0.0	·	·	·	4070.1	·	2715.5	2844.9	2715.5	2844.9	·	·
E19	9810	0.167%	9810.0	-16.4	4198.7	1579.4	4031.9	6665.9	0.60	2930.3	3221.5	2913.9	3205.1	0.72	0.79
E20	9810	0.300%	9810.0	-29.4	·	·	·	6665.9	·	·	·	·	·	·	·
E21	9810	0.300%	9810.0	-29.4	4159.4	1540.2	4110.3	6665.9	0.62	1985.6	2226.9	1956.2	2197.4	0.48	0.53
E22	9810	0.319%	9810.0	-31.3	5817.3	1402.8	2589.8	6665.9	0.39	1821.1	2095.1	1791.7	2065.6	0.69	0.80
E23	9810	0.129%	9810.0	-12.7	·	·	·	6665.9	·	·	·	·	·	·	·
E24	9810	-0.023%	9810.0	2.3	2776.2	1334.2	5699.6	6665.9	0.86	2964.1	3178.4	2966.4	3180.7	0.52	0.56
E25	9810	-0.441%	9809.9	43.3	3806.3	1599.0	4404.6	6665.9	0.66	2770.0	3001.9	2813.2	3045.1	0.64	0.69
E26	9810	-1.508%	9808.9	147.9	·	·	·	6665.9	·	3179.9	3416.6	3327.9	3564.5	·	·
E27	9810	-2.551%	9806.8	250.2	·	·	·	5800.7	·	·	·	·	·	·	·
E28	9810	-2.695%	9806.4	264.3	·	·	·	5800.7	·	2587.4	2746.8	2851.7	3011.1	·	·
E29	9810	-2.360%	9807.3	231.5	10182.8 Error	1138.0	·	6233.3	·	2054.2	2275.9	2285.7	2507.4	·	·
E30	9810	-2.207%	9807.6	-216.5	·	·	·	6233.3	·	1823.7	2001.2	1607.2	1784.8	·	·
E31	9810	-2.147%	9807.7	-210.6	5768.3	529.7	3509.7	6665.9	0.53	1599.8	2177.8	1389.2	1967.2	0.40	0.56
E32	9810	-2.671%	9806.5	-261.9	5120.8	686.7	3999.0	4935.4	0.81	2095.6	2376.7	1833.6	2114.7	0.46	0.53
E33	9810	-3.013%	9805.6	-295.4	2766.4	765.2	6274.0	4935.4	1.27	2861.9	3356.2	2566.5	3060.7	0.41	0.49

4 REALIGNMENT DATA AND ANALYSIS

4.1 The description of realignment

Two times of realignment had been made in HZMB tunnel, for element E4 and E5 respectively. Afterwards a free of realignment technique was developed, which will be discussed in a new paper. For element E4, E5, after installation, the go-through survey was carried out from inside to know the precise position of the new tunnel element, and jacks were set on the connection chamber between new element and existing element, see Figure 9. The jacks provided a pushing force F_1 (similar to that of Figure 2c). In addition to it, on the secondary end of the tunnel element, a line connecting to the immersion rig was used to pull the tunnel element provided a pulling force F_2, for the same purpose of realignment. In element E4 alignment work, another three jacks were set on the other side of the cross-section as a constraint to the opening of immersion joint (similar to that of Figure 2d).

4.2 Summary of the concerned data in the realignment operations

See Table 4.

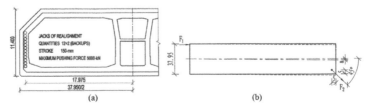

Figure 9. Realignment works in HZMB tunnel (a) cross-section; (b) plane layout.

Table 4. Operation record of the realignment of the new tunnel element, maximum jack-pushing force F_1 and line-pulling force F_2, corresponds to Figure 9b.

Tunnel element	Times	Vertical load (kN)	F_1 (kN)	F_2 (kN)	Displacement of tunnel element (mm)	Note
E4	1	47392	37425	-	0.8	Nearly moved
	2	27772	28253	-	26.8	
	3	14038	-	491	25.3	
	4	14715	19542	-	10.0	
E5	5	12753	24319	392	27.5	

5 OBSERVATION ON THE MOVEMENT OF ADJACENT TUNNEL ELEMENTS OF CLOSURE JOINT

5.1 Description

The closure joint was located between tunnel element E29 and E30. A new construction method of the deployable element was used (Lin, 2018a). After its installation, the water pressure on one end of the tunnel element E29, E30 vanished, causing a change of lateral force of around 96600 kN, see Figure 10. The vanished hydrostatic force must be replaced by the frictional force around tunnel element E29 and E30. The frictional force was from tunnel elements and their surroundings including the gravel bed and backfills. To ensure adequate static frictional force, the tunnel elements were loaded by backfilling on their roof, ballast concrete, and ballast water inside them. The total vertical loads were 360892 kN and 373180 kN on tunnel element E29 and E30 respectively. Supposing the averaged frictional coefficient is 0.4,

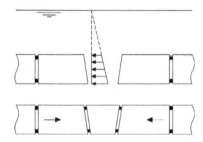

Figure 10. Closure joint installation before and after, longitudinal illustration.

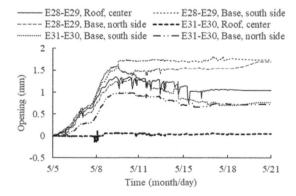

Figure 11. E30–E31 and E29–E29 immersion joint opening-time data.

the element will not move as the static friction can compensate for the unbalanced lateral force. However, we had observed the movement.

5.2 *Observations*

The closure joint installation was done on 4th May 2017. In the following days, we observed that the joint E29–E28 has an opening of 2 mm on average, while the bottom side of the joint of E30–E31 has an opening of 1 mm (Figure 11). One possibility is, under the action of waves, the gross lateral force of the tunnel element has conquered the frictional coefficient and then the tunnel element slowly move towards the closure joint.

6 CONCLUSION AND DISCUSSIONS

For the immersed tunnel installed by the vertical jack method, the bottom frictional force is small. Comparatively, element sitting on the gravel bed will have to face much larger frictional force. For the latter, we found that the observed data and test date were in general agreement, by comparing the test results in Section 2 and the monitored data from site work in Section 3 and Section 4. The friction coefficient in test ranges from 0.39 to 0.47 while the averaged friction coefficient from site observation ranges from 0.40 to 0.72. The difference may come from the inclined longitudinal slope and the scaling difference of the proto-type and the model.

Regarding the meaning of the test to the realignment works, in the rotational test, the frictional force showed some uncertainty, as previously discussed. Considering that the construction tolerance of the gravel bed, the sediments, and the variation of overload could lead to more uncertainty, to eliminate this uncertainty, we controlled the friction amount by means of controlling the lifting lines force.

Regarding the meaning of the test to the connection works, in the connection operation, to reduce the influence of element alignment deviation caused by the unevenness of friction, we controlled both the stroke and the force of the pulling-jacks. The tunnel element axis nevertheless deviates from the targeted alignment. The direction of deviation is at randomness, and the amount of the deviation depends on the frictional force and the centre of gravity of the tunnel element. However, the deviation of the curved tunnel element can be predicted by assuming the tunnel element as a rigid body, creating an equation of motion, and then solving it.

REFERENCES

Lin Ming, Lin Wei, Wang Qiang & Wang Xiaodong. 2018a. The deployable element, a new closure joint construction method for immersed tunnel. *Tunn. & Undergr. Sp. Tech.* 80(2018): 290–300.
Lin Ming & Lin Wei 2017. Hong Kong-Zhuhai-Macao Island and Tunnel Project. 2017. *Engineering* 3 (2017): 783–784.
Lin M., Lin W., Liu X., Yin H., Lu Y., Liang H. & Gao J. 2018b. Over and under. *Tunnel and Tunnelling International Edition* 2018 April: 38–48
Lin Wei, Zhang Zhigang, Liu Xiaodong & Lin Ming. 2017. Design for the Reliability of the Deepwater Immersed Tunnel of Hongkong-Zhuhai-Macau Bridge Project. *Proc 27th Int Offshore and Polar Eng Conf, San Francisco, ISOPE*, 1082–1086.
Shang Q. & Wang D. 2015. Reinforced concrete immersed tunnel element and foundation bed friction resistance test and study. *China Harbour Engineering* 2015-7: 46–48. (In Chinese)
Zhu Y., Lin M., Meng F., Liu X. & Lin W. 2018. Hong Kong-Zhuhai-Macao Bridge. *Engineering*: in press (doi: 10.1016/j.eng.2018.11.002)

*Tunnels and Underground Cities: Engineering and Innovation meet Archaeology,
Architecture and Art, Volume 7: Long and deep tunnels – Peila, Viggiani & Celestino (Eds)
© 2020 Taylor & Francis Group, London, ISBN 978-0-367-46872-9*

The Montcenis base tunnel: How to turn project environmental constraints into opportunities

S. Lione, L. Brino, P. Grieco & L. Pinchiaroglio
TELT sas, Turin, Italy

A. Mordasini
LOMBARDI SA, Minusio (TI), Switzerland

ABSTRACT: Security and safety concerns, mainly due to grassroots opposition in Italy, have led to the need to implement drastic changes in the Final Design for the construction phase of the new Turin-Lyon rail line. Alternative solutions in the design of the Montcenis Base tun-nel have included a reversal of the mechanized excavation direction in favour of a downhill path; a reduction of the space available for the main construction site, and a particularly com-plex logistical setup. Faced with these challenges, the classic design approach has been over-turned in order to convert constrains into opportunities for the project, especially for the local regions and the environment. These opportunities include a reduction in the use of soil, an in-crease in the proportion of the excavation conducted with mechanized means, and a wholly underground handling of "green" (asbestos-containing) rocks.

1 INTRODUCTION

The new Turin-Lyon Railway Line (NLTL) is an integral part of the "Mediterranean Corridor" which is, in turn, part of the European project known as the TEN-T (Trans-European Transport Network).

The TEN-T is a new European policy designed to encourage the movement of people and goods via rail, the most environmentally friendly mode of transportation. The goal is to decrease the use of road transport, which contributes to pollution and to greenhouse gas emis-sions. Within this network, the new Turin-Lyon rail connection is located at the crossroads of two large European communication axes, between north and south and between east and west. It is the central ring of the Mediterranean Corridor, serving 18% of the EU's population, in regions that produce about 17% of European GDP. The work will contribute to a necessary shift in transport habits, particularly in the sensitive Alpine region.

Completion of this line is therefore part of a development strategy that goes beyond the national interests of Italy and France to establish roots in the concertation between Alpine countries aimed at encouraging economic and social development in these areas, making sure that undesired traffic jams or transfers cannot threaten the economic feasibility of some routes.

Within the NLTL, the cross-border section extending from Bussoleno/Susa in Italy to Saint Jean de Maurienne in France, for an overall length of about 65km (57.5km of which are the Montcenis Base tunnel) has been the subject of studies and approval procedures since 1992.

The "Comitato Interministeriale per la Programmazione Economica" (CIPE) with reso-lution n. 19 dated 20 February 2015 declared the Public Utility of the cross-border section and approved the Definitive Design for the Italian side.

The prescriptive framework of resolution 19/2015, with provision n. 235, "*Construction site optimisation study*", reads: "*During the executive construction design, an alternative place-ment of the construction sites must be studied according to the security needs of the public*

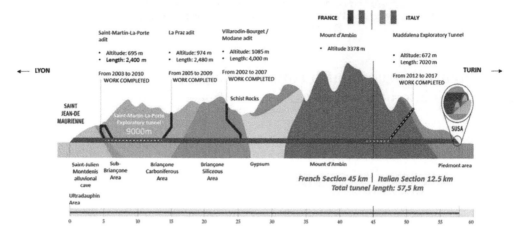

Figure 1. The Turin-Lyon base tunnel.

and with respect to the operational needs of the project... This study should also evaluate and quantify the resulting cost of classifying the aforementioned construction sites as strategic interest sites..."

Faced with this provision, TELT (Tunnel Euralpin Lyon Turin), the Public Promoter responsible for the creation and management of the international section of the NLTL, has entrusted a specialised consortium with the task of creating a study aimed at fully responding to the abovementioned provision. The result of this study has led to the need to draft a variation of the approved Definitive Design that involves the entirety of the project's construction site set-up.

This article, describing the Definitive Design as approved by CIPE in February 2015 and highlighting its strengths in relation to the application of the project's best practices, illustrates the approach followed to transform the constraints deriving from the new prescriptive framework into an opportunity for the local territory and the environment.

The usual and consolidated design approaches, adopted in the absence of constraints and public order requirements, have been completely revised. The result has led to the creation of an alternative Definitive Design, the so-called "Security Alternative" approved by CIPE with resolution n. 30 dated 21 March 2018, with innovative construction solutions that, with the same costs envisaged for the previous version, have further improved the environmental aspects without negatively impacting the efficiency of the industrial process used to carry out the works.

2 THE DEFINITIVE DESIGN

The original NLTL Definitive Design is the result of a decision-making process open to the territory and enriched by the contributions of interested parties and stakeholders that have actively collaborated in the works of the Observatory for the Turin-Lyon Railway.

That approach has been quite beneficial in guiding the planning of construction sites that, in addition to the criteria of consolidated organization and shared use for the creation of large works, has adopted the most advanced principles of innovation, sustainability and respect for the territory.

The environmental mitigation measures were intended to be works of sustainability, i.e. "planning actions" resulting from a set of inseparable best practices, respect for regulations, guidelines, and opportunities provided by the territory and its socio-economic context.

The general criteria for the choice and organization of the construction sites were based mainly on the following key points:

1. Conveyor belt
2. Building materials storage
3. Concrete aggregates storage
4. Plant for the valorisation of excavated materials
5. Concrete batching plant
6. Covered areas for temporary muck deposit
7. Precast segmental lining plant
8. Temporary deposit
9. Workshop + oil storage
10. Helipad
11. Offices
12. Visitor Building

Figure 2.   A rendering of the truck terminal.

- a reduction of the size of the construction areas, optimizing their functions;
- occupation of already-used areas and poor quality soils in order to reduce land consumption;
- elimination of temporary base camps for the workers in favour of lodging in hotels or similar structures already available in the area;
- transport of excavation materials using the railway as much as possible;
- completion of the main work processes in closed environments (to contain dust and noise), transforming work sites into true industrial establishments;
- use of the best technologies available (with the goal of optimum technical and energy efficiency), i.e., the adoption of recently applied or currently-being-applied best practices and best technologies at large alpine excavation work sites;
- earlier completion, i.e. already in the construction phase, of definitive measures aimed at environmental mitigation;
- optimization of the economic and occupational impact on the region, exploiting the opportunities provided by Piedmont regional law n. 4 of April 2011.

The main industrial area for the excavation of the Italian tract of the Montcenis Base Tunnel was located in Susa and it was to occupy a surface area of 12 hectares. It was designed to provide support to the outdoor construction sites and to the work areas, being the site of necessary supply plants (e.g. temporary muck deposits, a plant for the selection and valorisation of the excavation materials, a concrete batching plant, the precast concrete plan for segmental liners production, etc.). From this framework, the strategic nature of the Susa area in the original design of the industrial construction site, heart of the entire organization for the completion of the work on the Italian side, was clear.

The railway plant for the evacuation of the surplus excavation material was to be located in the same Susa area and it was designed to manage the railway equipment and systems after completion of the civil works.

Essentially, one might say that the industrial construction site of the Definitive Design was conceived as a linear construction site with a layout designed according to the production sequence.

3   THE WORK SITES SECURITY STUDY

In order to comply with the provisions of CIPE resolutions concerning the optimization of the construction site to ensure people's security, a study was drawn up by the National Inter-university Consortium for Transport and Logistics (NITEL); a Consortium of 19 prestigious Italian universities.

This study is based on a risk analysis of the different hypothetical ways the construction sites could be configured. The location of some work processes on different sites with respect to those established in the Definitive Design implied a different use of the individual areas, with their subsequent greater and/or lesser exposure to safety and security risks. Hence the need to carry out cumulative analyses of the various hypothetical solutions.

The level of risk exposure associated to each site was determined according to the relative level of sensitivity and impact. "Sensitivity" measures the degree to which a site may be subject to malicious action carried out by opponents, while "impact" is the measure of potential consequences that a malicious action carried out against a site may generate, assessed according to four points: impact on the population, on the workers, on work continuity and financial losses.

Therefore, the developers of the study, in close collaboration with the design team, have analysed the various technically possible options for construction site locations and, more specifically, have considered the different options for excavation sites: for the Base tunnel, muck evaluation and train loading, the ventilation central and the deposit sites.

Considering the technical restraints, it was possible to identify various configuration hypotheses, subsequently reduced to four, eliminating those with clear functional and/or security issues.

Once the risk associated with each of the four configurations had been analysed, in relation to the main Susa valley area, the security study then compared the exposure to the risk itself. The conclusion of the study led to the identification of the best solution, characterized by a rather different layout of the construction sites than in the previous plan.

Introducing a variable not traditionally considered in construction site design, the Security Study thus led to a change in terms of the organisational choices that are far from standard. The priorities and the design choices were dictated mainly by security requirements, which influenced the technical assessments. Except for the unchanged rail alignment, the construction site of the Base tunnel has been completely revised.

The designers thus had to grapple with new territorial, hydraulic and environmental constraints and they had to adapt the entire construction site system, which, like many work sites, is particularly complex, to the context identified due to security reasons.

4 CONSTRAINTS BECOME OPPORTUNITIES

The results of the construction site Security Study and the constraints imposed by the need to protect and ensure the security of the workers and the public have affected both the logistics of the construction phase and the changes to some final works.

The experience of the Maddalena exploration tunnel, a site already protected by law enforcement, has made it possible to identify the area as the best suited to host the main base tunnel boring construction site between the underground safety area of Clarea and Susa. Given the scarcity of available surfaces, at the confluence of the Clarea Stream and the Dora River, it was not possible to maintain the concept of an in-line unique construction site. Rather, it was necessary to identify other locations to contain all the necessary operations. For example, the evaluation and use of muck and its loading on trains for transport to deposit sites was located in Salbertrand, a new municipality that wasn't part of the Definitive Design, about a dozen km from the boring site. In order to carry out other logistical activities, other areas were identified adjacent to the existing construction site on the opposite side of the Clarea stream.

The identified layout also implies a revision of the tunnel advancement faces and the excavation methods for some types of lithology, leading to the better use of the TBM, as explained further in the following paragraph. The reversal of the boring direction from Chiomonte to Susa has greatly simplified the construction activities in the Susa area, with a reduction of environmental pressure factors, thanks to a simplification of the activities while work is under way. The shifting of the main Base tunnel boring activities from the Susa area to the area of the current construction site of Chiomonte has made it possible to predict the environmental impact, based on the environmental monitoring of excavation activities of the exploration tunnel in the last 5 years. In addition, in summer 2017, the Environment Minister, after having analysed the feedback from the exploration tunnel construction site, stated that *"the impact created on the environment by the construction site did not produce significant irreversible changes nor effects on any monitored environmental component"* and that *"we consider all the possible impact and effects on the environment of reference as tested, both in terms of the environment and management"*. Operating in a context proven in its transformations, makes a work site with lower environmental risks possible, where further techniques to control and reduce the impact on environmental components can be applied.

The relocation of the industrial construction site, originally located in the Susa area, has taken place in an environmentally degraded zone of the Salbertrand municipality, but close to other zones of great ecological importance. This location has led to the study and proposal of a "restoration ecology" design that aims to achieve more important objectives than simply restoring the sites to their original condition. The new design in fact allows for the ecological upgrading and recovery of the area involved, including a portion that has been heavily damaged from a natural and environmental point of view.

The new Maddalena underground interchange has created an important opportunity, making it possible to store underground green rocks that potentially contain asbestos. This green rocks storage method will better adhere to the principle of caution in the management of such materials, as it eliminates the risk of the fibres being dispersed into the environment. The material is managed entirely underground, in a tunnel section in confined environment conditions, without the need for open-air transport and handling.

One other positive direct consequence of the new construction complex was the elimination of the ventilation well and ventilation plant in the Clarea valley. The elimination of this final work thereby removes the effects generated by the plant's construction and operation, in a valley that is quite valuable environmentally, as well as in terms of its landscape.

Another important aspect was the chance to limit the underground 132 kV Susa-Venaus cable duct, with its environmental threats, to the land covering three municipalities. Thanks to the reversal of the excavation direction of the boring machines and the fact that their power supply will therefore come directly from Chiomonte instead of Susa, it will no longer be necessary to build the definitive connection as soon as possible in order to power the construction site. The overall length of the cable duct alignment after this modification will remain largely unchanged in respect to that of the approved Definitive Design (about 7.8 km). However, the most important thing to note is that only 1.4 km will still need to be buried near the surface,

Figure 3. The Maddalena reconnaissance tunnel construction site.

while the remaining tract will become part of the Base tunnel. This makes it possible to simplify the construction site and to significantly reduce the effects generated by drilling associated to the laying of cables, as well as the interference with local roads and public services, thereby also reducing completion times.

4.1 The Maddalena underground excavation site

The Montcenis Base Tunnel excavation site on the Italian side has been entirely revised. The excavation direction of the two tubes from Susa towards France, established in the Definitive Design as uphill, has been reversed completely, assuming a downhill approach. This choice, which is not standard in terms of normal excavation operations, has resulted in notable advantages in terms of the general layout.

The introduction of the Maddalena-2 tunnel has made it possible to bring the Base tunnel's advancement attack underground instead of out in the open, thereby greatly reducing potential environmental impact.

Below is a description of the plan for the new Maddalena excavation site, illustrated in Figure 4.

The Maddalena-1 tunnel coincides with the exploration tunnel, which has already been completed in order to gain a better understanding of the geology, geomechanics, and hydrogeology of the rock mass. From the portal, the tunnel leads to the Base tunnel, placed between its two tubes, in an overlying position, running parallel to them for about 4 km.

During the operational phase, this tunnel will be accessible to emergency vehicles for about the first 2.2 km, from the entrance to the starting point of connecting tunnel 1, along which they will proceed until the safety area/Base tunnel. The remaining portion of the Maddalena 1 tunnel will be completely closed in order to store the "green rocks", as described below.

Connecting tunnel 1, along with the first tract of the Maddalena-1 tunnel, will be the access way to the Base Tunnel for emergency vehicles. The considerable reduction of the route with respect to the approved Definitive Design will facilitate the timely access of vehicles during an emergency response.

The Maddalena 2 tunnel will be built to allow, as work in progress, the entry of the TBMs to be used to excavate towards Susa. Thanks to this tunnel, it will be possible to eliminate the works for the ventilation plant in the Clarea valley and its related shaft. The junction with Connection tunnel 2 will therefore have the function of ventilation in the safety area and of smoke extraction from the Base Tunnel in case of fire. Access will be limited exclusively to maintenance vehicles. The remaining portion of Maddalena-2 will be closed and it will also be used to store all the green rocks.

With the "Security Alternative", the new underground Maddalena construction site will thus become the main hub for underground boring management on all advancement faces of the tunnels excavated on the Italian side, helping to free the Susa Valley from the construction activities related to the execution of underground operations.

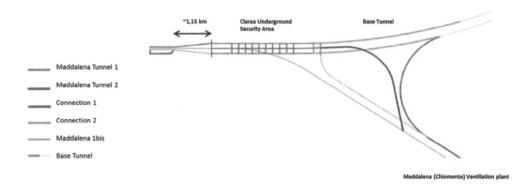

Figure 4. Tunnel layout in the Maddalena interchange.

The eastern portal of the Base Tunnel in the Susa area will become active only several years after the beginning of the works. It will be used only for the construction of the artificial tunnel and during the dismantling of the TBM, occupying a very small space and for only a short duration. Its function as a construction site for technological installations will, however, continue.

4.2 *The Salbertrand industrial area*

The new location for the temporary construction site, identified by the Security Study, is in an area suited to host the various features necessary for the operation of the construction site, including the transport of surplus excavation material by train, as it is adjacent to the existing railway line. It is located in the Municipality of Salbertrand, on the left bank of the Dora Riparia River, in an area that is currently subject to intense environmental pressure. Close to the Natura 2000 Site of the Gran Bosco di Salbertrand, the area is currently used for industrial activities and it is mainly occupied by soil piles and small contaminated areas.

The industrial construction site covers about 11 hectares.

The industrial area serving the underground construction sites has taken over the activities and functions originally planned for the Susa area: temporary muck deposits, a plant for the selection and valorisation of the excavation materials, a concrete batching plant, the precast concrete plant for segmental lining production and storage, a rail yard for transporting muck to final deposits, etc.

The zone where the new construction yard will be located is an area with hydraulic restrictions and constraints for construction works.

The limitations imposed by law to set up construction works in the aforementioned area and the constraints deriving from the unavailability of alternative locations (as declared by the Government Commissioner for the Turin-Lyon Railway) have forced the commissioning of a detailed hydraulic study finalised to ensure that site installations and the expected temporary activities do not affect the water tables (upstream and downstream) or other characteristics of particular importance for the natural ecosystem of the Dora Baltea River.

The permanence of the construction site for many years and the value of the landscape of the surrounding areas have led to the choice of a final restoration with nature as a priority, to allow for a marked improvement compared to current conditions.

The project includes the creation of wooded or shrub-covered areas, alternating with open spaces to improve environmental complexity and create diversified habitats for fauna and flora. A portion of the mitigation interventions will be carried out already during the construction phase, in order to make it possible to have meaningful clusters of native vegetation already established at the time of final restoration in the implementation phase. These cluster will be especially useful to accelerate the spontaneous recolonization of the industrial area used during the construction phase.

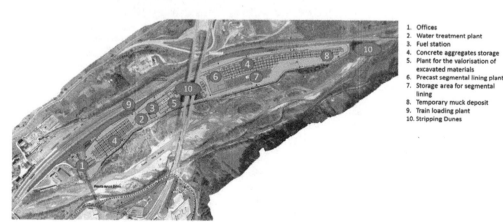

1. Offices
2. Water treatment plant
3. Fuel station
4. Concrete aggregates storage
5. Plant for the valorisation of excavated materials
6. Precast segmental lining plant
7. Storage area for segmental lining
8. Temporary muck deposit
9. Train loading plant
10. Stripping Dunes

Figure 5. The Salbertrand Industrial Construction Site.

The wooded and bushes communities will be heterogeneous, made up of native species with a local origin certificate. In particular, the restoration of wooded areas will take place through the use of species that form already part of the area's plant communities (phytocoenosis), alternating pioneering species with entities typical to more mature, stable developments.

In addition, restoration interventions will include the maintenance of broad grasslands, thereby facilitating their use by wildlife.

The environmental recovery design also proposes experimental interventions against invasive alien species such as *Buddleja davidii*, aimed at improving the ecological conditions of the pebble tracks along the Dora.

The field studies carried out for the project in this river area have produced a snapshot of an area of significant interest because of its ecosystems and of the presence of rare botanic species thanks to a very wide river bank freely shaped by the morphogenesis of the river.

Among the many plant communities surveyed in the study, there are single species of extreme interest for conservation, such as the *Carex alba* and the *Typha minima*. The project is a great opportunity for "ecological restoration" that, once the tunnel is completed, will allow the involved areas to move beyond their current degraded status, activating a process of natural restoration and ensuring a greater presence of wild fauna and reducing the current risk of damages to wildlife.

4.3 *The underground deposit of green rocks*

The area relative to the Mompantero zone, where the Base Tunnel portal for the Italian side is expected to be located, is characterised by the presence of ophiolitic rocks (basic and ultra-basic rocks) belonging to the tectono-metamorphic set of the Piedmont zone.

For a section of about 300–350 m up to the eastern portal of the Base Tunnel, the tunnel will be excavated through ophiolite, which are potentially asbestiform lithotypes part of the "green rocks" group.

Based on the available direct data and current knowledge, it is possible to estimate that the volume to be excavated in "green rocks" is about 80,000 m^3. Of this volume, however, only a limited portion could potentially contain asbestos and, in any case, only a small fraction will presumably be characterized by concentrations of asbestos over 1,000 mg/kg, which is the limit to distinguish a material as either:

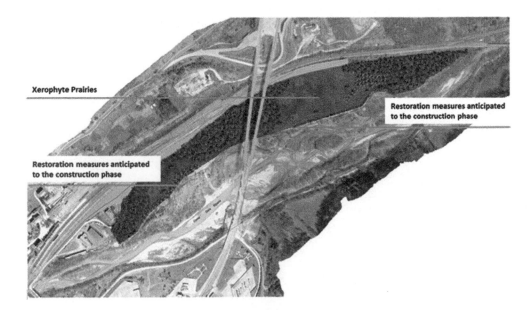

Figure 6.   Restoration activities and final development of the Salbertrand construction site.

- not dangerous, and potentially usable in construction works, according to art. 184 bis of Italian Legislative Decree 152/06 and subsequent amendments and additions, or
- dangerous waste, according to "CER code n. 17 05 03*", requiring permanent storage.

The passage from the calcareous schist formation to the ophiolite formation is not cut and dry and the available soil investigations in ophiolite formations show a scenario with a high variation of asbestos concentrations. An accurate estimate of the volume of material that might contain asbestos and, even more, a detailed evaluation of the volume of material classified as "dangerous waste" will become available only during the Construction Design phase. According to this approach, the Design has prescribed additional soil investigations and laboratory tests to be carried out during the excavation phase. The new data collection will increase the knowledge and will allow to determine the actual degree of presence of asbestiform rocks.

Currently, the "Security Alternative", inverting the Base Tunnel excavation direction from the Italian to the French side, as described above, will make it possible to excavate the section in "green rocks" entirely from within the mountain and not from outside. It also offers the opportunity to create an underground deposit in already-excavated temporary tunnels, as they will have no purpose once the railway begins operations.

Based on what has already been completed, an adapted version of the storage project for the ANAS underground deposit in Cesana Torinese (TO), this layout has been adopted for storing in an underground space all the 80,000 m^3 of "green rocks". The underground deposit design includes the burial of the "green rocks", the complete filling of the tunnel with cement mortar, and the sealing and waterproofing of the access closure with a reinforced concrete wall. This practice is borrowed from ancient techniques used in the industrial mining, such as the cultivation of the Fontane talc mine in Germanasca Valley, not far from Turin.

The project thereby provides for the material first to be packed into high-density polyethylene (HDPE) containers, and then to be transported and permanently stored in the underground deposit site.

The solution identified allows for the "green rocks" to be stored underground without any handling in open air. They will always remain within underground spaces, and no transport and disposal in other locations outside the site, with their related environmental impact, will be necessary.

The disposal methods and criteria, the provision of suitable equipment and of personnel and personal protective measures will be handled according to Italian Legislative Decree 36/2003 and to Italian Decree 27/09/2010.

It should be emphasized how the change in the excavation works makes it possible to tackle the excavation of the section in "green rocks" with a TBM instead of conventional excavation methods, leading to considerable protective benefits. The TBM can be considered a "travelling factory". It allows to mechanize and automate all the various operations, such as boring, clearing out, transport, and storage of the muck, reducing both the excavation time and the

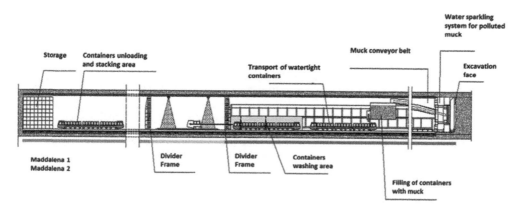

Figure 7. Diagram of the green rocks excavation and the underground deposit.

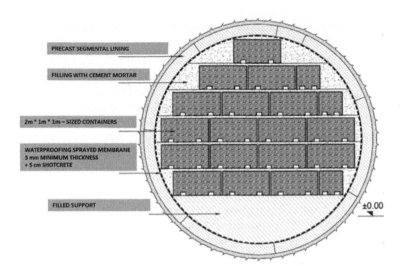

Figure 8. Storage of green rocks in the underground deposit (e.g., the Maddalena 2 tunnel).

number of workers necessary to manage the excavation phase, avoiding the creation of dust and the release of fibres in the environment.

The design choices have made it possible to optimize the reutilisation in totally safety conditions, both for the workers and for the environmental surroundings, of tunnels already excavated, as well as to achieve lower costs for the management of material that sees the presence of "green rocks".

5 CONCLUSIONS

The most important infrastructure projects, especially if in international scenarios, derive from the willingness to make plans with horizons of at least ten years. From the first feasibility studies to the completion of the project, through the authorization practices and necessary procedures, relatively long periods may pass, during which it is often possible to have important changes in social and political contexts, on local or international scale.

The true challenge is to create opportunities for the territory and the environment where these projects are implemented, subverting if necessary the usual design approaches and insisting, more and more, on sustainable designs, so that these projects can be seen and appreciated as a positive vehicle for change.

REFERENCES

Brino L. 2012. La Nuova Linea Torino Lione: gli aspetti tecnici di un grande progetto transfrontaliero, *Gallerie e Grandi Opere Sotterranee* 102: 11–15.
Brino L., Luchetti E., Chabert A. & Rettighieri M. 2013. Der Basistunnel Lyon-Turin: Technische Aspekte eines großen grenzübergreifenden Projekts, *Beton- und Stahlbetonbau Spezial – Europas Lange Tunnel*: 43–49.
Bufalini M, Dati G., Rocca M. & Scevaroli R. 2017. The Montcenis Base Tunnel, *Geomechanics and Tunnelling* 3: 246–25.
European Commission - Directorate General for Mobility and Transport 2014. *Mediterranean Core Network Corridor Study - Final Report 2014*. Brussels: European Union.
Virano M. 2016. Le scelte di buon uso del suolo e del sottosuolo negli attraversamenti alpini del nord ovest d'Italia lungo il Corridoio Mediterraneo. *Gallerie e grandi opere sotterranee* 117.
Virano M., 2017, Il tunnel di base del Moncenisio: le scelte di TELT all'interno di una strategia europea, *Strade & Autostrade*, 122: 18.

Tunnels and Underground Cities: Engineering and Innovation meet Archaeology,
Architecture and Art, Volume 7: Long and deep tunnels – Peila, Viggiani & Celestino (Eds)
© 2020 Taylor & Francis Group, London, ISBN 978-0-367-46872-9

Anisotropic convergence of tunnels in squeezing ground: The case of Saint-Martin-la-Porte survey gallery

Y. Liu & J. Sulem
Laboratoire Navier/CERMES, Ecole des Ponts ParisTech, IFSTTAR, CNRS, Université Paris-Est,
Marne-la-Vallée, France

D. Subrin
Centre d'études des tunnels (CETU), Bron, France

E. Humbert
Tunnel Euralpin Lyon Turin, Le Bourget du Lac cedex, France

ABSTRACT: As a part of the Lyon-Turin railway link project, the Saint-Martin-la-Porte (France) access gallery (SMP2) has been excavated through a highly squeezing Carboniferous formation. The anisotropic time-dependent behavior of the access gallery has been widely studied, particularly through convergence measurements during and after excavation. Recently, a new survey gallery (SMP4) started to be excavated along the direction and at the depth of the base tunnel. The excavation is accompanied by extensive monitoring including convergence measurements. Considering that the tunnel crosses the same Carboniferous formation, highly squeezing behavior was expected. In this paper, we test the predictive capabilities of the convergence law proposed by Vu et al. (2013) for SMP2 by analyzing the convergence data of SMP4 and by extrapolating the parameters calibrated on SMP2 to SMP4.

1 INTRODUCTION

In order to build up a more efficient connection of the European countries, the European Commission has launched the Trans-European Transport Network (Ten-T) project. The Lyon-Turin railway link is highly strategic as it is a key element in the Mediterranean corridor, which connects southwestern, central and eastern Europe. It contains a 57.5 km base tunnel as its main part, 45 km on the French side and 12 km on the Italian side. In the context of growing traffic between France and Italy, the construction of the Lyon-Turin railway link aims at reducing the road traffic and the carbon gas emission. Several excavation faces will be processing at the same time by using several intermediate accesses.

In Saint-Martin-la-Porte (France), some difficulties related to tunneling in squeezing ground have been encountered. Squeezing behavior is characterized by significant time-dependent and often anisotropic deformation during and after excavation because of the rock mass poor mechanical properties and the important field stress (Barla 2001). This typical ground behavior requires specific excavation and support methods as it may cause severe difficulties during tunneling works (Panet 1996).

In this paper, the geological context of the Saint-Martin-la-Porte survey project and the encountered difficulties are firstly presented. Then, the process to analyze the anisotropic deformation data proposed by Vu et al. (2013) is adopted in the case SMP4. Finally, the predictive capability of the convergence law is tested by extrapolating the model parameters obtained for SMP2 to predict the convergence evolution of SMP4.

2 SAINT-MARTIN-LA-PORTE SURVEY PROJECT

2.1 SMP2 Access Gallery

A survey project is in progress in Saint-Martin-la-Porte in order to study the geological context of one of the most complex section of the Lyon-Turin base tunnel. An access tunnel (SMP2) was firstly excavated in this site to reach the base tunnel level from Vallée de l'Arc. It offers intermediate front faces and provides ventilation during the excavation of the base tunnel. After achievement of the construction, it will serve as an emergency and maintenance access as well as for ventilation of the final structure. Tunneling works started in 2003 with an excavated profile of 77–125 m^2. They were completed in July 2010 with a final length of 2,329 m and a final internal profile of 54–63 m^2 up to a depth from 250 to 400 m. During excavation of this gallery, a Carboniferous formation was encountered on the western side of Mount Brequin. In this formation, highly tectonized productive Houiller was met, which exhibited a very stratified and highly fractured structure, consisting of schists and/or carboniferous schists (45–55 %), sandstone (40–50 %) and also a significant proportion of cataclastic rocks (up to 15 %). Several studies have been carried out on SMP2 (e.g. Barla et al. 2012, Bonini and Barla 2012, Vu et al. 2013, Tran-Manh et al. 2014), particularly for the strongly squeezing ground behavior due to the poor mechanical property of rock mass, and especially the anisotropic and time-dependent convergence of tunnel walls.

In SMP2, the squeezing behavior of the ground is characterized by large time-dependent deformation, associated with a decompressed zone of great extent around the gallery. Significant anisotropic response of the ground and ovalization of the tunnel section were observed (Figure 1a). The support profiles had to be adapted and optimized to these squeezing conditions (Bonini & Barla 2012). In the zones of very large convergence, a yield control support system was then adopted, including highly deformable concrete elements to stabilize the high convergence (Figure 1b).

2.2 SMP4 Survey Gallery

Recently, another survey gallery (SMP4) began to be excavated in Saint-Martin-la-Porte at the depth and along the orientation of the base tunnel (Figure 2). As it will cross the same Carboniferous formation as SMP2 but at a greater depth up to 600 m, a strongly squeezing behavior was expected. Tunneling works started with a full-face excavation with conventional excavation technique. The diameter of the tunnel is about 13 m and the rate of excavation is of 0.5 m/day - 0.7 m/day from West to East. The top side of the tunnel began to enter in the Houiller area at chainage 10267 m and the full section was excavated in the Houiller formation from chainage 10287 m. The

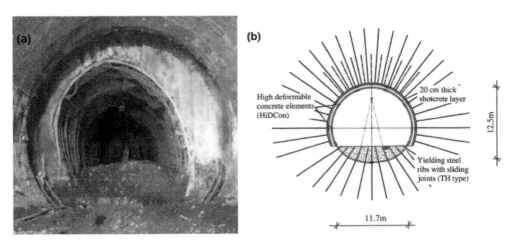

Figure 1. Anisotropic deformation (after Mathieu 2008) and support profiles of SMP2 access gallery (Bonini & Barla 2012).

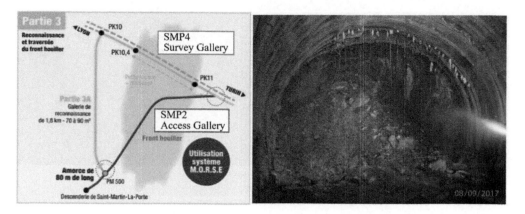

Figure 2. Position of SMP4 survey gallery and collapse at chainage 10303 m.

tunnel face collapsed on September 8, 2017 at chainage 10303 m because of the presence of a fault zone. After passing through the collapsed zones, excavation continued with a reduced section.

The excavation of SMP4 is associated with extensive geological survey and monitoring of the ground deformation (convergence and extensometer measurements, extrusion of tunnel face).

3 DEFORMATION OF TUNNEL WALLS

3.1 Geological units

As shown in the schematic representation of the geological units encountered in SMP4 (Figure 3), the rock formation is highly heterogeneous from chainage 10270 m to 10285 m. An inclined layer of anhydrite crosses the lower part of the tunnel section. A zone of carboniferous schists (Houiller formation) is encountered in the upper part of the tunnel and its extension is increasing from chainage 10267 to 10287 m as the tunnel advances. In the Houiller formation, coal and/or schists dominate, and several inclined bands of sandstone are embedded.

This heterogeneous character of the rock mass is clearly shown on the tunnel face (Figure 4). Due to the presence of the Houiller formation in the upper part of the tunnel, a highly squeezing behavior is observed in this zone whereas the lower part of the tunnel which is in the Anhydrite formation exhibits smaller deformation.

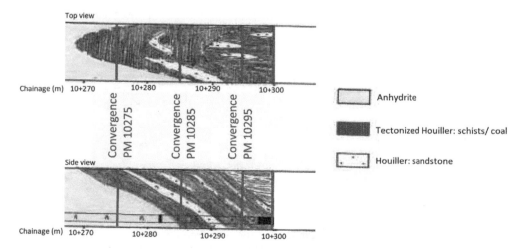

Figure 3. Distribution of geological units.

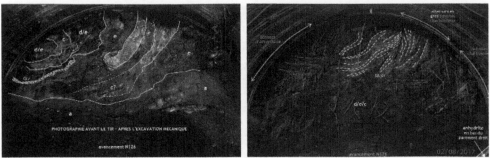

Lithology: a Anhydrite; c Sandstone; d Coal; e Carboniferous schists

Figure 4. Tunnel face at chainage 10281 and 10287 m.

Table 1. Position of monitoring points CT1-CT5.

Section	Upper part (Houiller)	Lower part (Anhydrite)
Chainage 10275 m	CT2; CT3; CT4	CT1; CT5
Chainage 10280 m	CT2; CT3; CT4	CT1; CT5
Chainage 10285 m	CT1; CT2; CT3; CT4	CT5
Chainage 10290 m	CT1; CT2; CT3; CT4; CT5	

3.2 Convergence monitoring data

Convergence measurements were carried out every 5 m by following the horizontal, vertical and axial displacements of 5 monitoring points on the tunnel wall. The convergence data recorded in several sections between chainage 10275 and 10290 m are analyzed in the following. As the monitoring points are located in different rock formations (Table 1), the convergence data of the upper part of the section where the deformation is more important are solely studied in the following.

Visual observation and convergence data show an elliptical deformation of the upper part of the section (Houiller formation). To better describe the anisotropy, the method proposed by Vu et al. (2013) is adopted here for geometrical processing of the measurement data. After projecting the points on the mean vertical plane of the considered cross-section, an elliptical

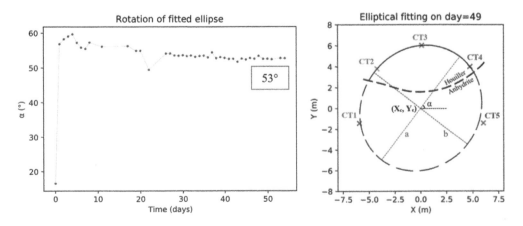

Figure 5. Example of convergence data processing for evaluation the section ovalization at chainage 10275 m (CT1 and CT5 are disregarded in the fitting procedure since they are both localized in the anhydritic lower part of the section).

approximation of the actual shape of the deformed section is performed by fitting the parameters of this ellipse. It is assumed that the coordinates of the ellipse center (X_c, Y_c) is fixed, so that only 3 parameters need to be fitted: the two semi-axes lengths a and b, the orientation α of ellipse. In every section under study, the orientation of the ellipse tends to stabilize as deformations develop because of the important convergence of the top left side of tunnel. Fixing the orientation at its final value, the evolution of the minor axis is calculated to represent the maximal convergence magnitude (Figure 5).

4 CALIBRATION OF THE CONVERGENCE LAW

4.1 *Parameters of convergence law*

The convergence of the minor axis of the fitted ellipse is approximated by using the convergence law proposed by Sulem (1983) and Sulem et al. (1987a, 1987b), where the convergence is expressed as a function of the distance x to the face and of the time t:

$$C(x,t) = C_{\infty x}\left[1 - \left(\frac{X}{x+X}\right)^2\right]\left\{1 + m\left[1 - \left(\frac{T}{t+T}\right)\right]^n\right\} \tag{1}$$

This convergence law depends on five parameters: T - characteristic time related to the time-dependent properties of the system; X - parameter related to the distance of influence of the face; $C_{\infty x}$ - instantaneous convergence obtained in the case of an infinite rate of face advance; m - parameter related to the ratio between the total convergence and the instantaneous convergence; n - constant which describes the form of the fitted curve, usually fixed as 0.3.

The interpretation of the measured convergence must take into account the displacement that has occurred between the opening of the section and the installation of the convergence targets (i.e. 'lost convergence'). The recorded convergence is thus (Guayacán-Carrillo et al. 2016):

$$\Delta C(x_i, t_i) = C(x_i, t_i) - C(x_0, t_0) \tag{2}$$

where x_0 is the face distance for the first record reading and t_0 is the time elapsed since the face crossed the considered section.

The predictive capability of the convergence model is tested by applying the parameters obtained in the study of SMP2 to SMP4. The average behavior in SMP2 has been characterized by the typical values of the parameters: T=20 days; X=15 m; m=18; n=0.3. These values of parameters are applied to SMP4 to fit the convergence law for the sections from chainage 10275 to 10290 m. The only parameter to evaluate for each section is $C_{\infty x}$.

The evaluation of the performance of the proposed convergence law is carried out in two stages. First, the parameters (X, T, m) of SMP2 are directly applied to SMP4 to examine if the measured data can be reproduced in an accurate way by fitting the single parameter $C_{\infty x}$ on all the recorded data. As proposed by Vu et al (2013), we test in a second stage the accuracy of the predicted long-term convergence by restricting the fitting of $C_{\infty x}$ to the first 20 days of measurements.

4.2 *Sections at chainage 10275 and 10280 m*

The calculation is firstly performed on the two sections at 10275 and 10280 m respectively. As shown in Figure 6, the recorded data are well reproduced by the convergence law when the fitting is performed on the complete set of data.

The instantaneous convergence $C_{\infty x}$ and the predicted total (long-term) convergence C_{∞} obtained from the two fitting procedures are very close as shown in Table 2. The parameters

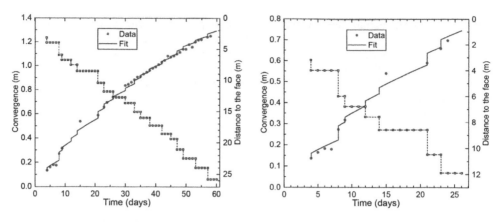

Chainage 10275 m (a) Fitting on all the measurements (b) Fitting on the first 20 days of measurements

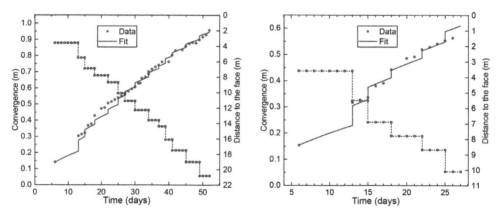

Figure 6. Calibration of the convergence law on sections at chainage 10275 and 10280 m.

Table 2. Fitted parameters with different fitting periods (chainages 10275 and 10280 m).

Section	All measurements		20 days of measurements	
	$C_{\infty x}$ (m)	C_{∞} (m)	$C_{\infty x}$ (m)	C_{∞} (m)
Chainage 10275 m	0.21	3.98	0.22	4.12
Chainage 10280 m	0.17	3.25	0.19	3.54

of the convergence law retrieved from SMP2 are applicable in SMP4 and only 20 days of continuous recording of the data are sufficient for accurate predictions of the convergence at least up to 50 days, corresponding to the last measure available.

4.3 Sections at chainage 10285 and 10290 m

The same process is applied to the two sections situated at chainage 10285 and 10290 m respectively. In this case, it was observed that the measured data cannot be reproduced by keeping the parameters of SMP2. A larger value of the characteristic time parameter T is obtained: $T=60$ days for the section at chainage 10285 m and $T=145$ days for the section at chainage 10290 m (Figure 7). This indicates that the time-dependent deformation of these two sections is significantly higher than the two previous ones. The values of C_{∞} (total

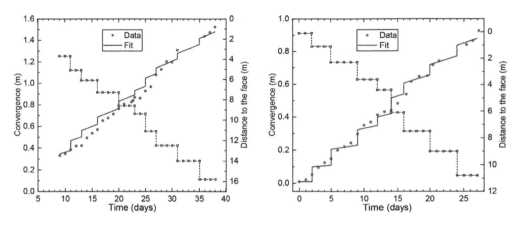

Figure 7. Calibration of convergence law at chainage 10285 (left) and 10290 m (right).

Table 3. Fitted parameters of sections at chainages 10285 and 10290 m.

Section	T (days)	$C_{\infty X}$ (m)	C_{∞} (m)
Chainage 10285 m	60	0.56	10.63
Chainage 10290 m	145	0.70	13.28

convergence) are very high for these two sections and correspond to full closure of the tunnel (Table 3). It reflects the highly damaged state of the rock mass and could be seen as an indicator of the collapse that occurred near chainage 10303 m.

5 CONCLUSION

On the basis of the extensive convergence monitoring of SMP survey projects, the predictive capability of the convergence law proposed for SMP2 (excavated at a depth from 250 to 400 m) (Vu et al. 2013, Tran Manh et al. 2015) is tested against the convergence data recorded for SMP4 (excavated along the future base tunnel of the Lyon Turin link at a depth of about 600 m).

For the studied sections of SMP4 situated between chainage 10275 and 10290 m, it is obtained that by extrapolating the parameters of the convergence law of SMP2, the convergence data can be accurately reproduced for the two first sections situated at chainage 10275 and 10280 m respectively. For the two others, it is necessary to consider a much higher value for the parameter T of the convergence law to correctly fit the data. This parameter controls the time-dependent response of the system. This reflects the highly degraded state of the rock mass for these two sections and could be seen as an indication of the collapse that has occurred at chainage 10303 m.

REFERENCES

Barla, G. 2001. Tunnelling under squeezing rock conditions. *Eurosummer-school in tunnel mechanics, Innsbruck*, 169–268.
Barla, G., Debernardi, D. & Sterpi, D. 2012. Time-dependent modeling of tunnels in squeezing conditions. *International Journal of Geomechanics*, 12(6),697–710.
Bonini, M. & Barla, G. 2012. The Saint Martin La Porte access adit (Lyon–Turin Base Tunnel) revisited. *Tunn Undergr Space Technol* 30:38–54.

Guayacán-Carrillo, L.-M., Sulem, J., Seyedi, D. M., Ghabezloo, S., Noiret, A. & Armand, G. 2016. Analysis of Long-Term Anisotropic Convergence in Drifts Excavated in Callovo-Oxfordian Claystone. *Rock Mechanics and Rock Engineering*, 49(1),97–114.

Mathieu, E. 2008. At the mercy of the mountain. *T & T international*, (OCT), 21–24.

Panet, M. 1996. Two case histories of tunnels through squeezing rocks. *Rock mechanics and rock engineering*, 29(3),155–164.

Sulem, J. 1983. Comportement différé des galeries profondes. *Thèse de l'Ecole Nationale des Ponts et Chaussées*.

Sulem, J., Panet, M. & Guenot, A. 1987a. Closure analysis of deep tunnels. *Int J Rock Mech Min Sci Geomech Abstr* 24(3):145–154.

Sulem, J., Panet, M. & Guenot, A. 1987b. Analytical solution for time- dependent displacements in a circular tunnel. *Int J Rock Mech Min Sci Geomech Abstr* 24(3):155–164.

Tran Manh, H., Sulem, J. & Subrin, D. 2016. Progressive degradation of rock properties and time-dependent behavior of deep tunnels. *Acta Geotechnica*, 11(3),693–711.

Tran Manh, H., Sulem, J., Subrin, D. & Billaux, D. (2015). Anisotropic Time-Dependent Modeling of Tunnel Excavation in Squeezing Ground. *Rock Mechanics and Rock Engineering*, 48(6),2301–2317.

Vu, T. M., Sulem, J., Subrin, D., Monin, N. & Lascols, J. 2013. Anisotropic closure in squeezing rocks: The example of Saint-Martin-la-Porte access gallery. *Rock Mechanics and Rock Engineering*, 46 (2),231–246.

Tunnels and Underground Cities: Engineering and Innovation meet Archaeology,
Architecture and Art, Volume 7: Long and deep tunnels – Peila, Viggiani & Celestino (Eds)
© 2020 Taylor & Francis Group, London, ISBN 978-0-367-46872-9

Overcoming extreme tunneling conditions on Vietnam's longest tunnel

S. Log, B. Li & P.N. Madhan
The Robbins Company, USA

ABSTRACT: Vietnam's Thuong Kon Tum Hydroelectric project is a 17.4 km headrace tunnel that will be the country's longest once complete. A section of the tunnel was excavated by a 4.5 m diameter Main Beam TBM in granitic rock up to 250 MPa UCS. Started in 2012, the project's original contractor left due to non-satisfactory performance. In 2016, the contract to refurbish the TBM and excavate the remaining 10.45 km of tunnel was awarded to a joint venture of Robbins and a local contractor. Robbins was fully responsible for the TBM operation, including supplying operational crews. The crew overcame massive granitic rock, fault zones gushing water at 600 l/s, and difficult conditions. In under two years, the TBM advanced from a standstill at 15 percent project completion to 85 percent complete. This paper will address the refurbishment of the TBM in the tunnel, the work to streamline operation, and challenges faced.

1 INTRODUCTION

The Thuong Kon Tum Hydroelectric Power Project is a 220 MW hydroelectric project located in the Kon Tum State, in central Vietnam. The power project consists, among other works of construction, of a 17.4 km headrace tunnel that will be the longest tunnel in Vietnam once completed. The project is located in the highlands of Vietnam, with very limited infrastructure and industrialization (see Figures 1–2).

The tunnel excavation started in 2012 with a section bored by a 4.5 m diameter Robbins Main Beam TBM designed for the granitic rock types expected (see Table 1). The remainder of the tunnel is being excavated conventionally.

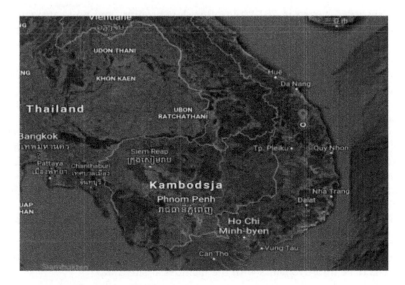

Figure 1. Location of Thuong Kon Tum project.

Figure 2. The Thuong Kon Tum job site.

Table 1. TBM specifications.

Machine Diameter	4,530 mm with new cutters
Cutters	17"
Number of Disc Cutters	30 (To be 30)
Maximum Recommended Individual Cutter Load	267 kN
Average Cutter Spacing	75.5 mm
Recommended Operating Cutterhead Thrust	8010 kN
Maximum Operating Cutterhead Thrust	13345 kN.
Cutterhead Drive	VFD Electric motors/safe sets, gear reducers
Cutterhead Power	1980 kW (6X 330 kW)
Cutterhead Speed	0 to 9.6 RPM

The geology of the project was described as a granitic rock type with strengths up to 120 MPa, however, the available geology of the project was very limited, due to the mountainous jungle above the tunnel which made pre-investigations complicated and expensive.

2 BACKGROUND

The tunnel excavation started in 2012, but after slow progress and non-satisfactory performance, the contract with the project's original contractor was cancelled in 2014. The TBM had then excavated 2636 m in two years in the hard and massive granitic rock, with some zones having significant stability problems.

In late 2014, Robbins was invited to the site to inspect the condition of the TBM and auxiliary equipment that had been left in the damp and warm tunnel for months after the original contractor left the project. After the initial inspection it was obvious that the equipment was not maintained in a proper fashion during operation, nor had any protective measures of the machine been taken before it was left in the tunnel in a tropical climate, with the temperatures in the tunnel being above 30°C and with a humidity above 90 percent. The combination of the previously stated conditions resulted in the TBM and its equipment needing considerable refurbishment before any boring could be restarted (Willis, 2018).

In addition, there was a severe lack of appropriate rock support in the already excavated tunnel, which needed to be properly supported (see Figure 3).

The condition of the tunnel and the TBM was summarized well in the inspection report written by Robbins Field Service Personnel in December 2014 (McNally, 2014):

"As reported during our two (2) previous visits on the 9th August & 16th September 2014. There needs to be improvements with the facilities at site/camp.

Figure 3. Example of inappropriate rock support.

- Conditions in the tunnel were substandard & unsafe
- Visual inspection of the TBM/Back-up showed total lack of maintenance
- No preventative actions were taken in the protection of the major components before closure of operations

> Therefore, I must stress our concern to the likely possibility of major/catastrophic problems we may encounter during the repair/refit of the TBM & associated equipment."

In 2016, more than two years after the original contractor left the site, a joint venture between The Robbins Company and the Vietnamese contractor, Construction Joint Stock Company No. 47 (VH47) was awarded a contract to refurbish the TBM and auxiliary equipment and excavate the remaining tunnel. Robbins was fully responsible for the refurbishment and TBM operation, including supplying operational crews and eventually rock support in the tunnel as well. Construction Joint Stock Company No. 47 was responsible for the site operation, logistics, and rock support in L3.

3 THE REFURBISHMENT OF THE TBM

The refurbishment of the equipment started in mid to late March 2016 and included, among other works, the following:

- Inspect all machine components
- Overhaul hydraulic system
- Repair/replace all electrical systems, including VFDs
- Repair tunnel conveyor and tunnel conveyor E-Stop
- Inspect and repair motors
- Repair and commission rock support system
- Recommission guidance system
- Recommission all systems, including VFDs

In addition, VH47 had to overhaul the tunnel logistic system, ventilation system, and install appropriate rock support in the already excavated tunnel, as the installed rock support was not appropriate. VH47 also made significant efforts to improve the living quarters and facilities at site.

The original proposed refurbishment plan called for four months of refurbishments; however, the owner's production plan insisted on shortening this significantly by instead starting excavation with some work left to be done during TBM operations. The works of the refurbishment followed a strict schedule, which allowed for about two months of refurbishment work before the start of boring. The work was being performed simultaneously with the overhaul of the rest of the tunnel, and this made the logistics and planning important. As an example, the ventilation system and emergency systems were being repaired in the same period as the TBM, resulting in downtime during the TBM refurbishment. The TBM started boring in June 2016 after two months of intense repair work.

4 COMMENCING BORING AND OPTIMIZATION OF OPERATION

The boring commenced in mid-June 2016. In addition to the general follow-up on the machine and the continuous refurbishment efforts, a thorough review of the operation conditions and what could be done to improve the performance and safety at site was implemented. Some of the topics that were identified include:

- Improvement of the structure and organization at site, especially between the different stakeholders
- The rock support already installed in the tunnel was not appropriate and needed to be improved to allow for efficient operation in the tunnel
- Inefficient rock support scheme and rock support methods required by the owner and consultants hampered the production and needed to be revised
- The process of geological mapping and decision on rock support needed to be improved
- Hot and damp working environment on the TBM and Back-up hampered production
- Inappropriate processes for Health, Safety & Environment (HSE)

Some of the measures implemented were purely physical, such as installing water curtains, installing air coolers, running the ventilation through the night to reduce temperatures, etc., but there was also a big focus on aspects of operational processes. A detailed methods statement for the remaining refurbishment and the machine operation was developed. This included, among other aspects, a clear distribution of the responsibilities down to the lowest levels of the project, principle machine operation instructions in different rock conditions, new rock support methodology, new HSE plan according to Robbins international procedures, clear procedures for any operations with safety concerns, and employing a JV geologist team to work with the owner's geologists to make decisions and install rock support as efficiently as possible.

One of the most important features employed was a detailed analysis of the rock support methodology on site. The rock support methodology was defined in the initial contract and was based on a conservative approach to the Q-system (NGI, 2015). This included systematical bolting and shotcrete in all rock classes. Upon evaluation of rock support classes an agreement of a new support methodology was agreed upon. This included quick conclusions on temporary rock support installed in the tunnel by a team of geologists and a revised support scheme based on a less conservative approach to the Q-System. The rock support developed is a guideline for rock support in the tunnel. It is important to mention that the installed support was decided in the tunnel by a team of geologists (see Table 2).

When employing the Q-system with rock support classes in a TBM tunnel it is imperative to consider that the system itself is based on empirical data from Drill and Blast tunnels. As it is very likely that some of the support installed in a D&B tunnel in hard rock is there to compensate for blasting damage, a direct use of the Q-system itself for TBM tunnels is, by nature, conservative. This is important to consider with great caution to avoid misuse of the system and an inappropriate support being installed. (NGI, 2015)

Table 2. Rock support scheme for Thuong Kon Tum.

Rock Class (Q-System)	Q-Values	Rock Bolt (Qty/m)	Mesh (m²/meter)	Ring Beam/m	Shotcrete thickness (cm)	NOTE
A	>100	0	0	0	0	Spot bolting very occasionally
B	>4	0 – 4	0	0	0	Spot bolting as needed
C	>0.1	3 – 6	6	Strap if necessary With McNally	2 to 6 if required	Rock Support needed (quantities might deviate)
D	>0.01	0 – 14	10	Ring Beam with McNally System	6 – 12 (crown portion)	Evaluated case by case
E	>0.001	0 – 14	As required	Ring Beam with McNally system	12 – 15 (crown portion)	Evaluated case by case

Note: McNally System will be adopted to increase safety in the Tunnel

5 GEOLOGY AND ROCK SUPPORT

The geological baseline of the project indicated the vast majority of the tunnel being bored in a granite biotite which was described as relatively fresh and massive, although, with a limited rock strength. In total, the initial geological report described the occurrence of three geological classes (see Table 3, Geological Baseline Report – Kon Tum).

The encountered geology in the tunnel is a combination of granitic rock with a varying degree of gneiss features. The rock is generally massive with limited fractures, resulting in a high Q value and limited rock support needed. There are several stretches with practically no fissures and fractures. The rock is generally experienced as hard, but brittle (see Figure 4).

During operation, UCS testing of the rock was performed at representative locations in the tunnel. This was sent to an internationally recognized laboratory and the results can be seen in Figure 5 (Dahl, 2016).

The results from the testing show a very high rock strength in the majority of the rock mass with strengths of up to 260 MPa. The average rock strength is 194 MPa.

Systematical mapping of the tunnel was also performed according to the Q-system, which confirmed the massive rock conditions (NGI, 2015). The homogenous and strong rock mass is illustrated in high Q value. There were several open water-leading joints in the tunnel causing

Table 3. Baseline geology.

Classification	I	II~III	IV ~ V
Percentage Estimated (%)	10~15	70~80	10~15
Engineering Geol. Description	IIB rock, monolithic. Thick overburden, mostly over 400m	IIB~IIA rock, sub-monolithic ~ massive, inlaid structure. Overburden 200~400m	IB~IA rock, block cracking~ cataclastic, and chipping. Two ends of the tunnel and the places with faults 20m wide as influenced belt for Class IV faults, 10m-wide as influenced belt for Class V faults
UCS (Dry)	95.00–105.00	70.00–90.00	60.00–70.00

Figure 4.   Example of typical rock conditions in the tunnel.

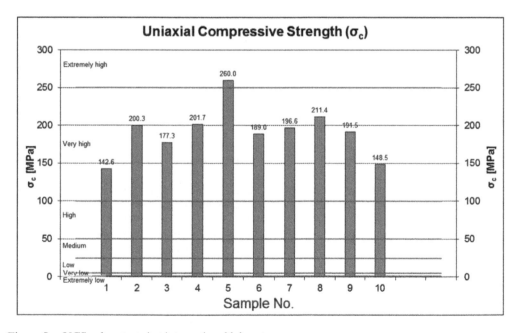

Figure 5.   UCS values tested at international laboratory.

a relative massive total water ingress in the tunnel over the length. The Q classes for a representative part of the tunnel are given in Figure 6.

The above charts show that the tunnel has been excavated in a very hard and massive granitic rock type, with more than 90 percent being in rock class A and B. This also gives a clear indication that the majority of the tunnel can be unsupported or supported only by spot bolting.

It is also worth mentioning that even though there have been ground conditions that have deviated from the baseline geology, the JV and the owner have been working efficiently to find a solution on how to handle the geological conditions during the project.

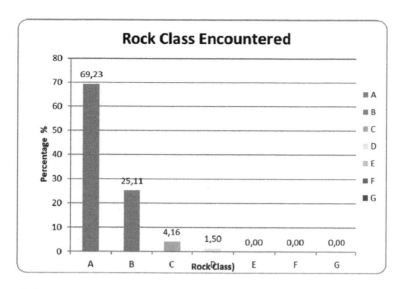

Figure 6. Rock class distribution as mapped by Robbins geologists.

This solution-driven approach has led to an absence of unsettled geological claims as the project is approaching its completion.

6 PRODUCTION AND PERFORMANCE

For the initial part of the project with the original contractor, the performance was not satisfactory with an average production rate of 120 m/month. In addition to the low production, there was a significant lack of maintenance. The lack of maintenance and the unprotected storage of the TBM and equipment in the tunnel caused damage to the machinery that limited the production on the project overall, due to more time being spent on repairs and maintenance than normal. This has especially been relevant to general maintenance, the cutters and cutterhead, and the conveyor system.

Based on the condition of the equipment and the geology, the site organization made a detailed plan for the excavation of the rest of the tunnel, with planned maintenance and down periods. This planning helped to achieve a steady production on the project, which seemed

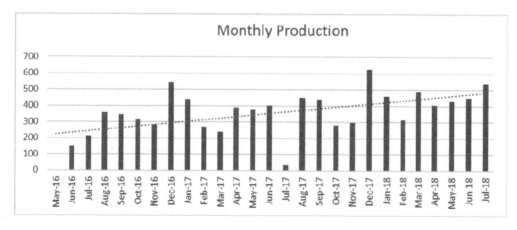

Figure 7. Monthly production.

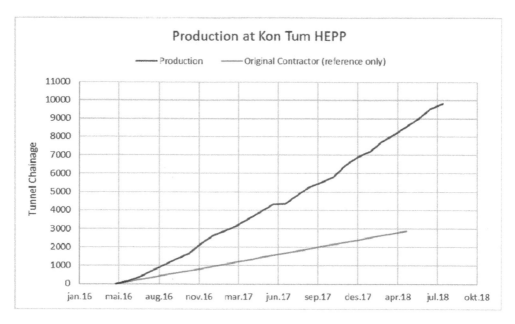

Figure 8.   Production chart and comparison to production in 2012–2014.

impossible at earlier stages of the project. Planning also included the remaining refurbishment of the system, which accounts for some of the lower production in the early months and the positive trendline in the production chart (see Figure 7).

The average production of the project after May 2016 until September 2018, with less than 500 m left to bore, has been about 380 m/month. This includes significant maintenance and advancing through adverse fault zones, which has required pre-excavation grouting. The steady production and vast improvement on the project is clearly visible in the production chart given in Figure 8.

7   DISCUSSION AND LESSONS LEARNED

The excavation of the headrace tunnel for the Thuong Kon Tum Hydroelectric project has been a very interesting project experience from an operational point of view. In the time before the operational contract with Robbins was signed, there were several discussions ongoing between the project owner, international consultants, and contractors on how to finalize the project within the time frame. Among the topics of discussion was the idea to employ a second TBM, replace the current TBM, go Drill and Blast, and/or change the cutterhead to 19"/20" cutters to better cope with the hard and massive rock mass.

After thorough inspection and evaluations, the JV committed to a production rate that would finalize the project in the given time frame. This included doing a thorough review of the operations at site, refurbishing the TBM and equipment, and replacing the current cutterhead with a 19"/20"-cutter compatible cutterhead. In hindsight, the performance proves that the correct decision was made at the time. When the operation restarted in June 2016 the performance was so satisfactory that a decision was made not to take the time to replace the cutterhead, even though the new cutterhead was already in production. This decision was made in consideration of the downtime it would take to replace the cutterhead that would have had a bigger effect on the completion date than the relatively lower production with the current cutterhead.

The experience from the project is a prime example of good and bad maintenance on a TBM and the effects this has, in addition to the importance of focusing on improving processes and making the TBM operate as much as possible. Considering the condition the equipment was in when Robbins took over operation, well-planned maintenance and a focus on a steady production has been imperative for the success of this project.

The results of the project give a proper reminder that tunnel excavation with a TBM is not a "push-the-button" solution. Even though TBMs have become more advanced and automated over the years there is still a big demand for skillful and experienced personnel. Proper training of the crews and supervision from experts at an early stage is still an important part of any TBM operation and it is not wise to forego this to save money.

One of the reasons why TBM tunneling is not a "push-the-button" solution is the material the machines are operating in. Geology varies in any rock mass and it is seldom exactly as expected. On this project, the geology was harder and more massive than what was known prior to the project. In hindsight, it is apparent that use of 19"/20" cutters should have been considered thoroughly due to the geology that was encountered. Especially in projects where the available geology is limited and over-design of the TBM is a cheap "insurance" if harder conditions are encountered.

Lastly, the rock support scheme on projects needs to be appropriately designed to be efficient with the technology of excavation. On some global projects there are set rock support schemes established as early as possible in the bidding phase. This is often based on some of the internationally recognized rock classification systems. In this case, the support scheme was based on a conservative approach to the Q-system, which specified a rock support scheme that was not optimized for TBM operation and made efficient operation challenging. It is important that the rock support methodology is considered with efficient operation in mind so that it can be an integrated part of the operation. The experience at Thuong Kon Tum also highlights the need for a standardized rock support system based on operation with a Main Beam TBM, as most of the current systems in use are based on D&B tunneling.

8 CONCLUSIONS

The experience from the excavation of the headrace tunnel for the Thuong Kon Tum Hydroelectric project reminds us of some obvious truths that sometimes are forgotten:

– A skilled contractor and knowledgeable crews are essential in any tunneling project
– Focus on planned activities and integration of maintenance in the planning stage is the best way to a steady production
– Rock support needs to be designed and customized to the excavation method and the specific TBM in cooperation with the contractor
– Improved processes and a clear focus on HSE has a clearly positive effect on production
– Even though the TBM is often blamed for low production on a project with sub-optimal performance, the TBM itself is rarely the cause

The project also shows us a few not so obvious reminders:

– Even TBM- related equipment that has been suffering from lack of maintenance and left in a damp tunnel in a tropical climate for two years can be put into operational shape
– It is possible to get a steady production from TBMs in very remote job sites with sufficient planning of activities

By September 2018 the tunneling machine had less than 500 m left before breakthrough. The vast improvement in performance has allowed the project to finish on time, which seemed like an impossible milestone less than 30 months ago.

REFERENCES

Dahl, F. (n.d.). 2016. SINTEF.

(n.d.). *Geological Baseline Report – Kon Tum.*

(n.d.). *Google Maps.*

McNally, J. (2014). *Report on site conditions at Kon Tum.* Internal.

NGI. (2015). *Using the Q-system – Handbook.*

Willis D. (2018). *A Gauntlet of Challenges.* Tunnelling Journal Apr/May 2018: 50–57.

*Tunnels and Underground Cities: Engineering and Innovation meet Archaeology,
Architecture and Art, Volume 7: Long and deep tunnels – Peila, Viggiani & Celestino (Eds)*
© 2020 Taylor & Francis Group, London, ISBN 978-0-367-46872-9

Maly Lubon road tunnel: The case of a full-face excavation of a very large tunnel in the Carpathian Flysch, Poland

A. Lombardi, A. De Pasquale & V. Capata
SGS Studio Geotecnico Strutturale S.r.l., Rome, Italy

ABSTRACT: The Maly Lubon Tunnel is the longest tunnel in Poland, as part of the expressway S7 Krakow–Rabka Zdroj investment. It is twin tube tunnel 2 km long with sections 220 m^2 large, excavated through Maly Lubon mountain by Astaldi company. The tunnel, still under construction, excavated in the Carpathians, consists of alternative layers of sandstone and shale with high variable degree of weathering. Due to the extreme variability of the rock mass behaviour, it was decided to change the excavation technology from the NATM to the Convergence-Confinement method. The big challenge of this change required a new design for the South portal, in the most weathered and heavily tectonized Flysch, already subject to large displacements during its geological history. Soil-structure interaction analysis were also carried out for a wider tunnel section which included emergency lanes and a vehicular cross passage intersecting with an additional technological tunnel. In order to study the complex geometry of the intersections between tunnels and the mutual effects potentially induced by the excavation of each tunnel, a specific 3D FEM analysis was carried out and provided a valid support to select suitable construction phases during the construction. The aim of the present paper is to describe the problems encountered during the excavation of a large road tunnel in Carpathian Flysch and to outlines the design solutions adopted to overcome the most critical sections of excavation.

1 INTRODUCTION

This paper focus on Maly Lubon Tunnel, the longest tunnel in Poland as a part of S7 from Kraków to Zakopane in the Lubień Rabka section. Each tunnel, between Naprawa and Skomielna Biala and under the Lubon Maly massif, are about 2 km long.

The original design has been heavily modified both in the tunnel excavation method and in the geometry of south access portal. At construction design stage, it was increased the mining tunnel length of about 87 m in order to reduce the slope height to about 21 m against 36 m according tender design. The proposed design solution follows a detailed evaluation of the topography and geotechnical context of the South portal area, thanks to the results got by additional ground investigations, carried out especially in this zone. This new design methodology does not involve any change of the Tender safety conditions, the main purpose has been only to anticipate the start of the excavation sequences by using full face excavation method and thereby avoid expensive works of soil improvements in extremely heterogeneous soils. The full face excavation method has been judged the most suitable to face the excavation of very large sections in the Carpathian Flysch. In this paper, we describe the main problems encountered in the excavation, highlighting the technical solutions adopted and the calculation methodologies.

2 TUNNEL DESCRIPTION

The tunnel consists of two parallel tubes S7 equipped with 2 lanes (Tajduś *et al* 2013), bands with the width of 2.50 m and 3.00 m and parking bays halfway along the tunnel. Both lines of the

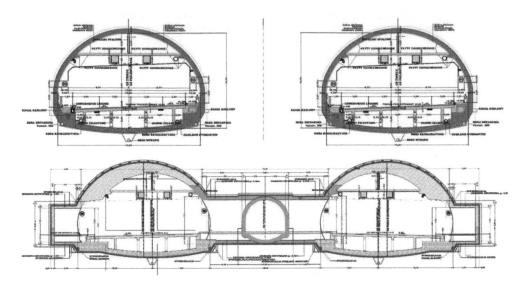

Figure 1. Typical cross section for main tunnel (above) and Lay-Bys (below).

tunnel will be connected to each other by pedestrian cross passages spaced every 172.5 m fulfilling the function of evacuation routes. The shape tunnel is polycentric closed at the bottom by an inverted arch. The section is fully waterproofed: the water captation is achieved through a drainage under the invert at the provisional stage during tunnel excavation and by means of two lateral drainage pipes (placed below the knees of final lining) at final stage. The final lining is executed in reinforced concrete with a constant thickness of 50 cm in case of support class 1 and 2 or variable thickness from 50 to 100 cm in case of support class 3. A full round waterproof system is installed made by a layer of pvc and a layer of geotextile. The design also includes signaling-alarm and teletechnical niches on the outer sides of each pipe of the tunnel spaced every about 172 m. Inside the tunnel, a vehicular cross passage for emergency is designed, as well as technical room ST3. This stretch of tunnel, called Lay-Bys, is characterized by an enlarged excavation section (over 200 m^2) to the presence of the parking lane (L = 2.50 m).

In the southern portal area, there is the technological building ST1 in which the air treatment plants are housed and which contributes to the stability of the retaining structures present on the sides of the portal.

The works started on March 2017 and the end of works it is fixed on end of 2020. The initial value of the tunnel it was estimated about 181.410.923 €.

3 GEOMORPHOLOGY AND CLIMATE OF THE AREA

The area of Maly Lubon Tunnel lies in the sub-province of Outer Western Carpathians, in the mesoregion of Island Beskids. A landscape feature characteristic of the Island Beskids is the occurrence of isolated peaks rising to heights of ca. 400–500 m above the level of the intermontane plain, up to the height of 850–1170 m a. s. l.. Mountain slopes are overgrown with a lower subalpine forest, and the intermontane plains are occupied by crops and residential buildings.

The region has a continental climate with cold winters from December to March. January temperatures average -1°C to -20°C. Summers, which extend from June to August, are usually warm, sunny and less humid than winter. July and August average temperatures range from 16°C to 19°C, though some days the temperature can easily reach even 35°C. The annual rainfall-snow can reach easily as much as 1300 mm/year. This rainy and snowy regime obviously influences the underground water circulation in a sensitive way especially in the first meters of depth.

4 GEOLOGICAL AND GEOTECHNICAL FRAMEWORK

4.1 *Geological framework*

The area of the designed works lies in the zone of the Magura nappe limited from the south with a zone of longitudinal strike-slip faults from the Pieniny rock belt, and from the north bordering on the Silesian nappe (Stupnicka, 2007). In Magura nappe, dominate thick-shoal sandstones with inclusions of conglomerates and shales. In their vicinity, there are belts of shales and thin-shoal sandstones, divided by faults, of the so called Hieroglyphic strata, classified in terms of age as the Middle Eocene. The southern section of the future road intersects with belts of hieroglyphic and sub-Magura strata, whereas the northern section for the designed tunnel, runs through thick-shoal sandstones and shales of the Magura strata, formed in the micaceous facies.

It is not possible to easily find a structure inside the rock mass, since the material occurs in most cases in the form of folded layers and with metrically variable discontinuity planes.

4.2 *Hydrogeological conditions*

In the area of study thicknesses of aquifers are low and do not exceed 5 m. Values of the filtration ratio are varied, depending on formation of deposits. The depth of water table is varied and strictly depends on precipitation inflow and the level of surface water. In the youngest river sediments, the free surface of water is shallow (from 0.2 to 1.0 m b.g.l.) and is closely connected with the level of water in rivers.

Hydrogeological parameters of the designed tunnel, where there are upper sections of watercourses, are low. This applies to both thickness and filtration parameters due to the often clayey nature of colluvium, in which filtration ratios drop to several m/day and, in extreme cases, even to approximately 1 m/day. Supply of water takes place through direct infiltration of precipitation, from side inflows and surface water.

In the cracked series of Carpathian Flysch, formed primarily as sandstones with shale inclusions, the aquifer is mostly composed of complexes of thick-shoal Magura sandstones with inclusions of silt-marl shales. The average thickness of the first (counting from the surface) aquifer is estimated to be about 15 m, whereas the average value of the filtration ratio of about 1.0 m/day. The best filtration parameters ($k = 10^{-5} - 10^{-6}$ m/s) occur down to the depth of 30–40 m b.g.l. Ground water occurs at different depths, from several meters in the bottoms of valleys to a dozen or so meters on slopes.

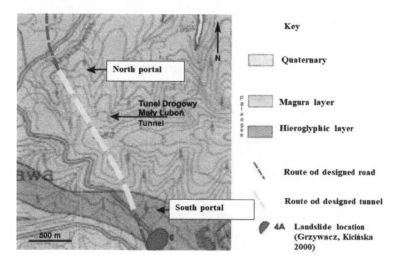

Figure 2. Contemporary geological situation in the mapped area.

4.3   Stratigraphy and geostructural properties

Starting from the surface there are the Quaternary deposits which consist of clayey silts with gravels and arenaceous boulders classified as Colluvium. The thickness of these deposits reaches a maximum of 10–12 m when depressions (such as in the area of the south portal) are present.

Below the Colluvium are the oldest formations, respectively the Flysch of the Hieroglyphic (Middle Miocene) and the Magura (Eocene-Oligocene).

These layers are present, below the Colluvium throughout the South area where are strongly disturbed from the tectonic point of view and anisotropic. It is not possible to easily find a structure in the rock mass since the material occurs in most cases in the form of folded layers and with metrically variable discontinuity planes.

Magura layers are marine Flysch sediments of the upper Eocene. The layers are composed of alternating layers of sandstone and slates. Sandstone series in Magura layers are fine or medium grained. Most commonly they take the form of medium to thick banks.

The average spacing can be assumed equal to 0.4 m in shales and 0.8 m in sandstone. Three different areas in the stratification, have been found. Homogeneous areas S1 and S2 are characterized by Magura layers, the homogeneous area S3 by Hieroglyphic layer systems. The table below summarize the geometry of these areas. In some of the boreholes, during examinations of the core, typical features of disturbances (dislocations) have been detected. These area were confirmed by the excavation of the tunnel, although in slightly different positions with respect to predictions. In some cases the excavation is progressing through strongly altered, fractured, deeply weakened and not homogeneous rocky materials, with stratification and faulting features changing in few meters - quite at each step, respect to the axis of the tunnel.

Table 1.   Geometric properties of main families of discontinuities for different layers.

Uniform area	boreholes	dip direction (°)	dip angle (°)
S1	BK1–BK8	175	25–60
S2	BK9–BK11	194	9–35
S3	BK12–BK14	194	29–61

Figure 3.   Hieroglyfic strata - typical borehole cores.

Figure 4.   Magura strata – good core of rock in main tunnel.

4.4 Geotechnical characteristics of soil/rock mass

For the geotechnical characterization of the area interested by the excavation of the Mały Luboń tunnel, test results provided in the Tender documents were interpreted and integrated with the results obtained by additional tests. For the measurement of strength properties have been used SPT, DPH and DS tests for colluvial and soft argillite/shale Flysch and PLT/UCS/Brazilian tests for rocky soils. Strain properties have been estimated by means of seismic refraction and MASW profiles. In the table below are the mechanical parameters for the different geotechnical units.

Table 2. Geomechanical parameters assumed for different layers.

Unit	γ (kN/m³)	GSI	σ'_f	c' (kPa)	φ (°)	E (MPa)
Colluvium	21	-	-	10–15	20–25	10
Hieroglyphic	23	15–35	0.5–15	20–120	24–26	100–300
Magura	23	45–55	15–35	300–430	44–52	1900–6600

The geotechnical profile along the tunnel was designed with the most relevant geotechnical units taken in account in the numerical analysis. The overburden above the tunnel goes from 15–25 m (on the portals) to 100 m, at the maximum overburden.

4.5 Water table conditions

In the boreholes, the piezometric level varies between 5.20 and 8.80 m below ground level, more or less at the colluvium/weathered shale transition level. It most likely represents a suspended

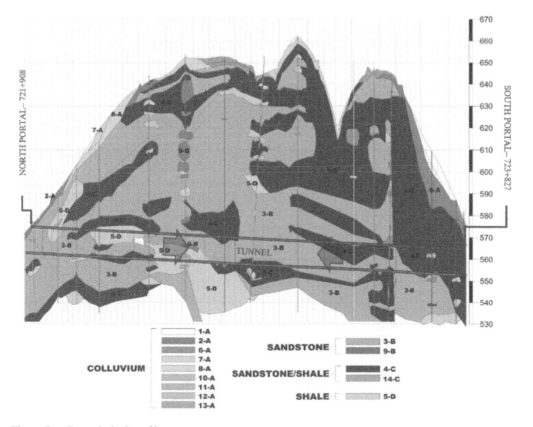

Figure 5. Geotechnical profile.

aquifer that does not affect the weathered rock material below the colluvium. This water, not at all easy to manage due to the extreme variability of the structure and the structure of the discontinuities, has been controlled with draining pipes arranged both along slopes/retaining walls at the portals and on the mining tunnel. In mining tunnel stretches, except in some cases, the flow rates flowing from the excavation face were of the order of few l/s.

5 EXCAVATION METHODOLOGY OF THE TUNNEL

The Maly Lubon tunnel is the largest and longest tunnel in Poland and it is the first one excavated with the full face method in this Country. The excavation method of the tunnel depends on the geomechanical properties of rock mass. There are defined five types of excavation sections called 1, 2, 3, NP, SP. For the sections 1 and 2, excavation can be performed with the use of explosives. For sections 3, NP and SP excavation has to be performed using mechanical means. In the part of tunnel where good rock mass exist, section type 1 or 2 are adopted. Temporary support lining consists of double steel ribs IPE180 at steps from 1.5 m to 2.00 meters and shotcrete 25 cm thick reinforced with steel welded mesh. The final lining has a thickness of 50 cm constant at the crown and in the tunnel invert. The section type 2 differs from the section type 1 in arrangement of radial rock bolts and the step of the steel ribs.

The section type 3 is characterized by forepoling crown support injected with cement grout. The truncated cone configuration is created to protect the excavation in advances. The pipes are arranged in an umbrella configuration around the excavation shape with n°45 steel tube, 15.00 m long, to be repeated after 8.65 meters of excavation steps.

The stabilization of the core face of the tunnel is performed by fiberglass pipes (n° = 46 – 100) 15.00 meters long. As for the forepoling, the fiberglass pipes have a minimum overlap of 6,00 m.

The primary lining consists of double ribs IPE 180 at step of 0.75–1 meter and shotcrete 25 cm thick reinforced with welded mesh. The final lining has a variable thickness for the crown from 50 to 100 cm and constant thickness of 50 cm for the invert arch. Before the final lining concreting, complete waterproofing of the section is provided.

The excavation is performed using mechanical means with step 1.00 m long. The excavation has a minimum area of about 172 m^2 at the first excavation and a maximum area of about 190 m^2 at the last excavation step.

6 MAIN ASPECTS AND PROBLEMS OF TUNNEL AND PORTALS DESIGN

One of the most relevant problems that have occurred during the execution of the tunnel was the soil displacements on South Portal slopes. In this case different geological and hydrogeological condition were encored compared to the design forecasts. In order to define a new geotechnical model the monitoring data were analysed. The monitoring instruments installed in that area were inclinometers, piezometers and targets. Furthermore, during the excavation of the slopes shallow landslides were observed as well. When the Berliner wall was completed and the excavation level reached the work tunnel level, the monitoring data

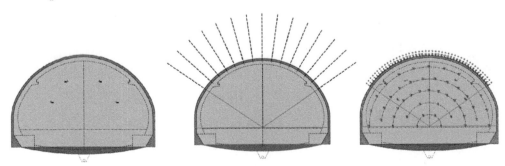

Figure 6.   Excavation section type 1 (on the left), 2 (on the middle) and 3 (on the right).

Figure 7.   Front face with sandstone and shale (on the left), shale (on the right).

showed an increasing of the deformation values both for the inclinometers dates and targets data. The first additional works performed on the slope, were the draining trenches in order to decrees the load of the suspended water table. Furthermore, even when the anchors on the Berliner wall were completed the horizontal deformation of the wall, measured on the optical target, reached maximum values up to 25–30 cm. These deformations of frontal Berliner wall were manifested as a result of a deep movement that also affected the future excavation section of at least one of the two tunnels. It was then necessary to take safety measures through the design of new additional works before star the tunnel excavation. There were designed and realized two important works that allowed the stop pf the deformation on the Berliner wall and on the slopes:

- reshaping on the slopes in order to remove gravity load on the head of the possible landslide;
- in situ casting canopy tunnel in order to load the foot of the possible landslide;

Through this two additional works, it was possible to stop the deformation and proceed with the tunnel excavation.

During the excavation of the tunnel the major difficulties encountered were related to the high heterogeneity of the excavated material. The geotechnics hypothesis adopted during the design stage were not fully confirmed during the construction stages. The front face showed in most of the logging present very difficult condition in term of discontinues and fracturing status due to the different litotypes observed and tectonic disturbance. Most of the front face consists in Flysch formation constituted by very thick sandstone beds with shale bench thin stratificated. Due to the high rock mass discontinuities, during the excavation of the tunnel especially on first meters, problems of face extrusion and extra excavation occurred. The worst condition of the front face was a front face totally composed of shale. In this condition the support class applied was the type 3 where fiberglass elements at the face were performed. In this material longitudinal strains were observed in the temporary lining during the excavation. Although the steel ribs were fixed at the foot by the invert, maximum displacements of 60–70 mm were measured in the upper part. The displacements have been exhausted once poured moreover in a very short time, the final RC lining.

7   MAIN TUNNEL ANALYSIS IN LAY-BYS STRECTH

Given the complex geometry of the intersections between the tunnels to be built in the Lay-Bys area, and the mutual effects induced by the excavation of each tunnel on the adjacent one, it was necessary to carry out three-dimensional numerical analyzes, the results of which provided a valid support in the choice of suitable executive phases.

In this case, the central tunnels (vehicular by-pass and ST3 technological tunnel), although much smaller than those of the lateral Lay-Bys, will be excavated in the portion of rock mass located between the two pipes of the main tunnel, in an area already disturbed and plasticized by the excavation of lateral main tunnel, also due to the reduced existing distance (about 14 m).

The FEM calculation code used for the numerical analyses is MIDAS GTS-NX 3D, version 2018 release 1.1, which allows to evaluate the stress-strain behavior of the soil-structures and their interaction, by simulating all construction phases of the tunnel.

In the analysis solid-tetrahedrycal elements are used made of 4 nodes classified as 3D sol-id shape. To the faces of the model the followings boundary conditions have been imposed:

– Upper surface: free;
– External side surfaces: fixed displacements in horizontal direction;
– Lower surface: fixed displacements in vertical and horizontal directions.

The model has been extended sufficiently so that to be able to hold negligible, in the zones of interest, the trouble effects due to the boundary conditions used. In this case we have laterally extended the model to a variable distance between. The model has been created using:

– About 460000 3D element (rock mass continuous space with properties of Magura strata);
– About 120000 2D elements (equivalent temporary supports and final lining as shell elements);
– About 2000 1D elements (rock bolts as beam elements).

For definition of the behavior of soil, a Mohr-Coulomb model has been used in FEM analysis, with plasticity associated with development of irreversible strains. A perfectly-plastic behavior is assumed for strain-stress calculation. The construction sequence must be the following one:

– Excavation of first lay-bys tunnel installing steel ribs and shotcrete;
– Rock bolts installation on cross passage area (rock bolts include both elements in the soil and for future support of cutted steel ribs at the beginning of the excavation of the vehicular cross passage);
– Pouring invert, knee and crown parts of first lay-bys tunnel except for the segment at the vehicular cross passage part;
– Implementation of steps 1 to 3 also for the second lay-bys tunnel;
– Vehicular cross-passage excavation by mean of steel ribs, steel wire mesh and shot-crete;
– Rock bolts installation at the technological tunnel ST3 intersection (rock bolts include both elements in the soil and for future support of cutted steel ribs at the be-ginning of the excavation of the ST3);
– Technological tunnel ST3 excavation installing steel ribs, steel wire mesh and shot-crete;
– Pouring of invert and knee parts of cross passage;
– Excavation of the niche on the opposite side of the vehicular cross-passage;
– Pouring of final lining for cross passage, ST3 tunnel, lay-bys tunnel crown and niches in one step.

The analyzes results show that the most critical phases are those that involving the excavation of the central by-pass and ST3 tunnels, as a consequence of a stress redistribution on the Lay-Bys tunnel as the excavation proceeds. The maximum stresses on the temporary lay-bys support occur precisely in these phases: this circumstance led to use a more robust steel ribs (2IPE220/0.75 m) for a total length of 12 m with midpoint on the axis of the vehicular by-pass.

In addition to this intervention, it was necessary, in order to reduce the stress values acting, to provide the invert and knees concreting for the entire length of two lay-bys tunnels and spring line/crown parts for the stretches not interested by vehicular by-pass. The temporary supports of by-pass tunnel has been calculated to bear the high forces generated by ST3 tunnel excavation, parallel to the main tunnel. The results of the numerical analyzes have finally allowed to verify all the tunnel linings structures both in temporary and final conditions and in exceptional conditions (in case of fire) in compliance with the requirements of the current reference standards. In the figures below are presented the main results of 3D FEM analysis.

Figure 8.   FEM model used to analyzed stress-strain state of Lay-Bys – ST3 stretch.

 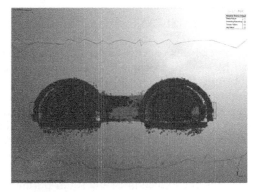

Figure 9.   Lay-Bys: plastic zones (red points) at the end of cross passage exacavation.

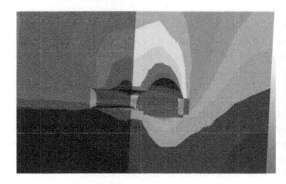 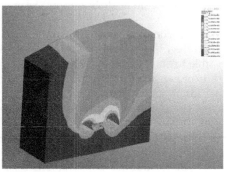

Figure 10.   Maximum total displacement on tunnels ate the end of technological tunnel ST3 excavation (max value 25 mm).

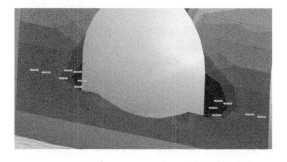 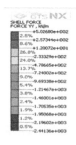

Figure 11.   Vehicular Cross Passage – Lay Bis tunnel intersection: axial force concentration on springlines.

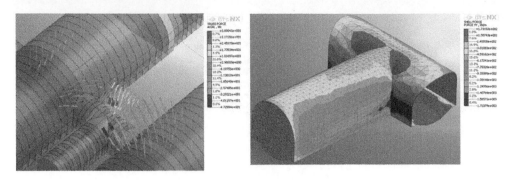

Figure 12. Vehicular Cross Passage – Lay Bys tunnel intersection: axial force on nails (on the left) axial force final lining (on the right).

8 CONCLUSIONS

The aim of this paper is to present the design of the Tunnel Maly Lubon still under construction by Astaldi Company. The main aspect described refer to the extreme variability of the rock mass behaviour, the big challenge of new design for the South portal and the soil-structure interaction analysis carried out through 3D FEM analysis. The main difficulties for the excavation of the tunnel regarding the extremely variability/anisotropy (Margielewski 2002) of the rock mass were faced with the support class variation. In the worst difficult material, the most heavy support class was applied, with fiberglass elements and umbrella steel pipes. The appropriate support class applied during the excavation made it possible to have convergences and forces in temporary support and final RC lining within the design values.

During the execution of the south portal area, big deformations due to different geological conditions occurred during the slope excavation made necessary the execution of additional works in order to increase the safety condition and in order to allow the excavation of the tunnel. Significant contribution has been given from different additional works as new drainage system on the slope that allowed to reduce the water load on the frontal Berliner wall; reinforced concrete canopy tunnel to load the foot of the landslide; reshaping of the slopes to unload the top of the landslide. After the additional works the deformations stopped and it was possible to start tunnel excavation.

All tunnel significant stretches have been designed by means of plane and three-dimensional finished models. In particular, in order to design the execution phases for the excavation of the Lay-bys (wider tunnel) at the intersection between the vehicular cross passage and technological tunnel, it was necessary to carry out 3D numerical analyses. The results of the numerical analysis allowed to verify all the tunnel linings structures both in temporary and final conditions, allowing to analyze the most critical areas, or those of intersection between the various tunnels, and to adopt executive steps suitable to perform the excavation in safe conditions.

9 ACKNOWLEDGEMENTS

Thanks to Eng. Giovanni Leuzzi for the support provided in numerical analysis. We thank our colleagues from Astaldi Spa for enabling designers to works in their offices to observe their daily operations and for their valuable technical support on this Project.

REFERENCES

Margielewski W. 2002. Geological control on the rocky landslides in the Polish Flysch Carpathians. *Folia Quaternaria* January 2002.
Stupnicka, 2007. Geologia regionalna Polski (Regional Geology of Poland). University of Warsaw (eds)
Tajduś, A., Tajduś K. 2013. The use of tunnels to development of transport in mountain areas. *Geomatics, Landmanagement and Landscape No.4, 2013*: 103–112.

Tunnels and Underground Cities: Engineering and Innovation meet Archaeology, Architecture and Art, Volume 7: Long and deep tunnels – Peila, Viggiani & Celestino (Eds)
© 2020 Taylor & Francis Group, London, ISBN 978-0-367-46872-9

Milan to Genoa high speed/capacity railway: The Italian section of the Rhine-Alpine corridor

P. Lunardi
Lunardi Geo-Engineering, Milan, Italy

G. Cassani & A. Bellocchio
Rocksoil S.p.A., Milan, Italy

N. Meistro
COCIV, Genoa, Italy

ABSTRACT: The new high speed/capacity Milan to Genoa rail line will improve railway connections between the Liguria port system with the main railway lines of Northern Italy and the rest of Europe. The project is part of the Rhine-Alpine Corridor, which is one of the corridors of the Trans-European Transport Network (TEN-T core network) connecting Europe's most populated and important industrial regions. The Rhine-Alpine Corridor constitutes one of the busiest freight routes of Europe, connecting the North Sea ports of Rotterdam and Antwerp to the Mediterranean basin in Genoa, via Switzerland and some of the major economic centres in the Rhein-Ruhr, the Rhein-Main-Neckar, regions and the urban agglomeration in Milan, Northern Italy. The new high speed/capacity rail line will be 53 km long, of which 39 km in tunnels. The new line will be connected to the existing line through four interconnections, 14 km long. The construction started in April 2012 and the completion of the six sections is scheduled for April 2021. The adopted excavation methods are conventional (ADECO RS) and mechanized. The paper will describe this complex project from a design point of view for both excavation methods, conventional and mechanized.

1 INTRODUCTION

The high speed railway Milan-Genoa is one of the 30 European Priority Projects adopted by the European Union on April 29, 2004 (Project n. 24 "Railway axis Lyon/Genova-Basel-Duisburg-Rotterdam/Antwerpen") as a new European project, the so-called "Bridge between Two Seas", a rail-link Genoa-Rotterdam. The line will improve the connection between the port of Genoa and the Po Valley inland and further with Northern Europe stimulating a significant increase in transport capacity, in particular in freight transportation, aiming to meet the growing traffic demand. The line runs along the Genoa-Milan route reaching Tortona, and proceeds along the Genova-Alessandria-Turin route up to Novi Ligure, crossing the provinces of Genoa and Alexandria. The new line will be connected to the South at Voltri and Bivio Fegino through interconnections with the railway facilities at Genoa hub and with dock basins of Voltri and Porto Storico, while connection to the North will be ensured by the existing railway lines Genoa-Torino and Tortona- Piacenza-Milan. The total length of the line will be approximately 53 km; the project requires the construction of 36 km of tunnels running through the Apennine mountains between Piedmont and Liguria. The overall scope of underground works, including dual tube single-track running tunnels, adit and interconnection tunnels, exceeds 90 km (Figure 1).

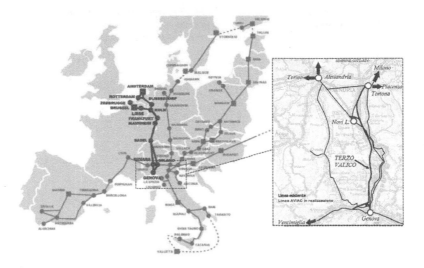

Figure 1.   Terzo Valico dei Giovi: general project layout.

The underground portion includes the approximately 700 m long Campasso Tunnel and the two interconnecting tunnels at Voltri, each with a length of approximately 2 km. Four intermediate access adits (Polcevera, Cravasco, Castagnola and Vallemme) are anticipated for the Valico Tunnel, both for structural and safety purposes. From the Serravalle Tunnel exit, the main line is predominantly above ground or in an artificial tunnel, until it joins the existing line in Tortona (en route to Milan), while a diverging branch line with a turnout speed limit of 160 km/h establishes the underground connection to and from Turin on the existing Genoa–Turin line. In terms of construction, the tunnels listed in Table 1 are the most significant underground works of the Terzo Valico. Tunnels excavations are designed and executed with conventional (61,7 km) and mechanized excavation (30,7 km) method with an overburden varying from 5,0 m to 600,0 m and cross sections area between 75 and 365 m^2. In compliance with the latest safety standards, the underground line consists of two single-track, side-by-side tunnels with cross-passages every 500 m which allow each tunnel to serve as a safe area for the other (Figure 2).

The General Contractor in charge of designing and building the Terzo Valico is the COCIV Consortium formed by the following major Italian construction companies: Salini Impregilo (64%), Società Italiana Condotte d'Acqua (31%) and CIV (5%).

Table 1.   Terzo Valico Tunnels.

Tunnels	Lunghezza	Section Type	Excavation method
Campasso	710 m	*single tube - dual track*	Conventional
Valico	27.032 m	*dual tube - single track*	Conventional & Mechanized
Serravalle	7.094 m	*dual tube - single track*	Mechanized
Volti Interconnection	3.023 m	*single tube - single track*	Conventional
Cravasco Adit Tunnel	1.260 m	*single tube - single track*	Conventional
Polcevera Adit Tunnel	1.763 m	*single tube - single track*	Mechanized
Castagnola Adit Tunnel	2.470 m	*single tube - single track*	Conventional
Val Lemme Adit Tunnel	1.590 m	*single tube - single track*	Conventional
Val Lemme Safety Area	750 m	*dual tube - single track*	Conventional
Novi Ligure Interconnection	2.860 m	*dual tube - single track*	Conventional
New route NV-01	1.010 m	*dual carriageway road*	Conventional
New route NV-02	306 m	*dual carriageway road*	Conventional
By-Pass	4.810 m	*pedestrian & vehicle way*	Conventional

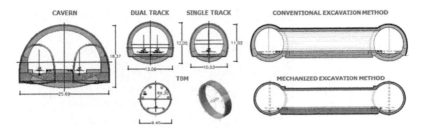

Figure 2. Terzo Valico dei Giovi: Cross-section of railway tunnels.

2 THE PROJECT

2.1 Characteristic of the new railway line

Design standards provide for a maximum speed of 250 km/h on the main line, 100–160 km/h for interconnections, a maximum gradient of 12.5 ‰, a DC power supply of 3 kV but with infrastructure that provides for 2 x 25 kV AC, and a Type 2 ERTMS signalling system.

A safety stop equipped for the evacuation of train passengers in the event of an accident or a significant failure is planned for inside the Valico Tunnel at the Vallemme adit tunnel.

The system involves the juxtaposition of the two railway tunnels with two other pedestrian tunnels for the evacuation of passengers; the tunnels are 750 m long and are linked together via a "transect" that passes over both tracks, reaching the Vallemme adit, which serves both as the emergency exit and as the emergency vehicle access point (Figure 3).

This overpass, along with a 15 + 15 by-pass, connects the two platforms with the two evacuation tunnels and affords the passengers of a damaged train safe passage to the opposite plat-form to board another train or, in extreme cases, route them to the safety exit at the Vallemme adit. The construction of a vehicular tunnel system that connects the adit tunnel with the odd track evacuation tunnel is also planned.

The railway line is crossed by means of a level passage. There are plans for a second safety area at Libarna, in the above-ground section of the line between the Valico and Serravalle Tun-nels; it will be equipped with a priority shelter track and will have the dual function of communications area and safety area.

2.2 The geological overview

The tunnel section of the "Terzo Valico" extended from Genoa to Tortona among two main geological units (Figure 4):

- from Genoa to geological contact zone with Tertiary Piedmontese Basin (TPB) (chainage 19+500), the layout in entirely within the Sestri – Voltaggio Zone (ZVS); particularly this zone is characterized by the "Argille a Palombini" Formation (aP), a sequence of argilloschists, claystones and limestone lenses; between chainage 8+500 and 12+500, rockmass is highly tectonized and squeezing because of tectonic Alpine evolution;

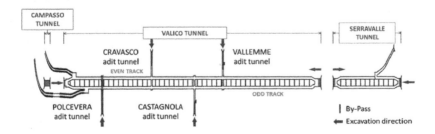

Figure 3. Terzo Valico dei Giovi: Running Tunnel layout.

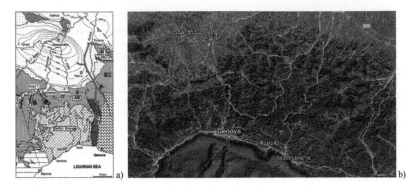

Figure 4. Terzo Valico dei Giovi: a) geo-structural setting; b) Satellite view.

– from change 19+500, tunnel stretch crosses Tertiary Piedmontese Basin Units (TPB), a sedimentary sequence constituted by conglomerates, sandstones, marls, claystones.

The ZSV represents an important tectonic area, and with the "Gruppo di Voltri" forms a complex geological context ("nodo collisionale ligure" di Laubscher at. al. 1992), interpreted as the West to East transition from Alpine rock sequences to Appennine rock sequences.

The ZSV is constituted by three different tectonic units, two of them ophiolitic (Cravasco-Voltaggio and Figogna) and the last one, Gazzo-Isoverde unit. During various stages alpine evolution, these three units experienced different temperature and pressure conditions (metamorphism) that determined the natural growth (by the original protoliths) of chrysotile, actinolite and tremolite.

The TPB represents the overlying tertiary sedimentary cover sequence of the ZVS. The Terzo Valico tunnel section crosses this important and very complicated geological, structural and lithlogical context, because of its involvement in the Alpine Evolution Phases. The tunnel stretch passes through the contact zone between The ZVS and TPB units, too. The most important lithologies along the alignment (Figure 5), from south (Genoa) to north (Milan) are listed below:

– claystone schists with limestone lenses (Palombini)-aP; this unit represents the predominant lithological unit (most than 60%);
– lenses of basalts in the claystones schists unit,
– sedimentary unit constituted by conglomerates, sandstones, marls and clays; this unit is the predominant lithological unit in the lowland area.

2.3 The Sestri Voltaggio Zone (ZVS)

The survey phase allow to identify "geomechanical groups" based on the following main factors:

– lithological criteria (petrographical and mineralogical composition, the degree of alteration, possible presence of water);
– structural criteria (characteristics of joints, RDQ index, foliation intensity, tectonisation intensity, such as the presence of folded structures including also microscale folding);
– lithomechanical criteria (with reference to the first assessment of physical properties, strength and deformability).

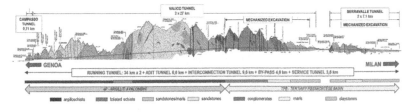

Figure 5. Terzo Valico dei Giovi: geological profile.

The evidences gathered during the excavation of exploratory tunnels, performed at the initial stage of the project, were complemented by interpretation of in situ investigations and the results of laboratory tests carried out for final design and detailed design stage. Classification of sampled material in terms of lithological composition, foliation pattern and recurring calcite veins has confirmed, also at the sample scale. The geomechanical parameters are summarized in Table 2.

2.4 The Tertiary Piedmontese Basin Units (TPB)

The Tertiary Piedmont Basin crops out in the southern part of the Piemonte region, northwestern Italy. Formed on a backvergent segment of the Alpine wedge after the meso-Alpine collisional events, the TPB and its highly deformed Alpine substratum became involved in the growth of the Apennines orogenic wedge from the Oligocene.

The section crossed by Terzo Valico is mainly constituted by conglomerates, sandstones, marls and clays. The geomechanical parameters are summarized in Table 3.

Table 2. Claystone schists.

Formation	γ [kN/m³]	υ [-]	E_{op} [GPa]	σ_c [Mpa]	m_i [-]	GSI [-]
GR1	27	0,25 - 0,3	3 ÷ 7,8	30 ÷ 40	15 ÷ 20	45 ÷ 55
GR2a	27	0,25 - 0,3	1,5 ÷ 2,0	10 ÷ 12	20 ÷ 25	40 ÷ 45
GR2b	27	0,25 - 0,3	1,0 ÷ 1,5		15 ÷ 20	35 ÷ 40
GR3a	26	0,3	1,0 ÷ 1,2	5 ÷ 7	19	30 ÷ 35
GR3b	26	0,3			19	25 ÷ 30

Table 3. Tertiary Piedmont Basin geomechanical parameters.

	Formation	γ [kN/m³]	υ [-]	UCS [MPa]	σ_t [Mpa]	m_i [-]	GSI [-]	E_{RM} [MPa]
Conglomerates of Molare	**Molare** *high cementation*						50÷55 (52)	3500÷5000 (4250)
	Molare *medium cementation*	26,00	0,25÷0,30	5÷30 (14)	1,60	18÷24 (21)	40÷50 (47)	1750÷3500 (2600)
	Fault Zone						35÷40 (37)	700÷1500 (1100)
	FMa	25,80	0,25÷0,30	10÷32 (22)	3,00	13÷21 (17)	35÷50 (45)	850÷2500 (1650)
Marls of Rigoroso	mR	25,50	0,25÷0,30	25÷45 (35)	2,50	5÷9 (7)	40÷55 (48)	1000÷3200 (1750)
	Fault Zone (mR)	25,50	0,25÷0,30	12.5÷25 (18)	2,50	5÷9 (7)	35÷40 (37)	300÷1000 (650)
	fR	25,60	0,25÷0,30	20÷40 (30)	2,70	10,75	35÷55 (43)	550÷2100 (1050)
	Fault Zone (fR)	25,60	0,25÷0,30	12÷20 (16)	2,70	10,75	30÷40 (35)	200÷1100 (580)
Costa Montada	uMa	23,50	0,30	7÷25 (14,50)	1,50	5÷9 (7)	45÷55 (50)	310÷2000 (1020)
	uMb	25,80	0,30	10÷32 (22)	3,00	13÷21 (17)	35÷50 (45)	850÷2500 (1650)
	uMc	24,50	0,30	7,5÷28 (17,50)	2,00	10,75	35÷50 (40)	300÷1200 (630)
Costa Areasa	fCa				≈ uMb			
	fC (FC1)	23,5	0,3	7÷25 (14,50)	1,5	5÷9 (7)	45÷55 (50)	310÷2000 (1020)

3 TUNNELS DESIGN

Conventional method will be used for the excavation of the Castagnola, Cravasco, Vallemme adit tunnels and for 20 km from South Entrance of Valico Tunnels while for its remaining length the excavation will be approach from the North Entrance using mechanized method. Serravalle Tunnels and Polcevera adit tunnel will be excavated by mechanized method.

3.1 Conventional Excavation Method

The survey phase allow to divide the rock mass into "geomechanical groups". Each group is characterized by a specific set of strength and deformability parameters that determine its stress-strain response to the excavation. In view of these assumptions, it became necessary at the design stage to determine a full set of section types, verified through all possible scenarios (depending on the overburden and on the variability of geomechanical parameters valid for statistically defined intervals). The main section types adopted, grouped according to the type of behaviour category, A, B and C. The 'variabilities' to be applied were designed for each section type for statistically probable conditions where, however, the precise location could not be predicted on the basis of the available data. It is essential to identify the variabilities for each tunnel section type that are admissible in relation to the actual response of the ground to excavation, which will in any case always be within the range of deformation predicted by the ADECO-RS approach. This is because it allows a high level of definition to be achieved in the design and also at the same time the flexibility needed to be able to adopt quality assurance systems during construction to advantage.

A design and contract instrument, called Guideline (Figure 6.a), has been defined for all tunnel sections (Figure 6.b), in order to manage the application of these section types. For each section type (Figure 6.c), a specific range of application was defined together with a variety of stabilization interventions assigned to it. In this way each section type, without changing the final structural characteristics of the tunnel, can fit to actual geomechanical conditions, the rock mass hydrogeological properties, tunnel face extrusion pattern and cavity deformation type.

The Guidelines, managing section type application, allowed to define the criteria that the designer can rely upon during the construction works in order to:

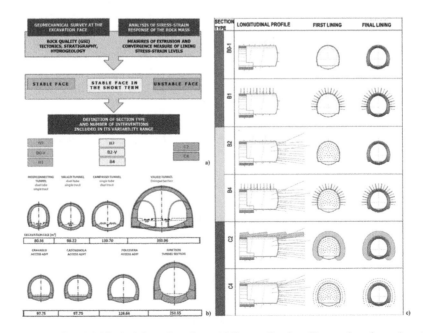

Figure 6.   Terzo Valico dei Giovi: a) flow chart for guideline application; b) tunnel sections; c) section type.

- confirm the most applicable section type, selecting it from among those already assigned to a particular tunnel stretch;
- define the section type most suitable for the actual geomechanical context, according to the designed variability;
- identify a different section type from those assigned to each particular stretch or anyway envisaged in the design for the same formation, in case actual conditions encountered during the excavation differ from those predicted.

During the design phase the following set of components were defined:

- section types each with specific interventions to control the ground deformation in the various geomechanical contexts specified at the survey phase;
- uniform sections of underground alignment, in terms of deformation response of rock mass to the excavation, together with a set of section types assigned to them; each section type is characterized by a percentage of its application defined by a special parametric analysis;
- alert and alarm threshold values set for rock mass and lining deformation parameters to be monitored during construction works by means of the instruments specified in the monitoring plan;
- three levels of variability established for the designed interventions (minimum, medium and maximum) defined for each section type, and introduction of Guidelines as a management tool operating with a complete set of different excavation sections.

3.2 *Mechanized Excavation Method*

The Serravalle Tunnel will be completed entirely with mechanized excavation, using two 9,73 m diameter EPB TBMs. The excavation of the Valico Tunnel will be carried out using both technologies: conventional excavation from the southern entrances and from the four access adits, and mechanized excavation, using two 9,77 m diameter EPB TBMs, from the northern entrances. Also the Polcevera tunnel, an adit one, was excavated by mechanized method. Table 4 summarize the characteristics of the used TBMs.

The Serravalle tunnel is a 7 km long tunnel, of which 6.3 km have a section with separate tubes, excavated by mechanized method. From the southern portal, the tunnel is in a context of transition between relief and plain (up to approximately pk 32+300). The overburden in this section vary from a minimum of 27 m (pk 30+500 approx.), up to a maximum of 130 m (pk 30+200 approx.).

Over pk 32+850, the tunnel pass through a lowland context, where the overburden gradually decrease reaching minimum value equal to 5–6 m, close to the northern portal. In the section between pk 32+850 and 33+500 the route is part of an urbanized area, included in particular a commercial area of the municipality of Serravalle. In this sector, the tunnels cross a geological context characterized by the presence of a palaeo-riverbed and a tectonized area.

The TBM - EPB has been selected for the excavation of Serravalle Tunnel considering the local condition. These machines are flexibly designed in terms of support and excavation methods. The tunnelling mode can be adapted to changing ground, requiring relatively short conversion times

Table 4. Terzo Valico Tunnel Boring Machines.

TBM CHARACTERISTICS	*Serravalle Tunnel*	*Valico Tunnel*	*Polcevera Tunnel*
Type		EPB Shield	
Excavation Diameter	9,730 [m]	9,770 [m]	9,790 [m]
Shield lenght	10,225 [m]	11,000 [m]	9,800 [m]
Power	4.000 [kW]	4.000 [kW]	4.680 [kW]
Bulkhead pressure	5,0 [bar]	5,0 [bar]	4,0 [bar]
Segmental Lining Outside ⌀	9,40 [m]	9,40 [m]	9,45 [m]
Segmental Lining Inner ⌀	8,60 [m]	8,60 [m]	8,65 [m]

This means that even tunnels with extremely varying geological and hydrogeological conditions can be constructed safely using the TBM-EPB. The following mode can be selected:

- Closed Mode with ground conditioning: excavation to be used if the tunnel face is not self-supporting (unstable) and/or when a groundwater pressure is to be balanced. The closed mode is typically known as EPB mode.
- Semi Closed Mode with compressed air: can be applied if the tunnel face is stable but if at the same time significant more water is flowing into the excavation chamber. Then, the water inflow could be prevented by application of compressed air.
- Open Mode: excavation can be applied if the tunnel face is stable and if just locally some disturbance or fissures are encountered. A groundwater pressure cannot be handled in this operation mode.

Figure 7 shows the correlation between the excavation mode and geological formation while the Table 5 summarizes the advantages and disadvantages of single excavation mode.

The north entrance of Valico Tunnel, from km 27+327,50 to km 19+700, will be characterized by dual tube single track excavated by mechanized method (from Radimeno Shaft). This section of Valico tunnels are inside the Tertiary Piedmont Basin and cross the following geological context:

- Costa Areasa Formation: flyschoid formation consisting of poorly cemented silty marl, cemented marl carbonates, poorly cemented sands and fine sandstones.
- Costa Montada Formation: arenaceous marls and medium to coarse sandstones.
- Marls of the Rigoroso Formation and the Costa Areasa Formation. Silty, clayey marl with intercalations of fine sandstone (Rigoroso Marls). Cemented marls and sandstones (Rigoroso Flysch).
- Conglomerates of the Molare Formation, consisting of polygenic conglomerates in benches and strata in an arenaceous matrix. The Molare Formation contains groundwater flow systems, making groundwater locally significant. In fact, the uppermost part of the Molare Formation is more highly cemented and contains an aquifer system which is widely used as a source of potable water. With regard to lithostratigraphic properties, excavation challenges are linked to the strong granulometric heterogeneity of the conglomerates,

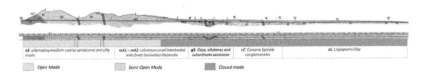

Figure 7.   Serravalle Tunnel: excavation mode related to the geological profile.

Table 5.   Advantages & Disadvantages of different excavation modes.

EXCAVATION MODE	ADVANTAGE	DISADVANTAGE
CLOSED MODE *Excavation by soil conditioning* *(foam and water)*	•Low wear of the TBM's mechanical parts •Production Increase •Good control of TBM	•The excavated material is too fluid and not shovelable •The excavation material must sediment in a settling tank
OPEN / SEMI-OPEN MODE *Excavation without* *soil conditioning* *(only water when necessary)*	•The excavated material is easy to handle •The excavated material can directly delivered to cleaning up site	•Reduction of Tunnel Production •TMB difficult driving •Very high wearing of cutterhead and screw

specifically to the occurrence of large stone blocks embedded in a poorly cemented fine matrix with poor cohesion.

Figure 8 shows the correlation between the excavation mode and geological formation.

The assessment of works impacts on aquifers is an aspect with important repercussions during tunnels design. A detailed hydrogeological studies for designing the underground works related to Terzo Valico has been carried out and presented in a previous paper (Lunardi, 2016) regarding the assessment and mitigation of the tunneling impacts on the existing aquifers.

Referring to the hydrogeological conditions the authors considering important to recall the following consideration applied to the Valico Tunnel. The tunnel excavation causes the generation of a plastic zone, with the resulting detension around the cavity, which could increase, even significantly, the degree of hydraulic interconnection inside the bored ground mass. The permeability increases in the immediate vicinity of the cavity for a distance at least equal to the radius of the plasticization band and the water distributed around the tunnel could transfers the corresponding hydraulic head to the tunnel lining. In particular, considering the waterproofing condition of the tunnel, it is possible to evaluate the evolution of hydraulic head around the cavity (Figure 9) which develops as follow:

- during the excavation phase, the initial natural hydraulic head determines a continuous flow from the saturated mass inside the tunnel (drain behavior of the tunnel and zero value hydraulic head);
- once the excavation is completed and the lining installed, the load starts to increases (timing depends on the overall permeability of the system);
- from short to long term, the hydraulic head on the tunnel is rebalanced and it will be definitively restored to a level equal to, or possibly lower than, the initial natural level, depending on the extent of irreversible disturbances induced to the entire system by the excavation;
- when the hydrogeological condition is completely rebalanced, the hydraulic head will not only act on the lining at the intersection with the individual aquifer discontinuities, but probably along a larger section of the tunnel, due to the creation of a permeable zone along the plastic band surrounding the entire tunnel, which extends parallel to the tunnel axis.

The rebalanced hydrogeological condition determines the drainage intervention.

The hydraulic head used as a reference value during the drainage intervention design was the one corresponding to the undisturbed condition. The permeability condition of Valico tunnel are summarized in Figure 10 and by the following category:

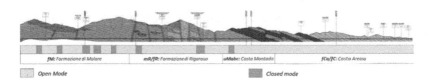

Figure 8.   Valico Tunnel: section tunnelled by EPB excavation mode related to the geological profile.

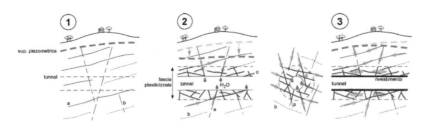

Figure 9.   Hydraulic head and water table evolution consequent to tunnel excavation: 1. Undisturbed condition; 2. Excavation phase & generation of a plastic zone; 3.Long term condition.

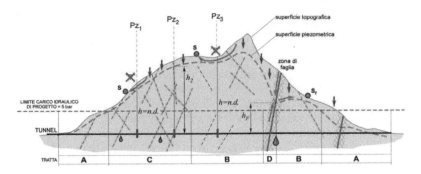

Figure 10.   Hydraulic head and water table.

A. hydraulic head < 0
B. hydraulic head > 50 m and rock mass permeability coefficient from 0 to extremely low value (typical condition for TPB's Marls);
C. hydraulic head > 50 m and rock mass permeability coefficient from low to medium value (typical condition for claystone and molare formation);
D. hydraulic head > 50 m and rock mass high permeability coefficient value.

The condition listed above required the definition of different type of intervention (summarise in Figure 11) related to the control of hydraulic head and water table level.

Nowadays Polcevera adit tunnel is completely excavated, it has been tunneled trough Argille a Palombini – aP. Figure 12 shows the geological profile of Polcevera adit tunnel. The ground mass consists of Claystone schists with limestone lenses («palombini»).

From a geological point of view, the aP is categorized as a lithostratigrafic complex composed of micaceous carbonate schists of dark grey colour, containing a very strong pervasive foliation and an abundance of intrafolial quartz and albite-bearing veins. The spacing of schistosity planes ranges from a few millimetres to several centimetres and locally the rock mass is strongly foliated. The aP contains diffused layers of very compact microcrystalline limestone, called "Palombini", characterized by massive texture ranging in thickness from centimetre to decimetre scale and interbedded with phyllite. The calcareous intercalations are heterogeneously and discontinuously distributed and, therefore, their location is not predictable. Schists may also contain lenticular basaltic bodies, often very fractured, which can occur associated with banded jaspers.

The experience gained during Polcevera Adit tunnel excavation, has been useful for the evaluation of the critical issues which affect the mechanized excavation through "Argille a

Hydraulic head	Permeability	Intervention Type		
		Drain Type	Drainage Section Step	Drains per Section
A < 50 m	-	NO INTERVENTION REQUIRED		
B > 50 m	0 ÷ low	Slotted Drain Ø 77 mm + geotextile	36 m every 20 rings	n° 2 minimum quantity
C > 50 m	low ÷ medium		27 m every 20 rings	
D > 50 m	medium ÷ high		1,8 m every rings	

Figure 11.   Intervention for the control of hydraulic head and water table level.

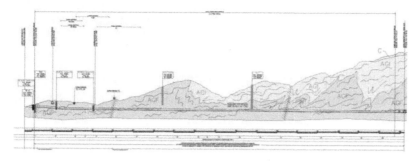

Figure 12. Polcevera adit tunnel geological profile.

Palombini" (i.e. monitoring parameter, such as: geological condition variation, tunnel convergence, TBM components wearing) and job site organization (i.e. conditioner material selection, excavated material treatment).

4 CONCLUSION

The Terzo Valico is a new high speed/ high capacity railway line that allows to strengthen the connections of the Ligurian harbour system with the main railway lines of Northern Italy and Europe. The project is part of the Rhine-Alps Corridor, one of the corridors of the strategic trans-European transport network (TEN-T core network) that connects the most densely populated and industrial regions of Europe.

The new railway line concerns a total of 53 km, 37 of which in tunnels, and involves 12 municipalities in the provinces of Genoa and Alessandria and the regions of Liguria and Piedmont. The Terzo Valico will be connected to the South, through the interconnection of Voltri and Bivio Fegino, with the railway network of the Genoa hub, for which important works of functional adaptation and enhancement are underway, as well as the port of Voltri and the historic port. To the north, from the Novi Ligure side, the route will connect with the existing Genoa-Turin lines (for traffic flows in the direction of Turin and Novara/Simplon) and to the Tortona-Piacenza line (for traffic flows in the direction of Milan-San Gottardo).

In line with the strategy of favouring eco-sustainable transport modes, the project will transfer significant amounts of freight traffic from road to rail, with advantages for the environment, safety and the economy, respecting the European prescription.

REFERENCES

Lunardi, P. 2008. *Design and Construction of Tunnels: Analysis of Controlled Deformations in Rock and Soils (ADECO-RS)*, 576. Berlin: Springer.

Pagani E., Cassani G., 2016. Terzo Valico dei Giovi: Milan-Genoa High Speed/High Capacity Line. *In: Proceedings of Swiss Tunnel Congress 2016*: 48–57. Luzern.

Capponi G., Crispini L., Cortesogno L., Gaggero L., Firpo M., Piccazzo M., Cabella R., Nosengo S., Bonci M.C., Vannucci G., Piazza M., Ramella A. & Perilli N. 2008. Note illustrative della carta geologica d'Italia alla scala 1:50.000, *foglio 213–230*. Genova.

Lunardi G., Cassani G., Bellocchio A., Pennino F., P. Perello, 2016. Studi Idrogeologici per la progettazione delle gallerie AV/AC Milano-Genova. Verifica e mitigazione degli impatti dello scavo sugli acquiferi esistenti. *Gallerie e Grandi Opere Sotterranee 117*: 17–24.

Hoek E., Brown E.T., 1980. *Underground excavations in rock*. London: Inst. Min. Metall.

Tunnels and Underground Cities: Engineering and Innovation meet Archaeology,
Architecture and Art, Volume 7: Long and deep tunnels – Peila, Viggiani & Celestino (Eds)
© 2020 Taylor & Francis Group, London, ISBN 978-0-367-46872-9

Cravasco jobsite optimization for the management of narrow spaces: Underground crushing plant and suspended conveyor belt

P. Mancuso, U. Russo, A. Fossati, O. Urbano, L. Russo & A. Filice
COCIV, Consorzio Collegamenti Integrati Veloci, Genova, Italy

ABSTRACT: The new high-speed railway line Milan-Genoa "Terzo Valico dei Giovi" includes the construction of the "Cravasco" adit, 1.260 m long, and its operational jobsite, which is aimed to serve the main tunnel lines, for a total length of 9.319 m, 4 excavation front faces and 2,5 million tons of tunnel muck. Orographic conditions, geographical position of the final disposal site and available narrow spaces of the operational jobsite (6.000 sqm) carried out an important design of the jobsite in order to overcome this condition. Due to narrow spaces have been designed and successfully realized an underground crushing plant, placed at the ending stretch of the adit within the Cravasco Cavern, and a complex system of 2 km conveyor belts capable of 800 t/h, with a relevant suspendend 260m long flyingbelt anchored to two large bases, which allow to transport and dump muck to the final disposal site.

1 PROJECT DESCRIPTION

The railway Milan–Genoa, part of the High Speed/High Capacity Italian system (Figure 1), is one of the 30 European priority projects approved by the European Union on April 29th 2014 (No. 24 "Railway axis between Lyon/Genoa – Basel – Duisburg – Rotterdam/Antwerp) as a new European project, so called "Bridge between two Seas" Genoa – Rotterdam. The new line will improve the connection from the port of Genoa with the hinterland of the Po Valley and northern Europe, with a significant increase in transport capacity, particularly cargo, to meet growing traffic demand.

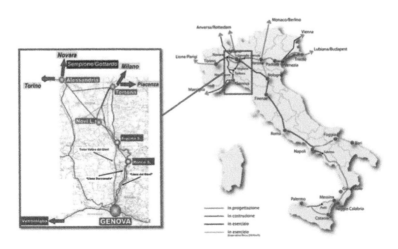

Figure 1. Italian HV/HC railway system.

The "Terzo Valico" project is 53 Km long and is challenging due to the presence of about 36 km of underground works in the complex chain of Appennini located between Piedmont and Liguria. In accordance with the most recent safety standards, the underground layout is formed by two single-track tunnels side by side with by-pass every 500 m, safer than one double-track tunnel in the remote event of an accident.

The layout crosses the provinces of Genoa and Alessandria, through the territory of 12 municipalities.

To the South, the new railway will be connected with the Genoa railway junction and the harbor basins of Voltri and the Historic Port by the "Voltri Interconnection" and the "Fegino Interconnection". To the North, in the Novi Ligure plain, the project connects existing Genoa- Turin rail line (for the traffic flows in the direction of Turin and Novara – Sempione) and Tortona – Piacenza –Milan rail line (for the traffic flows in the direction of Milan-Gotthard).

The project crosses Ligure Apennines with Valico tunnel, which is 27 km long, and exits outside in the municipality of Arquata Scrivia continuing towards the plain of Novi Ligure under passing, with the 7 km long Serravalle Tunnel, the territory of Serravalle Scrivia (Figure 2). The underground part includes Campasso tunnel, approximately 700 m long and the two "Voltri interconnection" twin tunnels, with a total length of approximately 6 km.

Valico tunnel includes four intermediate adits, both for constructive and safety reasons (Polcevera, Cravasco, Castagnola and Vallemme). After tunnel of Serravalle the main line runs outdoor in cut and cover tunnel, up to the junction to the existing line in Tortona (route to Milan); while a diverging branch line establishes the underground connection to and from Turin on the existing Genoa-Turin line.

From a construction point of view, the most significant works of the Terzo Valico are represented by the following tunnels:

- Campasso tunnel 716 m in length (single-tube double tracks)
- Voltri interconnection even tunnels 2021 m in length (single-tube single track)
- Voltri interconnection odd tunnels 3926 m in length (single-tube single track)
- Valico tunnel 27250 m in length (double tube single track)
- Serravalle tunnel 7094 m in length (double tube single track)
- Adits to the Valico tunnel 7200 m in length
- Cut and cover 2684 m in length
- Novi interconnection even tunnels 1206 m in length (single-tube single track)
- Novi interconnection odd tunnels 958 m in lenght (single-tube single track)

The project standards are: maximum speed on the main line of 250 km/h, a maximum gradient 12,5 ‰, track wheelbase 4,0 – 4,5 m, 3 kV DC power supply and a Type 2 ERTMS signalling system.

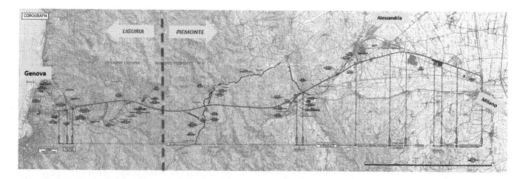

Figure 2.   Terzo Valico dei Giovi.

2 CRAVASCO BUILDING SITE

The relevant "Terzo Valico dei Giovi" underground development has entailed the realization of several adits and caverns which are directly connected to the Valico Tunnels through the up and down rail tracks. Moving from Genova to Northern Italy the railway line is linked to the following 4 access adits: Polcevera, Cravasco, Castagnola and Val Lemme. Topic of the present paper is the Cravasco building site and the management system of the tunnel muck hailing from main lines and carried to the final disposal area by means of conveyor belts.

Cravasco construction site (CSL2), is located along the left bank Verde stream in Campomorone district, near Cravasco hamlet and upstream the provincial road S.P. n°6. located in Genova territory. The jobsite area is situated within the Verde stream water catchment area, more specifically between Valletta brook and Verde stream itself (along the left bank). This area is configured as an urban margin campaign, not directly in contact with urban fabric, with rural dominance, which includes isolated nucleus, scattered and widespread buildings. The operational jobsite is articulated on a service area of reduced extensions, about 6,000 square meters, at an altitude of 293.50 m.s.l.m., and is connected to the provincial road S.P.n° 6 through a service track with a slope of about 11%.

The area, located inside a former quarry, belongs to dolomite and dolomitic limestone rock mass related to the geological formation of the "Dolomie del Monte Gazzo". In consideration of the fragile behavior of the aforementioned dolomites and of its relative geomechanical characteristics, it was necessary, prior to the construction activities, to carry out the activities of rocks loosening and securing the slope by means of protection nets and anchor bolts.

Once these operations were completed, it was possible to proceed with the site set-up stage and installation of the plants and equipment in order to support tunnel excavation activities.

Machinery and equipment such as vehicle maintenance workshop, fuel distribution site, weighbridge, offices, oxygen and acetylene storage tanks, electrical substation, electric generators, waste water treatment plant, tunnel muck temporary disposal area, manual and automated big bags packaging system, tunnel water treatment and filtration system and ventilation groups are all included within Cravasco CSL2 area.

It is important to underline that the discovery of asbestos throughout excavation of the adit led to significant changes and additions to the building site layout, generating a considerable reduction of available spaces as it was necessary to provide for the location of the plants suitable to manage the asbestos risk.

At the end of the realization of Terzo Valico, the abovementioned area will be restored to the original vegetational conditions through the green arrangement of the entire area, in compliance with the environmental context and the vocational tendencies of the area itself.

The main underground works included in Cravasco lot are the following:

a. Cravasco adit
   Total length of 1260 m, slope of 12.5% and excavation sections of the order of 95 square meters for a total amount of tunnel muck over 300,000 t. The main feature that

Figure 3.   Cravasco overall overview.

Figure 4. Cravasco Adit.

characterized this work was the excavation and management of over 30,000 t of tunnel muck containing asbestos in natural matrix.

b. Cravasco Cavern

Total length of 190 m and excavation sections of the order of 200 square meters for a total amount of tunnel muck over 100,000 t. As for the adit, even for the cavern it was necessary to restore plants for the management of asbestos.

c. Valico Tunnel

Total length of 9319 m and excavation sections of the order of 90 square meters for a total of over 2.5 million t of tunnel muck to be managed with potential presence of asbestos in natural matrix.

Currently the Valico Tunnel front face excavations (up and down rail track) along the two main directions, Genoa and Milan, is performed by traditional system.

The muck material which came from the excavation of adit and cavern, always managed in compliance with the requirements established by the law, has been transported and dumped to "Cava Castellaro – DP02" which is the final disposal area and environmental upgrading site. It is geographically located on the opposite side to the operational site at a distance of about 400 m and separated from it by a valley, Verde stream and Valletta brook.

The transport of this material to the final disposal area, which took place by road using dumpers, had an impact on the provincial road S.P.n°6 adjacent to the construction site, insisting on the existing road networks and the surrounding environment on a daily basis.

3   TUNNEL MUCK MANAGEMENT SYSTEM – CRUSHING PLANT AND CONVEYOR BELTS

In order to completely review and optimize construction site processes, a transporting tunnel muck system coming from Terzo Valico tunnels and directly connected to the final disposal area has been conceived, designed and implemented successfully.

It was therefore possible to minimize the interferences as the contemporaneity activities like dumping and all the other site activities, bypassing the road transport along the provincial road S.P.n°6, reducing times and costs of dumping at the final disposal area and therefore the environmental and acoustic impact.

The aforementioned plant system consists of the following elements:

1. underground crushing plant
2. traditional conveyor belt fixed to Cravasco adit
3. suspended conveyor "Flyingbelt"
4. traditional conveyor belt located in the final disposal area.

3.1 Underground crushing plant

The choice of locating the crushing plant in underground rather than the external site area provides an added value to the whole tunnel muck management system creating significant benefits both in terms of space optimization and in terms of reducing interference with site activities.

Furthermore, the concentration of vehicular traffic inside the tunnel means that gas/dust pollutants emissions are managed directly through the plants and equipment located in underground, thus reducing the impact in the living environment.

As can be seen from the following project excerpts reported below, dumpers proceed from the excavation front faces towards the plant, located inside the Cravasco cavern, which receives the large size tunnel muck and initiates a primary crushing process of it (grain size 0–250 mm) starting successively to load the traditional conveyor belt fixed to the adit final lining.

The plant has been designed in such a way as to minimize the risks for workers associated with the crushing process with specific reference to:

a. dispersion of asbestos fibers
b. presence of methane gas.

As regards the asbestos aspect, n°2 temporary gates has been positioned upfront the crushing plant in order to maintain a clear division of the areas within the tunnel. In this case, fully loaded dumpers coming from front faces are allowed to access to the crushing chamber and unload tunnel muck into the crusher hopper only after the first gate has completely closed.

In this way the asbestos fibers, potentially releasable into the working environment during the dumpers unloading operations, are not dispersed in the area upfront the crushing chamber, thus ensuring that this area is not contaminated. The structure of the decontamination

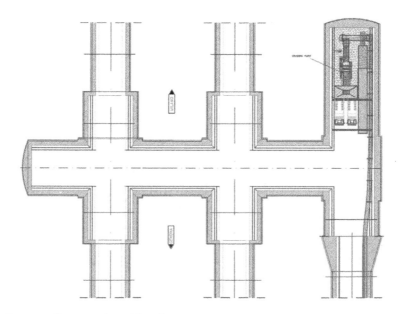

Figure 5. Cravasco Cavern and crushing plant.

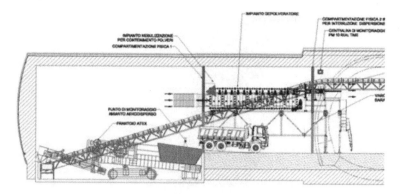

Figure 6. Underground crushing plant.

area is the following: on the top part of the crusher loading portal a series of spray nozzles in such a way as to ensure the removal of any powders, monitoring points for any airborne fibers inside the crushing chamber, aspirating ventilation system and finally an air filtration system extracted from the crushing chamber.

Regarding the risk inherent to the possible presence of methane gas discoveries during excavation operations, in consideration of the intrinsic risk that characterizes the rock mass, both the crusher system and the conveyor belt connected to it comply with the ATEX standards.

3.2 Traditional conveyor belt fixed to Cravasco adit

It has a length of about 1,400 ml and reaches the external construction area site following the same slope of the Cravasco adit, namely approximately 12.50%. The conveyor belt, directly loaded by the crushing system and fixed along the tunnel lining, allows to transport about 800 t/h of tunnel muck which come from the excavation faces, thus guaranteeing high standards of hourly production in the order to reach a maximum value of 5 m/h of front face excavation.

The tail part of the conveyor belt, belonging to the external site area, is at an altitude of +5 m respect to the walking surface in order to eliminate all possible interferences between the conveyance of the muck and the activities performed on the external area (moving machineries, movement of construction materials as steel ribs, fiberglass and so on).

The interface between the traditional belt and the next belt is regulated by a load chute which has the following characteristics:

1. passing grain size 0–250 mm
2. width of the mat 800 mm.
3. belt entering speed 3.2 m/s.
4. Inlet angle of the incoming belt, 5 ° in relation to the horizontal

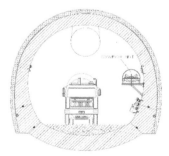

Figure 7. Traditional conveyor belt within Cravasco adit.

Figure 8. Loading chut.

5. Inclination angle of the incoming belt, in plan 30.4 ° (angle between traditional entering belt and flyingbelt at exit).

3.3 *Suspended conveyor "Flyingbelt"*

Starting from the load chute a suspended single-span conveyor "Flyingbelt" of 259 m was realized to reach the final disposal area "Cava Castellaro (DP02)" placed, as described previously on the opposite side respect to the CSL2 operational site overcoming an altitude difference of about 9m.

The main technical features that characterize the suspended "Flyingbelt" are the following:

a. Type of muck transport: unidirectional
b. Grain size: 0–250 mm
c. Belt width: 800 mm
d. Steelcord type belt resistance: 500 N/mm
e. Maximum belt speed: 3.2 m/s
f. Full capacity: 800 t/h
g. Belt motor drum power: 110 kW
h. Stopping time: 3.2 s
i. i) Safety factors (belt and carrying ropes): > 2.8 and > 6.67
j. Maximum wind pressure out of operation: 1.200 Pa

"Flyingbelt" profile view and the countryside plan are shown below.

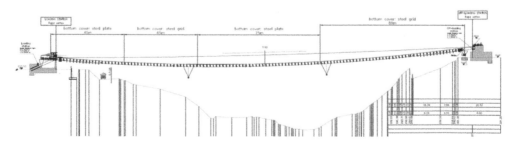

Figure 9. Flying belt profile view.

Such Flyingbelt has been designed and conceived in order to guarantee:

1. Adaptability to the layout and the morphology of the territory as the support structures of this conveyor belt are only the two supports at the initial and ending part. These supports, of a size equal to about 11m in elevation and 370 m3 of reinforced concrete, allow that external lands are total overflight.

2. Overcrossing the provincial road S.P.n°6 with a free vertical space for the transit of vehicles equal to 5.5 m in all loading conditions. Even during the preliminary assembly phases it was made that the impact on the existing road network was minimal, in fact the Flyingbelt was installed starting from "Cava Castellaro" by pushing pre-assembled frames (modular systems of about 2m) along the carrying ropes. The abovementioned system allowed the stationing of the belt above the provincial road with the minimum geometric footprint and for the shortest possible time.

3. Overcrossing of the medium voltage power line, only at the intersection at height with the suspended belt: the interfered operator has prescribed a minimum lateral margin of 2m with respect to the line support.

4. High degree of wind resistance in the transverse direction thanks to particular technical choices such as the use of Vackem belt, the configuration of the n°4 carrying ropes and the elimination of any suspended service catwalk.

5. Limited dispersion of powders into the atmosphere. This aspect is of crucial importance considering the area where the plant is located and the strategic importance of the work. A series of measures have been adopted with respect to the loading point, the line, the point of unloading and the roofing system.

   With regard to the loading point a dust removal system was provided combined to a water misting system immediately downstream of the loading point.

   As far as the line is concerned, the shape of the belt is able to satisfy the impact of the wind load on the transported material, in addition a complete tilting system has been adopted on the return sector.

   Regarding the discharge point, it has been composed by a steel plate structure with a first scraper, which cover the barrel and a Wash box equipped with double nozzles with a water flow capacity of 40 lt/s and 0.63 bars. The waste water which come from the mat cleansing activity is sent, by a pipe installed on the flyingbelt, to the waste water treatment located in the construction site.

   As regards the full coverage system, in order to restrict the material on the belt, it has been provided using welded steel mesh panels or stretched metal sheet. Furthermore, in order to avoid material which falling on delicate area such as the provincial road, Verde stream, Valletta brook or buildings, coverages in stretched sheet have been choose.

6. Minimum environmental impact in terms of noise impact and CO_2 emissions compared to dumpers transport.

Figure 10. Flying belt.

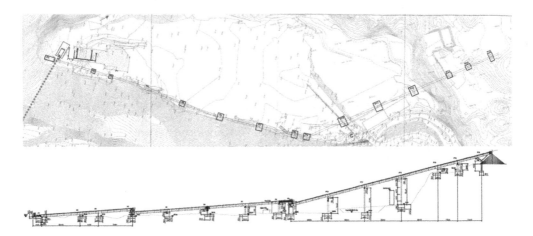

Figure 11. Traditional conveyor belt.

As follow a picture of the Flyingbelt in which is highlighted the overcrossing of the provincial road.

At the end of the construction of the Terzo Valico, the support structures will be demolished and the conveyor belt system will be completely removed, thus returning to the ante operam state.

3.4 Traditional conveyor belt located in the final disposal area

From the Flyingbelt, by additional unloading hoppers, tunnel mucks move through a sequence of traditional belts, of a length of 79mt, 143mt, 188mt, inside the internal disposal zone in Cava Castellaro. In its final part, the above mentioned system has to immediately unload the tunnel muck inside the final disposal area. Mandatory for the conveyor belt, a number of 15 variable dimension supports (the biggest one 10 mt tall) have been realized on the border area, in order to not interfere with the quarry works. Moreover, over the above mentioned foundations, support steel frameworks have been connected in order to sustain the conveyor belt. The highest one is 24mt high.

A layout plan and profile representation of the three conveyor belts follow.

Technical features such as the hourly capacity, medium material size, width, belt speed are the same of the flyingbelt. It is necessary to mention that, considering a 47mt difference in level among the starting stretch and final (within the disposal area), a growing power as 37kW, 55kW and 180kW is installed.

Regarding the environmental aspects, despite the fact that the material is completely humidified, on the transshipment zones, vaporized nozzles, in order to cut the dust down (15lt/min is the maximum vaporized water flow), transshipment point closing carters and superficial locking mechanism (movable for maintenance) have been provided.

4 CONCLUSIONS

Due to the local morphology and the narrow spaces usable as operative site, about 6.000sqm, a careful design of the external site area and the tunnel muck management, hailing from the four Valico tunnels, has been necessary.

A plant system has been designed and successfully realized. In particular, it is composed by: an underground crushing plant, characterized by a capacity of 800ton/hr, and a conveyor belts system (traditional and flyingbelt), long 2.069mt in total.

The system is able to carry all the material directly in the disposal area which is localized in the opposite zone of the valley respect the operational site.

Locating the crushing plant inside a confined area such as a tunnel and combine it with an aspirating ventilation system, prevent dust dispersions in the external living environment.

The traditional belts fixed to the final lining has been positioned in order to minimize every kind of interferences among the tunnel muck discharge activity and transit of machineries in both the tunnel and the external area.

The design and execution of a flyingbelt with a relevant span of 260mt, suspended and anchored to two support of a great dimension, provided to overcross an entire valley occupied by two water rivers and a provincial road in order to reach the final disposal site directly form the adit and avoiding to use dumpers on public roads.

Tunnels and Underground Cities: Engineering and Innovation meet Archaeology,
Architecture and Art, Volume 7: Long and deep tunnels – Peila, Viggiani & Celestino (Eds)
© 2020 Taylor & Francis Group, London, ISBN 978-0-367-46872-9

Brenner Base Tunnel, Italian Side: Mining methods stretches – design procedures

D. Marini & R. Zurlo
BBT SE, Bolzano, Italy

M. Rivoltini & E.M. Pizzarotti
Pro Iter, Milan, Italy

ABSTRACT: The Brenner Base Tunnel is mainly composed by two single track Railway Tunnels and a Service Tunnel; Lots Mules 1 and 2–3 involve a 22 km long stretch on the Italian side. They cross the South part of the mountainous dorsal between Austria and Italy, under overburdens up to 1850 m, consisting of rocks both of Southalpine and Australpine domains, separated by the major Periadriatic Fault. More than 30 km of tunnels must be carried out with Mining Methods, with average dimension ranging from 7 to 20 m and with a 200 years required service life. The paper describes the procedures implemented in the design: geological assessment; geomechanical characterization and selection of rock mass design parameters; analytic and numerical methods to forecast rock mass behaviour (varying from rock burst to squeezing); definition and sizing of rock enhancements, first phase and final linings, under a wide range of load conditions.

1 INTRODUCTION

The Brenner Base Tunnel (BBT) system mainly consists of two 55 km long parallel Railway Tunnels (at a distance varying from 40 to 70 m), and an Exploratory Tunnel. Lot Mules 1 was the continuation of the preliminary works for the construction on the Italian side; its key features were the creation of large cavities, the presence of several intersections among the cavities, and above all the Exploratory Tunnel crossing the Periadriatic Tectonic Lineament. Lot Mules 2–3 on Italian side includes 22 km of Railway Tunnels, from km 32 (Austrian border) to km 54 (close to the South access), and 17 km of Service Tunnel, from km 10 to km 27 (Figure 1a).

For most part of their length, Railway Tunnels only host one track each, but close to the adjacent lot 'Underground Crossing of the Isarco river' they become two-track and three-track. The Exploratory Tunnel is located between the two main ones, 12 m lower than the main axes (Figure 1b); only close to the South adit (from km 51+600 Eastern pipe) the Exploratory Tunnel deviates from its central position, and is no longer aligned with the Railway Tunnels up to its Aica South adit. Approximately at the Eastern tunnel chainage km 44+700 there will be a 530 m long Emergency Stop with connection tunnels for passenger evacuation and ventilation tunnels, which shall be used to manage emergency situations in case of fire. The Stop is connected to a ventilation and access tunnel that runs parallel to the line tunnel for approximately 4.5 km and subsequently inserts onto the 1720 m long Mules access tunnel, which connects the system to the outside. The tunnel system is completed by 69 cross passages (one approximately every 333 m) that provide an escape between the pipes and host technical rooms and firefighting water tanks. Vertical shafts connecting the cross tunnel with the underlying Exploratory Tunnel will be raised for plant connections and hydraulic purposes approximately every 2000 m. Close to the Mules access tunnel, there is an underground ventilation station, directly connected to the outside by means of a 60 m long vertical shaft. The intersection between Mules access tunnel

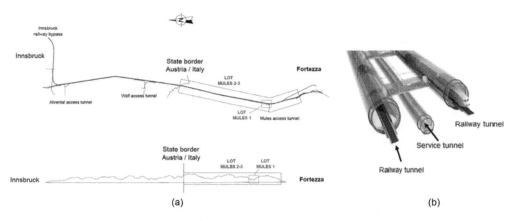

Figure 1. (a) BBT key plan with identification of Lots Mules 1 and 2–3. (b) General system configuration.

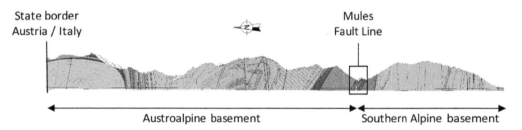

Figure 2. Geological section along tunnel.

and Exploratory Tunnel/Railway Tunnels hosts a system of tunnel and chambers which allow the underground transfer of most of construction logistics. From a geological point of view, the BBT develops across the main tectonic units forming the Alpine chain, with a maximum overburden of 1850 m. These units, which consist of several overlapping layers, are what remains of the collision between the European plate and the Adriatic (African) one; in the design area they form a dome, at the center of which it is possible to identify the Pennidic and Subpennidic units of the Tauern window, i.e. the deepest tectonic units that form the core of the Alps. Southward, the BBT crosses the fault zone that forms the Periadriatic Lineament, which separates the Austro-alpine basement from the Southern Alpine one (Figure 2). Several typical sections were designed, which vary based on both tunnel type (Railway Tunnels, Service Tunnel, Cross Tunnel, Emergency Stop and other logistic works), excavation method (in part mining methods, but mainly with TBM) and geological and geomechanical conditions.

2 GEOMECHANICAL CHARACTERIZATION

2.1 Analysis of geological and geomechanical model

The Detailed Design is based on the geological model completed with the geomechanical information drawn from surveys; these data were inferred from the critical examination of the documents developed during the previous phases of in-depth design study. An additional source of information is represented by the experience acquired during the excavation of the Service Tunnel, which anticipates the methodical excavation of other works and allows to reduce significantly the uncertainties caused by the lack of knowledge regarding rock mass that cannot be excluded during the design phase.

2.2 Definition of Geomechanical Homogeneous Stretches (GHS)

Geomechanical Homogeneous Stretches (GHS) were identified for characterizing the different crossed rock masses. For Pusteria and Mules Fault Lines the characterization stems partially from the evidence attained with the Exploratory Tunnel excavation; however, since at the time the project was drafted the Mules Fault Line had not been entirely crossed, the length of this critical area was defined based on reasonably cautious hypotheses. Generally speaking, GHS were identified providing a stratigraphic units' classification, divided into different rock mass types (basic homogeneous geomechanical unit), depending on the prevailing lithology in the case that the same stratigraphic unit contains several lithological types.

For each GHS, the following basic geomechanical parameters were defined for parameterization and used for the following design phases:

- Parameters of intact rock: natural unit weight of the intact rock γ; uniaxial matrix compression strength σ_{ci}; parameter of failure envelope m_i; modulus of deformation of intact rock E_i.
- Parameters of rock mass: Rock Mass Rating (*RMR* – Bieniawski 1989); Geological Strength Index (*GSI* – Hoek et al. 2002).

Given the extent of the planned area, for each of the GHS the parameters described above were given using the average value and the standard deviation acquired from basic documents and evaluated basing on the information taken from the geological data or obtained from the statistical analyses of data stemming from on-site surveys, from laboratory tests and from the excavation front surveys of already excavated works. Minor faults were identified based on basic geological data, in which faults are divided into a Core Zone (CZ) and a Damage Zone (DZ). The following procedure was implemented for the characterization of these sections:

1. Identification of the rock mass in which the fault falls in and choice of the characteristic geomechanical parameters of intact rock; for faults separating two rock masses, the parameters of the worst rock mass were conservatively taken.
2. Choice of rock mass index parameters to be linked with the fault: *GSI* = 30 for the DZ; *GSI* = 20 for the CZ. These are conservative values, as they reflect conditions that often don't correspond to the evidence obtained from excavations.

2.3 Geomechanical characterization

2.3.1 Peak parameters

Parameters of strength and deformability of rock masses used in the design analyses were determined according to the Hoek & Brown failure criterion (Hoek et al. 2002), expressed on the basis of maximum and minimum principal stress, uniaxial matrix compression strength and envelope curvature parameters m_b, s, a, depending on the lithological nature and on the rock mass fracturing condition and acquired starting from *GSI*, m_i and D (disturbance factor). Considering the non-linearity of failure envelopes, these have been rectified within the stress field of reference, thus obtaining friction angle and cohesion values in relation to the operating stress level.

The uniaxial rock mass compression strength, seen as the stress value corresponding to the start of the process of failures' propagation, has been evaluated putting $\sigma_3' = 0$ in the constitutive relation; for the global strength, the exceeding of which involves the rock mass collapse, the definition proposed in (Hoek et al. 2002) was taken.

Lastly, the deformation modulus of rock mass E_{rm} was derived from the matrix deformation modulus E_i, according to the indications of (Hoek & Diederichs 2006).

2.3.2 Post-peak parameters

For the post-peak strength parameters, the method described above was used choosing however a reduced *GSI* value called GSI_{res} (Cai et al. 2007):

$$GSI_{res} = GSI \cdot e^{-0.0134 \cdot GSI} \tag{1}$$

The dilation angle was defined depending on the difference between friction angles in peak and post-peak conditions, based on (Rowe 1962). Nevertheless, the increase in strength truly available was less than the theoretical one, due to phenomena of failure of rock joint asperity which were not considered in the theoretical formulation; considering also (Hoek and Brown 1997), the angle of dilation was calculated by dividing this difference by 1.5 rather than by 2.

2.3.3 Design values of rock mass parameters

In compliance with the NTC 2008 (Italian Technical Building Regulations) and with Eurocode 7, the design was developed using the characteristic values of geotechnical materials strength and deformability parameters, defined as a reasoned and precautionary estimate of the parameter value in the considered limit condition. As said, for the parameters of intact rock (σ_{ci}, m_i, E_i), each of the identified GHS encompassed the statistical variability expressed in terms of mean value and standard deviation; for these quantities, a conservative estimate of the average value was taken as a characteristic value, calculated by deducting half of the standard deviation from the average value. This approach allowed to implicitly consider the reduced geometric sizing of the sampled data in relation to the development of potential surfaces of failure. Statistical variability was available also with regard to *RMR* and *GSI*; however, in order to size the different excavation sections and assign them an implementation rate it was necessary to outline the distribution of values for homogeneous geomechanical quality classes; in the case at hand, the division into classes proposed by (Bieniawski 1989) was adopted according to the value of *RMR* index: Class I 81 – 100; Class II 61 – 80; Class III 41 – 60; Class IV 21 – 40; Class V 0 – 20. To estimate the distribution of the different classes within a GHS, each class was given a probability equal to the area under the curve of the probability density in the corresponding *RMR* interval (Figuer 3a), regardless of the existence of a class when its probability was lower than 5%.

In some cases, the area under the probability density curve within its validity range ($0 \leq RMR \leq 100$) was less than 100%; for example, the curve shown in Figure 3a has a probability of about 1% that *RMR* values higher than 100 will be reached (physically not significant). Because of this effect (caused by the fact that the data stock is not exactly normal), the design implementation rates were defined in a precautionary manner, reducing those of the highest classes and consequently increasing those of the lower classes. Once the geomechanical classes that characterize each section were identified (Figure 3b), in order to proceed with the sizing of the excavation sections it was necessary to assign an *RMR* and *GSI* reference value to each class:

- For *RMR* index, the minimum value corresponding to each class was precautionary used as a reference value.
- As far as the *GSI* index is concerned, it was not possible to create a direct link with the classes of behavior used in the design. However, as reported within the reference literature, there can be a linear link between *RMR* and *GSI*, net of hydraulic conditions and orientation of discontinuities, which are not considered by *GSI*. Assuming that within each GHS the hydraulic conditions and orientation of discontinuities are maintained approximately

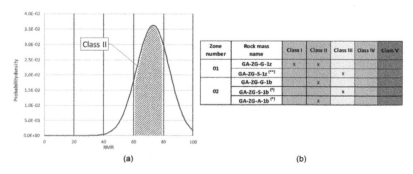

(a)   (b)

Figure 3. (a) Example of identification of the rates of classes implementation according to Bieniawski. (b) Distribution attained for some GHS.

constant, the difference between *RMR* and *GSI* in the stretch would also be constant; it was therefore decided to take the reference *GSI* values for each class starting from the difference between the average values of the two parameters (Eq. 2):

$$GSI_{rif} = RMR_{rif} + (GSI_m - RMR_m) \qquad (2)$$

where GSI_{rif} and RMR_{rif} = reference values and GSI_m and RMR_m = mean values for each stretch.

Starting from the characteristic values calculated as described, the characteristic values of rock mass parameters adopted in the sizing have been determined for each overburden using the equations of Hoek & Brown failure criterion (Hoek et al. 2002). Finally, to avoid making the analyses more demanding without achieving a real improvement in the quality of the design, for GHS with an expected length of less than 100 m it was decided to adopt a single value of *GSI* and *RMR* equal to the average one.

3 ROCK MASS BEHAVIOUR FORCAST

3.1 Empirical methods

The methods used for a quick and qualitative evaluation of squeezing or rock burst risks are briefly described below; the described analyses have been all carried out conservatively considering the maximum overburden of each GHS.

3.1.1 Risk of squeezing/front face instability

The criteria proposed in (Jehtwa and Singh 1984) and (Bhasin 1994) are based on the ratio between rock mass strength σ_{cm} and lithostatic pressure P_0 (Eq. 3 & 4), as well as on the method proposed in (Hoek & Marinos 2000), linking this ratio with the cavity deformations ε by means of the solutions of Carranza-Torres and Duncan-Fama (Eq. 5). Table 1 provides a summary of the different methods.

$$\text{Jehtwa}: \ N_c = \sigma_{cm}/P_0 \qquad (3)$$

$$\text{Bhasin}: \ N_r = 2 \cdot P_0 / \sigma_{cm} \qquad (4)$$

$$\text{Hoek}: \ \varepsilon = 0.2 \cdot (\sigma_{cm}/P_0)^{-2} \qquad (5)$$

The method described in (Panet 1995) is on the contrary based on the definitions of the critical deconfinement rate λ_e (Eq. 6) and stability factor N, identical to the N_t parameter referred to

Table 1. Squeezing risk evaluation (Jehtwa & Singh 1984; Bhasin 1994; Hoek & Marinos 2000).

Condition	Jehtwa	Bhasin	Hoek
No squeezing	$N_c > 2.0$	$N_t < 1.0$	$\varepsilon < 1.0\,\%$
Mildly squeezing	$N_c = 0.8 - 2.0$	$N_t = 1.0 - 5.0$	$\varepsilon = 1.0 - 2.5\,\%$
Moderately squeezing	$N_c = 0.4 - 0.8$		$\varepsilon = 2.5 - 5.0\,\%$
Highly squeezing	$N_c < 0.4$	$N_t > 5.0$	$\varepsilon > 5.0\,\%$

Table 2. Squeezing risk evaluation (Panet 1995).

Face behaviour		Face condition	
Elastic	$N < 2.0$	Stable	$\lambda_e = 0.6 - 1.0$
Partially plastic	$N = 2.0 - 5.0$	Stable (short term)	$\lambda_e = 0.3 - 0.6$
Plastic	$N > 5.0$	Unstable	$\lambda_e < 0.3$

in Equation 6. Table 2 shows the proposed behavioral limits and the conditions of the excavation front.

$$-\lambda_e = \frac{1}{4N}(\sqrt{m_b^2 + 8m_bN + 16s} - m_b) \tag{6}$$

3.1.2 Risk of rock burst

During the design phase, two methods were used to identify the possible risk of brittle failure (rock burst); these analyses were performed only for rock masses belonging to Bieniawski classes I and II ($RMR > 60$) and with uniaxial compression strengths over 100 MPa.

The criterion proposed in (Tao Z.Y. 1988) is based on the ratio between uniaxial compression strength of intact rock σ_{ci} and maximum principal stress in geostatic conditions σ_1 (in this case matching with the geostatic pressure P_0). The method proposed in (Hoek & Brown 1980) also links the risk of rock burst to the ratio between lithostatic pressure P_0 and uniaxial compression strength of intact rock σ_{ci}. Table 3 summarizes these two methods.

3.2 Convergence – Confinement method

The Convergence – Confinement method determines the rock mass behavior during excavation defining and analyzing five curves: 1) Radial stress vs Convergence; 2) Convergence vs Distance from the Front; 3) Radial stress vs Extension of the Plastic Zone; 4) Distance from the Front vs Extension of the Plastic Zone; 5) Distance from the Front vs Excavation Fictitious Forces. In the case at hand, curves 1) and 3) were drawn considering a Mohr-Coulomb elasto-plastic constitutive model type with softening and non-associated flow rule, as proposed in (Ribacchi & Riccioni 1977); on the contrary, curve 2) was obtained through a simplified analytical procedure using the ratio proposed in (Nguyen-Minh & Guo 1996), while curves 4) and 5) were derived from the other three by extrapolation. Table 4 shows the approach for estimating the behavior of the excavation front in the light of the results of the analyses obtained with the Convergence - Confinement method (Gamble 1971 & Sakurai 1997).

3.3 Results

Figures 4 and 5 show examples of the results achieved respectively with the empirical methods and the Convergence - Confinement method. The provided assessments, even if considered as preliminary values of indication and too conservative, were used for an initial definition of the geotechnical risk for each GHS (Figure 6).

Table 3. Rock burst risk evaluation (Tao Z.Y. 1988; Hoek & Brown 1980).

Condition	Tao Z.Y.	Hoek & Brown
No rock bursting	$\sigma_{ci}/\sigma_1 > 13.5$	$P_0/\sigma_{ci} = 0.1$
Low rock bursting activity	$\sigma_{ci}/\sigma_1 = 5.5 - 13.5$	$P_0/\sigma_{ci} = 0.2$
Moderate rock bursting activity	$\sigma_{ci}/\sigma_1 = 2.5 - 5.5$	$P_0/\sigma_{ci} = 0.3 - 0.4$
High rock bursting activity	$\sigma_{ci}/\sigma_1 < 2.5$	$P_0/\sigma_{ci} = 0.5$

Table 4. Criteria of stability derived from (Gamble 1971 & Sakurai 1997). c_f = convergence at the front; F_{pf} = extension of the plastic zone at the front; R = equivalent excavation radius

Face Condition	c_f/R	F_{pf}/R
Stable	$< 1\%$	$<< 1.0$
Stable – Short term	$1 - 2\%$	< 1.0
Tendency to instability	$2 - 3\%$	≥ 1.0
Unstable	$> 3\%$	$>> 1.0$

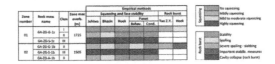

Figure 4. Example of the results of the empirical method.

Figure 5. Example of the results of the Convergence – Confinement method.

Figure 6. Example of the evaluation of the geotechnical risk level.

4 DEFINITION OF EXCAVATION TYPICAL SECTION

The analysis of geological, hydrogeological and geomechanical data, the estimates of geomechanical behavior obtained from the previously described methods and the experience gained during the previous excavations led the definition of a geotechnical risk matrix for each rock mass (Figure 6). In general, the expected behaviors are:

- Rock masses with high resistance and reduced fracturing: almost elastic behavior, with very modest convergences and equally modest plastic areas around the excavation. Rock bursts are possible for overburden of over 1000 m.
- Rock masses with medium-low resistance and moderate fracturing: medium elastoplastic behavior; for the poorest materials probable significant convergences linked with important thicknesses of plasticized zones around the excavation. In the presence of anhydrites and chalk, possible swelling behavior.
- Fault zones (with the exception of the Mules Fault Line): possible elasto-plastic behavior; significant convergences may occur linked with significant thicknesses of plastic zones around the excavation.
- Mules Fault Line: presence of significant rock mass deformation, with a strongly anisotropic rock mass behavior; a ratio of 1:10 between the maximum radial convergences, in the order of cm, and the maximum extrusions, in the order of dm, was indeed observed in the previous excavations (Fuoco et al., 2016).

The Excavation Sections were defined basing on the behaviors described above (Table 5), which application is essentially linked to three parameters: rock mass quality; overburden; lithology. Despite the in-depth examinations carried out, some significant variables on the responses of the rock mass to excavation can still exist during excavation. Therefore, a regular monitoring of the works was planned with the scope of adapting the support and consolidation actions (and the hypotheses that led to their definition), thus obtaining a more efficient application of the excavation sections during advancement. For this purpose, a number of key performance indicators, or KPI, were identified in the project to which the threshold values for the passage from one excavation section to the other were linked (Table 6). In particular, Attention Threshold was defined as the value of any of the indicators, at which all the means and materials must be set up to allow a timely passage to an excavation section heavier than the one currently applied according to the sequence T2/TRb, T3, T4, T5, T6. On the contrary,

Table 5. Support measures for mining methods of typical sections.

Typical section	Application	Support measures
T2	$RMR > 60$, Overburden < 1000 m	Radial bolting (90°); rigid 1st stage lining (shotcrete).
TRb	$RMR > 60$, Overburden > 1000 m	Radial bolting (210°); rigid 1st stage lining (shotcrete).
T3	$RMR = 41 - 60$	Radial bolting (120°); rigid 1st stage lining (shotcrete).
T4	$RMR < 41$ in Mules Fault Line (better lithologies)	Eventual ahead bolting on face/cavity; rigid 1st stage lining (shotcrete, steel ribs).
T5	$RMR < 41$ in Mules Fault Line (medium lithologies)	Ahead bolting on face/cavity; radial bolting (270°); rigid 1st stage lining (shotcrete, steel ribs).
T6	$RMR < 41$ in Mules Fault Line (worse lithologies) or swelling rock masses	Ahead bolting on face/cavity; radial bolting (360°); deformable 1st stage lining (shotcrete, yielding steel ribs).

Table 6. KPI and thresholds.

KPI	Attention threshold	Alarm threshold
RMR	See Table 5 for sections application fields	
Convergence	1 % of excavation radius	2 % of excavation radius
Extrusion *	1 % of excavation radius **	2 % of excavation radius **
Convergence rate ***	1.5 cm/m	2.0 cm/m
Stress/strain monitoring in steel ribs *	77% of material design strength	100% of material design strength
Acoustic emission monitoring ****	Acoustic emission not related to incipient fracture phenomena.	Acoustic emission related to incipient fracture phenomena.

* Only relevant for heavier sections (type T4–T6).
** Due to rock mass behaviour in previous excavations in Mules Fault Line (Fuoco et al. 2016), in this area extrusion attention and alarm threshold were set to 2 % and 4 % of excavation radius, respectively.
*** Only relevant for sec. T6; rate evaluated in a stretch of excavation of length equal to cavity diameter.
**** Only relevant for lighter sections (type T2, T3, TRb).

Alarm Threshold was defined as the value of any of the indicators, at which the Excavation Section heavier than the one currently applied must immediately be adopted according to the same sequence.

The data obtained from the Exploratory Tunnel crossing the Mules Fault Line have been of great importance for the Detailed Design. Through a careful monitoring activity, it was indeed possible to observe the conditions of the crossed rock mass as well as of the strains in lining. These data, acquired systematically during the advancement, allowed inter alia to evaluate the behavioral evolution over time also with regard to the excavations of other works (for example, in the fault stretches it was regularly registered an increase of the Exploratory Tunnel lining strains in the order of 30% at the passage of Railway Tunnels – Figure 7).

5 SIZING OF FIRST STAGE AND FINAL LINING

Following the classification of the crossed rock masses and the definition of the excavation sections, the sizing of linings was carried out using numerical methods. In this phase, particular relevance was given to the recommendation contained in the DD basic data, providing a service life of 200 years; the same data establish that the first-phase lining must not be considered in the sizing of the final lining, which then becomes alternative to the said first-stage lining. As said, for rock masses with high strength and reduced fracturing (Class I - II) the behavior is almost elastic; in these conditions, the loads on the linings were determined by analyzing potentially unstable blocks (Figure 8a). For rock masses characterized by lower

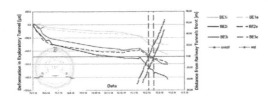

Figure 7.   Deformations measured in the lining of the Service Tunnel; noteworthy is the increase upon passing of the Railway Tunnels (vertical dotted lines).

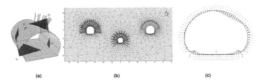

Figure 8.   (a) Analysis of blocks stability. (b) FEM (continuum). (c) FEM (hyperstatic reactions).

resistances and/or greater fracturing (Class IV - V), an FDM numerical modeling in plane strain was used to evaluate the interaction between rock mass and lining. These analyses were developed in order to monitor the load history and the stress-strain behavior of the rock mass during works implementation. The rock mass was modeled as a perfectly elasto-plastic continuous mass with Mohr-Coulomb failure criterion with softening and non-associated flow rule. In the intermediate situations (Class III) both analyses (blocks stability and FDM) have been carried out considering the most severe results for the sizing of the linings. In particular, in order to take into account the presence of rock mass enhancements (tunnel boundary and core nailing) analyses were carried out using analytical methods (Convergence - Confinement method) and FEM numerical analyses, in which the rock mass was considered as continuous and isotropic, with an ideal elasto-plastic behavior and Mohr Coulomb failure criterion (Figure 8b). In these analyses the rock mass enhancements at the front were schematized by an increase in core cohesion depending on the pressure conveyed to the excavation front. On the contrary, the enhancements on the contour were simulated with an increase in the rock mass mechanical properties at the cavity contour reckoned by standardizing the properties of each material proportionally to the area.

Finally, specific FEM analyses were carried out to determine the stresses in the final lining (Figure 8c), modeling the structure by means of beam elements on which the following loads were applied: loads caused by rock mass (obtained as described above), railway operations, differences in temperature, concrete viscosity and shrinkage, earthquakes, aerodynamic pressure due to train transit, impact and fire. The soil-structure interaction was simulated with linking elements at the model nodes that, when compressed, are able to send to the structure a reaction equal to the soil-structure contact pressure. The stiffness of the links was determined according to the characteristics of rock mass and waterproofing layer.

During the execution of the second part of the Lot Mules 1, which involved some Railway Tunnels sections in stretches already explored by the Service Tunnel, the provision of a back-analysis during works implementation has favored to enhance the design. The said back-analysis was based on the examination and interpretation of the Exploratory Tunnel monitoring data, which led to an update of the geomechanical-geological model and to the identification of the real geomechanical parameters. The Excavation Sections (in particular T4 and T5) were then optimized and sized comparing the strains expected in the design phase with those really measured; finally, the stretches for the application of the sections were redefined. For the SLS (Serviceability Limit State) and ULS (Ultimate Limit State) structural checks, the combinations of actions were considered in compliance with the requirements of NTC 2008. However, in order to contemplate a service life of 200 years, the partial coefficients on strength were increased in agreement with the following criteria:

- Concrete (reinforced or not), coefficient on concrete strength: standard conditions $\gamma_c = 1.60$ (instead of 1.50); exceptional conditions $\gamma_c = 1.20$ (rather than 1.00).
- Reinforced concrete, coefficient on steel strength: standard conditions $\gamma_s = 1.20$ (instead of 1.15); exceptional conditions $\gamma_c = 1.00$ (as per NTC 2008).
- Unreinforced concrete, reduction coefficient of strength for long duration: $\alpha_{cc} = 0.80$ (instead of 0.85).

6 CONCLUSIONS

The construction of a long railway tunnel provides the designer with a considerable amount of data for the design implementation and improvement. In particular, in this article we describe a procedure for managing the statistical variability of the basic geomechanical data of Lot Mules 2–3, the longest on the Italian side, with the aim of optimizing the design. This essay also describes the back-analysis carried out in Lot Mules 1 based on the Exploratory Tunnel excavation results, which allowed to further improve the design and approach the actual conditions.

REFERENCES

Bhasin R. 1994. Criteri rapidi ed economici per la previsione dei problemi di stabilità nelle gallerie costruite in argilla, roccia tenera e roccia dura. *Gallerie e grandi opere sotterranee.*
Bieniawski Z.T. 1989. *Engineering rock mass classifications.* New York: Wiley.
Cai M., Kaiser P.K., Tasaka Y., Minami M. 2007. Determination of residual strength parameters of jointed rock masses using the GSI system. *Int. Journ. of Rock Mechanics & Mining Sciences* (44): 247–256.
Correa, R., Merlini, D., Moja M., Pizzarotti E.M., Voza A. 2016. Scavo di gallerie con TBM scudate: rivestimenti in anelli prefabbricati in c.a. *Proceedings del Congresso "Evoluzione e sostenibilità delle strutture in calcestruzzo.* Roma.
Fuoco, S., Zurlo, R., Marini D., Pigorini A. 2016. Tunnel excavation solution in highly tectonized zones. *Proceedings of World Tunnelling Congress 2016.* San Francisco.
Gamble J.C. 1971. *Durability-plasticity classification of shales.* Ph. D. Thesis, Univ. of Illinois. USA.
Hoek E. & Brown E.T. 1980. *Undergr. Excav. in Rock.* London: Institution of Mining and Metallurgy.
Hoek E. & Brown E.T. 1997. Practical estimates of rock mass strength. *International Journal of Rock Mechanics & Mining Sciences* (34): 1165–1186.
Hoek E., Carranza Torres C., Corkum B. 2002. Hoek-Brown failure criterion – 2002 Edition. *Proceedings of NARMS-TAC Conference.* Toronto.
Hoek E. & Diederichs M.S. 2006. Empirical estimation of rock mass modulus. *International Journal of Rock Mechanics & Mining Sciences* (43): 203–215.
Hoek E. & Marinos P. 2000. Predicting squeezing. *Tunnels & Tunnelling International* (Nov-Dic).
Jehtwa J.L. & Singh B. 1984. Estimation of ultimate rock pressure for tunnel linings under squeezing rock conditions. *Proceedings of the ISRM Symposium.* Cambridge (UK).
Moja M. & Pizzarotti E.M. 2017. BBT lato Italia – Lotto Mules 2–3: le scelte di progetto dei rivestimenti definitivi. *Proc. of SIG Symposium on Great Alpine Railways Infrastructures under construction.*
Nguyen-Minh D. & Guo C. 1996. Recent progress in convergence confinement method. *Proceedings of ISRM International Symposium - EUROCK 96.* Torino.
Panet M. 1995. *Calcul des tunnels par la méthode convergence-confinement.* Paris: Presses de l'école nationale des ponts et chausses.
Ribacchi R. & Riccioni R. 1977. Stato di sforzo e di deformazione intorno ad una galleria circolare. *Gallerie e grandi opere sotterranee.*
Rowe P.W. 1962. The stress-dilatancy relation for static equilibrium of an assembly of particles in contact. *Proceedings of the Royal Society A* (269 1339): 500–527.
Sakurai S. 1997. Lessons Learned from Field Measurements in Tunnelling. *Tunnelling and Underground Space Technology.*
Tao Z.Y. 1988. Support design of tunnels subjected to rockbursting. *Proceedings of the Symposium on Rock Mechanics and Power Plants.* Madrid.

*Tunnels and Underground Cities: Engineering and Innovation meet Archaeology,
Architecture and Art, Volume 7: Long and deep tunnels – Peila, Viggiani & Celestino (Eds)*
© 2020 Taylor & Francis Group, London, ISBN 978-0-367-46872-9

Lessons from adverse geological occurrences during excavation in Rohtang tunnel, India

P. Mehra
Border Roads Organization, New Delhi, India

ABSTRACT: Rohtang Tunnel is being constructed in the Pir Panjal ranges of Himachal
Pradesh, India to provide all weather access between Manali and Lahaul on Manali-Sarchu-
Leh road which remains closed for six months in a year. Total length of Rohtang Tunnel is
8.802 Km. Drill & Blast technique for excavation is being used for the construction of Rohtang Tunnel. The tunnel is a very deep with a maximum overburden close to 2 kms. Rohtang
tunnel project is located within 'Central crystalline' litho-tectonic group of Himalayas. Main
rock types along the alignment are Phyllites, Quartzites, Mica schist, Migmatite and Gneiss.
During the excavation, the Tunnel has faced many geological challenges owing to the high
overburden and peculiar Himalayan geology. The project has been delayed by almost four
years after encountering a massive fault zone of almost 590 m which led to many muck flows
coupled with heavy ingress of water. These extraordinary Tunnelling conditions besides being
technically complex also posed many contractual challenges as the FIDIC contract agreement
in force could not provide any robust resolutions to payment related disputes. This paper
attempts to analyze the geological occurrences, reasons for rock mass failure, and recommend
necessary measures to ensure smooth handling of both technical and contractual issues.

1 INTRODUCTION

The government of India has planned to develop, the road from Manali to Leh connecting the
states of Himachal Pradesh to Jammu and Kashmir, into an all weather highway. In order to
achieve this all-weather connectivity a series of tunnels are required to be constructed across
the mountain passes. Rohtang Pass is the first pass on this axis from the Manali side and
hence a tunnel across Rohtang pass has been taken up. The Tunnel is being constructed in the
Pir-Panjal ranges of Himalayas as Manali- Leh road remains closed for six months in a year
due to Rohtang Pass being completely snow bound between Nov and Apr. It will connect
Manali to Lahaul & Spiti Valley throughout the year and will reduce the road length from
Manali to Leh by 46 km. Figure 1 depicts the alignment of the Rohtang Tunnel juxtaposed
with the existing Manali-Leh highway till Lahaul.

Rohtang tunnel project is located within 'Central crystalline' litho-tectonic group of Himalayas. The regional geological succession at the project site comprises the Tandi formation,
Batal formation, Salkhala group and the Rohtang Gneiss complex. The Rohtang tunnel alignment is mainly through Salkhala group (Precambrian). Main rock types along the alignment
are Phyllites, Quartzites, Mica schist, Migmatite and Gneiss. Major Geological structures in
the area are Seri Nala fault, Chandra-Kothi structure, Rohtang Ridge structure, Dhundi
structure, Palchan structure, Palchan fault, Sundar Nagar fault and Main central thrust fault.
The longitudinal cross-section of the total length of tunnel indicating the main fault structures
with indicative overburden heights is produced in Figure 2.

The South Portal of Rohtang Tunnel is located at a distance of 25 Km from Manali at an
altitude of 3060m, while the North Portal of the tunnel is located near village Teling, Sissu, in
Lahaul Valley at an altitude of 3071m. Total length of Rohtang Tunnel is 8.802 Km. Tunnel is

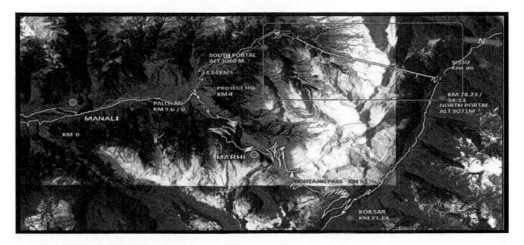

Figure 1. Alignment of Rohtang Tunnel.

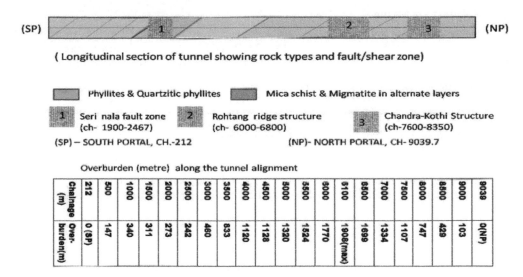

(SP) .... 1 .... 2 .... 3 .... (NP)

(Longitudinal section of tunnel showing rock types and fault/shear zone)

| | Phyllites & Quartzitic phyllites | | Mica schist & Migmatite in alternate layers |

| 1 | Seri nala fault zone (ch- 1900-2467) | 2 | Rohtang ridge structure (ch- 6000-6800) | 3 | Chandra-Kothi Structure (ch-7600-8350) |

(SP) – SOUTH PORTAL, CH.-212   (NP)- NORTH PORTAL, CH- 9039.7

Overburden (metre) along the tunnel alignment

Chainage (m)	212	500	1000	1500	2000	2500	3000	3500	4000	4500	5000	5500	6000	6100	6500	7000	7500	8000	8500	9000	9039
Over-burden(m)	0 (SP)	147	340	311	273	242	460	833	1120	1128	1320	1524	1770	1908(max)	1699	1334	1107	747	429	103	0(NP)

Figure 2. Overburden and Fault Profile of the Tunnel Alignment.

horse shoe shaped, single tube bi lane with 8.0m roadway. The representative cross section is produced at Figure 3. Drill & Blast technique for excavation coupled with New Austrian Tunnelling Method (NATM) philosophy is being used for the construction of Rohtang Tunnel. On completion, this would be one of the longest road tunnel in the world at an altitude above 3000m.

As per the final design submitted by designer the tunnel alignment consists of seven rock types based on NGI "Q" system and overburden depths. Each rock type has a unique support system including a combination of shotcrete, rock bolts and lattice girders. Special pre supports in the form of Fore-poling and Pipe roof Umbrellas have also been proposed for weaker rock mass. The supports were foreseen to be calibrated as per monitoring inputs through a comprehensive monitoring program based on 3D deformation bi reflex targets.

Post a protracted phase of mobilization owing to requirement of specialized imported machinery, first blast at South Portal of the Tunnel was taken on 29 Aug 2010 and the same at North Portal was achieved on 4th Oct 2010. The date of completion of civil works was assigned as 01 Feb 2015 based on anticipated rock type and contractor's method statement.

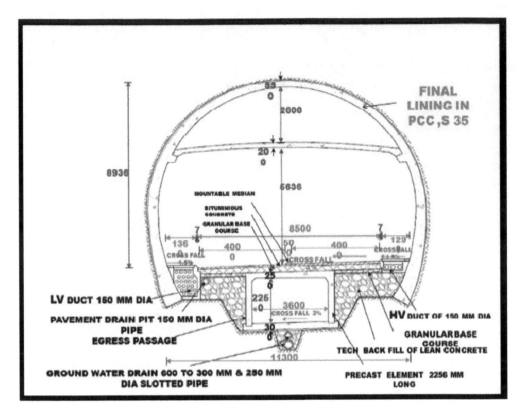

Figure 3.   Representative Cross sectional of Rohtang Tunnel.

The excavation from south portal continues through the year and but the North portal is inaccessible for any work between mid Nov to mid of May owing to closure of the pass.

2   SERI NALA FAULT ZONE (SNFZ)

After commencement of the work in the second half of 2010, the work from South Portal continued smoothly till Dec 2011. In Dec 2011 from chainage 1905 onward quality of rock started deteriorating leading to a retarded advance rate. Subsequently, at chainage 1950 the condition at the face became devoid of any rock and muck along with water started flowing from the face inside the tunnel due to the weathering caused by Seri Nala under which the tunnel was passing at that time. Heavy inflow of water at a rate of 60 – 70 litres per seconds was experienced. This stratum was much worse than the worst rock class defined in the design statement and hence the contract.

As per preliminary studies the Seri Nala fault was likely to be encountered between Chainage 2200 m to 2800 m. However, the zone was encountered 300 m prior and the geological state was found to be much worse than predicted in DPR. At the beginning around the chainage of 1905 m, the Seri Nala fault zone comprised of a mix Rock mass of Quartzite Schist on right hand side and faulted sediments in the form of Angular to Sub Angular Boulder gravels with pebbles of different rock types embedded in Silty-Sandy-Clayey material, on the left hand side. However, by the time heading excavation progressed to chainage 1949 m, loose strata along with flowing matter was encountered along the full face of the tunnel. The alignment of Seri Nala juxtaposed with tunnel is shown in Figure 4.

The following years proved to be extremely challenging as the tunnel experienced numerous muck and debris flows which included 20 major face flows and numerous cavity formations.

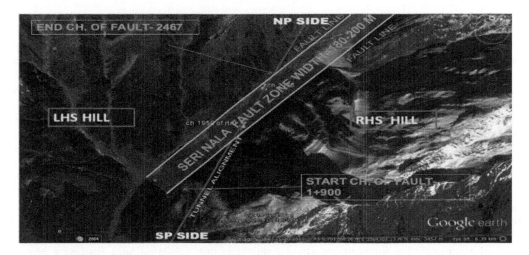

Figure 4.   Alignment of Seri Nala Fault Zone.

The tunnel experienced severe squeezing with radial convergences reaching almost a meter at certain instances. The Contractor, Engineer, Designer and Employer struggled hard to converge on to workable solutions to tackle the Seri Nala fault zone. In order to progress the excavation various techniques were adopted including improving the ground with Poly-Urethane grouting, attempts at Pilot Tunnels, partial excavation using side drifts but the deformations continued unabated and the pace of advance dropped to dismal levels of less than a five metres a week. Photographs showing cavity formations are produced as Figure 5 below.

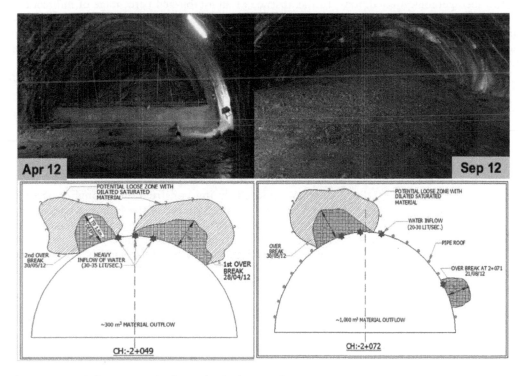

Figure 5.   Muck flow and cavity formation in the tunnel.

Various national level expert agencies visited the tunnel but, none of the agencies were able to give a viable proposal for further excavation in the tunnel owing to the extremely dynamic and progressively poor geological conditions being experienced. In fact, at one instance it was being considered to realign the Tunnel when a team led by BRO officials undertook a foot reconnaissance by trekking on top of the mountain, electronically mapped the Seri Nala and established that such a diversion would be counterproductive. This was proven to be a wise decision later when horizontal drilling confirmed the findings.

In the meantime, the site team kept scientifically graduating and calibrating the measures being taken. Finally, after detailed deliberations a workable methodology was evolved using partial excavation under a heavy pipe roof umbrella. This was further reinforced by carefully draining the water from the face and skilfully diverting the water away from the excavation area by using horizontally driven perforated pipes for drainage and intricate grouting mechanisms. Heavy pipe roof umbrellas using 114 mm diameter pipes instead of the initially designed 76 mm diameter pipes were used through a contractual variation. Excavation sequence was amended with reduced round lengths, 1m x 1m excavation panels with face support and a support core maintained behind face. Temporary Inverts were also used to close the support ring and obviate stress concentrations. Observational approach proved extremely effective in handling such poor geological conditions. The measures were calibrated based on observations.

Finally, on 3rd Jan 2016 at the chainage of 2462 m from south portal the heading excavation of Seri Nala fault zone was successfully tackled after having erected a total of 40 grouted pipe-roof umbrellas of 12–15 m length, most of which used 114 mm diameter pipes. This encompassed a number of variations from contract provisions and consumed more than 1200 days as compared to 90 days envisaged in the contract. This delayed the project by almost four years.

The challenge was not only restricted to geology related issues faced but extended to contractual complications as well. The contract was based on FIDIC red book and did not cater for any pragmatic mechanism to handle variations and effect of changed advance rate due to unforeseen geological conditions. In the absence of an established percentage of indirect and overhead costs the client and contractor struggled to establish payment rates for new items and compensate for the Time related costs due to prolonged project duration. Simplistic advance rates establish for foreseen ground conditions could not help establish contractual time for the unforeseen conditions.

Though through provisional payments the work has been progressed but many contractual disputes remain live even when the project is nearing physical completion.

3   LESSONS

During the excavation of the tunnel in the Seri Nala Fault Zone many lessons have emerged. Technically, it has been realized that a fault zone needs to analysed from a geomorphological angle, mapped in detail, further explored during construction through horizontal drilling and needs to be approached keeping in mind the science of stress development along a deep opening. Excavation and support sequence needs to be managed in a manner that prevents muck flows from the face and ensures a much reduced convergence around the tunnel boundary thereby ensuring least development of yielding/plasticity around the Tunnel openings.

A few steps proven effective in handling the fault zone are summarized below: -

a) Adopting an adequately stiff pre support i.e. a pipe roof umbrella with grouted pipes. If required, reinforcement bars may be inserted inside the pipes of the pipe-roof before grouting. A photo graph is produced as Figure 6.
b) Sequential excavation with partial excavation panels in annular rings with a support core to act as a buttress.
c) Face support with rock bolts.

Figure 6.   A Double Layered Grouted Pipe-Roof with Support Core.

d) Provision of temporary inverts.
e) Handling water through perforated pipes inserted before excavation with the pre-support pipe roof umbrella.
f) Close and frequent deformation monitoring to keep tunnel convergence in check.

Contractually it is extremely important that the indirect and overhead costs included in tendered BoQ rates and Time related cost of the project are made part of the tender itself. Underground sites especially in Himalayas will invariably encounter unpredicted ground behaviour which solicits new items not included in contract to be introduced to the site. Having established indirect and overhead costs beforehand will aid in arriving at a mutually agreeable price for such items. Also with an established TRC the contractor can be compensated for extensions of the contract period in absence of a contracted advance rate. Such provisions will help the client and the contractors to overcome extraordinary geological occurrences without disputes.

In this particular case of Rohtang tunnel the projection duration has been extended from the original period of 1915 days to more than 3500 days. The cost has also escalated due to introduction of new items and price escalation. The stake holders have arrived at a time related cost factor applicable for the period when project was going through the fault zone. The final payment modalities are still being worked out. This could have been avoided if the TRC and indirect, overhead costs were established during the tender stage itself.

4 CONCLUSION

In the particular case of Seri Nala, owing to extremely complex nature of the ground and inadequate design and contract provisions, the site executive's response could be graduated to the suitable excavation/support approach relatively slowly with excavation face having to idle several times. This lead to avoidable delays in the Projects and finally delaying the Project by more than 3.5 years. It is suggested that with this documented experience the other projects in the Himalayas can be best advised to promptly resort to partial excavation under a pre support system in case such severe squeezing is envisaged. It is pertinent that wherever such ground conditions are anticipated the necessary design and contractual provisions be catered ab initio.

The partial excavation approach based on understanding the mechanics of a deep underground openings has proven highly effective for excavation of Rohtang tunnel across one of the most challenging geology experienced by a highway Tunnel. Excavation of Rohtang tunnel across the Seri Nala has proven that NATM is capable of handling most ground conditions when coupled with a sound deformation monitoring practice and proper understanding of rock mechanics.

The contractual challenges being faced by the project could have been avoided if the indirect and overhead costs within BoQ rates and Time related cost of the project were established at the tender stage itself.

Tunnels and Underground Cities: Engineering and Innovation meet Archaeology,
Architecture and Art, Volume 7: Long and deep tunnels – Peila, Viggiani & Celestino (Eds)
© 2020 Taylor & Francis Group, London, ISBN 978-0-367-46872-9

Hydrogeological, environmental and logistical challenges for TBM excavation in the longest tunnel in the Italian territory

N. Meistro, S. Caruso, A. Mancarella, M. Ricci, L. Di Gati & A. Di Cara
Cociv (Consorzio Collegamenti Integrati Veloci), Genoa, Italy

ABSTRACT: With an overall length of approximately 27km, Valico Tunnel is the longest tunnel at present in Italy, and the main infrastructure along the new railway line "Terzo Valico dei Giovi", running from Genoa to Tortona. The first 7.7 km-long stretch in the northern side is being excavated by means of two 9.77m-diameter EPB-TBMs, crossing various geological formations at the junction between the southern part of the western Alps and western termination of the northern Apennines. Values of overburden up to 420 meters, the complexity of geology and high groundwater pressures around the lining, are some of the geotechnical challenges expected during the excavation. Furthermore, the TBM are expected to run into geological formations containing asbestos fibers and methane gas, requiring particular procedures that affects logistic of the jobsite. Such conditions make this tunnel one of the most challenging and complex project currently under construction.

1 INTRODUCTION

The railway Milan–Genoa, part of the High Speed/High Capacity Italian system (Figure 1), is one of the 30 European priority projects approved by the European Union on April 29th 2014 (No. 24 "Railway axis between Lyon/Genoa – Basel – Duisburg – Rotter-dam/Antwerp) as a new European project, so-called "Bridge between two Seas" Genoa – Rotterdam. The new line will improve the connection from the port of Genoa with the hinterland of the Po Valley and northern Europe, with a significant increase in transport capacity, particularly cargo, to meet growing traffic demand.

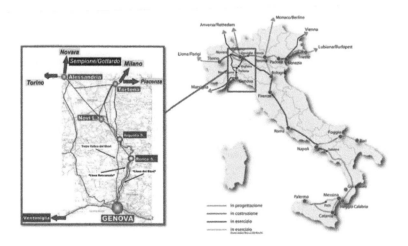

Figure 1. High-speed Italian system.

The "Terzo Valico" project is 53 Km long and is challenging due to the presence of about 36 km of underground works in the complex chain of Appennini located between Piedmont and Liguria. In accordance with the most recent safety standards, the under-ground layout is formed by two single-track tunnels side by side with by-pass every 500 m, safer than one double-track tunnel in the remote event of an accident.

The layout crosses the provinces of Genoa and Alessandria, through the territory of 12 municipalities.

To the South, the new railway will be connected with the Genoa railway junction and the harbor basins of Voltri and the Historic Port by the "Voltri Interconnection" and the "Fegino Interconnection". To the North, in the Novi Ligure plain, the project connects existing Genoa - Turin rail line (for the traffic flows in the direction of Turin and Novara – Sempione) and Tortona – Piacenza – Milan rail line (for the traffic flows in the direction of Milan - Gotthard).

The project crosses Ligure Apennines with Valico tunnel, which is 27 km long, and exits outside in the municipality of Arquata Scrivia continuing towards the plain of Novi Ligure under passing, with the 7 km long Serravalle Tunnel, the territory of Serravalle Scrivia (Figure 2). The underground part includes Campasso tunnel, approximately 700 m long and the two "Voltri interconnection" twin tunnels, with a total length of approximately 6 km.

Valico tunnel includes four intermediate adits, both for constructive and safety reasons (Polcevera, Cravasco, Castagnola and Vallemme). After tunnel of Serravalle the main line runs outdoor in cut and cover tunnel, up to the junction to the existing line in Tortona (route to Milan); while a diverging branch line establishes the underground connection to and from Turin on the existing Genoa - Turin line.

From a construction point of view, the most significant works of the Terzo Valico are represented by the following tunnels:

- Campasso tunnel 716 m in length (single-tube double tracks)
- Voltri interconnection even tunnels 2021 m in length (single-tube single track)
- Voltri interconnection odd tunnels 3926 m in length (single-tube single track)
- Valico tunnel 27250 m in length (double tube single track)
- Serravalle tunnel 7094 m in length (double tube single track)
- Adits to the Valico tunnel 7200 m in length
- Cut and cover 2684 m in length
- Novi interconnection even tunnels 1206 m in length (single-tube single track)
- Novi interconnection odd tunnels 958 m in lenght (single-tube single track)

The project standards are: maximum speed on the main line of 250 km/h, a maximum gradient 12,5 ‰, track wheelbase 4,0 – 4,5 m, 3 kV DC power supply and a Type 2 ERTMS signalling system.

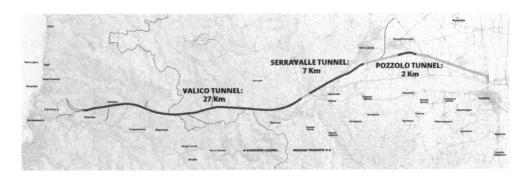

Figure 2.   Terzo Valico project.

2 TUNNELING CONTEXT

2.1 Geological features

The northern stretch of Valico tunnel, about 7.7km, crosses the sedimentary sequence of Tertiary Piedmontese Basin Unit (TPB) that crops out in the southern part of Piedmont and consisting of conglomerates, sandstones, marls and claystones. The complex TPB stratigraphic succession is the result of different depositional system modified during several deformation phases connected to growth of Alps and Appennines.

Moving from northern entrance towards Genoa, the tunnel section crosses the following geological formations:

- "Costa Areasa", consisting in alternating between silty marls and thin sandstones. The degree of fracturation occurring in this formation is very low.
- Formation of "Costa Montada" is composed by 3 main facies: silty marls, sandstones and an alternating between both. Locally conglomerates of quartz, gneiss, carbonates clasts are included in marls facies. Generally fracturing is limited to crossing zone between facies and in faults.
- "Marls of Rigoroso" is a lithotype composed by silty and clayey marls with intercalations of fine sandstones.
- "Molare" formation consisting of polygenic conglomerates in benches and is divided in 5 main litofacies: the *breccia facies* is composed by serpentinites and metaperidotites clasts in a very poor clays matrix but well cemented. The *ruditic facies* consists in conglomerates in a sandstones matrix; the clasts are metabasites, serpentinites, marble and scists from 10 to 20cm size even with the possibility to reach 100cm. The *cemented ruditic facies* is composed by conglomerates and polygenic breccias in a sandstone matrix heavy cemented; the clasts reaches 100cm size and consists in limestones, dolomites, basalts and metaophiolites. The heavy cementation is related to the carbonatic composition of this lithotype. *Silty – pelitic facies* consists in clayey marls, silty marls and fine arenites; this facies is not prevalent and it could be out from the crosses section. The *sandstone facies* is composed by fines and coarse sandstones locally fossiliferous.

In terms of geomechanics characteristics these formations are characterized by a good behavior. The GSI classification is between 45 and 55 on rock mass. Locally the section crosses few fault zones or transition zones between lithologies; in such areas the mechanics behavior decreases and the GSI is between 35 and 45. The average overburden in the first 4 km of the tunnel section is around 100m; then the coverage increases reaching 420m under Molare formation.

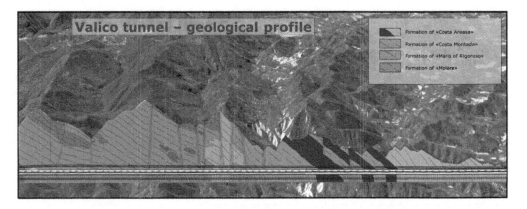

Figure 3. Geological profile.

2.2 Hydrogeological model

Rock mass within this area can be divided in various hydrogeological complex with different permeability. A large number of investigations have shown a very low conductivity: the main circulation is related to the presence of joint and fractures. On a large scale, two different contexts may be distinguished. The first context is predominant: permeability is related to the presence of a joint grid with low persistence or small fault. In tunnel scale the average permeability is very low because the limited connection of the joint. The second context is associated to the presence of main fault zones, relevant at the tunnel scale. In this case the average permeability could be much higher, because of large number of joint connections.

From the investigations that have been carried out, the cementation degree for the formation at issue is very high even in fault zones. Conductivity is expected between 10^{-7} and 10^{-8} m/s, reaching 10^{-6} m/s in fault zones.

The presence of carbonatic clasts in the cemented ruditic facies of Molare formation makes chemical dissolution possible. This occurrence may cause micro-karsism along fractures (around a meter sized) associated to increasing of conductivity.

The upper part of Molare formation contains an important aquifer system which is widely used as source of potable water. In order to avoid sterilization of this important reserves, specific studies have been carried out, by making use of 3D groundwater numerical models; these models allowed to determine possible connections between water basin and tunnel construction.

Groundwater level in the high-overburden stretch reaches 300-350 m, although the hydraulic pressure expected could be lower due the low hydraulic conductibility. During tunnel excavation, a plasticized zone with high permeability is expected around the bored area; in this zone, groundwater pressure could be considered as low as zero in short term conditions. In the middle-long term the groundwater pressure around the lining tend to restore to the initial condition.

2.3 Environmental issues

Geological complexity in Valico tunneling is not limited only to geomechanical and hydrogeological context. Because of his natural lythological growth, some kind of asbestos minerals may be found along the tunnel section in natural matrix.

A specific protocol for the management of Asbestos Risks, has been developed and approved by the local controlling agencies, with the purpose to define the asbestos occurrence during tunneling excavation, together with sampling and diagnostic methods for correct evaluations of airborne fibers and rock massive amount.

From a geological standpoint, a number of specific investigations have been executed in order to define a design tunnel profile that contains information on the Probability of Asbestos Minerals Occurrence (the so-called POMA classification).

The northern part of Valico tunnel has been divided in 2 main stretch:

- The first 3.4km of tunnel will be bored inside Costa Areasa, Costa Montada and Marne di Rigoroso formations; sampling test in these sedimentary rocks showed the presence of asbestos minerals although with concentration lesser than 1000 mg/kg.
- The last 4.3km of tunnel will be driven inside Molare formation; conglomerates and breccias may contains fiber clasts. Sampling analysis revealed that the asbestos concentration may be greater than 1000 mg/kg.

In addition to the asbestos issue, during the excavation is also expected the presence of limited quantity of methane gas. This gas is naturally present inside the sedimentary rock mass like those in question. On the basis of the geological and structural features and the mechanized tunneling method adopted, it was possible to define 2 main risk scenarios related to the presence of methane gas:

- poor presence or absence of methane in low pressure;
- possible and locally presence of methane gas inside faults or joints

According to these evaluations the TBMs were equipped with an advanced monitoring system designed to detect and quantify the presence of methane gas in the working areas of the machines. Finally, a series of procedures were developed to mitigate the risk and ensure the workers safety.

3 MAIN PROJECT FEATURES

3.1 Design specifications

According to design specifications, project consists of two single-track tunnel connected by by-pass to a distance of 500m each other.

Valico Tunnel has a total length of about 27 km. Starting from the construction site COP20 (Pozzo Radimero) toward Genoa, the first 7.7 km are going to be excavated by TBM, whereas the remaining 19.3 km are being excavated by conventional methods.

TBM excavation has been chosen as the best alternative for the northern stretch taking into consideration the following:

– Possibility to speed up tunnel construction time.
– General improvement in worker safety during tunnel construction.
– High complexity of geological and hydrogeological conditions.
– Probability to find asbestos minerals and methane gas inside the rock mass.

The two TBMs were named S-979 or "Paola", operating on the left tunnel, and S-1044 or "Daniela", operating on the right tunnel. Both have an excavation diameter of 9.77 meters, and were designed to excavate in Earth Pressure Balanced (EPB) mode. Face support is achieved by conditioning the excavated material using foam, polymers and water.

Project EPBs have been designed to apply a counter pressure up to 6 bars, in order to prevent risks associated to the geological and hydrogeological conditions. Final concrete lining allows to apply a maximum thrust of 128MN thanks to 19 couples of hydraulic cylinders. Revolution of the cutting head is given by 11 electric drives allowing a nominal torque of 18MNm.

Cutting head design takes into account excavation within rocks of medium resistance and consists on 47 single disc cutters, 6 double cutters in the center zone and 106 scrapers.

According to the design specifications, a support pressure is required ("close" mode) when crossing faults zones. Outside from fault zones it is possible to proceed without pressure control ("open" mode). Furthermore, the tunnel is designed to act a draining tube. Therefore, in order to reduce long term water pressure on concrete lining, dewatering has to be carried out by a certain number of spots along the tunnel.

Final lining of the tunnel makes use of the universal ring, consisting in 6 concrete segments plus the key element. Internal diameter of the ring is 8.6 m and its conicity allows a minimum curve radius of 300m. A continuous rubber gasket around each segment guarantees to seal the tunnel upon a pressure of at least 5 bars. In addition, it is possible to increase the water-proofing performance, by installing an expansive string coupled to the gaskets.

The annular gap between the excavated section and the ring extrados is filled during the advance by using a bicomponent backfilling, a mix of water, cement and bentonite injected directly from the end of the tail shield. Mix design of bicomponent ensure very quickly gelling time (about 10s), allowing to fill the anular gap and to support the ring during its way out from the tail. Moreover, the low permeability of the bicomponent reduces the risk of water inflow.

3.2 Jobsite overview

The construction site is located at the north entrance of the Valico Tunnel and is located in the municipality of Arquata Scrivia; jobsite covers an area about 25000 m^2 and includes all the supplies necessary for TBM excavation. The two tunnel entrances are located in a shaft about 13m deep with respect to the ground level. The shaft can be reached through metal stairs, lift or vehicle ramp.

Figure 4.   Jobsite aerial view.

The segments lining required for the advancing of the TBMs, are stored upon the line track of the tunnel. Two 80t cranes are dedicated to the storage of segments and the loading of the MSV (Multi Service Vehicle) down to the shaft. The storage area for both tunnel permit a maximum storage of 60 rings.

The site is equipped with two mixing and injection systems for the production of the backfilling mix. The bi-component injection occurs during the TBM excavation phase. This phase has an average duration of 60 min, and the necessary amount of the total bi-component per excavation is theoretically equal to 10 m^3. Each mixing and injection system has a capacity of 30m^3/h in order to fulfil the demand of both TBMs during the excavation phase.

The water entering the construction site is stored in two tanks used for tunnel excavation and for fire-fighting purpose. Tanks have 160 m^3 capacity each which 22 m^3 as fire-fighting reserve. From the storage tanks the water reaches two cooling towers and then is carried through the tunnel with steel pipes.

Chilled water is used to cool the TBM systems, to carry out the conditioning of the material in the excavation chamber and also to supply other various services, like TBM cleaning. At the end of the process, water comes out from the tunnel through 2 different pipes, and is pumped to the water treatment plant and reused within the tunnel. The surplus water is returned by a pipeline to a surface river (Rio Campora).

Water treatment plant is equipped with final ultrafiltration modules in order to clarify the water from the presence of asbestos fibers. Ultrafiltration is a separation process with a pressure membrane that separates the particulate from the water-soluble components.

Inside the jobsite area there are also an electrical - mechanical workshop, a warehouse for tools and spare parts, site offices and changing rooms for the personnel.

3.3   Logistic of the jobsite

The logistic organization of the jobsite is a main aspect to take into account for a successful advancing of the works. One of the key points is the muck management from the production in TBM to its final storage destination. Muck is carried from both TBMs to three temporary storage pits with a continuous conveyor belt. In order to extend the belt, a "storage belt" is located close to the tunnel entrance, allowing to store up 500m of belt.

Figure 5. Jobsite layout.

Muck is temporarily stored within three storage pits with a total capacity of almost 10000 m³, which allow to serve the TBMs for about 3 days. Here, a first sampling and testing of muck is performed in order to define if the material complies with the Italian standard law for reusing, or vice versa has to be managed as waste. In the first case, muck is carried by trucks from the storage pits to a temporary deposit where is laid down in order to facilitate the degradation process of the surfactant. Normally, we used an average of about 30 truck a day, which allow to carry about 3000 m³/day of material from the jobsite to the temporary disposal site.

Degradation time of the conditioned muck depends on the surfactant quantity and the weather conditions. In this case, it has been chosen to limit the tenside quantity for conditioning to the minimum to reach the EPB consistency and, at the same time, minimize the waiting time in the temporary area. Usually the degradation time is between 4 days in the best case until 2 weeks in the worst one.

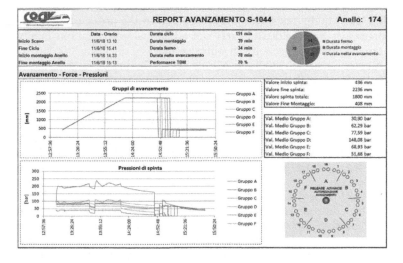

Figure 6. Data management report.

Finally, data management is a key aspect in mechanized tunneling. There are some of fundamental parameters for monitoring the process and controlling the excavation like the pressure inside the excavating chamber and along the screw conveyor, extracted quantities of material, volume and pressure of backfilling, torque, stroke, rotating speed of the cutter head, and TBM advancement speed, conditioning parameters, guidance system, production rate etc. This kind of data management is made by creating a detailed report for each advance that is constantly checked with the design engineer.

4 STATE OF WORKS AND MAIN CONSIDERATIONS

Mechanized tunneling works for realization of the Valico Tunnel started on September 2016. The first of two TBM's named "Paola" has begun his travel from "Radimero Shaft" towards Genoa in the "left tunnel". The excavation team works in 4 shifts composed by 15 workers in order to guarantee a 24h production on 7day a week. Furthermore, a daily shift of 18 workers was employed to fulfill TBM needing like mechanical and electrical workshops, belt conveyor systems, lining segment transfer, backfilling production, water treatment plant operation etc.

The initial part of the tunnel, first 1800m, was excavated inside "Costa Areasa" formation. After an initial "learning" period, the average production stood at 270m for each month, reaching the maximum of 387m in September 2017. During this initial part a total amount of 130 disc cutters was replaced during 20 maintenances with an incidence of about 0.001cut/m^3.

The second part of excavation crosses a several faults and connection zones inside "Costa Areasa", "Costa Montada" and "Rigoroso Marls" formations. The size composition of the crossed lithology is poor of fine fraction and the surfactant percentage has to be limited in order to guarantee reasonable time of degradation of the muck in the storage pit areas. These features and limitations cause high permeability inside the working chamber making inflow water risk possible. In September 2017, on the passage between "Costa Areasa" and "Costa Montada", a sudden and considerable water inflow has occurred. According to the project the machine was advancing in "open mode" without any earth pressure against the face. The event caused a sudden increase of the pressure inside the excavation chamber; in a few minutes pressure on the top of the chamber reached 2,5bar and over than 4bar in the bottom. The result was a series of advancing difficulties, above all during the extraction by the screw conveyor. In order to overcome this critical zone, it was decided to switch to "closed" mode by increasing the amount of material inside the excavation chamber with the purpose to reduce the permeability and avoid water inflow events.

The situation was rapidly resolved and, in addition to the probe drilling investigation laid down in the project, it was decided to carry out additional investigations in advance with the purpose to intercept potential pressurized water volumes.

The excavation has reached 3200m in March 2018. The presence of water was continuous but no more unexpected.

In order to check the actual hydraulic pressure in the back side of the lining, it was decided to install a series of electric piezometer by the internal side of the lining. The first of it was installed in correspondence of the first water inflow event. Water pressure data have been acquired for 10 months showing an excursion between 1 and 3 bar depending of the ground water level. Other instruments of this kind will be installed in order to assess the needing of dewatering spots and avoid the overloading of the final lining.

According to the Protocol for the Management of Asbestos Risks, which has been defined together with the controlling agencies, during the first 3200 m of advancement, both "airborne" and "massive" fiber contents were evaluated. The diagnostic analysis inside "massive" has never exceeded the limit value of 1000 mg/kg and only 4 times reaches the value of 300 mg/kg. The airborne evaluation has exceeded 3 times the limit value of 1 fiber per liter. In these conditions the above mentioned protocol provides for the switch from class LR-0 to

Table 1. Asbestos risk level.

Risk Level	Value of asbestos fibres in SEM	Materials characterisation	Risk
LR-0	0 ÷ ≤ 1 ff/l	absent or under the detection limit	No Risk
LR-1	> 1 ÷ ≤ 2 ff/l	under the threshold (1000 mg/kg)	Low
LR-2	higher than 2 ff/l	over 1000 mg/kg	High

class LR-1, requiring dedicated protective equipment. These results have been consistent to the design geological profile, where the local presence of asbestos was associated to these sedimentary formations.

The works for the second tunnel (the right tunnel) started in May 2017 with the realization of the pre-tunnel, excavated by conventional methods, and the delivery of the second TBM from the factory. The S-1044, named Giovanna, was assembled in 4 stages due to the limited space inside the shaft along the right alignment. Therefore, TBM was assembled up to the third backup gantry at first. Than it has been necessary to push ahead the machine in order to create enough space to assemble the remaining part.

In this temporary configuration some of services has been adjusted like power, water and bicomponent supply or the mucking system. In the provisional configuration, mucking has been carried out by using a chute installed in the end of the last assembled gantry. Finally, a series of trucks was employed in order to transfer the muck toward the storage pits.

Despite the mucking method was not continuously, it has been possible to reach – even in this provisional configuration – an advancement rate equal to approximately 6 rings per day, working 10h/day.

The definitive belt conveyor system was installed when the available space was enough for the assembling of the storage belt structure inside the tunnel, about 100m long. Once that the TBM was fully assembled, is was delivered to another contractor which carried on with the advancement. The second tube will be excavated in the same geological context of the first one with the advantage of the knowledge obtained during the left tunnel excavation. At present (august 2018), 900m of the right tunnel have been excavated.

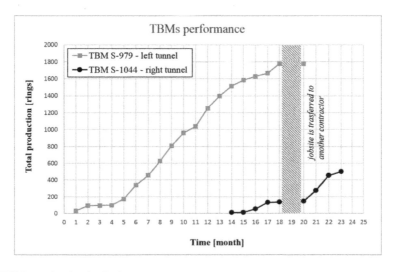

Figure 7. TBMs production in the last two years.

5 CONCLUSIONS

Valico tunnel is part of the new railway line "Terzo Valico dei Giovi", a key transit project running from Genova to Tortona, Italy. This 27 km-long tunnel, currently one of the longest in the world, will cross various geological formations, in a complex scenario that is often related to challenging geotechnical conditions (groundwater loads, high loads, fault zones) and environmental issues (presence of asbestos and of gas methane).

Experience has shown that, in addition to the technical problems that are related to the excavation of tunnel in difficult geological conditions, other problems related to the jobsite area, logistic and management have to be dealt and carefully evaluated, and shall not be underestimated.

The success of such a complex project depends on finding the compromise between the required construction performance, deadline, cost, safety, logistical constrains and working conditions, always taking into consideration the respect for environmental regulations.

The excavation of the left Valico tunnel has reached today the 30% of the total advance. The prosecution of the works will affect the crossing of Molare formation, a section that represent a really challenge on geological and hydrogeological point of view. First of all, the increasing of the overburden and the crossing of main faults related to the instability of the face. The excavation method adopted will allow to cross these zones event in presence of limited water inflow. The real challenge is represented by the potential crossing of pressured water volumes greater than the contrast capacity of the TBM. This issue will be dealt with ahead investigation from the machine like probe drilling, geoelectric e seismic exploration. According to the project, in order to drain water volumes, dewatering spots could be acted even ahead the TBM.

The second challenge in Molare formation will be the possible presence of Asbestos over than 1000mg/kg of content. In this case will be act all the procedures and technical updating to the machinery in order to comply the Protocol for the Management of Asbestos Risks.

Tunnels and Underground Cities: Engineering and Innovation meet Archaeology,
Architecture and Art, Volume 7: Long and deep tunnels – Peila, Viggiani & Celestino (Eds)
© 2020 Taylor & Francis Group, London, ISBN 978-0-367-46872-9

Surveying methods and geodetic control in long railway tunnels construction

N. Meistro, L. Surace, R. Maseroli, A. Mancarella & M. Vantini
COCIV Consorzio Collegamenti Integrati Veloci, Genoa, Italy

ABSTRACT: Today satellite survey techniques (GNSS) allow to design and construct any kind of engineering infrastructures with very high accuracy and relatively simple measurement operations. Such capabilities cannot be exploited during the construction of long railway tunnels, when very high accuracy is required. It is fundamental that the bearings of the excavations, usually driven from two or more entrances, are determined with the highest possible accuracy, because once the work is completed any axis correction would result in significant extra-costs. Conventional surveying instruments and methods are not sufficient for such goals, while a consistent aid can be provided by precision gyroscopic devices that allow independent checking of the azimuthal directions, provided that gathered data are correctly processed. This report deals with measurements, data processing techniques, relevant problems and specific solutions adopted to overcome and limit errors and their propagation in the construction of "Terzo Valico" tunnel.

1 INTRODUCTION

Today satellite surveying techniques allow achieving very high accuracy with relatively simple measurement operations. Such capabilities cannot be exploited during the construction of long railway tunnels, where moreover the elongated shape and the comparatively small diameter constitute a limit to the geometrical design of survey networks. Nevertheless it is fundamental that bearings of the excavations, usually driven simultaneously from two or more entrances, are determined with the highest possible accuracy, because once the work is completed any axis correction would result in significant extra-costs.

In tunnel constructions with TBM the required accuracy increases even more, because lining is performed only a few meters from the front of the tunnel and the surveyors are forced to be not far from the perfection to avoid *butterfly effect* in excavation.

Conventional surveying instruments and methods are not sufficient to meet the requirements, while a consistent aid can be provided by precision gyroscopic devices that allow independent checking of the azimuthal directions, provided that gathered data are correctly processed.

First, direction-finding in tunnels should take into account that terrestrial survey observables, but spatial distances, are related to the terrestrial gravity field, because survey instruments are set-up according to the local plumb line. Therefore the measurements are oriented with respect to the level surfaces, i.e. equipotential surfaces of the terrestrial gravity field. Such surfaces, of which the Geoid is the most important one, are not parallel to each other in a geometrical sense and they are impractical for survey computations. In the real world, the local direction of the plumb line is the result of physical forces acting on an object at a point. The functional model of survey observables processing requires a mathematically treatable surface, such as the bi-axial ellipsoid. Although the difference between the perpendicular to the geoid and the normal to the ellipsoid is not significant for many applications, it must be considered in special engineering surveying such as tunnel construction. Observed azimuths, angles and directions must be previously reduced to the reference ellipsoid in order to avoid significant and dangerous errors.

Secondly, the particular tunnel environment does not allow adopting the common model of propagation of electromagnetic waves and obliges to take into account the dangerous phenomenon of lateral refraction, responsible of significant azimuthal disorientations of underground traverses.

2 PROJECT DESCRIPTION

The railway Milan–Genoa, part of the High Speed/High Capacity Italian system (Figure 1), is one of the 30 European priority projects approved by the European Union on April 29th, 2014 (No. 24 "Railway axis between Lyon/Genoa – Basel – Duisburg – Rotterdam/Antwerp) as a new European project, so-called "Bridge between two Seas" Genoa – Rotterdam. The new line will improve the connection from the port of Genoa with the hinterland of the Po Valley and northern Europe, with a significant increase in transport capacity, particularly cargo, to meet growing traffic demand.

The "Terzo Valico" project is 53 km long and is challenging due to the presence of about 36 km of underground works in the complex chain of Appennini located between Piemonte and Liguria. In accordance with the most recent safety standards, the underground layout is formed by two single-track tunnels side by side with by-pass every 500 m, safer than one double-track tunnel in the remote event of an accident.

The layout crosses the provinces of Genoa and Alessandria, through the territory of 12 municipalities.

To the South, the new railway will be connected with the Genoa railway junction and the harbor basins of Voltri and the Historic Port by the "Voltri Interconnection" and the "Fegino Interconnection". To the North, in the Novi Ligure plain, the project connects existing Genoa - Turin railway (for the traffic flowing in the direction of Turin and Novara – Sempione) and Tortona – Piacenza –Milan railway (for the traffic flowing in the direction of Milan - Gotthard).

The project crosses Ligure Apennines with Valico tunnel, which is 27 km long, and exits outside in the municipality of Arquata Scrivia continuing towards the plain of Novi Ligure underpassing, with the 7 km long Serravalle Tunnel, the territory of Serravalle Scrivia (Figuer 2). The underground part then includes Campasso tunnel, approximately 700 m long and the two "Voltri interconnection" twin tunnels, with a total length of approximately 6 km.

Valico tunnel includes four intermediate adits, both for constructive and safety reasons (Polcevera, Cravasco, Castagnola and Vallemme). After tunnel of Serravalle the main line

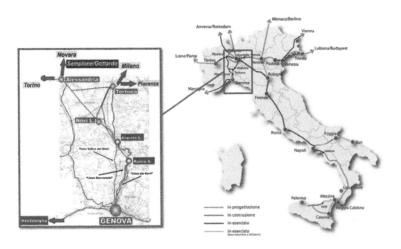

Figure 1.   The railway Milan–Genoa, part of the High Speed/High Capacity Italian system.

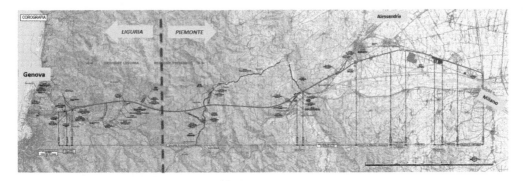

Figure 2. Layout of the new railway.

runs outdoor in cut and cover tunnel, up to the junction to the existing line in Tortona (route to Milan); while a diverging branch line establishes the underground connection to and from Turin on the existing Genoa-Turin line.

From a construction point of view, the most significant works of the Terzo Valico are represented by the following tunnels:

- Campasso tunnel 716 m in length (single tube double track)
- Voltri interconnection even tunnels 2021 m in length (single tube single track)
- Voltri interconnection odd tunnels 3926 m in length (single tube single track)
- Valico tunnel 27250 m in length (double tube single track)
- Serravalle tunnel 7094 m in length (double tube single track)
- Adits to the Valico tunnel 7200 m in length
- Artificial tunnels 2684 m in length
- Novi interconnection even tunnels 1206 m in length (single tube single track)
- Novi interconnection odd tunnels 958 m in lenght (single tube single track)

The project standards are: maximum main line speed 250 km/h, a maximum gradient 12,5 ‰, track wheelbase 4,0 – 4,5 m, 3 kV DC power supply and a Type 2 ERTMS signalling system.

3   COORDINATE SYSTEM IN DESIGNING RAILROAD TUNNELS

From a geodetic point of view, to correctly design railway tunnels a great attention should be paid to the surface network and to the underground one. The two networks play different roles: the former one is the base network, the latter one is the work/construction network.

Most national or regional geodetic networks have been designed, measured and adjusted with overall homogeneity in mind. Often this quality makes them unsuitable for controlling engineering projects, for which very high local accuracy is required. The accuracy of the underground network is much more critical than the accuracy of the surface network. Indeed the influence of the errors of the surface network is almost negligible in comparison with the influence of the underground traverse errors. The standard rules to improve any geodetic reference network are strengthening the geometric configuration and adding redundant surveys, but the possibilities to change configuration and to add extra surveys in a tunnel are very limited. The surface network is usually overdesigned and its accuracy requirements exaggerated while the underground control is often underestimated.

The reference surface for the base network is of course the geocentric ellipsoid to which the GNSS observations are referred, while a local plane coordinate system is the appropriate choice for an engineering survey requiring high relative positional accuracy.

The local system is based on the Transverse Mercator (TM) conformal map projection. TM projection produces a constant scale factor along an arbitrary meridian; therefore, taking into

4018

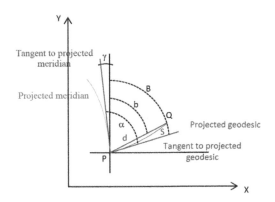

Figure 3. Geometric characteristics of the projection.

account also the topography and choosing suitable application conditions, it is possible to set up a "quasi-isometric" projection, i.e. a conformal projection where cartographic distances may be considered coincident within a defined tolerance with the observed distances reduced for the zenith angle.

The main geometric characteristics of the projection are shown in Figure 3, where S is the projected geodesic, d the chord, α the geodetic (ellipsoidal) azimuth, γ the meridian convergence, B the grid azimuth of the geodesic, b the grid azimuth of the chord and (B-b) is the arc-to-chord reduction.

Considering that tunnels are usually driven simultaneously from two or more opposite entrances, the most difficult part of the project, from a geodetic point of view, is to create a network several kilometres long and few metres wide, which should have a breakthrough error of less than a few tenths of mm. The geometry of tunnels doesn't allow flexibility in the design of the underground control network, so an open-ended traverse is usually the only practical choise and contemporarily the worst one from the point of view of error propagation and control. To some extent, speaking about the proper design of the underground control survey risks to remain an intellectual exercise.

4 A PRIOR ESTIMATION OF BREAKTHROUGH ERROR

A tunnel survey is an excellent problem to investigate when considering high accuracy requirements. Usually the logistic of the construction prevents to carry out the survey along the central part of the tunnel, so that open-ended traverses running along the walls often remain the only suitable option. The main goal of a survey control for the alignment of the tunnel axis is that the opposite headings meet at the breakthrough point without any need for additional excavations. Therefore the optimization of the breakthrough accuracy is the most important task to accomplish before surveying. The critical problem is to minimize the breakthrough error despite the weak configuration of the network and the particular environment conditions that may cause significant systematic errors.

One way of detecting and minimizing the influence of non-random errors, that will be examined later, is to carry out redundant and/or independent observations. Such observations, due to the particular geometry of a tunnel, are not specifically devoted to increase the accuracy of the survey, rather to individuate blunders and systematic errors. Assuming that these errors could be minimized in advance, it is useful to exploit the techniques of covariance propagation to determine the covariance matrix for each point in the open-end traverses and to estimate the expected breakthrough accuracy.

As far as the observation errors are concerned, standard a-priori errors of 0.3 mgon and of (0.8 mm + 1 ppm) have been assumed, respectively for directions and distances. Considering two

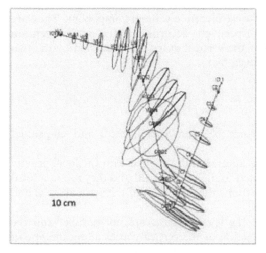

Figure 4. Propagation of TS observation errors.

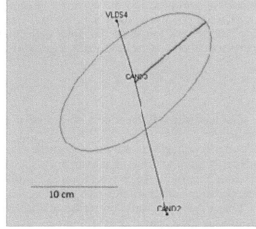

Figure 5.   Traverses breakthrough error with only TS.

traverses running from opposite entrances, each traverse should meet the other one in a theoretical point, according to the designed geometry. The unavoidable observation errors propagate as in Figure 4 and the theoretical breakthrough point may be considered as two separate points, so that a relative error ellipse can be calculated (Figure 5). The directional error can be computed as the component of the error curve (usually at 95% confidence level) in the direction perpendicular to the tunnel axis. The correlation between the two virtual points can be ignored, considering the amount of observations that commonly separate those points in the network.

The traverse 1 (red in Figure 4) has 14 legs for a total length of 3992 m and an estimated semimajor axis of the error ellipse at the breakthrough point of 55 mm. The traverse 2 (blue in Figure 4) has 12 legs for a total length 3253 m and a semimajor axis of the error ellipse at the breakthrough point of 90 mm. The estimated semimajor axis of the relative error ellipse at the breakthrough point is 106 mm (Figure 5).

5   ERROR ANALYSIS AND TREATMENT IN TUNNEL NETWORK SURVEYS

Considering the very high position accuracy required, all possible errors but random ones should be eliminated in advance, as open-ended traverses are carried out with very few or no over-determinations. In such a case the goal of checking surveys through calculations is quite unachievable. Anyhow, the more redundant surveys are, the higher possibility to detect blunders and systematic errors exists and therefore all measurable distances and directions must be measured. Actually short additional observations can be carried out very quickly and could be used as a first check. Much more dangerous than random errors are those sources of errors that may produce a systematic azimuthal deflection of the open-ended traverse.

The main problem to be considered is the error which will occur at the breakthrough of a tunnel in a mountain area if the effect of the gravity field is neglected. Moreover, equal attention should be paid to the effect of the atmospheric refraction inside the tunnel. Both systematic errors could have a big influence on the accuracy of the direction-finding inside the tunnel

5.1   *Vertical deflection influence and gravimetric reductions*

Deflection of the vertical at a point is the angle between the plumb line, i.e. the gravity vector, and the normal to the reference ellipsoid passing through the point. It is commonly separated into two orthogonal components, called ξ and η. The former one is the north-south or meridian component as the latter one is the east-west or prime vertical component. They are considered positive to the north and to the east respectively, according to the sign convention for latitude and longitude. The total deflection of the vertical in the plane defined by the two perpendiculars (astronomical and ellipsoidal) is given by:

$$\varepsilon = \sqrt{\xi^2 + \eta^2} \tag{1}$$

Geometrically it represents the angular distance between astronomical and ellipsoidal zenith.

After levelling a total station, its vertical axis is automatically aligned with the local gravity vector (Figure 6). Thus all measurements are related to the gravity field and they are in a local astronomic frame. To convert them to a geodetic frame the deflection of the vertical (ξ, η) and the perturbation in azimuth (ΔA) must be known.

Due to density variation inside of the Earth, the level surfaces are not parallel and the plumb line is a spatial curve (Figure 7). The deflection of the vertical varies along the line of force, as well as the direction and the module of the gravity vector, and depends on where it is measured.

Many tasks in geodesy require the deflection of the vertical at the geoid, other tasks need for the deflection of the vertical at the Earth's surface. The deflections of major interest in surveying are surface deflections rather than geoid deflections, since they are required to compute the reductions to the traditional observations. Surveying a tunnel network requires the knowledge of the deflection of the vertical at an intermediate equipotential surface passing through the underground traverse's points (Figure 7). This is an additional difficulty that makes tunnel construction a challenging work, even from the surveying point of view.

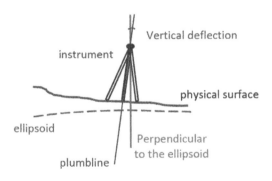

Figure 6.   Vertical deflection and TS set-up.

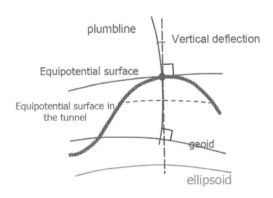

Figure 7.   Vertical deflection and level surfaces.

The determination of vertical deflection is an astronomical (or gravimetric) problem often beyond the competence of a surveyor, but today, with modern technologies and the increased observational accuracy, incorrect models can cause significant errors because they don't accurately describe the physical surface on which the measurements are carried out. It is therefore essential to define an appropriate functional model for the reductions of observations in order to take into account local variations of the Earth's gravity field. The deflection of the vertical is defined absolute if referred to a geocentric ellipsoid and relative if referred to a local ellipsoid. Depending on the choice of the ellipsoid, it can reach 20" in flat regions and up to 70" in rough areas. These are certainly not negligible quantities. The deflection of the vertical on the physical surface allows converting an astronomical azimuth A, such as the azimuth observed by a gyro device, into ellipsoidal azimuth through the Laplace equation of orientation:

$$\Delta A = A - \alpha = \eta \tan \varphi + (\xi \sin \alpha - \eta \cos \alpha) \cot \zeta \tag{2}$$

where φ is the geodetic (ellipsoidal) latitude, ζ is the ellipsoidal zenith angle to the observed point, ξ and η are the components of the vertical deflection.

The vertical deflection on the physical surface can be estimated with a mean accuracy between 0.1" and 0.5" by comparing astronomical (ϕ and Λ) and geodetic (φ and λ) coordinates of the same point. Assuming that the minor axis of the ellipsoid is parallel to the earth's mean rotation axis, defined by the CTP (Conventional Terrestrial Pole), the components of the deflection are:

$$\begin{aligned} \xi &= (\phi - \varphi) \\ \eta &= (\Lambda - \lambda) \cos \varphi \end{aligned} \tag{3}$$

The astrogeodetic method is the classical one and should provide the highest accuracy, but it is laborious and expensive. The deflection can be alternatively calculated from gravity observations with the well-known Vening-Meinesz formulae. The gravimetric method uses gravity anomalies to determine deflections on the geoid and has a mean accuracy of ± 2". To reduce them to the physical surface, it has to be considered that, at a local level, they are highly correlated with topography. The serious drawback of gravimetric methods is that density distribution of the Earth in the surroundings of the point should be known, but in very rugged terrain the topography has the predominant effect rather than the density distribution and a Digital Terrain Model allows to correctly compute the reduction.

5.2 *Atmospheric refraction and geometric corrections*

A dangerous source of errors in tunnel surveying is the atmospheric refraction. The thermal gradient between the layers of air close to the walls and those of the central part influences the azimuthal observations, causing systematic errors. Additionally a large temperature difference can be experienced at the tunnel entrance between outside and inside, that can result in errors on the first observations.

The effect of the refraction could be largely reduced carrying out the observations as closely as possible to the centre line and as far as possible from the walls. Unfortunately this is usually impossible, because the logistic of the construction prevents surveying along the central part of the tunnel, so that open-ended traverses running along and close to the walls remain the only suitable option.

The optical ray crossing zones at different temperatures (= different densities) is deviated and assumes a curvilinear pattern on the horizontal plane (Figure 8). The direction error due to the lateral refraction increases as the tunnel section decreases and thermal gradient increases; it accumulates in the traverses with many legs. According to the error propagation law, the longer the legs of the traverse are, the better the final accuracy, but when the leg length increases it is more difficult to avoid the influence of lateral refraction.

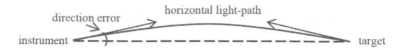

Figure 8.   Horizontal refraction.

6   GYROTHEODOLITE IN UNDERGROUND CONTROL NETWORKS

The errors caused by neglecting vertical deflection and lateral refraction propagate along the traverses and largely justify the use of independent gyro observations, which can significantly improve the breakthrough accuracy. The principle of the gyroscope is based on the fact that a quickly rotating mass with horizontal spin axis swings into the North direction due to the combined effects of the gyro spin, the gravity and the Earth rotation. An accuracy better than ±3" can be easily obtained from correct observation procedures.

As far as the influence of refraction is concerned, it is easy to verify that under similar atmospheric conditions the lateral refraction on a *n-leg* traverse surveyed with a gyro instrument is *n* times smaller than in the case of angle measurements.

The choice between carrying out few gyro observations as check and an extensive use of gyro should be investigated time by time and largely depends on local conditions.

With gyro it is possible to measure azimuths, which are independent from the computed azimuths of the traverse surveyed with traditional methods. In order to compare the observed gyro-azimuths (equivalent to astronomical azimuths) and the grid-azimuths of the traverse legs, the following reductions must be applied to the gyro observations:

– *Astronomical reduction*: the Earth rotates around an instantaneous axis that will not remain fixed to the Earth body and experiences a periodic movement (Polar motion, published periodically by the IERS) around a "mean" axis. Therefore it is necessary to reduce all observations to the CIO (Conventional International Origin), defined as the mean position of the instantaneous pole between 1902.0 and 1906.0. Usually the reduction could be neglected, but considering the duration of a long tunnel construction (many years), the reduction makes more consistent to one other the observations relevant to different epochs.

– *Gravimetric reduction*: gravity anomalies cause vertical deflection deserving special attention in mountainous regions like Alps, where it is suspected to be large and not negligible at all. The Laplace equation (2) allows a gyro azimuth (astronomical) to be converted to geodetic (ellipsoidal) azimuth. While the second part of the formula produces often small values (cot $\zeta \approx 0$), the first part significantly influences azimuth, especially in mountainous regions, where the value of η reaches even 20". Being multiplied by tan φ (≈ 1 in Northern Italy), it practically coincides with the reduction to be applied.

– *Cartographic reduction*: as the control survey is carried out in a local cartographic frame, the observed gyro-azimuth should be corrected for arc-to-chord reduction and for the convergence of the meridian (Figure 3). Both reductions relate to the projection system: the former one is obviously negligible for short traverse legs, while the latter one increases with increasing distance from the central meridian.

– *Instrumental reductions*: a misalignment does always exist between the heading of the gyroscope and the corresponding optical axis of the theodolite, depending on the mechanical coupling of the two instruments. Figure 9 shows the geometric relationships between the true azimuth of the gyro spin axis (G), the calibration value (E), the true azimuth of the geodesic PQ (A) and the TS reading (H). The calibration value, which corresponds to the angle between the spin axis of the gyro instrument and the zero of the graduation circle of the theodolite, is in theory the angle that the apparatus would provide when the theodolite's optical axis was oriented exactly to the astronomical North. It should be determined comparing the measured azimuth to a known value on a calibration baseline: therefore it includes the angle between theodolite's zero and gyro's zero, as well as the residual distortions of the base network which the baseline is linked to.

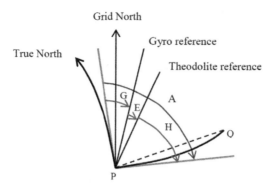

Figure 9. The calibration value E.

The value of E, determined by the manufacturer and provided with the calibration certificate isn't directly usable within the local frame, affected by local residual distortion. The local value allows the gyroscope to be used as "azimuth generator" consistent with the local geodetic context. For the above reasons, calibration baselines have been set up close to the entrances of the tunnels, adjusted within the surface network and materialized by a couple of permanent pillars.

$\Delta\alpha$ calculated with (2) shows that the contribution of the vertical deflection, both in terms of absolute value and variability, is never negligible and that it is necessary to consider a distinct value for each calibration base, assuming the constant of the nearest base at each measuring point

When using gyro device to check tunnel traverses a new problem arises due to the fact that the gravimetric reduction is position-dependent: $\Delta\alpha$ is in fact influenced by the local anomalies of the gravity field. The vertical deflection components ξ and η in the Laplace equation vary with position and $\Delta\alpha$ is not constant, but shows a significant variability from point to point, caused by the local topography and the density of the surrounding masses. To achieve the nominal accuracy, it is therefore essential to know the deflection of the vertical on all points of the calibration bases and of the traverses, in order to apply the corresponding correction to each measurement.

As gyro gives an independent azimuth, we can easily understand the advantage of using it in a tunnel traverse; surely the gyro observations reduce the propagation of the orientation errors and the influence of atmospheric refraction, with widely tested effectiveness.

Generally speaking, much less accurate gyro measurements will give the same positional accuracy as the much more precise angle measurements. As the number of stations increases,

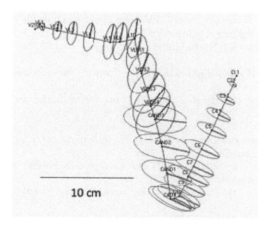

Figure 10. Propagation of Gyro errors.

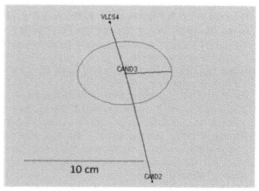

Figure 11. Gyro traverses breakthrough error.

the use of gyro instruments improves the results considerably. The same traverses as in Figures 5, 6 have been treated assuming fully gyro observations with a standard a-priori error of 0.8 mgon. The results (Figures. 10 and 11) show very clearly the improvement deriving from gyro: the estimated semimajor axis of the error ellipse at the breakthrough point is 35 mm for the traverse 1 and 25 mm for the traverse 2. The estimated semimajor axis of the relative error ellipse at the breakthrough point is 40 mm (Figure 11) instead of 106 mm estimated in the case of a traditional surveying.

7 CONCLUSIONS

Two dangerous sources of errors have to be taken into account when surveying a tunnel control network: they are vertical deflection and lateral refraction. High precision gyroscope observations, correctly processed, are an important method to both check and improve the accuracy of the underground network. The pre-analysis of the planned survey clearly shows that the breakthrough error of an open-ended traverse improves a lot using gyro observations. To correctly treat gyro observations the availability of the surface deflection values over the whole area with homogeneous density as well as the underground deflection values along the axis of the tunnel is crucial. It allows evaluating the position-dependent gyroscopic calibration values and the relevant area of application. The accuracy of the gyro-observations and the pre-analysis of error propagation suggest planning at least a gyro-measurement every fourth or fifth station and in any case when the estimated lateral deviation exceeds the value of 1 mgon at 95% confidence level. The spatial variability of the calibration values suggests periodic (yearly?) check for residual distortions of the base network, proceeding to an independent measurement of the geodetic (ellipsoidal) azimuth of each calibration base through long-lasting GNNS observations. The time variability of the calibration values must be monitored by periodic re-observation on the calibration bases. The periodic measure also serves to keep the physical characteristics of the apparatus under control, including the mechanical coupling between theodolite and gyroscope.

The influence of the refraction on a *n-leg* traverse surveyed with a gyro instrument is *n* times smaller than in the case of angle measurements, so the use of gyro-theodolites decreases error propagation in underground traverses, particularly when the influence of atmospheric refraction is suspected. In real working conditions, the use of such equipment is an investment strongly recommended not only to limit the propagation of the orientation error, but also to independently verify the presence of any defects of orientation and to plan the appropriate corrective actions.

REFERENCES

Heiskanen, W.A. & Vening Meinesz, F.A. 1958. *The earth and its gravity field*. New York: McGraw Hill
Brunner, F.K.(ed), 1984. *Geodetic refraction*. Berlin Heidelberg: Springer-Verlag
Vaniceck, P. & Krakiwsky, E.J. 1986. *Geodesy: The Concepts*. North-Holland
Torge, W. 1989. *Gravimetry*. Berlin: W. de Gruyter
Zanini, M. 1992. *Hochprazise Azimutbestimmung mit Vermessungskreisen*. Zurich: Institute fur Geodasie und Photogrammetrie
Reise, O. 1995. *Mesures gyroscopiques et deviation de la vertical*. Zurich: Institute fur Geodasie und Photogrammetrie
Torge, W. 2001. *Geodesy*. Berlin: W. de Gruyter
Featherstone, W. E. & Rueger, J. M. 2001. *The importance of using deviations of the vertical for the reduction of survey data to a geocentric datum*. The Australian Surveyor
Ceylan, A. 2009. *Determination of the deflection vertical components via GPS and levelling measurement: A case study of a GPS test network in Konya, Turkey*. Scientific Research and essay Vol. 4
Rezo, M. Markovinovic, D. & Sljivaric. 2014. *Influence of the Earth Topographic masses on vertical deflection*. Tehnicki vjesnik
Abele Balodis, M. Janpaule, J. Lasmane I. Rubans, A. & Zarins, A. 2012. *Digital zenith camera for vertical deflection determination*, Geodesy and Cartography, Vol. 38

Tunnels and Underground Cities: Engineering and Innovation meet Archaeology,
Architecture and Art, Volume 7: Long and deep tunnels – Peila, Viggiani & Celestino (Eds)
© 2020 Taylor & Francis Group, London, ISBN 978-0-367-46872-9

The Gronda Motorway by-pass at Genoa: Underground works east of the Polcevera River

P. Mele
Spea Engineering S.p.A, Milan, Italy

A. Raschillà
Autostrade per l'Italia S.p.A, Rome, Italy

L. Mancinelli, C. Silvestri, A. Damiani & M. Agosti
Lombardi Ingegneria srl, Milan, Italy

ABSTRACT: This paper describes the main aspects of the Detailed Design of a number of road tunnels to be built as part of this motorway bypass, which lies just north of Genoa in Italy. In association with a complementary project west of the Polcevera River, this project consists of connecting the three A7, A10, and A12 motorways. This will significantly improve vehicular circulation whilst also reducing urban and local traffic. The works consist of twelve tunnels to be excavated using the conventional method, a number of bypasses, and a mechanised safety tunnel. The Detailed Design updates and extends the Preliminary Project (2011) and incorporates the new geotechnical information that became available as a result of additional ground investigations. This paper provides an overview of the different methods of advancement proposed for the underground works in relation to the different layouts, which include tunnels whose internal cross-sections are designed to accommodate from 1 to 4 lanes; a specific description is also given of the deep and large-face tunnelling works. Following a general description of the geomechanical formations encountered and traversed is a section highlighting the issues that make work of this type so important for excavation, ground support, and the structural solutions adopted, as well as for managing multiple tunnelling works that are carried out simultaneously on one site.

1 INTRODUCTION

This article describes the site, the works, and activities that were performed as Detail Design stage for the tunnels of the road and motorway node at Genoa, in the area east of the Polcevera River.

The description that follows places the Gronda Motorway Bypass project in context. The project is designed as the upgrade to the stretches of the existing A10, A7, A12 and A26 motorways that lie within the municipal area of Genoa, by increasing the capacity of the A10 (the Gronda di Ponente) by constructing a new separated enlargement and partly with new separated enlargements of the A7 and A12 motorways.

2 THE GENOA-GRONDA MOTORWAY BYPASS

Although the motorway infrastructure in the Genoa area (the A10, A7, A12 and A26 motorways) was originally build to take traffic leaving the area, they also *de facto* serve as orbital motorways taking the urban traffic and traffic into and out of the city, and redistributing cross-town traffic. Today it has become clear that the current configuration of this motorway system - both because of its characteristics on plan and in section, which often do not meet minimum standards, and because of the traffic load, which includes a strongly commercial component - is no longer capable of performing both of those tasks.

The objective set for the Gronda di Genova project, also known as the "Gronda di Ponente", was therefore to reduce traffic on the stretch of the A10 motorway that has the most interconnections with the city of Genoa - i.e. the stretch between the toll gates at Genoa Ovest (Porto di Genova) and the urban area of Voltri - and transfer through traffic to a new motorway running alongside the existing A10 and, in effect, doubling it.

In this way, the traffic on the following roads:

* Milano - Ventimiglia (for departures and arrivals not attracted from the A26)
* Livorno - Porto di Voltri - Ventimiglia
* Genova Ovest (Porto) - Porto di Voltri - Ventimiglia

that does not need to access the urbanised parts of the city would move to the new motorway, thereby reducing the load on the A10, which would remain as a motorway that mainly serves the city of Genoa and its urban functions.

These works are to be made complete by:

* upgrading and connecting the A7 and A12 motorways;
* interconnecting the new Gronda, along with the existing A10, at the Morandi Viaduct and in the area of Voltri;
* a node at San Benigno to facilitate connections between the Genova Ovest junction and the port of Genoa.

The existing configuration of the territory on plan and in section, and the mountainous topography of Genoa, mean that the system "as existing" cannot be simply upgraded; thence the need for the new Gronda di Ponente motorway bypass, which along with the new A7 and the upgrade to the stretch of motorway between the connection to the A7 and the exit at Genova Est, must be constructed "alongside" the existing system.

This results in a fairly simple functional transport diagram: the new motorway system connects the existing access points distributed around the perimeter of the city (at Genova Est, Genova Ovest, and Bolzaneto), to the A26 at Voltri (taking advantage of the already existing connection to the junction for the port) and rejoins the A10 in the locality of Vesima.

The general layout will be made complete by the upgrades to the A7 between the junctions at Genova Ovest and Bolzaneto and the upgrade of the A12 between the entry point at Genova Est and the A7/A12 interchange: in both of these cases, the project has made provision for the construction of a new 3-lane carriageway in one direction and to re-allocate the two existing carriageways (i.e. 2+2 lanes) to take traffic going in the opposite direction.

Due to the very particular topography of the area, the new road system is almost completely below ground, and includes 25 tunnels of lengths that vary from 100 m to more than 6 km, for a total length of approximately 51 km of below-ground works.

The new Gronda only emerges into the open where it connects with the existing motorways. These, the only visible parts, will be concentrated at the points where the Gronda crosses river valleys, which will be achieved by building new bridges or viaducts, or widening those that already exist.

After the exits from the city at Genova Est and Genova Ovest, the road immediately goes underground. The tunnels converge towards the Val Torbella, where the new motorways overpass the A12 in the open, before returning underground until they reach Bolzaneto. Here, the different routes unify and continue westwards, crossing the Polcevera Valley on a viaduct.

The first long tunnel emerges into an open-air stretch to cross the Varenna Valley and then enters a second tunnel until it reaches the Leira Valley at Voltri, where viaducts carry it over the Leira and Cerusa streams.

These two river valleys are separated by a short tunnel passing through a mountain (on which the Sanctuary of the Madonna delle Grazie stands).

A final tunnel leads to the point where the project terminates, close to Vesima, where the new motorway reconnects back to the existing A10.

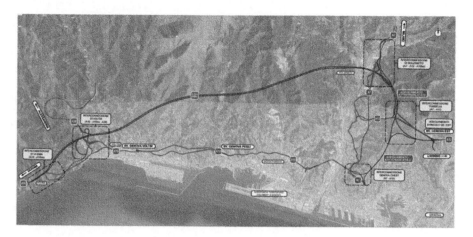

Figure 1. Map showing the major road and motorway node at Genoa.

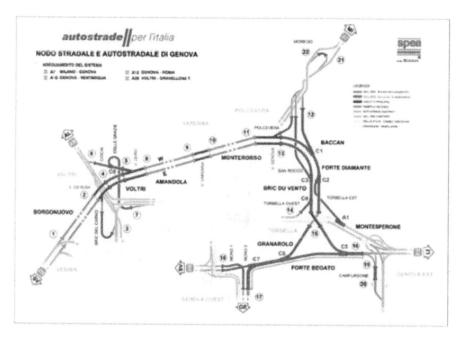

Figure 2. Functional diagram of the works showing the tunnels within the overall road plan.

The new coastal works, which are part of the upgrade to the Genoa motorway system, will be in the open and visible when completed as an artificial embankment crossing the Canale di Calma, a stretch of water that separates the airport from the sea.

Figure 1 showing a general overview of the proposed works.

Figure 2 is a general diagram locating the tunnels for the new Gronda Bypass within the overall road plan.

3 UNDERGROUND WORKS AND CONSTRUCTION TECHNOLOGY

The geometric and functional characteristics of the natural tunnels and the other underground works included in the project for the Gronda Bypass were defined on the basis of the current regulatory requirements (Ministerial Decree 5.11.2001, published in Issue 3 of the Official Gazette on 03/1/2002, and on Legislative Decree no. 26 of 45.10.2006).

In order to take into account the safety standards laid down by Ministerial Decree of 5.11.2001, which regulates the construction of new road infrastructure, the topography of the site meant that inevitably, most of the new roads would have to be in tunnels.

In fact more than 90% of the new system is in tunnels that vary in length from 100 m to over 6 km, for an overall total of approximately 51 km of tunnels. 29 km of these are to be mechanically excavated and 22 km traditionally excavated. Overall, the new Gronda Bypass requires the following 25 new tunnels:

- East of the Polcevera Valley: 14 traditionally excavated tunnels; Monte Sperone (L=2,011.9m), Granarolo (L=3,353.6m), Forte Begato (L=1,153.5m), Moro 1 (L=883.3m), Moro 2 (L=806.6m), Torbella Est+Alesaggio (L=1,530.3m), Torbella Ovest (L=415.0m), Forte Diamante (L=2,815.9m), Bric du Vento (L=2,514.0m), Baccan (L=1,136.7m), San Rocco (L=1,273.5m), Polcevera (L=551.2), Morego (L=145.0m), and Campursone (L=136.8m);
- West of the Polcevera Valley: 6 mechanically excavated tunnels (Monterosso, Amandola, and Borgonuovo, twin bore) and 5 traditionally excavated tunnels (in the Voltri area).

To these should be added the emergency tunnel that runs parallel to the intersection tunnels between GE Ovest and GE Est (Granarolo as far as Cavern 7 inclusive, Forte Begato and Montesperone from Cavern 5 inclusive). This bore, which has a smaller cross-section (Ø-$_{int}$=3.5m), is approximately 4,284m long and will be mechanically excavated (Hydro Shield).

The work schedule of East Polcevera's tunnels, defined by assigning average productions compatible with the planned works and with means and equipment of current use, has led to the work execution times approximately 8.5 years, at an estimated cost of 860 million Euros.

4 THE GEOTECHNICAL CONTEXT

The area to the east of the Polcevera River lies completely within the Apennine flysch domain. This includes four clayey and flyschoid tectonic and tectonometamorphic units (Mignanego, Montanesi, Ronco, and Antola) that are very homogeneous in terms of their lithology, with a progressively decreasing degree of metamorphism proceeding from W to E; left of the Polcevera River, the units in the hydrographic area can be considered as non-metamorphic. These are stacked with their vergence from E to W, and roughly occupy bands that are elongated from N-S along the Polcevera Valley.

The tectonic and tectonometamorphic units, listed above and showing in Figure 3, follow each other from the bottom towards the structural high and from W to E, as follows:

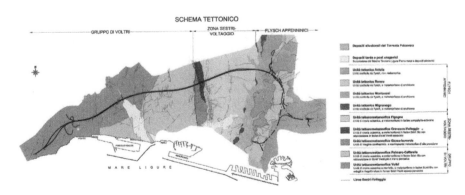

Figure 3. Tectonic diagram of the Gronda Motorway Bypass (from the CARG project as modified for the purposes of the present project).

- The Montanesi Tectonic Unit;
- The Ronco Tectonic Unit;
- The Antola Tectonic Unit.

These units are stacked and roughly occupy bands that are elongated from N-S along the Val Polcevera and are separated by planes that are mainly low-angle plunging towards the E, with their thrust vergence towards the W.

The Montanesi Formation (MTE) consists of thickly laminated argillites and silty argillites from blackish grey to black in colour, very rarely polychrome, and hazelnut in discoloration. They include occasional intercalations of fine laminated grey quartz sandstone, hazelnut in discoloration, in thin layers of millimetric to decimetric thickness. The sand/pelite ratio is on average 1:10. There is also a presence of numerous quartz veins, a few centimetres thick.

The laminating surfaces are smooth to very smooth, slightly altered, with surface oxidation and development of silty-clayey patinas. In places, the separation between the lamination surfaces is filled.

The stratification is generally disturbed; in the less disturbed parts, layers are identifiable that verge generally towards the E. A pseudostratification is observed locally, induced by squeezed isoclinal folds of small radius with interruptions in the continuity of the arenaceous layers, again with immersion towards the eastern quadrants and verging generally towards the W; these structures may have been induced by dragging phenomena associated with thrust planes sub-parallel to the stratification.

The Ronco (ROC) Formation (*upper Santonian-lower Campanian*) is of turbiditic origin and consists of greyish fine arenites, flat-parallel laminated, locally convoluted, and marly and argillitic siltites in layers of thicknesses from centimetric to decimetric with silty interlayers of thicknesses from millimetric to centimetric, although layers of multi-decimal thickness are present towards the roof of the formation. Again, the stratification is flat-parallel at the scale of the outcrop.

The disturbance is greater in the medium to low portion of the unit, where the ratio between the arenitic layers and the pelitic intercalations can oscillate between 1:1 and 1:2; in the medium to high part, which is characterised by significantly more massive layers with thin pelitic interlayers, the disturbances are less evident and the clastesis is concentrated in narrower bands. Samples taken along the route of the motorway show the presence of horizons of breccias to clasts from angular to subrounded, immersed in a pelitic-clay matrix.

The FAN Mt. Antola Formation (*upper Campanian*) is positioned on the roof of the Argillites of Montoggio and is formed by a sequence of carbonate turbidites, prevalently calcareous - marly, sometimes silty, with predominantly marly fine horizons; light grey calcarenites and marly calcarenites, whitish in alteration, marls and calcareous marl in planar layers, from decimetric up to plurimetric, intercalated with marly, silty, and to a lesser extent argillaceous layers from centimetric to decimetric thickness. The calcarenites/siltites ratio is much greater than 1.

5 THE ENGINEERING APPROACH

The Gronda is a challenging and extensive project; the motorway is of considerable length, and for the portion east of the Polcevera River alone (which is the subject of this paper) fourteen tunnels are planned (Morego, Polcevera, San Rocco, Bric du Vento, Baccan, Forte Diamante, Torbella Est, Torbella Ovest, Granarolo, Montesperone, Forte Begato, Moro 1, Moro 2, and Campursone). The works are completed by the safety system, which consists of several secondary tunnels (all walkable and accessible by vehicles) and the only mechanised tunnel (in the eastern portion) excavated with Hydro Shield technology, internal diameter 3.50 m and approximately 4.3 km long.

In terms of design, proceeding from north to south the scenarios change considerably. The different geological formations, low, high and very high covers, small and large cross-sections, intersections and interferences all make the design process a complex activity which, from an engineering point of view, requires strong synthesis. Figure 4 illustrates the design process as it was followed for the works described in this paper.

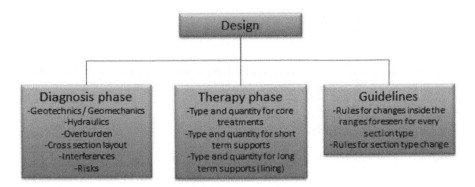

Figure 4. The design process.

5.1 New and updated elements at the Implementation Design stage

To a significant extent, the development of the Detailed Design (DD) was based on the Preliminary Design (PD) that had been developed since 2011 and was completed in 2016.

In this time interval, the project was submitted to preventive technical validation by ANAS, verification of environmental compatibility, authorization of excavated material managing plan and verification of planning compliance.

The Preliminary Design, adapted to the recommendations contained in the Final Acts of the approval procedures successfully concluded, was sent to the Ministry of Infrastructure and Transportation (MIT) in April 2016.

Since most of the important decisions had in fact already been taken, that made it possible to map out a clear process path for most situations. The project update at DD stage had to include the following main aspects:

- a new topographical survey;
- an updated construction assessment for the whole area of the works;
- a new geological, geotechnical, and geomechanical survey (updating the earlier information);
- updated road alignments;
- updated safety design.

It should be clear that these aspects had a significant effect on the outcomes, which in some situations was different from the original (PD) solution. To mention one or two simple examples, the new topographic morphology can influence the organisation of a tunnel working site or the best chainage for beginning underground activities; and new geotechnical/geomechanical conditions can change the type of structural protection required for external excavations, as well as the solutions for tunnel advancement.

5.2 The diagnosis stage

The diagnosis stage is the point at which the designer makes a synthesis of the fundamental tunnelling aspects, in order to finalise ideas and take decisions. These elements can be correctly represented as

- Tunnel alignment
  - Overburden (up to about 350m)
  - Interference between new tunnels
  - Interference between existing and new tunnels

- Geotechnical and geomechanical conditions;
- Tunnel cross-sectional layouts (dimensioning the different functional volumes that are required);

- Face conditions;
- Peculiar situations (parietal zones, fault zones, . . .);
- The construction programme;

Subsequently, using classic face stability methods (limit equilibrium method) and character-istic curves (face and cavity), the representative behaviour and the appropriate solutions for advancement were identified. It is worth noting that the excavation class assessment in each stretch is not a direct outcome of the analysis but involves the engineering judgment of the designer, who must consider all the available elements (e.g. uncertainties in the geotechnical-geomechanical information, possible interferences with other tunnels or slopes etc.).

The studies took into consideration:

- One, two and three lane cross-sections;
- Cross-sections that include a safety tunnel below the roadway;
- A 4-lane cross-section, which is present in the Forte Diamante Tunnel;
- A 3-lane cross-section with emergency stop lane, which is present in the Granarolo Tunnel;
- Caverns (at the intersection of two tunnels);
- Bypasses;
- Drainage tunnel;
- Safety tunnel (the only mechanised application in this project).

Figure 5, Figure 6 and Figure 7 showing an example of cross-sections designed.

5.3 The therapy phase: numerical models and structural concept

The purpose of the diagnosis phase is to identify scenarios with which the advancement solution can be associated. The therapy phase consists of defining an advancement solution that is identified by pre-supports (core treatments for face control), supports (for short-term cavity stability) and linings (for long-term cavity stability). The therapy is developed using numerical models which, in this case of the Gronda, were based on the finite elements method.

Several 2D FEM models were implemented using PLAXIS software to check tunnel behaviour and quantify the internal actions in the provisional and final lining. The excavation process was modelled using the subsequent relaxation phases of the initial geostatic stress. Since in this project the tunnel paths are fairly close together, the same tool was used to check the effect on an existing tunnel of excavating of a new tunnel close to it. Generally, the effect is an increase of the internal actions of the final lining of the existing tunnel.

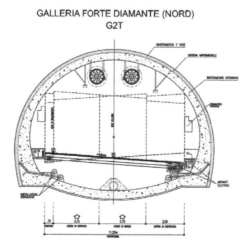

Figure 5.   - Forte Diamante Tunnel: 2-lane cross-section with emergency line.

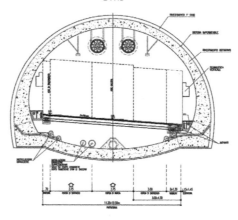

Figure 6.   - Forte Diamante Tunnel: 4-lane cross-section.

Figure 7.   – Safety Tunnel: Tunnel segmental lining - Universal Ring (4 segment + key).

Overall, the project includes tunnels with a concrete lining reinforced with steel bars as necessary. In terms of the excavation solutions, as per ADECO-RS method, the following cross- section types were used; each type is characterised by a particular provisional lining and by propping or consolidation ahead of the face:

– A: supported by steel bolts;
– B0: supported by steel ribs and shotcrete;
– B0V: supported by variable steel ribs, shotcrete and a tubular steel canopy;
– B2: supported by steel ribs, shotcrete and fibreglass elements at the face;
– B2V: supported by variable steel ribs, shotcrete and tubular steel canopy, with fibreglass elements at the face;
– C2: supported by steel ribs, with injection at the contour fibreglass elements at the face;

The application of typical sections is shown in the geomechanical profile as a percentage of each homogeneous segment. In accordance with the 'observational method', the advancement measures may be revised during the construction process by face inspections and analysis of the monitoring data.

Generally speaking, section type A refers to excavation in very good quality rock; minimal support measures are therefore allowed for and the excavation face is stable. Section B0 is designed for medium quality rock, using systematic steel ribs with shotcrete as provisional support (B2 includes the face consolidation). The conical sections B0V and B2V are placed in the areas where crown reinforcement is required, by propping with a steel canopy to give a safe excavation progress. In section C2 the tunnel face shows short-term stability or instability, requiring heavier steel ribs and systematic propping, with the final lining installed close to the excavation face.

In all of these situations, the availability of the long-term lining relates to the excavation conditions and is defined in terms of its distance from the face, for the invert and the crown. As is usual, shorter distances match with less stable situations, and thereafter with heavier works.

5.4 Geomechanical calculation

The geomechanical characterization has been developed in order to develop a rock mass site model, by means geometric and geomechanical parameters definition, and through any geostructural details that may influence the behavior of the rock mass conceived as a set of blocks and discontinuities.

Considering the intensity of the cracking and the state of alteration of the rock mass compared to the scale of the problem treated (ie the volume of rock that is affected by the effect of the "external disturbance"), the geomechanical characterization was performed considering the rock mass as a "continuous" medium with a behavior that is reasonably comparable an isotropic medium.

Under these assumptions, (rock mass considered homogeneous and isotropic), the estimates of the strength and deformation characteristics of rock masses, is performed in accordance with the Generalised Hoek-Brown criterion (Hoek et al., 2002), in which the shear strength is represented by a non-linear failure envelope determined from the rock mass quality index (GSI) and the basic parameters of the intact rock.

The Geological Strength Index (GSI), introduced by Hoek (1994) and Hoek, Kaiser and Bawden (1995) provides a number which, when combined with the intact rock properties, can be used for estimating the reduction in rock mass strength for different geological condition.

In relation to the lithological characteristics recognized, for the geomechanical characterization we have been taken into account the indications of Hoek and Marinos (2001) for the flysch constituted by alternations of different lithotypes. The parameters "m_i" and "σ_{ci}", related to the intact rock of the different lithotypes, are based on the results of laboratory tests (monoaxial and triaxial compression), carried out in the geological and geotechnical survey performed.

5.5 The importance of the geomechanical profiles

The outcome of the design activity is summarised in the geomechanical profile (i.e. Figure 8). This is a key document that includes, in summary form, a large amount of information along the length of the tunnel and relating to:

- geological formations;
- geomechanical parameters;
- tunnel cover;
- tunnel layout;
- advancement cross-section type;
- support works for short and long-term conditions;
- interferences;
- expected risks.

This geomechanical profile is thus a compact guide to the construction stage. In other words it is a document that results from the design process itself and will serve as a guide for the contractor throughout construction of the whole tunnel and, with the necessary amendments and updates, becomes a very valuable tool for future applications.

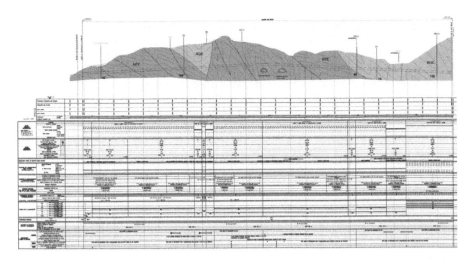

Figure 8. The geomechanical profile for the San Rocco tunnel.

5.6 *Outcomes in relation to cross-section types*

Following the diagnosis and therapy phases, all the engineering information was summarised. The outcomes, in relation to the application of the different cross-section types, are summarised in Figure 9. As per the ADECO-RS approach the indications A, B, and C represent respectively stability, short-term stability, and instability conditions for the cavity. The Detailed Design shows approx. 52% for A/B0/B0V sections, 33% for B2/B2V sections and 16% for the C2 sections. The total length of the tunnels on which the estimation was based is approx. 17 km (excluding bypasses and the mechanised safety tunnel). It is worth recalling that as the route passes through the Montanesi claystone, the Ronco formation, and then the Antola formation, the quality of the rock mass gradually improves from north to south. The heavier typical sections (C2, B2V) are used mainly in the northern tunnels whilst the lighter sections are used in the southern part.

5.7 *Waterproofing and drainage system*

The tunnel's waterproofing and drainage system has been studied to achieve the following results:

- to limit the hydraulic loads on the final lining, which are added to the geostatic loads;
- need to drain any accumulation areas present when the concrete lining filling between the formwork and the previously installed waterproofing layer is not perfectly made.

The designed waterproofing and drainage system consists of a layer of drainage geotextile and a waterproofing sheet placed at the extrados of the final lining (Figure 10). The drainage geotextile extends over the arch lining and unloads into a longitudinal collection pipe at the foot of arch; the invert is placed directly against the rock. This system is necessary to limit the overpressure at the boundary and, due to the permeability of rock mass, its effect is exclusively local.

The waterproofing system will also be equipped with special waterproof compartment joints positioned along the lining concrete casting joints and welded to PVC layer. These joints will be provided with re-injectible pipes for water-reactive resins in order to resolve any waterproofing defects that could occur (Figure 11).

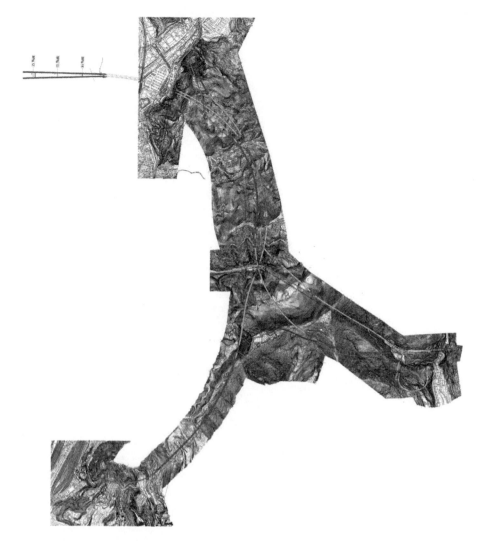

Figure 9. The tunnels east of the Polcevera River (plan view). The colours refer to the tunnel size (final clear section). The graph refers to the distribution of the typical excavation sections.

5.8 *Monitoring and guidelines*

The overall tunnel project was completed by a project for monitoring them - an absolutely essential prerequisite for proceeding with the works. Based on the expected values of convergences, extrusions, the load on the short-term supports and linings, and settlement, thresholds were defined with alert and alarm values. In association with these values, guidelines were defined for switching from one advancement solution to another. The excavation depth (the distance between the face and the first supported section), the distance between bolts or ribs, the number of treatments at the face, and the distance of the lining from the face, must all be appropriate for the actual geotechnical and geomechanical scenarios. Figure 12 describes the threshold checking procedure, with summary indications of the different activities expected.

Moreover, all of this must take place without any loss in terms of safety and works control. The system for monitoring during construction includes the following activities:

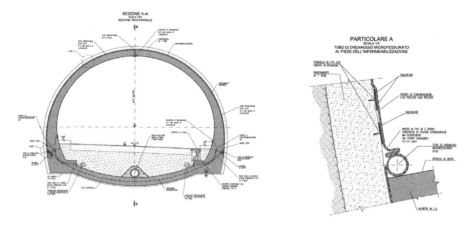

Figure 10.   Waterproofing and drainage system.

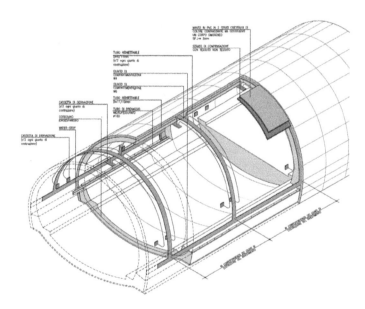

Figure 11.   Waterproof compartment system.

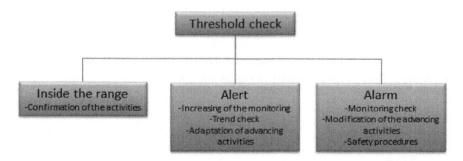

Figure 12.   The threshold checking procedure.

- Geomechanical survey at the tunnel face. The rock mass at the excavation face is assessed by carrying out frequent geological investigations;
- Boreholes at the tunnel face with perforation parameter recording (DAC - Test). Sudden variation in the perforation thrust force;
- Excavation profile closure (displacement of the provisional lining measured by carrying out topographical investigations);
- Longitudinal tunnel face deformation (in sensitive areas, by installing extrusometers ahead of the excavation face, or by means of topographical investigations if advancement has been suspended);
- Stress levels in the provisional lining: steel ribs and shotcrete (load/pressure sensors).

Monitoring during construction continues until the parameter of interest reaches a stable value. Long-term monitoring was also included in the design, in order to check the lining performance, which includes:

- Measuring final lining convergences or deformations using topographical investigations and MEMS technology;
- Measuring stress in the final concrete lining by means of pressure cells.
- Measuring water pressure around the tunnel using piezometers.

REFERENCES

Cassani G., Mancinelli L. 2007. The use of guidelines in tunneling. *Proceedings of ITA-AITES Congress.*

Hoek E., Carranza-Torres C.T., Corkum B. (2002) "Hoek-Brown failure criterion- 2002 edition", Proc. North American Rock Mechanics Society Meeting in Toronto, July, 267–273.

Hoek E. [2004]: "Numerical Modelling for Shallow Tunnels in Weak Rock" – Rocscience,

Hoek, E. (2001). Big tunnels in bad rock, The 36th Karl Terzaghi lecture. Journal of Geotechnical and Geo-environmental Engineering, A.S.C.E., 127(9), 726–740.

Panet M. & Guenot A. [1982]: "Analysis of convergence behind the face of a tunnel" – Tunnelling '82, Brighton, 197–204,

Panet M. [1995]: "Calcul des tunnels par la méthode convergence-confinement". Presses de l'école nationale des Ponts et Chaussées, Paris,

Ribacchi R. [1993]: "Recenti orientamenti nella progettazione statica delle gallerie", AGI, XVIII Convegno Nazionale di Geotecnica – Rimini,

Tamez E. [1984]: «Estabilidad de tuncles excavados en suelos»; Work presented upon joining the Mexican Engineering Academy, Mexico City,

Lunardi, P. 2000. Design and tunnel construction. ADECO-RS approach. *Tunnels & Tunnelling International special supplement.*

Cevasco, A. 2008. Relazioni fra alcune caratteristiche geomeccaniche delle formazioni argillose del genovesato and le problematiche evidenziate in tunnels storiche fra le valli Polcevera and Scrivia (Appennino ligure). *Geoingegneria Ambiente and Mineraria, Anno XLV, n. 2.*

Tunnels and Underground Cities: Engineering and Innovation meet Archaeology,
Architecture and Art, Volume 7: Long and deep tunnels – Peila, Viggiani & Celestino (Eds)
© 2020 Taylor & Francis Group, London, ISBN 978-0-367-46872-9

Ismailia: The first tunnels to be constructed under the Old and New Suez Canals – a case history

G.L. Menchini & M. Liti
CMC Cooperativa Muratori Cementisti, Ravenna, Italy.

ABSTRACT: The city of Ismailia is located in the Eastern part of Egypt near the Sinai Peninsula and close to the Suez Canal. In August 2014 the Egyptian Government's Ministry of Defense and the Engineering Authority of the Armed Forces, awarded several projects to enhance the development of the Sinai region (Eastern region of Egypt). The Ismailia Road Tunnel project consists of the construction of two new highway tunnels under both the old and new Suez Canals. A careful consideration of the tunneling technique to be used for the construction has been done before choosing a Tunnel Boring Machine (TBM) Mix Shield (slurry) for the construction of the 4.84 km long. The external diameter of the tunnel is13m and the internal is 11.4 m. The surrounding geological area is composed of a very dense sand with traces of clay and a high hydrostatic pressure.

1 INTRODUCTION

The Egyptian Government's Ministry of Defense and the Engineering Authority of the Armed Forces, awarded several projects to enhance the development of the Sinai region in the Eastern region of Egypt.

The Ismailia Road Tunnel project (Figure 1, Figure 2) is one of the investments made as part of the Suez Canal Corridor Development Project. In addition to the construction of the new Suez Canal, new rail and road tunnels will be constructed at three different locations, Port Said, Ismailia and Suez. The Ismailia Road Tunnel is located near the town of Ismailia halfway between Port Said and Suez alongside the Suez Canal.

The project is consisting of two parallel tunnel tubes that have been constructed using two 13.05 m diameter tunnel boring machines (TBMs), operating at maximum depth of approximately 67 meters below ground level. Each tunnel formed by the two TBMs' is around 4.8 kilometers long, excluding the connecting ramp structures.

The design adopted a tolerance of 100 mm. for tunnel misalignment and deformation of the lining and incorporated a 50mm gap for the installation of either a spayed lining or panels of fire proofing material. The invert of the tunnel, in addition with a service gallery, has filled with a sand cement mixture as a foundation for the asphalt roadbed.

To summarize, the basic tunnel data is listed below:

- Length: 4.84km;
- External diameter: 13.05m;
- Maximum gradient: 3.3%;
- Minimum radius: 500m;
- Rings: 2,400 per tunnel;
- Nominal width of ring segments: 2.00m;
- Thickness of segments: 600mm;
- Cross-passages: 10 no at 500m intervals.

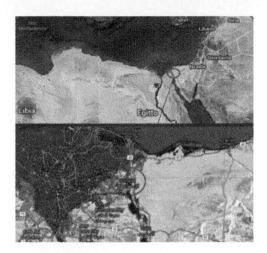

Figure 1. Suez Canal Region.

Figure 2. The Ismailia Road Tunnel.

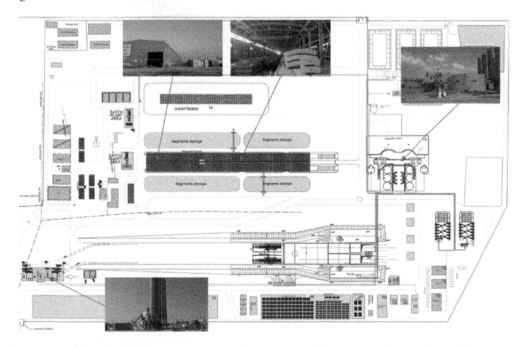

Figure 3. Jobsite plan: STP (Separation Plant), Grout Batching Plant, Segment Factory, Power Plant, Compressor Plant, Ventilation Plant, Water cooling system, Water supply system, Gantries, Workshop, Storage.

As shown in Figure 4 below, there is a culvert in the base of the tunnel beneath the road bed and that will be used for utilities.

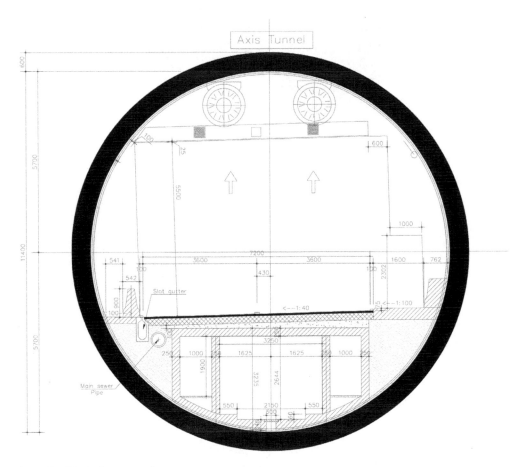

Figure 4. Typical cross section.

Figure 4 is showing the typical Cross Section; each tunnel has an internal diameter of 11.4 m, and contains a two-lane carriageway each lane being 3.65 m wide, and a 1.0m wide hard shoulder. Traffic barriers are constructed each side of the carriageway. A minimum of 5.50 m headroom was required.

2 SOIL CHARACTERISTIC

To help the determination of the soil stratigraphy and establish the best method of tunnel excavation, the contractor undertook ground investigations and detailed reviews of borehole logs taken along the alignment of the tunnel.

The Figure 5 above shows a cross section of the geological conditions discovered. The soil conditions can be can be briefly summarised as consisting predominantly of medium to fine sand (I), shallow silty clay (II), very dense sand (III), deep silty clay (IV) and sandstone/limestone (V). The density is increasing with depth from medium dense to very dense. Sandstone and limestone fragments, with sizes ranging from 5 cm to 40 cm, are present at random elevations within the sand formation. Random, discontinuous silty clay layers or variable thicknesses appear at different elevations within the soil profile in several locations.

Sandstone and limestone formations of various thickness are also observed at several locations and elevations with the cross section.

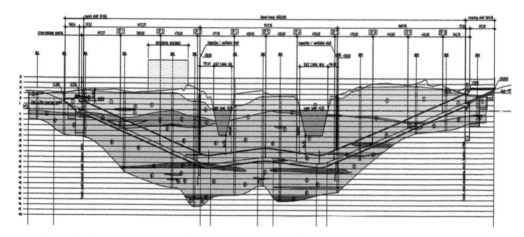

Figure 5. Stratigraphy of the soil.

3 TBM CHARACTERISTICS

The TBM selected for the project was a Mix shield (slurry) machine built by the renowned TBM manufacturer, Herrenknecht. The machine has the following technical specifications:

- Total length of TBM: 86m;
- Shield outer diameter: 13.05m;
- Shield length: 12.245m;
- Maximum advance speed: 65mm/min.;
- Maximum cutter head rotation speed: 3.2 rpm;
- Maximum working pressure: 7.5bar;
- Opening ration of the cutter head: 32%;
- Type of drive: electric;
- Number of drive motors installed: 10;
- Main drive free position: 5;
- Drive motor power: 350 kW;
- Nominal Thrust Force: 158000 kN;
- Nominal torque: 16779kNm;
- Overload torque: 23155kNm;
- Number of push cylinders: 50;
- 25 positions and 6 groups (A-F); stroke cylinders: 2800mm.
- Slurry circuit: line pipes: DN 500;
- Conveyance rate feed line: 2300 m3/h;
- Conveyance rate return: 2600 m3/h.

4 SLURRY TREATMENT PLANT (STP)

The main duties of the STP is to monitor the flow, density and yield of the slurry circuit as indicated in the Table 1 below.

Table 1. Slurry parameters in the different type of sections.

	Section name	Plug	A1	A1	A2	A3	A4	First Intermediate shaft	First Intermediate shaft	A4	A5	Second Intermediate shaft	Second Intermediate shaft	A5	A6	A7	A8	A9	Plug at receiving shaft
Sections characteristics	Ring Numbers	R-4 to 4	5	6-166	167-316	317-696	697-727	728-737	738	739-866	867-1601	1602-1611	1612	1613-1616	1617-1866	1867-2016	2017-2166	2167-2402	2403-2411
	Ground formation	Bentonite sand and cement	Medium to dense sand and shallow silty clay	Medium to dense sand and shallow silty clay	Shallow silty clay and very dense sand	Very dense sand	Very dense sand and deep silty clay	Concrete	Concrete	Very dense sand and deep silty clay	Very dense sand	Concrete	Concrete	Very dense sand	Very dense sand and deep silty clay	Very dense sand	Medium to dense sand and shallow silty clay	Medium to dense sand	Bentonite sand and cement
Main slurry parameters	Ground density	-	1.80	1.80	1.90	1.95	1.91			1.95	1.95			1.92	1.92	1.95	1.85	1.80	-
	Material excavated for each stroke (ton)	-	482	482	508	522	510			522	522			514	514	522	495	482	-
	Fresh bentonite concentration (g/l)	0	30	10	0	30	20	0	0	20	30	0	0	30	20	30	20	30	0
	Slurry - Density P1.1	1.10	1.15	1.10	1.10	1.15	1.15	1.10	1.10	1.15	1.15	1.10	1.10	1.15	1.10	1.15	1.10	1.15	1.10

The separation plant segregated the liquids from the solid part of the TBM slurry. Slurry is pumped by pipeline from TBM to STP. Only slurry discharge that has been cleaned and transformed into a specified grain size is recycled into the circuit. The fine contents remaining in the flushing circuit leads to a continuous increase in density of the slurry. When this density has reached a certain value, a portion of it must be cleared out and replaced with new slurry of a lower density. The waste slurry cannot be reused any more.

The designed flowrates of the slurry are between 0 and 2600 m3/h for the input flow and between 0 and 2800 m3/h for the output flow with a density varying between 1.10 (sand) to 1.30 (clay) maximum.

Depending on the characteristics of the soil, the yield value can be defined as a measurable quantity similar to the viscosity or it can be thought of as the initial resistance to flow under stress. A higher viscosity only slows down the rate of particle movement. The yield value is required to create a stable suspension. Before any hyperbaric intervention takes place, the cake is tested to see if it is brittle. If so, fresh bentonite should be added before testing the cake again. Should the cake still be brittle then whole slurry circuit should be replaced with a fresh bentonite with a high bentonite mix value in order to get a nice cake with a small filtrate value. The definition "high bentonite mix value" means a new slurry of fresh bentonite in order to ensure that a cake will form at the face and the air pressure will be sustainable.

Although it is important to have a cake during excavation, the quality of this cake is essential for hyperbaric interventions. The excavation process actually require fines in the ground with a balanced hydrostatic pressure to ensure the face stability. During Hyperbaric intervention the air alone is not sufficient enough to ensure the face stability, this is why the formation of a cake or a membrane with the help of compressed air can counterbalance the hydrostatic pressure and the lack of cohesion in some geological formation (i.e. sand).

4.1 Batching plant and grouting the annular gap between the soil and the tunnel

The gap created between the excavated ground and segmental lining (outside the rings) by the advancement and drive of the shield was filled with grout installed under pressure behind the shield's tail seal. The grout is delivered from the Grout Batching Plant to the TBM with dedicated Multi-Service Vehicles (MSV's), with two tanks by 11-ton each one. The injection of mortar grout was undertaken using four pumps and eight pipes crossing the tail skin with one pressure sensor in each service line. The grouting operation was carried out during the TBM excavation, using all the injection lines in order to ensure that that the void between the excavation and rings was filled with mortar as the TBM progressed.

The theoretical quantity of mortar required for each stroke was 18.13m3. In order to prevent any issues caused by mortar loss, a minimum quantity of 20m3 was prepared for each ring installed.

The intention of the grout injection was to create a bedding for the segmental ring that was made up of individual segments, whilst maintaining the natural state of the ground and minimizing any settlement. The bed also prevented any immediate contact between any aggressive soils and the concrete rings whilst improving waterproofing of the structure.

At least six lines have to be used at the same time during the TBM advancement.

5 DRIVE IN DIFFERENT SOIL CONDITIONS – CRITICAL OBSERVATIONS & PARAMETERS

During the excavation process, several types of soil were found. Principally sand and clay.

The main parameters that needed to be consider in these soil conditions were contact force, total thrust and advancement speed.

These parameters can change in function depending on the type of soil being driven through.

The different types of soil can be demonstrated with the pressure chamber parameter. In clay, the pressure in the bottom part of the CHD is instable (Figure 6). This inconstant pressure is a signal of the presence of clay. However when mining in sand (Figure 7) the pressure in the excavation chamber is very stable for both sensors.

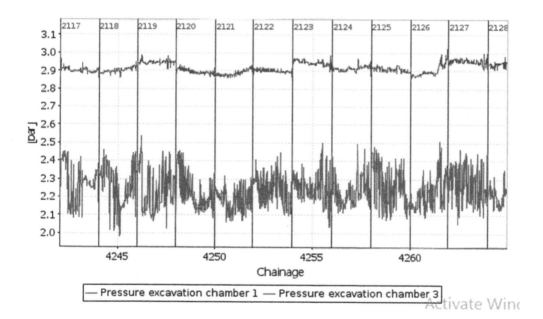

Figure 6.   Trend pressure excavation chamber in clay.

Whilst excavating in sand, an increase in the advancement speed was noted together with a decrease of contact force and thrust force.

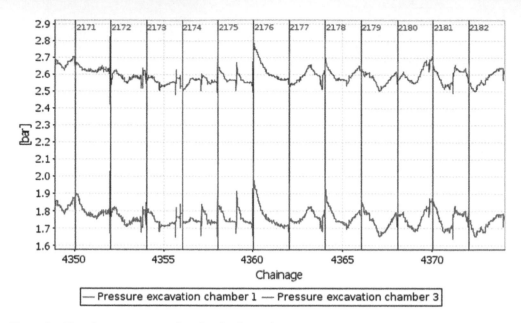

Figure 7.　Trend pressure excavation chamber in sand.

However, in clay (Figure 8) an increase of the contact force and thrust force with a decreasing of the advance speed was noted. In the Figure 8 below shows the trend of the main TBM parameters in both the type of soil.

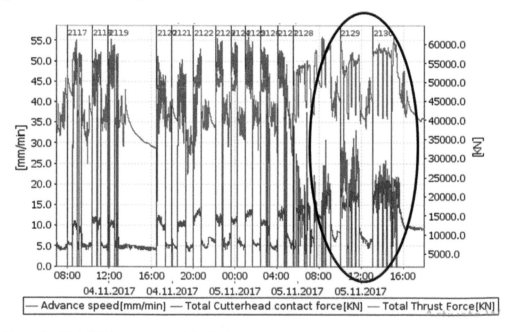

Figure 8.　Main TBM parameters mining in clay.

The red line represents the advancement speed, the blue line the contact force and the green line the thrust force. It is clearly shows the increase of the contact force and the decrease of the advancement speed. The thrust force increasing can be related more with the wear of the tools than the different soils. Regardless, the tool wear is higher in clay than in sand over equal distances.

SAND·SOIL¶

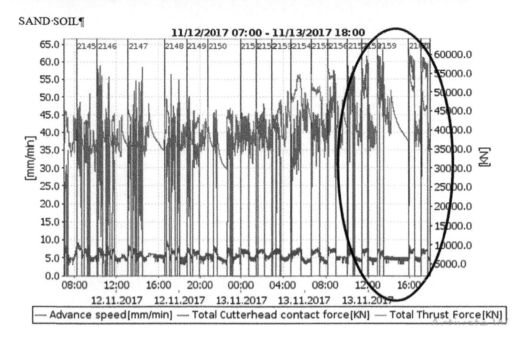

Figure 9. Trend main TBM parameters in sand.

In sand (Figure 9), a decrease of the contact force and the increase of advancement speed together with a small decrease of thrust force can be observed. These graphs are representative for the two types of soil found.

CLAY SOIL	SAND SOIL
Average CHD contact force: 13000 kN to 25000kN	Average CHD contact force: 5000 kN to 11000kN
Average Advance speed: 10 mm/min to 30 mm/min	Average Advance speed: 45 mm/min to 65 mm/min
Average Thrust Force: 55000 kN to 60000 kN	Average Thrust Force: 45000 kN to 55000 kN

In order to reduce clogging the CHD during pass through the clay layers and consequently the number of hyperbaric intervention to check it, the following procedures were applied. Mainly it consist to change completely the architecture of the cutter head. In fact, the initial configuration with disks cutter has been changed, replacing them by rippers. In that way we avoided the very frequent worn out of clocked disks, because of mud, that avoided them to tilt during the dig progress. Only on "the nose" the disks were left.

1. Increase the cutter-head rotation speed up to 2- 2.5 rpm.
2. Limit the advance speed to 15 mm/min.
3. Decrease the density of the slurry.
4. A flushing of the cutter-head helps to limit the clogging of the CHD. In any case the pressure in the excavation chamber must be monitored to prevent any over-excavation.
5. During the intervention it is important to clean the CHD thoroughly.
6. The number and the duration of the interventions depended on the stability of the front and from the availability of the divers.
7. The cutter-head intervention was stopped as soon as the front became unstable. After the front collapsed, at least two strokes must be done in order to overcome the collapsed area before a new cake filter could be created to allow a new cutter-head intervention.
8. Maintenance the replacement of rippers is much faster compared to the time employed to replace the discs cutter therefore saving time.

6 CONCLUSIONS

Specific parameters for the TBM advancement have to continually checked in order to take the correct countermeasure (interventions, changing the tools, etc.). In this instance, the rippers and cutter discs (single & double) have been be replaced in order to increase the performance of the TBM excavation and reduce the CHD inspection time. After completing the modifications to the CHD tools, the TBM performance increased considerably (Figure 10) and a reduction in the numbers of the interventions from every 60 rings to around 235 rings (2nd Suez Canal) was achieved. The time for the CHD maintenance was reduced by 20% as a result of the new tools installation. Due to the CHD's modifications the improved performance of the TBM, the excavation time (Figure 11), and the number of interventions has been reduced (Figure 12)

Upon completion of the excavation of the clay layer in November-December 2017 (Figure 13) the final 800m of excavation were done in sand and an increase in performance was noted to achieve the world's best monthly performance. The most important parameter that needs to be checked constantly is the contact force as cannot exceed 25000KN. If it does exceed 25000KN, an inspection must be undertaken after the completed stroke and ring installed.

The tunnel breakthrough was completed on 4th December 2017 as shown in the figures 14-15.

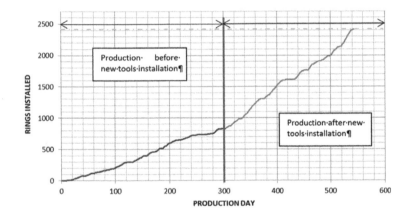

Figure 10.   Production trend TBM S-958.

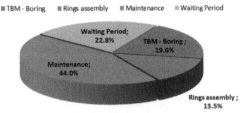

Figure 11.   Time activity TBM before Palmieri tools installation.

Table 2.   Comparison table number of rings/intervention.

From Ring	To Ring	CHD Status	Number interventions	Δrings/CHD interventions average
1	875	Before Changing Tools	79	10.4
875	2401	After Changing Tools	12	127.2

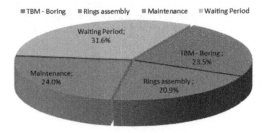

Figure 12.   Time activity TBM after the tools modification.

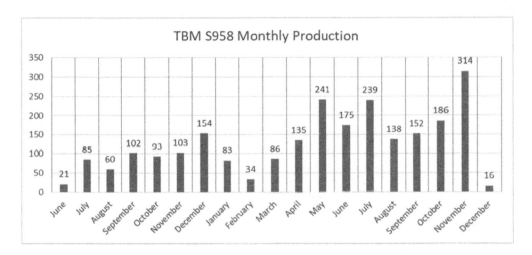

Figure 13.   The best monthly performance was achieved in November 2017 with 314 rings installed.

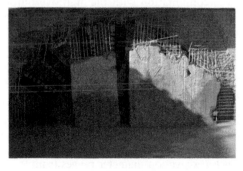

Figure 14-15.   Breakthrough 4th December 2017.

REFERENCES

Babenderende, T. & Berner, T. & Gohringer H. 2014. EPB Tunneling in hard rock conditions and transition zones.

Lunardi, G. & Cassani, G. & Canzonieri, A. & Carini, M. 2017. TBM excavation in difficult condition: the underpass of Vistula river.

Yang, Q. & Wu, Z. & Ju, Y. & Yang, Z. 2016. Super-large Diameter Slurry Shield Tunneling in Rock-soil mixed ground under 0.72 MPa pressure.

Popovic, I. 2014. Separation of used bentonite suspension in slurry shield tunneling.

Tunnels and Underground Cities: Engineering and Innovation meet Archaeology,
Architecture and Art, Volume 7: Long and deep tunnels – Peila, Viggiani & Celestino (Eds)
© 2020 Taylor & Francis Group, London, ISBN 978-0-367-46872-9

Brenner Base Tunnel, Lots Mules, 2–3 (Italy): The emergency stop in Trens

D. Merlini, M. Falanesca & D. Stocker
Pini Swiss Engineers, Lugano, Switzerland

A. Voza
BBT-SE, Bolzano, Italy

ABSTRACT: Safety is one of the crucial aspects considered in the conception of long railway tunnels. The present paper focuses on the design issues related to the construction of the emergency stops under high overburden planned for the 64 km long Brenner Base Tunnel (BBT). The project consists of two main tunnels, an exploratory tunnel and four adits.

The operational safety plan of the BBT includes the construction of cross-passages located every 333 m and 3 emergency stops spaced about 20 km. Cross-passages represent an escape route in case of emergency; emergency stops are connected to the external adits and allow the intervention of emergency vehicles and the extraction of smokes in case of fire.

The emergency stop of Trens consists in a 470 m long central tunnel connected to the main tunnels every 90 m. The overburden at the location of the emergency stop amounts to about 1100 m. The geology forecasted in the emergency stop area, mainly consists in Schist and, to a lower extent, dolomites, phyllites, quartzite, marls, serpentinites and anhydrites. In addition, two faults system are expected.

The main design challenges are represented by the complex interaction between underground structures located at 2 different elevations in combination with the high overburden and medium-low strength rock mass. The design involved high-end 2D and 3D analyses and included, in addition to the stress induced loading, swelling and creep.

1 INTRODUCTION

The Brenner Base Tunnel (BBT) is the high-speed rail link between Italy and Austria that underpass the Alps along the new European railway corridor called Scandinavian-Mediterranean (Bergmeister, 2015; Eckbauer et al, 2014). With its length of more than 55 km (64 km including all existing tunnel connections) this tunnel system will be amongst the world's longest traffic tunnels. The BBT project includes two twin single track tunnels (excavation diameter ranges from 9.7 to 10.8 m) and a service tunnel (excavation diameter ranges from 6.3 to 6.6 m) placed 12m below the main tunnels that is designed mainly to reduce the risk during construction, thus optimizing construction logistics, costs and scheduling. For operational and safety reasons, the main tunnels, spaced from 40 to 70 m, are connected every 333 m with bypasses. These distance were established, after detailed studies, to reduce the mutual interference. The tunnel system includes also 3 underground multifunctional stations and a connection tunnel with the Innsbruck existing underground. Cavern for ventilation plant and logistic, ventilation shaft, safety and service tunnels, complete the tunnel system. A general overview of the BBT global system is shown in Figure 1. The maximum cover is about 1800 m. The average overburden along the emergency stop of Trens is about 1100 m.

Different excavation methods are forecasted (Figure 2). Approximately 70% of the BBT is being built using TBMs (gripper and shield types). Some sections of the main tunnel and

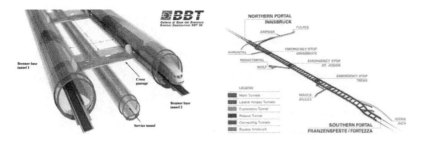

Figure 1.   Brenner Base Tunnel System.

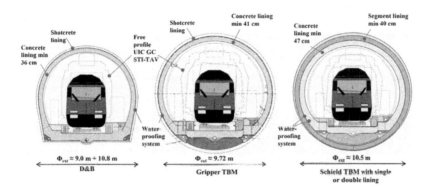

Figure 2.   Brenner Base Tunnel – typical cross sections.

exploratory tunnel, the access tunnels, the ventilation and logistics caverns, the connecting tunnels, the emergency stations and cross-ways are being excavated using conventional excavation method, designed with a double lining system. the temporary stability during construction, through the external ring, and long-term stability of the tunnel for the entire project life, through the internal ring are provided.

The final linings have been designed to guarantee a service life of 200 years. To ensure the durability, special attention was focused to the waterproofing and drainage system along the entire tunnel length. From the Italian portal the tunnels rise up to the Austrian border with a positive slope of 4 ‰ and then slope 6.7 ‰ down to the Austrian portal.

The project is financed by the Italian and Austrian governments (30% each), as well as by the European Union (40 %) and the overall costs until the scheme is completed amount to around 10 billion EUR (given an annual adjustment of 2.5 %). The costs, the provisions for chances, the risks and the construction times are updated on an annual basis. By the end of August 2018 some 88 km of a total tunnel network (26 km main tunnels, 34 km exploratory tunnels and 28 km other tunnels) of 230 km for the final project had been excavated. The design phase start since 1999 while the start of construction of the first section was in the year 2007. It is planned to complete the project in December 2026 with a subsequent daily average of 330 trains passages (approx. 80% cargo trains and 20% passenger trains) and a reduction of travel time between Fortezza and Innsbruck from the currently 80 minutes to 25 minutes.

2   THE MULES 2–3 CONSTRUCTION LOT

The Mules 2–3 construction lot is the main part of the BBT in Italian sector and started on September 2016 (Figure 3). Starting from the Mules construction site, the lateral access tunnel

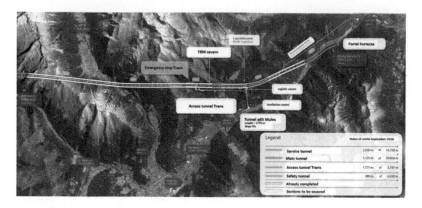

Figure 3. The mules 2–3 construction lot.

Figure 4. The Mules 2–3 construction lot during excavation phases.

was excavated from 2008. At about two kilometres distance, and in 200 meters depth, the level of the main tunnel tubes was reached. Since 2011, the various underground structures, such as the TBM assembly caverns, were prepared for the future construction activities.

Between 2011 and 2015, 3.8 km of main tubes and 2 km of exploratory tunnel were built, the latter connecting with the 10.5 km exploratory tunnel of the previous Aica-Mules lot. A total of nearly 20 km of tunnels were excavated by 2015 in the framework of the Aica-Mules and Peria-driatic Fault construction lots. After handing over the work to the new consortium, the first blasts of the construction lot Mules 2–3 were carried out at the end of 2016. On April 2017, with the opening of the sixth and last excavation front (Figure 4), the work entered full operation. Since then, four fronts are being excavated simultaneously to the north (the exploratory tunnel, the two main tubes and the access tunnel to the emergency stop) and two fronts to the south (the two main tubes). This lot extending for 23 km, includes a system of tunnels and cross passages driven through varying geological formations. Specifically, the construction lot includes the construction of about 44 km of main tunnels, partly forecasted driven by drill and blast (about 12 km) and partly using a TBM (about 28.5 km), about 15 km of exploratory tunnel, partly forecasted driven by drill and blast (about 0.6 km) and partly with a TBM (about 14 km), an access tunnel to the Trens emergency stop, 3.8 km long, to be excavated using drill and blast and a series of cross passages. At August 2018, a total of 9.2 km of tunnels have been driven. The 3,8 km access to emergency stop in Trens with a cross-section of approximately 80 m² will presumably be excavated by 2021. The end of construction is forecasted by August 2023 and total construction costs are equal to 993 million €.

3 SAFETY DURING DESIGN PHASES: THE EMERGENCY STOPS

The safety design of the project was originally based on the concept of the Gotthard Base Tunnel in Switzerland and during the review phase of the project a series of optimising technical layout and construction changes have been identified. The safety plan includes the construction of three

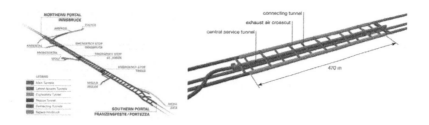

Figure 5. The emergency stops in BBT tunnel project.

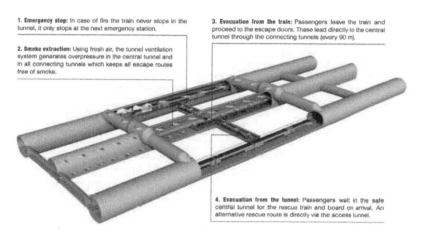

1. Emergency stop: In case of fire the train never stops in the tunnel, it only stops at the next emergency station.

2. Smoke extraction: Using fresh air, the tunnel ventilation system generates overpressure in the central tunnel and in all connecting tunnels which keeps all escape routes free of smoke.

3. Evacuation from the train: Passengers leave the train and proceed to the escape doors. These lead directly to the central tunnel through the connecting tunnels (every 90 m).

4. Evacuation from the tunnel: Passengers wait in the safe central tunnel for the rescue train and board on arrival. An alternative rescue route is directly via the access tunnel.

Figure 6. Layout of the emergency stop in BBT.

emergency stops, located south of Innsbruck, below St. Jodok and Trens, at a distance of approximately 20 kilometres from each other. All emergency stops are accessible through drivable tunnels in Ahrental (2.4 km in length), Wolf (4.0 km) and Trens (3.6 km) via Mauls (1.8 km) ad shown in Figure 5. The tunnel adits have a regular slope of 9–10%. The emergency stops are each 470 meters long and are connected to a central tunnel at 90-metre intervals. Extracted air connecting tunnels are offset by 45 metres, so that they are also located every 90 metres. The central tunnel is provided with an intermediate ceiling that divides the central tunnels into an upper and lower half. Between the two main tunnels and the central tunnel, exhaust air ducts crosscut with exhaust air dampers are located in the upper half of the central tunnel, whereas the lower half of the central tunnel holds connecting tunnels fitted with escape doors. Linking the emergency stops with the outside area allows fresh air intake and, if necessary, the generation of positive air pressure inside the tunnel, thus preventing smoke from spreading through the entire tunnel system. This ensures there is always fresh air in the crossways. The excavation diameter of the main tunnels ranges from 9.7 to 10.5 m (value inside the emergency stop), depending on the local position. The exploratory tunnel has an excavation diameter that ranges from 6.3 to 6.6 m. The central service tunnel has an excavation diameter of about 10.9 m that increase to 13.5 m at the crossing areas with the exhaust air crosscut. Polypropylene fibers inserted in inner lining are forecasted for fire safety.

4 GEOLOGY

The BBT is geographically positioned in the central eastern Alps. Geologically, the tunnel is crossing the collision area of the European and the African plate; the contact line between the two plates in this area is called Periadriatic Line. From 2000, are executed about 35 km of geognostic surveys since with a maximum depth of 1430 m (Figure 7).

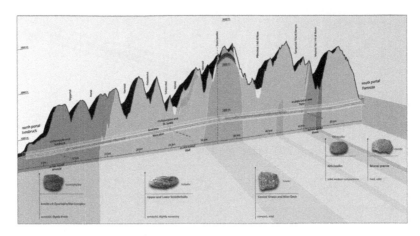

Figure 7. Longitudinal profile of Geology with main Lithological units with indication of Trens emergency stop sector.

The Northern sector from the Periadriatic line to the Austrian boundary is mainly constituted of metamorphic rocks, which include primarily Gneiss and Schists. Within these formations there are levels of Quartzites, Phyllites, Schists, and Marbles. South of the Periadriatic line the rock is constituted of Granite. Inside the Periadriatic Fault area, the rock masses are mainly constituted of Tonalite, Fillide, Black Schist, and Mica Schist. These rock masses are often strongly damaged by the tectonic action. Within the schistose rock masses, alternations of levels of Gneiss and Quartzite are also present.

The emergency stop of Trens (km 44+555 ÷ km 45+025) is located in the following lithological units:

- km 42.3 to 45.4: Bündnerschiefer of the Pfitscher Synform (changing geological conditions with phyllites, shists and dolomites)
- km 45,4 to 47,5: Cristalline of the eastern alps.

The geological information in which the emergency stop will be excavated is based on the results of the surface surveys and only one inclined borehole from surface that does not directly affect the area, as indicated in Figure 8.

Taking into account the following figure, the geomechanical units presents with an indication of the expected rock mass behaviour are summarized below:

1) Amphibolite (RMR = 70 ± 5): behaviour of rock mass from slightly-moderately squeezing and risk of tunnel face instability slightly-moderately.

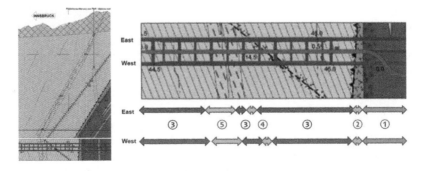

Figure 8. Longitudinal profile of Geology with main Lithological units.

2) Tauri fault system consisting of cataclasite, cachirite, fault gouge, phylonit rocks (RMR = 30 for the damage zone and RMR = 20 for the core zone): behavior of rock mass from moderately-high squeezing, high risk of rock-blocs instability and risk of tunnel face instability from moderately-high.

3) Bündnerschiefer consisting of schists, phyllites and dolomites (RMR = 60 ± 5): behaviour of rock mass from moderately-high squeezing and moderately risk of tunnel face instability.

4) Avens fault system consisting of fault gouge, cataclasite, cachirite rocks (RMR = 30 for the damage zone and RMR = 20 for the core zone): behaviour of rock mass from moderately-high squeezing, risk of rock-blocs instability and risk of tunnel face instability from moderately-high.

5) Schists with low carbonate, triassic and ophiolite (RMR = 60 ± 5): behaviour of rock mass from moderately-high squeezing, risk of rock-blocs instability and risk of tunnel face instability from moderately-high.

From the hydrogeological point of view in this area, the water ingress will limited to 3 l/s.

Despite the extensive investigations carried out (boreholes down to 700 m depth), significant geological and mechanical uncertainties inside the tectonized materials within this geological structure still remain. To reduce the level of uncertainty, the service tunnel has been designed to be excavated in advance at least 500 m before the excavation of the main tunnels. The data monitored in terms of convergences, extrusion of the tunnel face and loads on the support systems, provide important information to verify the suitability of the designed support systems and check the eventually optimizations for the future excavations.

Among the monitoring data collected during the excavation, an interesting point related to the mutual interaction in terms of increment of loads on the primary support system due to the passage of the above located main tunnels will be extrapolated (Fuoco et al. 2017).

5 CONSTRUCTIVE ASPECTS FOR THE EMERGENCY STOPS OF TRENS

The section types defined by the design are a function of rock mass quality and geomechanical analysis (Fuoco et al. 2016). These are the various excavation section types defined in the design executive phase for the main tunnels and the central service tunnel inside the emergency stops with the indication for applicability based on the different types of rock mass.

- T3 excavation section: applied to class III rock mass (41 ≤ RMR ≤ 60, forecasted inside the amphibolite). Excavation with blasting, 5 cm over-excavation, maximum drive length 3.00m, fibre-reinforced shotcrete C 30/37 both in the surrounding area (5 + 10 cm) and at the face (5cm), anchor bolts SuperSwellex Pm24 with yield strength Ny≥200kN (length 4.50m, p = 1.80m transv x 1.50m long).

- T4 excavation section: applied to class IV rock mass (21 ≤ RMR ≤ 40, mainly gneiss, paragneiss and quarzite), 10cm over-excavation, maximum drive length 1.50m, fibre-reinforced shotcrete C30/37, both in the surrounding area (5+25cm) and at the rock face (5cm), double 2 IPN180 ribs in S355JR steel with distance from 0.75 to 1.50m, radial consolidation with self-drilling R38N with length up to 6 m, consolidation around the cavity (if necessary) with R51N self-drilling cemented bolts (Ny≥630kN), length 12.00m, transversal interval = 0.75m and longitudinal interval 3.00m, stabilization of the rock face (if needed) with 32 R38N self-drilling cemented bolts, with yield strength Ny ≥ 400kN, length 15.00m, overlap 6.00m;

- T5 excavation section: applied to class IV rock mass (21 ≤ RMR ≤ 40, mainly schists and cataclastic rock), mechanized excavation, 10cm over-excavation, maximum drive length 1.50m, fibre-reinforced shotcrete, 30/37, both in the surrounding area (5+25cm) and at the rock face (5cm), double 2 IPN180 ribs in S355JR steel, radial consolidation with self-drilling R38N with length up to 8 m, consolidation around the cavity with R51N self-drilling cemented rods, length 12.00m, transv. interval 0.75m longitudinal interval 3.00m, stabilization of the rock face cemented self-drilling rods, length 15.00 m, overlap 6.00m.

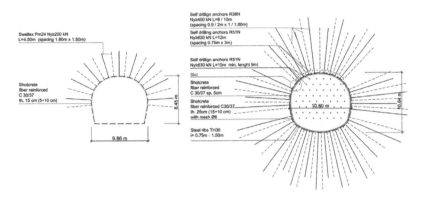

Figure 9.   T3 and T6 excavation sections of BBT inside the emergency stop.

- T6 excavation section: applied to class IV/V rock mass (0 ≤ RMR ≤ 40, mainly faults zones and rock mass with high squeezing behaviour) mechanized excavation, 30cm over-excavation, maximum drive length 1.00m, fibre-reinforced shotcrete, 30/37, both in the surrounding area (5+25cm) and at the rock face (10cm), TH36 ribs in S355JR steel, distance from 0.5 to 1.00m, radial consolidation with self-drilling R38N with length up to 10 m, consolidation around the cavity with R51N self-drilling cemented rods, length 12.00m, transv. interval 0.75m longitudinal interval 3.00m, stabilization of the rock face cemented self-drilling rods, length 15.00 m, overlap 6.00m

On the basis of forecasted geomechanical units along the tunnel alignment, percentages of application of excavation sections were considered as follow in order to take into account the presence of expected complex rock-mass behaviours: T3 (4%), T4 (28%), T5 (45%) and T6 (23%). For the sections under analysis, the cast in situ final inner lining consists of C30/37 reinforced concrete with minimum thickness, which take into account the tolerances in the construction phase, equal to 60 and 70 cm respectively in the tunnel crown and in the tunnel invert.

In the sector where the Trens emergency stop will be excavated, the exploratory tunnel is excavated by a shielded TBM. It has a radius of about 3.30 m (with a possible over-excavation of 10 cm) and is covered with a ring in precast concrete segments C50/C60 with a 30 cm thickness. Between the excavation radius and the extrados of the segments there is a gap of 10 cm (or 20 cm if provided over-excavation) which can be filled with Pea-gravel or expanded clay. A distance of installation of the segments from the excavation face equal to 12 m is foreseen. An in-situ cast inner lining of C30/37 reinforced concrete with minimum thickness 35 cm was forecasted in the design phase.

6   DESIGN MODIFICATION DURING CONSTRUCTION

During the construction, the consortium of contractors proposed a new design option, summarized below:

a) anticipation of the TBM-excavation of about 2.3 km through Fortezza. The mechanized excavation of the Trens emergency stop involves a TBM machine misalignment in the horizontal direction of about 46 cm, and an upward translation of 1.6cm with respect to the centers plan in order to guarantee the respect of the footprint limits and the definitive plants planned. For the emergency stop a double system lining is proposed consisting of a precast segments lining consisting of a ring composed of 6 + 1 trapezium prefabricated segments 45 cm thick and 1.75 m long and an internal lining in situ-cast complete with inverted arch, 23.5 cm thick in the tunnel crown and 163.4 cm in tunnel invert. The two lining are made structural coupling with metal plugs Φ22mm in steel S355 and between the two lining a spray waterproofing membrane (type Masterseal 345) with a thickness of 3 mm (Figure 10).

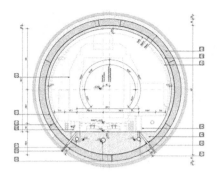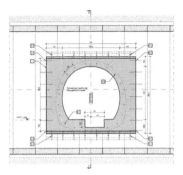

Figure 10. TBM cross section designed for the emergency stop and example of frame in reinforced concrete for the excavation of zone between main tunnel and connecting tunnel.

b) Use of expanded clay in place of the Pea-gravel on the main tunnels and the above exploratory tunnel along the emergency stop sector. The high deformability of this material guarantees the possibility of developing a part of its convergence, transmitting contained loads on the lining.
c) Execution of frame in reinforced concrete for the excavation of zone between main tunnel and connecting tunnel or ventilation tunnel with cut of segmental lining according with a rectangular shape.
d) The study of the interference described in the following paragraphs, highlighted the need to put down the Exploratory Tunnel of about 3m at the emergency stop.

7 NUMERICAL ANALYSIS

In order to predict the stress-strain behaviour of the rock mass and the effects of excavations between nearby tunnels, numerical calculation representative of the different types of rock mass encountered were implemented. The main design challenges are represented by the complex interaction between underground structures located at 2 different elevations in combination with the high overburden and medium-low strength rock mass. The design involved high-end 2D and 3D analyses and included, in addition to the stress induced loading, swelling and creep. The numerical simulations were developed using the 2D finite element code RocScience RS2 and also 3D finite element code MIDAS GTS NX. The analysis consider the different stages of excavation (1 – exploratory tunnel; 2/3 - main tunnels; 4 - central service tunnel) and the implementation of the safety measures. The geomechanical parameters were determined on the basis of geological investigations and the results of laboratory analyses and were averaged. A lithostatic stress field with a value of K_0 equal to 0.75 has been adopted and an overburden of 1100 m is considered. An elastic-plastic Mohr-Coulomb failure criterion is adopted. In the numerical analysis have been considered in the detail some innovative aspects in the modelling of the structural elements and their interaction with the surrounding rock mass. Detailed analyses were performed to design inner lining on the case of anhydrite swelling phenomena. A maximum load equal to 400 kPa was considered on the basis of laboratory test. Other analysis

Table 1. Geomechanical parameters considered for T4 geomechanical scenario.

	GSI [-]	γ [kN/m³]	E [MPa]	υ [-]	σ_{ci} [MPa]	K [MPa]	G [MPa]	Φ [°]	T [MPa]	C [MPa]	Ψ [°]
Schist Area	55	26.6	20.4	0.3	50	17	7900	33.8	1.1	3.06	0
consolidated	57	26.6	20.5	0.3	55	17.1	7910	35	1.1	3.20	0

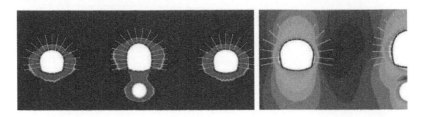

Figure 11. 2D RS2 analysis after the excavation are completed for T4 geomechanical parameters: (left) plasticized zone and (right) total displacements maximum 80 mm.

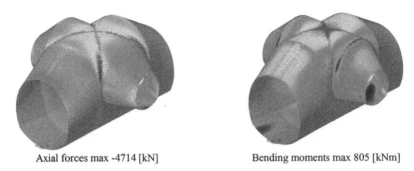

Axial forces max -4714 [kN]   Bending moments max 805 [kNm]

Figure 12. 3D analysis: connection zone between central service tunnel and exhaust air crosscut.

were also performed in order to analyse the influence of the number of segments, the type of connection, the filling conditions of the annular gap with different assumptions that strongly influence the internal actions inside the inner liner of service tunnel (Migliazza & Oreste, 2017). The geomechanical parameters for the analysis of T4 section are given in Table 1 and in the following are some results of the analysis for this geological scenario.

8   INVESTIGATION DURING THE EXCAVATION AND MONITORING

During the excavation, numerous sections equipped with measuring instruments were installed (Fuoco et al. 2017). The instruments installed (convergence sections, strain gauges on shotcrete or steel ribs, pressure cells, instrumented rock bolts sections, multi points extensometer, incremental extensometer at the tunnel face) allowed to both control stress and deformation phenomena during the excavation phase and to verify the effects on the support system during the excavation of the nearby tunnels. A diagram of a typical section equipped to measure stress and deformation of the support system for exploratory tunnel is shown in Figure 13.

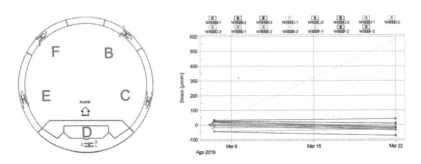

Figure 13.   Example of an instrument-equipped section installed in the exploratory tunnel.

9 STATUS OF THE WORKS

During August and September 2018, the Shield-TBM excavated the section of the exploratory tunnel where the emergency stop will be constructed. Good advance rate performance are recorded in this section, with an average value of 20 m/day and average thrust force of 4000 ÷ 5000 kN. Approximately 50 meters before the start of the emergency stop, due the cross of core zone of the Tauri fault system, TBM excavation is stopped for few working days for the execution of ground treatments ahead the face. The faults zone 4 and 5 (Figure 8) were excavated without particular difficulties. Average value of RMR index were recorded between 50 and 60. The excavation of the two base tunnels are forecasted during Spring 2019.

10 CONCLUSION

This paper focuses on the design issues related to the construction of the emergency stops of Trens that consists in a 470 m long central tunnel connected to the main tunnels every 90 m. The main design challenges are represented by the complex interaction between underground structures located at 2 different elevations in combination with the high overburden and medium-low strength rock mass. The design involved high-end 2D and 3D analyses and included, in addition to the stress induced loading, swelling and creep. The excavation will start in December 2018 and works are scheduled to be finished in 2024. The total constructive costs forecasted for only the emergency stop in Mules are equal to approximatively 70 M€.

REFERENCES

Eckbauer, W., Insam, R. & Zierl, D. 2014. Planning optimisation for the Brenner Base Tunnel considering both maintenance and sustainability. *Geomechanics and Tunnelling 7* (5): 601–609. Ernst & Sohn.

Bergmeister, K. 2015. Life cycle design and innovative construction technology. In FGU (ed.), *Swiss Tunnel Congress 2015; Proc. intern. Symp, Luzern.*

Fuoco S., Zurlo R., Marini D. & Pigorini A.: Tunnel Excavation Solution in Highly Tectonized Zones. In *WTC ITA AITES 2016; Proc. intern. Symp, S. Francisco.*

Fuoco S., Zurlo R. & Lanconelli M. 2017. Tunnel deformation limits and interaction with cavity support: The experience inside the exploratory tunnel of the Brenner Base Tunnel. In *WTC ITA AITES 2017 Proc. intern. Symp, Bergen.*

Migliazza M.R. & Oreste P.P.: Innovazione nella modellazione numerica di scavi sotterranei civili: il caso della galleria del Brennero. In *MIR 2017 Innovazioni nella progettazione, realizzazione e gestione delle opere in sotterraneo; Proc. intern. Symp, Torino.*

*Tunnels and Underground Cities: Engineering and Innovation meet Archaeology,
Architecture and Art, Volume 7: Long and deep tunnels – Peila, Viggiani & Celestino (Eds)
© 2020 Taylor & Francis Group, London, ISBN 978-0-367-46872-9*

Experiences of tunnels subject to earthquake in central Italy

A. Micheli, L. Cedrone & A. Andreacchio
Anas S.p.A., Rome, Italy

S. Pelizza
Politecnico di Torino, Turin, Italy

ABSTRACT: In tunnels design development, seismic action has always had a secondary role; past experience showed that, as structures with a circular shape in a continuous medium, confined and bound to the medium transmitting the action they are able to withstand the displacements and any induced stresses without suffering significant damage. The studies and knowledge in this field are relatively young and not codified as for classical buildings. In this work will be reported recent experience about seismic vulnerability of tunnels, gained on some works belonging to the Anas road network. Final purpose is to describe the conditions that determined damage, identifying elements that influenced the state of the damage detected and subsequently to introduce interventions to restore its safety and structural functionality. Specific relevance was recognized to seismic event characteristics, and to the peculiarities of the work itself, principally related to construction method and final quality obtained.

1 SEISMICITY IN ITALY

Italy is located on the convergent boundary between two tectonic plates, the African one and the Eurasian one, whose collision causes energy accumulation and deformation, which are released by earthquakes. Therefore, Italy is one of the countries, into the Mediterranean basin,with greatest seismic hazard.

Typically, tunnels are considered immune to the action produced by the earthquake, but we have memory of several cases in which, instead, damages were recorded. For the Italian experience we report the references of 4 seismic events that affected galeries of different types: the cavern of a hydroelectric power plant, a motorway tunnel under construction, a hydraulic tunnel and 2 road tunnels under Anas competence; only in some of these cases there were damages.

Table 1.   Italian seismic events in areas with tunnels considered in this paper.

| Earthquake | M_W | Types of underground works in the area | Epicentre | | Hypocentre |
			Site	Distance km	depth km
Friuli 1976	6.5	cavern of a hydroelectric power plant (Somplago)- motorway tunnel under construction (Lago di Cavazzo)	Gemona	7÷10	5.7
Irpinia 1980	6.9	hydraulic tunnel (Pavoncelli: Acquedotto Pugliese)	between Teora e Conza della Campania	In the epicenter area	12
L'Aquila 2009	6.3	road tunnel Monteluco (SS 17)	L'Aquila	3	8.8
Appennino Umbro-Marchigiano 2016	6.5	road tunnel San Benedetto and 9 others with minor damages (SS 685)	Amatrice - Accumuli	In the epicenter area	9

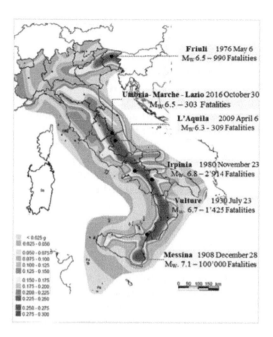

Figure 1.   Map of the seismicity in Italy with the most catastrofic recent events.

The events observed on the Italian territory have shown a concentration of the most significant earthquakes in correspondence with the Apennine arc.

On the basis of the data collected, a seismic hazard map has been drawn up, which shows a representative parameter of the most seismically dangerous areas in Italy (GdL MPS, 2004; ref. Ord. PCM 28.04.2006, n.3519, All. 1b, expressed in terms of horizontal acceleration of the soil, probability of excess of 10% in 50 years, referring to rigid soils Vs30>800 m/s; cat. A, point 3.2.1 of the D.M. of 14.09.2005); it is used as reference in the design phase of the works on the territory.

The Database of Individual Seismogenic Sources (DISS), developed by the National Institute of Geophysics and Volcanology, also identifies the sources capable of generating earthquakes of magnitude 5.5 and above:

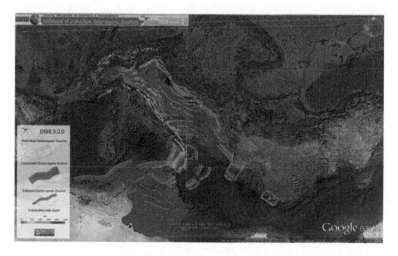

Figure 2.   DISS Working Group (2018). Database of Individual Seismogenic Sources (DISS), Version 3.2.1 A compilation of potential sources for earthquakes larger than M 5.5 in Italy and surrounding areas. http://diss.rm.ingv.it/diss/, Istituto Nazionale di Geofisica e Vulcanologia; DOI:10.6092/INGV.IT-DISS3.2.1.

2 THE ANAS ROAD NETWORK

Anas is the national company that designs, builds and controls more than 27,000 km of roads and motorways, that form a network throughout the country. In some of the areas with high seismic intensity (Fig.1) there are several tunnels at a relatively small distance from the epicentres of recorded seismic events. Some of these have been damaged during the most intense episodes. In fact, most of the natural tunnels are located along the Apennine arc, where the seismic hazard is higher and seismogenic areas are concentrated, over a total length of about 690 km (Fig.3).

It is interesting to note that not all tunnels exposed to the actions of recorded seismic events have suffered damage and, among the damaged ones, it is interesting to distinguish the type and severity of breakages found. The natural tunnels damaged by the recent earthquakes of 2009 (L'Aquila) and 2016 (Amatrice) are located between the regions of Umbria, Marche, Abruzzo and Lazio.

3 THE BEHAVIOUR OF TUNNELS IN EARTHQUAKE AREAS

First of all, two remarkable seismic events of the past are taken into consideration (Friuli and Irpinia); they are briefly described because they are already the subject of scientific-technical articles cited in the bibliography. Then the most recent episodes (L'Aquila and Appennino Umbro-Marchigiano) are reported, with a wider description of the events and the damage to tunnels in the area.

3.1 *Friuli 1976 earthquake*

This event was one of the worst Italian earthquakes for victims, damage and territorial extent (5500 km^2); the first shock was recorded on 6 May, with an intensity of M_W 6.5, followed by four other shocks, and with the epicenter located in the area of Gemona. In the area there were three major works: the hydroelectric power plant in Somplago cave (Capozza, F. 1981), the motorway tunnel of "Lago di Cavazzo" with two pipes under construction, at a distance

Site	Natural Tunnel	
	n°	km
Valle d'Aosta	9	2
Lombardia	145	115
Piemonte	20	16
Liguria	60	38
Friuli Venezia Giulia	13	14
Veneto	13	16
Emilia Romagna	31	23
Toscana	31	18
Umbria	64	67
Marche	82	57
Abruzzo	53	34
Molise	29	17
Puglia	13	11
Lazio	16	19
Campania	45	10
Basilicata	33	12
Calabria	115	54
Salerno – Reggio Calabria	138	116
Sicilia	21	7
Sardegna	50	39

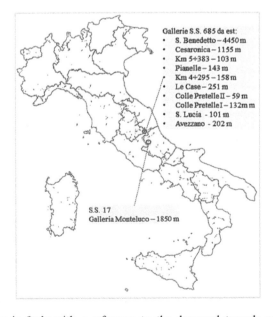

Figure 3. Natural Tunnels managed by Anas in Italy with a reference to the damaged tunnels on the SS 685.

of 7–9 km from the epicentre, built in coherent carbonatic formations, and which have not suffered any damage, and the large viaduct crossing the Tagliamento River, under construction at 3–4 km from the epicentre, of which we remember the cutting of the piers due to the fall of compact blocks of limestone ($5 \div 10 \mathrm{m}^3$) from the surrounding slopes.

3.2 Irpinia 1980 earthquake

This strong earthquake (M_W 6.9) hit a very large area (17,000 km^2) and was felt throughout southern Italy. The epicentral area is crossed by the initial stretch of the Apulian aqueduct (Sele-Calore), which runs from the sources of the Sele river to the west and Conza to the southeast, and consists of a tunnel of 12 km, with 9 m^2 of section and 6500 l/s of flow. The gallery, Pavoncelli, was built from 1906 to 1914 with brick or stone linings. The first section crossed the geological unit of the varicoloured clays for about 9 km, and was badly damaged so much so that it had to be abandoned (Cotecchia, V. 1986; Cotecchia, V. 1993; Corigliano, M. 2017) and replaced with a new tunnel. Immediately after the earthquake a temporary by-pass was built with an external pipe laid on the ground, while the new tunnel was recently built with a shield TBM using reinforced lining segments and the works were completed in October 2017.

3.3 L'Aquila 2009 earthquake

During the April 2009 seismic events, it was recorded that out of a total of 20,000 recognized tremors, 31 had an M_I magnitude between 3.5 and 5, and 3 more than 5:

Table 2. Seismic events in April 2009 in central Italy with Magnitude > 5.

Date	Magnitude		Depth km
	M_I	M_W	
06/04/2009	5.8	6.3	8
07/04/2009	5.3	5.6	15
09/04/2009	5.1	5.4	≈10

The area affected by the seismic sequence extends for more than 30 km in the NO-SE direction, parallel to the axis of the Apennine chain. During this period of intense seismic activity, and in particular following the shock recorded on 6 April, the Monteluco tunnel, on the SS 17, suffered damages to the final lining, with a widespread cracking pattern and concentrated displacements at the recasting between the segments of the final lining, which were disjointed with respect to the axis of the tunnel. The tunnel is located near the city of L'Aquila, about 3 km from the epicentre identified (Valente et al., 2010).

Figure 4. Crack on the final lining of the Monteluco tunnel.

Table 3. 2016 seismic events in central Italy with Magnitude > 5.

Date	Magnitude M_W	Depth km	Site
24/08/2016	6	8	Accumuli
24/08/2016	5.4	8	Norcia
26/10/2016	5.4	8	Castelsantagelo sul Nera
26/10/2016	5.9	10	Visso
30/10/2016	6.5	10	Norcia

3.4 Umbro-Marche Apennines 2016 earthquake

Since 24 August 2016, central Italy was affected by an anomalous seismic activity that recorded almost 45,000 earthquakes located in an area that stretches for about 80 km long and 20–25 km wide, straddling the regions of Lazio, Abruzzo, Umbria and Marche. In particular, during the last week of October 2016, the seismic sequence activated in August evolved with particularly destructive events (Tab. 3).

In this case, the distribution of the replicas in plan has a greater extension than the one of the L'Aquila event, but a similar spatial development. The October 30 event with an M_W magnitude of 6.5 was the strongest in the sequence.

During this period of seismic activity, already during the first shocks in August, the tunnels along the SS 685 suffered slight damage, in some cases immediately repaired, but they were significantly affected by the shocks of October. In particular, the San Benedetto tunnel, located on the border between Marche and Umbria and about 13 km from the epicentre of the earthquake of 30 October, located 5 km north of Norcia, has suffered damage such as to affect its viability.

On the other hand, just north of the areas affected by the events reported, there are, between the regions of Marche and Umbria, two important road axes in process of completion that connect the cities of Foligno and Civitanova Marche through the SS 77 and the cities of Perugia and Ancona through the SS 76 and SS 318 (Quadrilatero Marche-Umbria).

As part of this project, 45 natural tunnels were built for a total length of about 55 km, which did not suffer significant damage as a result of recorded seismic events, although they are located in the same geographical area as the damaged tunnels.

Following the shock of 30 October, in some sections of the SS 685 tunnels, only superficial damage was detected, such as not to compromise the structural integrity and functionality of the lining; in other areas, however, the lining was marked by the presence of deeper lesions, in some cases passing through the final lining, in others instead located at the structural joints or at the casting joint. In several sections the cracks were also found in the flooring, often in correspondence with the ones of the lining and in others, the action of the earthquake has led to

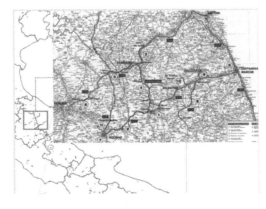

Figure 5. Tunnels of Quadrilatero Marche-Umbria design.

Figure 6. Damage between the casting joints in the Le Case and Senza Nome I tunnels, disarticulation of the joints and exposure of the steel reinforcement in the Colle Pretelle I tunnel.

Figure 7. St. Benedict's Gallery following the earthquake of 30 October 2016.

the disarticulation of some segments. In the areas where the lining was reinforced, the ejection of the steel cover, in some areas of the shell, occurred, with consequent exposure of the steel.

The San Benedetto tunnel, on the other hand, suffered localised damage at about 920 m from the Norcia side entrance, with the breakage, at several points, of the final lining and the lifting of the road platform by about 24 cm.

4 SOURCES OF DAMAGE

A good international bibliography that collects information on the effects of the earthquake on underground structures was created after the 70s, following earthquakes of high intensity that caused damage on tunnels of different uses. Several authors have contributed to identify some criteria for classifying the types of damage in relation to the characteristics of the specific seismic event, the particular context and the peculiarities of the structure (347 cases) (Powel et al., 1998, Corigliano, M. 2007). In order to determine the cause of the damage found, it is necessary to take into account the various factors that have characterized the specific seismic event, the peculiarities of the particular tunnel and the context in which it is located, and to identify the specifics of the type of damage found.

4.1 The Italian cases: Gallerie Pavoncelli, Monteluco and San Benedetto

In the case of the Pavoncelli tunnel (Irpinia 1980) there is a memory of four different types of damage. In proximity to consistent formations (limestone and sandstone) there were transversal breakages of the coating with relative dislocation of the two parts, in two different areas to the ends of the damaged section. In the only section of low cover (about 30m) there was instead a local breakage due to crushing of the lining in the crown, while over a long section

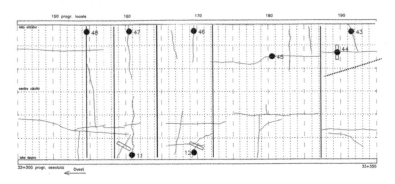

Figure 8.   Cracking and lesion pattern on the final lining of the Monteluco tunnel.

of about 80 m in the area of maximum cover (about 300 m) there was the lifting of the invert and detachment of the lining in the wall.

The types of damage found are strongly correlated with the particular geological context in which the work fits and in particular the type of the most intense damage occurred in the area of transition between the basic pelitic formation to the upper marly limestone. Moreover, there are reports in the literature, since the construction, that refer to a not easy geological context and episodes of surface collaps and damage during operation.

The Monteluco tunnel (L'Aquila 2009), on the other hand, crosses the basic limestone formation, with debris covers especially at the eastern portal these are debris-organogenic limestones, in layers of various thicknesses with the presence of intercalations of limestone breaches, in layers and banks, with alternating limestone marls. It has the typical lining of a natural tunnel, equipped with the invert.

The tunnel was particularly damaged for a section of 450 m, which was then investigated and monitored. In general, good structural conditions were found, even though in some points there was a decent frequency of fractures. During the monitoring period, the recorded seismic swarm did not give rise to neo-formation cracks or any indication of significant displacements along pre-existing cracks.

This event made it possible to detect that the seismic event determined a cracking pattern in the tunnel, particularly concentrated in the joints between the lining segments; the displacemente of the latter even though a few centimetres, have contained the development of the cracking pattern.

In the case of the San Benedetto tunnel, on the other hand, a more significant damage was found with the breakage, at several points, of the final lining and the lifting of the road platform by about 24 cm. From the original project (1986) it was inferred that the land is part of the stratigraphic succession of the Umbro-Marchigiana Series, in particular the excavation has involved rock formations of good quality, consisting of a stratigraphic succession of limestone: starting from the Norcia side portal, the escavation meets a stretch that crosses the Limestone Massif, then the Corniola with Megabrecce and finally the Corniola. Furthermore, in the section between the km. 550 and 720, more frequent faults and cutting areas were found, near the contact with the Corniola with Megabrecce.

5   SIGNIFICANT FACTORS HIGHLIGHTED BY THE SAN BENEDETTO CASE

The case of the San Benedetto tunnel can be considered particularly significant because there are available information that can reconstruct a fairly precise picture of the elements that caused the damage, which is related to the concomitance of specific factors.

In the area where the most significant damage occurred there is a heterogeneity of material due to the passage between two formations, so that the mechanical characteristics of the material behind the final coating are not homogeneous. In addition, geognostic investigations and work carried out during the restoration phase have revealed the presence of gaps between the lining and the rock that were not filled during construction. It can be deduced, therefore,

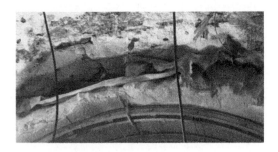

Figure 9. Presence of voids on the back of the final lining.

that an area of material with particular mechanical characteristics is crossed and where the confinement of the final coating by the surrounding rock is not effectively carried out.

The intensity of the seismic event is certainly a significant factor; in all the cases taken here into account it is a magnitude greater than 5, and in particular for the San Benedetto greater than 6, in accordance with the cases reported in the literature. The intensity of the action determined by the seismic event on this tunnel is also influenced by the distance of the work from the epicenter of the earthquake, which in the cases considered is significantly reduced, and by the depth of the epicenter, since these are always superficial earthquakes.

It is still necessary to consider the specificity of the structure determined by the absence of first-phase lining and by the particular characteristics of the final lining, which was unre-inforced, of a non-homogeneous thickness and not complete with the invert. In addition, there is a false ceiling slab for the aeration system. These elements influenced the distribution of the forces induced by the seismic action, causing the lower part of the lining to be crushed and the road surface to be lifted locally.

The distribution of the fractures detected and the results of the investigations carried out are reported:

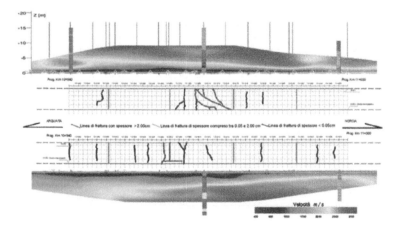

Figure 10. Cracking pattern and Sh waves velocity from REF investigations.

Considering the information described, a reference can be made to the classification available in the literature (Wang et al., 2001), in which some types of breakage have been identified in relation to the factors that determined it, in order to classify the type of damage and the factors that were decisive, as shown below:

Table 4. Type of cracks (Wang et al. 2001).

Type of cracks	Pattern	Type of cracks	Pattern
a) Sheared off lining		e) Inclined cracks	
b) Slope failure induced tunnel collapse		f) Extende cross cracks	
c) Longitudinal cracks		g) Wall deformation	
d) Trasverse cracks		h) Cracks nearby opening	

According to this classification the type of breakage detected is characterized by patterns of type d) Transverse cracks, e) Inclined cracks and g) Wall deformation. Analyzing the factors related to these types of cracks, the unfavourable ground conditions are identified as a strongly influential cause; among the secondary causes, the lack of reinforcement is common to the three types of cracks, while two of them are related to poor structural arrangements, such as the absence of the invert, the lack of reinforcement and the non-homogeneous resistant section (the proximity to the sliding surface was not taken into account because the phenomenon did not occur at low coverings).

Table 5. Type of cracks (Wang et al. 2001) for the San Benedetto tunnel.

Possible Factors	a	b	c	d	e	f	g	h
Passing through fault zones	◊							
Unfavorable ground condition				◊		◊		
Interface of hard-soft ground						◊		
Nearby slope surface or portal		◊		◊	◊	◊		
Collapse during construction			•	◊	•	•		
Lining cracks before earthquake			•	•				
Poor structural arrangements				•	•	•		◊
Unreinforced concrete lining	•	•		•	•	•	•	◊
Deteriorated lining material			•	•				
Cavity existed behind lining			◊		•			

◊ decisive link • weak link

It is also reported that through some numerical simulations, that are object of another work (Ricci, C. 2016), the behavior of the tunnel subjected to the action produced by the earthquake of October 30 was reconstructed, and the effects of increased stresses due to the dynamic component of the structure's response were found, observing a mechanism of rupture consistent with that observed in reality.

6 PRINCIPLES OF MITIGATION OF SEISMIC ACTION ON NATURAL TUNNELS

In literature there are works aimed at studying theoretically the type of measure to be taken in order to mitigate the seismic action on the linings of tunnels. Recent numerical studies, even 3D, have, for example, modelled the effects of the earthquake on final linings made for segments of reduced length compared to the usual size, obtaining a segmented lining, also taking into account the non-linear behavior of the ground (Fabozzi & Bilotta, 2016, Fabozzi, S. 2017). It has been observed that, by adopting this configuration, the stresses due to earthquakes acting on a circular tunnel are generally much lower than those reached in the case of continuous lining. This is due to the fact that the tunnel reacts to the seismic acceleration with relative displacements between the segments that generate rotation in the joints. Thiseffect can however lead to damage in the joint due to excessive rotation.

In accordance with this concept, the introduction of extended isolation of the structure from the surrounding ground is proposed (Kawashima, K. 2000); in this way, the deformable layer between the structure and the surrounding ground would absorb part of the seismic action, reducing that on the lining.

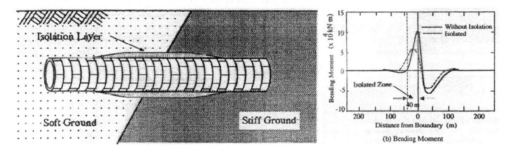

Figure 11. Isolation for a shield tunnel and effect of seismic isolation: bending moment decreasing (b) (Kawashima, 2000).

7 CONCLUSIONS

It has been observed that the damage of natural tunnels in case of an earthquake is affected by several factors related to:

1. characteristics of the seismic event (intensity of the earthquake, depth of epicentre);
2. position of the tunnel with respect to the earthquake (distance from the epicenter);
3. particular context in which the work is placed (characteristics of the rock);
4. characteristics of the work itself (geometry, characteristics of the linings).

In the case of the San Benedetto tunnel, the bad local characteristics of the rock crossed, in contact between two formations, the presence of voids on the back of the lining, the absence of the invert and reinforcement in the final lining have proved to be decisive.

In addition, due to the experience of the Monteluco tunnel, it was possible to deduce that the stresses and deformations accumulated as a result of the earthquake, if the structure does not present particular structural deficiencies, tend to concentrate in correspondence with the recasting between the segments of the final lining. Therefore, creating a system capable of conveying damage to defined areas, can avoid more significant damage.

According to what has been obtained from this experience and in accordance with what has been theoretically demonstrated in literature, the new definitive covering of the San Benedetto tunnel has been realized for successive segments of 3 m length, separated from each other by joints of width 1 cm, filled with a sheet of neoprene, realized both in the crown and on the piers and in the invert, so that each single segment is structurally detached from the previous and the following one. This solution allows each segment, in the presence of strong lateral thrusts due to the earthquake, to undergo transversal movements relative to those adjacent, avoiding breakage. Also the assembly of the reinforcement of the final covering has been provided for segments, so that the cage of each of them remains independent. The new lining has a reduced thickness (90 cm) compared to the original but it is reinforced on both the external and internal sides.

Since, due to the context in which the San Benedetto tunnel is located, new seismic events cannot be excluded, even if they are certainly not desirable; the observation of the behaviour of the new lining subjected to new seismic stresses may provide further material for study and additional information on the effectiveness of this type of intervention.

REFERENCES

Barla, M. & Di Donna, A. 2017. Nuovi conci prefabbricati per lo sfruttamento energetico delle gallerie. In MIR 2017, *Innovazioni nella progettazione e gestione delle opere in sotterra-neo. 16° ciclo di conferenze di meccanica e ingegneria delle rocce*: 249–262. 16–17 Febbraio, Torino.

Capozza, F. 1981. Effetti dei terremoti in superficie e in sotterraneo. Convegno su, *Il rischio sismico nelle industrie di processo e nei sistemi territoriali interconnessi*,29 giugno-1 luglio, San Giuliano Milanese, ENES-ISMES.

Cotecchia, V, Nuzzo, G., Salvemini, A. & Tafuni, N., 1986. "G. Pavoncelli" tunnel on the main canal of Apulia Water supply: geological and structural analysis of large underground structure damaged by the earthquake of November 23, 1980. Int. Symposium, *Engineering geology problems in seismic areas*. Bari.

Cotecchia, V. 1993. Opere in sotterraneo: rapporto con l'ambiente. *XVIII Convegno nazionale di geotecnica*. Rimini.

Corigliano, M. 2017. Progettazione di gallerie profonde in condizioni sismiche. In MIR 2017, *Innovazioni nella progettazione e gestione delle opere in sotterra-neo. 16° ciclo di conferenze di meccanica e ingegneria delle rocce*: 201–230. 16–17 Febbraio, Torino.

Fabozzi, S. 2017. Behaviour of segmental tunnel lining under static and dynamic loads, tesi di dottorato, Università di Napoli Federico II.

Fabozzi, S., Bilotta, E. 2017. Behaviour of a segmental tunnel lining under seismic actions. *Procedia Engineering*, 158, 230–235.

Guidoboni, E., Ferrari, G., Mariotti, D., Comastri, A., Tarabusi, G, Sgattoni G. & Valensise, G. 2018 - CFTI5Med, Catalogo dei Forti Terremoti in Italia (461 a.C.-1997) e nell'area Mediterranea (760 a.C.-1500). Istituto Nazionale di Geofisica e Vulcanologia (INGV). http://storing.ingv.it/cfti/cfti5/

Huang, T. H. et al. (1999), *Quick investigation and assessment on tunnel structures after earthquake, and the relevant reinforced methods*. Report for the Public Construction Commission, Taipei.

Kawashima, K 2000. Seismic design of underground structures in soft ground: A review. *Geotechnical Aspects of Underground Construction in Soft Ground*

Lanzano, G., Bilotta, E. & Russo, 2008. Tunnels under seismic loading: a review of damage case histories and protection methods. Strategy for Reduction of the Seismic Risk (Fabbrocino & Santucci de Magistris eds.).

Micheli, A., Cedrone, L. & Martino, M. 2018. Damaging of S. Benedetto tunnel after the quake of October 30, 2016: Study and repair. *Deep Foundations Institute* 2018. 65–74. ISBN 88-88102-15-3.

Ricci, C. 2017. Interpretation of the damage to the gallery San Benedetto following the earthquake of October 30, 2016. *Degree thesis*, supervisor L. Callisto, Sapienza University of Rome.

Tafuni, N. 1986. "G. Pavoncelli" tunnel on the main canal of Apulia water supply: geological and structural analysis of large underground structure damaged by the earthquake of November 23, 1980. *Egineering Geology problems in seismic areas*, Bari.

Valente, A., Micheli, A., Cedrone, L. & Bardani, C., 2010. La galleria Monteluco (L'Aquila): un caso di galleria soggetta a sisma. *Gallerie, 93*.

Wang, T.T., Wang, W.L., Su, J.J., Lin, C. H., Seng, C.R. & Huang, T.H. 2001. Assessment of damage in mountain tunnels due to the Taiwan Chi-Chi Earthquake. *Tunnelling and Underground Space Technology*. Volume 16, Issue 3, July 2001, Pages 133–150.

Wang, W. L. 2001. Assessment of Damages in Mountain Tunnels due to the Taiwan Chi-Chi Earthquake. *Tunneling and underground space technology*, 16, p.13–150.

Wang, Z. Z. & Zhang, Z. 2013. Seismic damage classification and risk assessment of mountain tunnels with a validation for the 2008 Wenchuan earthquake. *Soil Dynamics and Earthquake Engineering*, p. 45: 45 – 55.

Tunnels and Underground Cities: Engineering and Innovation meet Archaeology,
Architecture and Art, Volume 7: Long and deep tunnels – Peila, Viggiani & Celestino (Eds)
© 2020 Taylor & Francis Group, London, ISBN 978-0-367-46872-9

Tunneling challenges in Himalayas: A case study of Head Race Tunnel of 720 MW Mangdechhu Hydro-Electric Project, Bhutan

A.K. Mishra & I. Ahmed
Mangdechhu Hydroelectric Project Authority, Trongsa, Bhutan

ABSTRACT: Tunneling in Himalayas is a tough challenge attributed to complex geology brought by the complex orogenic processes, resulting in complicated folding, faulting, metamorphism and displacement which have brought the litho-units of different chronology and genesis in juxtaposition to each other giving rise to complex geology, where in unrelated rock masses of contrasting engineering properties has been brought in close association. The 6.5m diameter and 13.521km long Head Race Tunnel of Mangdechhu Hydro-Electric Project negotiated through the rocks of Higher Himalayan Crystalline comprising of strong granitic gneiss, quartzite and weak mica/biotite schist. The complex geological conditions, necessitated continuous modification and optimization of excavation methodology and support system to maintain a steady progress, which was achieved by implementing advanced excavation methodology comprising of pre-grouting, fore polling, pipe roofing, controlled blasting, adequate support, monitoring of rock mass behavior by geotechnical instruments and probe hole drilling to infer rock mass conditions ahead of the.

1 INTRODUCTION

Himalaya mountain system is a source of many large river systems having a huge hydro power potential, with a continuous rise in demand of electricity many hydroelectric projects are under construction along the mighty Himalayas in India, Nepal and Bhutan. Tapping of hydroelectric potential of the river often requires construction of long tunnels, which is one of the major challenges in developing a hydroelectric project. The tunneling challenges in Himalayan terrain are attributed to, complex geology, wide variation in rock mass properties, folding and faulting of the strata, and poor infrastructure which makes accessibility difficult during investigation stage.

The Mangdechhu Hydroelectric Project (MHEP) is one of the major project under construction in Bhutan, the project comprises of a 112m high concrete dam, an underground water conductor system comprising of 2 nos. of Desilting Chambers, Head Race Tunnel (HRT) and a Surge Shaft, and an underground Power House Complex. The 6.5m diameter, horse shoe shaped and 13.520km long HRT of the Mangdechhu Hydroelectric Project is located along the left bank of the Mangdechhu river. The invert level at the intake is at El. 1718m while that at the terminal end it is at El. 1650m with a total drop of 68m the gradient along the HRT is 1:198.86m which allows a design discharge of 118cumec to flow with a velocity of 3.71m/sec. To facilitate excavation of HRT five construction adits were provided to provide ten faces for excavation.

2 GEOLOGICAL SET UP OF THE AREA

The HRT alignment falls in Central Crystallines of Higher Himalaya (Ganser, 1964 and 1983) and stratigraphically is represented by two distinct suits of rocks namely the Thimpu Gneissic Complex and Chekha Formation; intruded by granitic batholith of Tertiary age. The Thimphu Gneissic Complex comprises of granite gneiss, augen gneiss with subordinate bands of garnetiferous mica schist and metasediments in form of schist, phyllite with lensoidal intrusions of granite and mafic rocks. The Chekha Formation comprises of micaceous quartzite and mica schist with

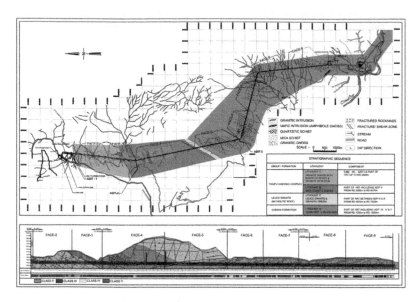

Figure 1. Plan and section showing encountered litho-zones along the HRT.

pegmatitic and amphibolitic intrusions (Nautiyal et.al., 1964, Guha Sarkar, 1979, Jangpangi, 1974 and 1978, Bhargava, 1995, Grujic et.al., 2002).The rocks of Thimphu Gneissic Complex were encountered upto RD 5394 of HRT including adit-1 and 2. Tertiary leuco-granitic batholith which on surface covers the central part of HRT between RD 4070 to 7000m was encountered only between RD 5394 to 5620 and was represented by granitic gneiss of contact zone area. The Chekha Formation dominates the downstream side of the HRT from RD 5620m onwards including adit-3, 4 and 5. The plan and section showing different litho zones as encountered along HRT are shown in figure 1.

3 TUNNELLING CHALLENGES AND SOLUTIONS

Major tunneling challenges attributed to complex geology of Himalayas faced during HRT excavation included enormous variation in predicted and encountered geological conditions, structural control on tunneling, high cover zones with superincumbent cover varying from 750 to 1000m, water charged strata where heavy ingress of water of upto 120 LPM were encountered, occurrences of previously un-envisaged shear zones and poor rock mass conditions, where excavation was carried out by continuous rib support along with strengthening crown by aid of pipe roofing, fore poling and grouting which were time consuming and resulted in slow progress as such delaying the overall construction schedule. To overcome these challenges and to maintain schedule and ensure stability of the structure continuous improvement and changes in excavation methodology and support system were implemented. The challenges faced during excavation of MHEP HRT and methodologies devised to counter them are discussed below.

3.1 *Problems along construction Adits*

Since construction adits are the means of approach to HRT for its excavation adverse conditions encountered along them directly affects the HRT excavation schedule. Among five Adits of MHEP HRT no major adverse conditions were encountered along Adit – 1, while during the development of the portal of Adit – 2 wide cracks of 50 to 100 cm opening developed along the slope. These cracks were passing just by the edge of the outer cutline of the portal at El. 1710m. Due to frequent downpours the overburden comprising of angular boulders embedded in soil and clay got water charged and started moving slowly under creep. The direction of creep was almost parallel to the adit alignment as such posed a threat to the structure therefore to ensure the stability of the structure the portal of the Adit was shifted away from the unstable area. From the new location excavation of the Adit was undertaken after developing portal. During the excavation of the Adit very poor rock mass was encountered along 16.9% of Adit length corresponding to110m while poor rock mass of Class-IV category was encountered along 12.68% of Adit length corresponding to 83m. From portal adit negotiated through adverse geology for first 168m (figure 2), this stretch was negotiated by strengthening the strata by aid of pre-grouting, fore polling/pipe roofing. Steel ribs support was also provided concurrent to excavation along this stretch.

Major problems in the form of poor geological strata and cavity formation were faced along Adit – 4 and 5. Along Adit – 4 though fair rock mass of class III category was encountered along 69% of the total Adit length, however very poor geological conditions prevailing along the initial 100m posed a major challenge in its excavation (figure 3), which was negotiated by multi drift method of excavation. Prior to excavation the crown of the Adit was strengthened by aid of pipe roofing and grouting. Two major cavities were formed in this very poor rock mass zone. The first cavity of 5 to 5.5 m height was formed between RD 3 to 9 m and second cavity was formed between RD 31 to 35, the extent of cavity resulted in day lighting of the Adit. The cavities were formed due to prevailing very poor rock mass conditions and water charged strata which resulted in flowing conditions. The adverse conditions were negotiated by multi drift method of excavation with the aid of 6m long, 89mm diameter pipe roofing and grouting.

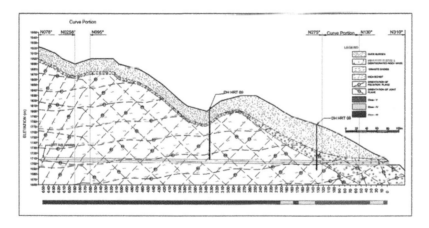

Figure 2. Geological section along Adit – 2.

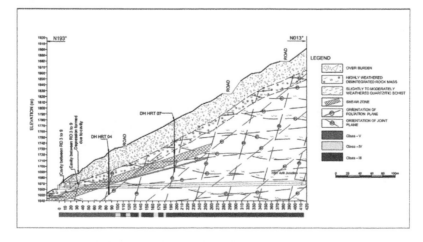

Figure 3. Geological section along Adit – 4.

4072

Similar to Adit – 4, very poor rock mass conditions were encountered along the initial 150m of the Adit – 5. This stretch was negotiated by mechanical means of excavation by multi drift method and strengthening crown by aid of pipe roofing and grouting. Apart from this, major length of the adit was negotiated through poor rock mass of Class-IV category which was encountered along 57% of the total adit length (figure 4). The excavation through poor rock mass was a slow and tedious process. Continuous over breaks to the tune of 2 to 3m beyond the adit profile occurred along the entire adit length. Prevailing poor rock mass conditions necessitated strengthening of the crown by aid of fore-poling and grouting followed by steel rib support. Fair rock mass of Class-III category was only encountered along 16% of the total adit length. However the RMR ratings were of the order of 41 to 43 only, this rock mass of lower Class-III category practically behaved as that of Class-IV category and as such steel ribs support was provided in major portion of Class-III rock mass also.

3.2 Variation in predicted and encountered rock mass conditions

Based on the anticipated geological as well as rock mass conditions as per DPR predictions resource planning, excavation methodology and construction schedule was planned. However considerable variation in encountered rock mass conditions with respect to DPR predictions (figure 5) necessitated re-planning and modification of excavation methodology to maintain the construction schedule. Apart from this variation other major challenges faced during HRT

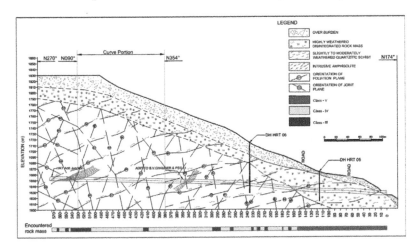

Figure 4.   Geological section along Adit – 5.

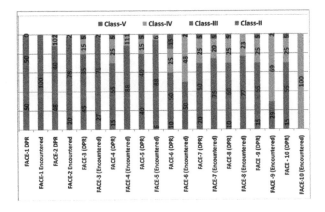

Figure 5.   Graph showing rock class comparison (predicted v/s encountered) along various faces of HRT.

excavation include, structural control on tunneling, tunnel squeezing and rock burst in high cover zones where superincumbent cover varied from 750 to 1000m, heavy ingress of water of upto 120 liters/minute, occurrences of previously un-envisaged shear zones and poor rock mass conditions, where excavation was carried out by continuous rib support along with strengthening crown by aid of pipe roofing, fore polling and grouting which were time consuming and resulted in slow progress as such delaying the overall construction schedule. To overcome these challenges and to maintain schedule and ensuring stability of the structure continuous improvement and changes in excavation methodology and support system were implemented. The geological and geotechnical challenges faced and methodology devised to face them is discussed below.

3.3 Structural control on tunneling

Tunneling and tunnel stability to a great extent is governed by the relation of tunnel alignment with the attitude of discontinuities. The structural control on tunnel determines the quantum of over breaks, stability and excavation rate. Along the HRT, structural control on tunneling was observed along face 6,7,8,9 and 10, which were excavated through gently dipping strata of Chekha Formation intersected by four major set of discontinuities. The mean attitude of the discontinuities recorded along these faces and their relation with respect to tunnel drive direction is tabulated in table 1.

Gently dipping and thinly foliated strata along Face-6, 7, 8, 9 and 10 induced slabbing along the crown. Along Face-6 and 7 tunnel alignment was oblique at high angle to the strike of foliation as such slabbing was less pronounced while along Face-8, 9 and 10 which were aligned almost parallel to the strike of foliation slabbing was more pronounced and governed the tunnel profile. Steeply dipping S2 joint set which also strikes parallel to tunnel alignment also contributed adversely in slabbing and also resulted in failure along the walls of the tunnel.

Intersection of discontinuities along Face-6 occasionally gave rise to wedge failure along the left side of the crown. However since foliation joint has a gentle dip the height of the unstable wedges was less generally of the order of 1 to 1.5m. Along Face-7 intersections of steeply dipping S2, S3 and S4 joint set resulted in formation of wedges of considerable height of 3 to 5m towards the right side of crown. Although intersection of steeply dipping joint set gave rise unstable

Table 1. Attitude of major discontinuities along different faces and their relation with tunnel alignment.

Face (drive direction)	Discontinuity set	S1	S2a	S2b	S3	S4
Face-6 (N104°)	Strike	030°-210°	167°-347°		070°-250°	150°-330°
	Dip direction/amount	300°/20°	077°/60°		160°/67°	240°/26°
	Relation/angle with drive direction	74°	63°		34°	46°
Face-7 (N284°)	Strike	000°-180°	173°-353°		090°-270°	065°-245°
	Dip direction/amount	270°/27°	083°/64°		180°/66°	335°/65°
	Relation/angle with drive direction	76°	69°		14°	39°
Face-8 (N174°)	Strike	160°-340°	175°-355°	005°-185°	084°-264°	070°-250°
	Dip direction/amount	250°/27°	085°/65°	275°/67°	174°/67°	340°65°
	Relation/angle with drive direction	14°	1°	11°	90°	76°
Face-9 (N354°)	Strike	010°-190°	001°-181°	002°-182°	089°-269°	077°-257°
	Dip direction/amount	280°/24°	091°/78°	272°/78°	179°/80°	347°/78°
	Relation/angle with drive direction	16°	7°	8°	85°	83°
Face-10 (N174°)	Strike	028°-208°	170°-350°	005°-185°	087°-267°	081°-261°
	Dip direction/amount	298°/25°	080°/61°	275°/81°	177°/78°	351°/84°
	Relation/angle with drive direction	34°	4°	11°	87°	87°

wedges of considerable height but genteelly dipping foliation joint which cuts across the wedges generally restricted the wedge formation to its theoretical apex height as such the height of the wedges formed was generally less than that expected with a flat end instead of a conical apex. Along Face- 8 and 9 intersection of steeply dipping joints gave rise to formation of cubical wedges with a flat apex formed by the dip surface of foliation, which cuts across the cubical wedges. Along Face-8 the wedges were formed along the right side of the crown while along Face-9 they were restricted to the left side of the crown.

3.4 Tunnel squeezing and rock burst in high cover zone

Tunnel squeezing/closure and mild rock bursts were observed between RD 4035 to 5950m along a total length of 1915m, covering part of Face-4 and 5 under high cover zone where superincumbent cover varied from 750 to 1000m. As per DPR, strong and massive granite, with good tunneling was anticipated along this high cover stretch. However during course of excavation granite was not encountered and tunnel negotiated through mica schist and quartz biotite gneiss. Fragile/ incompetent biotite schist was encountered along the stretch being excavated through Face-4, while comparatively competent quartzitic schist and gneiss was encountered along Face-5. High vertical stresses along Face-4 induced tunnel squeezing/closure and buckling of the steel ribs (figure 6). Along Face- mild rock bursts along with post excavation slabbing due to dilation of rock mass under high stress was observed where crown failure upto 1.5m depth were observed (figure 7). The most effected stretch of HRT where tunnel closure was observed lied between RD 1196 to 1428 of face -4, where convergence of more than 100m was generally observed, with a maximum convergence of 562mm recorded at RD 1358m.

To tackle the squeezing excavation was carried out by heading and benching method. Designed support system comprising of 100m Steel Fiber Reinforced Shotcrete (SFRS), 4m long 25mm dia. rock bolts was replaced by 150mm SFRS, 6m long 32mm dia. rock anchors with wider base plates (200 X 200mm) for better load distribution. 9 to 12 m long, 76mm dia. stress relief holes were

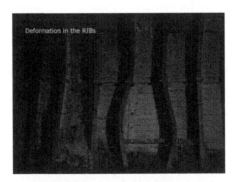

Figure 6.   Photograph of buckled ribs in high cover zone along Face 4.

Figure 7.   Post excavation dilation resulting in slabbing along Face – 5.

provided for release of stresses. In class-IV (Poor) rock mass the invert of the HRT was excavated in arc form instead of straight line as in D shaped. Invert struts, where squeezing continued during benching, were also provided. To maintain the tunnel profile, excavation was carried out by precision (line) drilling with low intensity blasting. Along Face-5 where mild rock bursts were observed/ anticipated excavation was carried out by controlled blasting with line drilling of dummy holes along the tunnel periphery to minimize the effect of blast induced vibrations on the rock mass along the tunnel periphery. SFRS was replaced by plain shotcrete with welded steel wire mesh. 4m long, 25mm dia. rock bolts were replaced by 6m long, 32mm dia. rock anchors. Stress relief holes were also provided.

3.5   *Tunneling through water charged strata*

Interception of ground water is one of the major challenges faced during the excavation of the tunnel. Ingress of ground water during tunneling may result in face and roof collapse hampering the progress (Sharma and Tiwari 2015). High pore pressure behind the tunnel periphery may also adversely affect the support system and cause its failure (Mauriya et. al., 2010). Along the HRT ground water with ingress of more than 100 liters/minute was intercepted along Face-8, 7 and 2, with some intermittent stretches along Face-3, 5, 7 and 9 (figure 6 and 7). Total stretch of water charged strata was 705m. The methodology adopted to negotiate through water charged reaches comprised of diverting water ahead of face by drilling 10 to 15m long drainage holes, inclined at an angle of 30 to 45° and radiating 15 to 20° outwards from the face (figure 10), followed by excavation and installation of support system as per rock class. Additional drainage holes were also drilled after providing primary and secondary support to ensure free flow of water from the supported reach to prevent accumulation of water and building up of pore water pressure. Flexible pipes were connected at the mouth of the drainage holes to channelize the water into the drainage network which comprised of interconnected system of sumps, drains and pipes with pumps attached for steady removal of water from the tunnel.

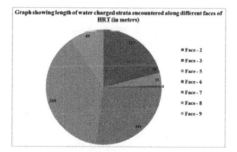

Figure 8.   Graph showing length of water charged strata along various faces.

Figure 9.   Photograph showing ingress of water from drainage holes drilled prior to *face advancement*.

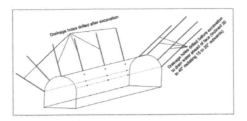

Figure 10.   Schematic Section showing configuration of drainage holes.

3.6 *Shear zones*

Shear zones are characterized by highly deformed, sheared/pulverized, water charged poor rock mass conditions is the biggest nightmare of tunnel engineer. Serious tunneling problems have been experienced when the rock mass is affected by shear zone. The most common problems associated with shear zones are, loose fall, muck flow, chimney formation, squeezing and heaving of the ground, face and crown collapse (Panthi, 2007, Maurya et. al., 2010, Sharma and Tiwari, 2015). These problems are attributed to less standup time of poor rock mass of shear zone where RMR value is less than 20 (Bieniawski 1989). Along the entire length of the HRT only two major shear zones were encountered. A foliation parallel shear zone of 7 to 8 m thickness was encountered at RD 287 of Face-7 and a 4 m thick at RD 1424 of Face-9. Both the shear zones comprised of rock fragments along with clay and rock flour. The encountered shear zones resulted in formation of 7 to 8 m high cavity between RD 287 to 293 of face-7 and RD 1424 to 1426m along Face-9, with continued loose fall. The shear zone along face-9 was accompanied with heavy ingress of water resulting in muck flow and face collapse. The continued loose fall was restricted by spraying many layers of shotcrete at a regular interval till falling of loose/sheared material stopped, followed by cautiously erecting steel ribs and back filling the cavity. After treatment of the cavity the extent of shear zones was probed by aid of drilling 15m long probe holes. Based on the probe hole data thickness of shear zone along Face-7 was estimated to be of the order of 7 to 8m and its extent along the tunnel was about 63m i.e. upto RD 350 (figure 11), while the thickness of shear zone along Face-9 was deciphered to be 4m and extending for a length of 43m along the face i.e. upto RD 1467 (figure 12).

Once the extent of the shear zone was estimated excavation was cautiously carried out by heading and benching method. Prior to excavation strata ahead of the face especially crown was strengthened by aid of pipe roofing, followed by grouting with micro-fine cement with a grout consistency varying from 3:1 to 1:1 (water: cement by weight) under pressure between 2 to 12 kg/cm2. The excavation in the shear zone area was carried out by heading and benching method by mechanical means and comprised of first excavating 1m span of tunnel upto spring level (spl), followed by erection of crown portion of the rib and back filling the void between

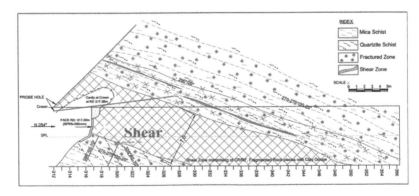

Figure 11.   L-Section along HRT Face-7showing disposition of shear zone between RD 287 to 350m.

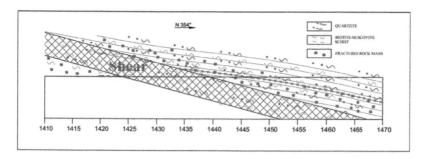

Figure 12. L-Section along HRT Face-9 showing disposition of shear zone between RD 1424 to 1467m.

rib crown and rock surface, this was continued till the heading was completed for 5 to 6m length. After heading for 5 to 6 m length and installation of support system along the crown, benching for this stretch was undertaken in stages of 1 to 2 m, followed by erection of columns. The same methodology of excavating and supporting crown to 5 to 6m length by heading in stages of 1 to 2 meter and then benching down to invert level in stages of 2m was continued till the entire shear zone strata was negotiated.

3.7 Poor rock mass conditions

The rocks of Himalayan orogen have undergone many faces of deformation, resulting in development of intense jointing and weakening of the rock mass, giving rise to adverse tunneling conditions. Adverse geological conditions in form of weak and fragile rock mass strata were encountered along entire Face-10, major part of Face-9, 7 and 6 along with some intermittent stretches along Face-2, 3, 4, 5 and 8. Length of poor rock mass condition along various faces is graphically shown in figure 13. Total length of poor rock mass conditions encountered along the

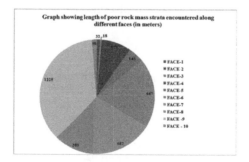

Figure 13. Graph showing length of poor rock mass strata along various faces.

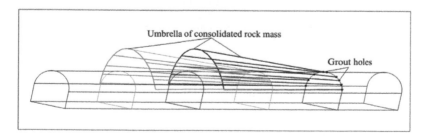

Figure 14. Schematic section showing formation of umbrella of consolidated rock mass around tunnel periphery by grouting.

HRT was 3387m corresponding to 25% of total tunnel length. The poor rock mass along Face-6, 7, 9 and 10 comprised of slight to moderately weathered, thinly foliated biotite schist and quartzite with bands of schist of low strength and intense jointing and characterized by presence of interfolial shear seams of thickness varying from 1 to 10cm. This low dipping stratum especially along Face-9 and 10 where the strike of foliation was almost parallel to tunnel drive direction resulted in continued slabbing and over breaks.

The excavation along poor rock mass, stretches was carried out by providing steel ribs support and lattice girders. Lattice girders were provided along faces where tunnel profile was not structurally controlled and over breaks were negligible especially in Face-4 and 5. Crown strengthening by pre-grouting, fore polling and pipe roofing to create an umbrella of consolidated rock mass around tunnel profile (figure 14) was undertaken to increasing stand up time so that adequate time is available for installation of primary and secondary support. The excavation methodology was continuously evolved so as to arrive at an optimum blasting pattern comprising of optimum drilling pattern, charging pattern and delays so that a desired pull with minimum over breaks be achieved. Apart from improving the rock mass strength by conventional method of grouting and strengthening crown by aid of fore polling and pipe roofing emphasis was given on reducing the cycle time so that the excavation of the HRT was completed as per schedule.

4 CONCLUSION

Tunneling in Himalayas is a tough challenge attributed to complexity of the geology brought by the complex orogenic processes. The complex geological conditions, along with enormous variation in rock mass conditions as projected in DPR necessitated continuous modification and optimization of excavation methodology and support system to maintain a steady progress, which was achieved by implementing advanced tunneling methodology comprising of ground improvement by grouting, crown strengthening by fore polling and pipe roofing, controlled blasting, installation of timely and adequate support system, monitoring of rock mass behavior by geotechnical instruments and probing strata ahead of face to infer the rock mass conditions for proper planning of excavation methodology, support requirement and resource planning in advance to ensure timely commissioning of the project, while also ensuring the stability of the structure.

REFERENCES

Bhargava, O.N. 1995. The Bhutan Himalaya: a geological account. Spec. Publ. Geol. Surv. India, Vol.39.
Bieniawski, Z.T. 1964. Engineering rock mass classification. John Willy and Sons Inc., New York.
Ganser, A., "The Geology of the Himalayas", Wiley Interscience, New York, 1964.
Ganser, A., "The Geology of the Bhutan Himalayas" Birkhavser, Basel, 1983.
Grujic, D., Holister, L.S., and Randall, R.P. 2002. Himalayan metamorphic sequence as an orogenic channel: Insight from Bhutan. Earth and Planetary Sciences Letters, Vol. 198.
Guha Sarkar, T.K. 1979. Geology of the Gaylephug-Trongsa-Byakar area, Central Bhutan. Publ. Geol. Surv. India Misc. Pub., Vol. 41(1), 1979.
Jangpangi, B.S. 1974. Stratigraphy and tectonics of parts of Eastern Bhutan. Himalayan Geology, Vol. 4.
Jangpangi, B.S. 1978. Stratigraphy and structure of Bhutan Himalayas. In Saklani, P.W.S., ed., Tectonic Geology of Himalaya, Todays' and Tomorrows' Publications, New Delhi.
Mangdechhu Hydroelectric Project (4X180 MW) Trongsa, Bhutan., "Detailed Project Report. Site investigation and geology", Volume IV, 2008.
Maurya, V.K., Yadav, P.K., Angra, V.K. 2010. Challenges and strategies for tunneling in the Himalaya region. Proceedings: Indian Geotechnical conference, GEOtrendz, Mumbai.
Nautiyal, S.P., Jangpangi, B.S., Singh, P., Guha Sarkar, T.K., Bhate, V.D., Raghavan, M.R., Sahai, T.N. 1964. A preliminary note on the geology of Bhutan Himalaya. Proc. 22nd Int. Geol. Cong., New Delhi, Vol.11.
Panthi, K.K. 2007. Underground spaces for infrastructure development and engineering geological challenges in tunnelling in the Himalaya. Hydro Nepal: Journal of Water, Energy and Environment, Vol. 1(1).
Sharma, H.R., Tiwari, A.N. 2015. Challenges of tunneling in adverse geology. Proceedings: EUROCK and 64th Geomechanics, Colloquium, Schubert (ed.), 2015.

Tunnels and Underground Cities: Engineering and Innovation meet Archaeology,
Architecture and Art, Volume 7: Long and deep tunnels – Peila, Viggiani & Celestino (Eds)
© 2020 Taylor & Francis Group, London, ISBN 978-0-367-46872-9

Study on cracks in concrete lining based on inspection records on tunnel

T. Miyaji, H. Hayashi & M. Shinji
Yamaguchi University, Ube, Japan

S. Kaise
Nippon Expressway Research Institute Company Limited, Tokyo, Japan

S. Morimoto
Dobocreate Corporation, Ube, Japan

ABSTRACT: Japan has over 10,000 road tunnels. There is something which consists of inspection A and B in Tunnel lining inspection system. Inspection A is carried out the desktop inspection system which extracts the priority inspection area through the desktop study of the photographing record by the lining inspection image photographing system. Based on the results of Inspection A, the detailed inspection with the proximity visual inspection is carried out by engineers (Inspection B). We compare the scored of both inspections, we found the gap in it. One of the reasons is the temperature of inspection.

In this paper, we compare the change of the width of the cracks records existing on the tunnel lining of both inspection by the temperature of the inspection. As the results, it was found that there was possibility of the change in crack width due to the material deterioration and other than change in temperature.

1 INTRODUCTION

Maintenance of civil engineering structures is an important issue in Japan, Currently, Japan has about 10,000 tunnels. Tunnels that are more than 50 years old since construction are 20% out of 10,000, the proportion is expected to increase in the future. Furthermore, Japan is concerned about the shortage of engineers due to the declining birthrate and aging population. Therefore, advancement of lining inspection is desired. On highway tunnels managed by East Nippon Expressway Company Limited, Central Nippon Expressway Company Limited and West Nippon Expressway Company Limited (NEXCO) lining detail inspection system consists of a combination of desktop inspection, which extracts the priority inspection points through the desktop study of the lining development image photographing record by the lining inspection image photographing system, and detailed inspection by engineers. To comprehend the characteristics of the inspection that NEXCO is doing, we compared the two inspection. NEXCO is conducting scoring at inspection. This score is called an evaluation point. Figure 1 shows the distribution of evaluation point inspection B and inspection A. The red line shown in Figure 1 is a straight line with the same number of evaluation point, which shows that there are variations in evaluation point at the time of inspection from Figure 1.

Therefore, in this paper, we focused on the inspection result of the highway tunnel. We examined the change of crack width due to temperature, which is considered to be one of factors of scattering of evaluation points.

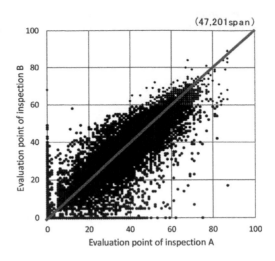

Figure 1.   Distribution of evaluation point inspection B and inspection A.

2   NEXCO'S TUNNEL INSPECTION

As mentioned earlier, NEXCO carries out two types of inspections as inspections A and B as regular inspections of the tunnel.

Inspection A is carried out the desktop inspection system which extracts the priority inspection area through the desktop study of the photographing record by the lining inspection image photographing system.

Based on the results of Inspection A, the detailed inspection with the proximity visual inspection is carried out by engineers (Inspection B). As in the case of inspection A, inspection B also calculates the evaluation points for the deformation from the crack index and observation items. Then, using the calculated evaluation points, soundness is evaluated for each span.

At the time of inspection of each of the inspection A and the inspection B, evaluation points are calculated using the detailed inspection A sheet (A sheet) and the detailed inspection B sheet (B sheet). From the crack map, NEXCO calculate the evaluation point of the external force of the evaluation points for the structural stability and the evaluation point of the evaluation point for the safety of the user. In principle, this sheet will be created in units of span, which is the length of the concrete lining. In addition, Tunnel-lining Crack Index (TCI) Points is used for the evaluation of two items "crack width, length, distribution" and "crack direction" of the A sheet and B sheet items. TCI is calculated using the width, length and angle of the crack

3   VARIATION OF RESULTS BY MULTIPLE TUNNEL INSPECTION

In this chapter, we tried to analyze and consider the inspection results in detail for the two tunnel objects that were inspected twice proximity visual inspection.

3.1   *Targeted tunnels*

In investigating cracks, in consideration of the inspection interval, temperature difference and tunnel extension in selecting the tunnel to be considered. Moreover, it is known that the value of TCI can fluctuate in inspection A and inspection B due to errors of lining image accuracy and interpretation from the past study. Furthermore, since the inspection A is a desktop inspection that inspects from the lining image, since the inspection B is an inspection that makes proximity visual inspection at the site, the tunnel which conducted the inspection B several times has the highest inspection accuracy. It can be judged. From this fact, in this paper, Tunnel P and tunnel Q in which inspection B was conducted twice were targeted. Table 1 shows outline of tunnels

Table 1. Outline of tunnels.

Tunnel	Length m	Number of spans	Construction year
P	316.7	31	1993
Q	245.1	24	1993

3.2 *Analysis of inspection results*

We extracted the span that changed at the evaluation point from the first B inspection to the second inspection from the B sheet of two times made with P, Tunnel Q. Then, we compare the B sheets of two extracted spans, and tracked the change in judgment on the deformation on the B sheet.

Figure 2 shows the items that changed as a result of tracking changes in judgments against variations and the number of spans. From this figure, it was found that items of "crack width, length, distribution," crack directionality "," crossing/branching "are changed in Tunnel P and tunnel Q. As an exception, the span No. 20 of the Tunnel Q also had items corresponding to "crack width, length, distribution" and "closed type" and the score changed. Also, in span No. 4, there were items corresponding to "crack width, length, distribution", "crossing/branching", "leakage", and the score changed.

Since the item of "crack width, length, distribution" calculated by using TCI in many spans among the spans where the evaluation point has changed has changed, the variation in the evaluation points in the inspection B in twice It is inferred that the change in the TCI value may have a large effect. Figure 3 shows a comparison of evaluation point distributions at two inspections B. The straight line (red) is a straight line with which the evaluation points coincide. From this figure, it can be seen that there are some cases that the second evaluation point becomes high overall, and in some cases the result is as large as 20 points.

Considering that the above-mentioned variation is greatly influenced by fluctuation of TCI value, the following factors can be mentioned as fluctuation factors of TCI value from reference.

1) Progress of crack injury
2) Overestimation of crack damage
3) Underestimation of crack damage
4) Implement repair for crack damage
5) Fluctuation of crack width due to temperature change

In literature, a temperature change of 19.7 ° C. was confirmed from the time of the previous inspection, but it is considered that the main factor of the increase in the crack width is progress of crack damage. Therefore, in the next chapter, we will investigate whether the effect of temperature is small even when the target tunnel changes, in the tunnel targeted in this paper.

4 CHANGES IN CRACK WIDTH DUE TO CHANGES IN TEMPERATURE

Table 2 shows the temperature during inspection of tunnel P and Tunnel Q and the interval between inspection days. Since the temperature on the day of inspection was not measured

Table 2. Temperature during inspection of tunnel P and Tunnel Q and the inspection interval.

Tunnel	1st Inspection Date	2nd Inspection Date	1st Inspection temperature °C	2nd Inspection temperature °C	Inspection Interval days
P	5/9/2013	3/8/2016	23.9	28.7	1063
Q	28/8/2013	15/7/2016	27.3	24.8	1052

data of the temperature inside the tunnel, the nearest daily average temperature was adopted for the tunnel. In order to confirm the change in the crack width due to the change in temperature, crack maps made in two inspections B were used to compare crack widths in the first inspection B and the second inspection B. The inspection interval in this chapter is the number of days from the first inspection B to the second inspection B.

4.1 *Comparison method*

For the crack map, the one obtained by two inspections B was used for the comparison in the tunnel P and Tunnel Q. Crack map Comparison was made by visual comparison, the one with changed crack width was extracted and the number was counted. Here, the judgment was made based on the crack recorded at the time of the second inspection. Figure 4 show examples of changes in crack width. The difference in temperature at the time of inspection was calculated from (the temperature at the time of the second inspection) – (the temperature at the time of the first inspection).

4.2 *Comparison result and consideration*

When the temperature difference is positive The temperature is higher at the time of the second inspection B and the lumber at the time of the second inspection B is expanded compared to the lining at the time of the first inspection B so that the crack width It is easily thought that it becomes narrow. Conversely, if the difference in temperature is negative, the

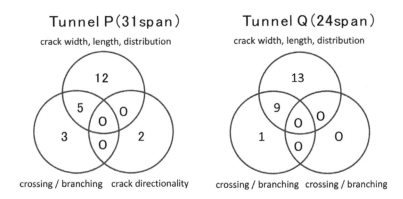

Figure 2.   Changed item and its number of spans.

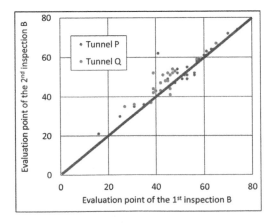

Figure 3.   Distribution of evaluation point tunnel P and tunnel Q.

4083

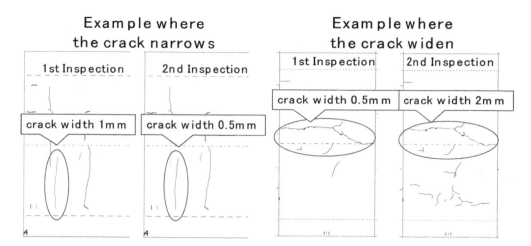

Figure 4.   Examples of changes in crack width.

Table 3.   Investigation results on the change in crack width.

Tunnel	Crack width change	Number of crack width changed	Temperature difference °C
P	Narrow	9	4.9
	Wide	2	
Q	Narrow	14	2.5
	Wide	2	

lining at the time of the second inspection B is shrinking compared with the lining at the time of the first inspection B, so the crack width becomes wider.

Table 3 shows the investigation results on the change in crack width. As shown in this table, in the Tunnel Q, it was found that the case where the temperature drops decreases seven times as much as the case where the crack width narrows. From this, it became clear that the influence on the crack width due to the temperature might be. However, when looking at the temperature difference of the Tunnel P, it is thought that the crack width itself narrows because the temperature is rising at the second inspection B. However, from this table, the case where the case where the crack width widens becomes narrow becomes 4.5 it turns out that it exists twice. From this, there is a possibility that the crack width changed due to external factors other than the influence on the crack due to temperature difference and progress of material degradation. However, since the acquired temperature is not directly measured inside the tunnel pit, we think that it is necessary to acquire the temperature inside the tunnel and investigate again in the future.

5   CONCLUSION AND FUTURE TASK

In this paper, from the inspection record for the tunnel which was carried out the inspection B twice, it was confirmed which variation on the B sheet the variation of the inspection result changed. In addition, crack map were compared, considering the possibility that the factors of the variation in the results of the two inspections are caused by the change in crack width due to the change in temperature.

As a result, in the Tunnel P and tunnel Q, it was found that the scores of the inspections B twice were mainly due to the fact that items calculated using TCI changed. In the comparison of crack maps, it was found that there was the possibility that the crack width due to material deterioration and external factors changed in addition to the difference in temperature.

In this paper, we analyzed the dispersion of inspection results of two times of two tunnels of Tunnel P and tunnel Q. Variations occurred in the comparison of inspection B, and from this fact, we assume that variations may occur due to various factors. In the future, we will continue to analyze this variation in tunnels other than Tunnel P and tunnel Q. In order to prove the relation between the temperature and the crack width, we think that it is necessary to measure the temperature and the crack width in the actual tunnel.

REFERENCES

Kaise, S., Ito, T., Yagi, H., Mizno, M., Maeda, K., Shinji, M. 2018. Evaluation method of crack propagation of tunnel lining validation *Journal of Japan Society of Civil Engineers, Ser. F1 (Tunnel Engineering)*, vol.73: I_10-I_20. Tokyo: Japan Society of Civil Engineers.

Kaise, S., Ito, T., Yagi, H., Mizno, M., Maeda, K., Shinji, M. 2018. Verification on quantitative health monitoring method of tunnel concrete lining *Journal of Japan Society of Civil Engineers, Ser. F1 (Tunnel Engineering)*, vol.74: 1–14. Tokyo: Japan Society of Civil Engineers.

East Nippon Expressway Company Limited. & Central Nippon Expressway Company Limited. & West Nippon Expressway Company Limited. 2017. *Maintenance Inspection Procedure Structure edition.* Tokyo: Nippon Expressway Research Institute Company Limited

*Tunnels and Underground Cities: Engineering and Innovation meet Archaeology,
Architecture and Art, Volume 7: Long and deep tunnels – Peila, Viggiani & Celestino (Eds)*
© 2020 Taylor & Francis Group, London, ISBN 978-0-367-46872-9

Study on deformation and failure behavior of mountain tunnel linings which consist of various materials

M. Mizutani & K. Yashiro
Railway technical research institute, Kokubunji, Japan

ABSTRACT: This paper presents conclusions drawn from model tests of mountain tunnel linings consisting of various materials. Plain concrete linings did not show any decrease in load, indicating good deformability, whereas compressive cracks and spalling occurred. Brick linings demonstrated lower structural stiffness and bearing capacity than plain concrete linings, and interlayer cracks occurred. Short-fiber-reinforced concrete linings showed good anti-spalling performance, whereas structural stiffness and bearing capacity were almost the same as in plain concrete linings. Reinforced concrete linings had the highest structural stiffness and bearing capacity, but displayed frequent shear failures.

1 INTRODUCTION

Mountain tunnels are subject to ground pressure and seismic motion depending on the geological and other conditions of where they are built. Studies have shown that mountain tunnel linings can withstand significant deformation, however, these studies do not fully take into account the characteristics of arch-structured mountain tunnels. In practice, alternative parameters are used to evaluate safety in the design process, such as tensile cracking, yielding of reinforcing bars, etc., and crack width, convergence rate, etc. for maintenance management. The use of these parameters can be explained by the lack of a clear definition of what the critical state is for mountain tunnel linings. There are a number of possible reasons why defining a critical state for tunnel linings, is difficult. Firstly, a mountain tunnel is an arched structure encased in the ground and the axial forces and bending moments intensify as deformation increases, leading to complex deformation and failure behaviors. Secondly, the quality of the tunnel lining itself which depends on the quality of construction: poor work can lead to deficiencies such as insufficient tunnel lining thickness and cavities behind the tunnel lining, which are quite common. Thirdly, the wide range of materials used for linings. Figure 1 shows the materials used for mountain tunnel lining at different times in history. Figure 1 shows that a variety of materials have been used, including bricks, concrete blocks and cast-in-place concrete. In the Meiji Era, linings were made with bricks and stones. In the Taisho Era, cast-in-place concrete was gradually introduced. At that time, concrete mixing and pouring was done manually. For a certain time, concrete blocks were used to achieve the design thickness of linings. In the Showa Era, cast-in-place concrete became the norm.

Depending on geological and topographical conditions tunnel linings may need to be reinforced. Figure 2 shows the different places where plain concrete and reinforced concrete are used for reinforcement in mountain tunnels. Plain concrete is normally used on stable ground, whereas reinforced concrete is used for tunnel entrances, shallow-overburdened sections and for fractured ground sections in anticipation of loads which will be exerted on the tunnel. More recently, fiber-reinforced concrete has in some cases been used.

Given this background, experiments were conducted using 1/5 scale models of tunnel linings made with different materials including plain concrete, brick, short-fiber-reinforced concrete and reinforced concrete in order to observe their deformation and failure behavior and thus clarify

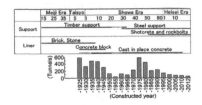

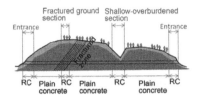

Figure 1. Materials used through history up to the present day for mountain tunnel linings.

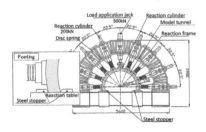

Figure 2. Typical distribution of plain concrete and reinforced concrete use in a mountain tunnel.

what the critical state is for mountain tunnel linings thereby facilitating related safety evaluation. This paper discusses the results of these experiments.

2 EXPERIMENTS

2.1 Experimental device

The experiments were conducted using an experimental device that included scale models of tunnel linings (Figure 3). The experimental device was a 1/5 scale model (with dimensions corresponding to a standard section of Shinkansen double-track tunnel). The device was composed of a hydraulic load application jack, hydraulic reaction cylinders, reaction frames, etc. Disc springs were arranged across the outer circumference of the model lining to simulate interaction between tunnel linings and the ground. Table 1 shows the specifications of the experimental device. Using a reduced 1/5 scale is less onerous than conducting full scale experiments and at the same time enables use of materials that are actually employed in tunnels, making it possible observe failure conditions similar to those found in real tunnels.

2.2 Experimental device

Five experimental cases, shown in Table 2, were used to identify any difference between the model lining materials. Figure 4 shows the dimensions of the model linings used. The model linings had a standard thickness of 150 mm, based on the thickness of 70 cm for Shinkansen tunnels used in the conventional method of the 1960s, and a longitudinal length of 300 mm.

Figure 3. The experimental device [3].

Table 1. Specifications of the experimental device.

Element	Item	Specification
Hydraulic load application jack	Maximum load	500kN
	Maximum pressure	5.6MPa
	Stroke	250mm
Hydraulic reaction cylinders	Stroke	200mm
Disc spring	Spring coefficient	3,000kN/m
	Modulus of subgrade reaction	16MN/m^3

Table 2. Experimental cases.

No.	Designation	Material	Specification
1	Plain concrete	Concrete	Plain concrete
2	Brick	Brick	Brick 2 layers
3	PPF	Concrete	Polypropylene fiber, 0.05vol% (Left half)
			Polypropylene fiber, 0.1vol% (Right half)
4	SF	Concrete	Steel fiber, 0.5vol%
5	RC	Mortar	Reinforced concrete

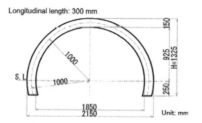

Figure 4. Dimensions of the model linings.

In Case 2, a brick lining was used. The bricks used were of the kind that are freely available on the market and had similar properties to those employed for tunnel linings in the Meiji Era. As shown in Figure 5, the bricks were arranged in a staggered pattern lengthways and stacked two layers high in the radial direction using mortar joints.

Cases 3 and 4 focused on short-fiber-reinforced concrete. In Case 3, short polypropylene fibers (PPF) were mixed with concrete according to the specifications for reinforced concrete structures designed to prevent the cover concrete from spalling to verify the spalling prevention performance when used in mountain tunnel linings. As the PPF mixture ratio of Case 3 was considered to be rather low and therefore have little effect on the load bearing performance of the model, a second ratio was also used to detect any difference between them, so the model was made with two different PPF mixture ratios, one for the right half and the other for the left half. In Case 4, steel fibers (SF) were mixed with concrete at a mixture ratio of 0.5 vol%, a ratio expected to improve flexural ductility, for comparison with Case 3.

Case 5 was designed to simulate reinforced concrete linings. Mortar was used instead. There were no design standards for reinforced concrete linings of tunnels constructed according to the conventional method. Consequently, reinforcing bars were arranged in a double reinforced arrangement as shown in Figure 6 by referring to examples (D22, bars approximately 150 mm apart) from actual railway tunnels. Shear reinforcement bars are round, therefore a reinforcement ratio of 0.22% was chosen based on the standards for reinforced concrete structures. Table 3 shows specifications for the materials used.

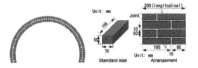

Figure 5.   Arrangement of bricks (Case 2).

Table 3.   Specifications for the materials.

Item	Specification
Concrete	Maximum size of coarse aggregate: 20mm
	Strength: 24N/mm^2 (28 day)
Mortar	1:5Mortar, Strength: 22N/mm^2 (28 day)
Brick	Size: 195 - 50 - 70mm, Joint: 1:3 mortar
Steel fiber	Size: φ0.7mm - 43mm, Mixture ratio: 0.5vol%
Polypropylene	Size: φ64.8μm - 12mm
fiber	Mixture ratio: 0.05vol% and 0.1vol%
Reinforcing	Double reinforcing, SD345
bar	Longitudinal bar: D6@27.2mm (Reinforcement ratio 1.55%)
	Stirrup: φ3 - 5 @60mm (Reinforcement ratio 0.22%)

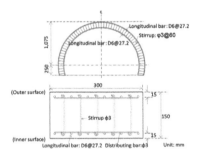

Figure 6.   Arrangement of reinforcing bars (Case 5).

Experiments were conducted by applying a load perpendicularly onto the crown using a hydraulic load application jack until a controlled displacement of 50 mm (40 mm in Case 2) was reached.

Figure 7 shows the measurements that were considered. While loads were being applied, the following measurements were taken: the load being applied, the displacement of the crown where the load was being applied and normal displacements (at nine angles) on the inner surface of the lining. In Case 5 (reinforced concrete), the strain on the reinforcing bars was measured at the angles of the lining where normal displacement, mentioned above, was measured.

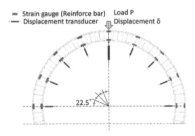

Figure 7.   Measurement items.

3 EXPERIMENTS RESULTS OF EXPERIMENTS

3.1 *Failure modes*

There were three main failure modes. After the appearance of some tensile cracks caused by bending, the following was observed: bending compression failure and spalling in Case 1 (plain concrete), Case 3 (PPF) and Case 4 (SF); interlayer cracking in Case 2 (brick); and inclined cracking in Case 5 (reinforced concrete).

With Case 1 (plain concrete), Case 3 (PPF) and Case 4 (SF), all of which used concrete, when δ was about 2 mm, tensile cracks appeared on the inner side of the crown and on the outer side of the lining at the left and right shoulders, loading the lining into a three-hinged arch shape, and when δ was about 20 mm compressive failure occurred on the inner side of the lining at the left and right shoulders. Cracking progressed in a similar manner with all of the materials, with and without fibers. However, with the compressive failure observed at δ = 50 mm (Figure 8), Case 1 (plain concrete) saw a lump as large as a hand coming off and falling, while Case 3 (PPF) and Case 4 (SF) saw pieces no larger than those of coarse aggregate (about 20 mm) coming off and falling. There were no major differences between the PPF mixture ratios or the materials. The lining spalling preventive effect was observed even with the lower PPF mixture ratio of 0.05 vol%. This is considered to be because mortar-jointed fibers limited the spalling of small pieces of concrete even with closed cracks.

Similarly, in Case 2 (brick), when δ was about 2 mm, tensile cracks appeared on the inner side of the crown and on the outer side of the lining at the left and right shoulders, pushing the lining into a three-hinged arch shape. When δ was about 20 mm, cracks appeared at the crown, in the joint between the upper and lower brick layers. As loading continued, compressive failure did not appear on the inner side of the shoulders. When δ was about 32 mm and onward, however, cracks at the crown between the brick layers spread rapidly to the shoulders on both sides (Figure 8). In both Case 1 (plain concrete) and Case 2 (brick), the lining took on a three-hinged arch shape, despite different materials being used. However, in Case 2 the brick joints made the lining structurally uneven, and the difference in curvature between the upper and lower layers caused by deformation is thought to have led to shear force being applied to the interlayer joint as the lining deformed and consequently led to the joint splitting.

In Case 5 (reinforced concrete), when δ was about 3 mm, tensile cracking as well as inclined cracking appeared at a number of locations on the inner side of the crown. Then, tensile cracking appeared in a number of locations on the outer side between the arch shoulders and spring lines, and the inclined cracking gradually grew in width. When δ was about 25 mm and onward, the inclined cracking widened rapidly. On the other hand, the tensile cracking grew little on the inner side of the crown and in the areas between the arch shoulders and spring lines on the outer side, and shearing failure occurred as a tunnel structure (Figure 8). Compared with Case 1 (plain concrete), tensile cracks at the crown and shoulders were relatively narrow and shallow, and no compressive failure was observed on the inner side of the shoulders. This is considered to be because longitudinal reinforcement bars were provided, making the lining more resistant to bending.

3.2 *Load-displacement curve*

Figure 9 shows the relationship between load P and displacement δ (hereafter "the load-displacement curve") for each of Cases.

In Case 1 (plain concrete), Case 3 (PPF) and Case 4 (SF) in which the lining was made of concrete, when δ was about 2 mm tensile cracking appeared and lining became a three-hinged arch shaped, and the load temporarily weakened. There were no significant differences in the curve between the different types of fiber and PPF mixture ratios. With Case 4 (SF), however, the decline in load after the lining became shaped like a three-hinged arch was less than Case 1 and Case 3 and the load was larger than Case 1 and Case 3 when δ was less than 20 mm. This was probably because the decline in bending rigidity after the appearance of the tensile cracks was limited as a result of the comprehensive impact of the difference in fiber material, fiber mixture ratio, fiber geometry and fiber length. When deformation progressed up to a point

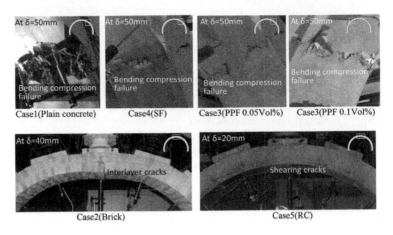

Figure 8. Failure situation.

and δ was about 20 mm and above, the load became almost the same for all Cases. After the appearance of compressive failure at the crown, rigidity declined slightly.

With Case 2 (brick), the load was lighter than the other Cases. This was considered to be due to discontinuity in the lining because of the numerous joints and that the modulus of elasticity of each single brick was smaller than the concrete, making the brick lining less rigid than the other materials.

Case 5 (reinforced concrete) displayed the highest rigidity and the largest maximum load among the materials. No significant decline in rigidity caused by tensile cracking was observed after δ reached about 2 mm, as was seen in Cases 1, 3 and 4. The rigidity declined when δ was about 3 mm and again about 14 mm. The load peaked when δ was about 22 mm and started declining rapidly when δ was about 27 mm. With Case 5, the load on the legs of the lining increased to such an extent that the steel stopper for the legs broke, resulting in both legs moving outward by about 10 mm by the time δ reached about 20 mm. The legs did not move after δ reached about 20 mm.

Figure 10 shows the deformation profiles when δ was 10 mm for Cases 1, 2 and 5. Cases 3 and 4 showed similar deformation profiles to Case 1 and therefore were omitted. Generally, the lining was displaced toward the inner space in the crown area directly below the load application point and toward the disc springs on the left and right shoulders. The outward displacement of the shoulders from highest to lowest was: Case 1 (plain concrete), Case 2 (brick) and Case 5 (reinforced concrete). Case 1 (plain concrete) saw the largest outward displacement of the shoulders probably because the location of the shoulders roughly corresponded to two hinges in the three-hinged arch where bending rigidity declined significantly. In Case 2 (brick), the lining had joints, which were absent in the plain concrete of Case 1, making the structure less rigid. It is therefore considered that the displacement caused by load application was absorbed in the area just at and around the load application point, resulting in the

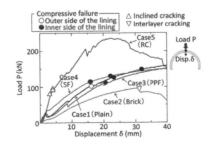

Figure 9. Load-displacement curve.

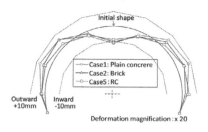

Figure 10.   Illustration of deformation (At δ=10mm).

outward displacement of the shoulders being less than that for plain concrete. Case 5 (reinforced concrete) saw the smallest displacement of all the materials. This is probably because the reinforcement bars enhanced the bending rigidity of the structure.

3.3 Strain of the reinforcing bars

In Case 5 (reinforced concrete), Figure 11 shows distribution of the strain on the longitudinal reinforcement bars when δ was 10 mm and Figure 12 shows the relationship between displacement and strain on the reinforcing bars. Figure 11 shows the reinforcement bars having a tendency to deform more at the crown and both shoulders. Figure 12 shows tensile yield at A (inner reinforcement bars at the crown) when δ was 3 mm, compressive yield at B (outer reinforcement bars at the crown) when δ was 8 mm, tensile yield at D (outer reinforcement bars at the shoulders) when δ was 12 mm and compressive yield at C (inner reinforcement bars at the shoulders) when δ was 22 mm. Figure 9 shows decline in rigidity when δ was about 3 mm and again about 14 mm. The former decline appears to have been caused by yielding at A (inner reinforcement bars at the crown) while the latter likely resulted from yielding at D (outer reinforcement bars at the shoulders). In addition, the load stopped increasing when δ was 22 mm, which is probably because the lining was pushed into a three-hinged arch shape after yielding at C (inner reinforcement bars at the shoulders).

3.4 Load bearing capacity and deformation performance

The load bearing capacity and deformation performance observed in each Case are summarized below.

Figure 13 summarizes the maximum load P_{max} (load bearing capacity) and the displacement at failure δ_d (deformation performance) for each Case. As shown in Figure 9, no decline in load was observed in Case 1, Case 3 or Case 4 even when δ reached 40 mm (corresponding to 200 mm in real size), a magnitude of deformation rarely occurring in real tunnel, making it impossible to define structural failure appropriately. Failure was therefore defined from a safety point of view, when spalling (of pieces larger than those of coarse aggregate) occurred on the inner surface of the lining as a result of compressive failure. With Case 2 (brick), failure was defined as having occurred when interlayer cracking appeared. With Case 5 (reinforced concrete), failure was

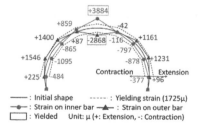

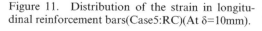

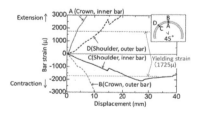

Figure 11.   Distribution of the strain in longitudinal reinforcement bars(Case5:RC)(At δ=10mm).

Figure 12.   Relationship between displacement and reinforcing bar strain (Case5:RC).

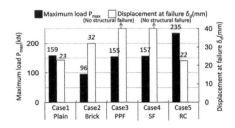

Figure 13. Load bearing capacity and deformation performance.

defined as having occurred when the load rapidly declined following the appearance of shearing cracks.

Figure 13 indicates the following: in Case 2 (brick), P_{max} (load bearing capacity) was smaller but δ_d (deformation performance) was greater than in Case 1 (plain concrete). In Case 3 (PPF) and Case 4 (SF), P_{max} (load bearing capacity) was roughly the same as in Case 1 (plain concrete) and deformation performance was also considered high as no major spalling was observed as shown in Figure 8. In Case 5 (reinforced concrete), P_{max} (load bearing capacity) was high but δ_d was nearly the same as in Case 1 (plain concrete) as shear failure occurred and the load declined relatively early in this experiment. Based on the above, reinforced concrete was clearly not an option to significantly improve deformation performance. This appears to be because tensile cracking to a certain extent poses no problem for arch-shaped linings in mountain tunnels, offsetting the advantage of increased ductility that reinforced concrete could bring.

4 DIFFERENCE IN LOAD BEARING CAPACITY AND DEFORMATION ERFORMANCE BETWEEN THE MATERIALS

Assuming that the materials tested above are adoptable as mountain tunnel linings, this chapter will discuss differences in load bearing capacity and deformation performance between the materials. Mountain tunnels are constructed with a variety of section profiles and structures to meet various ground conditions. In addition, tunnel linings are often found to have cavities, or insufficient thickness. The subsequent section will discuss tendencies of tunnels satisfying the following assumptions:

– The ground is either sediment or low-strength soft rock. The lining has a sectional force that is generated due to moderate bending moments and axial forces.
– The tunnel lining does not have cavities and is sufficiently thick.

(1) Plain concrete lining
As deformation progresses, cracking occurs, reducing rigidity, while the load being applied holds until the deformation becomes relatively large. These are some of the benefits of the arch structure. That said, compressive failure occurs on the inner side of the lining before the load peaks.

Plain concrete with reasonable levels of load bearing capacity and deformation performance is considered to be appropriate as the material for the lining of mountain tunnels being constructed in ordinary ground, which requires some margin and does not require application of specific loads. Nevertheless, as deformation progresses, compressive failure and spalling can occur. Therefore, a tunnel with cracks and deformation needs to be monitored regularly and retrofitted with anti-spalling and other measures suited to the amount of deformation to prevent compressive failure and spalling.

(2) Brick lining
Brick linings have lower rigidity and load bearing capacity than concrete, and develop interlayer cracks as deformation progresses. Having low bending rigidity, brick linings contain deformation within a small area around the load application point.

Given the many hinges in the brick lining structure of a mountain tunnel, brick linings have lower bending rigidity and load bearing capacity and are more vulnerable to deformation than plain concrete linings. Being a layered structure, a brick lining needs inspection to detect interlayer separation if the tunnel is deformed.

(3) Short-fiber-reinforced concrete lining

Ordinary short-fiber-reinforced concrete linings can have different mixture ratios and has roughly the same levels of rigidity and load bearing capacity as plain concrete linings. The decline in load with steel-fiber-reinforced concrete linings is less that for plain concrete linings when the lining becomes shaped like a three-hinged arch as long as deformation is moderate. Spalling occurs less often, and this still holds with lower mixture ratios of fiber.

For large deformation, fiber-reinforced concrete linings has the same level of load bearing capacity as plain concrete linings but is effective in spalling prevention and has higher deformation performance. Therefore, it can be used as a measure to prevent spalling for a tunnel constructed in fractured ground sections and other adverse conditions where deformation caused by earthquakes etc. is expected. For small deformation, fiber-reinforced concrete linings with appropriate types of fiber and mixture ratios can help limit the opening of tensile cracks and deformation.

(4) Reinforced concrete lining

Reinforcing bars greatly improve rigidity and load bearing capacity over plain concrete linings. Load increases even after reinforcing bars yield, which is the arch effect. On the other hand, bending reinforcement and resultant increase in bending strength can induce a shear failure mode.

Reinforced concrete linings has highest rigidity and load bearing capacity and is most appropriate where high load bearing performance is needed such as at tunnel entrances and in shallow-overburdened sections. In that context, however, shear reinforcement also needs to be considered.

5 CONCLUSION

Experiments were conducted using 1/5 scale models of tunnel linings made of various materials, to observe their deformation and failure, with a view to helping clarify critical states for mountain tunnel linings and facilitate related safety evaluations. Assuming that the materials experimented above were adoptable as mountain tunnel linings, differences in their load bearing capacity and deformation performance were examined. These exercises clarified, applicable to limited conditions though, critical states, load bearing capacity and deformation performance of mountain tunnels. Mountain tunnels are constructed with a variety of section profiles and structures to meet various local geological conditions. In addition, deficiencies such as insufficient tunnel lining thickness and cavities behind the lining are seen in many tunnels. It must be noted that what is presented in this paper may not be applicable to certain conditions. Further study is needed to be undertaken to clarify the deformation and failure behaviors of tunnel linings in an even wider range of conditions.

REFERENCES

Asakura, T. and Sato, Y. 1992. How does tunnel lining get broken, *QR of RTRI*, Vol.33(3):169–172
Railway technical research institute. 2007.*Digests of design and maintenance standards for railway structures and commentaries.*
Okano, N. 2007. Development of a testing machine with a large tunnel lining model, *Railway technology avalanche*, No.19: 2.

Tunnels and Underground Cities: Engineering and Innovation meet Archaeology,
Architecture and Art, Volume 7: Long and deep tunnels – Peila, Viggiani & Celestino (Eds)
© 2020 Taylor & Francis Group, London, ISBN 978-0-367-46872-9

Developing cross passages and safety niches in a rationalized way using remote controlled demolition robots

C. Montorfano & R. Giti Ruberto
Brokk Italia, Italy

ABSTRACT: Tunneling applications present a wide range of opportunities for high powered and compact remote controlled demolition robots (RCDRs). Robots are utilized in utilities, cables, water, railways, metro and road tunnels for excavation of tunnel faces, cross passages, substations, safety niches, shaft sinking and lining renovation in addition to other duties. RCDRs substitute for or work together with traditional equipment and working methods generally intended as drill & blast (D&B) procedure or the use of diesel-powered excavators. RCDRs combine power and accessibility where space limitations make it impossible or impractical to use large machines or where traditional methods slow down the workflow or are forbidden for safety reasons. RCDRs provide remote-controlled operation, small size, flexibility, reach and capability in confined spaces. RCDRs can be equipped with different types of attachments and can handle tasks such as drilling, rock breaking, excavating, scaling, rock splitting, removal of debris, and shot-creting. In this paper three case studies are presented where robots substituted for or coordinated with traditional working methods in cross passages and safety niche development.

1 INTRODUCTION

In railway and road tunnels, cross passages and safety niches are provided at regular intervals as a safe means of egress in the event of an emergency and for maintenance. A cross passage is a connecting structure between twin-tube tunnels or between tunnel and shaft structure. Cross passages are generally built by cutting an opening in the completed bored tunnel lining by dismantling the tunnel segment and mining through this opening from one tunnel to the opposite tunnel. It should be noted that, especially in case of soft bedrock, the excavation of an underground cross passage is a risky operation; under the threat of high water and earth pressure, an opening must be cut in the steel lining segment of the main tunnel. For these reasons, cross passage development is an important process in tunnel construction that requires dedicated equipment and working methods.

In the following paragraphs three case studies are presented, detailing jobsite specifics, geological conditions, working methods, equipment used, logistics, main selecting features for the equipment and results achieved. All these parameters and the results are then discussed and compared in a matrix in Table 3, to reconstruct the common path that brought to the decision of using RCDRs. The authors intend to report as much in detail as possible the data and logic behind the decision of using RCDRs but an exhaustive market review of the available equipment is outside of its scope.

For clarity, it is necessary to define that traditional equipment and working methods, refers to drill & blast operations executed by means of drill jumbos, while diesel-powered excavators, or just excavators in the text, are hydraulic heavy construction machines generally consisting of a two part boom carrying bucket or hydraulic hammer attachments and a cab on a rotating platform on top of an undercarriage with tracks or wheels. RCDR refers to an electric or diesel engine-powered hydraulic machine, radio remote controlled, with a

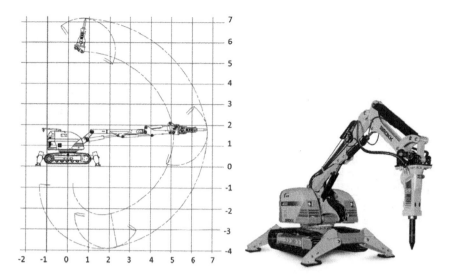

Figure 1. remote controlled demolition robot model Brokk B400, used in the Legacy way case study. On the left: arm reach in meters, on the right a picture of the machine. B400 has a weight of approximately 5.100 kg without attachment, an electric motor rated 30 kW powering a hydraulic pump with 175 bars and 130 l/min capacity.

three segment arm able to carry different types of attachments, mounted on a rotating platform on top of a tracked undercarriage (Figure 1).

2 CASE STUDY 1: PONTREMOLESE DOUBLING RAILWAY TUNNEL PROJECT, ITALY

In the Pontremolese doubling of Parma-La Spezia railway jobsite was deployed one RCDR unit model Brokk B800 equipped with a hydraulic hammer model Atlas Copco 1200 MB E and a HEB350-240 driller. The ground conditions included strong marly limestone up to 66,4 MPa UCS and average 2,05 MPa of Is50. The robot was selected among traditional working methods and analysis were done for monitoring noise, induced vibrations, excavated volumes, demolition performance and operation time reported in Table 1, with the scope of collecting data for further and larger scale RCDR deployments. The average measured rock demolition productivity was 5,23 m3/hr on net excavation time, excluding other operations while the calculated total time for execution of a complete safety niche was 9,97 working hours (Table 1). During rock demolition with the hydraulic hammer, induced vibrations to the ground ranged from 1,270 mm/s to 2,413 mm/s at 256,0 Hz while the noise level at 5 m meters distance was approximately 110 dB.

3 CASE STUDY 2: LEGACY WAY PROJECT, AUSTRALIA

The Legacy Way project in Brisbane, Australia included the construction of two parallel highway tunnels of approximately 4,3 km length each and 12,4 m outer diameter. The main tunnel excavation was carried on with tunnel boring machine (TBM) while the 35 cross passages with a size of 4,5 x 5 x 10 m, were excavated with other methods, mainly RCDRs (Kenyon, 2013 and Mucoli, 2014). The ground conditions included Bunya phyllite and Neranleigh Fernvale Beds metamorphic rocks. The Bunya phyllite strength, measured in the jobsite, ranged from 4 MPa to 160 MPa (including point load testing) and it was noted that even though there was a consistent foliation throughout the rock mass, high strength phyllite offered little or no

Table 1. Results of the performance test, conducted at the Pontremolese doubling of Parma-La Spezia railway jobsite with remote controlled demolition robot model Brokk B800 and hydraulic hammer model Atlas Copco 1200 MB E.

Test parameters		Measurement
-	Rock geotechnical characteristics	• Marly limestone • UCS (2 tests), 62,9 MPa and 66,4 MPa • Point load test average of 5 tests: Is50 2,05 MPa
-	Deployed RCBD model	Brokk B800 equipped with Atlas 1200 MB E hydraulic hammer
a	Transversal section of the niche	13,97 m2
b	Excavation depth of niche	2,58 m
c	Niche volume (a*b)	36,01 m3
d	Over-excavation volume coefficient	1,05
e	Niche volume including the over-excavation coefficient (d*c)	39,61 m3
f	Excavated volume during the test	12,79 m3
g	Excavated volume including over excavation coefficient "d"	13,43 m3
	Measured operations time	
h	Demolition of main tunnel lining	56 min
i	Cut of main tunnel beam (type 2IPN200)	88 min
l	Excavation time (g)	140 min
m	Time coefficient for shaping of niche profile	1,1
n	Total working time including coefficient m (l*m)	154 min
	Performance	
o	Excavation performance (g/n*60)	5,23 m3/hr
p	Estimated time for excavation of the niche (e/o)	7,57 hours
q	Estimated total time for excavation of niche and demolition of main tunnel lining ((h+i)/60+p)	9,97 hours

Measured vibration	measure 1	2,5m distance measure 2	measure 3	5m distance measure 4	measure 5
Vector sum mm/s	1,651	2,413	1,27	0,508	0,889
Frequency of the main axis in Hz	256,0	256,0	256,0	256,0	256,0

Notes:

• measures above were taken in the Pontremolese jobsite during rock demolition, with hydraulic hammer working at different angles in respect to the ground and therefore they are unique and not repeatable

• the frequencies measured along the three main axes (radial, vertical and transversal) are not less than 170,6 Hz

Figure 2. Left: remote controlled demolition robots (RCDRs) unit Brokk model B800 equipped with hydraulic hammer demolishing the tunnel lining segments with the necessary precision. Right: Brokk B800 shaping and refining tunnel walls.

planes of lower strength that could benefit mechanical excavation (Bennet and Norbert, 2014). Two RCDR units model Brokk B800 with 1.200 kg hydraulic hammers and two units model Brokk B400 with 550 kg hydraulic hammers were deployed in this project. The B800 was used for the cross passage tunnel face excavation while the B400 was used for shaping the tunnel walls and edges. Robots worked in minimum space without interfering with other operations ongoing in the tunnel, in support of the TBM and gave minimum vibration to the bedrock not interfering with building in surface. Two working methods were used: RCDRs alone in weak to moderately strong rock conditions and a combination of drill & blast and RCDRs in strong rock conditions. Where RCDRs were deployed after drill & blast operations, they executed the scaling and the final shaping to the requested tunnel profile corresponding to approximately 20% of the excavated volume. RCDRs then executed the finishing of the cross passages including the removal of excavated material, shotcreting and bolting. The productivity registered for the B400 robot during rock excavation was an average 2,5 m^3/hr net, excluding standby time and reached peaks of 3,5 m^3/hr.

4 CASE STUDY 3: DOHA METRO PROJECT, QATAR

Qatar is upgrading its mass transportation infrastructure, developing the Doha Metro project, that includes four railway lines named Red Line South, Red Line North, Green Line and Gold Line and the completion of the work is expected in 2019. Several RCDRs were deployed in the Doha Metro and working method and details for Red Line South and Gold Line are reported in the following paragraphs, synthetized in Table 2 and described in the next two subchapters (Ruberto, 2017). Cross passages, in Red Line and Green Line, were developed after completing the main tunnels excavation, while in the Gold Line they were excavated during TBM operations. RCDRs were selected for their high power to size ratio, their ability to handle several tasks, electrical motors with zero emissions, small size that required minimal working space and the three-part arm system that allowed for precise positioning of attachments, including bucket, breaker, rock drill and scabbler. Two different models were selected by four contractors: the Brokk B260 and the Brokk B160 with respectively about six and five meters of arm reach.

4.1 The Red Line:

The Red Line consists of two approximately 30 km long parallel tunnels with 34 cross-passages and 6 dewatering sub-pits in the lower parts of the tunnel. The cross-passages have diameters of approximately 5 m and lengths ranging from 10 m to 40 m. One RCDR model

Table 2. remote controlled demolition robots (RCDRs) models selected and deployed in the Doha Metro project for each line and working method summary.

Doha Metro line	Number of machines deployed and model	Excavation of cross passage	Working method and operations executed by RCDRs
Green Line	1 unit Brokk B260	AFTER tunnel boring machine operations	1. Consolidation phase: secure the cross passage with drilling rig attachment
Red Line North	1 unit Brokk B260		2. Demolition phase: mining the passage with rock breaker attachment for strong rock conditions or scabbler attachment for weaker rock conditions, from opposite sides, to meet in the center
Red Line South	1 unit Brokk B160 1 unit Brokk B260		3. Finishing phase: trimming excavation edges, shotcreting and removing debris
Gold Line	6 units Brokk B160	DURING tunnel boring machine operations	1. Excavation phase: mining cross passages with hydraulic hammer attachment 2. Refining phase: trimming of tunnel walls 3. Finishing phase: spoil removal and primary supports installation.

Brokk B260 and one model Brokk B160 were used in the Red Line south portion (Murrow, 2017). The working method included an initial excavation phase, placing the robots at the opposite ends of the cross passages and mining the tunnel to meet up in the middle. The subsequent refining phase included trimming the edges of the profile. Finishing phase included spoil removal and primary support installation. Excavation performance was approximately 25 m³/day per machine, in limestone and shale with compressive strength ranging from 10 to 25 MPa. In the Red Line North and Green Line in the finishing phase scabbler attachments were also used. Scabblers are hydraulic powered attachments with two rotating drums furnished with conical carbide picks allowing cutting of weak to strong rock, specially employed for trimming, shaping of tunnel walls and excavation of tunnel face through weaker rock conditions. The scabbler model ER100 has a cutting head 370mm wide that generated a 65mm cutting depth for each passage.

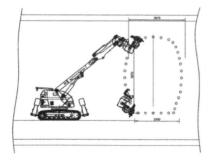 

Figure 3. Left: working scheme for remote controlled demolition robots (RCDRs) indicating drilling points in main tunnel lining for developing of cross passage with drilling rig attachment, before demolishing the lining with the hydraulic hammer and proceeding with the mining operation. Right: RCDR model Brokk B260 in Doha Metro Red Line's, mining the cross passage face.

4.2 The Gold Line:

The Gold line consists of two tunnels approximately 32 km in length, with 10 stations, 23 cross passages and 7 sub-pits for dewatering. Ground conditions included weak to moderately strong marly limestone and shale, strong to very strong limestone and dolomitic limestone up to 120

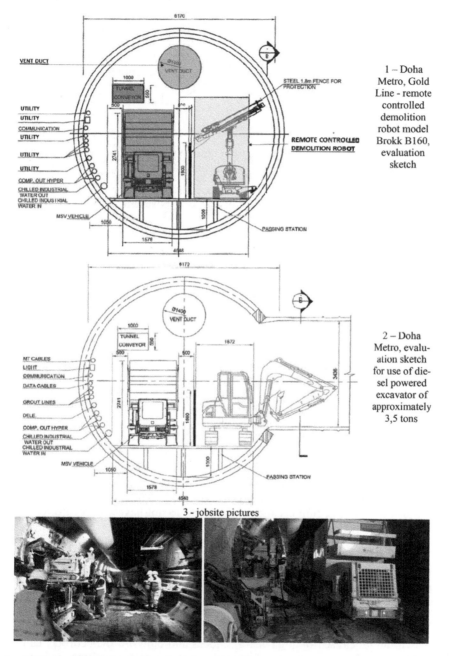

1 – Doha Metro, Gold Line - remote controlled demolition robot model Brokk B160, evaluation sketch

2 – Doha Metro, evaluation sketch for use of diesel powered excavator of approximately 3,5 tons

3 - jobsite pictures

Figure 4.   Equipment deployment layout of the Doha Metro Gold Line showing: 1) sizes and deployment of remote controlled demolition robot model Brokk B160 on the right side (yellow) operating drilling rig attachment without interfering with the multi-service vehicles (blue), the tunnel conveyor system (purple) and the vent duct (green). 2) diesel-powered excavators scenario, discarded during evaluation phase. 3) jobsite pictures of robot at work.

MPa and water saturated areas. Cross passage development had to be carried on simultaneously while the TBM was operating without interfering, therefore the available space was approximately 3 meters or about half of the main tunnel diameter (Figure 4). Cross passage excavation equipment had to be operated in tight spaces while keeping the road open and safe for maintenance vehicles, transportation of excavated material and prefabricated lining (Figure 4). Six RCDR units Brokk model B160 were deployed in the Gold Line with TEI-drilling rigs, hydraulic breakers and shotcrete attachments. The robots' working method included a consolidation phase to secure the cross passage using the TEI-drilling rigs to execute approximately 40 holes of 3" diameter and 9 m depth. The average measured time to drill one hole was approximately 15 minutes. The subsequent demolition phase included mining the cross passages with rock breaker attachment. Finishing phase included RCDRs trimming the excavation edges, shotcreting the exposed rock and removing debris.

5   DISCUSSION

During the evaluation process, for selecting working methods and equipment in the three presented case studies, the following nine main parameters, summarized in Table 3, were considered:

1) Operations shall be mechanized to avoid blasting. Blasting for cross passages and safety niches, in an already completed tunnel, causes risk exposure to utilities, structures, personnel and other ongoing operations.
2) Development of cross passages and safety niches is carried out in the narrow spaces of the main tunnel, and require dismantling of tunnel lining and beams, de facto removing the tunnel supports and therefore exposing nearby spaces to collapses of bedrock and water infiltration. Robots allow operators to maintain the necessary safe distances whether they are a few meters away or more, adapting to most situations and positioning for the best angle of view.
3) Ground conditions may vary within a project, especially rock strength, rock fracturation and presence of water; therefore, equipment shall be able to adapt as quick as possible to maximize productivity. Drill & blast method is generally carried on with dedicated drill jumbos, while diesel excavators and RCDRs offer more flexibility being able to change digging attachments; for instance, hydraulic hammer for strong rock and scabbler for weak and moderately strong rock. RCDRs offer the possibility to change the attachment in situ, without intervention of a dedicated technician, which represents an advantage. Robots and excavators can also use rock splitting attachments for very strong to extremely strong rock.
4) Equipment shall operate inside cross passages and safety niches, which are generally narrow working spaces, for example approximately 3 m wide and between 4 to 5 m in height. This requirement is particularly limiting large drill jumbos and large diesel-powered excavators. RCDRs are specifically designed for operating in narrow spaces, with very low ceilings, inside buildings, tunnels and mines. They do not have an operator cabin and they have three part arms, thus reducing the operating and traveling height. Nonetheless the small size of robots allows designers to minimize cross passages sections, bringing time and costs savings (Figure 1).
5) When operating in narrow confined spaces such as tunnels, exhaust and noise emissions are important health risk factors for workers. Exposure can be reduced and limited, but the best solution is removing their sources, for example by utilizing electric motors which generate no emissions and operate with minimum noise levels. The majority of robots, to minimize size (Figure 1) while keeping maximum torque, are engineered to utilize electrical motors.
6) Induced vibration, caused by high energy shocks and blows to the bedrock, can reach surface buildings and houses, and are a factor of major concern in underground projects. Generally, the origin of induced vibrations are hydraulic hammers, but especially blasting with its high power release. For cross passages and safety niche excavation, for example, one of the solutions to reduce vibration is to use lower weight hydraulic hammers, meaning low impact power, but with a high blow frequency. Nonetheless it must be maintained a

Table 3. Cross passages & safety niche development equipment comparison matrix with the main parameters evaluated in the three discussed case studies, comparing traditional methods intended as drill & blast, diesel-powered excavators and remote controlled demolition robots (RCDRs).

		(Traditional methods)		
#	Facts	Drill & blast	Diesel-powered excavators	Remote controlled demolition robots
1	Blasting may affect or damage tunnel structures, utilities, bedrock and other operations	Avoid blasting if possible	-	-
2	Safety distance for operators	Need to evacuate nearby areas	Operator always in the cabin	Remote controlled. Operator can freely move in the best suited position
3	Adaptability to changing ground conditions	Minimum possibility of changing attachments	Can use scabbler, hydraulic hammer and rock splitter attachments	Can use scabbler, hydraulic hammer and rock splitter and rock driller attachments
4	Ability to work in cross passages and safety niches narrow space	Require small drill jumbo	Cabin on board and two parts arms require tall working spaces	No cabin onboard and three segments harm allow low height
5	Exhausts and noise emissions	Blasting emissions and possible diesel exhaust	Diesel exhausts	Electric motors have no exhausts and minimal noise emissions
6	Induced vibration to bedrock	High power release during blasting	Hydraulic hammer attachments cause lower vibrations	Hydraulic hammer attachments cause lower vibrations
7	Possibility to execute different operations or tasks and use many attachments	Minimal	Medium. Two part arm limiting maneuverability and possible attachments	High. Three part arm for best maneuverability and reach allowing many attachments
8	Engine fuel	Diesel or electric	Diesel	Electric supplied with cable reel
9	Effects on the tunnel workflow	Require stopping all operations during blasting	Require larger working space. Not specifically designed for tunnel application	Require smaller working space. Specifically designed for tunnel application and compatible with other ongoing operations

balance between the size of the hydraulic hammer, its carrier, the tunnel size and an acceptable productivity. Robots, with the advantage of their compact size, balance an adequate hydraulic hammer with the possibility to work in narrow space while maintaining high productivity (Table 1 and Figure 1).

7) During development of cross passages and safety niches it is necessary to perform different tasks in a relatively short amount of time. For this reason, equipment able to change tools quickly and without specialized technicians are preferable. Drill jumbos are specialized equipment and have minimum possibility to change digging attachments. Excavators and RCDRs, on the contrary, can mount several types of attachments. Robots furthermore have the compactness and the three part arm advantage, which allows them to accomplish complex tool movements. For example, robots are able to shotcrete in tight environments, such as cross passages, replacing hand work.

8) In tunnel jobsites with electrical power supply networks in place (most of them), it is preferable to use machinery that can draw power from this utility avoiding managing diesel supply and delivery and diesel exhaust. Electrical power is widely available and easily

accessible while the power cable is managed with a cable reel. As mentioned at point 5 above, most robots are powered by electric motors.

9) Cross passages and safety niches developed are one of the several operations carried on simultaneously during tunnel construction. For this reason, in and out transportation of material and personnel along the main shaft shall not be interrupted or limited, so as not to affect the tunnel workflow efficiency. For example, operations of removal of excavated material, supply of tunnel support material and air venting are continuously carried on and they also require large portions of the main shaft space. Drill & blast method for cross passages and safety niches requires interruption of most operation in the tunnel therefore impacting efficiency the most. Excavators and robots do not require interrupting operations, but they may impact on the main shaft viability if their size is not adequately dimensioned. Demolition robots, as explained at point 5 above and shown in Figure 4, offer efficient size to power combination. With robots it is possible to work in minimal working space while maintaining satisfactory productivity.

6 CONCLUSIONS

The three presented case studies are located in far flung countries: Italy, Qatar and Australia. They have different ground conditions including shale, marly limestone, limestone, dolomitic limestone and metamorphic rocks; they are dedicated for different utilities, railway, metro and highway but they share similar safety, technical and design requirements. When selecting the equipment for developing cross passages and safety niches, the three contractors had to choose between drill & blast, excavators or robots, and all three came to the same conclusion of using robots. They based their decision on the main facts outlined in the previous section and summarized in Table 3. The comparison matrix of Table 3 is therefore proposed as a rationalized guideline in tunneling projects for selecting the appropriate equipment and working method for developing cross passages and safety niches. Robots demonstrated their ability to work in parallel with traditional working methods and side by side with other equipment, complementing and optimizing some tasks, or completing the cross passages and safety niches on their own. Demolition robots are a relatively young option in tunneling works, and in their technological evolution they have increased their performance while remaining compact, safe and environmentally friendly: they are well positioned to be able to face the future challenges in modern tunnel projects and should be evaluated as a viable, effective alternative solution.

7 AKNOWLEDGMENTS

Special acknowledgment to Mr Haitham Gouda of Brokk Middle East, Mr Will Visser of Brokk Australia, Mr Nasko Cepilov, Mr Marcello Montorfano and Mr John Broumis (Plant & Equipment Manager, Doha Metro project).

REFERENCES

Bennett C. and Norbert M. 2014. Tunneling within the Bunya Phyllite of legacy way, Brisbane, Queensland. *15th Australasian Tunneling Conference 2014: Underground Space - Solutions for the Future. Barton, ACT: Engineers Australia and Australasian Institute of Mining and Metallurgy*: pp 131-140.

Kenyon P. 2013 April 23, Rapid excavation breaks through in Brisbane, *https://tunneltalk.com/Australia-25Apr13-Brisbane-Legacy-Way-traffic-tunnel-first-breakthrough.php*

Mucoli M. 2014, Grande successo in Australia, *quarry & construction September 2014*, pp 35-37.

Murrow R. 2017, Breaking bad underground, https://www.tunneltalk.com/TunnelTECH-March2017-Breakers-breading-into-the-underground.php

Ruberto R. 2017, La metro di Doha in Qatar, *Strade & Autostrade*, 3–2017, pp 56-58

Tunnels and Underground Cities: Engineering and Innovation meet Archaeology, Architecture and Art, Volume 7: Long and deep tunnels – Peila, Viggiani & Celestino (Eds)
© 2020 Taylor & Francis Group, London, ISBN 978-0-367-46872-9

Mixed Ground TBM tunneling

R.M. Myrlund, H.I. Frostad & P.D. Jakobsen
Norwegian University of Science and Technology (NTNU), Trondheim, Norway

F.J. Macias
JMConsulting-Rock Engineering AS, Oslo, Norway

ABSTRACT: This paper investigates the machine performance and cutter life in a mixed ground Tunnel Boring Machine (TBM) project. The paper studies correlations between TBM operating parameters and the corresponding cutter life. The research involves a 2.2 km long tunnel stretch, with extremely complex geological conditions, changing between basalt (UCS up to 95 MPa), soft ground (silt and clay) and mixed face. The length of the zones are varying from less than a hundred meters to about 900 meters making the boring process very varying. The cutter changes reports have been used to back-calculate the instantaneous cutter life along the different geological zones, making this the basis for a machine parameter study. The main machine operating parameters focused on in this study are the thrust, torque, RPM and performance parameter penetration rate, which together with the geological pre-investigation and laboratory-testing, provide a wide range of aspects connected to the boring process and ground conditions.

1 INTRODUCTION

TBM tunneling project in mixed ground and complex geological conditions is challenging. Geological pre-investigations and lab-tests gives important information to the project management in selecting TBM machine type and excavation tools. Machine performance and cutter life in a mixed ground TBM project is investigated and the paper studies correlations between TBM operating parameters and corresponding cutter life. The main machine operating parameters focused on in this study are the thrust, torque, RPM and performance parameter penetration rate.

1.1 *Geological definitions*

Based on the scope of this study and the level of geological description the definitions by the International society for Rock Mechanics (ISRM) are used. The terms of ISRM defines ground conditions soil, soft rock and hard rock by their uniaxial compressive strength (UCS) (Barton, 1978). ISRM divided originally soil in six different grades and rock in seven. To fit the scope in this study some ground conditions are being merged (Table 1).

The ground condition with UCS between 0.25 MPa and 0.5 MPa is in a transition zone between extremely weak rock and hard soil (Table 1). In this project the geological zones vary from less than a hundred meters to about 900 meters and changes between basalt (UCS up to 95 MPa), soft ground (silt and clay) and mixed face.

Table 1. Proposed soil UCS grades.

Parameter	UCS (MPa)
Soil	< 0.5
Extremely weak to weak rock	0.25 – 25
Medium strong rock	25 – 50
Strong rock	> 50

1.2 Mixed ground conditions

From the literature on the topic of TBM boring in mixed ground conditions the term "mixed ground", there is no clear definition and can be considered different dependent on country and project. The most used definition in the industry is probably that the geological properties of two ground types needs to have a difference in uniaxial compressive strength (UCS) of 1:10 or more (Zhang, 2010, Steingrimsson, 2002, Tóth, 2013).

The term mixed ground is in this study used for the presence of two or more geological conditions with clear difference in ground properties along a planned tunnel path and mixed-face will be used for a presence of two or more geological condition in the same tunnel cross-section. The initial definition related to mixed ground that two different ground type's need a difference in UCS of 1:10 would also be the basis for mixed-face.

1.3 TBM challenges in mixed ground and mixed face.

One of the challenges in mixed-face boring is the impact loads on the cutterhead and cutter tools that often causes time delay and extra costs. Regarding to cutter tools and composition it is often necessary to make compromises. In hard abrasive rock types is hard steel alloys often preferred, but this also makes the steel very brittle. If the steel alloys is too soft will the abrasive wear be unnecessary high.

Dependent of the rock strength in mixed-face TBM boring the steel alloy could be somewhat designed for the present ground condition. At the same time, the composition of the cutting tools must be able to excavate the strongest rock one reckons to encounter. If the cutting tools are too soft, the tool wear would be unnecessary high and the TBM performance would sink to a minimum.

The steel alloys and type of cutting tools are not the only parameters influencing mixed-face TBM tunneling. The TBM operation parameters are also of great importance dealing with mixed-face issues. To keep the machine and cutter wear at minimum and at the same time keep the production as high as possible requires a lot from the TBM operator.

If the cutterhead speed is too high the constant impacts on the cutters hitting the hard rock would cause disproportionally many broken cutters due to cracking and chipping. Boring with a low RPM often results in a low production due to the low degree of rock breaking, but reduce the broken cutters.

Table 2 shows typical wear pattern and reasons for failure for TBM boring in hard rock with disc cutters.

Table 2. Typical wear pattern and reasons for failure in hard rock.

Typical wear pattern
• Normal abrasive wear mainly on the cutter tip.
Reasons for cutter failure
• Cutter ring changed due to extensive chipping.
• Cutter ring changed due to smaller scale of chipping.
• Cutter ring stopped caused by blocked hub, resulting in flat spots on the cutter ring.
• Cutter ring worn down by normal abrasive wear.
• Cutter ring changed when the wear limit is reached.

Table 3. Typical wear pattern and reasons for failure in soft ground.

Typical wear pattern
• Pyramid form on the cutter ring.
Reasons for cutter failure
• Exposed to wear on the ring tip and on the sides.

Table 4. Typical wear pattern and reasons for failure in mixed-face.

Typical wear pattern
• Cutter ring broken due to radial fractures and ring damaged by chipping.
• Broken cutter ring sealing which causes material to penetrate the inside of the hub.
Reasons for cutter failure in mixed-face
• Failure caused by extreme stresses induced going from soft ground and hitting a hard surface.
• Cutter ring stopped caused by blocked hub, resulting in flat spots on the cutter ring.
• Damage/ broken seals causes soil to get into the hub and bearings.
• Chipped/ broken cutter rings.

Table 3 shows typical wear pattern and reasons for failure for TBM boring in soft ground. Table 4 shows typical wear pattern and reasons for failure for TBM boring in mixed-face.

2 ANALYSES AND RESULTS

In the analyzing process the results from the software TBM Instantaneous Cutter Consumption Database (TBM ICCD version 1.2) (Frostad, 2013) has been put together with some of the TBM parameters logged in the PLC system (Computer based data logger). The software TBM ICCD version 1.2 is based on the NTNU instantaneous consumption model (Bruland, 2000).

While working with the huge amount of data from the PLC-logs some data manipulations has been done. For instance, has small local extreme values been deleted or let out of the analysis since they interrupt and reduce the resolution of the graphs. This is regarded to have no significate impact on the final results since only a few values has been changed of more than a 100,000 unique data sets. Many of the extreme values are also of such character that they are believed to be wrong.

2.1 Cutter life

Figure 1 shows the estimated cutter life between shaft "1" and shaft "2" and the calculated cutter life is also a part of the basis in the following TBM operation parameter study. The cutter life is calculated by using the program TBM ICCD version 1.2. Close to a breakthrough into a shaft the results from the software program must be used with some caution. After the breakthrough in shaft "1" all the cutters on the cutterhead were changed before the boring process started up again, even if it is unlikely that all cutters were worn out/broken.

Delayed results

Since the cutters are changed when they are broken or worn down to the wear limit, and not when the geological conditions are changing, there would be a delay between the results from the software program and the geological sections.

Especially the face cutters placed in the most favorable positions with the least wear could be kept on the cutterhead through several different geological zones. The amount of wear on these cutters would therefore be hard to connect to one specific geological zone.

Thus, there is a margin of error which is not possible to eliminate since the amount of wear is not logged in each intervention. The amount of interventions is increasing significantly from around chainage 13±000 and correlates well with the reduced cutter life. It is possible to

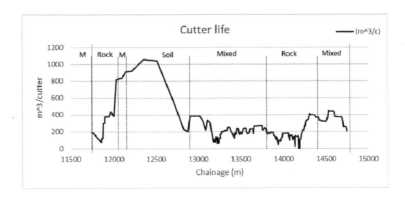

Figure 1. Cutter life between shaft 1 and shaft 2.

estimate approximately when a cutterhead intervention is necessary based on previous experience from similar ground conditions, TBM operation parameters and the geological profile.

2.2 Cutter life in the different geological zones

In Table 5 one section of the three different geological zones, mixed-face ground, soft ground and rock is compared. These three sections have been chosen because they are the longest continuous zones in each ground condition and are valuable for comparison since they have quite similar lengths. One should note that the number of interventions in the soft ground zones is less than 1/10 of the other two zones. Due to the considerable difference in tool life, this is a reasonable result. However, due to the low number of interventions, drawing any generalized conclusion from these results could be dangerous.

2.3 Type of cutter failure

The three most common cutter failure types between shaft "1" and shaft "2" are listed below, after their frequency of occurrence.

1. Cracked and/or chipped cutters
2. Blocked cutters
3. Abrasive wear

2.4 TBM parameter study

The TBM operation parameters have been analyzed for the 1473 first rings, which is equal to chainage 13 875. The cutter life is calculated all the way to shaft "2" at chainage 14 777. Thus, because of the delay between the TBM operation parameters and the tool life has the following graphs been plotted until chainage 14 200. After this is the cutter life of little interest to this part of the study.

2.5 Penetration rate

The penetration rate is calculated by dividing the boring length on the time used. In the PLC logs the boring length and corresponding boring time is logged every 20 mm. Thus, the

Table 5. Cutter life in the different geological zones.

Zone	Zone length (m)	Tool life (m³/cutter)	Number of interventions
Soil	750	667,5	6
Mixed	900	211,0	70
Rock	600	194,5	79

calculated penetration rate is close to an instantaneous penetration rate and not averaged per stroke or per 10 meter etc. Due to irregularities during boring, as for example small push-backs and subsequent boring through already excavated space, the calculated penetration rate sometimes shows extreme values e.g. penetration rate over 1000 mm/min or negative. Since these values are not of interest in this study, the data have been corrected manually.

All negative values and penetration rates above 100 mm/min has been averaged with the two neighbor values. This has been conducted to less than ten values of the more than 100 000 unique datasets. Penetration rate and cutter life in the longest soft ground and mixed-face section (Figure 2).

2.6 *RPM*

RPM and cutter life in the longest soft ground and mixed-face section (Figure 3).

2.7 *Torque*

The Torque and cutter life in the longest soft ground section and mixed-face section (Figure 4).

2.8 *Thrust*

The Thrust and cutter life in the longest soft ground section and mixed-face section (Figure 5).

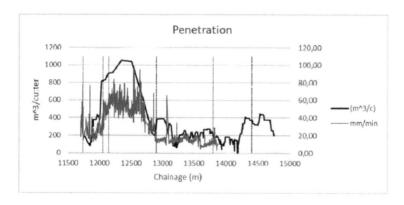

Figure 2.   Net penetration rate and cutter life.

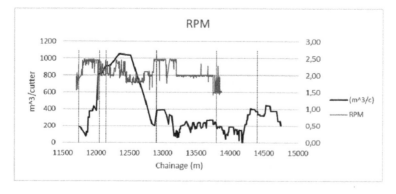

Figure 3.   RPM and cutter life.

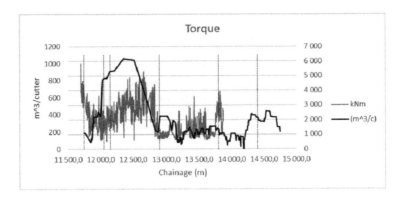

Figure 4. Torque and cutter life.

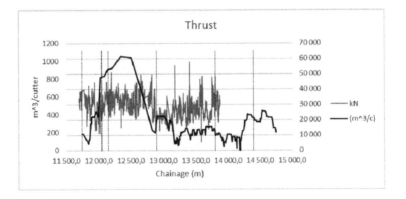

Figure 5. Thrust and cutter life.

Table 6. Average TBM operation parameters in different geological zones.

Zone	Chainage [m] rate	Zone length [m]	Average RPM	Average Torque [kNm]	Average Thrusth [kN]	Average Penetration [mm/min]
Rock	11 740–12 050	310	2,40	1 917	28 193	24,7
Mixed	12 050–12 150	100	2,11	1 866	31 708	50,1
Soil	12 150–12 900	750	2,05	2 763	32 653	50,5
Mixed	12 900–13 800	900	2,11	1 239	25 597	14,1

2.9 Average values

The rock zone between chainage 13 800 and 14 400 are included where the TBM operation parameters has not been analyzed (Table 6). Ideally, this rock zone should have been included in the TBM operation parameter study due to the extensive length, but the PLC-logs was in a format unfit for this study.

3 DISCUSSION

Strengths and weaknesses with both the methods and the results, and potential margin of errors are identified. The "human factor" is also given attention since this after all has crucial influence on TBM boring in any condition.

3.1 Cutter life

The graph in Figure 1 shows huge variations in the calculated cutter life between shaft "1" and shaft "2". There are sections of hundreds of meters' length with an average cutter life above 800 m^3/cutter followed by another section of more than a kilometer with average cutter life less than 200 m^3/cutter. This indicates large variations in ground conditions. What should be noted is that the three kilometer stretch between shaft "1" and shaft "2" is the first section. The TBM launched on this job was brand new and even if the personnel had experience with TBM tunneling from beforehand, some deployment time is necessary to get a sufficient routine.

Boring in mixed ground with the potential of ground water also requires extra attention and skills from the TBM personnel. Based on this some of the variations, especially in the beginning, could be a result more connected to operating the TBM rather than the ground conditions.

For the subsequent discussion on the difference in cutter life in the different geological sections the rock zone is used as basis since the TBM is equipped with hard rock disc cutters. The discussion is based on the results in Table 5. From this table one could see that the cutter life in the mixed-face zone is actually 8.5% higher than in the rock zone. This is not quite as expected. In a defined rock-face the wear pattern and main reason for cutter change is abrasive wear.

Problems with destroyed cutters due to extensive chipping, radial fractures and blocking should be minimized. However, one has to note if the TBM operation parameters are note adjusted to the current ground condition here is no problem to destroy cutters and reduce the cutter life to a minimum even in an optimum "homogenous" rock-face zone. The reason for the relatively high cutter life in the mixed-face zone, especially compared to the defined rock-face zones, could be connected to several factors.

For instance, the rock-face and mixed-face zones defined in Table 5 are, not totally corresponding with the geological sections. If the geological sections are close to what has been described this means that some of the cutter life in the rock-face zones is actually included in the mixed-face cutter life. Another possible reason for this unexpected result could be that the zone (s) defined as rock could be more of a mixed-face ground. All dependent on how fare from the pre-investigation drill holes one is boring the reliability and accuracy of the geological map decreases.

The basalt rock is also fractured. This ground property together with the possibility of some deep ground weathering could cause some variations in geology that is not included. It is possible that these results are correct. The cutter life could actually be somewhat higher in the mixed-face ground than in the rock. For instance, if the transition zones from rock to soft ground is consisting of mostly weathered rock, the extreme stresses induced into the cutter going from soft to hard ground would be somewhat reduced. In this case the cutters would benefit from the soft ground which, as shown in Table 5, causes a significantly higher cutter life than the rock-face.

In the end, the high cutter life in the soft ground zone of the TBM cross section would over-compensate for the issues connected to mixed-face TBM boring and result in a mixed-face cutter life slightly higher than the rock-face cutter life. In the soft ground zone, the cutter life is in average 243.5% higher than in the rock zone. This is a huge difference, but not unlikely due to the mechanisms causing the tool wear. If the ground has the necessary strength to make the cutter spin, the soft ground would pack around the cutter causing abrasive wear on the cutter ring tip, the cutter ring sides, as well as the hub. If the ground is too soft, the cutter would stop spinning and be worn flat on one or more sides.

Even if the cutter life is high, compared to the rock-face and mixed-face zones, disc cutters are not the preferable tool in soft ground. Here rippers and scrapers would be the favorable cutter tools. Rippers and scrapers are designed to excavate soft ground. The ground treatment, like, the injection of air, water and foam, would be more optimal and result in a reduced number of interventions in the end causing higher advance rates.

3.2 *TBM operation parameters*

In this part of chapter 4 the TBM operation parameters influence on the cutter life will be discussed. One of the main objectives with this paper was too reveal if one or more of the TBM operation parameters influence the cutter life to such a degree that an optimum set of parameters could be proposed. The RPM was individually one of the sub goals since this parameter is thought to have considerable influence on the cutter life in mixed-face TBM boring.

3.3 *Penetration rate*

In hard rock tunneling with TBM one adjusts the RPM, thrust and torque to get as high penetration rate as possible. When it comes to soft ground TBM boring the ground is much easier to excavate. The penetration rate is therefore optional. One tries to achieve as high penetration rate as possible without causing a change in the surrounding ground conditions. The key is to balance the excavated material, injected soil, water etc. and the forward movement of the machine. The penetration rate could therefore be seen as a semi-TBM operating parameter.

In this case the penetration rate is used to partly validate the cutter life and therefore its relevance. The huge difference in cutter life between shaft "1" and shaft "2" could be related to more than the geological conditions, for instance the "human factor". An unusual high cutter life for a period could for instance be caused by too few cutterhead interventions. The cutter tools would then be on for too long but since the ground is so soft this would just partly interrupt the boring process and/or damage the machine. This is however related to how long the interventions have been postponed. As a result, the cutter life would be artificially high.

Figure 2 shows that the penetration rate curve is correlation well with the cutter life curve. In sections with high cutter life, for instance from chainage 12150 to 12900 the penetration rate is significantly higher than in the following sections. Based on these results one could argue that the reliability of the cutter life curve is high and could be used for the further TBM operation parameter study.

3.4 *Torque*

Soft ground TBM boring is characterized by low RPM and relatively high torque where the high torque is caused by the weight of the muck inside the chamber while boring. Based on dialogue with experienced TBM personnel the applied torque could to some extent be used as a guide to determine the overall shape/condition of the cutter tools.

The practical knowledge is that a significant increase in applied torque, without any considerable change in ground condition, could often be caused by worn and/or broken cutters and an intervention is necessary.

Because the worn cutters are not excavating the ground in a satisfactory way, more workload is transferred to a reduced number of cutters and to maintain the approximately same penetration rate the applied torque increases.

In Figure 6 four cases where high torque is followed by a reduction in cutter life is highlighted. The blue rings are indicating tops in the applied torque while the following green ring is marking a local low in cutter life.

As mentioned earlier in this chapter some deployment time is often needed in the start of a TBM project. In this period the variations in both TBM operating parameters and cutter life could be more connected to human factors than the actual ground conditions.

Based on this there is a risk that case number one in Figure 6 (number one from the left) is related to issues with mixed ground TBM boring and not the overall condition of the cutter tools. Case number two from the right is also a bit uncertain.

The applied torque is significantly higher just for a short period and could be connected to other factors than the condition of the cutter tools. In general, the two cases, number to and four from the left, are the most valid examples. In these two cases the torque is increasing gradually and are followed by a relatively clear local low in the cutter life.

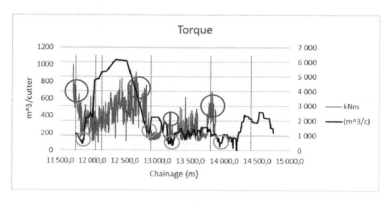

Figure 6. Possible connection between high torque and subsequent reduction in cutter life.

3.5 *RPM and thrust*

When it comes to RPM and thrust very few correlations between cutter life, geology and operating parameters are found. One of the few correlations found is that the applied thrust is varying some between the different geological zones. Especially when going from soil to mixed-face ground at approximately chainage 12900 the thrust is decreasing significantly. In both the geological zones enlarged in Figure 5, the cutter life and applied thrust are varying without any direct correlation. As for the thrust there are marginal correlating results between the RPM, cutter life and geology.

From a physical point of view there are good indications that RPM should influence the cutter life. As described in chapter 3.4 cracked and chipped cutter rings was the number one failure reason in this part of the project. Cracking and chipping is a typical brittle failure and is often connected to the extreme stresses induced when the cutters hits a hard surface like rock. Based on this a reduced RPM, which reduces the extreme stresses, should influence the cutter life in a positive way.

The fact that the RPM and cutter life curves are showing little or no correlation could be connected to several reasons. One of the things that have the potential to influence the analysis in such a way that a correlation could be hard to find is the geological sectioning.

As discussed in chapter 4.1 there are possibilities that the geological sectioning done is not completely corresponding with the actual situation. This is partly based on the experiences gained from the TBM personnel. If this is the case some of the adjustments of RPM, which are probably done in accordance with the geological map, are not done in the right place and would not give the influence on the cutter life it was intendent. Another possible reason is that a variation of RPM between 1.8 to 2.5 is too little to influence the cutter life in any significant way. In the end of the analyzed stretch the RPM is also lowered down to around 1.5. This is maybe what is needed to actually see a positive influence on the cutter life.

4 CONCLUSION

The calculations show huge variations in cutter life. This is reasonable considering the challenging and varying geological conditions and the analysis of the penetration rate indicates that the reliability of the calculated cutter life is high and is valid for further analysis. Even with some geological experiences that in some cases are not corresponding completely with the geological mapping, the results are quite good.

The cutter life in the soft ground zone is significantly higher (2–3 times) than in rock-face and mixed-face zones. To get a good division between the rock-face and mixed-face cutter life was proven harder. The mixed-face cutter life was actually 8.5 % higher than the rock-face cutter life. This difference is very little and is based on too little data to make any clear conclusion. There are also high possibilities that the results are influenced by some uncertainties connected to the boundaries of the geological zones.

In fact, the experience is that rock bodies formed like horizontal lenses could be hard to divide in different geological zones. If the pre-investigation is not very thorough it could be hard to determine with certainty how much of the tunnel cross-section is consisting of soft ground and how much is hard ground. In the transition zone between soft ground and rock, the rock could also be somewhat weathered.

One observation during the cutter life study, is that geological zones of short length should be neglected. At least the results should be used with caution. In relatively soft grounds the cutter life is so high that the cutters are kept on through several zones. The amount of wear is therefore difficult to connect to any specific section.

The TBM operation parameter study shows very little correlation between TBM operation parameters and cutter life. In the case of applied thrust the only correlation found was between the geological condition and the thrust, not the cutter life. In the RPM study there was big hopes of revealing relations between RPM and cutter life in mixed-face TBM boring. Surprisingly there was no evidence in the results that indicated that a variation in RPM between 1.8 and 2.5 causes any significant variations in cutter life.

However, the results are suffering a bit from ground conditions that are not mapped or described thoroughly enough. For instance, there are examples of geological experiences that are not corresponding completely with some of the geological mapping. Partly because of this it was hard to divide the different geological zones and know if the results were based on the correct assumptions.

Anyhow, there are still, from a physical point of view, good reasons to believe that especially the applied RPM should influence the cutter life in mixed-face TBM boring. In future studies longer sections of mixed-face ground should be investigated and a wider specter of RPM should be analyzed.

Based on the examples shown in chapter 4.4 and dialogue with experienced TBM personnel there are results that indicate that the applied torque could be connected to the overall condition of the cutter tools. Based on only a few examples the evidence is too limited to draw any conclusion. The most notable case is if the torque is increasing significantly in the same geological ground conditions. If this happens there is a high possibility that the cutter tools are worn and should be replaced. It is of great importance that the variations in ground conditions must be kept at a minimum. If not the variations in TBM operation parameters, including the torque, could be connected to something else than the cutter wear.

REFERENCES

Barton, N. 1978. Suggested methods for the quantitative description of discontinuities in rock masses. *ISRM, International Journal of Rock Mechanics and Mining Sciences & Geomechanics Abstracts*, 15.

Bruland, A. 2000. *Hard rock tunnel boring : Vol. 6 : Performance data and back-mapping*, Trondheim, Norwegian University of Science and Technology, Department of Building and Construction Engineering.

Frostad, H.-I. 2013. TBM Kutterslitasje database. *MSc Thesis. Norwegian University of Science and Technology (NTNU), Trondheim, Norway (2013), (in Norwegian)*.

Steingrimsson, J., Grv, E. & Nilsen, B. 2002. The significance of mixed-face conditions for TBM performance. Mixed Face TBM Performance. *Sydney: World Tunnelling*, 435–441.

TÓTh, Á., Gong, Q. & Zhao, J. 2013. Case studies of TBM tunneling performance in rock–soil interface mixed ground. *Tunnelling and Underground Space Technology*, 38, 140–150.

Zhang, K., Yu, H., Liu, Z. & Lai, X. 2010. Dynamic characteristic analysis of TBM tunnelling in mixed-face conditions. *Simulation Modelling Practice and Theory*, 18, 1019–1031.

Tunnels and Underground Cities: Engineering and Innovation meet Archaeology,
Architecture and Art, Volume 7: Long and deep tunnels – Peila, Viggiani & Celestino (Eds)
© *2020 Taylor & Francis Group, London, ISBN 978-0-367-46872-9*

Design of countermeasure for squeezing and swelling in mountain tunnel

Y. Okui, S. Kunimura & H. Ota
OYO corporation, Saitama-City, Japan

K. Maegawa, T. Ito & S. Kaise
NEXCO Research Institute Japan, Machida-City, Japan

H. Yagi
Central Nippon Express Company Limited, Nagoya-City, Japan

K. Nishimura & N. Isago
Tokyo Metropolitan University, Hachioji-City, Japan

ABSTRACT: The squeezing and swelling deformation causes serious damage to tunnel. For the design stage and countermeasures during construction and in service, it is necessary to predict tunnel deformation. It means that the mechanical behavior of squeezing and swelling of ground is needed to be clarified. In this paper, the relationship between swelling strains, confinement pressure and time dependency is configured. And the strength reduction should be considered. In addition, above relationships is incorporated into elasto-plastic analysis by finite element method. The results of a numerical back calculation of measured deformation in the ground and floor was identified by this method using field observation and laboratory tests. Finally, design concept of the countermeasure for residual heaving in mountain tunnel is introduced.

1 INTRODUCTION

When constructing a tunnel in swelling and squeezing ground, it is necessary to ensure not only deformation behavior during excavation but also long-term stability of a tunnel. Indeed, it is important to clarify the mechanical behavior of squeezing and swelling rock, to predict the tunnel deformation (by simple and practical method) and to establish engineering design concept (tunnel, countermeasure).

However, as mentioned later, deformation cause of swelling and squeezing often occur after tunnel construction. In this case, it is sometimes difficult to predict the long-term behavior of the tunnel, in spite of many studies that continue from the past to the present.

As for swelling, various causes have been reported from the past research, 'Mechanical', 'Osmotic', 'Intracrystalline' swelling (cray)/hydration (anhydrite), and it is briefly described by Einstein (1996) as follows: "Time dependent volume increase of the ground, leading to inward movement of the tunnel perimeter. This can be compared to a corresponding definition of squeezing: Time dependent shearing of the ground, leading to inward movement of the tunnel perimeter. And Swelling and squeezing are said to occur simultaneously."

In short, it is necessary to combine both squeezing and swelling to predict the deformation of the tunnel. Therefore, we attempt to formulate the following three ideas in combination.

1) Specific formulation of stress relief such as tunnel excavation as a decrease of mean stress.
2) Generation of volumetric swelling strain due to decrease mean stress.
3) Shear failure due to reduction of strength of ground cause of occurrence of swelling strain.

2 SQUEEZING AND SWELLING BEHAVIOR

2.1 *Typical case histories*

Some typical case histories of squeezing and swelling behavior in Japan are shown as follows.

Figure 1 shows the large deformation of tunnel section during construction. This tunnel is famous as the first railway tunnel where conventional method was applied in Japan. The conventional method was adopted instead of the timber method due to the large deformation as shown in this figure.

Figure 2 shows the long lasting deformation. In this picture, we can see the deformation of side groove and scratches on road surface by cutting as temporary measures against road heaving.

Figure 3 shows the damaged invert under road surface. There are many cracks on the surface under inverted arch by squeezing pressure. Almost cracks were caused by shear failure. When 17 years after tunnel completion, the sudden and huge heaving of road surface at the

Figure 1. Sudden squeezing in construction.

Figure 2. Long continuous deformation of pavement (Okui et al. 2010).

Figure 3. Sudden squeezing in 17 years after operation (Okui et al. 2009).

Figure 4. Typical restoration work.

center of the tunnel occurred unexpectedly, and the traffic of this tunnel has been heavily interrupted. Approximately 400mm as heaving of road surface was measured in ten days.

Figure 4 shows the situation under re-construction. The existing invert was yielded by squeezing pressure. Restoration work was the excavation of the sheared invert and install of shotcrete, inverted steel support, and casting new concrete invert of new curvature. At the result of the measurement has been applied this tunnel, the maximum heaving deformation was measured 950mm as compared with the designed cross section.

2.2 Case studies for the period up to countermeasure

Figure 5 shows the relationship between the time period up to the countermeasure is applied and the deformation velocity. A relationship between deformation velocity and timing to apply countermeasures in damaged mountain tunnels was studied in squeezing or swelling ground. In this figure, each point plotted on the graph shows the maximum value of the deformation velocity of individual tunnels and the period to countermeasure at the corresponding cross section.

Generally, when the deformation velocity is fast, the time period up to the countermeasure is short. On the contrary, when the deformation velocity is slow, it is understood that the time period up to the countermeasure is long. In Figure 5, three curves of total displacement assuming that deformation occurs from service of the tunnel are described. According this figure total displacement is in the range of 30mm to 120mm.

Therefore, it is presumed that countermeasures are being implemented by observing the state of damage and displacement of the lining.

3 CONCEPT OF PROPOSED ANALYSIS AND PROCEDURE

3.1 Concept of proposed analysis

To clarify the mechanical behavior of swelling rock, swelling test should be adopted. Figure 6 shows the apparatus of swelling test, called *oedometer* test. The consolidation testing machine is applied for this test. A rock sample with a circular rigid steel ring is mounted into an *oedometer* and is loaded with the normal stress σ.

Figure 7 shows a typical curve of time dependence of swelling strain. Upper curve is under no confinement stress, and lower curve is under confinement stress applied. On the other hand, Figure 8 shows typical relation between swelling strain and confinement stress proposed by Grob (1972).

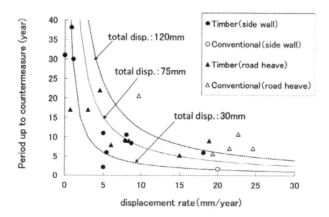

Figure 5. Period up to countermeasure vs. displacement rate(Case studies, one dot corresponds to one tunnel) (Okui et al. 2016).

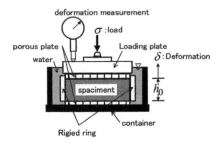

Figure 6.   Swelling test image.

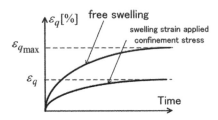

Figure 7.   Swelling vs. time relation(Swelling test results).

It is said that the swelling vs. confinement stress relationship proposed becomes a straight line as shown in Figure 8 when the constraint pressure is logarithmic as shown in Figure 9. We propose the relationship with respect to the time dependency of swelling-confinement stress as shown. Upper line shows free swelling line which applied no confinement stress. And this line implies a relationship in the final time of swelling-confinement stress. And lower line

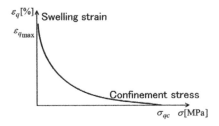

Figure 8.   Grob's swelling law.

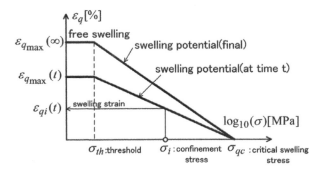

Figure 9.   Time dependent swelling law.

shows swelling potential at a certain time t. When the confinement stress σ_i is applied, swelling strain generated is $\varepsilon_{qi}(t)$. We think that this line changes to $\varepsilon_{qmax}(\infty)$ with time.

Furthermore, in order to link swelling and squeezing process, the following three ideas are integrate.

First, ISRM (1999) and Okui et al. (2018) proposed that the first invariant of the total stress controls the volumetric swelling deformations as shown in Figure 10.

Second, the relationship between strength and swelling strain presented by Uemoto et al. (1988), shows rock strength vs. swelling strain as shown in Figure 11. This figure shows one of the test results which show the relationship between strength of the rock and swelling strain. As shown in the figure, cohesion and friction angle sharply decrease due to occurrence of swelling strain. Furthermore, as seen in the slaking test and the like, it is considered that all the physical property values (for example, deformation coefficient) are reduced.

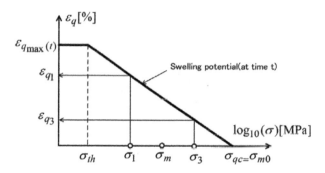

Figure 10. Specifically calculation procedure for swelling strain.

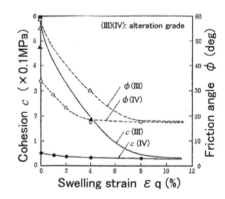

Figure 11. Swelling strain vs. strength.

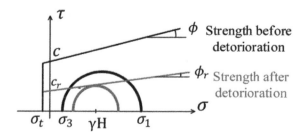

Figure 12. Strength reduction method(Phi-c reduction) by mohr-Coulomb model.

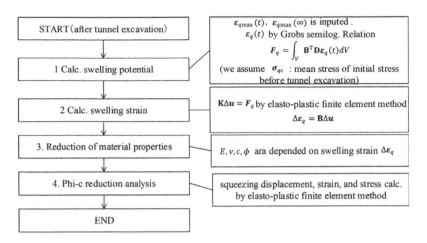

Figure 13. Procedure of numerical analysis.

Third, applying the above relationship, the Phi-c reduction method by finite element method is applied.

By integration above concept, swelling strain due to the decrease in mean stress and the reduction of strength of the rock due to the occurrence of swelling strain can be combined. In addition, it is possible to simultaneously consider plastic deformation due to decrease strength of rock.

3.2 Procedure of proposed numerical analysis

Procedure of proposed numerical analysis is shown in Figure 13. The calculation procedures aim to model the construction process. After tunnel excavation and lining installation, the first process is the calculation of swelling potential (final and time t) by the same relationship as shown Figure 9 corresponding swelling pressure. The second process is the calculation of finite element stiffness equation corresponding displacement and strain. The third process is to reduce the strength of the ground according to corresponding generated swelling strain. And the forth process is that the strength reduction method (Phi-c reduction) applied.

4 CASE STUDY FOR DEFORMED TUNNEL

4.1 Summary of deformed tunnel

The proposed method in the previous section is used to analyze one of the tunnels in swelling rocks in Japan. This tunnel was built in 1993, total length of tunnels is about 3km, and the maximum overburden is about 300m.

Figure 14 shows the heaving on road surface for 16 years after construction (okui et al. 2010). Geology of the tunnel consists of sandstone, the alternation layer of mudstone, sandstone and Tuffy conglomerate of the Tertiary. All layers contents the expansive clayey mineral (smectite).

The heaving of the road surface occurred at the alternation layer of mudstone and Tuffy conglomerate. And the vertical displacement was about 160mm at the maximum in 2009, and this phenomenon is continuing now. In this section, the overburden is 100m.

Level on road surface and underground displacement were measured with extensometer meter. Radial stress on the lining was measured by over coring method.

Vertical displacement in the ground occurred down to a depth of about 6 m, and a displacement of about 90 mm was measured at the head. Radial stresses on the lining measured about 3.5MPa at tunnel arch, and about 2MPa at side wall.

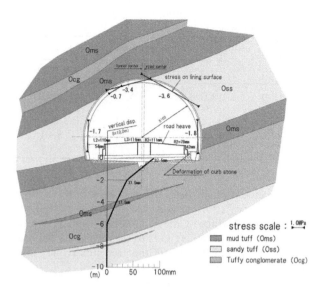

Figure 14.   Result of the measurements on the geological cross section map.

Result of a level at near the center of road surface measured as a corrected value from 2,000 to 2,010 as shown in Figure 15.

Residual displacements are predicted by curve fitting as also shown in Figure 15. As a result of prediction, the vertical displacements of center of road surface converge 266mm.

4.2   Cases of numerical analysis

Cases of numerical analysis are set as shown in Table 1. Here, case (a) is a typical value assumed from the support system applied at deformed zone, case (b) is a case where the property of the ground is lowered by one rank in view of the occurrence of a deformation, and case (c) and residual value are assumed from laboratory test.

4.3   Results of back calculation

Results of back calculation to reproduce the measurements are shown in Table 2 and Table 3. Here, Table 2 shows displacement, yield zone and underground displacement by extensometer, and shows calculated and measured value. And Table 3 shows stress distribution at tunnel lining and calculated and measured value.

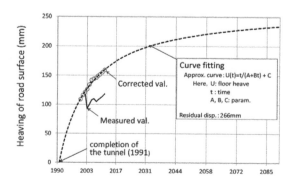

Figure 15.   The result of the level at the center of road surface and prediction of residual displacement.

Table 1. Calculation cases and input data.

Item	Initial value			Residual value
	(a)CII	(b)DI	(c)DI(ϕ15)	
Deformation modulus E (MPa)	1,000.0	500.0	500.0	30.0
Poisson's ratio v	0.30	0.35	0.35	0.48
Cohesion c (MPa)	1.0	0.4	0.4	0.1
Friction angle $\phi(°)$	40.0	35.0	15.0	4.0
Tensile strength σ_t (MPa)	0.2	0.08	0.08	0.02

In three cases, the displacement of the head of the extensometer is the same.

Underground displacements of case (a) and case (b) show a similar tendency, a displacement distribution near the road surface of the ground is dominant. On the other hand, in case (c), a displacement from the depth of 6 m similar to the measured displacement occurs.

Regarding the above differences, it is a result that depends largely on friction angle. Therefore, it is important to set friction angle.

The lining stress shown in Table 3 is represented a stress distribution reflecting the displacement of sidewall of the tunnel shown in Table 2. In particular, the case (a) shows the tendency of the sidewall of the tunnel to spread outward, which is opposite to the measurement.

From Table 1 and 2, case(c) is most closely agreement with the measurements.

Table 2. Displacement, yield zone and underground displacement.

case	(a) CII	(b) DI	(c) DI(ϕ 15)
swelling param.	$\varepsilon_{qmax}(2009) = 13.5\%$ $\varepsilon_{qmax}(\infty) = 40.5\%$	$\varepsilon_{qmax}(2009) = 7.0\%$ $\varepsilon_{qmax}(\infty) = 19.0\%$	$\varepsilon_{qmax}(2009) = 4.0\%$ $\varepsilon_{qmax}(\infty) = 11.5\%$
disp. vec. and disp. below road			

Table 3. Stress distribution at tunnel lining.

case	(a) CII	(b) DI	(c) DI(ϕ 15)
swelling param.	$\varepsilon_{qmax}(2009) = 13.5\%$ $\varepsilon_{qmax}(\infty) = 40.5\%$	$\varepsilon_{qmax}(2009) = 7.0\%$ $\varepsilon_{qmax}(\infty) = 19.0\%$	$\varepsilon_{qmax}(2009) = 4.0\%$ $\varepsilon_{qmax}(\infty) = 11.5\%$
hoop stress at lining surface	+1.1MPa +0.6MPa (-3.4) (-3.6) -3.5MPa (-1.7)	-1.2MPa -1.4MPa (-3.4) (-3.6) -1.9MPa (-1.7)	-3.0MPa -4.2MPa (-3.4) (-3.6) -1.0MPa (-1.7) ($\cdot$) measured

5 MODELLING THE TIME DEPENDENT ANALYSIS

5.1 *Modelling the time dependent analysis*

The constitutive equation reproduced above is limited to the stationary state (final and specified time t), i.e. it relates the developed swelling strains to the stress condition that is present at time $t = t$ and $t = \infty$.

Subsequently, the relationship shall be extended to the time scale. To achieve this, we start with a formal viscous approach. The rate of swelling strains is defined as following equation.

$$\dot{\varepsilon}_q(\sigma(t), t) = \langle \Phi \rangle \cdot \frac{\Psi(\sigma(t), \varepsilon_q(\sigma(t), t))}{\eta(t)} \tag{1}$$

with the retardation time η as a viscosity parameter and the vector valued creep function Ψ. This equation denotes the visco-elasto-plastic swelling strains that have developed at the considered time t. In Equation (1),

$$\Phi(t) = \frac{\sigma_m(t) - \sigma_{m0}}{\sigma_{m0}} \tag{2}$$

here,

$$\langle \Phi \rangle - \begin{Bmatrix} \Phi, & (\Phi > 0) \\ 0, & \Phi \leq 0 \end{Bmatrix} \tag{3}$$

and the vector valued creep function Ψ is

$$\Psi(\sigma(t), \varepsilon_q(\sigma(t), t)) - (\varepsilon_q(\sigma(t_0), \infty) - \varepsilon_q(\sigma(t), t)) \tag{4}$$

5.2 *Tunnel simulation*

Figure 16 shows the result of above time dependency analysis and the vertical displacement of road surface with time. In this analysis, when after the tunnel lining is installed (time t = 0), a swelling analysis is performed. This Figure shows the comparison between measured values (dot) and calculated one (dotted line). In this Figure, the upper line is the case where do not implement countermeasures. In contrast, the lower line shows the case where countermeasures were taken 20 years after the start of tunnel service.

When no countermeasure is applied, the vertical displacement of road surface became 266mm. On the other hand, the vertical displacement of road surface became 196 mm.

It can be pointed out that displacement coursed by swelling and squeezing can be suppressed by applying invert.

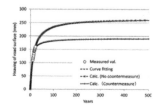

Figure 16.   Floor heave of road surface after countermeasure.

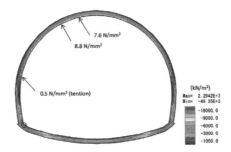

Figure 17.   Stress distributions in a lining.

Next, Figure 17 shows stress distribution of the tunnel lining. Due to the restraint closed by invert, compression stress at arch and tensile stress at side wall increase. And maximum stress emerges at a joint part between lining and invert.

6   CONCLUSIONS

In this paper, an alternative procedure for the simulation of swelling effects is proposed. The procedure is based on a stationary formulation and an explicit continuum constitutive formulation within a viscous framework.

With this approach, an algorithmic formulation that can be adopted within a finite element solution process is derived.

The results obtained in this study are summarized below.

1) A numerical analysis incorporated three mechanisms, stress relief, swelling and shear failure by strength reduction, into finite element method was proposed.
2) The proposed method is applied to an actual deformed tunnel, which shows that heaving of road surface and extrusion of the side wall can be reproduced well.
3) Design concept of the countermeasure for residual deformation in mountain tunnel is introduced.

REFERENCES

Einstein H. H. 1996. Tunnelling in difficult ground -Swelling behaviour and identification of swelling rocks. Rock mech. rock engng (29–3): 113–124.
Grob, H. 1972. Schwelldruck im Belchentunnel. Berichte des Int. Symposiums für Untertagbau, Luzern.
ISRM 1994. Commission on Swelling Rock. Comments and recommendation on design and analysis procedures for structures in argillaceous swelling rock, Pergamon Press, Oxford.
Okui, Y. Tsuruhara, T. Ota, H. Sakuma, S. & Nakata, C. 2009. Analysis of heaving behavior in Sakazukiyama road tunnel under use. Proceedings of tunnel engineering (19): 173–180, JSCE, (in Japanese).
Okui, Y. Ota, H, Hayakawa Y. & ISaji, S. 2010. Funamental study of tunnel stability based on elasto-plastic model for a road surface heaving. Proceedings of tunnel engineering (20): 85–92, JSCE, (in Japanese).
Okui, Y. Ota, H. Kunimura, S. Kitamura, H. & Nishimura, K. 2016. Case study of establish timing to apply countermeasures for damaged tunnels in squeezing rock. Proceedings of the Symposium on Rock Mechanics (44): 283–288, JSCE, (in Japanese).
Okui, Y. Nishimura, K. 2018. A model for swelling and squeezing Ground in tunneling. Journal of Japan society of civil engng. Ser. F1, Tunnel engng. (74–1): 40–51, JSCE, (in Japanese).
Uemoto, N. Otsuka, Y. & Mitsu, H. 1988. Deformation behavior of swelling-rock mass and the effects of countermeasure. Soil mechanics and foundation engnng. (36–5): 43–48, Japanese geotechnical society, (in Japanese).

Tunnels and Underground Cities: Engineering and Innovation meet Archaeology, Architecture and Art, Volume 7: Long and deep tunnels – Peila, Viggiani & Celestino (Eds)
© 2020 Taylor & Francis Group, London, ISBN 978-0-367-46872-9

Tunnel T-48: A challenge in India

A. Panciera
Lombardi Engineering Ltd., Locarno, Switzerland

ABSTRACT: India is developing its railway network, linking the peripheral regions to the national network. In Kashmir, a new line extends from Jammu, to Srinagar and Baramulla (USBRL), in Kashmir, 270 km long: 50% is underground with 4 tunnels longer than 10 km. Indian Government assigned the general project management to IRCON, a subsidiary of the national railway ministry. Long of 10'185 m, T-48 is the third longest railway tunnel in India, after T-80 and T-49 on the same line. It is single lane with lateral escape tunnel (excavation 53-74 resp. 27-34 m^2). Excavated D&B through carbonaceous and quarzitic phyllite, it has a maximum overburden of 1'150 m. At present 45% of the excavation is done. Major difficulties are the underground works execution and the logistics. The initial contractor was terminated. A new contract was awarded after 6 months. Completion is now foreseen end 2020 (3 years delay).

1 INTRODUCTION AND GENERAL OVERVIEW

The Indian government undertook a general campaign for approaching the peripheral regions to the pulsing central core of this huge nation, defined subcontinent not by chance. The railway line penetrating the core of Kashmir meets this program. By this project, the present network in the Uttar Pradesh lowland, with Delhi at its center, reaching Jammu and Udhampur in the south of the state of Jammu and Kashmir, is linked to Srinagar and Baramulla in the north of this state: Udhampur-Srinagar-Baramulla Railway Line (USBRL). Similar projects are also in progress in the states of Himachal Pradesh, Uttarakhand, Sikkim and Manipur. This new line falls in the network of the Northern Railways (the Owner), active in this region and in the neighboring Indian states. In the whole North of India, this branch of the India Railways manages a total of approx. 7'000 km of railway lines. The project, design and construction management are deputed to IRCON International Ltd (the Engineer, and Client for the Designers), a subsidiary company of the Railway Ministry. The design is assigned to international design companies, especially regarding this line with a significant underground portion. The works are awarded by public tender to Indian major contractors, specialized in public works.

The topography of the region requires a sinuous and rather steep alignment. At the border of the Uttar Pradesh lowland, the first raises of the Pir Panjal range are the core of the lower Himalaya, between the lowland and the Srinagar plateau, in the hearth of Kashmir. The railway line makes a total of approx. 270 km starting from Udhampur (755 m a.s.l.) and climbs along approx. 155 km in the Chenab valley up to Banihal (1'666 m a.s.l.), with an average gradient of 0.6% and a maximum of approx. 1.1%. Along this section, approx. 144 km are underground. Once passed the main mountain chain between Banihal and Qazigund, the railway leads to Srinagar and then Baramulla on a gentle alignment along the remaining approx. 115 km. The final portion (be-tween Banihal and Baramulla) was already completed 6 years back and it is already operating. For doing so, the whole railway equipment was transported along the national road, by winning a mountain pass slightly more than 2'000 m a.s.l. and another with a top tunnel at approx. 2'200 m a.s.l..

2 CHARACTERISTICS OF THE PROJECT

Tunnel T-48 is approx. 40 km far from the top point of the line, in Banihal, and it is the third longest railway tunnel in India once its neighboring T-49 will be completed. The main characteristics of the tunnel are (Figures 1 and 2):

- Length: 10'185 m (10'085 m underground)
- Scheme: Railway tunnel, single lane with lateral escape tunnel at 25 m distance axis to axis. Escape bypass every 375 m, 18 m long. The crossing of the trains occurs mostly in the stations, scattered along the whole line with nearly constant distances.
- Internal free section: Main (railway) tunnel 37.7 m^2
                           Escape tunnel, for the safety and maintenance equipment 18.3 m^2 (schedule 3.0 x 3.6 m) (see Figure 2)
- Excavation section: Main tunnel from 53.4 to (variable conical section) 73.9 m^2
                           Escape tunnel 26.9 to (in case of variable conical section) 33.7 m^2
- Maximum overburden: 1'150 m (see longitudinal profile in Figure 3)
- Present situation: Excavation in progress, realized 58% of the total (21'372 m).

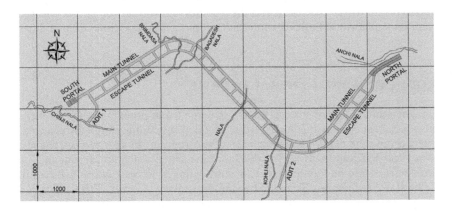

Figure 1.   Planimetric scheme of the tunnel T-48.

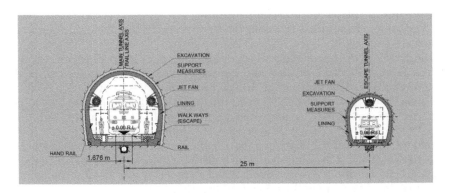

Figure 2.   Typical sections of the Main (railway) Tunnel and of the Escape Tunnel.

3 GEOLOGY

The project is located on the left side of the Chinjy Nala, a right affluent of the Chenab river and lies completely in the Ramban formation made, in this region, of mostly carbonaceous phyllite with intercalations of quarzitic phyllite, and quarzitic phyllite (see Figure 3). This zone is in in contact with the neighboring Murree formation, overlapping it with a tectonic contact analogue to the Himalayan Main Boundary Thrust. This contact lies at depth greater than 200 up to 400 m in the project area, with a minimum at the South portal just less than 100 m, increasing towards the North portal.

Frequent quarzitic veins of centimeter to several centimeters thickness cross the rock beds, particularly towards the South portal nearer to the tectonic contact. The bedding is against the valley slope and it is crossed by the tunnel in its strike, therefore with generally transversal/diagonal orientation in the excavation face (see Figure 4).

The North portal is located within a stable slope debris, residual of an ancient landslide (see Figure 5).

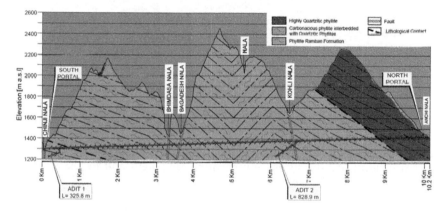

Figure 3. Geological profile of the railway tunnel.

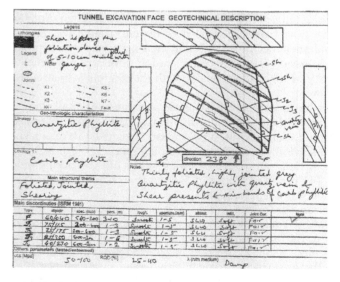

Figure 4. Face mapping (typical).

Figure 5. Slope cut at the North portal.

4 EXCAVATION AND SUPPORT MEASURES (OBSERVATIONAL METHOD)

The excavation is done by drill and blast. The support measures are defined on the base of hazard scenarios identified during the investigation campaigns and by the experience of the designer, developed respecting the Eurocodes and the Swiss codes for the design of underground structures.

These scenarios consider the rock formation, the overburden, the presence of water and relevant effect on the rock mass characteristics and behavior as well as particular aspects such as the brittle behavior of the rock foliation or the swelling potential.

The identified scenarios allowed to adopt rather typical solutions: rock bolts, steel ribs and shotcrete. These are completed by drainages in various configurations, pre-support of the tunnel face by glass fiber bars, pre-support of the crown arch by micropile umbrella (pipe-roofing) or self-drill anchors (forepoling) and consolidation or waterproofing grouting, with or without invert, generally by full face excavation, with a crown-bench solution in case of severe face instability or high convergences, with the possibility to strengthen it by a temporary invert in the crown.

Because of the high overburden, solutions are also foreseen for allowing larger convergences neither losing the tunnel shape, nor bringing the instability to part of it, keeping the necessary strength for providing a significant support also during the development of these deformations. In order to solve these conditions avoiding solutions such as the tunnel re-profiling, expensive in terms of financial point of view as well as of the working program, steel ribs are adopted by using TH profiles (Toussaint-Heintzmann) with sliding joints (see Figure 6). At present, this solution was not yet adopted.

The support measures are applied on the base of the Observational Method, applying a multicriteria process, which considers the rock class defined by RMR and GSI as well as other aspects concurring to the behavior of the opening, according to the identified hazard scenarios. This process takes place at the excavation face, with an initial geo-mapping and the relevant classification plus the geo-structural and hydrological aspects participating into defining the mentioned hazard scenarios. The solution to be adopted is henceforth evaluated and adopted.

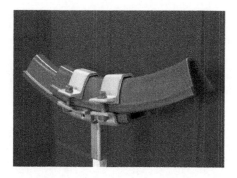

Figure 6.   Sliding joint of a TH profile (Toussaint-Heintzmann).

The behavior of the tunnel is constantly monitored by instrumented sections, generally every 50 m or in other specific positions as directed by the work supervision team. Instrumented sections completed by multi-rod extensometers and piezometers are located every 500 m or in other specific positions as directed by the work supervision team.

Based on the relevant feedback, the project can be considered well calibrated, with solutions that until now showed to be suitable for assuring the tunnel stability. An adjustment is presently ongoing of some design parameters. Indeed, the experience up to 850 m overburden shows lesser convergences than expected, letting suppose a better rock mass condition it its general nature, in the meantime subject to breaky behavior in the more relaxed zone near the excavation. This is likely due to the quarzitic bands and banks also in the zones where the rock seems mostly carbonaceous, while the lithologic structure at the small scale suffers the stress release by the use of explosive and the support time in general. Because of the schist texture of the rock, this leads to caving. This effect is limited, compared to the excavation volume, but a handicap for the generated over-excavation and the relevant necessity of backfilling by significant shotcrete volume or later by final lining concrete volume.

5   SAFETY, VENTILATION AND ELECTROMECHANICAL EQUIPMENT

A challenge itself is the electromechanical design, related to the concept of safety and ventilation. These tunnels, especially the ones longer than 5 km, are the first ones realized in India (Pir Panjal tunnel T-80, 11'215 m operated since 2013 and Karbude, 6'500 m operated since 1998).

Despite the effort spent by the Owner and the Client, the contractual design organization divided by single tunnel makes difficult the adoption of a common uniform concept along the whole line. Maintenance and operation would profit of a smoother coordination in this regard.

Nowadays diesel traction is foreseen, to shift in future to electric traction. This requires a significant hygienic ventilation for assuring acceptable air quality in the tunnels. The accident and fire conditions are considered by a 50 MW fire power and the EUREKA temperature curve. The piston effect and the steep gradient favoring the chimenea effect facilitate the air flow to become turbulent. The tunnel ventilation is deputed to control the smoke propagation and stratified distribution, for limiting the relevant expansion in the whole tunnel section because of excessive turbulence. The system is longitudinal with groups of 880 mm diameter axial jet fans, located near the transformer and command rooms. In case of fire, the fire detection cable allows to identify the position of the same and the nearby fans are excluded for avoiding the perturbance of the aerodynamic flow. The passenger escape is foreseen through pressurized bypasses, equipped by double doors in order to exclude the smoke propagation to the escape tunnel (see Figures 7 and 8).

The power distribution in low tension is managed through 5 transformer stations 11kV/ 450V and relevant command rooms distributed along the tunnel every approx. 2'500 m, 3 of them underground (in specific caverns) and 2 at the portals. At the North portal and

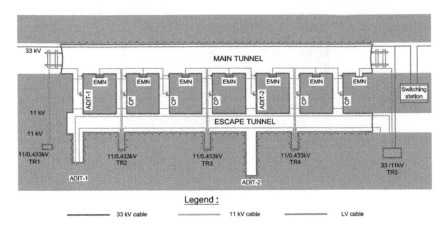

Figure 7.   Power supply scheme and escape ways.

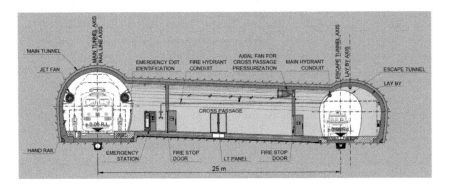

Figure 8.   Bypass with double doors and pressurization, with axial jet fans in the Main Tunnel.

downstream the previous next tunnel (T-47), two switch points 33/11kV are planned. The 33kV main line lies along the entire railway alignment. It is placed in the Main Tunnel opposite to the escape ways, fully independent to the tunnel power supply. The 11kV main power supply lines lie in the Escape Tunnel, simplifying the circuits to/from the transformers. The distribution in 450V and the signal/command network is managed by clusters of 187.5 m, starting from the bypass and from the electromechanical distribution cabins located in the middle point between two subsequent bypasses, in the Main Tunnel.

The electromechanical design follows the civil works one. This limits the manageability of the various line and interface positioning. The coordination and the strict cooperation with the final Owner (the Northern Railways) allows the requirements of the railway line to be met with a homogeneous setup so far as possible. Specific solutions remain limitedly applied, generally inspired to a common approach to the problem solving along other tunnels of this line.

This process is demanding, but largely appreciated by the Owner and the Client. Despite the management of approx. 7'000 km of railway lines, the experience is still young in project management of large underground opening such as the tunnel T-48.

6   WORK EXECUTION

The project is presently in the middle of the civil works execution. The two adits, 6'063 m (60%) of the Main Tunnel excavation and support, 5'702 m (59%) of the Escape Tunnel and 10 bypasses are realized for a total of 13'125 m over 21'372 m (61%). The concrete final lining is completed over only 1′073 m in the Main Tunnel (10%) and 388 in the Escape Tunnel (3%). this activity is delayed because of non-optimized method and execution cycle (approval of the formwork and of the methodology and sequence, unsuitable traffic scheme management face/lining location/outside, unsuitable local work sequence for a chain work and linear progress). Recent high water inflows are a significant concern for the working sequence, needing a powerful dewatering until the breakthrough is provided.

The major difficulties are nevertheless in the logistics and organization. Kashmir suffers of weak and fragile connections. The roads are not on a suitable standard for the traffic intensity and the heavy vehicle are particularly slow. The driving style is also far from the European safety standards. In the valleys, it is usual to consider an average transport speed of 30 km/h. Furthermore, despite Kashmir climate is generally continental, in the Chenab valley the monsoon still gives sign of its presence. The heavy and sometime violent rainfalls in this late-summer period give way to frequent landslides cutting the communication and delaying the supply up to various days and blocking the people movement with urgent roaming works. This cause the traffic to accumulate and requires a one-way traffic management over alternated days along whole valleys (see Figure 9).

Winter is typically continental-alpine, bringing heavy snowfalls, which oblige the traffic to be blocked, then to be drained by one-way circulation in alternated days after the road reinstatement (see Figure 10).

These conditions should lead the designers and the clients to the formulation of specifications regarding the availability of material, equipment and spare parts and for a logistic support and supply line suitable to the needs, not letting the stocks unattended. Measures for enforcing the

Figure 9.   Bad weather and landslides. Traffic jam, waiting for the restart of the opposite one-way flow.

Figure 10.   Winter conditions 06.11.2018.

relevant application should also be part of it. The firsts are becoming an actual part of the contractual formulation in this market essentially targeting the financial profitability, thanks to the experience of international design companies bringing proposals of optimized working methods and cycles for the actual improvement of the production rates. The latter, instead, still find a hard way to become actual in their formulation and application. The budget offered by the Contractors are sometimes such to make inapplicable any form of pressure or negotiation.

Because the equipment is rarely optimized for the desired production (e.g. automatic jumbos, loaders with lateral tilting bucket, Toro loaders, dumpers with double drive direction) and the construction material is sometimes scarce at site, contingency measures and flexibility of application are a frequent strategy that the work supervision teams are called to practice in order to avoid the stop of the construction works. The contactors' budgets, very competitive as mentioned, generally do not allow a pre-emptive capitalization such to face the bottlenecks and the consequent clogging of several days along the supply line. Not in any case the principle for a rapid execution favoring the final capitalization finds its realization within construction companies with limited self-financing capability compared to the actual aspects to be faced. The parties are often facing a continuous negotiation over the construction financing modalities.

7 PRESENT SITUATION

The construction contract with Hindustan Construction Company, working on the project until July 2017, was terminated.

The balance of works was retendered and awarded to GAMMON. The complete processing of this phase took only three and a half months (100 days). The excavation works resumed middle of November 2017. The commissioning was postponed from end 2017 to end 2020.

8 CONCLUSIONS

This experience might lead some conclusions, good for further progresses by future projects of this league, be these railway or roadway.

Performant equipment has to be preferred, with high modernization. Adequate marking during the evaluation have to be assigned, as opposite to an old, inappropriate and imprecise one.

Particular solutions and details have to be specified more than usual (e.g. steel ribs and LG footing, proper grouting including relevant equipment, the use and the how of TH ribs or LSC).

A positive synergy with the Client is required, for targeting the correct implementation of design and execution concepts not yet completely entered in the common use.

The mechanism hazards - consequences - risks is fundamental for increasing the safety.

Don't compromise to bad attitudes, difficult to correct afterwards, because of a silent consent.

On the other hand, so far as possible, always consider the existing equipment, material and human resources by designing a tunnel and its construction.

Always discuss extensively beforehand with the Client the introduction of new or more modern techniques and approaches, including the cost-benefit aspects. Keep insisting on the required specifications to apply for assuring the required result. Missing this, the possible backfire is the loss of Client's trust in the apparently ineffective proposals.

Always consider a suitable support to the Client all along the construction. Tunnel design is more and more linked to practical efficiency and the unsuitable implementation of several concepts and details can easily be the killer factor against innovation.

Always defend the deontological difference of safety (including factors of safety) as opposite to oversizing.

Don't be or feel constrained by standardized solutions such as NATM, NTM, ADECO-RS or other "coded" approaches and methods. Rather follow the observational method and the hazard scenario approach defined (first between others) by the Swiss codes (Gefährdungsbildermethode).

Tunnels and Underground Cities: Engineering and Innovation meet Archaeology,
Architecture and Art, Volume 7: Long and deep tunnels – Peila, Viggiani & Celestino (Eds)
© 2020 Taylor & Francis Group, London, ISBN 978-0-367-46872-9

Assessments on the TBM performance at the Minas-San Francisco Hydropower Project in Ecuador

K.K. Panthi & J. Encalada
Norwegian University of Science and Technology (NTNU), Trondheim, Norway

ABSTRACT: The fossil energy is impacting global environment considerably. Therefore, focus on the development of renewable energy is of priority in world. Hydropower is among the favored renewal sources which many developing nations have prioritized. One such example is recently constructed Minas-San Francisco Hydropower Project in Ecuador where both drill & blast and TBM methods of tunneling were used. This manuscript highlights engineering geological conditions of headrace tunnel and evaluates TBM performance. Influence by rock mass quality in TBM penetration rate is assessed, and has been emphasized that knowledge of rock mass quality parameters and cutter technology are key factors that influence on the estimation of net penetration of TBM. Hard rock mass of high abrasivity encountered in the project resulted in very low penetration rate than that was predicted during design. Analyses further shows that throughout TBM boring the penetration mainly depended on the mechanical strength of the rock mass.

1 INTRODUCTION

For the construction of long tunnels, TBM gives the advantage of shortening the excavation time and reducing the costs, compared to the drill and blast method. In addition, TBM produces a smooth excavation contour that helps to minimize the effect of hydraulic head loss. Mechanical breaking of rock using disc cutters involves penetration and crack propagation. In the first place, high thrust generates grooves of 1-15 mm as the cutter passes. Later on, tensional cracks are developed when the grooves are deep enough, causing spalling and consequently chipping of the rock (Nilsen & Thindemann, 1993). The development of tensile fractures will depend on the toughness of the rock. According to Whittaker et al (1992), for disc cutters, cracks will develop in a mixture of tensional (mode I) and edge-sliding (mode II). It is important to understand that fracture toughness, which is known as the critical stress intensity factor, is the resistance to microscopic separation and cracking (Shen et al, 2014). Whittaker et al. (1992) also shows that values of toughness are correlated to the uniaxial compressive strength. Similarly, Atkinson (2015) associated toughness to the fracturing process of materials where the effect of loads is either static or dynamic (percussive). According to Bruland (2000), the net penetration rate of a TBM cutter head is a function of the machine and rock parameters. Machine parameters are cutter thrust, spacing, shape and size, cutter head rpm and installed power. Geological parameters are fracture frequency and orientation, drilling rate index (DRI) and strength of rock mass.

Nilsen & Thindemann (1993) stated that the advance rate will be less dependent on the rock mass strength for a high degree of jointing with continuous joints rather than thin fissures. While Bruland (2000) emphasized that the joints with negligible or no shear strength are indeed involved in TBM boring. The degree of fracturing is represented by a fracture class, which is classified between 0 and VI where class category 0-II are frequently observed in rock mass such as quartzite and gabbro. Macfarlane et al (2008) reported very low penetration rate for igneous rocks characterized by a fracture class 0. On the other hand, cutter penetration in metamorphic rocks analyzed by Yagiz (2008) produced results that are insensitive of the fracture class described by Bruland (2000). Regarding fracture orientation, a small angle between tunnel axis and fracture planes is the most favorable condition for boring (Nilsen & Thindemann, 1993). The advance rate is also affected by abrasiveness of minerals such as quartz, feldspar and epidote, which can also influence on the usable life of cutters. To tackle problems of abrasion, 19-inch cutters are used in order to apply high thrusts and extend the time of the cutter change (Dammyr, 2017).

In regard to the character of the rock, it has been found that penetration increases as DRI and thrust per cutter increase (Movinkel & Johansen, 1986). This relationship has become fundamental for the development of prognosis models such as developed at NTNU (Bruland, 2000). Other models such as the Q-TBM (Barton, 2002) incorporates the mechanical strength and the rock mass quality. It is acknowledged by Barton (2009) that Q values exceeding100 can represent difficult ground conditions for a TBM; a situation that make the penetration mode dependent on the available power.

The effect of the machine parameters under diverse geological conditions is represented by the penetration index (PI), and it has been widely analyzed by using records of mechanical excavation by Hassanpour et al. (2011). Bilgin et al. (2006) reported good correlations of PI with the mechanical strength of the rock. In their study, penetration increases as the UCS and DRI increase. The same behavior is reported by Zare Naghadehi & Ramezanzadeh (2017), who highlighted the need for high thrusts to achieve a reasonable level of penetration in massive and strong rocks. In addition to penetration index (PI), the specific energy (SE) measured in kW/m3 will give an idea of the capability of the equipment. The advance rate is linked to specific energy and parameters such as the rotational speed (rpm), the geometry of the cutters and the cutting power determines the level of specific energy. Cutter spacing plays a role in optimizing the level of energy. A small spacing between the cutters will compromise the cutting tools due to excessive abrasion and a wide distance between the cutters will cause a lack of interaction and poor fragmentation. Thus, reaching an optimum energy depends on the configuration of cutting tools as discussed by Harrison & Hudson (2000a). Curves of SE for a variety of rock specimens reported by Bilgin et al (2006) show a relation between the optimum SE and the uniaxial compressive strength (UCS) of the rocks indicating that high level of energy is needed for strong rocks.

This manuscript evaluates the impact of geological and machine parameters during the excavation of the TBM tunnel. Important relations between geological and machine parameters are obtained so that they can be used as support for the planning of new TBM tunnels under similar geological conditions.

2 THE MINAS SAN-FRANCISCO HYDROPOWER PROJECT

The concept of firm energy is relevant in Latin-American countries, especially in Ecuador. This geographic location is characterized by high annual runoff and a difference in topographic elevation. As a result, large hydropower systems have been and are being developed across the Andean region. The Minas San Francisco Hydropower Project (MSF HPP) was recently implemented at the southeast region of Ecuador. The project has an installed capacity of 275 MW and produces 1290 GWh energy annually (MEER, 2018). As shown in Figure 1, the main features of the project consists of an RCC dam of 77 m height (measured from the foundation), a 12 km low pressure headrace tunnel, a penstock shaft, and an underground power station equipped with three pelton turbine units. The implementation of the MSF HPP started in 2012 and was extended until February 2018 (Torres & Vega, 2011).

In the context of large hydro power projects, three critical components are always in the focus of analysis; i.e. headworks, waterway system and the underground power station. In the case of waterway system, tunneling techniques are influenced by the rock mass conditions and the extent of the excavation. Thus, cost-effective solutions in long tunnels include the use of tunnel boring machines (TBMs). For the MSF HPP, a 9 km of the headrace tunnel was

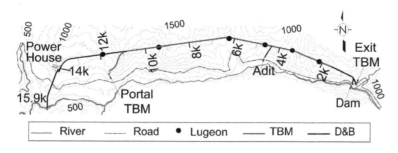

Figure 1. Main features of the MSF HPP (reproduced from Torres & Vega, 2011; Bedoya, 2018; Jimenez, 2018).

excavated by using a double shield TBM. An interesting aspect of the excavation was the low advance rate experienced, at this project despite in-depth knowledge on the character of the rock mass in the design stage.

In the area of the reservoir, rocks of volcanic origin with some intrusions of quartz-monzonite are common with their average strength of about 50MPa, and GSI values ranging from 60 to 80 (Riemmer et al, 2013). The segment of the headrace tunnel on the other hand is distinguished by zones of volcanic and plutonic rocks. The TBM excavation with a cross section diameter of 5.76 m took place in both rock types between chainages 10.8 km and 1.6 km (Torres & Vega, 2011). The location of adits and portals used for the excavation are shown in Figure 1. Rocks such as gabbro and andesite prevailed along the segments excavated by the TBM. Only last segment of excavation from chainage 2 km is composed of tuff and breccia of volcanic sediments. A low performance was experienced, especially when excavating intrusive rocks. In the case of the MSF HPP, 19-inch cutters allowed the application of loads up to 300 kN per cutter, reaching the nominal capacity of steel (Della Valle, 2015).

3 EXCAVATION WITH TBM

Daily records of parameters such as UCS, water inflow, average cutter thrust, and rotational speed (rpm) are used in this analysis. Useful results from the feasibility stage are taken as to complement the present work. It is already known that the uniaxial compressive strength of the rock varied widely along the tunnel alignment. In the same way, previous analysis shows that most sections of the TBM tunnel satisfy the Norwegian criteria for minimum rock cover (Riemmer et al, 2013). In regards to potential inflow in the tunnel, it was observed that the ground water table is relatively deep with no as such water ingress observed in the tunnel excluding in boreholes located at the reservoir area (Suescún & Jiménez, 2017).

3.1 Types of rocks and their character

The rocks excavated were andesite, hornfels, gabbro, diorite, volcanic breccia and alternated layers of tuff and shale (Figure 2). Andesitic dykes-oriented N-E and dipping 70° NW appeared sporadically. The Cerchar Abrasivity Index (CAI) for rock specimens tested during excavation show that andesite was abrasive to extremely abrasive, gabbro very abrasive, and the sedimentary rocks were abrasive (Suescún & Jiménez, 2017). Hornfels was observed sporadically and no CAI was available for this rock.

It was seen that the abrasiveness determined during construction was not different from values reported in feasibility studies; neither do DRI values show discrepancy.

Furthermore, the rocks described above are characterized by having porosities below 1% having dominated minerals of quartz and epidote.

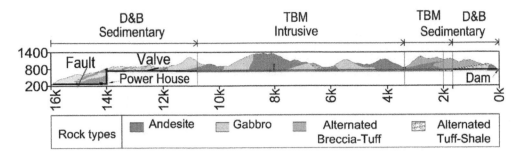

Figure 2. Main geological features along the tunnel alignment (reproduced from Torres & Vega, 2011; Bedoya, 2018; Jimenez, 2018).

3.2 *Weakness zones and discontinuities*

A single shear zone of 20 cm thickness was met at chainage 6.0 km belonging to the hydro-thermally altered zone, a shear zone of low cover (Suescún & Jiménez, 2017). This finding greatly differs from the idea of thrust faults with altered bands of 3 m thickness described in the feasibility study and detailed design. Otherwise, rock mass along the tunnel alignment was found to be with joint surfaces having rough, undulating and relatively tight. More than half of the joints were found to be filled with quartz particles, while remaining joints were found to be filled with some staining or without infilling but very tight. In addition, the weathering was of very limited with a weathering grade of fresh rocks (W1) according to ISRM (1978). This condition is important because weathering can affect the mechanical strength of the rock according to Panthi (2006). Suescún & Jiménez (2017) also found that borehole logs and reports during excavation classify most of the rock as fresh rocks. The jointing intensity and their orientation is presented in Figure 3. The records of registered jointing condition along the TBM tunnel provides possibility to characterize tunnel segments of interest. As one can see in the Figure 3 (right), there are mainly three system of joint sets; i.e. J1, J2 and J3. The joint systems are strongly dipping along the whole tunnel, a situation that is considered not favorable for the advance rate.

It is also important to highlight here that the degree of fracturing has been analyzed in terms of RQD rather than only fractures per meter. The problem is that classification based on the mean spacing value (fracture frequency) is limited and the concept of RQD in solving this problem has been used. Effectively, this study made use of a probability density function (pdf) that replicates the natural clustering of fractures according to Harrison & Hudson (2000a) that considers small spacing values occurring more frequent than the one with large spacing (Figure 3, left).

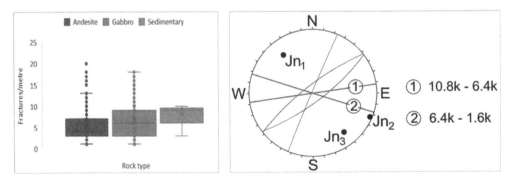

Figure 3. Fracture frequency and orientation along the TBM tunnel (plotted based on Torres & Vega 2011, Suescún & Jiménez 2017).

3.3 Compressive strength

The strength of the intrusive hard to very hard rock varies greatly depending upon weathering condition and grain size distribution. The UCS records of 216 tested specimens from along the tunnel is presented in Figure 4. Hornfels exhibit an extremely high resistance to compression. As discussed previously, Hornfels and Diorite appeared sporadically at chainages 9.9 km and 10.1 km in segments of 20 m and 85 m, respectively.

For volcanic and sedimentary rock formations, the compressive strength can be inferred from tests reported in the reservoir area where typical mean value was found to be around 50 MPa (Suescún & Jiménez, 2017). This observation is correct as high penetration rates were achieved in these rock formations.

3.4 Rock mass quality

After geological mapping, the rock quality results was found to be superior to the values reported during feasibility and detailed design studies. Figure 5 gives the distribution of predicted and actual rock mass class according to RMR. Since double shield TBM with continuous concrete segment lining was used, it was not necessary to use support optimization using rock mass class.

As one can see in Figure 5, much better quality of rock mass was registered along the tunnel alignment than that was predicted. The high rock mass quality class experienced has direct influence on the production rate of the TBM.

3.5 Cutter thrust

For the analysis of the advance rate, 29 single cutters and 4 twin-cutters contained in the cutter head were studied independently with a consideration on the analysis of average thrust per cutter. The simplified version of the prognosis model suggested by Barton (2009) and Bruland (2000) was utilized for this purpose. This was also justified with field observations of the performance of disc cutters. The cutter rings sent to the garbage store showed signs of failure of the steel, indicating working conditions close to the nominal capacity of 300 kN. Della Valle (2015) noted that the whole thrust was concentrated mainly on the frontal cutters. Therefore, load on peripheral cutters was disregarded for the evaluation of average thrust per cutter.

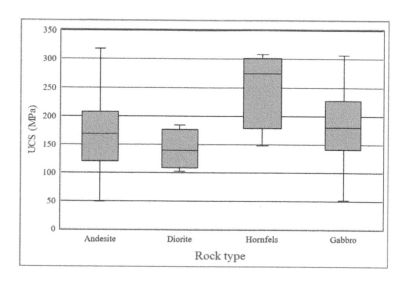

Figure 4.   Compressive strength for the rocks along the TBM tunnel (based on Suescún & Jiménez 2017).

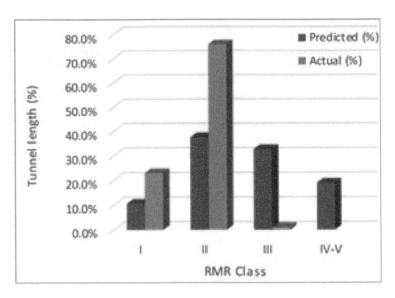

Figure 5. Actual and predicted rock mass class based on mapped RMR values.

3.6 *Operational capability*

In TBM excavation, about 1 km tunnel including part of the adit TBM can in general be considered as to gain experience. It can be seen in Figure 6, despite a rotational speed of 6-7 rpm and cutter thrust of 300kN, the penetration significantly dropped by April 2015 at chainage 10 km

Since the beginning of the excavation in December 2014 until April 2015, the production dropped gradually from 544 to 41 m3/cutter. In the zone of low production, the rocks consisting of dykes of hornfels and gabbro were encountered. The change of rings was frequent, which mostly indicated failure on the steel. The average consumption of the cutter ring was approximately 54 per month according to Della Valle (2015). The drop in the production was

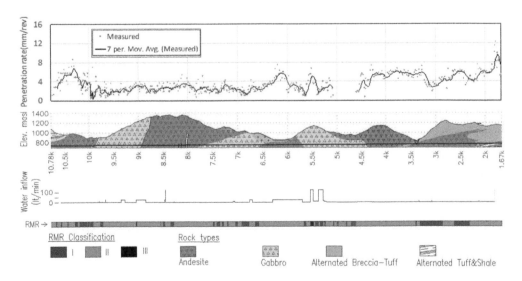

Figure 6. Penetration and rock mass conditions (based on Torres & Vega 2011, Suescún & Jiménez 2017).

caused by failure of a twin cutter housing, representing large fissures on the steel, which was finally solved in August 2015 after reaching the chainage 6 km.

3.7 *Behavior of penetration under varied geological conditions*

The penetration rate recorded is shown Figure 6. Two extremes can be observed in Figure 6; i.e. massive and relatively fresh gabbro at chainage 9.6 km and slightly weathered volcanic sedimentary rock formations at chainage 1.7 km. It is clear that the excavation process is subject to inaccuracies in operation and sampling data. It was also found that the strength of rock had high variability locally causing fluctuation on the global tendency.

The effect of mechanical failure in between the chainage interval of 10.1 km - 9.02 km can be easily observed and it is reasonable to exclude this data for further analyses. Based on reliable records, three aspects of TBM performance can be pointed out. First, the penetration is to some extent related to the depth of overburden due to the degree of fracturing, which is in general higher at shallow depths. Secondly, the rock mass quality is related to the penetration rate as observed at chainages 3 km and 5.5 km having a low performance for rock mass class category I and a noticeable increase in penetration rate for rock mass class category III. Thirdly, the change in advance rate from igneous to sedimentary rocks is also noticeable, which can be attributed to the lower mechanical strength of the sedimentary rock formations. The effect of the mechanical strength on the penetration can also be seen in chips product of excavation (Figure 7), revealing the high level of toughness of the intact rock.

According to Suescún & Jiménez (2017), most chips were of 1-3 cm thickness, 5-10 cm wide, and 10-20 cm in length. The specimen shown in the figure is no more than 15 cm in length and represents a zone of favorable borability. However, throughout the excavation, most chips presented small, planar and angular shapes, indicating a low cutting power compared to the exceptional characteristics of the rock encountered.

3.8 *Inflow and leakage*

It was observed in Figure 6 that a relatively high inflow of 25 - 125 ltr/min was met in two occasions. The first one, near to zones of rock type III at chainage 5.5 km and the second one at 8.5 km. Overall inflow and leakage conditions of the tunnel are reflected on the mean permeability observed after several lugeon tests. Along the tunnel, permeability varied between 1.6×10^{-7} and 4.2×10^{-8}. Inflow measurement during construction was not required because of the wide knowledge of the permeability beforehand. Measurement of water leakage also reveals a very low leakage with a mean value of 1.2 ltr/min/m and a maximum leakage of 3.6 ltr/min/m.

Figure 7.   Chip product of excavation at chainage 10.14 k (Suescún & Jiménez 2017).

4 ANALSYIS AND DISCUSSION

Data from the penetration rate (mm/rev) was analyzed in terms of the parameters discussed above. An attempt was made to find out a correlation between the TBM penetration rate and degree of fracturing represented by RQD (Figure 8, left). As one can see in the analysis plot, there exist very poor correlation. This is quite logical since the RQD is a value represented by length of cores longer than 10 cm in a one meter drilled core, which may sometime give false information on the jointing condition in the rock mass. For example, a rock mass having joint system spaced slightly over 10 cm can give a RQD value exceeding 80. On the other hand, the same rock mass having joint spacing slightly lower than 10 cm will end with RQD value below 20. Hence, only RQD is not enough parameter to estimate TBM penetration rate. Similarly, a correlation between penetration rate and UCS was also assessed, which resulted an improvement on the correlation (Figure 8, right). However, the improvement is not that significant either.

Further attempt was made to find a correlation between the ratio of thrust and penetration rate, which represents Penetration Index (PI) with the multiplication of UCS and RQD (Figure 9, left). Both lab tested UCS results of 216 specimens, estimated UCS at the tunnel and mapped RQD values were used for this purpose. As one can see, a fairly good correlation was found. Similarly, it was found that there exist a good correlation between Specific Energy (SE) with the multiplication of UCS and RQD values (Figure 9, right).

The findings of Figure 9 have significant importance in estimating Penetration Index (PI) and Specific Energy (SE) for the tunnels passing through similar geological conditions. The relationship achieved here can be further enhanced and improved using data and information of the TBM performance of other TBM tunnel projects built in other geological environment. In addition, the effect of abrasiveness can also be seen in the form of gross advance rate (Figure 10).

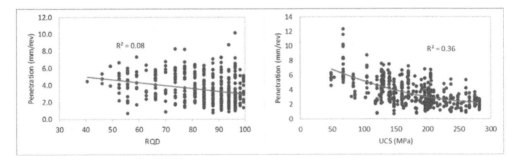

Figure 8.   Penetration rate in relation with RQD (left) and UCS (right).

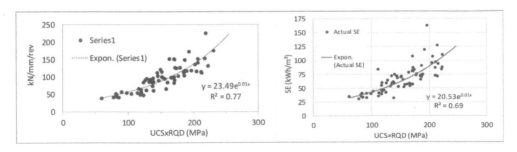

Figure 9.   Penetration Index (PI) (left) and Specific Energy (right) in relation with RQD multiplied by UCS.

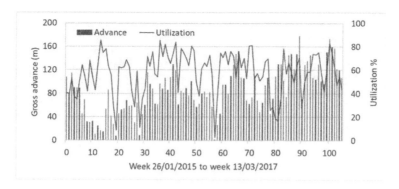

Figure 10. Weekly utilization and gross advance rate (based on data from Suescún & Jiménez, 2017).

The damage produced to the twin cutter host affected the gross advance from Week 8 to Week 30. During this time lapse, the advance was low in spite of high utilization of the machine. As indicated in Figure 10, a change of disc cutter took place during March 2016 (Weeks 57-58) at chainage 6+800, which clearly indicates noticeable drop in the advance rate. It can also be seen in the figure that the production was considerably improved for the same utilization time when the excavation approached sedimentary rock formations with enhanced borability characteristic.

Despite the favorable mechanical properties of the rock mass along the TBM tunnel, permanent concrete lining was used at this project based on the recommendation made in the design and contract documents where double shielded TBM was recommended. No considerations were made about the inflow and leakage observed, neither the quality of rock mass to select the final rock support. Experience on the use of unlined high-pressure shafts and tunnels in Norway (Benson 1989, Buen & Palmstrom 1982, Broch 1982, Panthi 2014 and Basnet & Panthi 2018) suggests that permeability and rock quality similar to the present case should have been very suitable for the implementation of unlined pressure tunnel concept. In the present study, most segments of the tunnel that was mechanically excavated showed a great advantage in terms of stand-up time (high rock mass quality) and low permeability. A second project downstream of the MSF HPP is to be developed and part of it will follow the rock formation dominated by granitic gneiss and quartzite. This condition should be considered as a possible choice of the implementation of unlined low-pressure headrace tunnel.

5 CONCLUSIONS

The engineering geological parameters in relation with both TBM Penetration Index (PI), Specific Energy (SE) and the advance rate of TBM at MSF Hydropower Project have been analyzed. The analysis concludes that there exists fairly good correlation between Penetration Index (PI) and Specific Energy (SE) with the multiplication of RQD and UCS. The proposed correlation can be enhanced by using TBM performance data and information from other TBM projects having different geological environment. However, it is emphasized here that a care should be taken while using the proposed correlations since they represent from only one single TBM project.

REFERENCES

Alber, M. 1996. Prediction of Penetration and Utilization for Hard Rock TBMs. *EUROCK 96, the ISRM International Symposium*, Turin, Italy.
Atkinson, B. K. 2015. Fracture Mechanics of Rrock. *Elsevier*.
Barton, N. 2002. Some New Q-value Correlations to Assist in Site Characterization and Tunnel Design. *International Journal of Rock Mechanics and Mining Sciences* (39): 185–216.
Barton, N. 2009. TBM Prognoses in Hard Rock with Faults using Q TBM Methods. *IoM3 Institute of Materials, Minerals & Mining*. Hong Kong.

Basnet, C. B. & Panthi, K. K. 2018. Analysis of Unlined Pressure Shafts and Tunnels of Selected Norwegian Hydropower Projects. *Journal of Rock Mechanics and Geotechnical Engineering* (10): 486–512.

Benson, R. P. 1989. Design of unlined and lined pressure tunnels. *Tunneling and Underground Space Technology* (4): 155–170.

Bieniawski, Z. T. 1989. Engineering Rock Mass Classifications: A Complete Manual for Engineers and Geologists in Mining, Civil, and Petroleum Engineering. *Wiley.*

Bilgin, N., Demircin, M. A., Copur, H., Balci, C., Tuncdemir, H. & Akcin, N. 2006. Dominant rock properties affecting the performance of conical picks and the comparison of some experimental and theoretical results. *International Journal of Rock Mechanics and Mining Sciences* (43): 139–156.

Broch, E. 1982. The Development of Unlined Pressure Shafts and Tunnels in Norway. *ISRM International Symposium.* Aachen, Germany.

Bruland, A. 2000. Hard Rock Tunnel Boring. *Advance Rate and Cutter Wear* (3).

Buen, B. & Palmstrom, A. 1982. Design and Supervision of Unlined Hydro Power Shafts and Tunnels with Head up to 590 meters. *ISRM International Symposium.* Aachen, Germany.

Dammyr, Ø. 2017. Evaluation of the potential for TBM use in future Norwegian Tunneling Projects. *PhD thesis.* Norwegian University of Science and Technology (NTNU).

Deere, D. U. 1963. Technical Description of Rock Cores for Engineering Purpose. *Rock Mechanics and Engineering Geology* (1): 22.

Della Valle, N. 2015. Informe de la visita realizada por el Dr. *Nicola Della Valle.* CELECEP.

Harrison, J. P. & Hudson, J. A. 2000a. Engineering Rock Mechanics: An Introduction to the Principles. *Elsevier Science.* Burlington.

Harrison, J. P. & Hudson, J. A. 2000b. Engineering Rock Mechanics (Part 2: Illustrative worked examples). *Elsevier.* Oxford.

Hassanpour, J., Rostami, J. & Zhao, J. 2011. A New Hard Rock TBM Performance Prediction Model for Project Planning. *Tunneling and Underground Space Technology* (26): 595–603.

Macfarlane, D. F., Watts, C. R. & Nilsen, B. 2008. Field Application of NTH Fracture Classification at the Second Panapouri Tailrace Tunnel. *New Zealand.*

MEER. 2018. *Proyecto Hidroelectrico Minas – San Francisco.* Ministreio de Electricidad y Energia Renovable. Ecuador.

Movinkel, T. & Johansen, O. 1986. Geological Parameters for Hard Rock Tunnel Boring. *Tunnels and Tunneling.*

Palmström, A. 2009. Combining the RMR, Q and RMi Classification Systems. *Tunneling and Underground Space Technology* (24): 491–492.

Panthi, K. K. 2006. Analysis of Engineering Geological Uncertainties Related to Tunneling in Himalayan Rock Mass Conditions. *PhD Thesis.* Norwegian University of Science and Technology. Trondheim.

Panthi K. K. 2014. Norwegian Design Principle for High Pressure Tunnels and Shafts: It's Applicability in the Himalaya. *Hydro Nepal* (14): 36–40.

Riemmer, W., Forbes, B. & Villegas, F. 2013. Proyecto Hidroelectrico Minas-San Francisco. *Informe1-Panel de Consultores.* Enerjubones.

Shen, B., Stephansson, O. & Rinne, M. 2014. Modelling Rock Fracturing Processes: A Fracture Mechanics Approach using FRACOD.

Suescún, C. & Jiménez, C. 2017. Geologia Tunnel de Conduccion. *463-IT-017-A.* CELECEP.

Torres, L. & Vega, G. 2011. Proyecto Minas – San Francisco-Informe de Geologia y Geotecnia – Evento 7, Estudio Geológico Definitivo. *Enerjubones Generadora Hidroelectrica.* Ecuador.

Whittaker, B. N., Singh, R. N. & Sun, G. 1992. Rock Fracture Mechanics: Principles, Design and Applications. *Elsevier.* Amsterdam.

Yagiz, S. 2008. Utilizing Rock Mass Properties for Predicting TBM Performance in Hard Rock Condition.

Zare Naghadehi, M. & Ramezanzadeh, A. 2017. Models for Estimation of TBM Performance in Granitic and Mica Gneiss Hard Rocks in a Hydropower Tunnel. *Bulletin of Engineering Geology and the Environment* (76): 1627–1641.

Tunnels and Underground Cities: Engineering and Innovation meet Archaeology,
Architecture and Art, Volume 7: Long and deep tunnels – Peila, Viggiani & Celestino (Eds)
© 2020 Taylor & Francis Group, London, ISBN 978-0-367-46872-9

Shotcrete lining installed from an open gripper TBM remedial works

G. Peach, B. Ashcroft, R. Amici & J. Mierzejewski
Multiconsult Norge AS, Oslo, Norway

ABSTRACT: Twin 8.5 m diameter, 10 km long parallel headrace tunnels under overburdens of up to 1870 m were excavated for the Neelum Jhelum Hydroelectric Project located in the Azad Kashmir region of northeast Pakistan using two open gripper Tunnel Boring Machines (TBMs). Both TBMs started headrace tunnel excavation in early 2013 with completion in May 2017.

This paper briefly outlines the shotcrete application procedure in different zones on the TBM, and the issues encountered during construction that led to the need for remedial works to the shotcrete lining. The paper then details the sequence for implementing the remedial works, which was complicated by limited access, and describes the various methods employed to carry out the finishing works, such as grouting operations and repair of construction joints. Details of quantities used for the Neelum Jhelum project are provided for use as a guideline for similar projects in the future.

1 PROJECT DESCRIPTION AND OVERVIEW

The Neelum Jhelum hydropower project is located in the Muzaffarabad District of Azad Jammu & Kashmir (AJK), in northeastern Pakistan within the Himalayan foothill zone known as the Sub-Himalayan Range. The terrain is rugged with ground elevations that range from 600 m to 3200 m above sea level.

The project is a run-of-river scheme, employing 28.6 km of headrace and 3.6 km of tailrace tunnels that bypass a major loop in the river system, transferring waters from the Neelum River into the Jhelum River, for a total static head gain of 420 m (Figure 1). The headrace tunnels include both single bore (31%) and twin bores (69%). The tailrace tunnel consists of a single tunnel. Design capacity of the waterway system is 283 m³/s. The project, which was completed in 2018, has an installed capacity of 969 MW, generated by four Francis turbines located in an underground powerhouse.

Tunneling commenced in 2008 using conventional drill & blast techniques. It soon became apparent that a 13.5-kilometer-long section of the twin headrace tunnels (under high mountainous overburden that precluded construction of additional access adits) would take too long to excavate. The contract was amended to incorporate two 8.5 m diameter open gripper hard rock TBMs to excavate approximately 10.5 km of twin headrace tunnels (Figure 1).

The gripper design offered flexibility for the expected conditions: possible squeezing ground given the relatively weak rock mass and overburdens up to 1870 meters, and the potential for rockbursts in the stronger beds.

1.1 *Geological Settings*

The entire project was excavated in the sedimentary rocks of the Murree Formation, which is of Eocene to Miocene age, and comprises closely interbedded sandstones, siltstones and mudstones. The TBM tunnels were constructed through a zone bounded by two major Himalayan faults that trend sub-perpendicular to the tunnels: the Main Boundary Thrust, and the subsidiary Muzaffarabad reverse/thrust fault (Figure 1).

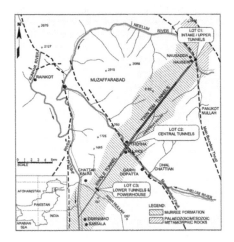

Figure 1. Neelum Jhelum project layout showing TBM Twin tunnels (in bold), major faults (dashed) and alignment geology.

2 OPEN GRIPPER TBM CONFIGURATIONS FOR SUPPORT INSTALLATION

Open gripper TBMs are designed for hard, competent rock where little to no support is required. The disc cutters mounted on the cutterhead are thrust against the rock, cutting concentric tracks in the rock mass as the cutterhead rotates. As the tracks get deeper, the rock between adjacent tracks spalls, and the spalled rock fragments are scooped up by buckets in the cutterhead and transferred to a belt conveyor.

The required thrust of the cutterhead is provided by thrust cylinders attached to grippers, which are braced against the tunnel wall. Once the thrust cylinders reach the end of their stroke, the rear cutterhead support is lowered and the grippers and cylinders are pulled in. They are then repositioned for the next boring cycle. The grippers then re-engage the tunnel wall, the rear cutterhead support is raised and the next cycle starts.

Initial support is applied at the 'L1 zone' just behind the TBM shield and any further support is applied at the 'L2 zone'. Figure 2 shows a typical schematic of an open gripper TBM.

2.1 L1 zone

The L1 zone is where the initial support is placed, primarily to secure any loose rock and to limit the convergence of the rock mass. Initial tunnel support installed at the L1 zone is generally incorporated into the permanent support. The principle is to install sufficient support to stabilize the surrounding rock surface and provide a safe working environment. In the case of Neelum Jhelum, the majority of the permanent support elements, with the exception of shotcrete, was installed in the L1 zone.

On the project, support elements employed at the L1 zone included 3.85 m long rock bolts, wire mesh (6, 8 or 10 mm diameter bars), mining straps (upper 180° of tunnel), full circular steel ring beams and, where essential, shotcrete. Full circular steel ring beams required a special erector for installation.

Shotcrete application is generally avoided at the L1 zone because overspray and rebound fouls the TBM's electrical and hydraulic equipment, and because the clamping pressures imposed by the grippers unavoidably damage green shotcrete. Consequently, shotcrete application is delayed where possible until the L2 zone, where dedicated robots mounted behind a shield can operate freely in a less congested environment. However, shotcrete application at the L1 zone was sometimes necessary where poor ground conditions were encountered.

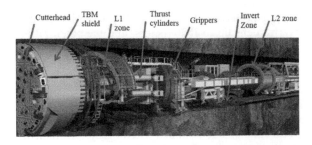

Figure 2. Schematic of gripper TBM (indicative only).

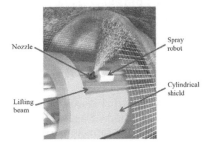

Figure 3. L2 zone shotcrete installation.

2.2 Invert Installation Zone

On an open gripper TBM, the invert is typically installed behind the rear cutterhead support and in front of the L2 zone (i.e. between the area where the 'feet' of the TBM are placed during re-gripping at the end of a stroke, and where the rail track is laid).

On the Neelum Jhelum project the invert installation zone was located 20 m behind the L1 zone. The tunnel was cleaned in this zone and small precast concrete blocks were placed at specified spacings to support the railway sleepers for the rails along which the TBM, back up gantries and locomotives travel. Shotcrete was then sprayed from a mobile applicator that could travel up to 12 m longitudinally along the tunnel, embedding the concrete blocks and completing the permanent invert, which comprised the lower 100° of the tunnel circumference. A construction joint was formed where the invert joins with the crown portion of the lining, installed subsequently at the L2 zone.

2.3 L2 zone

The L2 zone was located 65 m behind the cutterhead. Here, shotcrete was sprayed systematically over the initial support that had been installed at the L1 zone, to complete the permanent lining. The thickness of the shotcrete sprayed at the L2 zone typically varied between 125 mm and 250 mm and was dependent upon the quality of the rock mass.

The shotcrete spraying system itself consisted of two spray robots mounted on lifting beams, one on each side of the TBM. The robots were able move 7 m longitudinally, parallel to the tunnel axis, on these beams (Figure 3). The lifting beams themselves moved radially, perpendicular to the tunnel axis, with each lifting beam able to cover 135° on its own side, as well as overlapping by 45° past the vertical with its opposite number, for a total coverage of 270°.

The spray robots operated outside of a cylindrical shield (to which they were attached) that protected workers and equipment from overspray and rebound, and allowed unhindered access to the cutterhead along the upper deck of the TBM. The shield itself could move 3 m relative to the TBM, giving a total shotcrete reach of 10 m of tunnel length.

3 FACTORS INFLUENCING THE REQUIREMENT FOR REMEDIAL WORKS

As tunnel excavation advanced, every effort was made to complete the tunnel support and the shotcrete lining installation in one pass to avoid the need for any remedial works. Nevertheless, a number of unavoidable factors were encountered during construction that resulted in a less-than-perfect lining, necessitating remedial works being required for some parts of the tunnels. The following section discusses these factors in more detail, and how they pertained to the Neelum Jhelum project.

3.1 Support Element Configuration

The tunnel support design relies on the majority of support elements being installed with full contact against the excavated tunnel profile. The presence of overbreak, which was at times difficult or impossible to control, particularly where sub-horizontal stresses were high and the rock mass weaker than average, precluded such a full contact from being achieved. Any overbreak necessitated additional support such as wire mesh, mining straps and full circular steel rings, each of which potentially prevented full contact to some degree.

3.2 Shadowing

A phenomenon related to the support element issue is shadowing, which on the project was the largest contributor to creating voids within or behind the installed shotcrete lining on the project. Shadowing occurs when an item (such as wire mesh) obstructs the free propagation of shotcrete creating an area of missing or poorly compacted shotcrete. Subsequent remedial work is then required in the form of removal and replacement of shotcrete, or additional grouting to fill the voids. Other factors that can cause this is limited access to the tunnel periphery or poor application by the tunnel personnel.

3.3 *Rebound*

The application of shotcrete with differing sequences, as necessitated by variable ground conditions, sometimes precluded use of a systematic application scheme, leading to shotcrete of reduced quality. For example, ad hoc shotcrete application at the L1 zone, if not carefully monitored, could result in rebound falling onto the sidewalls or invert. If this were not removed or cleaned before subsequent layers of shotcrete were sprayed in the invert or at the L2 zone, this rebound could become incorporated into the lining, with significant voids or honeycombing resulting. Of course, observation of such behaviour by supervision staff, and its immediate rectification, was generally implemented, but odd examples occasionally slipped through, particularly early in the excavation before QA/QC procedures had been optimized.

3.4 *Water ingress*

The inflow of water can be particularly problematic for a shotcrete lining as it affects the adhesion of shotcrete to the rock profile, particularly if the rock is prone to softening when wetted. Shotcrete applied under these conditions can result in the development of voids behind the lining. Fortunately, little water ingress was encountered in the project's TBM tunnels.

3.5 *Specific Deleterious Geological Conditions*

Squeezing ground conditions present a significant challenge to the shotcrete lining, particularly if ground movements occur over many months. The TBM itself will usually give an early indication of squeezing ground by increased pressures experienced by the roof and side supports of the TBM shield, and the tunnel profile will become noticeably tighter over a distance equivalent to several tunnel diameters. However, time-dependant creep can occur long after excavation and convergence stations are used to measure such conditions. Squeezing ground will result in cracked and delaminated shotcrete which will require remedial works at a later stage.

Though anticipated, little squeezing ground was encountered on the project, although localized squeezing did result in damaged shotcrete that needed to be repaired.

A rockburst is a sudden and spontaneous release of strain energy resulting from stresses due to high overburden or tectonic forces exceeding rock strength. This relatively unpredictable event can range in severity from very minor to major events that can inflict extensive damage on the shotcrete lining, ranging from small cracks to total destruction. On the Neelum Jhelum project there were approximately 1700 rockbursts recorded in both TBM tunnels, many of which resulted in damage to the shotcrete lining.

3.6 *Additional Factors*

The following additional factors may result in remedial works to shotcrete linings being required.

- TBM gantries coming into contact with and deeply gouging the lining as the TBM advances, due to rockbursts or squeezing ground.
- The accumulation of dust or fines on the rock surface causing poor adhesion of subsequently applied shotcrete layers.
- Poor TBM operation resulting in a poor tunnel profile, for example causing excessive overbreak, if boring parameters are poorly selected for the ground conditions being excavated.

Again, observing and fixing or avoiding such issues is clearly part of good tunnel management. However, in the harsh environment of the Neelum Jhelum project and unanticipated stress conditions, it sometimes took time to identify these problems and establish appropriate solutions.

4 STRATEGIES TO ELIMINATE OR MINIMIZE THE REQUIREMENT FOR REMEDIAL WORKS

The strategies needed to eliminate or at least mitigate the issues listed above are well understood in tunnelling. Firstly, strict application of a Quality Assurance/Quality Control (QA/QC) system is essential to ensure that the design requirements are achieved and that the shotcrete delivered to the TBM is to the required specification. This should include an appropriate QA/QC plan for the materials being used.

Secondly, analysis of the application of shotcrete relative to the planned support elements to be installed should examine how the shotcrete can be applied around individual support elements whilst maintaining the required shotcrete nozzle to tunnel periphery distance, remaining cognizant of the spatial constraints within a TBM.

In difficult geological conditions, a permanent shotcrete lining in a tunnel excavated by TBM has a high potential for remedial works because of the factors outlined in section 3. Therefore, before this type of lining is selected, an initial assessment should be carried out to identify all factors that could lead to sub-standard shotcrete necessitating remedial works, and how they can best be mitigated.

5 TIMING, LOCATION AND IMPLEMENTATION OF REMEDIAL WORKS

5.1 Introduction

If ground conditions are optimal, tunnelling teams dedicated and highly trained, and the TBM is perfectly suited to the conditions, no remedial works will be needed. However, in real-world conditions, particularly in harsh environments and difficult rock stress conditions such as those encountered on Neelum Jhelum, some level of remedial works is unavoidable. Having established that remedial works are necessary, determining how delays can be minimized becomes the crucial factor. The different phases of tunnelling when remedial works could be carried out are outlined below.

Behind the TBM - TBM still excavating - This provides greatly improved access to the shotcrete tunnel lining (as compared to from the TBM), although full access is masked by services still required by the TBM (air, electricity, water, ventilation, conveyor). However, it is logistically the most complex approach, since its implementation cannot conflict with the TBM railway supply system, although water and compressed air supply being provided to the TBM remain available. Importantly, if these works can be completed before the TBM finishes excavation, then the time penalty to the programme is minimal.

After Completion of the TBM Excavation - TBM operations completed, track in place - This provides maximum access to the shotcrete tunnel lining. This approach requires dedicated electricity, water and compressed air supply and its own railway requirements, but since these are retained from the TBM's system, the logistics are usually not complex. However, the major disadvantage of this approach is the time penalty, since the programme completion date needs to be put back to accommodate these works.

After Completion of the TBM Excavation - TBM operations completed, track removed - This type of operation is similar to the one above, with a similar time penalty, except that the operation is directed primarily at the invert.

The following sections discuss the TBM remedial works that were undertaken for the Neelum Jhelum project.

Determining the Need for Remedial Works

Two requirements need to be fulfilled to fully document shotcrete defects, and thus the requirement for remedial works.

Construction Records - Provision should be made in the TBM shift report to enable recording of shotcrete defects. This allows a remedial works register to be established that records the locations of where remedial works will be required. On the Neelum Jhelum project, a dedicated section for possible shotcrete remedial works was incorporated within the TBM shift

Table 1. Example of remedial work item on TBM shift report.

Cause	Tunnel Chainage	Location	Area	Other details
Rockburst	9+743	11-2 o'clock	1.2 m x 1.1 m	0.32 m deep

reports that described the nature of the defect and its extent. An example in Table 1 has been extracted from the TBM shift reports.

Inspection of Completed Lining – While this action may be initiated due to Contract requirements (such as testing for drummy shotcrete), QA/QC procedures or randomly due to special occurrence, simple due diligence requires that such a detailed inspection be performed regardless.

The potential remedial work item should be documented in the format of the remedial works register, which will then define the scope of the works and also identify trends and patterns to allow efficient resource and equipment deployment.

5.2 *Implementing Remedial Works*

In practice on the Neelum Jhelum project, no remediation of damaged shotcrete was possible on the TBM itself, and nearly all the works were carried out after passage of the TBM, both while it was still excavating (termed 'Stage 1' below), after excavation had been completed but track was in place ('Stage 2'), and after the track had been removed ('Stage 3').

5.2.1 *Stage 1*

The first stage of carrying out shotcrete remedial works had reduced access due to the presence of TBM services such as ventilation, conveyor, power, water and compressed air etc. plus the railway line itself. Thus, the access gantries and equipment had to be designed to fit within the existing tunnel configuration, allowing safe access to the exposed sections of shotcrete lining for repairs. The general configuration of this equipment is shown in Figure 4.

As shown in Figure 4, the shotcrete remedial works were carried out from two 12 m long two-level access gantries equipped with a generator, air compressor, drilling equipment, a grout mixer and an injection pump. The TBM excavated tunnels had a single railway track with short double-track sections at 3 km intervals to allow passage of trains in opposite directions.

In order to allow uninterrupted train movement in the tunnel, the remedial work gantries were required to run on a separate track outside of the TBM railway track. These gantries were advanced on this dedicated track and progress was maintained by lifting the track at the rear of the gantries and transporting it forward to extend these rails at the front, allowing the gantries to advance to the next location. Figure 4 (a) shows a cross-section of stage 1 remedial works. The circumference of the tunnel has been divided into eight sections that relate to accessible sections during different stages of the remedial works operations. In stage 1, the accessible sections were 1, 3, 5 and 7 (shown in grey in Figure 4 (a)). The remaining sections were inaccessible due to the presence of the TBM support services, such as ventilation, rails, conveyor etc.

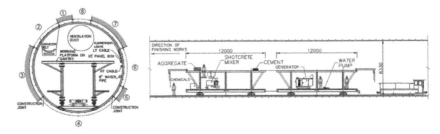

Figure 4. (a) Cross-section of remedial works equipment (available sections of tunnel periphery during stage 1 shaded in grey) (b) Longitudinal section of remedial works equipment.

Figure 5. Remedial works gantry within the tunnel.

The second stage of remedial works was carried out when the TBM had completed excavation, whilst the TBM and support services were being removed from the tunnel. This stage allowed remedial works to be carried out in previously inaccessible areas (shown in grey in Figure 6 (a)). The general arrangement of stage 2 remedial works is shown in Figure 6. Here the long section has the remedial work equipment removed in order to show the rail track removal process.

The third stage of remedial works was carried out whilst the TBM railway tracks and supporting sleepers were removed. During this stage debris from the invert of the tunnel was removed as this

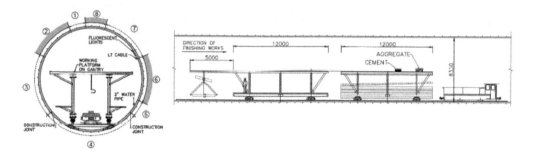

Figure 6. (a) Cross-section of remedial works equipment (available sections of tunnel periphery during stage 2 shaded in grey) (b) Longitudinal section of remedial works equipment Stage 3.

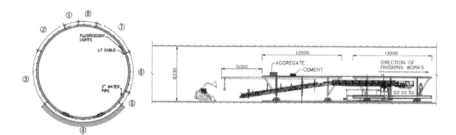

Figure 7. (a) Cross-section of remedial works equipment (available sections of tunnel periphery during stage 3 shaded in grey) (b) Longitudinal section of remedial works equipment.

area was previously inaccessible (shown in grey in Figure 7 (a)). The general arrangement of stage 3 remedial works is shown in Figure 6.

6 REMEDIAL WORKS PROCEDURES AND RESULTS

6.1 *Contact Grouting*

Intimate contact between the initial support and the installed lining is essential to maintain the integrity of the final support of the tunnel. It was therefore vital that there were no voids between the lining and surrounding rock. The remedial works register was used to assist in identifying potential voids locations, although testing was carried out along the entire tunnel. The remedial works gantries allowed direct access to the full periphery of the tunnel to carry out sounding inspections of suspected void locations, by tapping a geological hammer at regular intervals against the shotcrete surface to test for drumminess. The resulting sound allowed either a potential void or delamination to be distinguished from a competent contact between the lining and the surrounding ground.

Once the remedial area had been identified and marked with spray paint, the primary course of action was to contact grout the void or delaminated area. This grouting operation was undertaken at low pressures (300 kPa) and with different grout design mixes depending upon the presence of groundwater. Table 2 shows that the amount of remedial contact grouting undertaken was approximately the same for both left and right tunnels. The data shows that from stage 1 to stage 3, the area affected and the amount of grout consumed decreased with each stage in both tunnels. This indicates that the first two stages are the most effective and should be commenced as soon as possible

6.2 Consolidation Grouting of the tunnel periphery in zones of high rockburst activity

Rockbursts result in overbreak and delamination of the surrounding ground adjacent to the location of the rockburst. An average overbreak of 480 mm was recorded for each rockburst. Consequently, all the consolidation grouting holes were drilled to a depth of 1000 mm to account for both the overbreak and delamination resulting from the rockburst. The location of the holes was in sections 1 and 7. Table 3 shows that the amount of remedial consolidation grouting was broadly the same for both tunnels

6.3 Construction Joint repairs

There are two radial construction joints within the shotcrete lining, one each side of the tunnel, Table 4 shows that a total of 2073 m (9.9%) and 2039 m (10.3%) of the installed construction joint required remedial work in the left and right tunnels respectively. The values are similar, indicating a similar learning curve, standard of workmanship and quality control was applied in both tunnels. The 'other causes' listed in Table 4 primarily relate to inadequate removal of shotcrete rebound.

6.4 Repair of Detached Shotcrete

Larger areas of shotcrete occasionally detached or were deemed suspect due to geological ground conditions (e.g. fault zones and severe rockbursts) where a simple repair was insufficient, and instead additional lining was required. There were two types of additional lining installed on the TBM shotcrete lined tunnels: an additional internal shotcrete lining 250 mm

Table 2. Contact grouting details for both tunnels.

	Left Tunnel (Length of 10 498 m)		Right Tunnel (Length of 9 893 m)	
Remedial Work Stage	Proportion of shotcrete surfacecross-sectional area where remedial works were undertaken (%)	Average quantity of grout injected per square metre of affected shotcrete surface (litres)	Proportion of shotcrete surfacecross-sectional area where remedial works were undertaken (%)	Average quantity of grout injected per square metre of affected shotcrete surface (litres)
1	2.11	23	2.77	22
2	1.78	21	1.80	19
3	1.3	18.9	1.35	18.4
Total Tunnel contact grouting remedial works	1.78%		2.08%	

Table 3. Consolidation grouting details for both tunnels (only carried out during stage 1).

Remedial Work Stage 1	Left Tunnel (Length of 10 498 m)		Right Tunnel (Length of 9 893 m)	
	Proportion of shotcrete surfacecross-sectional area where remedial works were undertaken (%)	Average quantity of grout injected per square metre of shotcrete surface (litres)	Proportion of shotcrete surfacecross-sectional area where remedial works were undertaken (%)	Average quantity of grout injected per square metre of shotcrete surface (litres)
Total Tunnel contact grouting remedial works	2.31 2.31%	34.4	2.95 2.95%	29

Table 4. Construction joint details for twin tunnels.

	Left Tunnel (Length of 10 498 m)				Right Tunnel (Length of 9 893 m)			
	LHS construction joint		RHS construction joint		LHS construction joint		RHS construction joint	
	Damage due to rockburst (m)	Damage due to other causes (m)	Damage due to rockburst (m)	Damage due to other causes (m)	Damage due to rockburst (m)	Damage due to other causes (m)	Damage due to rockburst (m)	Damage due to other causes (m)
Total	107	1125	102	739	109	922	123	885

Table 5. Additional lining details for twin tunnels.

Type of additional lining required	Left Tunnel (Length of 10 498 m)		Right Tunnel (Length of 9 893 m)	
	Additional length required (m)	Proportion of tunnel where additional lining required (%)	Additional length required (m)	Proportion of tunnel where additional lining required (%)
Shotcrete	51	0.5	64	0.6
Concrete	307	2.9	275	2.8

Figure 8. Marking the drummy shotcrete area and drilling for grout holes.

thick and a concrete lining within the existing shotcrete lining. Table 5 shows the lengths over which additional lining was required.

The concrete lining was installed in areas of very poor/faulted ground conditions. In both cases an important design feature was the connection between the existing shotcrete and the additional lining. This was a tapered section, with a camber that allowed for smooth water flow over the transition. Figure 9 shows the design detail of this connection.

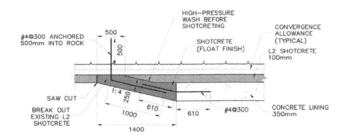

Figure 9. Connection detail between existing shotcrete and additional lining – upstream transition.

6.5 *Contour Checking*

The contract specified that the longitudinal contour profile should be smoothened to a maximum angle of 1:12 looking downstream and 1:4 looking upstream with smooth transitions. Contouring is important in water tunnels since abrupt changes in cross sectional area of the tunnel cause head losses and may contribute to cavitation, which in turn can cause serious damage to the lining. Measuring arms were installed on the finishing gantries in the same position at opposite ends of the gantry. A string line was then installed between opposite pairs of measuring arms and tensioned. A dip survey was then carried out along the 12 m meter long gantry. The non-compliant sections of the lining were identified and excessive peaks in the shotcrete lining were removed. Excessive troughs were filled in with shotcrete and the area was smoothed over so that it was compliant with the specification.

6.6 *Drainage holes*

There was a requirement for drainage holes to be drilled longitudinally every 3 m longitudinally and equally spaced radially around the tunnel. Drainage holes were required for the reduction of differential water pressures across the lining, which could occur whilst tunnels are filled, in operation or dewatered. Since the remedial works gantries were equipped to drill the holes required for grouting, this equipment was reused to drill the drainage holes after grouting was complete.

7 CONCLUSIONS

The management of the remedial works associated with a shotcrete lining installed by a TBM may not be as glamorous as the excavation and lining installation operations, but it nevertheless forms an critical part of completing the tunnel works. The selection of a shotcrete permanent tunnel lining offers programme benefits over placing a cast in-situ concrete lining, but these anticipated time savings can easily be lost if excessive shotcrete defects require extensive repair, particularly if the remedial works are poorly planned. In difficult conditions, a permanent shotcrete lining in a tunnel excavated by TBM has a high potential for requiring remedial works. Therefore, an initial assessment should be carried out to identify all suitable mitigation measures that will result in shotcrete of optimal quality, thereby reducing the need for remedial works to the extent possible.

The quantities given in Tables 2, 3, 4 and 5 are specific to the Neelum Jhelum project but nevertheless can be used as a general guide when planning similar operations with a final shotcrete lining in a tunnel excavated by an open gripper TBM.

Tunnels and Underground Cities: Engineering and Innovation meet Archaeology,
Architecture and Art, Volume 7: Long and deep tunnels – Peila, Viggiani & Celestino (Eds)
© 2020 Taylor & Francis Group, London, ISBN 978-0-367-46872-9

Long and deep hard rock double shield TBM under the Alps. Base Brenner Tunnel Mules 2-3

P. Pediconi, M. Maffucci & G. Giacomin
Ghella S.p.A., Rome, Italy

ABSTRACT: Once completed, the Brenner Base Tunnel (BBT) will constitute the world's longest railway tunnel. The paper describes the construction of the Italian stretch of the tunnel comprised between the Isarco river and the Italian-Austrian border. The project includes 16km of twin tunnels excavated with two large diameter TBMs; 16km excavated with a smaller TBM and 12km excavated by conventional method. The complex logistics and the challenging underground activities will be presented, with special consideration of the mechanized tunnelling works. The geology is extremely complex with different geological formations such as granite, gneiss and schist, interested by the "Periadriatic Fault", an extended regional fault crossing the eastern alps. Rock strengths up to 250MPa and extreme abrasiveness, squeezing grounds and aggressive waters pose severe challenges to the TBM advance. Furthermore the configuration and design of the tunnel boring machines for tunnel excavation is explained, including all the special plants and equipment required, as well as the study of the articulated conveyor system, ventilation and railway selected for optimizing the complex logistics of the jobsite in order to achieve the ambitious goal.

1 INTRODUCTION

The Brenner Base Tunnel Project will be the world's longest railway tunnel, connecting Innsbruck to Fortezza (55 km). The BBT is part of the SCAN-MED corridor, one of the nine trans-European communication corridors. The object of the paper is based on the Italian Mules 2-3 Lot, the largest currently under construction, assigned to the BTC consortium of Astaldi S.p.A, Ghella S.p.A., Obcrosler Cav Pietro, Cogeis S.p.A. and PAC S.p.A for a total work amount of 1 billion €. The Mules 2-3 Lot is the last stretch on Italian grounds, extending up to the Austrian border and will cover more than 17km of the overall 64km of railway line under the Alps. The work, whose completion is scheduled for 2023, will include the excavation of approximately 68 km of tunnels divided as follows:

- 40.3 km of main tunnels;
- 14.7 km of pilot tunnel;
- 85 Cross Passages each 333m;
- 1 emergency stop in Trens;
- 4 km of access tunnel;

The main access point for the work activities is from the decline adit in Mules, connecting to the principal logistic underground hub composed of big caverns and connecting tunnels.

The first part of the excavations and the main tunnels towards the south are planned with drill and blast method. The remaining part of the tunnel will be excavated with mechanized method through the use of three TBMs Double Shield: two for the main tunnel and one for the pilot tunnel. The pilot tunnel TBM has a boring diameter of 6,85m while the two TBMs for the main tunnels have a boring diameter of 10,71m.

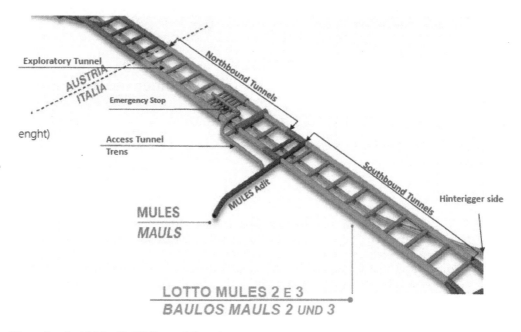

Figure 1.   Lot Mules II–III General Overview.

2   GEOLOGY

The Brenner Base Tunnel crosses the major tectonic units of the Alps: these units, with the shape of multiple overlapping strata, represent the remains of the collision area between the European and Adriatic Plates constituting the separation zone between the European and African plates. Considering the extent of the works, their geographic location and geological variability, rock masses with different resistance and mechanical behavior are expected.

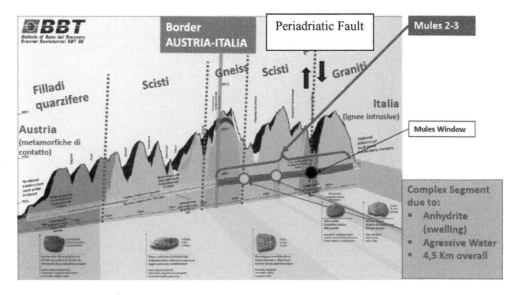

Figure 2.   Geological Profile Brenner Base Tunnel.

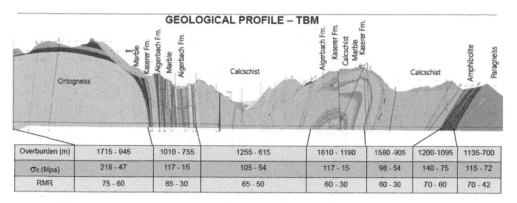

GEOLOGICAL PROFILE – TBM

Overburden (m)	1715 - 945	1010 - 755	1255 - 615	1610 - 1190	1580 - 905	1200 - 1095	1135 - 700
σc (Mpa)	218 - 47	117 - 15	105 - 54	117 - 15	98 - 54	140 - 75	115 - 72
RMR	75 - 60	65 - 30	65 - 50	60 - 30	60 - 30	70 - 60	70 - 42

Figure 3. TBM Tunnel Geological Profile.

The Periadriatic Fault is the contact zone between European and African plates, moving towards each other, causing the closing of Tetide sea and the Alps origin. The rocks, due to subduction, show the Alps plastic metamorphosis, while the Southalps formation have fragile behavior.

In the TBM excavated tunnel the geological formations are Kaserer Formation (phyllite, phyllite-quartzite, mica-schist, quartzite, meta-arkose, subordinate calcareous mica-schist) and Aigerbach Formation (chloritoid phyllite and chloritoid calcareous schist, quartzite, micaceous quartzite, dolomitic marble, meta-arkose and meta-conglomerate, phyllite-quartzite and dark-phyllite, anhydrite, gypsum, cargneuls). In the D&B tunnels the geological formations are the Periadriatic Fault, phyllite and mica-schist, cataclasite, quartzite.

The aim of the pilot tunnel is to serve as an exploration tunnel, to give the geotechnical and geological information to the main tunnels. For this reason, throughout its excavation, an extensive geotechnical monitoring is executed with probe drillings, extensometers, geo-seismic surveys, convergence sections, etc.

The average overburden of the tunnel is in the range of 1.000 1.500m with peaks up to 1.715m. Laboratory tests show that the rock mass is hard and abrasive. The expected UCS varies within a range of 100 MPa to 300 MPa.

Ground water inflow from the rock mass is generally restricted to fissures and weakness zones with the hydraulic conductivity is expected to be low; locations with higher permeability values are expected in correspondence of the big faults.

3   SITE LOGISTICS

Supply to the 3 TBMs is managed from a unique access where the TBMs are assembled and from a second access point from where the train feeds the TBMs through very complex logistic schemes.

There are different site installations for precast production, spoil management in Hinterrigger and tunnelling in Mules. This last area accommodates also the main offices for the site management, the camp and canteen facilities to host 300 tunnel and precast workers (see figure 4). This set-up guarantees high logistical efficiency with minimum impact on the local community due to most of the traffic being internal.

A total of approximately 20.000m^2 of shed structures have been installed to cover precast factories, workshops, warehouses, water treatment and filter press plants, spoil discharge and handling facilities, crushing plants, aggregate distribution facilities, concrete and grout batching plants.

A total of 1.720.000m^3 of rock will be excavated with the D&B method for the main tunnel to the South.

Road based transports for the material supply to the TBMs have been reduced to a minimum. Mucking is managed by conveyor systems. Each TBM has an independent conveyor in

Figure 4. Aerial photo of the site set-up.

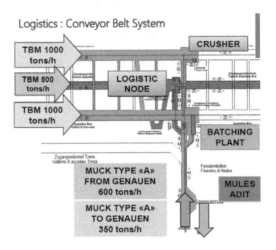

Figure 5. Conveyor Belt System.

the tunnel; in the main caverns all conveyors discharge in the same pit from which another conveyor transports all the muck up to the final area through the existing Aica tunnel, 8km long. The material is discharged in a temporary spoil area where part is used as aggregate for the precast segments or for the concrete for the lining, part for the ballast. The remaining parts are transported by dumpers to the final deposit (see Figure 5).

Part of the backfilling grout of the annular gap is pea-gravel transported by train to the TBMs (120° in the lower part) while the remaining backfilling is bi-component grout directly pumped via pipelines from the grout plant to the TBM backups.

All the transport of material, segments and personnel from the main cavern to the TBMs is performed by train: in order to manage all the tunnel at its peak, the JV estimates to use a configuration of No. 20 trains, between 180 Tons for the pilot tunnel and up to 380 Tons for the main tunnels, for a total railway length of 110km.

Segment handling on surface is managed with 3 gantry cranes with 32 Tons capacity and 20.5 m span.

4 PRECAST SEGMENTS

The single-shell, watertight precast concrete segmental lining for the main tunnels consists of 6 trapezoidal and rhombic segments plus 1 key stone, plus the invert segment. The pilot tunnel

Figure 6. Precast segments on the Mules II-III.

sees instead a 5 trapezoidal and rhombic segment plus 1 key stone, plus the special shape of the invert segment. For the main tunnels, the segment ring has an internal diameter of 9.27 m, a thickness of 45 cm and a length of 1.75 m while the pilot tunnel segment ring has an internal diameter of 5.82 m, a thickness of 30 cm and a length of 1.50 m (see figure 6).

The segment design includes shear connectors in the transversal joint and guiding rods in the longitudinal joint which assist in a quick ring erection with minimum lips and steps.

The waterproof EPDM gasket has been engineered and tested for a design water load over the 100-year design life. The JV has chosen to utilize an anchored gasket solution and is so far very satisfied with the result.

The precast elements are produced within a precast factory in Hinterrigger, with capacity to manufacture both types of segments with stationary 6 sets of molds, a dedicated concrete batch plant and controlled curing after demolding.

5   TBM TUNNELING

The JV has procured the two Ø 10.71 m Double Shield TBMs and one Ø 6.85 m Double Shield TBMs from Herrenknecht (see figure 7); the Tunnel Division Department of Ghella S. p.A. has paid particular attention in preparing the TBMs. The TBM shield and cutterhead used for the main tunnels, with a weight of 250 Tons, are on the very heavy range compared to other hard rock TBMs of this size. The main bearing diameter of 6.0m is specifically designed for this kind of rock. The cutterhead is fitted with 62 x 19" cutters rated 315 kN.

The TBM is further equipped with sealing systems for telescopic shield, gripper windows and a muck ring which can be activated in case of emergency to stop uncontrolled water ingress and prevent environmental and structural damages.

The length of the back-up of the main TBMs is approximately 210m, with a nominal thrust force of 95.000kN and a total power installed of 7.1MW. The average estimated penetration rate is 30-50mm/min. The length of the back-up of the pilot TBM is approximately 290m, with a nominal thrust force of 42.750kN and a total power installed of 4.9MW. The average estimated penetration rate is 30-50mm/min.

The contract requires close geological monitoring to be performed on all TBMs. This includes probe drilling for an exceptional length of 150m, face mapping and core drillings which JV performs systematically on a daily basis during the maintenance shift or each 100m.

The probe holes are normally drilled to around 150 m length with the help of two drill rigs which are installed on the erector ring of the TBM.

Figure 7. Follo Line TBM.

For the moment only the excavation of the pilot Tunnel has started, in May 2018. The TBM has excavated 12% of the total tunnel length (status as of end of August 2018) with an average penetration rate of 6 to 7 mm/rev.

The cutter lifetime has so far settled on an overall average of 350 m³/cutter. The TBM is operating 7 days per week, 24 hours a day. In the third month of production, with a very good and fast learning curve, the production has achieved the good result of 480m. For the moment the production is above expectations.

6   VENTILATION SYSTEM

The ventilation system of the project is very complex due to the different tunnels and connections existing underground. Through the Mules' adit and Aica, the JV supplies fresh air. The air flow exhausts are in the Mules adit and in the ventilation shaft. In each tunnel there is a dedicated ventilation duct with fresh air connected directly to the excavation face. In case of fire, the system allows to manage the smoke. To guarantee the good air quality, No. 4 fans of 800kW each are installed, with a power of 3.200kW and working pressure of 6.000Pa. The entire system is managed through an automatic remote ventilation control: the JV will analyze different scenarios throughout 18 different phases of the construction activity along the 7 years of construction. The overall air consumption has been determined to allow a maximum flow rate consumption.

7   CONCLUSION

The Base Brenner Tunnel Project is a challenging and exciting large scale Hard Rock Double Shield TBM project with precast segmental lining in very complex geological environment. The project will allow an increased experience in hard rock geology characterized by very high overbound. The main challenge of the project consists in the study, execution and control of a perfect clockwork logistical system; the first two years of the jobsite have proved hard work resulting in a continuously unified teamwork aimed at achieving all goals and milestones.

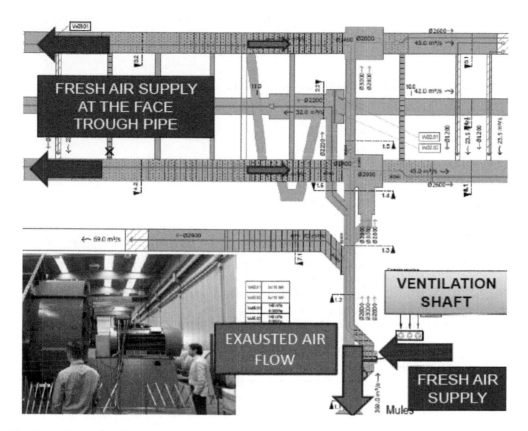

Figure 8. Ventilation System Design.

REFERENCES

A. Menozzi, G. Cimino, F. Maltese, B. Tiberi (SWS Engineering spa), E. Campa, D. Buttafoco, J. Debenedetti BTC (Brennero Tunnel Construction), Bim Implementation – Brenner Base Tunnel Project;

F. Gasbarrone, A. Oss, L. Ziller (SWS Engineering SpA), E. Campa, D. Buttafoco, J. Debenedetti BTC (Brennero Tunnel Construction), Use of expanded clay as annular gap filling. Design and application at the Brenner Base Tunnel.

Tunnels and Underground Cities: Engineering and Innovation meet Archaeology,
Architecture and Art, Volume 7: Long and deep tunnels – Peila, Viggiani & Celestino (Eds)
© 2020 Taylor & Francis Group, London, ISBN 978-0-367-46872-9

Empirical and numerical analysis of the blast - induced structural damage in rock tunnels

P. Perazzelli
Pini Swiss Engineers, Zurich, Switzerland

C. Soli & D. Boldini
DICAM, University of Bologna, Bologna, Italy

ABSTRACT: One of the main hazards in rock tunneling by using the drill and blast method is the damage of the nearby pre-existing underground structures due to ground vibrations generated by the explosion. A simplified empirical approach is commonly adopted for the assessment of blast - induced structural damage, based on the peak particle velocity (PPV) and related empirical threshold values. A more sophisticated approach consists in numerical analysis of rock – support interaction under the dynamic load induced by blasting. This allows evaluate ground vibrations and internal forces inside the underground structures. Both approaches are applied in the present paper for evaluating the feasibility of a drill and blast tunnel excavation in the vicinity of a pre-existing tunnel. Based on the presented results, it is possible to conclude that the simplified empirical approach significantly overestimates the risk of the blast - induced structural damage compare to the numerical stress analyses.

1   INTRODUCTION

In rock tunnelling a widely used method of excavation is the drill and blast technique (Fig. 1), because of its adaptability to different geological conditions, relatively low capital cost tied to the equipment, convenience in short, non-circular and non-linear tunnels compare to the TBM excavation.

Drilling jumbos are used to drill holes in the excavation face (Fig. 1). Then holes are filled with explosives and detonated (Fig. 1). Rock breaking near the hole is caused both by the shock wave and the high pressure. Once the breakout material is removed, tunnel support is applied in the form of shotcrete, anchors, lattice girders, steel meshes or steel sets.

The detonation causes the propagation of elastic waves through the surrounding rock; they are initially mainly compression waves, while at larger distances, due to reflection and

Figure 1.   Images of tunnelling by the drill and blast method.

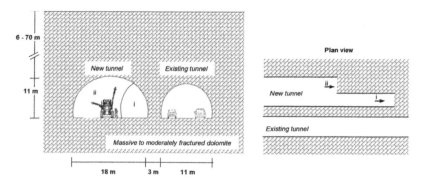

Figure 2. Road 60 Long tunnel project.

refraction phenomena, also shear and surface waves generate. Ground vibration and air over-pressure may induce damage on existing structures, both underground or above ground. Ground vibrations pose the greatest risks, and so blast design is approached as a compromise between achieving the maximum tunnel production and maintaining damaging effects of blast-induced vibrations below specific limits.

Due to the complexity of the mechanisms involved, a simplified empirical approach is commonly adopted for the assessment of blast - induced structural damage. This adopts the peak particle velocity (PPV), i.e., the maximum velocity of the ground particle during the transient phenomenon, as a measure of disturbance level accordingly to empirical thresholds values.

A more sophisticated approach consists in numerical analysis of rock – support interaction under the dynamic load induced by blasting. This allows evaluate ground vibrations and internal forces inside the underground structures, i.e., normal forces, shear forces and bending moments that may be used for structural verifications.

Only few research works investigated numerically the effects of blasting on the surrounding rock-mass (e.g., Cho & Kaneko 2004, Ma & An 2010, Jommi & Pandolfi 2010, Sazid & Singh 2013, Yan et al. 2013). In these works the blasting is simulated by applying a pressure decay law at the boundary of the blast hole. Different nonlinear rock models were adopted for simulating cracking of the nearby rock-mass (e.g., a cohesive crack model was used by Jommi & Pandolfi (2010) and a damage model was used by Ma & An (2010)). Only one example of a design analysis for a specific tunnel construction site using a numerical model was found in literature (Rosengren et al. 2003). More specifically, this study investigated the effect of tunnel blasting on the stresses and deformations on the surrounding rock-mass and an on the primary lining of an existing nearby tunnel.

The present paper focuses on the risk of damage in a pre-existing tunnel induced by the drill and blast excavation of a close proximity parallel tunnel. The case study (Fig. 2) is presented in Section 2 and consists in the Road 60 Long tunnel project (Western Asia). Both empirical and numerical approaches are applied for this purpose. Estimations of the PPV based on empirical laws are presented in Section 3, while dynamic numerical stress analyses using the Finite Element Method (FEM) are presented in Section 4.

2 CASE STUDY

2.1 Overview

The Road 60 Long tunnel (Western Asia) will have a length of about 880 m and it will run almost parallel to an existing tunnel (Fig. 2). The existing tunnel accommodates 2 traffic lanes (bi-directional traffic) and 2 emergency sidewalks (one per each side) and its lining consists in the primary support only (shotcrete and steel ribs or bolts). The new tunnel will accommodate 3 traffic lanes (one of which used for bus transit only) and 2 emergency pedestrian lanes (safety walks). The width and the height of the existing tunnel are about equal to 11 m and 8 m, respectively (Fig. 2), with a maximum overburden of 70 m. The corresponding dimensions of the new tunnel are instead about equal to 18 m and 11 m (Fig. 2). The thickness of the pillar between the new and existing tunnel varies between 3 and 7 m.

The main challenges of the project are related to the close vicinity between the new and the existing tunnels and to the fact that during the construction of the new tunnel the existing tunnel will be kept in operation (Fig. 2).

2.2 Geological and geotechnical conditions

The tunnels cross massive to moderately fractured dolomite (Fig. 3a) in the northern half the tunnel, highly fractured dolomite (Fig. 3b) and blocks of dolomite embedded in marl or in pockets of sandy dolomite (Fig. 3c) in the southern half the tunnel. Karst phenomena (open fractures filled with clay and rubble) may randomly occur (Fig. 3d). The tunnels are located above the water table. Table 1 shows the geotechnical parameters of the main rock unit (i.e., dolomite) in the area of the project.

Partial excavation with one side drift will be adopted in the northern half the tunnel (i.e., in good rock mass conditions) (Fig. 2), while partial excavation with two side drifts and closed arches will be adopted in the southern half the tunnel (i.e., in poor rock mass conditions).

Table 1. Geotechnical parameters of the dolomite (intact rock): Young's modulus E, Poisson ratio v; unit weight γ; uniaxial compressive strength UCS, tensile strength f_t.

E [GPa]	v [-]	γ [kN/m³]	UCS [MPa]	f_t [MPa]
3– 8*, 56**	0.2 – 0.3	23 - 27	20 - 100	5 - 20

* from laboratory compressive tests,
** from a seismic in situ test

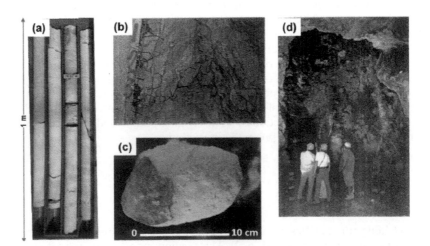

Figure 3. Geology in the area of the project: (a) moderately to fractured dolomite, (b) highly fractured dolomite encountered during the excavation of the existing tunnel, (c) sample of sandy dolomite, (d) karstic cavity encountered during the excavation of the existing tunnel.

3 EMPIRICAL ANALYSIS

The PPV is considered as an index of damage to structures. According to the DIN4150-3 standard, the peak velocity limit for blasting-induced vibration in shotcrete-lined tunnels is fixed at 80 mm/s. The limit applies to high vibration frequencies (i.e. small distances between considered object and detonation point).

According to the literature, the PPV depends on the charge per delay Q, distance from charge point D, rock-mass characteristics (e.g., stiffness, unit weight, fracturation degree), blast hole conditions, presence of water, explosive type and blasting pattern. A number of empirical equations exist in literature for estimating PPV; a review of different equations can be found in Kumar et al. (2016). The equations generally read as follow:

$$PPV = K(D^a/Q^c)^{-n} \qquad (1)$$

where K, a, c and n can be determined by the interpretation of in situ measurements. According to the literature (e.g., Kumar et al. 2016), the parameter c is generally assumed equal to 0.5 or 1, while the parameter a equal to 1 or 1.5, n value varies typically between 0.5 and 1.5, while K value between 100 and 1000.

A graphical representation of the most common empirical equations is given in Figure 4. Figure 4 shows the peak particle velocity as a function of the distance from charge point D for a specific value of charge per delay ($Q = 1$ kg). Each curve of Figure 4 may be used to evaluate the risk of structural damage in the shotcrete lining of a tunnel by comparing the PPV value given by the curve and the assumed limit value of 80 mm/s (i.e., points of the curves above the dashed line in Figure 4 thus indicate risk of damage).

According to Figure 4, there is no risk of blast - induced structural damage in the existing Road 60 long tunnel if the distance from charge point D is larger than 6 m (i.e., if the thickness of the pillar between new and existing Road 60 long tunnels is larger than 6 m). Al lower distances D, the risk may exist depending on the considered model.

Empirical curves shown in Figure 4 were determined in site conditions different from those of the project and as such they cannot be considered strictly valid for the present case-history. For more accurate predictions, in situ tests need to be executed to determine a specific empirical law for the Road 60 project.

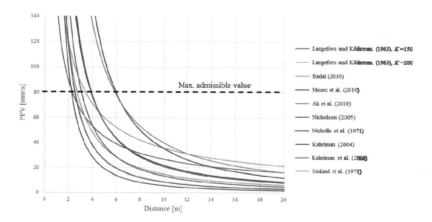

Figure 4. PPV (peak particle velocity) as a function of the distance from charge point according to different empirical models (charge per delay $Q = 1$ kg).

4 NUMERICAL STRESS ANALYSIS

4.1 Basic assumptions

Another approach for predicting blasting effects consists in the dynamic numerical stress analyses. Only few works have been found in literature on this topic (Section 1). The most uncertainties of this approach concern the modelling of the rock-mass behaviour and the simulation of blasting.

Blasting induces a dynamic loading and unloading of the rock-mass, characterized by shear failure (crushing) and tensile failure (cracking) in the so-called near-field region (i.e. in a small region adjacent to the blast hole) and elastic deformations far from the blast hole, in the so-called far-field region. In consideration of the aim of the study, focused on the analysis of far-field conditions, only a very simple model for the rock, consisting in a continuum medium with linear visco-elastic behavior, was adopted.

Energy liberated by initiation of the explosive charge during blasting manifests itself in two forms: as a shock pulse and as energy of the expansion of explosive product. The shock pulse energy expands like an elastic wave, spreading radially in all directions starting from the blast hole. The shock pulse is here is modelled using the pressure decay function (Fig. 5) proposed by Duvall (1953) and modified by Cho & Kaneko (2004):

$$P = P_p \xi (e^{-\alpha t} - e^{-\beta t}) \tag{2}$$

where:

$$\xi = 1/(e^{-\alpha t0} - e^{-\beta t0}) \tag{3}$$

$$t_0 = (1/(\beta - \alpha))\ln(\beta/\alpha) \tag{4}$$

P is the pressure at time t, P_P is the peak pressure, t_0 is the time at which the pressure peak is reached and α, β are constants. The present study assumes t_0 equal to 0.0001 (Cho & Kaneko 2004) and the ratio β/α equal to 1.5. Taking into account these assumptions, the Equations (3) and (4) lead to the following parameter values: $\alpha = 8109$, $\beta = 12164$ and $\xi = 6.65$.

The peak of pressure P_p is assumed to coincide with the denotation pressure. This latter was computed according to the National Highway Institute (Konya & Walter 1991) by using the following equation:

$$P_p = 449.93 SG_e(VOD^2)/(1 + 0.8SG_e) \tag{5}$$

where P_P is the detonation pressure (Pa), SG_e is the density of the explosive (g/cm^3) and VOD is the detonation velocity of the explosive (m/s). The present study assumes a density of the explosive equal to 0,8 g/cm^3, a velocity of detonation equal to 3500 m/s and a corresponding detonation pressure of 2689 MPa. Figure 5 shows the pressure decay function considered in the present study.

In the reality several blast holes are executed on the face, their density is typically bigger than 1 hole/m^2. The blast holes may be detonated at the same time, with standard delay (0.25–0.5 sec) or with micro delay (0.02–0.03 sec).

The micro and standard delays of detonation are typically high enough compare to the duration of the blasting perturbation (dependent on the pressure decay and rock stiffness, order of magnitude of 0.001 sec) to not cause the overlap of the perturbation of different blast holes.

The present study assumes that each blast hole is delayed or micro delayed compare to the other holes. This condition together with the simplified assumption of visco-elastic behaviour of the rock-mass makes it possible to simplify significantly the numerical model. As shown in Figure 6, the numerical model adopted in the present study accounts for a single blast hole; the modelling of all blast holes is not considered as necessary, since the effects of the different hole

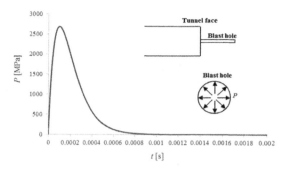

Figure 5. Pressure decay function.

detonation do not overlap and the rock-mass properties remain constant after each blasting (i.e., rock damage is neglected).

Dynamic plain strain analyses were performed using the Finite Element Method (FEM) and the commercial code Plaxis 2D. Rayleigh damping was considered in the equation of motion; this is a viscous damping which is calculated as a linear relationship between the mass and stiffness of the material:

$$\mathbf{C} = \alpha_R \mathbf{M} + \beta_R \mathbf{K} \tag{6}$$

where $\mathbf{C}$, $\mathbf{M}$ and $\mathbf{K}$ are respectively the damping, mass and stiffness matrix of the system, α_R and β_R are the so called Rayleigh damping coefficients. These coefficients were calibrated according to the procedure suggested by Amorosi et al. (2010). This procedure required the computation of the Fourier's spectrum of the pressure decay function for evaluating the frequencies containing the interval with the high energy content (Soli 2018).

4.2 Computational models

.Two types of analyses were performed.

The first analyses investigated the stresses and the deformations of the rock-mass induced by the blasting load under green field conditions (i.e., without considering the presence of the existing tunnel).

The second analysis focused on the internal forces in the primary lining of the existing tunnel induced by the blasting load. Both analyses applied the numerical model presented in Figure 6. The following computational stages were considered: initialization of the stresses in the rock-mass assuming a lithostatic stress distribution; excavation and support of the existing tunnel (in the analysis of the existing tunnel only); excavation of the blast hole; application of the blasting load.

Different boundary conditions were applied for the static stages (i.e., the base of the model is fixed, the upper face is let free and the two vertical sides are normally fixed) and the dynamic stage.

In fact, special boundary conditions are required in case of dynamic loading in order to take into account properly the far-field behaviour and avoid spurious reflections of waves on the model boundaries. The so-called viscous boundaries were adopted for the lateral sides while a compliant base was adopted for the bottom of the model (Plaxis 2017). Table 2 summarizes the model parameters for the rock-mass. The largest Young's modulus refers to a rock-mass composed by a massive dolomite while the lowest value refers to a highly fractured dolomite. The primary lining consisting in a reinforced shotcrete layer with thickness of 0.25 m was modelled by means of elastic beam elements with axial stiffness EA of 8×10^6 kN/m, bending stiffness EI of 41667 kNm^2/m, weight w of 5 kN/m/m. Dry conditions were assumed in the analyses consistently with the in situ hydraulic conditions.

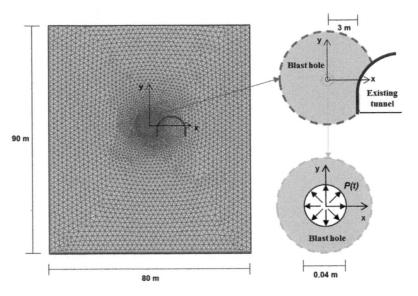

Figure 6. Numerical model.

Table 2. Rock-mass parameters in the numerical analyses: Young's modulus E, Poisson ratio v; unit weight γ; coefficient of earth pressure at rest K_0, Rayleigh damping coefficients α_R and β_R.

E [GPa]	v [-]	γ [kN/m^3]	K_0 [MPa]	α_R [-]	β_R [-]
1-56	0.3	25	1	94.25	0.0265x10^{-3}

4.3 Results

Figure 7 shows as example the radial displacement, the radial velocity and radial stress as a function of the time at three different distances from the blast hole (i.e., $x = 2$, 5 and 20 m) under green-field conditions and adopting $E = 56$ GPa. The blasting load (Fig. 5) induces a ground oscillation and a temporary variation of the stresses in the ground (Fig. 7). The amplitude and the frequency of the oscillation decreases with the distance from the blast hole (the frequency is around 50 Hz for $x = 2$ m and 35 Hz for $x = 20$ m in the example of Figure 7). The peaks of the displacement, velocity and of the radial stress decreases no linearly with the distance from the blast hole, their velocity of propagation coincides with the velocity of P waves (around 5000 m/s in the example of Figure 7).

Figure 8 shows the peak velocity as a function of the distance from the blast hole for different rock-mass stiffness under green-field conditions. The same figure reports for comparison the maximum admissible vibration as assumed in Section 3. The trend and the magnitude of the peak velocity are comparable with those of the empirical equations reported in Figure 4. As expected the peak velocity in a specific point increases with decreasing the Young's modulus of the rock-mass. At distance of 3 m from the blast hole, the peak velocity is equal to 45 mm/s for a Young's modulus of 20 GPa while it is between 190 and 360 mm/s for Young's modulus between 1 and 3 GPa. Similarly to Figure 4, the results of Figure 8 indicate that in the existing Road 60 the risk of blasting-induced structural damage exists and depends on the specific site conditions. This risk is high in case of rock-mass with low stiffness (i.e., in case of highly fractured dolomite). This case was investigated with a more sophisticated model including the presence of the existing tunnel. The results are shown in Figures 9 and 10.

Figure 9 shows the contours of the ground velocity at different times. It is interesting to see that the existing tunnel protects the rock-mass at the opposite side of the tunnel from the

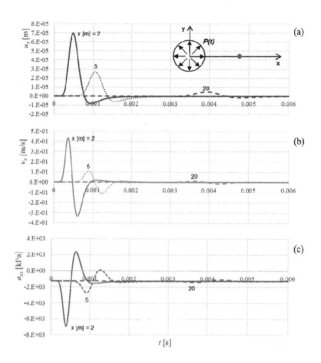

Figure 7.   (a) Horizontal displacement, (b) horizontal velocity, (c) horizontal stress as a function of the time in the rock-mass at different distance from the blasting hole (see the sketch inside the figure, x = 2, 5 and 20 m) (green-field analyses, E = 56 GPa, other parameters according to Table 2).

blasting perturbation. The peak velocity at a distance of 3 m from the blast hole in proximity of the existing tunnel is equal to 155 mm/s (Soli 2018); it is 20% smaller than the one computed under green-field conditions (190 mm/s according to Figure 8) but still larger than the admissible one according to the DIN regulations (i.e. 80 mm/s).

Finally Figure 10 shows the distribution envelope of the axial forces and bending moments in the primary lining of the existing tunnel induced by the blasting load only. The maximum axial force induced by the blasting load is 143 kN/m (Fig. 10) and corresponds to 7% of the maximum axial force under static conditions (2000 kNm/m according to Soli 2018). The

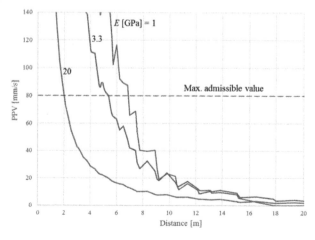

Figure 8.   PPV (peak particle velocity) as a function of the distance from charge point according to the numerical stress analyses for different Young's modulus (green-field analyses, E = 1, 3.3 and 20 GPa, other parameters according to Table 2).

4166

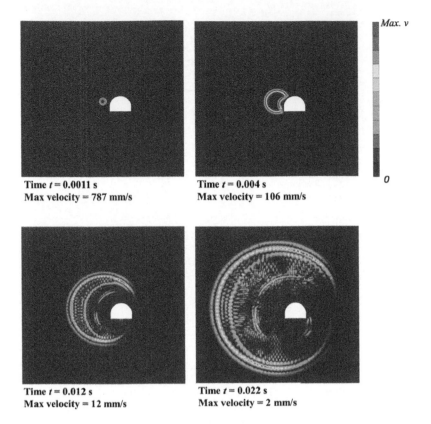

Time *t* = 0.0011 s
Max velocity = 787 mm/s

Time *t* = 0.004 s
Max velocity = 106 mm/s

Time *t* = 0.012 s
Max velocity = 12 mm/s

Time *t* = 0.022 s
Max velocity = 2 mm/s

Figure 9.   Contours of the velocity for different times (E = 3.3 GPa, other parameters according to Table 2).

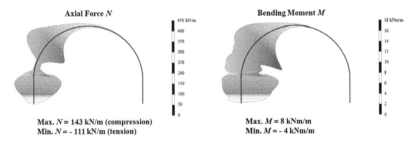

Axial Force *N*

Max. *N* = 143 kN/m (compression)
Min. *N* = - 111 kN/m (tension)

Bending Moment *M*

Max. *M* = 8 kNm/m
Min. *M* = - 4 kNm/m

Figure 10.   Axial force and bending moment distribution envelope (due to the blasting load only) (E = 3.3 GPa, other parameters according to Table 2).

maximum bending moment induced by the blasting load is 8 kNm/m (Fig. 10) and corresponds to 13% of the maximum bending moment under static conditions (60 kNm/m according to Soli 2018). Soli (2018) showed that the ultimate limit states verifications are fulfilled for the both static and dynamic conditions.

Finally, for the considered case study it is shown that in points of the ground in the vicinity of the existing tunnel the peak ground velocity exceeds significantly the regulations limits. Axial forces and bending moment on the structure, however, are not so large to justify damage.

5   CONCLUSIONS

The present paper describes a numerical approach for predicting the blast - induced structural damage in rock tunnels. This approach consists in the determination of the peak ground velocity and of the internal forces in the underground structures during blasting by dynamic FEM analyses. The blasting is simulated with a time dependent pressure function applied at the boundary of a blasting hole. The method is applied for the analyses of a real case-study, i.e. the Road 60 Long tunnel project (Western Asia). The risk of blast - induced structural damage is particularly relevant for this project because a new tunnel will be excavated in the proximity of an existing tunnel (thickness of the pillar between 3 - 7 m) and because during the construction of the new tunnel the existing tunnel will be kept in operation.

For the considered case study it is shown that in points of the ground in the vicinity of the existing tunnel the peak ground velocity exceeds significantly the regulations limits in the worst case conditions of thickness of the pillar equal to 3 m and relative low Young's modulus of the rock mass ($E = 3.3$ GPa). Axial forces and bending moment on the structure, however, are not so large to justify damage. Based on the presented results, it is possible to conclude that the currently adopted regulations limits of the peak ground velocity for assessing the risk of blast - induced structural damage are largely conservative compared to the numerical stress analyses considering the ground – structure interaction.

REFERENCES

Ak, H., Iphar, M., Yavuz, M. & Konuk, A. 2009. Evaluation of ground vibration effect of blasting operations in a magnesite mine. *Soil Dynamics and Earthquake Engineering*; 29(4):669–676.

Amorosi, A., Boldini, D. & Elia, G. 2010. Parametric study on seismic ground response by finite element modelling. *Computers and Geotechnics* 37(4): 515–528.

Badal, K.K. 2010. *Blast vibration studies in surface mines*. BS Thesis. National Institute of Technology Rourkela, 2010.

Cho, S.H. & Kaneko, K. 2004. Influence of the applied pressure waveform on the dynamic fracture processes in rock. *International Journal of Rock Mechanics & Mining Sciences* 41: 771–784.

DIN 4150 – 3 standard. Vibrations in buildings.

Jommi, C. & Pandolfi, A. 2008. Vibrations induced by blasting in rock: a numerical approach. *Rivista Italiana di Geotecnica* 2: 77–94.

Kahariman, A. 2004. Analysis of parameters of ground vibration produced from bench blasting at a limestone quarry. *Soil Dynamics and Earthquake Engineering*, 24(11):887–892.

Kahariman, A., Ozer, U., Aksoy, M., Karadogan, A. & Tuncer, G. 2006. Environmental impacts of bench blasting at Hisarcik Boron open pit mine in Turkey. *Environmental Geology*; 50(7):1015–1023.

Konya, C.J. & Walter, E.J. 1991. *Rock blasting and overbreak control*. Publication No. FHWA-HI-92-001.

Langefors, U. & Kihlstrom, B. 1963. *The modern technique of rock blasting*. John Wiley and Sons; 1963.

Mesec, J., Kovač, I. & Soldo, B. 2010. Estimation of particle velocity based on blast event measurements at different rock units. *Soil Dynamics and Earthquake Engineering*; 30(10):1004–1009.

Ma, G.W. & An, X.M. 2008. Numerical simulation of blasting-induced rock fractures. *International Journal of Rock Mechanics & Mining Sciences* 45: 966–975.

Nicholson, R.F. 2005. *Determination of blast vibrations using peak particle velocity at Bengal Quarry, in St Ann, Jamaica*. MS Thesis. Lulea University of Technology, 2005.

Nicholls, H.R., Charles, F.J. & Duvall, W.I. 1971. *Blasting vibrations and their effects on structures*. U.S. Department of the Interior, Bureau of Mines.

PLAXIS 2D 2017. Reference Manual, Delft, The Netherlands.

Rosengren, L., Brandshaug, T., Andersson, P. & Lundman, P. 2003. Modeling effects of accidental explosions in rock tunnels. In Matthew Handley & Dick Stacey (eds.), *Technology Roadmap for Rock Mechanics; Proc. intern. symp., Johannesburg, 8-12 September 2013*. Ohannesburg: South African Institute of Mining and Metallurgy.

Sazid, M. & Singh, T.N. 2013. Two-dimensional dynamic finite element simulation of rock blasting. *Arabian Journal of Geosciences* 6: 3703–3708.

Siskind, D.E., Stagg, M.S., Kopp, J.W. & Dowding C.H. 1971. *Structure response and damage produced by ground vibration from surface mine blasting*. U.S. Department of the Interior, Bureau of Mines.

Soli, C. 2018. *Evaluation of damage induced by drill and blast excavation in tunnelling on existing structures*. Master Thesis, University of Bologna, 2017.

Yan, B., Zeng, X. & Li Y. 2015. Subsection forward modelling method of blasting stress wave underground. *Mathematical Problems in Engineering*, Article ID 678468, 9 pages.

Tunnels and Underground Cities: Engineering and Innovation meet Archaeology,
Architecture and Art, Volume 7: Long and deep tunnels – Peila, Viggiani & Celestino (Eds)
© 2020 Taylor & Francis Group, London, ISBN 978-0-367-46872-9

BBT, Lot Mules 2–3. Management of data gained by the pilot tunnel drive for the twin main tubes

M. Pescara, M. Spanò & N. Della Valle
Tunnelconsult Engineering SL, Sant Cugat del Vallés, Barcelona, Spain

R. Sorge
Astaldi SpA, Rome, Italy

G. Giacomin
Ghella SpA, Rome, Italy

D. Buttafoco & M. Masci
BTC Brenner Tunnel Construction, Rome, Italy

E. Nuzzo
SWS Engineering SpA, Trento, Italy

ABSTRACT: Along Lot Mules 2–3 of the Brenner Basis Tunnel the Client BBT choose to improve geological knowledge along the alignment by a pilot tunnel to be excavated contemporarily but in advance to the main twin tubes. BTC (JV from Astaldi-Ghella-Pac-Cogeis) was awarded this project based on a tender where all the three tubes will be excavated using a DS TBM. The management of the data provided by the complex system of investigation and monitoring through the pilot tunnel ahead is a key point of the tunnel construction with the aim (1) to give instruction for the excavation of the pilot tunnel to explore what was never directly investigated and (2) to prepare the "engineered as built profile" of the pilot to use as a guide for the two bigger machines. The paper shows how this system is going to be organized and which parameters and correlations seems to be relevant for the purpose.

1 INTRODUCTION

Lot Mules 2–3 of BBT is linking the area of the Isarco underpass with the underground Italy-Austria State border for a length of the main tunnels of 21+927 m (East tube). The main access to the underground construction site is done by the Mules adit that connects the National Road SS 12 to the underground logistic area from where the 5 advancing faces start:

- two fronts are used to excavate the main twin tunnels direction South
- two fronts are used to excavate the main twin tunnels direction North
- one front is used to excavate the pilot tunnel direction North (CE in the following, started in May 2018 by TBM)

From the same logistic area one more front has been designed to connect the Mules Adit to the node of Trens emergency stop as well as to connect the level of the main tunnels with the level of the pilot tunnel placed 11 m underneath.

The pilot tunnel reaching this point from South is already excavated being the well-known tunnel Aicha-Mules connecting directly with the open air logistic area of Unterplatter.

Figure 1 and Figure 2 show the schematic layout of this construction Lot and the typical transversal section of the three running tunnels in correspondence of a cross passage.

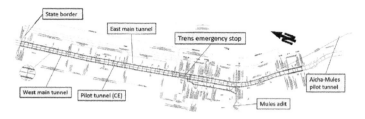

Figure 1. Layout of Lot Mules 2–3.

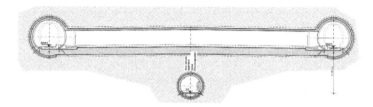

Figure 2. Transversal section of the running tunnels along a cross passage.

The pilot tunnel and the twin tubes are excavated at the same time with a distance between the pilot and the more advanced of the main tunnel which cannot be less than 500 m.

The geological investigation performed during the design phase was done by several long and even inclined boreholes, but the relevant length of these tunnels still left uncovered many portions of the alignment, thus the pilot tunnel has the main function to explore in advance what has been never investigated before. To assist the construction of the three tubes and specifically to properly organize all the possible information that can be obtained by the drive of the pilot tunnel, a thorough investigation and monitoring plan has been organized.

The purpose of this paper is to illustrate the main elements of the investigation and monitoring plan, how they are linked together, and which results have been obtained so far.

2 THE GEOLOGICAL CONTEXT OF LOT MULES 2–3

2.1 General overview

The Brenner Basis Tunnel crosses the central part of the Eastern Alps and the main tectonic nappes involved in the collision zone between the European and the Adriatic (African) plates, thanks to the presence of the huge antiform fold coinciding with the Tauern Window.

The Tauern Window, indeed divided into two different nuclei, the Tux at North and the Tinnertal at South, is constituted by the uplifted southern limit of the European continent, the complex of the Sub-Penninic nappes, namely the so-called Central Gneiss with its Mesozoic cover sediments of the lower Schiefe-rhülle.

Upward, the Central Gneiss and its cover are wrapped by the complex of the Penninic nappes, the upper Schiefe-rhülle, mainly calcschists (Bündner Schiefer) and ophiolites, the rocks of oceanic origin which overthrusted above the subpenninic nappes at the moment of the subduction phase. At the edge of the Tauern window, either on the north on the south side, the Austroalpine nappes are present, once constituting the Adriatic (African) continental border.

On the south side of the Periadrial line fault system, the Southern Alps nappes, on the contrary of all the so far mentioned tectonic units, have not been involved in the subduction process, are not characterized by alpine metamorphism and show only brittle alpine deformation (Frisch 1976, 1979, Ratschbacher et al., 1991, Fügenschuh et al., 1997 Lammerer & Weger, 1998).

According to the geological setting described here above, the excavation of the Brenner Base Tunnel and specifically the sector corresponding to the Lot Mules 2–3, crosses the rocks/ tectonic units summarized in the following table.

The overburden is ranging between 595 and 1715 m.

Table 1. Tectonics units and lithologies along the alignment of Lot Mules 2–3.

Tectonic units	Lithological units	Sector Lgth (m)
Upper Austoalpine Crystalline basement	Paragneiss	1510
Upper Austoalpine Crystalline basement	Amphibolites	400
Penninic upper Schieferhülle	Calcschists, Ophiolites, Triassic horizons, Marbles	2290
Penninic lower Schieferhülle	Kaserer Form. with triassic rock interbedding	350
Penninic upper Schieferhülle	Triassic rock, Calceschist	390
Penninic upper Schieferhülle	Kaserer Form. With triassic rock interbedding 26	775
Penninic upper Schieferhülle	Triassic at the base of Vizze nappe 27	105
Penninic upper Schieferhülle	Kaserer Form. with triassic rock interbedding 26	40
Penninic upper Schieferhülle	Triassic rock, Carbonate quarzites and carbonate calcschists, Marbles, Prasinites, evaporitic sequence, silicoclastic sequence	4875
Penninic upper Schieferhülle	Triassic rock, marbles, calcareous marbles	295
Sub-Penninic lower Schieferhülle	Central gneiss and pre-granitic basement	3902

Table 2. Main hazards and uncertainties.

Typical hazard	Where it is expected
Waters infiltrations and interstitial pressure: sudden water incoming and possible invasions of material	Along main fault zones and/or highly fractured horizons
Excavation in mixed lithotypes: inhomogeneous rock mass behavior, transition zones with different permeability or gas presence	Within lower and upper Penninic tectonic units, especially close to Tux and the Tinnertal antiforms nuclei, where several different kinds of rocks are present.
Swelling conditions	With huge overburden and rock mass poor geomechanical conditions
Rock burst	With huge overburden and rock mass very good geomechanical conditions
Interferences between neighboring cavities	Within karst conditions, within marbles and calcareous rocks
High temperatures	With huge overburden and at main fault zones crossing along which hydrothermal waters can raise from deep crustal sectors.
Impact on water resources	Within karst conditions, within marbles and calcareous rocks

From the hydrogeological point of view, the carried studies foresee water incomings flows, for 10 m of tunnel, of 0,4 to 2 l/s within little or no calcareous rocks, 0,4 to 10 l/s/10m within calcareous rocks and larger than 10 l/s/10m within fault zones.

2.2 *The main hazards and uncertainties*

Referring specifically to the portion of the CE alignment excavated using the DS-TBMs (i.e. from Pk 13+075,95 to State border at pk 27+217 m) and considering the geological setting described with the corresponding overburden/state o stress, the main hazards and uncertainties along the alignment can be summarized as follows in Table 2.

3 CHARACTERISTICS OF THE SELECTED TBMS

Given the picture described in chapter 2, the TBMs have been designed and equipped to face most of the risks coming from the highlighted hazards, starting from the basic choice to use a Double-Shield TBM for both the pilot tunnel (CE) and the main tunnels (GLEN and GLON) direction North.

The principal focus points to further customize the machines were:

* minimum length of the shields
* overcutting capacity
* conicity of the shields
* thrust force for the principal and auxiliary cylinders
* torque
* number of drilling positions through the cutter-head and the shields
* equipment for geotechnical monitoring

The TBMs are equipped to install a universal ring made of:

* CE: 5+1, 30cm thick, 1500mm long, 5820mm ID segments with single gasket at extrados
* GLEN/GLON: 6+1, 45cm thick, 1750mm long, 9270mm ID segments with double gasket at intrados and extrados

The following table gives a synthesis of the main strategic characteristics of the three machines.

Table 3. DS-TBM for the excavation of the three tunnels.

CHARACTERISTICS	HK TBM – S1054 - CE	HK TBM – S1071/72 - GL
Machine type	Double Shield TBM	Double Shield TBM
Installed power	approx. 4 900 kW	approx. 4 200 kW
Length of shield and cutterhead	12195 mm	12480 mm
Conicity (cutterhead-tail shield)	220 mm (difference in diameter)	290 mm (difference in diameter)
Cutterhead Nominal boring diameter	6850 mm	10710 mm
Overcutting (tool at crown)	224 mm	224 mm
Torque (nominal to breakaway, kNm)	3619 to 14013	13600 to 30636
N. of main cylinders and max thrust	n. 10 – 42750 kN at 420 bar	n. 18 – 95000 kN at 420 bar
N. of auxiliary cylinder and max thrust	n. 16 – 97000 kN at 600 bar	n. 38 – 212700 kN at 550 bar
N. of drillings through the cutterhead	n.22 diam 125 mm	n. 22 diam 125 mm
N. of drillings through the shield	n. 22 diam 125 mm and n. 2 diam 152mm	n. 10 diam 125 mm
Auxiliary geotechnical equipment	n. 5 fontimeters, n. 3 pressure cells	n. 1 fontimeters, n. 3 pressure cells

4 THE INVESTIGATION AND MONITORING SYSTEM

Although the DS-TBMs have characteristics that allow to face a wide range of geotechnical behavior while tunneling, the need to investigate ahead remains a main point to excavate long and deep tunnel where the possibility of a wide and extensive direct investigation during design phase is anyhow limited.

Therefore, systematic investigations are carried out aimed to detail and verify the information provided by the geotechnical profile while tunneling as per the following list:

- probe drilling (with possibility of core recovery if required);
- seismic geophysical investigation, using Tunnel Seismic While Drilling (TSWD) and Tunnel Seismic Prediction (TSP) techniques;
- geo-electric investigations, using BEAM system+;
- acoustic emissions measurement by means of geophones (only for main tubes)
- recording and analysis of main TBM excavation parameters;
- gas and radiation monitoring;
- water flow measurements.

The philosophy of the system is to check with one direct punctual investigation (the probe drilling) the general condition far ahead of the excavation face to detect well in advance zones with potential relevant hazards (poor geotechnical conditions associated with faults area or water inflow, ecc..) and to further detail the mechanical characteristics close to the TBM cutterhead while advancing with indirect investigations. The data obtained are checked against the performance of the TBM so that the overall system can learn while advancing.

4.1 *Probe drilling*

Core destruction boreholes are generally carried out every 100 m for a depth of roughly 150 m, so, with a roughly 50 m overlap between followings boreholes.

The probe drillings are carried out by an Eurodrill RH11–6 rotary percussive drilling machine (with DTH hammer - Wassara W70 or W100 model, respectively, with Φ 82 or 115 mm bit) installed on the TBM bridge, which allows an inclination up to 8°-9° and a 120° contour distribution; the use of preventer connected to the shield and sealed into the rock is also foreseen.

The overall drilling parameters are recorded so that the performance and hence the characteristics of the rock mass is investigated. The main information that allow the interpretation of the probe drill are the followings from which the specific energy is derived:

- water flow (l/min)
- percussion pressure (bar).
- rotation speed (g/min);
- rotation pressure (bar).
- penetration speed (m/h);
- thrust (kN);
- torque (kNm)

Eventually, it would be possible to carry out direct and indirect tests within the borehole to better characterize the rock mass (geophysical tests, temperature measurements, pressure gauges and/or transducers), as well as to verify the rock mass in situ stress conditions or deformations by means of borehole extensometers.

Moreover, in case of crossing of particularly critical zones (main fault zones, zones characterized by main uncertainties, etc.) and in accordance with the supervision responsible, it is foreseen the possibility to carry out a borehole with core recovery, radial core destruction boreholes, hydrogeological survey and rock mass deformation measurements by means of n° 3 radial multi-base extensometers.

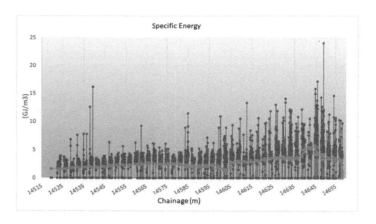

Figure 3. typical example of Specific Energy graph, for probe drill n. 15.

4.2 *Geophysical investigation*

The systematic geophysical investigations performed within the CE are the TSWD and TSP methodologies, this last being normally executed during boreholes drilling.

4.2.1 *TSP*

The TSP methodology, aims for determining any variations in the rock mass quality, namely discontinuities (faults, pervasive joint systems etc.) beyond TBM face with a test done in correspondence of the stops for the probe-drilling.

For this purpose, the seismic source signal is generated by detonation of explosive micro-charges into dedicated short holes drilled through the segmental lining (or by means of a hitting hammer. The seismic signal, generated by the detonations or by the hammer hits and reflected by the rock mass discontinuities are recorded by 2 pairs of triaxial geophones, inserted as well into dedicated holes. The acquired data are afterward processed by means of a specific software, separating from the acquired 3D axes signals, the P (compressional) and S (shear) waves as well as the converted SH and SV.

From seismic waves velocities and their related ratio, other rock mass parameters are evaluated/derived, such as Poisson coefficient, static and dynamic Young modulus, as well as the shear modulus.

4.2.2 *TSWD*

The TSWD methodology can derive the same parameters of the TSP but with a continuous acquisition since it uses as source signal the elastic waves generated by the cutter-head vibrations during excavation. Such continuous signal is transformed into a classic seismogram thanks to the use of a pilot signal directly measured at the source.

The TSWD procedure avoids downtime since no explosive is used and with the cutter-head vibrations allow to get a high frequency signal, reflected by geological discontinuities related

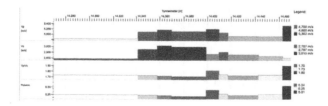

Figure 4. Example of executed TSP investigation by Akron (software TSP303 from Amber SA).

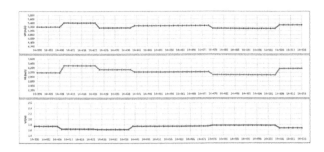

Figure 5. Example of one of last executed TSWD report by Akron Srl.

to faults, competence variations, voids, highly fractured horizons, etc., which can be easily extracted from background noise.

The TSWD methodology allows to get spatial and temporal continuity information from reflecting horizons because the investigation is carried out permanently moving forward the measuring stations and keeping continuously the same distance from the cutter head: in this way an overlap between following reports is kept allowing continuous updating of the information.

The comparison between the two methodologies TSP and TSWD allows a calibration of the two systems.

4.2.3 BEAM

The BEAM (Bore tunneling Electrical Ahead Monitoring) is a geophysical investigation methodology, consisting to induct polarization, at different frequencies, through the rock mass, to measure related resistivity. The BEAM configuration is constituted by 3-electrodes, being:

- A0+ can be the cutter-head itself, one of the single elements in contact with ground,
- A1+, the tail shield or one of the safety constructional components,
- returning one B (-), positioned on the rings (or at the portal).

The generated static electric field leads to the identification of rock mass volumes characterized by a roughly homogeneous conductivity capacity and thus it is possible to highlight the presence of underground waters and/or poor geomechanical conditions.

4.3 TBM parameters analysis

The overall parameters recorded by the PLC of the TBM are stored in the Tunneling Process Control (TPC) software, which allows to further plot, analyze and correlate all of them, filtering by time, ring, chainages (pk), and time. Routinely, main TBM key performance parameters are considered and examined to verify interpretations got from drilled boreholes and performed geophysical investigations, always with the aim to verify rock mass conditions foreseen in the geotechnical profile and to improve the interpretation ability to steer CE excavation and the knowledge along the alignment to facilitate the drive of the two big TBMs. The TBM key performances parameters are:

- Cutter-head rotation speed (rpm);
- Cutter-head advancement speed (mm/min)
- Cutter-head penetration (mm/rev);
- Cutter-head rotation torque (kNm);
- Main cylinders thrust (kN);
- Specific excavation energy [kWh/m^3];
- Excavated volumes (m^3);
- Volumes (bi-component grout) injected at shield bottom tail (m^3);
- Volumes (pea-gravel or expanded clay) injected behind ring segments (m^3);

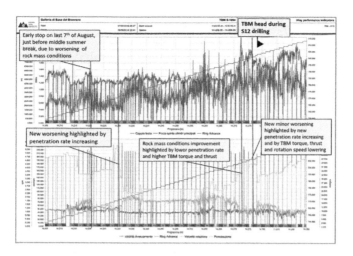

Figure 6.   Example of interpreted TBM key performances parameters chart.

4.4   *Convergence monitoring*

The CE convergence monitoring plan in intended to check after the excavation the effective stability of the tunnel and supporting system; this is obtained by implementing different instruments with different aims:

- Convergence monitoring by TBM shield.
- Monitoring annular gap behind ring extrados;
- Convergence measures with optical targets and with AWCS system.

Monitoring by TBM Shield

The rock mass tendencies to close against the shield is checked to monitor the risk of shield blockage using:

- N° 5 extendable jacks (fontimeters) for the automatic reading of the annular gap above the shield extrados, n° 3 of them placed on the frontal and n° 2 on the gripper shield
- N° 6 pressure cells, n° 3 of them placed on the frontal shield and n° 3 on the gripper shield.

Monitoring annular gap behind ring extrados

The annular void filling behind the rings is performed by a bi-component grout in the invert and pea-gravel or expanded clay on crown and sidewalls; as knonw this system typical of a DS.TBM operation has the intrinsic risk of leaving empty volumes, which need to be detected and filled to assure the long-term stability of the rings. This check is carried out using a

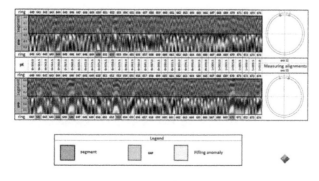

Figure 7.   Example of georadar report at 11 and at 13 o'clock from Akron Srl.

4176

georadar with double frequency antenna (200 and 600 MHz) along two 50m long profiles placed at 11 and at 13 o'clock.

<u>Segments convergence measures with optical targets and with AWCS system</u>

Eventual movements of the segmental lining are monitored by the classical topographical monitoring using optical targets specifically placed at in sensitive areas.

Moreover, also the AWCS (Automatic Wireless Convergence System) system is implemented with the aim to follow the movement of the rings just after the installation inside the back-up up to stabilization in an area where optical target is ineffective for a TBM drive.

5 THE TYPICAL FLOW OF INFORMATION FROM THE TUNNEL

The information is flowing from two main sources:

- from TPC software
- from deliverables done by specialized subcontractors

In the frame of such huge quantity of available data, the management of those concerned to rock mass geological conditions interpretation, have been so far organized according to what detailed hereunder.

<u>Daily analysis:</u>

- analysis of TBM advancement key parameters from TPC
- analysis of possible over-excavations/convergence from TPC such as excavated volumes, fontimeters, shield pressure cells, etc.

Just after every probe drilling (almost weekly):

- drilling parameters analysis and interpretation
- geophysical investigations analysis and interpretation (TSP/TSWD)

Weekly, all the recorded, analyzed and interpreted data, are presented and discussed with the Engineer team.

6 THE CONSTRUCTION OF THE ENGINEERED AS BUILT PROFILE

The main recorded and analyzed data concerning investigations, as-built excavation parameters etc., are finally condensed in a summarizing engineered as-built profile drawn up by means of excel software.

Within the engineered as-built profile the following basic information are collected:

- Ring installation data: number, key segment location, date of installation, initial and final pk, relevant TBE head pk and CE absolute pk

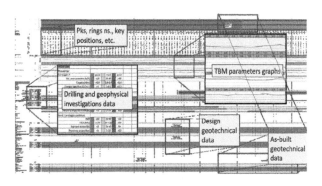

Figure 8.   Extract from the final summarizing engineered as-built profile.

- Main TBM excavation parameters: average for ring values of cutter-head rotation torque, Main cylinders thrust, Auxiliary cylinders thrust, cutter-head penetration and rotation speed;
- Investigation results highlighted, by means of colored legends, for probe drilling and geophysical investigations (for probe drilling, advancing speed, torque, Specific Energy and estimated water incomings flow.
- Simplified geology as-built which is actually the result of all the above-mentioned interpretation

7 CONCLUSIONS

While writing the paper (September 14th, 2018) the CE TBM reached to excavate the 11,2% of its drive at pk 14+653 (ring 1052) well inside the calcschists tectonic unit, while the following investigations has been carried out:

- n° 15 core destruction boreholes have been drilled
- n° 12 TSP investigation have been executed
- n° 26 TSWD investigation reports are already produced with a usual overlapping, between closing reports of 40–50 m.
- BEAM reports are produced, on average, every advancing meter and cover an investigation depth of 20 in front of TBM head.
- n° 8 instrumented rings have been installed

Along this first stretch of alignment the DS-TBM never faced really critical situations, not even along the passage between Austroalpine paraschists and amphibolites to the calcschists tectonic unit, whose relevant contacts were supposed to be faulted.

On the other hand, some limited problems have been related to the crossing of Afens fault system, approximately between PK 14+210 and 14+270 m, approached just before the middle August stop and completely crossed just after.

Such fault system was detected by probe drills n. 10 and 11, which allowed to advice the crew well in advance thus being prepared to this condition.

Before this, only one significant water incoming has been crossed, at rough 70 m depth (about pk 13+840 m) during the drilling of core destruction boreholes S7, for an approximate flow of 10–20 l/s (which, at the end of borehole drilling was already significantly reduced).

The short story recorded so far and the possibility to detect in advance the few critical points crossed is considered a good achievement and a signal that the overall system of investigation is organized to react and put in place the proper counter-measures.

REFERENCES

Frisch, W., 1976. Ein Modell zur alpidischen Evolution und Orogenese des Tauernfensters. Geologische Rundschau, 65: 375–392. Berlin:Springer Nature

Frisch, W., 1979. Tectonic progradation and plate tectonic evolution of the Alps. Tectonophysics, 60: 121–139. Amsterdam: Elsevier

Fügenschuh, B., Seward, D. & Mancktelow, N.S., 1997. Exhumation in a convergent orogen: the western Tauern window. Terra Nova, 9 (5–6):213–217. Hoboken NJ USA: Wiley

Lammerer, B. & Weger, M., 1998. Footwall Uplift in an Orogenic Wedge - The Tauern Window in the Eastern Alps of Europe. Tectonophysics, 285 (3–4):213–230. Amsterdam: Elsevier

Ratschbacher, L., Frisch, W., Neubauer, F., Schmid, S.M. & Neugebauer, J., 1989. Extension in compressional orogenic belts: the Eastern Alps. Geology, 17: 404–407. McLean, VA, USA: Geological Society of America

Ratschbacher, L., Frisch, W., Linzer, H.-G. & Merle, O., 1991. Lateral extrusion in the Eastern Alps, Part 2: Structural Analysis. Tectonics, 10 (2):257–271. Hoboken NJ USA: Wiley

Selverstone, J., 1988. Evidence for east-west crustal extension in the eastern Alps: implications for the unroofing history of the Tauern Window. Tectonics, 7 (1):87–105. Hoboken NJ USA: Wiley

*Tunnels and Underground Cities: Engineering and Innovation meet Archaeology,
Architecture and Art, Volume 7: Long and deep tunnels – Peila, Viggiani & Celestino (Eds)*
© 2020 Taylor & Francis Group, London, ISBN 978-0-367-46872-9

Measurements of long time lining strain in creeping rock mass and modelling with viscoplastic laws

R. Plassart
EDF Centre d'Ingénierie Hydraulique, Le Bourget-du-Lac, France

O. Gay & A. Rochat
EGIS Géotechnique, Seyssins, France

A. Saïtta
EGIS Tunnel, Pringy, France

F. Martin
BG Ingénieurs conseils, Lyon, France

ABSTRACT: To improve our understanding of long term behaviour of large galleries, this paper details the experience of the road tunnel of Chamoise (France), including an advanced processing of the available data (rock and lining displacements, linings yielding) recorded since the tunnel excavation, more than 20 years ago. Besides, calculations have been carried on until 100 years, improving the demonstration of feasibility of a large gallery in claystone for the challenging Cigéo project, the French project for nuclear waste underground repository.

1 INTRODUCTION

The French project of deep geological storage for nuclear waste (called Cigéo and managed by ANDRA) gives rise to a lot of innovative issues for underground space technologies. The nuclear activity of the intermediate level (ILW) and high level (HLW) radioactive waste generates temperature effects on both the lining and the surrounding rock mass. In order to avoid these unfavorable effects, the number of waste containers is controlled in the gallery section. The galleries section has consequently been ranged from a diameter of 4 to 6 meters.

However, for intermediate level waste (ILW), due to the lower temperature, the number of containers in a cross section of a gallery can safely be increased. Therefore, a largest diameter gallery (typically 10 meters excavated) could be more cost effective.

The difficulty for the design of the linings is due to the 500 meters overburden combined with the soft nature of the rock mass, which is Callovo-Oxfordian (COx) argillite, a claystone with a creeping behaviour.

The challenge is so now to demonstrate the technical feasibility of a big gallery in a context of high displacements. Moreover, these displacements have to be managed for 100 years at least, which corresponds to the activity and reversibility period of the storage. Some theoretical calculations have been carried out (Plassart et al. 2016) but without the possibility to confront the results with *in situ* recordings on real and in use tunnels, especially on a long time, *i.e.* over a few years.

To improve our understanding of long term behaviour of large galleries, this paper details the experience of the road tunnel of Chamoise, in France, between Lyon and Geneva.

The diameter of this tunnel is over to 10 meters and the geology is locally composed of Effingen's marls, a layer geologically and mechanically very close to COx argillite. In spite of a lighter overburden (400m versus 520m), an anisotropic initial stress state and a variable rate of carbonates, the highway tunnel of Chamoise is so quite similar to the planned gallery for intermediate-level nuclear waste, with a geology very close to the Cigéo's one in some sections. The paper proposes a review and an advanced processing of the available data (rock and lining displacements, linings yielding) recorded since the tunnel excavation, more than 20 years ago (measurements made daily).

Besides, a numerical modelling of the existing tunnel has been carried out, using two constitutive laws accounting for a viscoplastic mechanism: L&K developed by EDF (Kleine et al. 2006) and H&B-Lemaitre recommended by ANDRA (Souley et al. 2011). The elasto-plasticity of both laws is characterized by a positive hardening in pre-peak and a softening behaviour beyond resistance. Concerning viscoplasticity, L&K is based on the Perzyna's theory while H&B-Lemaitre include a modified Lemaitre model in which the viscoplastic strain rate is directly related to damage.

The simulations results have then been discussed and compared with the processed records on 20 years. Finally, the calculations have been carried on until 100 years, improving the demonstration of feasibility of a large gallery in claystone for the challenging Cigéo project.

2 TUNNEL OF CHAMOISE (FRENCH JURA)

2.1 Presentation of the tunnel

The highway tunnel of Chamoise is constituted of two large tunnels of 3300m long with an excavated section of 110 to 130 m² (diameter about 12m). The geology alternates west-dipping layers of quite stiff limestone and marls with variable content in calcium carbonate. The softest marls are Effingen's one (close to Callovo-Oxfordian argillite of the site of the CIGEO project), in the center part of the tunnel (Figure 1).

2.2 Excavation history

The northern tube has been excavated first, between 1981 and 1986, thanks to road header machines in three sections. Crossing the Effingen's marls, several over-excavations and spalling were encountered, especially in the bench (Figures 2-3).

Thanks to this experience, the excavation of the southern tube between 1992 and 1995 has been carried out by Drill & Blast method in two steps, with a large top bench and a small curved invert (Figure 4).

The design of the immediate support consists in long bolts (4 to 5 meters long for a density of 1 to 2 bolts per m²), light ribs and a shotcrete layer of 15cm. Bolts were added in the invert for the southern tube. For the two tubes, the final concrete lining is about 50cm thick, reinforced in invert.

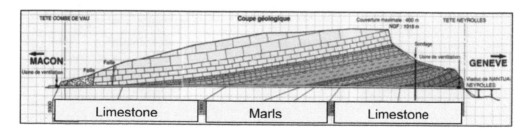

Figure 1. Longitudinal profile of the tunnel of Chamoise.

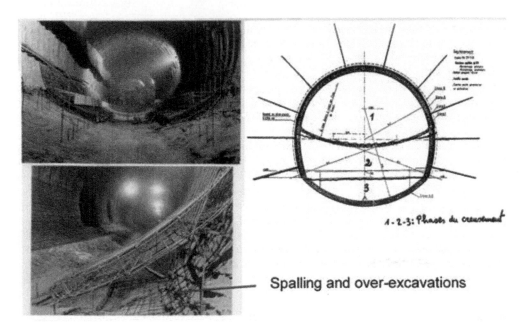

Figure 2. North tube of the tunnel of Chamoise – Pictures of excavation works (left) and support system in cross section (right).

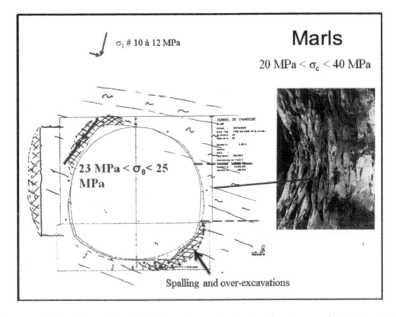

Figure 3. Interpretation of spalling in Effingen's marls, due to the stress anisotropy – North tube of Chamoise tunnel.

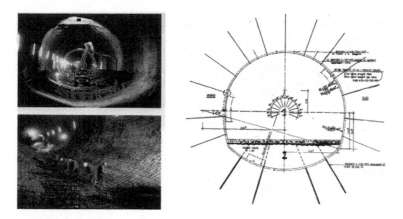

Figure 4. South tube of the tunnel of Chamoise – Pictures of excavation works (left) and support system in cross section (right).

3 MONITORING OVER 20 YEARS

3.1 *Monitoring principles*

Due to the mechanical particularities of the Effingen's marls, a specific monitoring has historically been set up, especially in the south tube, in order to measure the delayed effects on.

- The strain of the rock mass;
- The stresses in the concrete annulus.

The monitoring is composed of four different tools (Figure 5). Convergences are measured thanks to invar alloy wires in different directions inside the tunnel two times a year, with a precision of 0.1mm. Displacements in the heart of the rock mass (until 10m) are daily provided by borehole extensometers. "Distofor" are electric multipoint extensometers, bottom

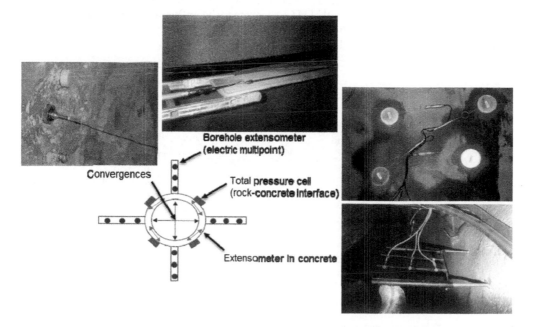

Figure 5. Different monitoring tools set up in the south tunnel of Chamoise since 1995.

anchored (point assumed to be fixed) and with a precision of 0.05mm. At the rock-concrete interface, total pressure cells have been disposed all around the lining in order to measure stresses. These tools consist in hydraulic cells with a pressure sensor based on vibrating strings, with a precision of 5 kPa. The last tool is the extensometer in concrete, drown at different angles in the lining. The precision is about 5 mm/m.

3.2 State after 20 years

Before the present project, a lot of data provided by all the monitoring tools described above were not really exploited. A first analysis has consisted in identifying tools still in use or not. Among operating tools, some of achieved data are not relevant and have to be removed from the analysis. At least, 20-25% of the expected data have to be excluded. The fact that almost eighty percent of all the monitoring tools give relevant data after 20 years is already in itself a considerable result.

3.3 Data processing on selected sections

For the comparison with the Cigéo project, two cross sections have been selected in the south tube: S1501 and S1605. These sections are located in the Effingen's marls, with 400m overburden, and with a large monitoring still operating after 20 years, generating redundant data.

The logging has started on December 1995, 25 months after the end of the excavation. It can be assumed that the immediate effects of the excavation are not considered, and only the creeping effect of the rock mass or of the concrete lining is recorded.

All the sensors have been connected to automatic dataloggers were data are recorded and stored on a dedicated PC. For this project, the selected data have been extracted and processed. The most important process deals with clearing the effect of the external temperature. As a well-ventilated road tunnel, in a mountain area, season effects are indeed notable. The Figures 6–8 present the extracted data including the temperature effect for the section 1501 from one borehole extensometer in crown (Figure 6), one pressure cell (Figure 7) and one extensometer in concrete (Figure 8) in invert.

In order to clear the temperature effect, the different data have been processed by the HST model, a model used by EDF-DTG to extract the influence of hydrostatic pressure (H), season (S) and drift with time (T). The processing of season effects gives so relevant results.

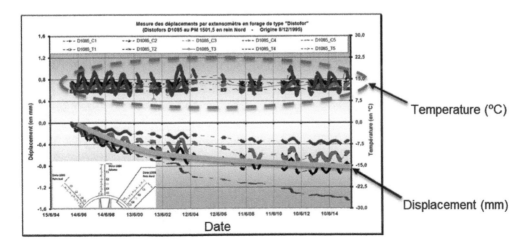

Figure 6.   Borehole extensometer (South tube of Chamoise: S1501) – No temperature processing.

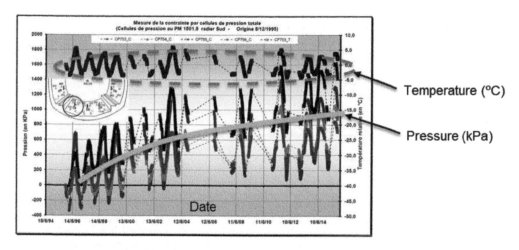

Figure 7.   Total pressure cells (South tube of Chamoise: S1501) – No temperature processing.

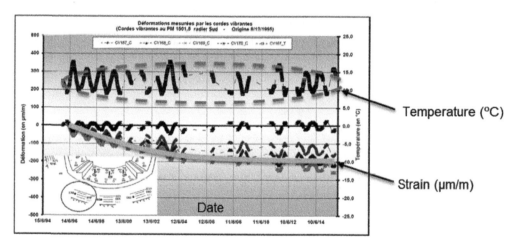

Figure 8.   Extensometer in concrete (South tube of Chamoise: S1501) – No temperature processing.

The second purpose of the processing is to extrapolate the curves beyond 20 years, until 100 years. The application of the HST model is more disappointing in this case, due to difficulties to calibrate the drift from 20 to 100 years. The equations of Soulem et al. (1987), specifically established for underground works, give more suitable results. Coupling HST and Sulem models, the Figure 9 presents an example of extrapolation until 100 years for the strain in an extensometer in concrete.

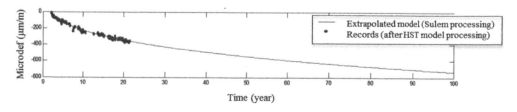

Figure 9.   Comparison between extrapolated model (Sulem) and records for strain in lining.

4 NUMERICAL MODELLING ATTEMPTS

Previous numerical modelling have been carried out comparing two creeping models applied to the Cigéo project (Plassart et al. 2016). These calculations were consistent but theoretical. In this section, the aim is to validate the methodology for the Cigéo project thanks to real recorded data in Chamoise.

4.1 Loading conditions and support description

The excavation is simulated in one step by the convergence-confinement method (Vlachopoulos & Diederichs 2009): a confinement rate is decreased from 1 to 0 according to the real time-dependent excavation progress (4m per day). After the end of the excavation, the calculation is extended to 100 years. The gallery axis is oriented according to the observed stress state (Figure 3).

The primary support (4m bolts spaced of 1.5m and 15cm shotcrete with a Young's modulus of 10GPa) is installed immediately after excavation, whereas the lining (50cm with a Young's modulus of 12GPa) is set one year later.

4.2 Constitutive laws including a viscoplastic mechanism

Modelling is carried out with FLAC software (Itasca 2011) and the following behavior laws for the rock mass including a viscoplastic mechanism: L&K or H&B-Lemaitre. Equations and parameters of both models are not detailed here, but they are available in references (Kleine et al. 2006, Kleine 2007, Souley et al. 2011, Plassart et al. 2013).

4.2.1 L&K model

EDF-CIH has developed a constitutive model, called L&K (Kleine et al. 2006), which is able to describe delayed mechanical effects of rocks thanks to a viscoplastic mechanism based on the overstress theory of Perzyna (Perzyna 1966). This viscoplastic mechanism is fully coupled with the plastic mechanism, which is an extended variation of the Hoek & Brown model. Besides, this model is highly dependent on variation of the dilatancy which governs the volumetric strain of the material during a solicitation.

4.2.2 HB-Lemaitre model

The HB-Lemaitre model (Souley et al. 2011) is the association of the Hoek & Brown model with softening, and the creep model of Lemaitre. There is no full coupling between the two mechanisms: creep is activated only after the excavation and after reaching instantaneous equilibrium.

4.2.3 Models calibration

Unfortunately, due to the lack of available data, it has been impossible to properly calibrate both models for the Effingen's marls of Chamoise. The set of parameters has so been provided for each model thanks to Cigéo data (Cox argillite), allowed by the large similarity of the two rock masses. The main properties are listed in Table 1.

For the L&K model, the constitutive law parameters have been calibrated thanks to instantaneous and delayed tests on COx argillite samples only (Kleine et al. 2006).

Table 1. Main properties for Effingen's Marls or Cox argillite.

Parameter	L&K model	HB-L model
Young's modulus E	4 GPa	5 GPa
Poisson's coefficient ν	0.12	0.3
Volumetric mass	2400 kg/m3	2400 kg/m3
Compression strength σc	12 MPa	12 MPa

The HB-Lemaitre constitutive law parameters definition is twofold: by curve-fitting of galleries convergences measured in the French URL, and by creeping laboratory tests results. For further details, please consult ANDRA's articles (*e.g.* Souley et al. 2011).

4.3 *Main results*

In comparison to previous calculations (Plassart et al. 2016), the tunnel of Chamoise presents an inclination of its principal stresses, leading to a local concentration of stresses and strain. The asymmetric strain of the concrete lining and the rock mass is observed both in numerical modelling and in records (Figure 10). The anisotropy of the behavior is however slightly underestimated by calculations, for the two constitutive laws.

Several records, extrapolated until 100 years thanks to the HST and Sulem methods, have been compared with the results of the numerical modelling carried out with the two reference laws, L&K and HB-Lemaitre. Correlations are generally relevant, even if some gaps remain, especially for pressure cells. This monitoring tool often presents higher variability and drift with time than the other's one. The comparison for the strain of an extensometer in crown of the concrete lining is proposed in Figure 11. It can be seen the measurement extrapolation of strain is properly framed by the two simulations until 100 years and probably beyond. Both models have yet to be improved in their calibration if we want the converging curves to be more tightened. An *in situ* tests campaign is so planned within the coming months.

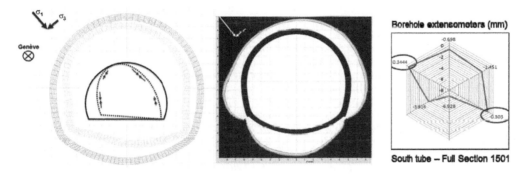

Figure 10. Strain of the tunnel of Chamoise – Comparison of calculations and records (20 years).

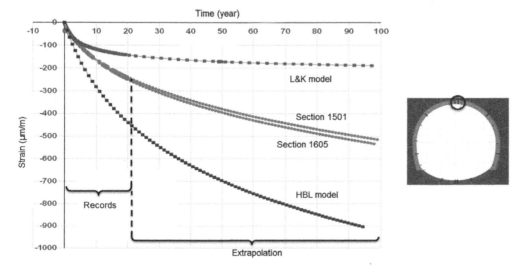

Figure 11. Comparison between extrapolated records and numerical modelling for strain in lining (crown).

5 CONCLUSIONS AND PROSPECTS

The road tunnel of Chamoise (France) is a rare example of tunnel in a creeping rock mass monitored for more than 20 years, with monitoring tools still operating and providing still evolving data. This paper has talked about some methods (HST and Sulem) to properly extract the data and with the aim to extrapolate the evolution until 100 years.

After data processing and numerical modelling with viscoplastic models (L&K and HB-Lemaitre), extrapolated records and calculations agree relatively well, in spite of all the uncertainties bounded to this type of study. In the perspective of the challenging Cigéo project, this result enhances the idea of the long term sustainability of a large gallery in claystone, with stresses and strain widely under control.

Thanks to tests on rock mass material, a better calibration of the numerical models on Effingen's marls would probably improve the results for both the models. Another way of improvement of the study would be to better characterize the current mechanical state of the concrete lining. The *in situ* tests campaign planned within the coming months aims at collecting concrete samples and so evaluating the current properties of the lining.

REFERENCES

Itasca Consulting Group, Inc. 2015. *FLAC — Fast Lagrangian Analysis of Continua*, Ver. 8.0. Minneapolis: Itasca.

Kleine, A., Laigle, F., & Giraud, A. 2006. An elastoplastic strain-softening constitutive model for deep rock – three-dimensional applications for nuclear waste repository problem. *4th, International FLAC symposium; FLAC and numerical modeling in geomechanics*; Madrid.

Kleine, A. 2007. Modélisation numérique du comportement des ouvrages souterrains par une approche viscoplastique. PhD Thesis. INP Lorraine.

Perzyna, P. 1966. Fundamental problems in viscoplasticity. In Advances in applied mechanics. Edited by Academic Press New York and London. 9:243–377.

Plassart, R., Fernandes, R., Giraud, A., Hoxha, D., & Laigle, F. 2013. Hydromechanical modelling of an excavation in an underground research laboratory with an elastoviscoplastic behaviour law and regularization by second gradient of dilation. *Int. J. of Rock Mech. and Mining Sciences* 58:23–33.

Plassart, R., Laigle, F., Saïtta, A., & Martin, F. 2016. Design of a large diameter gallery for a nuclear waste storage project. Application to the French repository project (Cigéo). *World Tunneling Congress; Proc. intern. symp., San Francisco, 22–28 April 2016*.

Soulem, J., Panet, M., & Guenot, A. 1987. An analytical solution for time-dependent displacements in a circular tunnel. *Int. J. of Rock Mech. and Mining Sciences* 24(3):155–164.

Souley, M., Armand, G., Su, K., & Ghoreychi, M. 2011. Modeling the viscoplastic and damage behavior in deep argillaceous rocks. *Physics and Chemistry of the Earth* 36:1949–1959.

Vlachopoulos, N., & Diederichs, M. S. 2009. Improved longitudinal displacement profiles for convergence-confinement analysis of deep tunnels. *Rock Mechanics and Rock Engineering* 42:131–146.

Tunnels and Underground Cities: Engineering and Innovation meet Archaeology,
Architecture and Art, Volume 7: Long and deep tunnels – Peila, Viggiani & Celestino (Eds)
© 2020 Taylor & Francis Group, London, ISBN 978-0-367-46872-9

Polcevera Adit – getting unstuck on EPB-TBM in squeezing rock

F. Poma, U. Russo, F. Ruggiero & L. Lampiano
COCIV, Consorzio Collegamenti Integrati Veloci, Genova, Italy

ABSTRACT: The construction of the new high-speed freight railway line between Milan and Genoa, the "Terzo Valico dei Giovi", required the excavation of 4 adits which will provide access to the excavation faces during the construction phase as well as to the railway line for mainten-ance and safety purposes during the operation phase. The Polcevera adit has been realized by an EPB TBM named "Giulia". However, the TBM got stuck after driving about 1.300 m inside the tunnel, although capable to deliver a maximum thrust force of 88.000 kN (in normal conditions the applied thrust force is equal to 20.000 kN). Soon after the TBM getting stuck, a number of measures were carried out in order to increase the TBM's thrust force and at the same time to decrease friction between the rock mass and the TBM's shield and tail.

1   TERZO VALICO DEI GIOVI

The railway Milan–Genoa, part of the High Speed/High Capacity Italian system (Figure 1), is one of the 30 European priority projects approved by the European Union on April 29th 2014 (No. 24 "Railway axis between Lyon/Genoa – Basel – Duisburg – Rotterdam/Antwerp) as a new European project, so-called "Bridge between two Seas" Genoa – Rotterdam. The new line will improve the connection from the port of Genoa with the hinterland of the Po Valley and northern Europe, with a significant increase in transport capacity, particularly cargo, to meet growing traffic demand.

The "Terzo Valico" project is 53 Km long and is challenging due to the presence of about 36 km of underground works in the complex chain of Appennini located between Piedmont and Liguria. In accordance with the most recent safety standards, the under-ground layout is

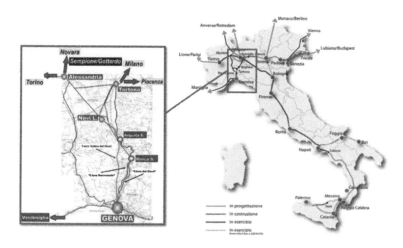

Figure 1.   High-speed Italian system.

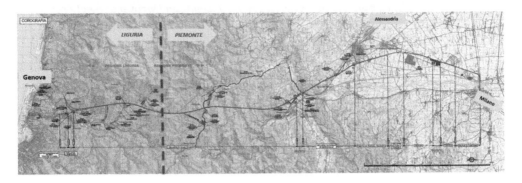

Figure 2. Terzo Valico project.

formed by two single-track tunnels side by side with by-pass every 500 m, safer than one double-track tunnel in the remote event of an accident.

The layout crosses the provinces of Genoa and Alessandria, through the territory of 12 municipalities.

To the South, the new railway will be connected with the Genoa railway junction and the harbor basins of Voltri and the Historic Port by the "Voltri Interconnection" and the "Fegino Interconnection". To the North, in the Novi Ligure plain, the project connects ex-isting Genoa-Turin rail line (for the traffic flows in the direction of Turin and Novara – Sempione) and Tortona – Piacenza –Milan rail line (for the traffic flows in the direction of Milan- Gotthard).

The project crosses Ligure Apennines with Valico tunnel, which is 27 km long, and ex-its outside in the municipality of Arquata Scrivia continuing towards the plain of Novi Ligure under passing, with the 7 km long Serravalle Tunnel, the territory of Serravalle Scrivia (Figure 2). The underground part includes Campasso tunnel, approximately 700 m long and the two "Voltri interconnection" twin tunnels, with a total length of approximate-ly 6 km.

Valico tunnel includes four intermediate adits, both for constructive and safety reasons (Polcevera, Cravasco, Castagnola and Vallemme). After tunnel of Serravalle the main line runs outdoor in cut and cover tunnel, up to the junction to the existing line in Tortona (route to Milan); while a diverging branch line establishes the underground connection to and from Turin on the existing Genoa-Turin line.

From a construction point of view, the most significant works of the Terzo Valico are repre-sented by the following tunnels:

- Campasso tunnel 716 m in length (single-tube double tracks)
- Voltri interconnection even tunnels 2021 m in length (single-tube single track)
- Voltri interconnection odd tunnels 3926 m in length (single-tube single track)
- Valico tunnel 27250 m in length (double tube single track)
- Serravalle tunnel 7094 m in length (double tube single track)
- Adits to the Valico tunnel 7200 m in length
- Cut and cover 2684 m in length
- Novi interconnection even tunnels 1206 m in length (single-tube single track)
- Novi interconnection odd tunnels 958 m in lenght (single-tube single track)

The project standards are: maximum speed on the main line of 250 km/h, a maximum gradient 12,5 ‰, track wheelbase 4,0 – 4,5 m, 3 kV DC power supply and a Type 2 ERTMS signalling system.

2   POLCEVERA ADIT

Polcevera is the southernmost intermediate access tunnel to the basis tunnel, it is about 2000 m and crosses through the geological formation of Argilliti a Palombini. This adit allows to achieve 4 single track tunnels, in order to excavate them during the excavation phase, and in order to provide access to the railway line for maintenance and safety during the operation period. The adit was excavated by an EPB/TBM S-914 named "Giulia", with an excavation diameter of about 9.79 m and a length of 67 m. Because of the smallnest of the construction site area, it was necessary to excavate the first 25 m of the tunnel by traditional methods in order to place the first part of the TBM.

The tunnel was divided into two sectors in order to face the different overburden: low overburden for the first 450 m and high overburden for the rest of the tunnel where there could be tool wear.

The final lining has been realized laying down precast elements rings, composed by 7 reinforced concrete precast elements and 1 ashlar key. Final lining rings are 40 cm wide, an external diameter of 9.45 cm and each ring has 1.40 m length.

3 TBM/EPB GIULIA

The TBM Giulia is an EPB (Earth Pressure Balance) type,, which allows to support the excavation front face against the earth pressure in the excavation chamber, in case loose materials are encountered. The TBM foresees the possibility to work effectively either in open or close configuration that allows the TBM to face different geomechanical conditions from the very good quality rock mass to loose material.

The shield's operative principle of "Earth Pressure Balanced Shield" (EPBS) is based on the effect coming from the material excavated for the retaining of the front face excavation while the cutter head's purpose is only related directly to the excavation. The shattered material coming from the excavation process passing through the cutter head flows into a chamber (excavation chamber) located behind the cutter head where it is kept in pressure through a diaphragm placed behind the material and using the shield jack system. The system transfers the pressure throughout the material located in the excavation chamber, and thus is transferred to the front face excavation.

The debris coming directly from the excavation is taken out from the excavation chamber through a screw conveyor system and then taken to a temporary disposal area. The screw conveyor is regulated to ensure that the amount of material taken out from the excavation chamber match the amount coming from the excavation. Therefore, the amount of material contained in the excavation chamber is kept constant and therefore the earth balancing pressure.

In order to reduce the size of the excavated material, EPB shields inject a free flowing agent through some specific nozzles located on the cutter head, in the excavation chamber and directly in the screw conveyor. The injection of this free flowing agent gives the excavated loose material the behaviour of a paste and thus transferring the pressure within the material from the jacks system as it was a fluid.

The main features of the machine are shown in the following chart:

Table 1. TBM features.

TUNNEL SEGMENT LINING		
Number of pieces	qty.	7+1 (7 x 51,43° + 1 x 11,66°)
Outer segment diameter	m	9.45
Inner segment diameter	m	8.65
Segment thickness	m	0.4
Segment lenght	m	1.4
Weight of the heaviest segment	kg	approx. 5810

(Continued)

Table 1. (*Continued*)

GENERAL TBM DATA

Type		EPB
Excavation diameter	m	9790
Nominal overcut on the shield	mm	17.5
Total Installed power	kW	approx. 4680
Total Weight (TBM complete)	t	approx. 1300
Weight of cutterhead without cutting tools	t	approx. 120
Weight of forward shield	t	approx. 400
Weight of main bearing assembly with drives	t	approx. 130
Weight of intermediate shield (complete)	t	-
Weight of tail shield	t	approx. 60
Weight of segment erector	t	approx. 40
Weight of screw conveyor	t	approx. 45

SHIELD STRUCTURES

Diameter	mm	9755
Bulkhead pressure (working)	bar	4
Bulkhead pressure (design)	bar	4
Static mixing paddles (fixed to bulkhead)	qty.	4
Active mixing paddles (fixed to cutterhead)	qty.	4
Soil injection within the bulkhead	qty.	8 possible
Bentonite injection within the bulkhead	qty.	2
Bentonite injection out of the shield	qty.	3 rows
Horizontal probe drilling lines	qty.	1
Inclined probe drilling lines	qty.	28
Valve to remove air cushion	yes/no	yes
Personnel lock	yes/no	yes
EPB cells	qty.	6
Hydraulic stabilizer	qty.	2 (in the top)
Bulkhead gate for screw	qty.	1
	sealing pressure (bar)	not pressure tight
	stroke monitor	no
Additional wear plates at screw invert	yes/no	yes
Access doors to the Man airlock	qty.	1

THRUST SYSTEM

Power	kW	132
Number of cylinder pairs	qty.	14
Number of cylinders	qty.	28
Cylinder dimension	mm	320/260
Propulsion stroke	mm	2000
Pressure	bar	400
Thrust force @ 400 bar	kN	90.076
Thrust force @ 400 bar/per cylinder	kN	3217
Maximum Excentricity of Thrust	mm	25
Distance from edge of pad to edge of segment	mm	according to design
Variable Pressure Groups	yes/no	yes
Independent Control	yes/no	yes
Maximum extension speed (all cylinders)	mn/min	80
Exensometer	qty.	4

During the excavation, even though the thrust force recorded by the TBM was 88.000 KN, an higher value than the ordinary ones, the Tunnel Boring Machine got stuck.

The reason of the machine getting stuck can be associated to the rock mass squeezing behavior. Once the initial stress condition of the rock mass is disturbed, and there is a stress release, the rock, absorbing the water from the air flowing along the exposed rock, squeezed and this led to an increase of the containment of the rock surrounding the machine and therefore increasing the friction between the TBM and the surrounding rock mass.

During the excavation, there were some abnormal values recorded by the machine: even though thrust values recorded on hydrauliuc jacks located between the shield and the tail reached the pressure upper limit the TBM, the advancement speed was zero and the TBM was stuck for two entire weeks.

In the following days, several attempts were made to release the TBM.

The first action undertaken was dismantling some tail rings and remove all the rock blocks fitted in between the shield and the tail in order to free the mechanism and reduce the thrust force needed to make the machine moving. After this action, nevertheless, the thrust force carried out by the machine was close to the upper limit, the advancement was still zero.

In the time in which the machine was stuck, the rock mass has swelled, because of its clay composition; the stress release due to the excavation induced in this kind of material the activation of negative overpressure that involved itself the activation of a water flow from the air to the rock mass, increasing the rock mass volume and therefore the friction between the rock and the TBM.

On the basis of the first analysis, some possible intervention have been identified in order to solve the problem.

The phenomenon could be easily described physically by the friction relation: the very high thrust force needed to get the machine restart excavation is due to the interaction between the contrast force and the friction degree.

The phenomenon can be described qualitatively by the following friction formula:

$$S_{max} = N_C tan\varphi \tag{1}$$

where S_{max} is the thrust force needed to overcome the friction and make the TBM run again.

Nc represents the contrasting force due to the rock squeezing, and φ represents the friction degree of the surface between the shields and tails and the rock.

In order to overcome the problem, the solution has been searched to reduce the friction and increasing the thrust force simultaneously. Operating in this way, there was the possibility to take action on two phenomena and therefore obtain a more effective result that eventually freed the machine and allowed to start again the excavation just after 2 weeks.

Some supplementary plants and equipment have been installed to realize some of the activities described previously giving to the machine such a different configuration able to exploit a bigger thrust power.

The main actions identified during the first analysis can be listed as the followings:

- Reduction of friction between the rock mass and the tail/shield set, through punctual hydrodemolition interventions with the adoption of high pressure rotating nozzles. For this intervention, some existing holes present on the shield and on the tail were used as well as some holes realized specifically for the intervention (Figure 4). In order to reduce the friction, injections of a lubricant polymer (CONDAT TFA 34) has been adopted too. These activities significantly reduced significantly the friction necessary to make the machine run again. In order to carry out polymer injections, it was necessary to prepare a specific plant using the machine grease pumping system, the oil pumping system and a water pumping system. The position of the above mentioned pumping systems were chosen in such a way so as not to affect other operations on the tail and the shield, considering narrow spaces available for carrying out other operations.

- Increasing the thrust force of the TBM, obtained by installing a supplementary system of jacks applied on the tail's joints and an empowerment of the shield pushing system. The additional plant was realized by installing 18 additional hydraulic jacks ENERPAC along tail joints and connected to a dedicated hydraulic control unit, composed of twelve 1470 KN hydraulic jacks and six 2450 KN hydraulic jacks, for a total capacity of about 32.000 KN. Furthermore, the original machine pushing system has been enhanced by introducing an hydraulic control unit Herrenknecht. The new control unit is able to increase the cylinders' pressure up to 520 bar (original value was 400 bar) and it was connected to 10 out of 14 couples of pistons of the machine. Valves with higher performance were also installed, in order to make the pipe system able to transfer the higher pressure coming from the empowered system. The capacity of the improved pushing system was increased to a total value of the thrust of 110.000 KN, 30.000 KN more than the machine original system.

The reasoning leading to determine the action to adopt to unlock the machine can be schematized in the following chart, which helped to identify a right and clear procedure that eventually has been adopted:

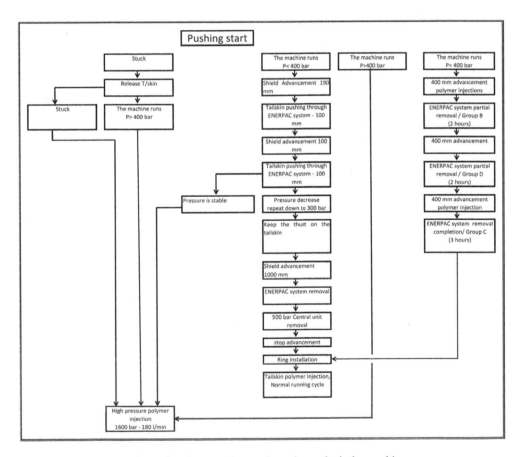

Figure 3. Flow chart to determine the procedures adopted to unlock the machine.

A brief narrative of action adopted to unlock the machine is given below.

4.1 Installation of hydraulic jacks pushing system

The first adjustment adopted was increasing the thrust force of the Tunnel Boring Machine, by installing a supplementary pushing system. The original machine pushing system has been

Figure 4. Supplemetary hydraulic jack system.

increased by installing a supplementary system. The pushing system is composed by a group of hydraulic jacks which push on the last final lining ring. The resultant force applied by all the hydraulic jacks is the force that allows the advancement of the machine. The default hydraulic jacks system was improved by introducing additional jacks, dedicated to the advancement of the tail. Subsequently, the machine's central unit was connected to the jacks system, and then it was possible to carry out some other pushing tests. The first test was unsuccessful. Despite the increase in the thrust force, after an initial advancement of 25 mm of the tail, the system was still blocked.

4.2 520 bar HKT pushing system installation

During this stage, it was decided to install a new pushing central unit for 10 out of 14 hydraulic jacks groups. Thanks to the power exploitable from this central unit, the new pushing system was capable to exert an higher pressure on the jacks hydraulic circuit. The pressure in the system was increased up to a value of about 500 bar (this value is close to the system's maximum operating pressure) and this induced an advancement of about 35 mm. With this pressure the total thrust force was about 105.000 KN. The result of this test was still unsuccessful: the machine stopped moving after the initial movement but the tail kept being stuck even though the shield was advancing. Eventually it was noted that this value was close to the one necessary to make the machine run again, however, it was still not enough to make the tail follow the shield.

4.3 Hydro demolition

A further increase in the thrust force, due to technological and financial limits, was not enough to make the machine run on its ordinary condition. It was therefore necessary to act on the friction degree between the rock mass and the tail by undertaking hydro demolition intervention on the contour of the tunnel surrounding directly in contact with the machine, in such a way as to reduce the contrast between the squeezing rock and the machine, and thus reducing the thrust force needed to make the machine moving. In order to reach all the areas directly interested by hydro demolition, it was necessary to undertake some intervention on the TBM tailskin articulation area. To make rotating high pressure nozzles reach the intervention area, it was necessary to drill some holes and buttonholes on the skin.

Figure 5. Holes and Buttonholes needed for hydro demolition intervention.

Once all the preparatory works were carried out, it was possible to perform the proper hydro demolition for all those rock parts touching directly the tailskin, reducing significantly the containment due to the roughness and to the squeezing of the rock mass around on the tailskin. The evidence of the effectiveness of this activity is that after the hydro demolition processes, the following pushing test the tailskin advanced of about 13 cm. The advancement in this case is a magnitude greater than ones recorded during previous tests. Since hydro demolition led us to excellent results, all the procedures adopted on the tailskin were applied on the shield as well. In order to reach the area around the shield, in the cutter-head direction, there were drilled some inflages starting from holes made for the previous work. The hydro demolition around the shield induced a reduction of the stress around the machine, but the thrust force needed to overcome the friction was still greater than the machine capacity.

4.4 Polymer injections and pushing tests

To reduce the amount of necessary thrust force, the hydro demolition was extended to other elements on the contour and a lubricant polymer has been injected on high pressure. The polymer injection was done around the tailskin and around the shield in two successive stages.

The lubrication has been done injecting first the lubricant polymer CONDAT TFA 34 and then injecting hydraulic oil.

Once the injections were done through holes specifically made for the activity, acting almost on the entire contact surface between the machine and the rock, some more pushing tests have been performed following some different configurations recording their reuslts.

The initial thrust force was 105.000 KN (a value corresponding to a hydraulic jack pressure of about 500 bar, close to their maximum capacity) made the shield advance of 400 mm. After the initial pushing test, the thrust force stood around a constant value of 87.000 KN (corresponding to an hydraulic jacks pressure of 390 bar, close to exercise conditions).

These results confirmed the effectiveness of both hydro demolition and polymer injections that actually reduced the thrust needed to run the TBM in its normal condition. Despite the reduction of the thrust value, the tailskin kept suffering the contrast due to the rock mass, resulting unable to proceed by itself.

Hence, during all the thrust stage, the tailskin has never followed the shield advancement. After any shield advancement, it was necessary to push the tailsking by using the ENERPAC additional hydraulic jack system. Thrust values recorded on the tailskin, as well as those related to the shield were higher in the beginning, but after a couple of pushing tests, these values decreased down to a value of about 12.000 KN.

Because of the effectiveness of the method so far adopted, the next stage was to repeat again the same activities, hence more lubricant injections were performed, improving the results. The needed thrust value kept decreasing and it stabilized around a value of 80.000 KN (corresponding to a pressure of about 340 bar, which was closer to normal values). Good results were obtained in terms of advancement as well. The problem was still linked to the tailskin that kept remaining behind and after each pushing test it required to be helped by the ENERPAC jack system to follow the shield. In this stage, the value of the thrust force exerted by additional jacks was still high, comparable to that registered in the previous stages.

In order to overcome the critical area, even though the advancement values were still low and it was necessary in each step to recover the tailskin by using the supplementary system, there were performed several pushing tests. By summing all the advancement obtained during the pushing tests it was possible to take the machine out of the critical zone and get back to the ordinary conditions. The thrust force needed to get the shield advance became compatible with the capacity exertable by the machine in its original set, hence the central unit adopted for the shield jacks system was removed and it was connected to the original central unit since it was able to operate in a faster mode. The above-mentioned processes were further applied only for the tailskin until the critical zone was overcome.

4.5 TBM unlocking

On 28th of July 2015, the TBM advancement reached its normal performance, in its original configuration. From this stage on, values of thrust force came back to those before the stuck: about 20.000 KN. Once the critical area was passed over, the TBM kept advancing without further problems, until it reached the end of works.

5 CONCLUSIONS

All the adjustments adopted to unlock the machine aimed to satisfy the equation (1).

The initial value of the thrust force needed was 105.000 KN, but after all the punctual hydro demolition interventions directly on the rock, affecting shield and tail; and thanks to the adoption of the lubricant polymer CONDAT TFA this value decreased down to 80.000 KN. Furthermore, analyzing all the results coming from each intervention, it is possible to determine the individual contribution of the hydro demolition and the lubricant polymer.

Before the hydro demolition interventions were done, the thrust force was greater than 105.000 KN, while after its execution thrust force decreased up to about 88.000 KN. Following the lubricant polymer injection, the thrust force reached a value of about 80.000 KN. When the machine came over the critical area, the thrust force value was once more compatible with the machine capacity.

It should be highlighted that the TBM's original pushing system capacity was not sufficient to exert the necessary thrust force even after action undertake to reduce the friction. Therefore increasing this value was necessary in order to benefit from the results coming from hydro demolition interventions and lubrication, which was achieved by adopting an additional pushing system. Furthermore, the additional hydraulic jacks system was necessary to guarantee that the thrust force was enough for the advancement of the tailskin in each stage.

Tunnels and Underground Cities: Engineering and Innovation meet Archaeology,
Architecture and Art, Volume 7: Long and deep tunnels – Peila, Viggiani & Celestino (Eds)
© 2020 Taylor & Francis Group, London, ISBN 978-0-367-46872-9

Sealing of huge water ingress in headrace tunnel of Uma Oya Project, Sri Lanka

A. Rahbar Farshbar & A.H. Hosseini
Farab Co., Tehran, Iran

ABSTRACT: Uma Oya Multipurpose Development Project is under construction in Sri Lanka. The headrace Tunnel (HRT) as a water transfer tunnel in this project has already passed 11.5km milestone and is mined in hard rock by TBM. The basic strategy for avoiding damage and negative influence on the surface (environmental problems) due to the water ingress into the tunnel relies on the execution of pre-excavation grouting. This paper reviews, the operation of TBM in headrace tunnel in more detail, focusing on the challenges of excavating through very strong and abrasive quartzite and gneiss, as well as encountering some high inflow of water along the tunnel. Furthermore, the water control procedures including Pre-Excavation Grouting (PEG) and Post Grouting (PG) methods to reduce the ground water ingress done in HRT are described.

1 INTRODUCTION

Uma Oya Multipurpose Development Project (UOMDP) as a hydro mechanical project in Sri Lanka, targeted generating hydro-electric power, transferring water for irrigation purposes and controlling seasonal devastating flood. This project lies in the south-eastern part of the central highland region of Sri Lanka [Rahbar, A. & Rostami, J. 2016].

This project involves 2 Rolled Concrete Core (RCC) Dams connected through a tunnel currently under construction by Drill and Blast method. The water is transferred from first diversion dam to the 2nd regulatory dam and from which, it will be transferred through a 15.4 km headrace tunnel to the top of a 650 m deep drop shaft that feeds the high pressure water to an underground powerhouse and turbine chamber for generation of 120 MW of electricity. The Headrace and Tailrace tunnels is being excavated by two 4.3 m diameter double shield TBMs.

Shielded TBMs with segmental lining are hereby facing severe challenges in sealing heavy water ingress, due to the constraints of access to the face for pre grouting ahead of the face as well as the available space for the installation of equipment for later post grouting. The shortcomings for pre excavation grouting often necessitate later more complex and time-consuming post-grouting measures in case of the necessity for sealing long term water inflow.

Post grouting in the tunnel behind the TBM with large amount of water inflow require complex measures and workflow to seal the water bearing ground to an acceptable water ingress level. The grouting design must consider technical and economic aspects, including long-term stability aspects of the lining design, accessibility and space limitations.

Excavation by DS-TBM has become a standard method for long small to medium sized pressure tunnels. Benefits comprise of the immediate support, the high flexibility of the TBM and the lining which can be utilized as long-term support (Figure 1).

The bedding of the segmental lining is established by pea gravel filling of the annulus gap between the segment and the rock at the side wall and crown segments immediately after ring building, while the invert segment is grouted by mortar.

Upon completion of the excavation this pea gravel is filled up with a standard cement based grout mix. High rock mass groundwater pressure requires a certain pressure decrease in the vicinity of the tunnel by systematic rock mass grouting, together with water relief through the

Figure 1.   Hexagonal segmental lining type.

segmental lining, which is achieved by a porous mortar within the segment joints and in case of a high external water pressure by small drainage holes drilled 30 to 50 cm into the rock mass.

This paper aims to review the significant 1000 lit/sec water ingress in 4.3m mechanized tunneling and all the activities and plans due to Pre-Excavation Grouting (PEG) and Post Grouting (PG) to reduction of the ground water ingress and other sealing recently done in Headrace Tunnel (HRT). Technical panel discussed all the residual ingress after excavation and installation of concrete segments, as far as possible backfilled with pea gravel, has dropped to about 500 L/s after roughly 3 months or less than 50% of the initial water inrush along this section of tunnel. Currently, practically all the ingress is concentrated in the invert between R6367 and R6396 (or about 35m tunnel length, in the following called the 'wet section'). This is average about 685 L/min/m for these 35 m.

2   GEOLOGY

The Highland series rocks cover the whole project area. They include pre-Cambrian rocks formed under high grade metamorphic conditions and are composed of two main types of rocks namely metasediments and charnockites or Charnochitic-gneisses. The metasedimentary rocks are metamorphosed sedimentary rocks and consist of garnet sillimanite gneisses or Khondalites, quartzites, quartz feldspar granulites, garnet gneisses, marble and impure crystalline limestone. Charnochitic gneisses are the most common rock types of the Highland series. Quartzite's are very hard rocks and difficult to drill through. Marble and impure crystalline limestone may ran into a problem when encountered in tunnels, shafts and dam foundations or across reservoir periphery, especially when associated with faults and karstic zones. The project area is folded into a series of large domes, basins, anticline and syncline structures [Rahbar, A. & Rostami, J. 2016].

3   HEADRACE TUNNEL (HRT)

Headrace tunnel, designed to excavate in 4.3m diameter and 15.4Km length with mechanized excavation method. According to the time schedule and also the geological condition Double Shield TBM proposed and supplied from Herrenknecht Germany under the code of M-1684

& M-1685. TBM M-1685 started with excavation of 500m adit from the portal and then excavated HRT from outlet in upward alignment. But according to the land acquisition issues in the job site, the other one excavated the Tailrace Tunnel and then after mobilization and getting ready the Inlet portal, TBM M-1684 re-assembled and started the excavation from HRT Inlet adit in 17 of July 2016 in downward alignment.

There are the design features separating this tunnel from other similar cases. First, this tunnel is pressure tunnel, meaning that it will operate under hydraulic pressure in excess of 20 bars at the intake shaft location with water up to 200m water column. Second unique feature of HRT in this project is the various adverse geological conditions including hard abrasive rocks, undetected faults containing high pressure water and mud, as well as the possibility of encountering karst. Finally, there is no middle access in ~15Km of HRT. Four different ground support types (GST) have been selected for the HRT based on the anticipated ground conditions. They include No support (invert segment only), Spot bolts where required (with invert segment), Systematic bolt pattern (with invert segment), Full ring segmental lining [Rahbar, A. & Rostami, J. 2016].

4 PROBING AND SITE INVESTIGATIONS

TBM's designed and later modified to drill the probe hole through the shield ports. The end of probe holes must always be at least 4m ahead of the tunnel face and maximum 6m outside of the future tunnel contour. For borehole positions with 7°, 11° and 15° outlook angles, the maximum length of probe hole are measured from the rear end of the drilling guide pipe installed in the TBM (Figure 2).

5 WATER INGRESS HISTORY

First excessive water ingress into the tunnel on Dec. 24th 2014 at Chainage 11+160 also caused extended delays. The high water inflow to the tone of up to 400L/sec has nearly flooded the cutterhead and heading and required additional provisions to allow working in the heading [Wannenmacher, H. et al. 2016]. Next major water ingress encountered on April 17 in the downstream heading of the HRT at ring no. R6342 (Chainage 7+317.5) of about 1000 L/s. This incident caused more than 4 months stoppage of excavation and came in addition to previous and still existing problems of large water ingress for this heading.

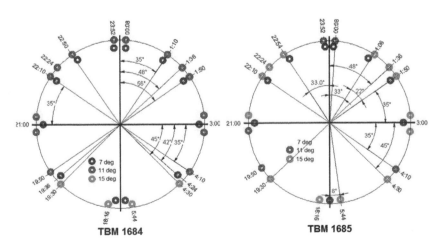

Figure 2.   Position of probe holes.

The features originally producing the ingress during tunnel advance into this section started as primarily local highly conductive channels in hard rock, with virgin ground water head of about 20 bar. As the face moved forward from R6320 until R6342 (7+317.5), where the major inrush was hit in April, the rock got gradually and increasingly more broken and water conducting. From R6342 to about R6388, the rock is generally highly fractured, so much so that significant over break developed in the tunnel crown from about R6368 to R6387 (about 23 m) with fallout to maximum 4–5 m above tunnel crown. This is basically where most of the 500 L/s ingress can now be observed in the invert, with some further ingress upstream to about R6396 (Figure 3).

When attempting to seal off such high ingress rate, the main problem was linked to the concentrated and high flow rate through highly broken ground combined with likely rapid pressure build up following any attempted flow restriction. Direct grouting easily lead to grout material getting flushed back into the tunnel through the layer of pea gravel backfill and the joints between concrete segments.

It must be noted that successful sealing-off of water ingress cannot be done by just sealing the joints between concrete segments in combination with grouting of the pea gravel layer alone (even if it were possible), since this would produce full water pressure load onto the concrete segments, for which they are not designed. A major part of the water pressure drop must consequently take place within the surrounding rock mass to prevent overloading of the segments. The high conductivity of the broken rock mass must therefore be grouted and sealed within the last few meters of rock outside of the tunnel circumference.

To get around the above mentioned main problem, a combination of temporary drainage to reduce flow rate and pressure build-up in the pea gravel layer while grouting off this layer has to be implemented. The strategy is simple in principle, but depends on the sufficient number of drainage holes drilled into the surrounding rock to provide enough conductivity to drain and divert a major part of the total ingress. By placing pipes from the tunnel, through segments and pea gravel into rock, the intersected and diverted part of the water flow can be conducted to the tunnel passing the pea gravel within the pipe. The remaining problem will be to decide when sufficient diversion has been achieved to successfully execute grouting of the pea gravel without too much pressure build-up and grout washout back into the tunnel. The combination of these two aspects of the approach can only be established by common sense and practical experience in the tunnel. Conditions will typically vary within quite wide limits along the tunnel section being considered and necessary adaptations should be expected.

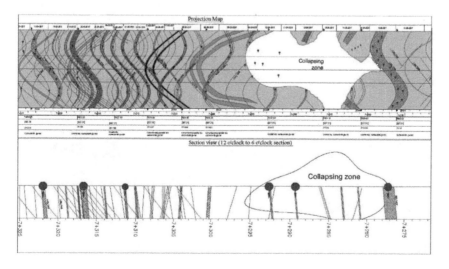

Figure 3.   Geological as built of water ingress area.

Figure 4. Water ingress of 7+317 area.

6 FINAL DECISION

Due to water inrush at Ch. 7+318, TBM advance was set on hold. It was appointed to assess the situation of HRT by a technical commitment for making decision whether excavation should be continued or other measures should be considered. Based on the current situation, all the possible options were assessed in a decision making matrix in order to propose the only one design solution deemed useful to face in the short term as follows:

Scenario 1 - Long holes from behind the face for drainage and grouting of the tunnel periphery and core.

Scenario 2 - Niches closely behind the shield and drilling for grouting of tunnel periphery.

Scenario 3 - Sealing only behind TBM backup where the access is guaranteed.

Scenario 4 - Sealing in the TBM backup by taking TBM backup apart for access.

Scenario 5 - Excavation without installing rings until the joint is exposed behind the shield and sealing by steel plates & grouting closely behind the shield.

Scenario 6 - Access to the front after pulling back of TBM and grouting from the face.

Scenario 7 - Bypass along the shield and grouting from there.

Scenario 8 - Bypass out far from the shield and grouting from there

Scenario 9 - Bypass in the crown or side wall, concrete wall at the face (plug) and grouting.

Scenario 10 - Drilling a ring of many holes by modified probe drill machine and grouting a ring around the TBM in the joint/fractured area.

Finally, two scenarios 3 and 9 were chosen as the top rated options but the amount of water was reduced in a short time and consequently, the risk of water inrush on manpower workability at shield area and electrical hazard (due to excavation continuation) were mitigated hence it was decided to recommence the excavation (Scenario 3) by considering some remedial measures for the enhancement of excavation procedure (Figure 5).

In this way, the excavation recommenced on 26 May 2017 until where entering into a dry rock mass and all TBM length was located in a safe and dry conditions. Then the excavation was stopped and post grouting activities were implemented in this huge water ingress area.

Figure 5. I 7+317 area.

7 POST GROUTING 7+317

After about 200m excavation and pass the wet zone, TBM parked in a good rock with min water ingress to start the Post Grouting. Then 2 main bulkheads made in downstream and upstream of the wet zone. With completion of the contact grouting in these dry zone, 4–6m deep drainage holes drilled through the lower wall segment erector holes. With installed split-pipes reaching about 1.5 m into rock, ground water drained and conveyed through the pea gravel fill and the concrete segment into the tunnel.

Wet zone contact grouting started through the invert holes while allowing the wall drainage holes to stay open for water diversion. After the grout has set, test the result by closing the valves on wall segment drain holes drainage holes to check for leakages. However, when the invert erector holes closed, pressure built up and water diverted to the wall drain holes (Figure 6).

Testing the efficiency of drain holes must be executed by closing valves in the invert for pressure build-up and the evaluation of how many wall drain holes would be required. If local areas are leaking more than acceptable, all the drain pipes are to be opened and drill and grout activities are carried out again as required. Then the pressure build up test is repeated. Cement grouting of the surrounding rock by pumping on the drainage holes. After the fulfilment of post grouting activities and grouting of all drainage holes, the water ingress measured by Parshall flume at portal reduce to less than 250l/s (Figure 7).

8 PRE-EXCAVATION METHOD STATEMENT AFTER 7+317

Regarding the complicated ground condition and social effect, it was advised to systematically keep the specified number of probe holes ahead of the tunnel face for minimum separate the 2 probe holes 180° in Outlet and minimum separate the 3 probe holes 120° in Inlet. The end of the probe holes must be kept at least 4 m ahead of the tunnel face and maximum 6 m outside

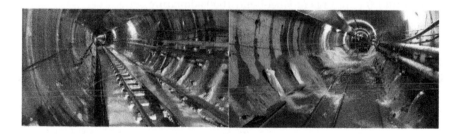

Figure 6.   Post Grouting of 7+317 area, before and after the post grouting.

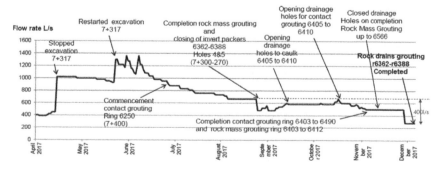

Figure 7.   The fluctuation of water flow amount measured at the tunnel portal.

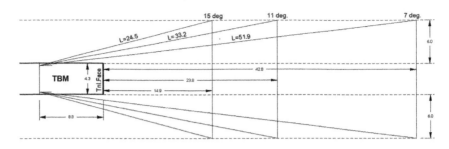

Figure 8.   Longitudinal section showing drilling ahead of the face.

of the future tunnel contour in difference length depends on positions with 7°, 11° and 15° outlook angles as illustrated in Figure 8 (Garshol, K. 2017 & 2018a).

In Dry Ground Condition, considered to use only 7° holes for probing and to drill the holes to the maximum length of 51.9 m from the rear end of the drilling guide pipe in the TBM. Under such conditions, the minimum of 4m overlap could be acceptable, which may allow the maximum TBM advance of 38.8 m before next probing ahead station.

When probe hole ingress results are triggering grouting, additional holes drilled forward of the face. The first stage of grouting requires 8 holes in. The 8 holes need to be distributed around the tunnel periphery with reasonably even spacing. The end of the holes should stay within 6 m of the tunnel contour.

Disposable packers with a non-return valve, expanded by the grout pressure allow retrieval of the insertion pipes immediately upon finished pumping of grout, which can then be easily cleaned by flushing with water. Packers placed in the borehole minimum 2 m passed the tunnel face.

Inflatable and mechanical packers used for temporary drainage and grouting. Base mix MFC, w/c-ratio 0.8 by weight with 1.0 % dosage of TamCem 28 admixture. Accelerator may be added through the Y-piece at the packer at dosage rates between 1% and 10% by weight of accelerator product against weight of cement in the mix design. Base mix OPC, w/c-ratio 0.5 by weight with 1.0 % dosage of Rheobuild 1000 admixture.

9   EQUIPMENT

Both TBMs must be equipped with modern el-hydraulic grouting units containing colloidal mixer, agitator and pump that can deliver maximum grout pressure of 100 bar. The units are professional grade modern devices, that allow to pre-set the maximum pressure, which automatically reduce pumping rate to what flow rate can be accepted by the ground at that pressure. The pumps have dosing units for (diluted) accelerator, to be added at the packer in dosage range between 1% and 10% by weight of cement.

For use of quick-foaming PU, suitable 2-component pump are available. Hoses, couplings, valves, pipes and fittings subjected to grout pressure must be rated for at least 100 bar.

10   POST GROUTING AFTER 7+317 IN FULL RING

Contact grouting was typically carried out in all sections; some locations had bulkheads injected beneath the selected invert segments. Also an invert bulkhead was in place in both ends.

Next step was drilling oversize Ø64 mm holes for the placement of standpipe to 1.5 m into rock in the selected hole-positions. Drill 4–6 m into rock through all the installed standpipes. The target of these holes is to cross water. The purpose of this step was to seal off leakage paths that exist through the contact grouted zone, as well as the first meter or two of the rock mass. This was to ensure that the main conductive water channels could be filled without losing the grout because of backflow to the tunnel. Packers placed in the segments. All grouting done by OPC with mix design tested and approved with w/c-ratio 0.5 including 1.0 % dosage of Rheobuild 1000 admixture. Stop criteria on single hook-up: Reaching pressure of 12 bar at < 1.0 L/min. flow rate. Or reaching grout quantity of 500 L for packers placed in segment (Garshol, K. 2017 & 2018B).

After ensuring that the contact zone and the near field rock mass had been sufficiently sealed to prevent grout backflow to the tunnel, rock mass grouting executed. Drilling did through available segment erector hole-positions. Because of the provided pressure release and water diversion through the open drains, Ø41 mm boreholes drilled through the decided segment erector holes. Holes drilled to 4 to 6 m depth depending on where and how much water that is encountered in each hole. The decided starting pattern could typically include less than 50% of the segment erector holes as considered necessary in the first stage of drilling and grouting. Grouting to be done by OPC.

After the 3rd Step grouting has been allowed sufficient setting time, all the drain-hole valves (stand pipes) were closed to allow water pressure build-up as a test of achieved water leakage cut-off. It took some hours or a day of pressure build-up to identify any leakage spots or other areas that considered for potential additional treatment. Such additional treatment consist of extra holes drilled through segment joints, to 1.0 m depth with packer in the segment joint, or to 4–6 m depth with packer at 1.5 m depth depending on where the water was encountered during drilling. Grouting executed as described above for Second Step or Third Step depending on where the packer is placed. The grouting process started with all valves on the drainage holes fully open.

Grouting commenced in the upstream end of the section by first pumping on any drains through invert segments, working direction downstream and then moving to wall drains and any roof drains also from upstream to downstream. Pumping was kept on the current position until one of the stop criteria was reached.

11   POST GROUTING AFTER 7+317 IN INVERT SEGMENT ONLY

Post Grouting of these conditions were completely based on the water ingress conditions in the tunnel. Drainage holes to be drilled to minimum depth 3 m and up to 6 m, until significant water was detected. Ø41 mm or Ø64 mm bits used to intersect the main water conducting joints to temporarily drain the area. Ø41 mm or Ø64 mm holes drilled to a depth of 3.0 m into the rock mass at a radial pattern as given by the local Instruction, placing the packers at 1 m depth. The pattern started with 3 m between radial fans and about 3 m distance between holes within a fan. Grouting started at the bottom and proceeded up to the crown. Grout the drain holes as the last step, started with the holes of lowest ingress and ending with the largest ingress holes. In a case of significant grout backflow anywhere through cracks and joints, the pump was slowed down and dosing accelerator was started at the packer and/or switch to PU if required.

12   PROGRESS HISTORY

Figure 9 presents the length of probe drilling versus the monthly tunnel excavation advance. Probe drilling has increased significantly and excavation rate decreased since Sep 2016. Since 7/9/17 about 2000~3000m drilling per month for 17~135m advance (44m drilling/m excavation) which is ten times the amount used between December 2015 and March 2017. Since then, the advance rate is about 2m/day and the cement consumption rate for PEG has been estimated about 2790 kg/m excavation.

13   CONCLUSION

Shielded TBMs with segmental lining are hereby facing severe challenges in sealing heavy water ingress, due to the constraints of access to the face for pre excavation grouting ahead of the face as well as the available space for the installation of equipment for later post grouting.

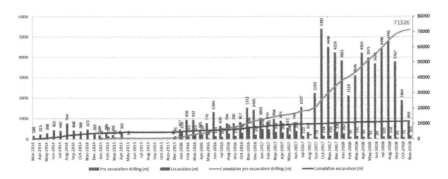

Figure 9. Probe Drilling ver. monthly tunnel advance rates in HRT.

The shortcomings for pre excavation grouting often necessitate later more complex and time-consuming post-grouting measures in case of the necessity for sealing long term water inflow.

The main hazard along HRT alignment is the unpredictable\unforeseeable adverse geological conditions and the water ingress\inrush which is encountered ahead of the face.

New, strict procedures of grouting works have been being followed in HRT. These activities have caused the tunnel advance rate, intensively reduced. Vice versa, PEG and probe drilling activities as well as the cement consumption have been considerably increased.

At the tunnel face, the water inflow seems to be under control, due to the proper PEG. Hence no major water ingress is envisaged to be likely encountered in future and the risk of encountering such water inrushes from the face similar to the previous experiences in HRT has been considerably mitigated. Approximately, 1200 m long has been excavated since Sep 2017. No new damaged buildings have been reported and there has been an increase in useable wells since Sep 2017. Consequently, Strategy and preventive measures which are currently pursuing, can be summarized as follows:

- Systematic probe drilling and TSP for the investigation of geological condition in front of the face
- Due to the environmental problems on the surface and side impact of tunnel excavation, only limited water ingress into the tunnel is allowed to have no effect on the surface. Hence sealing of tunnel by pre-excavation grouting is paramount in comparison to the excavation continuation and post-excavation grouting activities.
- Post grouting activities (Parallel to excavation) as the second priority in the limited and local zones where the residual water ingress are still more than the allowable ingress amounts.

AKNOWLEDGEMENTS

Authors would like to acknowledge their gratitude to Water and Irrigation Authority of Sri Lanka and Dr. Sunil De Silva. Also authors would like to thank colleagues at Farab Co. as well as subcontractor for TBM operation, Parhoon-Tarh Co., Mr. Hakimzadeh and Mr. Norouzi for their help and support and sharing of data.

REFERENCES

Rahbar, A. & Rostami, J. 2016. Construction of Headrace Tunnel of Uma Oya Water Conveyance Project Sri Lanka, World Tunneling Congress 2016. Proc. intern. symp. San Francisco, USA, 22–28 April 2016.

Wannenmacher, H. & Hosseini, A.H., Wenner, D., Rahbar, A. & Shahrokhi, Z. 2016, Water Ingress and reduction measures in the headrace tunnel at the Uma Oya multipurpose project, Hydro 2016. Proc. intern. symp. Montreux, Switzerland, 10–12 April 2016.

Garshol, K. 2017 & 2018a. Pre-Excavation Grouting (PEG) Method Statement, Rev. 01 to 06.

Garshol, K. 2017 & 2018b. Segment Lined HRT – Post Grouting Method Statement, Rev 01 to 05.

Tunnels and Underground Cities: Engineering and Innovation meet Archaeology,
Architecture and Art, Volume 7: Long and deep tunnels – Peila, Viggiani & Celestino (Eds)
© 2020 Taylor & Francis Group, London, ISBN 978-0-367-46872-9

Challenges of designing a Tunnel Boring Machine (TBM) for development of underground structures on the Moon

J. Rostami & C. Dreyer
Colorado School of Mines, Golden, USA

R. Duhme
Herrenknecht Asia Headquarters, Singapore

B. Khorshidi
McNally Construction Inc., Toronto, Canada,

ABSTRACT: Current and future goals for human activities on the Moon could be achieved by the creation of underground spaces. Locating a human habitat several meters below the lunar surface by tunnelling would provide greater protection from hazards than a surface habitat covered in regolith while also providing a more stable thermal environment than the surface. A Lunar Tunnel Boring Machines (LTBM) would be designed for the lunar environment and boring in lunar material. A LTBM could be used to access riles (lava tubes) at a distance from collapsed sections or skylights, thus reducing the hazard of rock fall when exploring these regions. LTBM could also be adapted for the frozen regolith in the permanently shadowed craters at the lunar poles, which may be an attractive option for access to lunar ice resources. This paper will discuss lunar base design concepts and requirements and possibility of application of a LTBM.

1 INTRODUCTION

Reaching for deep space is no longer just a scientific fantasy merely out of curiosity, but rather a possibility for commercial applications, and perhaps a necessity as the resources on the Earth seem to be stretched with the increase in population and our way of lavish living. These, among other reasons have fostered renewed interest in the space activities and discussions of landing man on the moon again and going beyond the moon for commercial utilization of space resources. With the advent of developing new reusable rockets which can reduce the cost of launching drastically, the establishment of the bases on the Moon and Mars has got much attention in last couple of years. *"Any attempt to establish a continuously staffed base or permanent settlement on the Moon must safely meet the challenges posed by the Moon's surface environment. This environment is drastically different from the Earth's, and radiation and meteoroids are significant hazards to human safety. These dangers may be mitigated through the use of underground habitats, the piling up of lunar material as shielding, and the use of tele-operated devices for surface operations"* (Lewis 1992).

Lunar base illustrations were typically depicted with large underground openings as shown in Figure 1. The benefits of underground access on the moon were recognized then, as now, and include protection from meteor impact, natural radiation shielding, and thermal stability. While this might be sufficient for a small base, there are studies to examine the feasibility of using Lava tubes for larger habitats, that could encompass areas the size of a large city. Obviously, the connection between these cities should be through underground excavations as well. There are other designs than encompass the use of underground space as residences,

Figure 1. Schematic drawings of large underground lunar base from National Geographic, 1969 (McKay, 1992).

workshops, and perhaps farms, with the use of sky light system, based on periscope with special filters to mimic the lights on the earth. (Figure 2,3).

The concept of a lunar tunnel boring machine has been proposed previously and an early conceptual design was developed (Allen, 1988). New discoveries about the Moon warrant a re-examination of the value that secure underground access may have for the development of a safe future lunar infrastructure. In addition, recent technological advancements in tunnel boring machines (TBM) and related technology suggest expanding the set of mining methods

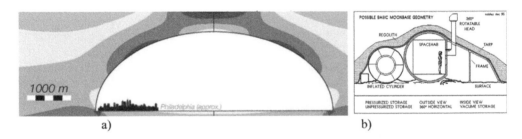

a)   b)

Figure 2. a) Lunar lava tubes that can host a large scale city, b) and Schematic of an underground space with skylights (priescope) for lighting (right) (Robinson et. Al 2012, and The Artemis Project).

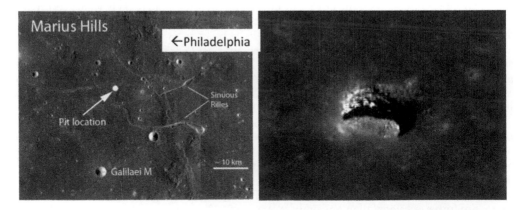

Figure 3. Pictures of Marius Hills on the Moon and the skylights indicating lava tubes (Walker 2016).

that should be considered in order to build man-made structures on the surface as well as under the Moon's surface.

New discoveries and opportunities for underground space include lava tubes and the polar shadowed regions. While both have been known for some time, recent discoveries suggest an increase in their importance for lunar exploration. Lava tubes, or rilles, are an ancient lava flow that formed into a hollow structure as the outer layers of lava cooled. These structures may possess unique environments on the lunar surface and thus are of interest for future science driven missions. Intact lava tubes are natural underground spaces that could be well suited for situating a manned habitat or someday a Moon colony. Recent work showed that under the unique lunar conditions lava tubes on the Moon could be very large (Blair, 2017). The unique lighting conditions of the permanently shadowed polar regions have been recognized for some time and the possibility that resources may be present in these regions that would not last long on other areas of the Moon. Limited exploration efforts have shown that water ice, carbon dioxide, and several other volatiles are present in and near the permanently shadowed craters of the lunar north and south poles. Tunneling and creation of underground space presents expanded opportunities for access to and extraction of resources and base construction in this unique environment.

Numerous plans for establishing bases and living quarters on Moon have been introduced. While some components of a lunar base must be placed on the surface, given the extreme environment of the Moon, it is likely that the most sought-after options for long term habitation of work or living space will be those that are created underground. There are various mining machines and methods for excavating soil and rock on the Earth that one can imagine using to develop underground space on the Moon. Yet, none of the mechanical excavation machines used on the Earth to date are more diverse and promising than Tunnel Boring Machines (TBMs). These systems now excavate pretty much any medium, from soft soil and flowing saturated sand to the hardest of rocks that are found on Earth, and everything in between. The worst cutting conditions known to industry are mixed ground that includes soft and unstable flowing soils combined with boulders and/or large sections of hard rock at the face. Yet, there are TBMs working on this exact combination every day in every part of the world. Another notable capability of these machines is to work under pressurized conditions (as someone put it - they work like submarines, but in soil and rock). This refers to the ability of the machine to work under substantial differential pressure. Such flexibility can come in handy in order to allow the operation to proceed in vacuum, while the tunneling operation can proceed at atmospheric pressure or higher inside the tunnel.

This paper will examine the need for the development of underground openings, followed by the list of many challenges for mining and construction activities on the moon. The use of TBMs and their great potential for use on the Moon for a variety of applications will be discussed. However, while the concept of the LTBM machine is sound, the actual operation and component design of TBMs for use on the Moon requires a number of important changes from the terrestrial baseline. Specific ways to approach machine design and rearranging the operational issues that meet the restrictions and constraints of the lunar environment will also be discussed in this paper.

2 UNDERGROUND SPACE FOR HUMAN HABITAT

To understand the need for underground space on the Moon, one only needs to list the conditions on the Lunar surface. This includes the vacuum and lack of breathable atmosphere, surface temperature variations, impacts by micrometeorites, and radiation. As such, situating a manned base underground on the Moon would provide several advantages over a lunar surface base. These advantages have long been recognized so we will only briefly enumerate them here.

Providing atmosphere: Granted that the lack of atmosphere means near absolute vacuum on the moon, pressurizing the underground space to 100 bar pressure, which is fairly modest, and maintaining the temperature and moisture content to near what it is on the earth facilitates normal living and working conditions of the crew. This can be rather easily achieved by using the underground structure for support and use of natural impermeable material or installation of a membrane for keeping the air pressure.

Thermal stability: The surface temperature varies from roughly -150 to +150°C at the lunar equator between day and night. This temperature difference reduces somewhat at higher latitudes, however a structure on the surface at high latitudes near the poles would at times be exposed to nearly horizontal sunlight on one side and radiate to deep space or a cold lunar surface on the opposite side. The need for thermal management for surface habitats is well recognized (Hertzberg, 1992). The swing from low to high temperature decreases to ±3K at just 30 cm beneath the lunar surface and there is essentially no day/night variation below 700 mm depth (Vaniman, 1991).

Radiation shielding: Radiation possesses a significant risk to human life in space from solar flares and galactic cosmic rays. Using regolith for radiation shielding has been recognized as a potential solution to this problem. Silberberg et al. (1985) recommend that lunar residents should spend most of the time protected by about 2 m of densely packed regolith, while at times of intense solar flares the regolith depth should be nearly doubled. This means that for long term use, the depth to the crown of >4 m would provide a safe living environment.

Micrometeorite protection: The lunar surface is constantly bombarded with meteorites from large crater forming impacts to micron sized dust grains moving at high velocity (micrometeorites) that garden the lunar surface. An impact of a large meteorite into a manned habitat could be a devastating event, but even micrometeorite impact would slowly erode the habitat surface if not protected. Covering a lunar habitat with regolith would protect from micrometeorites, while situating in an underground space would protect from larger meteorite impacts with more protection afforded the deeper the base is situated.

It is worth considering the burial of a base versus situating of a base in an underground space. In the first instance, a base could be protected from many of the hazards mentioned above by placement on the surface and burial in some amount of regolith. While situating in a true underground space formed by a tunnel boring machine would provide significantly more protection. A TBM has advantages beyond base situating, such as for access to lava tube and resource acquisition at the lunar poles. Lava tubes have been observed in lunar images, such as shown in Figure 2. Concepts for access to lava tubes have often focused on entry through collapsed sections, also known as skylights. However, it in these sections the lava tube may be most prone to collapse and thus pose a risk to exploration or base emplacement. Entry to a lava tube where it has not collapsed could be accomplished with a LTBM and allow access to a more stable section than near a skylight and potentially a space that has been sealed from the lunar surface environment for a long time. Situating a human habitat in a lava tube may also be advantageous because it provides an underground space with all the benefits and protections without the need to excavate the entire underground space.

Water ice and other volatiles have been captured by the low temperatures at the permanently shadowed regions (PSR) of the lunar poles. The PSRs present a significant opportunity for exploration of the moon and for development of human activity in space due to the availability of water in the form of frozen regolith. The PSRs activities require work in extreme low temperatures (as low as 40K in the deepest craters) and lack of lighting. A TBM could be used to tunnel into the PSR from a sunlit region and the product, water, to be transported to the portal. Situating a manned base underground in this region may also be advantageous.

3   LUNAR MATERIAL AND GEOTECHNICAL CONSIDERATIONS

The anticipated geology on the Moon's surface is a 10-meter layer of regolith, which becomes highly compacted within the first 10–20 cm of depth. Depending upon local impact history, this would grade into boulders then fractured rock and deeper into intact bedrock, which would be basaltic in the Mare regions and anorthositic in the Highlands. Frozen regolith at the PSR would combine fine grained through coarse grained rock including boulders of anorthosite mixed with icy volatiles, creating a permafrost region with variable mechanical strength.

Until underground openings are excavated at such depth that become fully basalt or other bedrock, the cutting conditions at the face will be that of the "mixed conditions." One should keep in mind that frozen regolith is the most valued material on the Moon since it is considered to be the primary source of much needed water for support of life and as fuel. Excavating such a mix from the surface is much more challenging since cutting of frozen regolith, boulders, and bedrock requires high cutting forces that cannot be provided by the weight of the machines. High cutting forces to penetrate harder material requires some sort of anchoring which limits the manoeuvre of the machines, or by gripping on to the excavated walls, which is exactly what TBMs are designed to do.

4 TUNNEL BORING MACHINES (TBM) AND THEIR APPLICATION

TBMs have more or less become the primary choice of tunnelling, despite their limited geometry to be circular, as compared to other geometries preferred in different applications. The widespread use of TBMs expands from hard rock TBM to various full-face shield machines for use in soft ground tunnelling under pressurized face conditions. This includes earth pressure balance (EPB) and slurry shields that are the most commonly used machines for soft ground tunnelling today. Figure 4 shows the classification of the TBMs (Rostami, 2016). Machines are specific to ground types, but there are also hybrid machines that can bore through various ground conditions. Ideally, machines are designed for the anticipated ground conditions and are equipped to cope with various types of rocks, soil, and mixed ground conditions. In general, there has been a desire to design and manufacture a "universal TBM" that can theoretically adopt to and work in variable ground conditions, and do so quickly. This would perhaps be one of the hybrid machine types.

The available machines range from microtunneling machines that are operated remotely (without man entry to the tunnel) and range from 250 mm to 3m in diameter to normal TBMs (man entry) with diameters around 2m and goes up to 14.5m in rock and 17.6m in soft ground tunnels. Machines can negotiate horizontal and vertical curves with radius as tight as 300m while there are designs for machines that can turn a 50m radius. There are shielded machines that are designed to work under groundwater pressure of up to 17 bar. The pressurized face machines show the capability of the machines to operate in differential pressures between the tunnel face, which is the ambient pressure of the ground and atmospheric pressure inside the tunnel. Given the capabilities of the TBMs and their potentials for use in new applications,

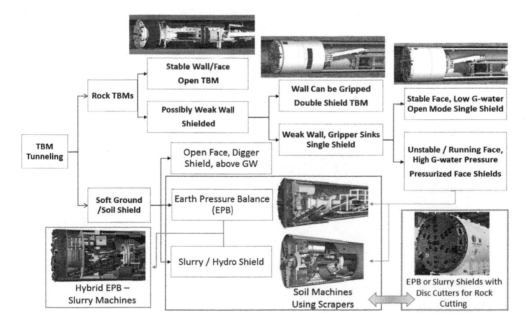

Figure 4. General classification of TBMs for various ground conditions (Rostami, 2016).

the authors strongly feel that their use in excavation of underground facilities and structures on the Moon for living quarters or simply for production of targeted materials are feasible. Obviously, there needs to be many changes in the way we approach certain aspect of operation. Following are some of the thoughts on the challenges and possible solutions for using TBM in mining of subsurface structures on the moon.

5 CHALLENGES FOR DEVELOPMENT OF LTBM

Cutting rock or soil underground is going to be largely the same whether done on the Moon or on the Earth. This means that a critical threshold of cutting force is needed to penetrate into the rock or soil face and fragment it to the extent that it can be separated and hauled away. This is based on the experience in rock/soil excavation on Earth where variations in the conditions have a minor impact on the excavation process itself. However, the very same conditions can have amplified impact on operations, including tunneling-related processes behind the machine. This includes ground support (reinforcement of soft overhead strata), materials handling, management of pressure differentials including, liquid drainage and exfiltration of gas, handling of temperature, permafrost conditions, machine maintenance, repairs and modifications, surveying, etc. Given the accelerated development of novel sensors and artificial intelligent systems, machines can now be operated in fully automated mode and optimized in real time as they excavate. Yet, TBMs are still being run by operators today. Meanwhile, most of the machines come equipped with various self-diagnostic systems that increasingly facilitate adaptive maintenance and repair. There is still no remedy or robots that can replace the regular or scheduled maintenance or occasional repairs as yet, but may have to be developed for the use of LTBM. Thus, man entry into the tunnel are still a part of the process on the Earth, while this may not be a workable solution for using TBMs on the Moon. Following are some of the challenges that one can imagine as part of using TBMs for development of underground Lunar bases or mining operations.

Weight Limits: A 2–3 m diameter TBM can weight over 100 tons, including the backup systems and trailing gear. This is primarily due to reliance on low-cost steel as the basis for the primary support structure of the machine and shields. Custom hardened steel alloys of higher value are also utilized for the gearbox, cutterhead, and cutting tools (typically disc cutters). Obviously, some steel alloys with specific properties are needed in critical parts of the machine, including the hardened steel for disc cutters in order to maximize wear life. While the dominant cutter type for cutting rock is disc cutters (Figure 5) that are heavy, there might be some alternative tools or ways of reducing the weight on these cutters. Maintenance will be an issue in lunar conditions irrespective of their location. The possibility also exists that a large percentage of the primary support structure could be manufactured from local lunar resources, including cast basalt or compressed sintered regolith reinforced with metallic members or basaltic tensile fibers.

It may also be possible to use lighter materials with higher hardness to replace the cutting tools, although this approach will increase total costs. The same should be considered for

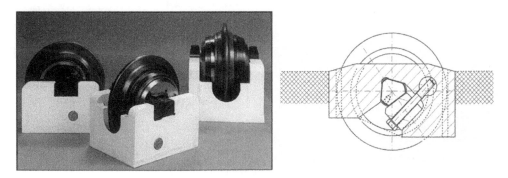

Figure 5.   Picture of disc cutters, and back loading assembly. (Robbins Co and Palmeri).

structural elements of the machine to switch to carbon fiber composite parts and lighter composites or alloys for various parts of the machine. Switching materials has some implications on repairs and maintenance such as welding of the parts that should be considered in the design. The added complexity is that the behaviour of material will be different at very low temperatures, if they are encountered. This refers to the brittle nature of most metals at low temperatures of around 50°K. As such, extensive research is needed to characterize the material properties at temperatures anticipated at the face. In this case, it is assumed that only the head will be exposed to such extreme temperatures. Furthermore, given the working conditions on the moon, the materials for seals that are very critical in success of the disc cutters, as well as lubricants for the disc cutters should be revisited. The use of graphite could be a possible solution as a lubricant in various parts of the machine.

Power and Energy Consumption: TBMs are power-hungry machines. Typical 2–3 m diameter TBM requires 500–700 kWe of input power. Variable Frequency or VFD drive systems are typically used on TBMs with feed cables carrying 5–25kV power from the portal or shafts and then transforming the input voltage to 480, 660, or 1000V for the drive units. Such power ratings are not common in space applications, and this issue will need to be visited by both machine designers and space operators. However, comparing different excavation on earth shows that to reach low specific energy consumption

per excavated volume, high forces and thus high power are necessary. Figure 6 is showing a comparison of different rock excavation methods specific energy consumption. On the contrary, "low force" excavation methods such as thermal spalling or microwave excavation require amounts of specific energy several orders of magnitude higher than the most efficient mechanical methods. In an environment, where on one hand power and weight levels have to be kept low, on the other hand energy must be conserved, this dilemma has different implications for the suitability of a solution than on earth. Thus, lunar TBMs may well adopt excavation techniques, which are not economical on earth. High efficiency components should be employed wherever possible in order to use every bit of power that is available and minimize sources of energy loss in the machine. Low friction bearings, high efficiency drive system, and preventing power losses from electrical to mechanical will all be a part of the equation to

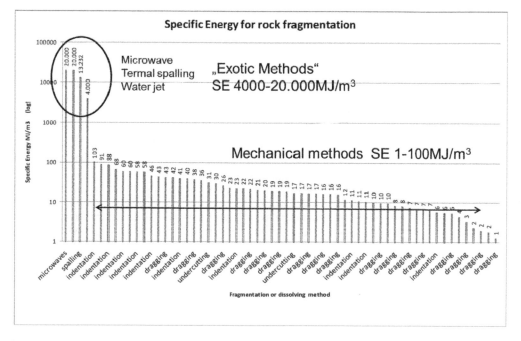

Figure 6.   Evaluation of specific energy for rock cutting in literature.

make LTBM a reality. The use of nuclear or solar power could meet the demand for a typical TBM, so this target would seem to be within reach.

Abrasivity: The regolith and bedrock of basalt encountered on Moon's surface are known to be very abrasive. This is due to the mineralogy of the material being from harder silicates that are present in the mineral composition of the rock, and hence in the regolith that is generated by the meteorites impacting the surface. Also, the regolith is reported to be jagged and having sharp edges due to the process of fragmentation being impact as opposed to the typical erosional process that happens on the Earth. The implication of highly abrasive materials to be expected on the machine is that the cutting tools wear at much higher rates and will need replacement, also the inner components of the machine may also be subject to higher wear than normal. The Appollo program has for example shown the detoriating impact of abrasive lunar dust on all kinds of seals. Consequently, disc cutters may wear faster and should be monitored more carefully. The use of conditioning media such as foam in TBM tunnelling reduces machine torque and wear of tools and other machine components, similar approaches may be envisioned for lunar applications to minimize abrasive wear on machine components and cutting tools.

Vacuum: Lack of atmosphere and working in vacuum pose significant challenges. Putting a system in place where pressure can be built up inside the tunnel will be important, and will certainly be easier once operations are entirely in hard rock. An effort to seal the entry and design of the tunnel to support pressure will have three benefits. First it would facilitate the capture of any valuable volatiles created during the tunnelling process. Second, it would enable the later use of tunnels to contain an atmosphere for human habitation. Third, it would allow for use of 'Earth-normal' methods and operations which would simplify the LTBM design process and subsequently, the operation. However, a higher standard of designing for vacuum operations will be adopted as a design requirement. This is especially critical for starting the tunnel. One important element of vacuum operations is the lack of convective heat transfer. This could result in excessive heating of the discs or cutting tools and components of the TBM drive system as well as other thermal management issues. Another issue is the very high and very low temperatures at the surface and limited possibility for heat exchange to maintain an ambient and acceptable temperature within the workings, although thermal conditions will stabilize once the TBM is beneath the surface.

The need for routine maintenance and repairs on the LTBM may require human intervention, and while there is the possibility to use space suits, wearing such suits will limit the ability of the workers to do their job, especially in the confined spaces that are available within the envelope of the excavated tunnel and with the presence of heavy equipment. This will put special emphasis on the clever use of automated systems and robotics to the greatest extent possible.

Lack of flushing/cooling material: Lack of atmosphere and fluids on the Moon surface means the lack of flushing material that is often used for muck transportation in smaller holes drilled for ground support. Drilling is often an integral part of TBM tunnelling for probing ahead of the machine or for installation of roof bolts. Also, the motors and VFD drive systems for cutterhead rotation as well as the motors used for various pumps, cooling systems of the transformers, and all other heat generating units need some form heat dissipation that is usually implemented with a fluid and this could be an issue for using TBM or any other heavy-duty machines on the Moon.

Implications of extreme temperature: As noted before, the extreme temperatures of roughly +/- 150°C should be expected on Moon surface. The higher temperatures require cooling to allow for proper functioning of the equipment, and possibly the work environment for workers. The low temperatures that are reported on the Moon and in the location of lunar ice, may need to be mined and present permafrost challenges that are unique to deal with even on the Earth. These extremes are less critical at higher depth, but granted that the mined-out materials should be transferred to the surface for various uses and applications, they still pose critical challenges to the operations.

Ground stability and support: Ground support is installed to assure safe use of the created underground space. The support type is selected based on the ground conditions and strength of the material surrounding the tunnel and to prevent loose blocks from falling in, soil from destabilizing and failure, and in some cases where the ground could creep and over time incur plastic deformation, or become squeezing. In this context, frozen regolith, if encountered, can show time dependent behavior and need to be contained. The most common ground support type is the use of rock bolts/dowels that is often steel rebar that can be placed in a drilled hole. If resin or grout is used, the bolt can be anchored and even put under tension by using a plate and nut at tunnel surface to apply a load and press the blocks against each other and thus increase the ground strength. Other ground support measures such as wire mesh and sprayed concrete are also commonly used in tunnelling operations.

For Lunar tunneling, one might think of Kevlar or carbon fiber mesh along with light polymer or resin based cements to be sprayed on the surface of the bored tunnel. This allows for containing the air pressure in the opening as well. There needs to be a system for installation of the ground support that works hand in hand with the machine to contain the pressure and allow the operation to continue in atmospheric pressure. One possibility is to cast special segments from light materials and build air tight rings in the back of a shielded machine and extrude the support as it happens in single shield TBM operations. This is similar to the machines that work in submerged conditions and can easily keep 7–8 bar of water pressure at bay. These materials could be formed by 3D printing rock powder produced by the TBM or regolith.

Maintaining air pressure in the opening: It is critical to maintain atmospheric pressures inside the opening to allow the crew to work in normal conditions. Tunnelling operations under high groundwater pressure is a good example for such conditions where shielded machines can be used under air tight support, installed within the tail shield. This requires a close coordination of the operation and development of a light weight, air tight support system. At this stage, the lunar tunnels are anticipated to be at depth of less than 100m, most likely no more than 10m and the ground pressure is not very high, meaning that the rock, even compacted soil can easily contain the stresses. If the tunnel is in regolith or granulated materials, a thin membrane can be used to contain the air, while the load is transferred to the ground support measures installed around the tunnel to react to the local stresses. This is commonly done by injecting grout into the soil behind the shield in the terrestrial application. However, there needs to be some studies of material types and development of new cements to react to the special mineral content of the regolith. Cast segments made of sintered 3D printed rock powder or regolith are options to explore.

Material transportation: Removal of the material at the face is typically done by buckets if there is no need to operate the machine in closed mode to contain the pressure. This is the easiest form of mucking in TBM tunnelling in rock. However, given that there is the need to operate the machine in closed mode, the excavated material can be mixed with a medium, perhaps a mud, or a polymer based thick fluid, or foam and extracted from the face, while maintaining the pressure at the face or vice versa. The use of this medium allows for transfer of the muck from the cutting chamber and plenum to the space behind the bulkhead where a differential pressure is maintained. Once the muck is transported to the work space within the tunnel, then it should be separated from the carrying medium. The medium can be recirculated to the head and the muck can be transported out via rail, trackless vehicles, or some type of conveyor to the portal or to the launch shaft. Perhaps a set of airlocks can allow the transition of transported materials in and out of the tunnel between vacuum and atmospheric pressure. If the machine is launched from a shaft, then it needs to be lifted out and dumped in pre-designated areas, possibly the mounds around the launch pads. The same is true for the supplies that go to the heading and the same mode of transportation can be used to move supplies to the heading. The muck could also be formed as a fluidized particle flow by a gas supply on the LTBM.

Utilities: TBM is indeed a tunnelling factory. As such, it does require some support and utilities to be provided for the continuous operation of the machine. The first and most important item is the electricity which is transferred via insulated high voltage cables. These cables are extended at every 300–500 meter of tunnel advance. Perhaps in this context, the compressed air lines are the second priority. The compressed air can be used for supplying the atmospheric pressure in the tunnel as well as being the power source for some of the equipment on the back-up system. While on the earth installation of water lines for cooling and dust suppression is essential, their use in lunar tunnelling with TBM does not seem to be as critical. Ventilation of the heading can be done using specialized air filtering and cleaning system to re-supply the oxygen used at the heading by workers since there is no source of air contamination such as diesel equipment in the operation. An air filtering system similar to what is currently used in space station can function as ventilation system, while the heat exchangers will make up for lack of heat transfer of ventilation in terrestrial tunnelling. Monitoring and communication systems are anticipated to function the same way on the Moon as they do in normal tunnelling operations on the earth within the tunnel, but they should be shielded against radiation at the portal. Rail system does not seem to be essential, as the transportation of men and material can be done by trackless vehicles.

Other Issues: Fluids, specially the use of hydraulic oil that is very common in the machines on the earth should be reevaluated since they are relatively heavy and should be tested and maintained on a regular basis. A remedy for the hydraulic systems is to use mechanical systems for rams and movement of the parts that are normally moved by hydraulic cylinders. At this stage, the tunnelling operations are not anticipated to be under cryogenic or elevated temperatures. Meanwhile, if the components are sensitive to temperature, special provisions should be taken to insulate them for transfer from support vehicles to the site for assembly. The same is true for electronic parts that work in the tunnel and that radiation is unlikely to impact them during operation in the tunnel but special provisions may be required for their transition to the site.

Detection of the geological features ahead of the machine and along the tunnel alignment is a very critical issue for lunar TBM operations. At this time, the detection of various features

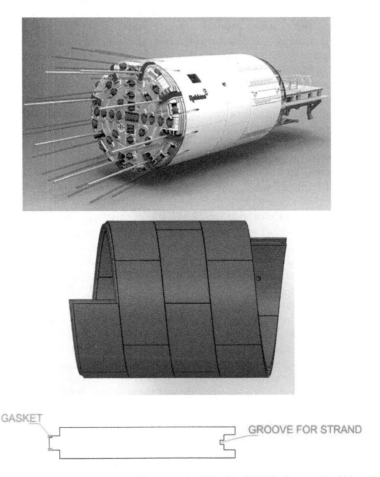

Figure 7. Schematic drawing of Probe drilling ahead of Shielded TBM (Source: Robbins Co on the top left), and Helical Segmental lining (top right) and sealing system (bottom).

ahead of the TBM is still an unsolved issue in terrestrial applications but there are multiple groups working to resolve this issue. One of the promising technologies could be the ground penetrating radar (GPR) and incorporating GPR into machine operation. An alternative is the use of various geophysical methods. For example, geophones and similar sensors can be deployed ahead of the machine on the surface and use the cutterhead as the source to identify certain features ahead of the TBM using micro-seismic techniques. The purpose of this exercise is to understand the nature of the ground ahead of the machine and identify hard and soft formations, perhaps voids and joints, by identifying high and low-pressure wave velocities in various media. Another alternative is the use of smart probe drilling systems for material characterization ahead of the TBM as shown in Figure 7. (Kahraman et. al. 2016).

In parallel, there are recent development of a new tunnel lining installation using helical segmental lining system that allows continuous operation of the TBM while supporting the ground and sealing the working for pressurized working condition. This is accomplished by using an integrated gasket in the tongue and groove to prevent water leakage into the tunnel in terrestrial application and preventing air leakage outside the working on the lunar application (Khorshidi 2018).

6 CONCLUSIONS

Use of underground space for establishing bases on the Moon seem to be a logical choice, perhaps an inevitable option. These spaces can protect the future lunar inhabitants and equipment from the harsh environment at the surface, including impact by meteorites, radiation, harsh temperatures, storms, and other unknown conditions that is an integral part of long term activities on the Moon surface. The use of underground space also allows for living in atmospheric pressure and conducting normal activities for the inhabitants. In addition, the ability to excavate underground space also lends itself to effective mining of some target material for various space activities. Among the various equipment and methods for underground construction and tunneling, TBM seem to be most promising method of excavation for application on the Moon. These machines have a very good track record of activities in tunneling on the earth and are capable of offering high performance and are the easiest systems to automate and operate remotely. This paper offered some initial thoughts about the use of these machines and necessary components that are required to meet the challenges of operating such equipment in the lunar environment and offer a safe and efficient operation of TBMs.

REFERENCES

Allen, Christopher S., et al., 1988, "Proposal for a lunar tunnel-boring machine," NASA-CR-184746, NAS 1.26:184746.

Blair, D. M., Chappaz, L, Sood, R, Milbury, C, Bobet, A, Melosh, H. J., Howell, K. C., and Freed, A. M., 2017, "The structural stability of lunar lava tubes". Icarus 282, 47–55.

Hertzberg, Abraham, 1992, "Thermal management in space," in Space resources. Volume 2: Energy, Power, and Transport.

Hunt D.V.L., Nash D., Rogers D. F., 2014, "Sustainable utility placement via Multi-Utility Tunnels," Tunneling and Underground Space Technology, Volume 39, January 2014, Pages 15–26, DOI: 10.1016/j.tust.2012.02.001

Kahraman S, Rostami J, Naeimipour A., 2016, "Review of Ground Characterization by Using Instrumented Drills for Underground Mining and Construction," Rock Mech Rock Eng, 49:585–602, DOI 10.1007/s00603-015-0756-4

Khorshidi, Behzad, 2018. "Helical Lining", Tunnel and Tunneling, Sept. 2018. https://secure.viewer.zmags.com/publication/f4738e6e#/f4738e6e/12

Lewis Hr 1992, "Human Safety in the Lunar Environment". NASA SP-509, vol 4, Social Concerns, Editors: Mary Fae McKay, David S. McKay, and Michael B. Duke, Lyndon B. Johnson Space Center Houston, Texas. www.nss.org/settlement/nasa/spaceresvol4/human.html

McKay, Mary Fae, David S. McKay, and Michael B. Duke, 1992, Space resources. Volume 4: Social Concerns.

Robinson M.S., Ashley J.W., Boyd A.K., Wagner R.V., Speyerer E.J., Ray Hawke B., Hiesinger H., van der Bogert C.H., 2012 "Confirmation of sublunarean voids and thin layer in mare deposits", Planetary and Space Science, 69 (2012), pp. 18–27

Rostami J., 2016, "Performance prediction of Hard rock Tunnel Boring Machines (TBM) in Difficult Grounds," Tunneling and Underground Space Technology (TUST), 10.1016/j.tust.2016.01.009.

Silberberg, R.; C. H. Tsao; J. H. Adams, Jr.; and John R. Letaw, 1985, "Radiation Transport of Cosmic Ray Nuclei in Lunar Material and Radiation Doses," in Lunar Bases and Space Activities of the 21st Century, ed. W. W. Mendell, 663–669, Houston: Lunar & Planetary Inst.

The Artimis Proejct; http://www.asi.org/adb/04/02/03/periscope-anecdote.html

The Robbins Company: http://www.therobbinscompany.com/category/blog/.

Vaniman, D., Reedy, R., Heiken, G., Olhoeft, G., & Mendell, W., 1991, "The lunar environment," The lunar Sourcebook, CUP, 27–60.

Walker, R. 2016, "An Astronaut Gardener On The Moon - Summits Of Sunlight And Vast Lunar Caves In Low Gravity", Science 2.0, October 9th 2016. http://www.science20.com

Tunnels and Underground Cities: Engineering and Innovation meet Archaeology,
Architecture and Art, Volume 7: Long and deep tunnels – Peila, Viggiani & Celestino (Eds)
© 2020 Taylor & Francis Group, London, ISBN 978-0-367-46872-9

Problems associated with an EPB-TBM in a complex geology with serpentinites and peridotites in Turkey

M. Sakalli & D. Talu
Dogus Construction and Trade Inc., Istanbul, Turkey
N. Bilgin & I.H. Aksoy
Istanbul Technical University, Turkey

ABSTRACT: Belpinar irrigation Tunnel of 5450 km in length was opened with an EPB-TBM having a diameter of 6.82 m. The tunnel has a complex geology with serpentinites and peridotites, highly fractured in most cases with water ingress reaching up to 450 l/s. A mean daily advance rate of 8.04 m with stoppages and 9.93 m without stoppages were obtained with a maximum daily advance of 21 m. The mean ratio of compressive strength to tensile strength ratio of the intact rock is extraordinary low changing between 3.8 and 4.1. This enabled to compare thrust index, torque index, penetration and specific energy values obtained in different tunnels in Turkey and classifying the excavability of the rocks according the ratio of compressive strength to tensile strength ratios. The complexity of the rock mass in Belpinar was divided in five parts. The Machine utilization, repairs of the machine, stoppages and disc consumptions were discussed for each part. Some recommendations were made for more efficient excavation for similar complex geologic formations.

1 INTRODUCTION

The excavation of 5,450 meters long Belpinar Tunnel of 6.82 meters in diameter which connects Saglik Plain to Emen Plain, was started on December 24th 2014 by an EPB TBM and completed on November 6th 2016 which was within the scope of construction of 48,564 meters long Kilavuzlu Irrigation Main Channel. Irrigation water will be provided to a total area of 55,536 ha in the cities of Kahramanmaras and Gaziantep and an additional area of 33,400 ha in Amik Plain. The completion of the project will lead to a rural development by increasing agricultural productivity in Saglik, Emen and Amik plains.

Tunnel line crossed ultrabasic rocks, peridotites and serpentinites with different geotechnical properties. Typical view of peridotites around the tunnel portal is seen in Figure 1. EPB TBM performance parameters such as torque, thrust force, specific energy penetration values of the EPB-TBM were evaluated in different ground conditions. It is believed that data provided within this paper will provide a more accurate planning and scheduling of the project in similar ground conditions (Sakalli, 2017).

2 GEOLOGY AND GEOTECHNICAL PROPERTIES OF THE GROUND

Peridotites, serpentinites and peridotites-serpentinites are found along the tunnel as described in Table 1. Peridotite is a magmatic rock. Serpentinite is a metamorphic rock and hydrothermally transformed case of peridotites. Peridotite-serpentinite is a metamorphic rock and half-hydrothermally transformed case of peridotites. These rock formations may be very problematic in most cases in tunneling (Marinos et al. 2006, Clark & Chorley 2014, Glawe & Upreti 2014).

Figure 1. Typical view of peridotites in the tunnel portal and TBM.

Table 1. Ground characteristics along the tunnel route.

Chainage	Description of the ground	RMR	GSI
65+098–65+309	Serpentinite, fractured in some places with water ingress	20–28	30–38
65+309–67+090	Peridotite –Serpentinite, fractured in some places with water ingress	22–33	32–43
67+090–70+548	Peridotite, massif and fractured in some places with water ingress	27–49	38–62

Table 2. Physical and mechanical properties of peridotites and serpentinites in Belpinar Tunnel, in chromite mines and in Greece.

Rock	Density, (g/cm³)	U. Compressive Strength, σ_c (MPa)	Tensile Strength, σ_t (MPa)	σ_c/σ_t	Elastic Modulus, (GPa)	Reference
Peridotite, In Belpinar	2.78	77	18.70	4.10	0.8	DSI, 1999
Serpentinite, in Belpinar	2.50	39	10.3	3.80	0.6	DSI, 1999
Serpentinite, in Chromite Mines	2.49	38	5.70	6.70	2.3	Bilgin et.al, 2006
Peridotites / Serpentinite, in Greece	2.6–2.7	87–375	10.8–25.2	3.1–12.5	1.3–2.5	Coumantakis, 1982
Serpentinite, in Greece	2.5–2.6	40–135	39–134	3.1–12.5	0.6–1.7	Coumantakis, 1982

Some physical and mechanical properties of the rocks found around the tunnel are given in Table 2. The values obtained from the literature are also given in the same table for comparative reasons. One important point emerging from this table is that for peridotites and serpentinites the ratio of compressive strength to tensile strength may go down to 4 even 3. As it will be explained in the following section, this ratio plays a very important role in the performance of TBMs.

3 PERFORMANCE OF TBM IN BELPINAR TUNNEL

3.1 *TBM design parameters and general excavation performance*

With major stoppages a mean daily advance rate of 8.04 m and mean monthly advance rate of 241.23 m were obtained; these values went up to 9.93 m and 297 m if the major stoppages of 128 days were excluded. Design parameters of EPB-TBM are shown in Table 3.

Table 3.   Design parameters of EPB-TBM.

TBM excavation diameter	6,820 mm
Shield diameter	6,790 mm
Number of single discs	39
Number of double discs	4
Disc diameter	432 mm
Maximum cutter head speed	3.5 rpm
Maximum torque	8,307 kNm
Maximum thrust force	48,657 kN
Maximum power	2,785 kW

Table 4.   TBM performance in different ground conditions.

Formation	Thrust Index, kN/ (mm/rev) +/–s.d	Torque Index, kNm/ (mm/rev) +/– s.d	FT/T	Mean Penetration, mm/ rev	Time spent to excavate one ring of 1.5 m min +/– s.d	SE kWh/m³ +/– s.d
Peridotite, dry	865 +/– 487	380 +/– 191	2.50	11.7	59.4 +/– 30.1	18.3 +/– 9.2
Peridotite, water ingress of 0–6 l/s	649 +/– 184	375 +/– 112	1.75	12.3	56.0 +/– 5.0	18 +/– 5.4
Peridotite, highly fractured water ingress 6–220 l/s, RQD 0–30	884 +/– 638	401 +/– 228	2.16	12.3	60.7 +/– 26	19.3 +/– 11
Peridotite–Serpentinite, highly fractured dry, RQD 0–30	754 +/– 287	402 +/– 97	1.86	11.6	55.1 +/– 22.5	19.3 +/– 4.7
Peridotite–serpentinites, highly fractured water ingress 6–220 l/s RQD 0–30%	1115 +/– 928	478 +/– 262	2.17	10.3	60.8 +/– 46.6	23 +/– 12.6
Serpentinite, dry,	707 +/– 170	442 +/– 75	1.61	10.6	57.8 +/– 17.0	21.2 +/– 3.6
Serpentinite, water ingress	633 +/– 87	424 +/– 73	1.53	10.3	59.0 +/– 30	20.4 +/– 3.5
Mean	801 +/– 397	415 +/– 148	1.9	11.3	58.4 +/– 25	19.9 +/– 7.1

3.2 TBM Performance in different ground conditions

Geotechnical parameters effecting TBM performance are given in Table 4. For each penetration (in mm/rev) of the cutters, thrust and torque values of TBM are different, so normalized values of thrust and torque are used in (kN/(mm/rev)) and in (kNm/(mm/rev)) for comparative reasons in different ground conditions. These normalized values are called thrust index and torque index. Specific energy is the energy spent to excavate a unit volume of rock and used to compare the excavation efficiency, for the prediction of instantaneous advance rates of mechanical excavators and for the classification of the excavability of different rocks (Rostami et al. 1994, Copur et al. 2001).

The important points emerging from Table 4 are as follow. In dry conditions, the mean penetration rate of TBM in peridotite is 10% higher than serpentinite. However, the existence of water ingress this difference increases up to 20%. In highly fractured peridotite-serpentinite, water ingress causes the clogging of disc cutters and excessive wear due to the clogging as seen in Figure 2. Consequently, the torque index increase up to 20% compared to dry condition. Specific energy changes between 18 and 23 kWh/m^3, which is almost double, compared to brittle rocks (Bilgin et al. 2014). The highest thrust index is obtained in highly fractured peridotites-serpentinites in the existence of highly water ingress to the tunnel. This value is almost 40% higher than of dry condition.

Breakdowns in tunneling operation obtained in different ground conditions are given in Table 5. The highest mechanical breakdown of 28.8% and time spent for cleaning the tunnel of 39.3% are obtained in highly fractured Peridotites-Serpentinites where the water ingress is highly. This might due to that; this rock formation becomes muddy in the existence of water. Peridotite is highly abrasive compared to serpentine, and breakdown for changing the disc cutters is higher in dry condition compared to the rocks and ground condition.

3.3 Comparison of TBM performance data with those obtained in different tunneling projects

The comparative performance values are given in Table 6. The ratio of compressive strength is one of the most important factor related to the brittleness of the rocks and it is reported by different authors that the performance of TBMs and drilling rigs are closely related to this

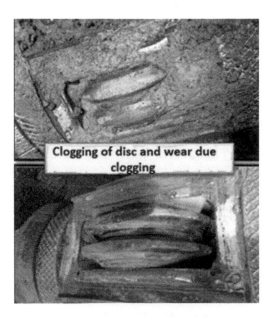

Figure 2.   Clogging of discs and excessive cutter wear.

Table 5. Breakdowns in different ground conditions.

Ground conditions	Mechanical breakdown (%)	Time spent for changing the cutters, (%)	Time spent for cleaning the tunnel, (%)
Peridotite, dry	26.7	26.8	24.1
Peridotite, water ingress	0.00	0.00	0.00
Peridotite highly fractured	26.3	25.8	25.1
Peridotite-Serpentinite dry	6.3	15.3	1.3
Peridotites -Serpentinites, highly fractured and water ingress	28.8	21.5	39.3
Serpentinite, dry	11.9	4.9	10.2
Serpentinite with water ingress	0.00	5.7	0.00
Sum	100.00	100.00	100.00

Table 6. The comparison of TBM performance data from Belpinar Tunnel with those obtained in different tunneling projects.

Project Formation	6c/6t	TBM Diameter m	Thrust Index kN/(mm/rev) +/− s.d	Torque Index kNm/(mm/rev) +/− s.d	FT/T	Mean Penetration, (mm/rev)	SE, (kWh/ m³)
Belpınar Ayazaga Cayirbasi Trakya	58/14.5 = 4	6.82	801 +/−397	415+/−148	1.9	11.3	19.9
	100/9 = 11.1	3.12	2267.1 +/− 436	222 +/− 42.7	10.2	5.2	9.4
Otogar Bagcilar Fossilated Lim	31.9/3.5=9.1	6.52	990.3	157.9	6.3	13	8.3
Melen Kartal Limestone	85/10.2 = 8.3	6.15	840 +/−112	164.1 +/−21	5.1	9.3	10.4
Ido Kirazli	46/4.8 = 9.5	6.15	12009 +/− 804	2258 +/− 217	5.3	14	9.5
Kartal Kadikoy	105/8.8 = 11.8	6.50	5485 +/− 435.3	1714 +/− 136	3.3	12.6	7.8
Mahmutbey Trakya 1	120/15 = 8	6.15	545.2	147.5	3.7	18.9	11.4
Mahmutbey Hard Clay	45/3.5 = 12	6.15	427+/−47	132.3 ı/15	3.2	16.5	7.4
Nurdagi	213/22.4 = 9.5	7.8	12500 +/− 1666.7	1700 +− 226.6	7.4	7.5	9.5

ratio (Kahraman 2002, Altindag 2010). In brittle rocks this ratio is higher than 10. As seen from table this ratio is at least 2 times less than any ratio obtained for eight tunneling projects in Turkey where EPB TBMs are used except Nurdagi Railway Project. Figure 3 gives the relation between specific energy values and compressive strength ratios obtained in eight different tunneling projects in Turkey. As seen from this Figure the relationships between dependent variables are highly significant with a high correlation coefficient of 0.94.

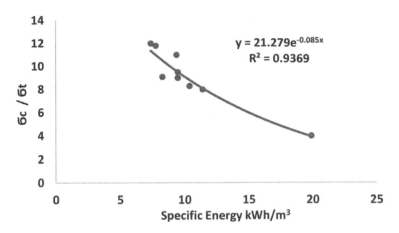

Figure 3. The relation between specific energy values and compressive strength ratios obtained in different tunneling projects in Turkey.

After Figure 3 and the experiences gained from the past projects, as a first approach the rock formations may be classified according to their excavability based on the rock brittleness as given below.

If the brittleness of the rock is less than five, the rock is very hard to excavate. If the brittleness of the rock is between five and eight the rock is hard to excavate. If the brittleness is between eight and 12 the rock is easy to excavate and if it is higher than 12 the rock is very easy to excavate.

4 CONCLUSIONS

Belpinar tunnel was excavated in peridotites, in peridotites-serpentinites and serpentinites. These rock formations may be very problematic in most cases in tunneling (Marinos et al. 2006, Clark & Chorley 2014, Glawe & Upreti 2014). The main conclusions may be summarized as follows.

With major stoppages a mean daily advance rate of 8.04 m and mean monthly advance rate of 241.23 m were obtained, these values went up to 9.93 m and 297 m if the major stoppages of 128 days were excluded in difficult ground conditions.

In dry conditions, the mean penetration rate of TBM in peridotite is 10% higher than serpentinite. However, the existence of water ingress this difference increases up to 20%. In highly fractured peridotite-serpentinite, water ingress causes the clogging of disc cutters and excessive wear due to the clogging. Consequently, the torque index increase up up to 20% compared to dry condition. Specific energy changes between 18 and 23 kWh/m^3, which is almost double, compared to brittle rocks (Bilgin et al., 2014). The highest thrust index is obtained in highly fractured peridotites-serpentinites in the existence of high water ingress to the tunnel. This value is almost 40% higher than of dry condition.

The ratio between compressive strength and tensile strength values is a good indicator of rock brittleness and the excavability of the rock. In brittle rock is usually higher than 10. In Belpinar Tunnel, this ratio for peridotites and serpentinites is around 4, showing the reason of difficulties in excavation.

It is believed that data provided within this paper will provide a more accurate planning and scheduling of the project in similar ground conditions.

REFERENCES

Altındag, R. 2010. Assessment of some brittleness indexes in rock drilling efficiency. *Rock Mech Rock Eng* 43: 361–370.

Bilgin, N., Demircin, M.A., Copur, H., Balci, C., Tuncdemir, H. & Akcin, N. 2006. Dominant rock properties affecting the performance of conical picks and the comparison of some experimental and theoretical results. *Int J Rock Mech Min Sci* 43 (1): 139–156.

Bilgin, N., Copur, H. & Balci, C. 2014. *Mechanical Excavation in Mining and Civil Industries*, CRC Press-Taylor and Francis Group, London.

Clark, J. & Chorley, S. 2014. The Greatest challenges in TBM tunneling: experiences from the field. In: Proceedings of North American Tunnelling Conference, pp. 101–108.

Copur, H., Tuncdemir, H., Bilgin, N. & Dincer, T. 2001. Specific energy as a criterion for the use of rapid excavation systems in Turkish Mines. *The Institution of Mining and Metallurgy, Transactions Section-A Mining Technology*, Sept.-Dec., (110): A149–157.

Coumantakis, J. 1982. Behaviour of peridotites and serpentinites of Greece rom the point of view of engineering geology and their mechanical and physical properties. *Bulletin of the International Association of Engineering Geology* No: 25, pp. 53–60.

DSI (State Water Authority XX Region). 2017. Geological and geotechnical report for Kilavuzlu Irrigation Project Belpinar Tunnel.

Glawe, U. & Upreti, B.N. 2014. Better understanding the strengths of serpentinite bimrock and homogeneous serpentinite, *Felsbau* 22 (5): 53–60.

Kahraman, S. 2002. Correlation of TBM and drilling machine performances with rock brittleness. *Eng Geol* 65(4): 269–283.

Marinos, P., Hoek, E. & Marinos, V. 2006. Variability of the engineering properties of rock masses quantified by the geological strength index: the case of ophiolites with special emphasis on tunneling, *Bull Eng Geol Env* 65: 129–142.

Rostami, J., Ozdemir, L. & Neil, D.M. 1994. Performance prediction: A key issue in mechanical hard rock mining. *Mining Engineering*. 11: 1263–1267.

Sakalli, M. 2017. *Geological and geotechnical factors effecting the performance of a TBM in Belpinar Tunnel*. MSc Thesis, Istanbul Technical University, p.68.

Tunnels and Underground Cities: Engineering and Innovation meet Archaeology,
Architecture and Art, Volume 7: Long and deep tunnels – Peila, Viggiani & Celestino (Eds)
© 2020 Taylor & Francis Group, London, ISBN 978-0-367-46872-9

A new underground laboratory for exploring the deep universe: The design of a third generation of a gravitational waves observatory

L. Schiavinato & P. Mazzalai
SWS Engineering S.p.A., Trento, Italy

G. Gemme[a], G. Losurdo[b] & M. Punturo[c]
INFN, Istituto Nazionale di Fisica Nucleare, ([a]Genova, [b]Pisa, [c]Perugia), Italy

A. Paoli
EGO European Gravitational Observatory, Cascina (PI), Italy

F. Ricci
INFN and Università di Roma "La Sapienza", Rome, Italy

E. Calloni
INFN and Università degli Studi di Napoli "Federico II", Naples, Italy

G. Oggiano
Università degli Studi di Sassari, Sassari, Italy

M. Carpinelli
INFN and Università degli Studi di Sassari, Sassari, Italy

ABSTRACT: The gravitational waves are space-time ripples caused by very violent phenomena, such as black holes' collisions, supernovae's explosions or the relic signal of the Big Bang that gave rise to the universe. In 2015, LIGO and the Virgo projects announced in parallel in Washington and in Cascina at the EGO site (European Gravitational Observatory) the first direct observation of a gravitational wave signal, announcing the birth of a new way to explore the Universe. At present, the realization of a third generation Gravitational Wave observatory (Einstein Telescope ET) is being pursued. After the Conceptual Design Study (CD) released by the original partners, a new phase is now open, with submission of the site proposal for a ET detector hosted in a large underground infrastructure. The feasibility study for the candidate site of Lula (Sardinia) was carried out through a CD's scientific revision and a deep and strong interaction with the geomechanical needs.

1 INTRODUCTION

1.1 *The Gravitational Waves*

The existence of gravitational waves (GWs) was first predicted by Albert Einstein in 1916. According to Einstein's theory of general relativity, gravity results from how mass warps the fabric of space and time. Under certain conditions, the accelerated motion of large and compact masses generates GWs that travel at the speed of light, stretching and squeezing space-time along the way.

The detectable effects of GWs are extraordinarily weak, and a century of studies, model developments, technological progresses and large infrastructures construction has been necessary to arrive to the direct detection of a GW signal. On 14 September 2015, the first observation of gravitational waves was made and it was announced by the LIGO and Virgo collaborations on 11 February 2016. Previously GWs had only been inferred indirectly, via the radiotelescope observations of the orbital periods of pulsars in binary star systems.

Measuring the splitted laser that travels each of the 4 kilometres interferometer arms, infinitesimal variations (of the order of 10^{-19} m) were detected, allowing to reconstruct for the first time a cosmic catastrophe: two black holes approached each other, rotating around the center of mass of that binary system in a spiral in which their speed has reached 150.000 kilometres per second, equal to about one half that of the light speed: there was a frightening energy release, as if three suns were gone in a flash of gravitational energy. This work earned three scientists the 2017 Nobel Prize in physics in October 2017. After that, several new detentions were made by LIGO and VIRGO.

1.2 Scientific impact of GW sciences

The detection of GW has been a huge scientific achievement, result of a century of efforts, but actually it is the beginning of a new era in the observation of the Universe. A broad community is relying on detection of gravitational waves for the following studies:

– Fundamental physics: access to dynamic strong field regime, new tests of General Relativity; Black hole science, etc.;
– Astrophysics: first observation for binary neutron star merger, relation to sGRB, Evidence for a kilonova, explanation for creation of elements heavier than iron;
– Astronomy: start of gravitational wave astronomy, population studies, formation of progenitors;
– Cosmology: Dark Matter and Dark Energy;
– Nuclear physics: tidal interactions between neutron stars get imprinted on gravitational waves; Access to equation of state.

1.3 The Virgo Observatory

Virgo is an interferometric gravitational-wave antenna located in Cascina, near Pisa, Italy. Virgo is a European collaboration with about 300 members, with the participation of scientists from France, Italy, The Netherlands, Poland, Hungary, Spain and Germany.

The structure consists of two 3-kilometre-long arms, which house the various machinery required to form a laser interferometer. A beam-splitter divides a laser beam into two equal components, which are subsequently sent into the two interferometer arms. In each arm, a two-mirror Fabry-Perot resonant cavity extends the optical length from 3 kilometres to approximately 100 km, because of multiple reflections that occur within each arms and the consequently amplification of the distance variation caused by a gravitational wave. The instrumentation is able to detect difference in the order of 10^{-18} m, about 1000 times less than the diameter of a proton.

Figure 1. The Virgo site. Plant view.

1.4 *The 3rd generation GW Observatory*

In order to substantially improve the sensitivity of the current detectors, new infrastructures, allowing the observation of GW sources at cosmological distance, are needed. A new global network of 3rd generation GW observatories is expected to be operative in the 2030+ decade. Einstein Telescope is the pioneer of this new network and the design and realisation of its infrastructure is one of the most important challenges in this research field.

2 THE ET CONCEPTUAL DESIGN

2.1 *Introduction*

The conceptual design of ET was realised with the support of the European Community's Seventh Framework Programme (FP7/2007–2013) under grant agreement n 211743 (2008–2011) and the Conceptual Design Report (CDR) has been delivered in the document coded "ET-0106A-10" publicly available in the ET site http://www.et-gw.eu/. This document describes the scientific case, the fundamental design options, the site requirements, the main technological solutions, a rough evaluation of the costs and a schematic time plan. These evaluations are still the reference framework for the ET project, but a deeper and updated study is now necessary to arrive to the production of the ET Technical Design Report (TDR) in the next years. This study presents a preliminary evaluation of the possible options that are currently under evaluation.

2.2 *The site selection*

The gravitational wave detectors are large and complex plants, and the selection of the site location is an issue of great importance. The selected site should allow the highest possible level of scientific productivity in connection with a reasonable cost of construction and operation and at minimum risk. The requirements for the definition of the site location are specified in order to obtain a certain detector sensitivity, considering specific models that translate the environmental disturbances into sensitivity limits. These requirements are:

– low seismic motion, which is addressed according to source frequency, including ambient seismic background, microseismicity, meteorologically generated seismic noise, and cultural seismicity from anthropogenic activity (local infrastructure, population density, etc.);
– low Newtonian noise originates from fluctuations in the surround geologic and atmospheric density, causing a variation in the Newtonian gravitational field.

It must be noted that most of the anthropogenic generated seismic noise propagates along the earth's surface and diminish with depth. Currently operating GW detectors have spent much effort developing seismic filter chains to support the main optics of the interferometer, preventing seismic induced vibrations to pass through and affect the detector sensitivity.

The site selection of a large research infrastructure like ET is always a difficult task, as the problems are not only of a technical nature. Currently, The ET "monitored" sites are:

– G-B-N Site, in the Meuse-Rhine region between The Netherlands, Belgium and Germany;
– Mátra Mountain Site, in the northern Hungary;
– Lula Site, in Sardinia, Italy.

The aim for 2019 is to submit ET as project to be included in the ESFRI (European Strategy Forum on Research Infrastructures) roadmap listing the potential sites and describing the procedure for the selection. A final decision about the location is expected to be done in 2021–2022.

2.3 *The underground infrastructures*

According to the CDR the underground infrastructure is composed by the following parts:

- three corner stations connected by 10 km long tunnels;
- each corner station has 2 caverns with a diameter of 30 m and a height of 30 m and a main cavern with a diameter of about 65 m and a height of 30 m;
- the connection tunnels contain the interferometer arms and have an inner diameter of 5.5 m.

The CDR identifies the reference dimensions for the underground works planned, which for the corner stations and the shaft appear significantly large. It is worth to note that the CDR describes some examples of excavation of caverns, but significantly smaller than the one expected for the new interferometer. The required dimensions, approximately 65m of diameter and 30m of height, are comparable to the Gjøvik Mountain Sports Hall, characterized by transversal dimensions 61m x 24m (Barton et al., 1992). Its excavation was possible only thanks to the concomitance of numerous favourable and maybe unique factors.

3 FROM THE CONCEPTUAL TO THE TECHNICAL DESIGN: IMPLEMENTATION EVALUATIONS

3.1 *Review of technological systems and related optimization*

Some of the key points of the CDR on which the present analysis is focused are presented following:

- increase and optimize the space inside the tunnel in relation to the planned equipment;
- assure the stability to the towers through a solid and enough large basement, avoiding conflicts with the pipes;
- critical shape and dimensions for the main cavern at vertex.

In particular we studied the case of the detector construction in the Sardinia site.

3.2 *The Lula (Sardinia, Italy) site*

The idea of having ET in Sardinia starts from the peculiar characteristics of the area:

- seismically quiet;
- not urbanized/industrialized (one of the least populated areas in Europe);
- absence of subsidence effects;
- long mining history in the area, but that do not interfere with the new planned works.

From a geological point of view, the area is affected by the following rocks:

- orthogneiss and granodiorite, with good intact rock parameters (UCS = 61 ÷ 92 MPa), suitable for the construction of large excavations;
- micaschist, paragneiss and quartzites, which present a great variability of behavior (UCS = 9 ÷ 68 MPa), inside which the Lula SOS Enattos mine is excavated.

A specific insitu and laboratory tests campaign has not been carried out (it is expected in the following design phases), and the current design data are derived from the mining site tests and the available data from the University of Sassari.

Figure 2. Graphic visualization (credits ET Design team) of the key elements of the CDR evaluated in this work.

3.2.1 *The Design Review*

A deep revision of the CD design was carried out, in order to progress in the key aspects mentioned above and, at the same time, take into account the points related to geomechanics and assess the feasibility of the underground works. This led to a possible revision of the layout of the internal systems, maintaining the triangular scheme of the CDR. The geomechanical aspect has been studied in more detail, through the localization of the vertex in the Lula site within the rock formations characterized by greater resistance.

At the same time, a study of a possible alternative solution was carried out, considering that the ideal perspective is to have a 3G network and a global governance of a 3G effort. The study was focused on an "L" scheme that offers some constructive advantages, some cost optimisation but important potential scientific drawbacks.

4 UNDERGROUND STRUCTURE

4.1 *Design of underground structure*

A feasibility study of the underground structure has been performed in order to evaluate the stability of the excavations and to have a first estimation of the necessary stabilization measures.

4.1.1 *Vertex caverns*

Particular attention was paid to the design aspects related to the caverns, which represent the most critical elements in the project works. As already seen above, the interaction between the various design components led to the definition of two layouts (triangular and "L") characterized by a smaller size compared to those provided in the CDR. However the dimension of the works is still relevant and interests different rock formations such as the "Granodiorite of Bitti" and the "Orthogneiss of Lodè".

Based on the current knowledge, the geotechnical model is assumed as an "equivalent continuum", starting from the information related to the characteristics of the intact rock, the parameters of rock mass have been defined through the criterion of Hoek and Brown for the rock masses, on the basis of values of the parameter m_i equal to 30–28 and a quality of the rock mass equal to RMR = $50 \div 60$/GSI = $45 \div 55$.

As previously mentioned, the evaluation of the excavation behaviour involves the assessment of the stress state around the excavation areas. Figure 5 and Figure 6 show the geological profiles for the planned works according to the geometrical configurations defined. The reference condition for the stress state is defined according to the overburden along the alignment. The following preliminary assessment have been carried out:

- Panet stability number N_s, given by the ratio between the theoretical circumferential tension at the contour of a circular excavation, and the resistance of the rock mass;
- Ground Reaction Curve (according an isotropic stress state and circular shape excavation by the equivalent radius of excavation). These assessments are necessary for a first definition of the excavation behaviour in intrinsic conditions.

After preliminary evaluations indicated that the condition of the rock masses was relatively favourable to the excavation conditions, the excavation behaviour of the real transversal shape and size was carried out through bi-dimensional numerical methods. The initial vertical

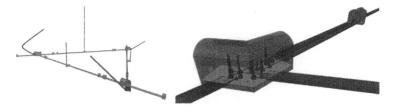

Figure 3.   Geometric scheme and vertex detail according to the Triangular scheme (Heights are not to scale).

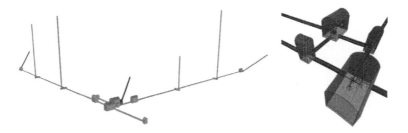

Figure 4. Geometric scheme and vertex detail according to the "L" scheme (Heights are not to scale).

Table 1. Large excavation basic assessments. Panet number.

Lithology	Granodiorite		Orthogneiss	
Overburden	480m	200m	480m	200m
RMR = 60	Ns = 1.76	Ns = 1.07	Ns = 2.11	Ns = 1.23
RMR = 50	Ns = 2.31	Ns = 1.38	Ns = 2.64	Ns = 1.58

stress has been set equal to the gravity load and the horizontal component variable between 0.8 and 1.2 times the vertical stress. In order to consider the excavation method (drill & blast) a layer of 2m is taken in account with residual parameters to model the blasting effects on the excavation boundary (Cai, 2007) assuming a disturbance parameter D=0.70. The results of the evaluations performed are shown in Table 2. As can be seen, the analyses confirm that what is proposed by the CD design is not stable. Instead, the excavations according to the revised layout are stable, thanks to the reduction of their size.

To complete the evaluation of the behaviour presented above, some three-dimensional models were made in order to validate the beneficial effect of the shape modification also in the third dimension and to assess the interference between the different excavations.

The 3D simulations allowed to investigate in detail the behaviour of the excavation in the connection point between the different tunnels and caverns. The same assumption given for the previously presented two-dimensional models have been considered also in this case, in particular regarding the simulation of the presence of disturbed material around the excavation boundary, characterized by a reduction of the mechanical properties due to the use of the D&B excavation method.

The obtained results are in general consistent with the ones obtained with the 2D analyses presented above, while it has been possible to gain addition information for the connection areas and for the case of the triangular scheme.

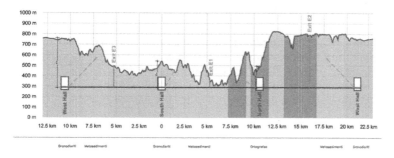

Figure 5. Schematic works profile for triangular scheme (the colors correspond to the lithology lower described).

4230

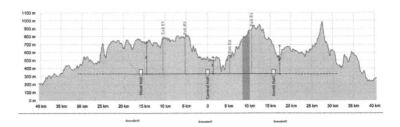

Figure 6.   Schematic works profile for L scheme (the colors correspond to the lithology lower described).

Table 2.   Numerical 2D simulation. Comparison results for the West hall for RMR = 60.

Parameter	CD scheme	Triangular scheme	"L" scheme
Numerical convergence	Not reached	Reached	Reached
Maximum displacements	70 cm (crown)	6 cm (walls)	2.5 cm (walls)
Plastic zones thickness- crown	> 50 m	4 m	2 m
Plastic zones thickness- walls	9 m	8 m	4 m

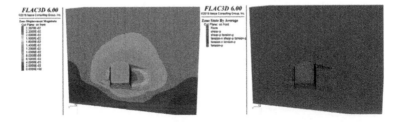

Figure 7.   3D numerical models generated for the vertex: Triangular scheme (left), "L" scheme (right).

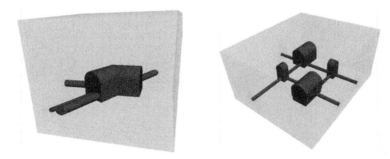

Figure 8.   Example of results obtained for the Triangular scheme: displacements (left), plastic zones (right).

The calculations proved the feasibility of the excavation of the works in both the planned geometric configurations. Subsequently, according to literature indications, the support interventions were defined, aimed to limit the deterioration of the rock mass properties due to the reduction of confinement and to prevent the mobilization of kinematically unstable blocks. This mainly to obtain a reasonable definition of the excavation costs.

4.1.2 Connection tunnels

The connection tunnels will be excavate in different rock mass types. Based on the available information, the excavation is expected to be mechanized using a rock TBM. This choice has been made after an overall assessment, considering the criticality of the mechanized excavation compared to the D&B method (for example used into the 3km underground interferometer

Figure 9.   Example of results obtained for the "L" scheme: displacements (left), plastic zones (right).

Table 3.   Connection tunnels: segmental liners properties.

Parameter	Triangular scheme	"L" scheme
External Diameter	11.0 m	6.3 m
Thickness	40 cm	30 cm
Segments length	1.75 m	1.50 m

with cryogenic sapphire test masses "KAGRA project" in Japan). The productivity of the mechanized excavation is much higher compared to an excavation with the traditional method, but the latter is more flexible and can take advantage of intermediate access adits.

In order to reduce the involved risks a shield TBM is expected to be used (Single shield TBM or Double shield TBM). If the geological survey would show the presence of areas with poor (soil-like) mechanical properties, a Dual Mode TBM or MixShield could be used so in order to safely cross these weakened areas. Modern TBMs, in any case, are equipped for the advance recognition of the variability of the geology through sonic or seismic systems, and for the execution of rock mass improvements from the head in progress.

Further critical aspects are linked to the logistic aspect, where long lengths determine very large times for mucking transportation, lining supply, material transfer, etc., with a consequent increase in the risks associated with possible inconveniences: logistics will be optimized using intermediate accesses in order to reduce supply distances, or to distribute services on two lines.

The advantages of the proposed solution are extended to the experimental phase since the mechanized excavation is characterized by the installation of a high quality lining and a limited production of dusts. This last aspect is particularly important for the foreseen measurements.

4.1.3 Access tunnels
The excavation size of the access tunnels must allow the passage of the TBM components and the technical installations parts:

– for the TBM, the larger non-removable part is represented by the TBM main bearing, with a size equal to 0.7 * Diameter Cutterhead TBM;
– for the technical/experimental parts, the largest dimensions are relative to the HF (4 x 4 x 3.0 m) and LF foundation block (5 x 5 x 3.5 m).

For the triangular scheme the size of the access tunnels is ruled by the TBM dimension, and the excavation size for the tunnel is around 35% greater than the one expected for the "L"

Table 4.   Reference dimension for access tunnels.

Part	"L" scheme	Triangular scheme
TBM Main Bearing	4.62	7.98
Excavation height	10.3 m	9.60 m
Excavation width	15.3 m	12.0 m
Excavation Area	136 m^2	100 m^2

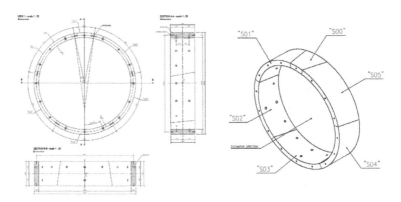

Figure 10. Connection tunnels. Example of tunnel liner formworks for "L" scheme.

Table 5. Excavation volume estimation. Summary table (in Millions of cubic meters).

Part	ET book (p.313) (no excavation factor)	Triangular scheme (excavation factor=1.3)	"L" scheme (excavation factor=1.3)
Tunnels	1.081	4.458	1.482
Caverns	0.426	0.831	0.654
Accesses	0.236	1.786	1.057
Total	1.742	7.075	3.193

scheme solution. In this case, the size of the access tunnels is ruled by the dimension of the technical/experimental parts.

4.2 Excavation muck and environmental issues

The excavated rock could be employed in the recovery of the nearby quarry sites. The quarried surfaced in the Buddusò District (granite extraction) covers ~2 mm^2. The muck produced by the excavations could be easily used for landscape rehabilitation.

4.3 Final evaluations

In conclusion, a comparison in term of excavation amount of the three solutions (basic, triangular scheme and "L" scheme) is presented.

The volumes associated with the planned excavations are shown below. The comparison is carried out adopting an excavation volume increase factor of 1.30. It has to be noted that this condition was not considered in the CD.

5 FUTURE DEVELOPMENTS

Future developments include an investigation campaign focused on defining the detailed characteristics of the rock mass in the area, through boreholes and specific geophysical surveys. In a systematic way, geomechanical structural survey will be carried out on outcrops in the external areas, in order to be able to develop in detail also the tunnel portal areas. At the same time, specific tests will be carried out also in relation to the excavation and the abrasivity of the rock formations.

Through these investigations, the conceptual model of the rock mass will be refined, in order to evaluate also the discontinuous geomechanical approach model, and define a detailed design based on the analysis of potential risks in the area and the aspects related to maintenance.

6 CONCLUSIONS

This paper presents the requirements and the preliminary study of the underground infrastructure for a 3rd generation detector of gravitational waves.

About 100 years after the theoretical formulation of Einstein, the existence of gravitational waves has been proven. Their study can provide important elements for understanding how the universe works. The current challenge is to create new infrastructures able to allow measurements with detail, quality and precision superior to those currently possible, aiming to a continuous improvement of the technological and scientific progress. To this end, the scientific community has come together to define the guidelines for implementing the third generation of GW observatory in the subsoil, with the advantages of low seismic noise, which not only improves the sensitivity in the observation band but also increases the stability of the detector so that the requirements on the control system are eased, and low gravity-gradient noise, which is one of the largest obstacles in improving the low-frequency sensitivity of next-generation detectors.

The proposed Italian site of Lula (Sardinia) has optimal characteristics to accommodate these instruments, being characterized by a good geological stability, a limited seismic disturbance, absence of subsidence phenomena and located in a less anthropized area, this site meets all the requirements for hosting the new detector.

A preliminary study connected to the feasibility of this infrastructure was carried out going beyond the CD's initial evaluations, through a deep investigation of the geo-mechanical needs. Beside the reference triangular geometry, described in the CD, the technical feasibility study of underground works as been addressed also to a possible alternative L-shaped geometry.

The conceptual design drawn up by the scientific community was developed through the technical and economical evaluation of the aspects related to the measuring instruments and from the point of view of the realization of the underground works. This has been an iterative process, which has involved the technical and plant engineering aspects in relation to the real possibility to realize the works, and that led to the definition of an original solution, which on the Lula site finds an optimization based on the possibility of realization of vertex caverns in favourable geomechanical conditions. The complexity of the work required a series of detailed evaluations from the geomechanical point of view, which were carried out through a process of subsequent studies, up to the development of three-dimensional numerical analyses in order to evaluate the effects of the excavation and the mutual interferences between works. Despite the preliminary planning phase, these tools allowed us to provide important elements in the definition of the functional aspects of the work and in the overall layout of the proposed solutions.

REFERENCES

European Commission FP7, 2011. ET-0106A-10. Einstein gravitational wave Telescope conceptual design study.

Losurdo, G., Paoli, A., Paoli, L., Oggiano, G., Cuccuru, S., Carpinelli, M., Mazzalai, P., Schiavinato, L., Cabriolu M., Loddo L., 2018. Some considerations on the et infrastructure the case for a Sardinian site. 9th Einstein Telescope Symposium, Cascina (Italy)

Barton, N., By, T.L., Chryssanthakis, P., Tunbridge, L., Kristiansen, J., Løset, F., Bhasin, R.K., Westerdahl, H. & Vik, G. 1994. Predicted and measured performance of the 62m span Norwegian Olympic Ice Hockey Cavern at Gjøvik. Int. J. Rock Mech, Min. Sci. & Geomech. Abstr. 31:6: 617–641. Pergamon.

Grandori, R. 2017. Vantaggi e limiti dello scavo meccanizzato: quando con la macchina e quando senza. Gallerie e grandi opere sotterranee (124) 39–44.

Cai, M., Kaiser, P. K., Tasaka, Y., Minami, M. 2007. Determination of Residual Strength Parameters of Jointed Rock Masses Using the GSI System. International Journal of Rock Mechanics and Mining Sciences 44(2):247–265

Hoek, E., Carranza-Torres, C., Corkum, B. 2002 Hoek-Brown failure criterion –2002 Edition. In R. Hammah, W. Bawden, J. Curran, and M. Telesnicki (Eds.), Proceedings of NARMS-TAC 2002, Mining Innovation and Technology. Toronto – 10 July 2002, pp. 267–273. University of Toronto

Tunnels and Underground Cities: Engineering and Innovation meet Archaeology,
Architecture and Art, Volume 7: Long and deep tunnels – Peila, Viggiani & Celestino (Eds)
© 2020 Taylor & Francis Group, London, ISBN 978-0-367-46872-9

Cost analysis of mine roadways driven by drilling and blasting method and a roadheader

O. Su
Department of Mining and Mineral Extraction, Bülent Ecevit University, Zonguldak, Turkey

M. Akkaş
Hattat Energy and Mining Co., Bartin, Turkey

ABSTRACT: To ensure mining activities safely and save production costs, it is necessary to select optimum excavation method. Both the drilling and blasting and mechanized excavation methods are commonly applied in mining and civil industries. Feasibility analyses lead to determine appropriate method selection. In this context, initial investments and running costs also play an important role. The main objective of this paper is to analyze the running cost of driving roadways by two methods. Cost analyses of drill & blast method and excavation of roadheader were evaluated based on the data obtained from a mine colliery in Turkey. The results show that the running costs per one meter of progress in drilling and blasting is 47% more expensive than that of roadheader excavation cost. In order to reduce the costs more, daily and monthly advance rates of roadheader should be increased. In addition, it is suggested that a reliable estimation of the running costs should be conducted before starting development works in a roadway and buying a machine.

1   INTRODUCTION

Drilling and blasting (D&B) has been applied as a conventional method since the beginning of 1900s for excavating the short distances of tunnels. However, it can be applied for driving long distances of mine roadways if the conditions are not suitable for mechanical excavation. D&B is suitable for most cases due to its flexibility, low investment cost and not requiring high technology (Jafari et al. 2011). Various types of tunnels, mine drifts and roadways can be advanced by this method. It generally becomes more economic in hard and abrasive rock where high percentage of quartz content is included. In conventional method, holes are drilled by using handheld or jumbo drills even though their penetration rates are quite low.

On the other hand, mechanized excavation machines such as roadheaders are capable of cutting soft and very hard rock. High advance rates can be achieved when optimum cutting conditions are provided. The roadheaders are usually driven in mine drifts, roadways and tunnels. They are able to excavate the face selectively and classified according to their power and weight produced in low, medium, and heavy duty. The geological and rock mass properties lead to select the proper type of roadheader. However, larger cross sectional is necessary to install the machine. Abrasive rocks increase the bit consumption and this method can sometimes be uneconomical. In addition, machine utilization is a significant factor that should be taken into account to monitor the machine performance. The advance rates, which depend upon the machine performance, can be obtained by Equation 1. The machine utilization is generally between 30% and 35% for the best cutting conditions.

$$AR = PR \ U \qquad (1)$$

where AR = advance rate; PR = penetration rate; and U = machine utilization factor.

The advance rate of a roadheader comparing to the D&B method may be faster, but its applicability is limited from soft to medium-strength rocks as well as abrasive mineral content of rocks. Excavation time, costs, rock support, and risk evaluation would be guiding to choose most appropriate method with certain ground conditions and performance specifications (Zare et al. 2016).

Acaroglu & Ergin (2006) pointed out that mechanized systems are more advantageous than conventional methods. Brino et al. (2013) compared technic and economic aspects of excavation by D&B method and by roadheader in gypsum quarries. Ocak & Bilgin (2010) highlighted that that drill and blast method is very efficient in high strength rocks and roadheader can be more productive than impact hammers in terms of production rate and machine utilization time. Zare et al. (2016) evaluated drill and blast method versus TBM tunnelling in terms of advance rate and excavation costs.

In this study, cost analyses of drill & blast and mechanical excavation methods were compared based on the data obtained from a mine roadway in Amasra colliery in Turkey. The roadway was initially developed by D&B method in which the blastholes were drilled by a jumbo drill using button bits. Then, a transverse type roadheader was bought by Hattat Co. and it was advanced at another roadway. Based on the monthly advance rates, an estimation of the total costs including initial investments was calculated in light of operating, equipment, and labor costs.

2 COST ANALYSES

Total energy (power), rate of penetration, bit wear and the total costs are considered when evaluating the performance of any excavation system or any drilling rig. Energy consumption and bit wear should be in minimum level in order to provide lower drilling costs. Penetration and advance rate are also the main concerns in the excavation process.

Hema company in Amasra, Turkey has opened a number of roadways in the scope of underground development headings. The roadways started to be developed by drilling and blasting method in claystone formations in the cross-section area of 28 m^2. For this purpose, the company rented a rotary-percussive type jumbo drill and drilled the blastholes in the diameter of 42 mm. Later, the company purchased a transverse type roadheader, 70 tons in weight, and started to develop another roadway in the same area. This study aims to analyze the running costs of two methods applied in the past and at the present. However, it should be noticed that prices and cost analyses given below would change according to the geological conditions.

2.1 Drilling and blasting costs

Rock excavation by drill and blast method consists of several stages. These are drilling, blasting, loading of muck removal, hauling, and supporting. Within these stages, drilling is the most critical since it affects the total cost of advances per meter. Because, the number of bits consumed by rock drill and its oil can be changed depending on the geological and environmental conditions. However, the number of explosive does not change frequently in blasting unless the cross-section area of the face and the charge ratio of the holes are changed. Muck is also removed by using a loader. The amount of the muck is essential at this stage. The factors affecting the rock drill costs can be listed as follows and they can vary with the operational conditions of the tunnels and mine collieries (Table 1). (Kantarci et al. 2013).

A successful drilling process can be performed by selecting the optimum bit according to the rock mass properties and also optimizing the drilling parameters such as thrust, rotation and weight on the bit. Thus, an optimum penetration and advance rate can be obtained. Besides, operator experience is quite important since he controls the drill rig and monitors the performance of the rock drill.

In the scope of this study, the performance of HL510 T jumbo drill operated in underground roadway of Amasra colliery was monitored. The impact frequency and the power (P) of the rotary-percussive type drill were 42 Hz and 20 kW, respectively. During the drilling process in the field, the average penetration rate of 2.85 m/min was measured. Other drilling parameters are listed in Table 2.

Table 1. The basic parameters related to the rock drill (Kantarci et al. 2013).

Rock drill supplies	Range
Oil consumption (lt/h)	15–20
The life of extension rod (m)	4000–6000
The life of shank adaptor (hour)	8000–10000
The life of a rock drill (hour)	400–600
Drilling performance (m/h)	150–200

Table 2. Drilling parameters applied in the mine roadway.

Drilling parameters	Value
Length of hole (m)	2.5
Average drilling time (sec)	54
Hole diameter (m)	0.042
Cross-section area of the face (m²)	30
Total number of holes	120
Total drilled hole length per blasting (m)	300
Number of blasting in a day	2

The capital costs in drilling and blasting method mainly include the price of drilling machine, the price of the power (compressor or electricity generator), and also other equipment such as crusher, pumps, etc. On the other hand, the running costs are drilling equipment, maintenance and repair of drill rig, cost of compressed air, electricity, water, salaries of operators, and operating costs such as preparing the drilling site, laying the pipe lines and cables etc. (Tamrock 1978). Based on these parameters, the running costs of a jumbo drill can be summarized as in Table 3. Calculations are performed in light of renting a rock drill. If the company would buy the rock drill, then the amortization period should be taken into consideration.

It is also possible to calculate the running costs of excavation per meter. If the total distance excavated by a jumbo drill is accepted to be 18000 m/month by considering 120 holes having the length of 2.5 m for each hole in 2 shift and 30 days, the rock drill cost would be 0.44 $/m. However, another option is to check total advance rates in the roadway. If it is accepted to be around 50 m/month, the cost of renting would be 160$/m. accordingly, energy requirement (E) of the rock drill, whose power is 20 kW, is found to be 0.31 kWh based on Equation 2. Average drilling period of one blasthole in the field was also measured to be 55 sec.

$$E = Pt \tag{2}$$

Table 3. The running costs of a jumbo drill.

Rock drilling expenses	Cost ($/month)
Rent of a rock drill	8000
Energy	500
Oil & Water	1500
Operators and mine workers	62600
Average drilling supplies	1000
Maintenance & spare parts	3600
Total	77200

where E = energy; P = power of drill; and t = drilling time.

If the unit price of the electricity is assumed to be 0.17 $/kWh; then the drilling cost of each hole would be $0.053. When 120 holes were drilled at the face, the cost due to percussion power of the drill is calculated to be $6.36 for each blasting. However, hydraulic system and electrical system should also be taken into account during calculation. In this context, jumbo drill draws the current of 60 A and the voltage of 380 V. The drilling process takes 2 hours in a shift. Thus, a total energy of 45.6 kWh is consumed by the machine during 120 minutes. The daily energy cost would be 91.2 kWh/day and 2736 kWh/month. Then, the running cost of total energy in accordance with 50 m/month estimated advance rate would be 9.3 $/m for opening the blasting holes.

The button bit used during drilling is approximately $50 and its can be employed for drilling nearly 500 m in rock. Thus, the cost of bit would be 0.1 $/m. The cost of an extension rod is about $120. If the average life is accepted to be 5000 m, then its cost would be 0.024 $/m.

Based on the data given in Table 3, oil and water consumption of the rock drill cost is accepted to be 25 $/shift while the cost of maintenance, repair and spare parts are assumed to be 60 $/shift. Since 50 m/month of the advance rate is derived from drilling and blasting method, 90 $/m is allocated for those expenses of rock drill.

During the drilling process in a day, the mine operated three 8 hours shift. However, drilling is conducted two shifts with 2 operators ($3500 × 2), 2 mechanics ($3300 × 2), 2 electricians ($3300 × 2) and 2 firers ($3200 × 2) and 12 development miner for supporting ($3000 × 12). If the total distance excavated by jumbo drill is calculated to be 15600 m/month, the labor cost would be 0.90 $/m when it is rented. Furthermore, if the advance rate by conventional method is accepted to be 50 m/month, the labor costs would be 1252 $/m. As a result, the running costs in one meter of advance rate can be calculated as summarized in Table 4.

2.2 Excavation costs of a roadheader

Mechanized cutting systems eliminate negative effects of drilling and blasting methods due to handling, storage, transportation of explosives, etc. However, when cutting a face with a roadheader, significant problems such as wear on the bits, handling with the dust, electric cables, water, etc. are encountered. In this context, there are few trials of driving roadheaders in the Zonguldak hardcoal region. The reason is that the coal measure rocks generally include high percentage of quartz. It basically leads to wear the picks mounted on the cutterhead of roadheaders. However, in Amasra colliery, there are claystone, siltstone, and sandstone zones which do not cover any quartz. Therefore, the mine roadways in this colliery could be excavated by roadheaders.

Roadheaders generally use conical shaped picks which are able to self-sharpening. Essentially, wear of the picks is the main problem since it increases the unit costs of excavation. It is the fact that bit consumption should be lower than 0.5 cutter/m^3. If it is exceeded, cutter brakes, wearing increase rapidly and forces acting on the tool increases tremendously. If it is between 0.2–0.5 cutter/m^3, the mining operations are considered to be at a very critical level.

Table 4. The running costs per meter of roadway progress by using drilling and blasting method.

Expenses	Cost ($/m)
Rent of jumbo drill	160
Energy	9.30
Bit consumption	0.10
Extension rod consumption	0.02
Water & oil consumption	30
Maintenance, repair & spare parts	60
Labor	1252
Explosive	600
Total	2111.42

However, if the consumption rate is lower than 0.2 cutter/m^3, the excavation would be very economical (Bilgin et al. 2014).

Wear can be determined from tool forces and machine vibration. As the tool forces increase, it indicates that they are worn. The greater force is, the worse the potential abrasion (Wu et al. 2013). Besides, when the tools are worn, the machine starts to vibrate gradually. Then, the picks on the cutterhead should be checked as soon as possible and changed with the new one. In this sense, economic analyses should definitely be performed based on the rock mass properties before buying or renting a roadheader.

When the roadway is developed by a roadheader, it is inevitable that there will be high capital and running costs. A case study from Amasra colliery, where a Deilmann-Haniel R60T roadheader is driven in the cross-sectional area of 28 m^2, is carried out as follows.

Roadheader was purchased for $2100000. If the amortization period is accepted to be 8 years, the cost would be 262500 $/year and 21875 $/month. The average advance rate of the same roadheader is accepted to be about 70 m/month. Thus, the excavation cost can be accounted for 312.5 $/m.

The machine worked at the average current of 100 A and at the voltage of 1100 V. Thus, the average cutting power of the machine was calculated to be 110 kW although it had the installed cutting power of 160 kW. However, depending on the geology, the current could be increased up to 130 A. According to the problems encountered in a shift, the utilization time of the machines changes between 20% and 30%. By considering the average utilization rate of the machine to be 25% during a day, the energy requirement would be 25740 kWh/month. In this sense, if the average advance rate and unit electricity price is assumed to be 70 m/month and 0.17 $/kWh respectively, the cost of energy in light of monthly advance rates is calculated to be 62.5 $/m.

An estimated value of pick consumption so far is assumed to be around 0.10 cutter/m^3. A point attack pick is approximately $15.2 and its life relies on the geology and abrasive mineral content. The average advance rate of roadheader is about 70 m/month with the cross-sectional area of 28 m^2. Thus, the excavated volume of muck would be 1960 m^3/month. Then, average pick consumption would be approximately196 cutter/month. In this case, pick consumption for 70 m of advance in a month would be 2.8 cutter/m and the cost of conical pick would be 42.5 $/m.

Although the machine is supposed to be working 30 days in a month, it can never be achieved due to the problems encountered on the dust suppression unit, electric panel, oil tank, water hose, pipe lines, belt conveyor, etc. In order to overcome those problems, 150 $/day, which equals to 4500 $/month, is allocated for maintenance, repair, and also spare parts. Accordingly, 65 $/m is spent for the extra machine expenses when the monthly advance rate is assumed to be 70 m. However, it is a fact that average advance rate of a roadheader in a month should be higher than 100 m for an economic and efficient cutting. Therefore, the company management and the operators with development miners should gain more experience with respect roadheader use according to variable geologic conditions.

During the excavation, roadheader is driven three shifts with 3 operators ($3700 x 3), 3 mechanics ($3250 x 3) and 3 electricians ($3250 x 3) and 12 development miners for supporting ($3000 x 12). Then, total amount of labor costs would be 66600 $/month. In this context, the labor cost would be 950 $/m with the average advance rate of 70 m/month. In conclusion, the running costs of per meter of advance rate can be given as in Table 5.

Table 5. Total cost per meter of roadway driven by roadheader.

Expenses	Cost ($/m)
Roadheader	312.5
Energy	62.5
Cutter consumption	42.5
Maintenance, repair & spare parts	65
Labor	950
Total	1432.5

3 CONCLUSIONS

The cost of the overall progress, including the capital and operating costs, should be estimated prior to starting an excavation project. The progress rate in conventional methods range from 5 to 40 m/week whereas it is about 15–60 m/week for a roadheader. Although the rates can vary relying on the geological conditions, the initial investments and running costs would be very high in roadheader excavation.

In this study, the running costs of unit advance ($/m) was calculated and the results were compared with the running cost of advance rate arising from drilling and blasting in Amasra colliery. The results indicated that the cost of driving a roadheader is 47% cheaper than the conventional method. It also has the advantage of faster advance rates.

As can be seen from the tables above, the initial investments in both excavation methods are very high. However, if the blast holes would be drilled by hand held drills, the cost would decrease as low as possible. When kilometers of roadways considered to be developed, the fact is that renting a machine could not be very economical. In that case, it is definitely better to buy a jumbo drill or a roadheader for a mining company.

The other major concern during the excavation also consists of labor and energy costs. Since labor costs are very high within the total cost, it is always suggested to employ as low as possible number of workers. Moreover, the cost of energy depends mainly on the machine weight and cutterhead power. Since the roadheader is fully electrically powered, its cost is much greater than a jumbo drill.

Besides, the number of cutter cost essentially influence the running costs. Since the claystone formation did not include higher amount of quartz, we did not calculate the wear costs in the overall study.

It is clear that the advance rate of a roadheader is higher than conventional method. As pointed out above, the calculations were based on average advance rate of 70 m in one month which cannot be validated within the acceptable ranges for a roadheader. This low value increases the running costs. The reason of higher costs of driving roadheader is that the machine has installed underground recently and the crew does not have enough experience to operate it. In order to maximize productivity and minimize the costs, the advance rates of the roadheader should definitely be increased to more than that of 100 m/month for an economic excavation. The basic advantages of the roadheader are that it provides high safety with little disturbance to surrounding rock mass and low manpower even though the running cost are very high.

If the entire calculations for a roadheader would be performed with an average advance rate of 100 m, the running cost would be almost half of the cost of drill and blast method.

ACKNOWLEDGEMENT

Scientific Research Project of Bulent Ecevit University numbered 2014-29011448-02 is gratefully acknowledged for the financial support.

REFERENCES

Acaroglu, O. & Ergin, H. 2006. A new method to evaluate roadheader operational stability. *Tunnelling and Underground Space Technology*, 21: 172–179.

Bilgin N., Copur, H. & Balci, C. 2014. *Mechanical Excavation in Mining and Civil Industries*, 1st edn. CRC Press, Taylor and Francis Group.

Brino, G., Cardu, M., Gennaro, S. & Gianotti, A. 2013. Technical-economical comparison between excavation by D&B and by roadheader in two underground gypsum quarries. *23rd International Mining Congress & Exhibition of Turkey*. Antalya, pp. 813–824.

Jafari, A., Hossaini, M.F. & Alipour, A. 2009. Prediction of specific charge in tunnel blasting using ANNs. *Proc. of Rock Characterisation, Modelling & Engineering Design Methods*. Hong Kong, pp. 786–790.

Kantarci, E.A., Ergener, B. & Buyurgan, G. 2013. Positive effects of developments in drilling machines on tool life and drilling costs. *Proc. of VII. Drilling and Blasting Symposium*, Istanbul, pp. 235–238.

Ocak, I. & Bilgin, N. 2010. Comparative studies on the performance of a roadheader, impact hammer and drilling and blasting method in the excavation of metro station tunnels in Istanbul. *Tunnelling and Underground Space Technology*, 25: 181–187.

Tamrock 1978. *Handbook of Surface Drilling and Blasting*. Finland, 236 p.

Wu, L., Guan, T. & Lei, L. 2013. Discrete element model for performance analysis of cutterhead excavation system of EPB machine. *Tunnelling and Underground Space Technology*, 37: 37–44.

Zare, S., Bruland, A. & Rostami, J. 2016. Evaluating D&B and TBM tunnelling using NTNU prediction model. *Tunnelling and Underground Space Technology*, 59: 55–64.

Tunnels and Underground Cities: Engineering and Innovation meet Archaeology,
Architecture and Art, Volume 7: Long and deep tunnels – Peila, Viggiani & Celestino (Eds)
© 2020 Taylor & Francis Group, London, ISBN 978-0-367-46872-9

A softening damage-based model for the failure zone around deep tunnels in quasi-brittle claystone

E. Trivellato
Laboratoire Navier, École des Ponts ParisTech, Cité Descartes, Champs-sur-Marne, France
Andra, R&D Direction, Châtenay-Malabry, France

A. Pouya
Laboratoire Navier, École des Ponts ParisTech, Cité Descartes, Champs-sur-Marne, France

M.-N. Vu & D.M. Seyedi
Andra, R&D Direction, Châtenay-Malabry, France

ABSTRACT: The excavation of galleries in deep rocks induces the formation of failure zones around these structures. For semi-brittle rocks, accounting for softening damage phenomena seems necessary for a correct description of the short term failure around galleries perimeter. In this work, an elastic-damage softening model with anisotropic shear strength is employed to analyze the short term extension of the excavation-induced failure zone as well as the displacements of the tunnels walls. No presence of support syste is considered in the performed numerical analyses.

1 INTRODUCTION

The excavation of deep galleries in quasi-brittle rocks generates a fractured zone around these underground structures, which shape and extension constitutes crucial aspects for their design and future employment. This type of structures becomes relevant for the technique of long-term geological repository to the radioactive waste disposal. Andra, the French National Radioactive Waste Management Agency, is studying the impact of a high-level and intermediate-level long lived waste disposal in the Callovo-Oxfordian (COx) claystone formation. Feasibility studies are conducted in the Andra Underground Research Laboratory (URL), located in the Meuse Haute-Marne department, France. At the main level (-490 m), Andra URL is composed by a network of horizontal drifts, following, in general, the main horizontal stress components. To distinguish the major from the minor component, they are usually indicated, respectively, σ_H and σ_h. The vertical stress component at the depth considered is denoted by σ_v. Table 1 reports the estimated in-situ values for each component. Among the investigations performed in the URL, the evaluation of the short-term failure zone, developed in the claystone around the drifts, constitutes an important aspect of research advances in rock mechanics related to tunneling. Around the cavities, one can distinguish a zone of interconnected fractures, in proximity of the walls, and an adjacent zone, characterized by diffuse failure. According to the definition already proposed by Emsley et al. (1997), they are called, respectively, *Excavation Damaged Zone* (EDZ) and *Excavation Disturbed Zone* (EdZ). A conceptual model for the failure zone and the excavation-induced fractures is reported in Figure 1, for two instrumented drifts in the URL. While one follows the direction of σ_H (Figure 1a), the other one is excavated along σ_h (Figure 1b).

Geological surveys suggest that softening damage is a leading phenomenon for the short-term response of the excavated rock. Numerical FEM simulations already showed that a softening elastic-damage model may analyze correctly the physical short-term degradation of a brittle rock around drifts (Pouya et al. 2016, Trivellato et al. 2018). In this work, evidences of the failure zone for the drifts shown in Figure 1 are considered to validate a damage-based softening

Table 1. In-situ stress state at the main level of the URL (Armand et al, 2013; Wiliveau et al, 2007).

σ_v [MPa]	σ_H [MPa]	σ_h [MPa]
12.7	16.2	12.4

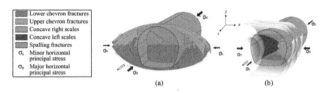

Figure 1. Conceptual model of the EDZ-EdZ system around the analyzed drifts, following σ_H (a) and σ_h (b) at the Andra Underground Research Laboratory (Armand et al., 2013).

model accounting for the anisotropy of the failure criterion. This can be a principal cause for failure localization either on the horizontal or the vertical direction with respect to the cross section of the drifts studied, together with anisotropies of the initial stress, if present. Around the drifts wall, a transition from a brittle to a ductile post-peak response of the COx claystone - due to a radial increase of the mean stress acting on the material - is accounted.

2 ANALYSIS APPROACH

Finite Element numerical simulations were performed with the code *POROFIS* (Pouya, 2015). Excavation-related analyses are, typically, three-dimensional problems (Figure 1). Here, a simplified procedure with a two-dimensional approach, in plane deformations ($\varepsilon_{zz} = 0$), has been chosen. Figure 2a and 2b illustrate, respectively, the failure system formed by EDZ and EdZ, for the drifts shown in Figure 1a and 1b, around their cross-section plane. Shape and extension of the zones reported in Figure 2 constitute the references to validate the model proposed in this work. Similarly, the displacement fields calculated around galleries are compared to in-situ measurements, focusing on values at their perimeter (convergences). The drift in Figure 1a and 2a, excavated along σ_H, is identified as GCS. The drift in Figure 1b and 2b is identified as GED and is excavated along σ_h (Armand et al. 2013).

The basic damage model includes isotropic linear elasticity with a scalar damage variable, D, defined from 0 to 1, and a failure criterion detailed in the following section. The constitutive law is written according to Equation 1 (Pouya 2015, Pouya et al. 2016). C is the elastic stiffness tensor, defined by the Young's modulus, E, and Poisson's ratio, v. The internal variable D reduces progressively the material's stiffness. From the basic constitutive law in Equation 1, an anisotropic elastic-damage model has been developed and employed for numerical simulations.

$$\boldsymbol{\sigma} = (1 - D)\mathbf{C} : \boldsymbol{\varepsilon} \tag{1}$$

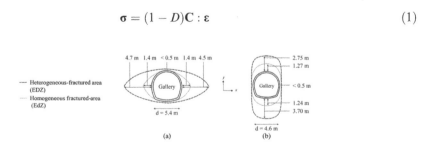

Figure 2. Bi-dimensional schemes of the extension of the system EDZ-EdZ on the cross-sections perpendicular to the two principal drifts directions (Armand et al., 2014).

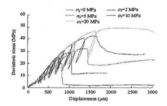

Figure 3. Triaxial compression tests on Callovo-Oxfordian claystone specimens for increasing lateral confinements (Armand et al., 2013).

As mentioned, numerical simulations consider the transition from a fragile behavior to an increasing ductility, shown by the COx claystone at yield, after the resistance peak. This is clearly noticed in laboratory triaxial tests, as reported in Figure 3. In general, an increased lateral confinement on the material corresponds to a reduced difference among the residual and the peak strength. It is worth defining their ratio, named, for this work's purposes, *stress softening ratio* and denoted η_r. It will be recalled later, when the mathematical constitutive modelling is discussed. At the underground structure scale, this brittle-ductile transition corresponds to a less fragile behavior after failure, when the radial distance from the tunnel perimeter increases and in-situ (i.e. initial) stress conditions are progressively restored.

3 METHODS

3.1 *Mathematical modelling*

The mathematical formulation of the elastic-damage softening model, available in *POR-OFIS* and used with some modifications in this work, is discussed hereby. This model produces the uniaxial stress-strain response shown in Figure 4. It is underlined that the resistance peak σ_0 coincides with the elastic limit, i.e. $f \leq 0$, where f indicates the failure criterion. Here, σ_0 is also defined as uniaxial compressive strength (*UCS*). A linear softening phase describes the strength decrease before the residual constant value, σ_r, is attained. According to Figure 4, the softening stress ratio mentioned previously is defined here as $\eta_r = \sigma_r/\sigma_0$.

In the following, a first modification of the stiffness tensor **C** is introduced. It accounts for an anisotropy of the damage evolution, based on the principal stress directions. For sake of simplicity, the compliance matrix $\mathbf{S} = \mathbf{C}^{-1}$ is employed to present this modification. In plane strain analyses, the formulation of the modified compliance matrix, $\tilde{\mathbf{S}}$, writes:

$$\tilde{\mathbf{C}}^{-1} = \tilde{\mathbf{S}} = \begin{bmatrix} \frac{1}{(1-D)E} & \frac{-\nu}{(1-D)E} & \frac{-\nu}{E} & 0 \\ \frac{-\nu}{(1-D)E} & \frac{1}{(1-D)E} & \frac{-\nu}{E} & 0 \\ \frac{-\nu}{E} & \frac{-\nu}{E} & \frac{1}{E} & 0 \\ 0 & 0 & 0 & \frac{2(1+\nu)}{(1-D)E} \end{bmatrix} \tag{2}$$

The inversion of $\tilde{\mathbf{S}}$ provides the correspondent stiffness matrix. According to Equation 2, damage reduces only the stiffness acting along the directions x and y. They corresponds to the axes identifying the plane of galleries cross section. Thus, a damage evolution characterized by

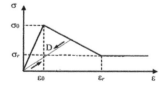

Figure 4. Softening model of the stress-strain curve in uniaxial compression.

transverse anisotropy has been implemented the internal variable D does not affect the axial (or *out-of-plane*) stiffness, perpendicular to the cross section. This choice is due to the fact that the main model's purpose is the prediction of damage evolution within galleries cross section only (Figure 2).

Then, a second phenomenon is considered to define, eventually, the elastic-damage stiffness tensor, employed for numerical simulations. Based on physical hypothesis and evidences of claystone brittleness after failure, the idea is a reduction of the cohesive properties of the damage material. On the contrary, the friction that it can develop after failure remains constant. From the physical point of view, in the cross section plane $(x; y)$, the material, initially cohesive, becomes purely granular when D increases. The correspondent strain response requires the material bulk modulus, K, not to be affected by damage. Only the shear modulus, in the cross section plane, is degraded by the internal variable D. From the mathematical point of view, if the spherical part of the stiffness tensor is discerned by the deviatoric part, the elastic constitutive law $\boldsymbol{\sigma} = \mathbf{C}: \boldsymbol{\varepsilon}$ according to the following form:

$$\boldsymbol{\sigma} = \frac{1}{3}\boldsymbol{\delta} \cdot tr\{\boldsymbol{\sigma}\} + \mathbf{s} = 3\mathrm{K}\frac{1}{3}\boldsymbol{\delta} \cdot tr\{\boldsymbol{\varepsilon}\} + 2\mu\left(\boldsymbol{\varepsilon} - \frac{1}{3}\boldsymbol{\delta} \cdot tr\{\boldsymbol{\varepsilon}\}\right) \tag{3}$$

To analyze this strain response, the damage factor $(1-D)$ in Equation 1 multiplies only the stiffness component of the deviatoric part of $\mathbf{C}$, i.e. 2μ.

Concerning the failure criterion modelling, this initial formulation is proposed:

$$f(\boldsymbol{\sigma}, D) = \sqrt{\left(\sigma_{xx} - \sigma_{yy}\right)^2 + 4\sigma_{xy}{}^2} + \sin\phi\left(\sigma_{xx} + \sigma_{yy}\right) - g(D)K \tag{4}$$

where φ is the friction angle and K is the resistance parameter depending also on cohesion (Alejano & Bobet, 2012). In Equation 4, the function $g(D)$ corresponds to the damage law, providing the linear plus residual softening behavior, at failure, illustrated in Figure 4. (Pouya 2015). According to several cases, reported in literature, of failure analysis around tunnels (e.g. Reed, 1986, Reed, 1987), the role of the axial (or *out-of-plane*) stress component, σ_{zz}, has been neglected. Even if its evolution is calculated in FEM analyses, it offers no contribution in the failure threshold during drifts excavation. At the same time, the anisotropy of the material strength was formalized and implemented. In particular, the inclination of the claystone stratification with respect to the principal loading direction is accounted (e.g. Guayacán-Carrillo et al. 2016, Mánica et al. 2016). The reference system in Figure 5, described locally by the unit vectors $\mathbf{n}$ and $\mathbf{m}$, is defined to consider the strength anisotropy. The angle between the unit vector $\mathbf{n}$, normal to the material stratification, and the horizontal (x) is denoted by ω. An anisotropic fabric tensor $\mathbf{H}$ is defined according to Equation 5:

$$\mathbf{H} = h_n\left(\mathbf{n} \otimes \mathbf{n}^T\right) \otimes \left(\mathbf{n} \otimes \mathbf{n}^T\right) + \left(\mathbf{m} \otimes \mathbf{m}^T\right) \otimes \left(\mathbf{m} \otimes \mathbf{m}^T\right) + \\ + \frac{h_s}{2}\left[\left(\mathbf{m} \otimes \mathbf{n}^T + \mathbf{n} \otimes \mathbf{m}^T\right) \otimes \left(\mathbf{m} \otimes \mathbf{n}^T + \mathbf{n} \otimes \mathbf{m}^T\right)\right] \tag{5}$$

The independent parameters h_n and h_s in Equation 5 allow to adjust the variation of the shear strength anisotropy, in uniaxial or triaxial conditions, for whatever interval of ω, between $[0; 2\pi]$. Adopting the tensor $\mathbf{H}$ introduced previously, a *generalized* stress tensor is defined in two dimensions, according to Equation 6. Each of its components replaces the correspondent one in the initial failure criterion (Eq. 4). The modified criterion, depending implicitly on ω, is

Figure 5. Reference system for the anisotropic strength definition in the galleries cross section $(x; y)$.

reported in Equation 7. It has been eventually employed for the failure calculation at the elastic limit in the numerical analyses of galleries excavation further presented.

$$\tilde{\sigma} = \mathbf{H}^T \cdot \begin{bmatrix} \sigma_{xx} & \sigma_{xy} \\ \sigma_{xy} & \sigma_{yy} \end{bmatrix} \cdot \mathbf{H} \tag{6}$$

$$f(\tilde{\sigma}, D) = \sqrt{(\tilde{\sigma}_{xx} - \tilde{\sigma}_{yy})^2 + 4\tilde{\sigma}_{xy}{}^2} + \sin\varphi(\tilde{\sigma}_{xx} + \tilde{\sigma}_{yy}) - g(D)K \tag{7}$$

The circular (radar) diagram in Figure 6 shows an example of the variation of the material strength with ω in uniaxial loading conditions: in particular, according to Equation 6, only the vertical stress component is not null, i.e. $\sigma_{yy} > 0$. The choice of numerical values of h_n and h_s induces, as shown, a minimum of the uniaxial compressive strength (UCS) at $\omega = (\pi/4) + n \cdot (\pi/2)$, with $n = 1, 2, 3$. In Figure 6, the circumference formed by the angular coordinate ω can be imagined as the perimeter of a circular gallery section: here, it is possible to deduce that failure will be firstly attained in correspondence of these minimal values.

3.2 Numerical modelling

It is presented hereby the Finite Element mesh adopted for numerical analyses of the failure zone around GCS and GED drifts (Figure 1 and Figure 2). According to cylindrical symmetry, it was possible to focus only on a quart of the entire domain. Geometry and boundary conditions are shown on the mesh and are reported in Figure 7. The radius r varies from GCS to GED analyses, respectively equal to 2.6 and 2.3 m (e.g. Guayacán-Carrillo et al. 2016). Concerning stress initial and boundary conditions, the in-situ geostatic state refers on the vertical and horizontal components reported in Table 1. The principal horizontal component, σ_{xx}, varies according to the drift analyzed: respectively, $\sigma_{xx} = \sigma_h$ for GCS and $\sigma_{xx} = \sigma_H$ for GED. Before excavation, it is immediately clear that the second drift is characterized by a stronger anisotropy of the initial stress state, on its cross section. Initial stress components were imposed, on the domain in Figure 7, with zero initial displacements, according to values in Table 2. In both cases, the stress state in the main reference system has been imposed accounting for the circular geometry by calculating the local inclination α at each finite element at the

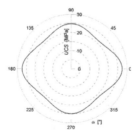

Figure 6. Radar diagram of the uniaxial compressive strength variation with $0 < \omega < 2\pi$, for assigned values of resistance parameters.

Table 2. Initial principal stress components in the cross-section of the drifts analyzed.

Case-study drift	σ_{xx} [MPa]	σ_{yy} [MPa]
GCS (parallel to σ_H)	12.4	12.7
GCS (parallel to σ_H)	16.2	12.7

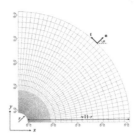

Figure 7.   Geometry, FE mesh and boundary conditions employed for plane strain numerical analyses of drifts in Figure 2.

internal and external boundaries. Hence, based on principal stresses in Table 2, the local normal and tangential components, σ_n and τ, have been imposed.

According to the local reference system formed by unit vectors **n** and **t**, with the inclination α to the main system (x; y) in the cross section the local normal and tangential components has been calculated as follows:

$$\sigma_n = \mathbf{n}^t \cdot \boldsymbol{\sigma} \cdot \mathbf{n} = (\cos\alpha \quad \sin\alpha) \cdot \begin{pmatrix} \sigma_{xx} & 0 \\ 0 & \sigma_{yy} \end{pmatrix} \cdot \begin{pmatrix} \cos\alpha \\ \sin\alpha \end{pmatrix} \tag{8}$$

$$\tau = \mathbf{t}^t \cdot \boldsymbol{\sigma} \cdot \mathbf{n} = (-\sin\alpha \quad \cos\alpha) \cdot \begin{pmatrix} \sigma_{xx} & 0 \\ 0 & \sigma_{yy} \end{pmatrix} \cdot \begin{pmatrix} \cos\alpha \\ \sin\alpha \end{pmatrix} \tag{9}$$

Equations 8 and 9 apply the principle of Cauchy stress tensor. Starting from the geostatic stress state, the excavation is performed in steady state, applying a tensile stress state at the internal circular boundary in Figure 6 (the drift perimeter) to nullify the initial one.

The initial and boundary axial stress component is not null and its value is derived according to the plane strain condition ($\varepsilon_{zz} = 0$), meaning:

$$\sigma_{zz} = \nu(\sigma_{xx} + \sigma_{yy}) \tag{10}$$

where the Poisson ratio $\nu = 0.25$ (Armand et al. 2014, Guayacán-Carrillo et al. 2016). In the framework of short-term behavior analyses, a value $E = 6000$ MPa, correspondent to an undrained response, has been chosen for the Young modulus, based on in-situ borehole data (Armand et al. 2017).

The calibration of parameters for the failure criterion comprehends both the resistance parameters (Eq. 4 and 7) as well as values for h_n and h_s (Eq. 5). According to literature data (Armand et al. 2013), a value $\varphi = 20°$ was chosen for the friction angle. The calibration of the other parameters is based on laboratory data of deviator stress values under triaxial compression at different confinements and inclinations ω of the specimens (Andra, pers. comm.). An example of the calibration performed to analyze the deviator variability with ω according to Equations 5-7 is reported in Figure 8. Failure is attained under a lateral confinement equal to 12 MPa.

Eventually, the numerical modelling of the brittle-ductile transition of the claystone is presented. In particular, it appeared necessary the definition of different zones, assumed circular and concentrically distributed around the drifts, where the material post-failure behavior changes. In this work, these zones are called *transition zones*, denoted as *t.z.* To approximate the transition occurring in the claystone softening, different *transition zones* has been characterized by different values of the softening stress ratio $\eta_r = \sigma_r/\sigma_0$. In Figure 7, six *transition zones* have been included in the area of possible damage evolution (according to the models in Figure 2). Each *t.z.* covers an annular concentric area with a 1 m thickness.

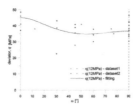

Figure 8. Values of deviator at failure for COx specimens confined at 12 MPa (Andra, pers. comm.) and calibration of anisotropic strength parameters with $\varphi = 20°$.

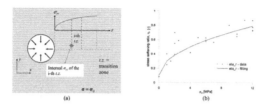

Figure 9. Scheme of the approach adopted to consider numerically the brittle-ductile transition of the post-peak behavior of the COx claystone with the increase of the radial stress.

Figure 9 describes the approach adopted to consider the increase of η_r along whatever radial distance from the drift wall, simulating the brittle-ductile transition. Considering a circular cavity after a complete unload, the elastic solution of the radial stress component σ_{rr} is traced (Figure 9a). In parallel, a certain quantity of concentric *transition zones* is distributed around the drift. In this work, the elastic solution of σ_{rr}, evaluated at the internal side of each *t.z.*, is assumed as the confining pressure correspondent to a certain value of η_r. Triaxial compression data on COx claystone, for different confining pressure, were reviewed (Belmokhtar et al. 2018, Bésuelle et al. 2006, Hu et al. 2004, Zhang et al. 2010). They confirm the post-peak softening response characterized by increasing ductility shown in Figure 3. For every triaxial stress-strain path, the softening stress ratio η_r was calculated, and plotted as function of the confining pressure. Its approximation with the elastic solution of the radial stress σ_{rr}, for a completely excavated gallery, allowed to trace the graphic in Figure 2b. Thus, it was possible to assign, for each transition zone imposed in the numerical model (Figure 7), a reference value for η_r characterizing the local post-peak material response. A function relying η_r with σ_{rr}, based on Figure 9b, was written according to Equation 11. The evaluation of its parameters is based on the least squares method.

$$\eta_r(\sigma_{rr}) = 0.1 + 0.2\sigma_{rr}^{0.5} \tag{11}$$

4 RESULTS DISCUSSION

4.1 Damaged failure zone

In Figure 10, the distribution of the damage variable D at the end of the excavation is reported. Figure 10a compares the calculated result with the model of EDZ-EdZ for the drift GCS, while Figure 10b shows the same result and comparison for GED. The contour plots are illustrated in their deformed configurations, according to the computed displacements, discussed later. For both drifts, damage at the elastic limit is attained at the internal perimeter, neither along the horizontal direction (x) nor along the vertical (y). Due to the anisotropic failure criterion, which calibration is shown in Figure 8, damage firstly develops along inclined directions between the horizontal and vertical. For both drifts, these directions are superposed to the zone of the localization of deformations. Then, damage evolves rejoining the horizontal and

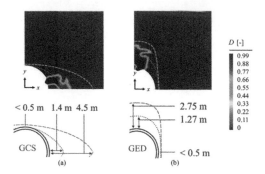

Figure 10. Distribution of the damage variable D for a complete unloading of the GCS (a) and GED (b) drifts.

vertical directions, respectively in GCS and GED drift. At the end of the excavation, its extension, in both cases, is included in the area between the red and green curves, in Figure 10a and Figure 10b, delimiting the EDZ and EdZ, respectively. Compared to GCS, the shape and extension of D in GED approximate better the bi-dimensional conceptual model of the failure zone estimated in-situ. It is possible to notice an area, included between the damage zones and the drifts wall, where elastic conditions are still maintained. This area is more extended for GCS drift. Physically, it can be assumed to the formation of a claystone block potentially subjected to a rigid translation inside the galleries. Its existence recalls in-situ observations showing locally a less damaged area, at the galleries perimeter, surrounded by a diffusely fractured medium, in the EDZ.

4.2 Displacements analyses

Contour plots of the displacements field are reported in Figure 11a and Figure 11b, respectively, for GCS and GED drift. For each node of the FE numerical model, the displacement is traced as the norm of the local horizontal and vertical components, along the principal directions, U_x and U_y. Thus, values reported in Figure 11 are plotted according to Equation 12:

$$\|U\| = \sqrt{U_x^2 + U_y^2} \tag{12}$$

Spatial distributions of displacements confirm the deformed configurations already shown in Figure 10, with an almost horizontal localizations of strains for the drift GCS and a vertical one

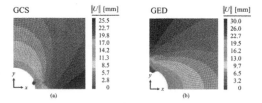

Figure 11. Displacements norm fields calculated for the GCS (a) and GED drifts (b).

Table 3. Comparison between short-term displacements, estimated at the drift walls (Armand et al. 2013, Guayacán-Carrillo et al. 2016), and correspondent values calculated.

	U_x, GCS [mm]	U_y, GCS [mm]	U_x, GED [mm]	U_y, GED [mm]
In-situ measurements	~ 20.0	~ 10.0	~ 10.0	~ 40.0
Max. calculated values	25.5	11.2	14.4	30.1

for GED. Focusing on the conditions predicted at the drifts internal perimeter, a comparison between estimations from in-situ geotechnical surveys and computed values is reported in Table 3.

5 CONCLUSIONS

This work presented a predictive model for the short-term analyses of the failure induced by deep galleries excavated in the Callovo-Oxfordian claystone. This material, in general, can be considered representative for several geological formations corresponding to sedimentary brittle rocks. In this model, damage is considered the only irreversible phenomenon of energy dissipation occurring in the instantaneous material response, after the elastic limit. Different yielding behaviors, from very brittle to more ductile softening conditions, were considered. Mathematically, damage is the internal variable reducing the material stiffness, when failure conditions are attained. Both failure conditions and damage evolutions include anisotropies correspondent to experimental evidences, at the laboratory scale, as well as specific purposes for numerical simulations (i.e. plane strain conditions).

Observations of the failure zones around two instrumented drifts at the Andra Underground Rock Laboratory (URL) were considered to validate the model proposed for different undisturbed stress states on their cross section. Shape and extension of the damaged area calculated around the two galleries develop according the main directions suggested by in-situ measurements, even if it does not predict the entire zone illustrated in conceptual models. In one numerical analysis (drift GCS), damage remains more or less confined around the drift wall. In general, damage gradient is much reduced, showing rapid transitions from a fully damaged zone to still elastic conditions. Concerning displacements analyses, they are, in general, consistent with in-situ measurements. In some cases, under- or over-estimations of horizontal and vertical convergences are obtained. As done for the failure criterion, considering an anisotropic elasticity, due to the claystone bedding, may improve the displacements prediction. In particular, a transverse isotropy, increasing the horizontal stiffness parallel to the stratification, seems suitable and physically consistent with the material microstructure.

In the context of deep galleries analyses, supervised by Andra, several works already proposed constitutive models accounting for dissipation phenomena in plasticity or visco-plasticity. Coupling a fragile phenomenon as damage with these types of failure may constitute an interesting perspective to this work. This could lead to a predictive model accounting, at the same time, the instantaneous fragile behavior around galleries and the hardening response of the loaded material, deeper in the claystone formation, together with time-dependent effects.

REFERENCES

Alejano, L. R. & Bobet, A. 2012. Drucker–Prager Criterion. *Rock Mechanics and Rock Engineering* 45(6): 995–999.

Armand, G., Conil, N., Talandier, J. & Seyedi, D. M. 2017. Fundamental aspects of the hydromechanical behaviour of Callovo-Oxfordian claystone: From experimental studies to model calibration and validation. *Computers and Geotechnics* 87: 277–286.

Armand, G., Leveau, F., Nussbaum, C., De La Vaissiere, R., Noiret, A., Jaeggi, D., Landrein, P. & Righini, C. 2014. Geometry and Properties of the Excavation-Induced Fractures at the Meuse/Haute-Marne URL Drifts. *Rock Mechanics and Rock Engineering* 47(1): 21–41.

Armand, G., Noiret, A., Zghondi, J. & Seyedi, D.M. 2013. Short- and long-term behaviors of drifts in the Callovo-Oxfordian claystone at the Meuse/Haute-Marne Underground Research Laboratory. *Journal of Rock Mechanics and Geotechnical Engineering* 5(3): 221–230.

Belmokhtar, M., Delage, P., Ghabezloo, S. & Conil, N. 2018. Drained Triaxial Tests in Low-Permeability Shales: Application to the Callovo-Oxfordian Claystone. *Rock Mechanics and Rock Engineering*, 51(7): 1979–1993.

Bésuelle, P., Viggiani, G., Lenoir, N., Desrues, J. & Bornert, M. 2006. X-ray Micro CT for Studying Strain Localization in Clay Rocks under Triaxial Compression. In Desrues, J., Viggiani, G. & Bésuelle, P. (Eds.), *Advances in X-ray Tomography for Geomaterials*, ISTE Ltd.

Emsley, S., Olsson, O., Stenberg, L., Alheid, H-J. & Falls, S. 1997. ZEDEX - A study of damage and disturbance from tunnel excavation by blasting and tunnel boring. *Swedish Nuclear Fuel and Waste Management Co.*, Stockholm, Sweden.

Guayacán-Carrillo, L.M., Ghabezloo, S., Sulem, J., Seyedi, D.M. & Armand G. 2016a. Effect of anisotropy and hydro-mechanical couplings on pore pressure evolution during tunnel excavation in low-permeability ground. *Rock Mechanics and Rock Engineering* 49(1): 97–114.

Guayacán-Carrillo, L. M., Sulem, J., Seyedi, D. M., Ghabezloo, S., Noiret, A. & Armand, G. 2016b. Analysis of Long-Term Anisotropic Convergence in Drifts Excavated in Callovo-Oxfordian Claystone. *Rock Mechanics and Rock Engineering* 49(1): 97–114.

Hu, D.W., Zhang, F., & Shao, J.F. 2014. Experimental study of poromechanical behaviour of saturated claystone under triaxial compression. *Acta Geotechnica*, 9(2): 207–214.

Mánica, M., Gens, A., Vaunat, J. & Ruiz, D. F. 2016. A cross-anisotropic formulation for elasto-plastic models. *Géotechnique Letters* 6: 156–162.

Pouya, A. 2015a. A finite element method for modelling coupled flow and deformation in porous fractured media. *International Journal for Numerical and Analytical Methods in Geomechanics* 39(16): 1836–1852.

Pouya, A. 2015b. Manuel d'utilisateur de POROFIS. *FRACSIMA - Fracture Simulation in Materials* (Copyright), Arcueil, France.

Pouya, A., Trivellato, E., Seyedi, D.M. & Vu, M-N. 2016. Apport des modèles d'endommagement sur la géométrie de la zone de rupture autour des ouvrages profonds dans des roches quasi-fragiles. *Actes des 8èmes Journées Nationales de Géotechnique et de Géologie de l'Ingénieur, 6-8 juillet 2016* (online). Nancy, France.

Reed, M. B. 1986. Stresses and displacements around a cylindrical cavity in soft rock. *IMA Journal of Applied Mathematics*, 36(3): 223–245.

Reed, M. B. 1987. The Influence of Out-Of-Plane Stress on a Plane Strain Problem in Rock Mechanics. Technical Report. *Department of Mathematics and Statistics, Brunel University Uxbridge*, United Kingdom (UK).

Seyedi, D. M., Vu, M-N., Armand, G. & Noiret, A. 2015. Numerical modeling of damage patterns around drifts in the Meuse/Haute-Marne URL. *The 13th International ISRM Congress, 10-13 May 2015*. Montréal, Québec, Canada.

Trivellato, E., Pouya, A., Vu M-N. & Seyedi D.M. 2018. Modélisation en endommagement radoucissant de la zone de rupture autour des ouvrages profonds dans des roches anisotropes quasi-fragiles. In Delage, P., Chevalier, C., Cui, Y.J. & Semblat J.F. (Eds.), *Actes des 9èmes Journées Nationales de Géotechnique et de Géologie de l'Ingénieur, 13-15 juin 2018* (online). IFSTTAR-ENPC, Marne-la-Vallée, France.

Wileveau, Y., Cornet, F.H., Desroches, J. & Blumling, P. 2007. Complete in situ stress determination in an argillite sedimentary formation. *Physics and Chemistry of the Earth* 32: 866–878.

Zhang, C.L., Czaikowski, O. & Rothfuchs, T. 2010. Thermo-Hydro-Mechanical Behaviour of the Callovo-Oxfordian Clay Rock. *Gesellschaft für Anlagen- und Reaktorsicherheit (GRS) mbH*. Technical report GRS - 266 ISBN 978-3-939355-42-7

Tunnels and Underground Cities: Engineering and Innovation meet Archaeology,
Architecture and Art, Volume 7: Long and deep tunnels – Peila, Viggiani & Celestino (Eds)
© 2020 Taylor & Francis Group, London, ISBN 978-0-367-46872-9

The use of Integrated Big Data software management systems on a complex TBM project. The Follo Line Project data analysis

F. Vara Ortiz de la Torre
Acciona Ghella Joint Venture Project Director, Oslo, Norway

ABSTRACT: The Follo Line project is a unique project and currently the largest TBM project in Norway. It uses four TBM's and will when completed provide the country with the longest rail tunnel in any of the Nordic countries. The project employs four TBM's, all of which have been deployed from the same starting point. Together they will be jointly responsible for drilling 36 km of tunnels between Oslo and Ski. The object of this abstract is to highlight how Innovation has been implemented through different in-house developed software systems, such as Cutter Disc Management systems and Connected Operational Intelligence Systems (COI). The developed software is classified as a specific Big Data-management system and it is based on live information taken from various sources, such as, geographical input, TBMs, concrete precast reports and cutter disc consumption studies. It provides updated information about TBM performance, availability, downtime, pregrouting, traffic management, cutter consumption and quality mapping from the segment factory. In summary, it represents an innovative up to date more efficient control system and a useful Big Data management tool for future project analysis.

1 INTRODUCTION

1.1 *Big Data Management in construction project. Smart cities*

The Follo Line Project is currently the largest transport project in Norway. The project consists of a new 22 km double track railway line between Oslo Central Station and the new station at Ski, south of Oslo. Acciona Ghella Joint Venture will construct the main part of the 20 km long tunnel, which will be Scandinavia's longest when it is finalized at the end of 2021.

One of the main challenges of the project is (in addition to dealing with the extremely hard Norwegian rock and 36 km. of tunnel excavation) the excavation of the four TBMs commencing from the same location. This means complex logistics must be considered and an appropriate design of the area where the works are placed must be produced. Production, maintenance and logistics must be perfectly defined in order to develop an installation area that is operational 24/7, ready for winter and summer and that has no impact on the surrounding area in terms of noise and dust.

In fact, The Follo Line EPC TBM project has built at Åsland what could be considered to be akin to a small city, where around 1,000 habitants have developed and carry out different activities in which coordination and efficiency is key in order to achieve the right productivity and safety standards.

Smart cities have become a reality in our world, almost all of our daily activities are starting to be controlled and managed by different software and Apps, (parking, driving, improved efficiencies in lighting consumption, traffic management, etc.). Everyday use and control of these Apps has started to become a normal part of our everyday life. Apps on our mobile phones help us to control such activities, helping us to improve our time management on a daily basis.

This concept has been applied on the Follo Line EPC TBM project, transforming the job site into a "small smart city", where production, traffic control, safety and cost control are all integrated into a digital platform allowing us to better manage and understand the millions of data changes and transfers that take place on site on a daily basis.

This paper will provide a brief summary of the capabilities of two different tools that have been specifically developed by Acciona for the Follo Line project:

1- The cutter monitoring system
2- Connected Operational Intelligence (COI).

1.2 *Project facts*

The Follo Line will form the core part of InterCity development south of Oslo and is commissioned by the Norwegian National Railway Administration (Bane NOR). The project consists of five EPC-contracts, in which the construction of the tunnels with TBMs is the most extensive. Installation of railway systems is also a part of the project, which will be completed in 2021. The tunnel contract has an approx. value of 8.7 billion NOK (1 billion euro). Key project facts:

- 37,000 ml TBM tunnel.
- 9 M tons of rock
- 140,000 concrete segments
- 1 M m³ of concrete
- 350 km of auxiliary pipes
- 1,000 workers peak
- More than 25 nationalities

For the excavation of the 36 km. of tunnels, Acciona Ghella Joint Venture procured four brand new tunnel-boring machines (TBMs) from Herrenknecht, a worldwide German TBM specialized company. The main characteristics of the TBMs are:

- Adapted to high rock strength. Bane Nor input based on Norwegian experience
- 71 Cutter rings 19" on 70 tracks
- Heavy structure, stiff support
- Diameter machine/tunnel: 9.96 m/8.75m
- Length of machines: 150 m
- Weight: 2,400 tons
- Installed Power: 6,200 kW.
- Main bearing size increased to ø 6.6 m

2 PROJECT CONFIGURATION

To better understand the needs and capabilities of the software to integrate the activities at the job site, a brief description on how the site has been developed is required.

The configuration of the area where the project was settled is approximately 200,000 m² of platform including the spoil area. It is close to exit 26 of the E-6 Highway that connects Oslo with Stockholm. As part of Acciona Ghella's Joint Venture scope, it was necessary to excavate the access tunnels, transport and auxiliary tunnels and assembly caverns where the machines were going to be assembled. This activity was contemporary with the execution of the civil works for the factories, auxiliary installations, and TBM assembly.

The first complex task was to define and integrate all factories and installations that were needed, while at the same time recognizing that the time factor was critical (contract was signed on 23rd March 2015, the land was provided by Bane NOR on November 2015 and start of TBMs excavation was planned for September 2016).

Finally, the main logistics areas were split in two: one for the three precast factories required to produce 140,000 precast concrete segments (including the spoil area), this being placed at level 169. The second area is the adit tunnel entrance, the access to both the main tunnels being placed at level 150. Finally, the design of the lower level was completed and allowed for the integration of facilities within the confined space that was created by the

Figure 1. 3D design of Auxiliary, transport, adit, rescue tunnels and caverns where TBMs are assembled.

excavation of the portal area, in addition to this the tunnels access and necessary wide logistical roads were also constructed. The Acciona Ghella Joint Venture then placed in this area the grouting plants, water treatment plant, substations, workshops and warehouses.

2.1 *Precast factories*

The project needed to have a minimum storage capacity on site of around 2,500 precast concrete rings; therefore, the precast factories had to be assembled in order to start segment production well in advance of the TBMs.

AGJV has built a large concrete segment factory at the construction site to ensure the continuous supply of precast concrete elements to the TBMs. The entire area is about 20,000 m^2 and consists of three factories with three production lines, batching plants and other auxiliary installations.

Figure 2. Aerial picture of the lower area.

Seven concrete segments are needed in order to assemble one complete tunnel ring. The factories were designed to produce nearly 20,000 complete rings, or 140,000 concrete elements, in addition to 20,000 invert elements.

2.2 Tunnel facilities and logistic configuration

In the lower area, different plant and equipment have been assembled that will feed the necessities of the four TBMs. The main facilities are:

- Crushing plant
- Conveyor belts
- Grouting plants
- Water treatment plants
- Ventilation system
- Workshops and warehouse
- High and low voltage transformers

Another key element for this type of continuous production site, and one that is just as important as the design of the whole site, is how the site functions logistically on a day-to-day basis. Amongst other activities, the logistics team is responsible for the coordination of tasks such as daily internal and external deliveries on and into the site, the coordination of warehouse operations and the management of workers shifts and the continuous rotation of such. One of the main activities undertaken by the logistics team is the delivery to the TBMs of workers, materials and segments.

The integration and coordination of all the equipment installed into the main areas previously described above has been one of the key elements and success factors on the EPC TBM Follo Line project. On the top of the described areas, a common "construction green" area, that includes offices, barracks for 450 workers and a 24 h canteen has been assembled close to the site for the white and blue-collar employees.

The integration of all these facilities and activities has been the main purpose of the two in-house developed software tools: The Cutter Management System and the COI (Connected Operational Intelligence system).

3 CUTTER MANAGEMENT SYSTEM

The cutter discs are a key element of any tunnel that is bored with the use of Tunnel Boring Machines.

Cutter discs are steel tools integrated into the front part of the machine (cutter head), and are the main elements responsible for the excavation of the TBMs. Efficient cutter disc maintenance is key on any TBM project, as when cutters must be changed in the middle of a tunnel, the costs of downtime as well as the cost of refurbishing or replacing the cutters can be huge.

The number of cutter changes in the Follo Line EPC Project was expected to be in the order of 1,000 cutters for the four TBMs each month; this was across a 24 month planned excavation schedule. Therefore, 24,000, 200 kg cutter changes had to be undertaken. This called for the best possible design to ensure cutter changes could be performed as quickly and efficiently as possible so that TBM production interference was reduced to a minimum. The improvement to the number of changes and the control and maintenance of this activity was critical for the project and was the main purpose of the development of the Cutter Management System, Acciona invests around 200 M € on a yearly basis in R&D, and the development of this software is part of this investment. The cutter monitoring system is a tool developed by Acciona's IT department for the control and management of cutter changes. The main purpose of the tool is to facilitate the daily activities of cutter wear control and cutter changes.

Previously and under the old system TBM cutter control was managed through a classic paper-based system. The TBM worker got access into the cutter head chamber and after a visual check of the cutter's wear status, decided how many cutters need to be changed. A

Figure 3.   Tablets used for cutter disk control.

second operator collected the data and updated it into a daily paper sheet, during the day this data was manually submitted to the main office where a site TBM engineer introduced all the data into an excel file. This process has served a purpose for many years, but it was clear improvements were needed in order to reduce occurrences of:

- The need for continuous access to check the cutter status, along with the safety risk this operation presents,
- Paper is the main part of the process. Clear risk of losing the data, lost paper, uncertainties in the reporting.
- Number of resources and timing of the process. It can be a long time until the information is processed.
- Decisions taken without the availability of the right data or information

With the Cutter Management System, some of the above constraints are avoided. The tool is based upon and works via a tablet device concept, which is robust enough to allow for use by TBM workers on a daily basis. The system needs to operate in a very aggressive environment, and the TBM cutter head provides a very unforgiving environment in which to operate. Each device is assigned to each TBM and can be used by different users. The main persons responsible for the management of the software and the tablet is the cutter head leading hand, it is he who controls and decides the workers who enter the cutter head and his is the final decision in terms of whether the cutter disc has to be changed or not.

The data that is generated in the cutter head chamber feeds both the system and the reporting tool in the same way as the previous paper-based system did. However, the Wi-Fi connectivity of the TBM and the digital platform now allows for the instantaneous transfer of data. In addition to this, all tablets and devices allow for real time access to data. This is a clear improvement on the whole process and creates a much more robust and useful platform for control and analysis.

The two main tools that are used in the application are wearing forecast and cutter head operation.

3.1   Wearing Forecast

The wearing forecast option allows users to estimate the status of the cutters without the need to access the TBM cutter head chamber.

It takes as the starting point for analysis from the last revision made, the application then makes the wearing forecast based on the last rings excavated. It is therefore key to correctly report the extent of the wearing as of the last ring so the model can be accurate.

The software produces an image of the cutter head in graphic form on which the real position of each cutter is shown. The wearing is estimated based on a color coding system that

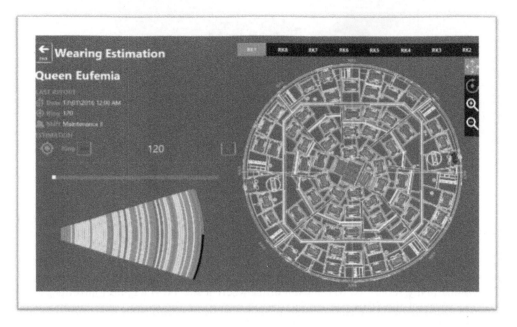

Figure 4. Wearing forecast menu. Screen shot.

provides the status of each cutter and the potential for damage during the subsequent advances of the TBM. This then allows the operator and engineers to make the decision whether or not to change any of the cutters in the short term. The information as introduced by the cutter head worker then flows immediately to all the teams, so that analysis and discussion can take place immediately allowing a decision to be taken in real time thereby introducing increased cost efficiency into the overall process by reducing the time, efforts and risks of making wrong decisions by mistake

3.2 Cutter head operation

The second main tool of the software is the cutter head operation option. The cutter operation mode gives the opportunity to the cutter head operator to make real time analysis on the status of the cutter head, checking the historic reporting of each cutter including how many times it has been changed. This can be done under 3 options: full cutter head view, detailed view and the cutter profile option.

Based on the information provided by the system, the time for review, inspection and the preventive measures to take in advance are all optimized; the process is therefore improved in terms of time and cost.

3.3 Conclusions

The implementation of the Cutter Monitoring System on the Follo Line EPC TBM project has been a huge step forward in the cutter analysis and maintenance field. The main advantages and improvements of the new system are:

- It is user friendly, so it speeds up the whole process.
- Reduction in the use of paper, as the system is fully integrated onto a digital platform.
- Reduced manual steps, so reducing the potential for human error.
- One single resource manages the whole process, clearly reducing the cost and resources involved in the process

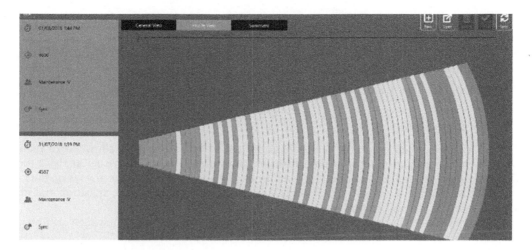

Figure 5. Cutter head operation. Cutter profile mode.

• All the project information is shared, allowing common knowledge to be shared amongst the team.

4   CONNECTED OPERATIONAL INTELLIGENCE SYSTEM (COI)

The Follo Line EPC TBM project is especially unique due to all four TBMs starting from the same point. This is very unusual and generates from a logistics point of view a challenge in terms of the coordination of activities. Any improvement in this area is of great advantage from a cost efficiency perspective.

Until now, the construction industry has evolved through incremental improvements, being generally slow or reluctant to embrace new technologies. Acciona Infrastructure is focused on improving (with the help of new technologies) how to better manage their projects. One of the innovative tools that has been implemented for the Follo Line project jointly by Acciona´s Digital Transformation Department, and technology company Worldsensing, is the connected operational intelligence system (COI).

The implemented solution is a GIS real-time and historical decision support system, that through four specific use cases provides enhanced data visualization, real time monitoring and historic data access. The system also provides on-site personnel with a single tool with which to perform status checks and record historic data of the project for future reference.

The main characteristics of the COI are:

• Support the user (Site Manager), with real-time information and analytics to help with quick decision-making.
• Real-time info regarding Health & Safety and Environment conditions.
• Simplify and automate reporting requirements.
• Be intuitive and simple to learn and implement by users.

After analysis and discussion with the site team about what areas were more suitable for the use of the COI system, four were identified:

• TBM Productivity: to automatize the TBM data collection process and analyze it in real-time to deliver tailor-made reports for quick decision making.
• Tunnel Precast Segments Production Monitoring: monitor production progress in real-time for the three production lines. Quality control analysis with tailor-made reporting.

- SCADA and CCTV: integration of the SCADA system and cameras into the tool, to provide visual information and real-time status of sensors.
- Tunnel Traffic Monitoring: plan and control tunnel traffic based on width and height allowances.

The first two will be described in this paper.

4.1 *TBM Productivity*

The TBM productivity is one of the tools under the COI platform that allows us to reduce paper dependency and eliminates data transfer onto Excel. It provides visual analysis tools that increases stakeholder's visibility of operations and also provides online always-on data. It creates and manages a massive big data archive for legacy analysis, data acquisition and provides easy access to pdf reports.

Previously the modus operandi for TBM reporting was based on the following workflow:

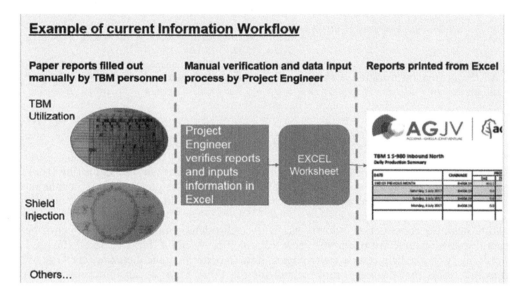

Figure 6.   Previous TBM production report workflow.

This reporting methodology is very time consuming in terms of resources, as all the key data and parameters that the TBM produces has to be manually transferred to excel files suited for specific analysis. This is done in different steps: the TBM operator fills the TBM production report in the TBM cabin and when he finishes his production shift, gives it to the site engineer. The engineer when back from the TBM at the office, transfers all the data as registered in the paper report to the excel file. It is then that reports can be produced and different key aspects of the TBM daily production (TBM performance, ring build time, pressures and consumes in the shield injection, pregrouting, etc.) can be analyzed and studied by the managers to help with improvement, adjustments and cost analysis, etc. With the use of COI and the TBM production tool, the data is managed as follows:

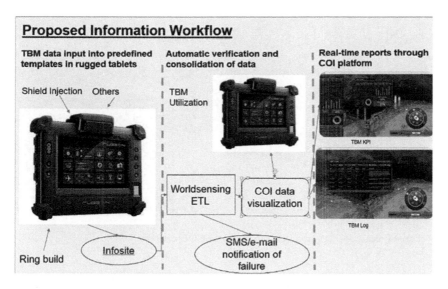

Figure 7. COI TBM production report workflow.

Under this new methodology, the TBM operator introduces all the data in robust tablets, all data is then registered by the software that can then be made instantly available to the team. Different kinds of reports can then be made available for analysis; these are provided in a much more user-friendly style and are totally integrated and linked to the overall process with the use of just one resource and in much shorter time.

This allows the management team to report and review the daily TBM data and share the information in a much better way than before.

4.2 Tunnel Precast Segments Production Monitoring

As previously mentioned in this paper, 120,000 Precast Segments are produced on site in three concrete precast factories. As with any industrial precast factory, every second or minute that can be saved through improvement in the industrial production process can translate into savings in terms of time and cost. With the COI Tunnel Precast Segments Production Monitoring tool operating under a similar concept as the TBM Productivity menu, the data of the daily production reports (previously based on paper reporting and posterior excel file transfers) are completed with the use of tablets by the precast operators and foreman. The three factories are integrated under the same platform and the software integrates all the dates related to

Figure 8. Some examples of daily TBM production reports.

4260

Figure 9.   Three Precast daily production report under COI platform.

concrete production, segment defects, segment repairs, consumables, timing of curing, transport, etc. under a full three factories integrated report. A much more comprehensive review using real data analysis can then be done by the managers, with decisions and reports being optimized with much less resources and therefore becoming less time consuming.

4.3   *Summary*

The COI is a unique project tool that integrates a very large amount of data being produced by both the TBMs and precast factories. It provides data access via a single platform for analysis and reporting purposes, with the use of much less resources than previously utilized whilst at the same time avoiding potential mistakes and loss of information. In summary, it provides:

- Automatic data cleanup and business rules
- Easy correction of imported data
- Elimination of data mangling in Excel
- Visual analysis tools covering all KPIs
- Production traceability
- Increases stakeholder's visibility of operations
- Online 'always-on' data
- Archive for historical analysis

5   CONCLUSION

The Follo Line EPC TBM project is a very complex and iconic infrastructure project. The use and integration of innovation and new technologies into daily production activities has changed the way of managing big data and reporting, as a result huge improvements of the whole production cycle and cost and time improvements have been achieved. This is clearly the way that future infrastructure and construction projects must be oriented, with the use and integration of new technologies allowing continuous improvement of the cost, safety and efficiency of the whole production chain.

ACKNOWLEDGEMENTS

Special thanks to Miguel Angel Heras (Acciona´s Digital Transformation Director) and his team, Ivan Lazaro (Acciona IT Director) and his team, Jacobo Arnanz (AGJV Construction Director), Worldsensing team and all AGJV team for their fantastic and professional work they have done at the Follo Line EPC TBM Project.

Tunnels and Underground Cities: Engineering and Innovation meet Archaeology,
Architecture and Art, Volume 7: Long and deep tunnels – Peila, Viggiani & Celestino (Eds)
© 2020 Taylor & Francis Group, London, ISBN 978-0-367-46872-9

Brenner Base Tunnel, Italian Side, Lot Mules 2–3: Risk management procedures

A. Voza & R. Zurlo
BBT SE, Bolzano, Italy

S. Bellini & E.M. Pizzarotti
Pro Iter, Milan, Italy

ABSTRACT: The Brenner Base Tunnel is mainly composed by two single track Railway Tunnels and an Exploratory Tunnel; Lot Mules 2–3 concerns a 22 km long stretch on the Italian side. It crosses the South part of mountainous dorsal between Austria and Italy, under overburdens up to 1850 m, consisting of rocks both of the Southalpine and Australpine domains, separated by the major Periadriatic Fault. The tunnels have to be carried out both with Mining and Mechanized (TBM) Methods, with average dimension ranging from 7 m (Exploratory Tunnel) to 20 m (Logistics Caverns). The forecasted rock mass behaviour varies from rock burst to squeezing phenomenon. The paper illustrates the risk management procedures planned at the design stage and the first experiences of their application: monitoring, surveys and tests during the advancement, guidelines for the application of excavation sections, design and operational risk identification, assessment & analysis, ownership, mitigation & control phases.

1 INTRODUCTION

The Brenner Base Tunnel (BBT) system mainly consists of two 55 km long parallel Railway Tunnels (at a distance varying from 40 to 70 m), and an Exploratory Tunnel. The Lot Mules 2–3 on Italian side includes 22 km of Railway Tunnels, from km 32 (Austrian border) to km 54 (close to the South access), and 17 km of Exploratory Tunnel, from km 10 to km 27 (Figure 1a).

For most part of their length Railway Tunnels only host one track each, but close to the adjacent lot 'Underground Crossing of the Isarco river' they become two-track and three-track.

The Exploratory Tunnel is located between the two main ones, 12 m lower than the main axes (Figure 1b); only close to the South adit (from km 51+600 Eastern pipe) the Exploratory Tunnel deviates from its central position, and is no longer aligned with the Railway Tunnels up to its Aica South adit.

Approximately at Eastern tunnel chainage km 44+700 there will be a 530 m long Emergency Stop with connection tunnels for passenger evacuation and ventilation tunnels, which shall be used to manage emergency situations in case of fire. The Stop is connected to a ventilation and access tunnel that runs parallel to the line tunnel for approximately 4.5 km and subsequently inserts onto the 1720 m long Mules access tunnel, which connects the system to the outside. The tunnel system is completed by 69 cross passages (one approximately every 333 m) that provide an escape between the pipes and host technical rooms and firefighting water tanks. Vertical shafts connecting the cross tunnel with the underlying Exploratory Tunnel will be raised for plant connections and hydraulic purposes approximately every 2000 m. Close to the Mules access tunnel there is an underground ventilation station, directly connected to the outside by means of a 60 m long vertical shaft. The intersection between Mules access tunnel and Exploratory Tunnel/Railway Tunnels hosts a system of tunnel and chambers which allow to transfer underground most of construction logistics.

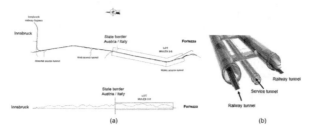

(a)   (b)

Figure 1.   (a) BBT key plan with identification of Lot Mules 2–3. (b) General system configuration.

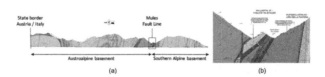

(a)   (b)

Figure 2.   (a) Geological section along tunnel. (b) Detail of Mules Fault Line.

From a geological point of view, the BBT develops across the main tectonic units forming the Alpine chain, with a maximum overburden of 1850 m. These units, which consist of several overlapping layers, are what remains of the collision between the European plate and the Adriatic (African) one; in the design area they form a dome, at the center of which it is possible to identify the Pennidic and Subpennidic units of the Tauern window, i.e. the deepest tectonic units that form the core of the Alps. Southward, the BBT crosses the fault zone that forms the Periadriatic Lineament, which separates the Austroalpine basement from the Southern Alpine one (Figure 2).

Several typical sections were designed, which vary based on both tunnel type (Railway Tunnels, Exploratory Tunnel, Cross Tunnel, Emergency Stop and other logistic works), excavation method (in part mining methods, but mainly with TBM) and geological and geomechanical conditions.

2   RISK MANAGEMENT IN DESIGN PHASE

As known, risk is the product of the measure of the probability of occurrence of a certain phenomenon and of the value of its consequences, and the geotechnical risk management is the global and systematic process of geotechnical risk identification and of its mitigation and control during construction.

The phases of risk management, dealt with in sequence consistently with the progress of the project and its implementation, as further discussed below, are: risk assessment, risk analysis, risk management.

2.1   Risk assessment

The first phase consists in the evaluation or identification of the risk, encompassing the recognition of hazards (seen as a situation or physical condition that has the potential to generate harm, or, in any case, to give rise to unintended consequences) and in the evaluation of the corresponding probability of occurrence and of the specific consequences.

Key steps in the risk assessment phase, which in general can be performed either by means of quantity-, semi-quantity-, or quality-based procedures (as in the present case), are listed below:

1. Identification of hazards related to the specific project (PRS - Potential Risk Situation).
2. Evaluation of consequences and impacts.
3. Estimate of the probability of occurrence.
4. Risk quantification.

5. Risk classification.

2.1.1 *Identification of hazards*
The risk assessment or identification began with the definition of dangerous hazards (Hazard) significant for the works concerned, meaning with Hazard (or Potential Risk Situation - PRS) the phenomena that have a potential impact in terms of negative consequences, i.e. they are the source of the resulting risk.

The possible hazards examined for the project are listed in Table 1.

2.1.2 *Classification of risks*
All PRS considered for the project can essentially lead to the three risk categories listed below:

a. Conditions such as to block the regular excavation process (e.g. shield entrapment for excavations with shielded TBMs). These conditions can be found in truly deteriorating geomechanical situations.
b. Collapse or significant damage of the coating. These are less extreme conditions (therefore more probable!) than those referred to in point a), but more subtle and difficult to highlight and predict; these conditions can still be just as important in terms of value of the harm.
c. Impacts on environment and other structures.

Table 1 shows the association between risk categories and Hazard.

2.2 *Risk analysis*

Following the identification and classification of the possible risk sources, the analysis phase allowed to organize the risks in terms of probability of occurrence and impact. It should be noted that, where necessary, the risk management process also provides for the introduction of new risk elements after the process has started by rechecking the classification of the items and the recalculated risk values.

Starting from the organized list, the following phases consisted in the planning of mitigation and management actions. Finally, the risk acceptance criteria were applied defining and comparing the most appropriate strategies and the most suitable design solutions, the most suitable equipment to be used, the implementation of procedures deemed appropriate and the contingent actions.

To summarize, the performed phases of risk analysis are:

1. Drafting and/or updating of the risk register
2. Application of risk acceptance criteria
3. Identification and analysis of possible mitigation measures to conveniently limit the calculated risks.

The identification of the mitigation actions for the expected Hazards is reported in the aforementioned Table 1; noteworthy in this regard, is that the special measures Type 1, 2, 3 and 4 consist in:

- Type 1: if required, installation of ahead drainages; execution of cavity reinforcement (grouted fiberglass bars) from TBM shield and head.
- Type 2: installation of ahead drainages; consolidation and reduction of rock mass permeability on the front face and boundary by means of MPSP (Multi Packer Sleeved Pipe) cement mix injection from fiberglass bars.
- Type 3: stop of Exploratory Tunnel TBM and excavation of first by-pass (mining methods) to reach the damaged area; execution of drainages and MPSP (Multi Packer Sleeved Pipe) cement mix injection from fiberglass bars, from by-pass to Exploratory Tunnel; restart of Exploratory Tunnel TBM; excavation of a second by-pass and execution of drainages and injections from by-passes to Railway Tunnels.
- Type 4: execution of radial perforation around Exploratory Tunnel and Railway Tunnels, followed by resin injections to reduce rock mass permeability.

Table 1. Considered hazards, risks and mitigation measures.

Hazard	Risk Cat.	Mitigation measures Mining methods	Mitigation measures TBM
Face instability	a	Introduction/increase of face reinforcement.	Implementation of specific measures (Type 1, 2, 3).
Squeezing	a, b	Moving to a heavier typical section/ with deformable lining.	
Asymmetrical stresses and strains	b	Moving to a heavier typical section; increasing of bolting in unstable areas.	
Cave in *	a, c		Excavation interruption; geological, hydrogeological and geomechanical investigation; implementation of specific measures (Type 1, 2, 3).
Spalling and crumbling	b	Increasing of bolting in unstable areas.	
Faults (filling instability, bad general conditions, water or gas inflow)	a, b	Thin fault lines: moving to a heavier typical section. Thick fault lines: geological, hydro-geological and geomechanical investigation, then definition of proper mitigation measures.	
Water inflow and/or water pressure	a	Introduction/increase of ahead drainages.	Implementation of specific measures (Type 1, 2, 3) with ahead drainages.
Dissolution or transport of finer particles due to water inflow	b	Introduction/increase of ahead drainages; introduction of face reinforcement (cement paste, resin).	
Presence of gas	a	Strengthening of ventilation system.	
Mixed face lithology	b	Moving to a heavier typical section; increasing of bolting in unstable areas.	Excavation interruption; geological and geomechanical investigation; implementation of specific measures (Type 1, 2, 3).
Swelling	a, b	Reduction of water use; reduction of surface exposure time; if possible, adoption of rigid first stage lining; if necessary, moving to a typical section with deformable lining and bolting; appropriate final lining sizing.	Excavation interruption; geological, hydrogeological and geomechanical investigation followed by the definition of specific measures; reduction of water use; appropriate final lining sizing.
Rock blocks instability	a, b	Moving to a heavier typical section; increasing of bolting in unstable areas.	Excavation interruption; geological and geomechanical investigation followed by the definition of specific measures
Rock burst	a	Moving to a -TRb typical section; excavation slowing; implementation of in ahead drillings to allow stress reduction.	Excavation interruption; geological and geomechanical investigation followed by the definition of specific measures; excavation slowing; implementation of ahead drillings to allow stress reduction.
Interference with other cavities	b	Reinforcement of rock mass between the cavities.	
Instability of structures during multi stage excavation	a	Appropriate connection between structures (e.g. steel mesh overlapping, anchored with bolts).	***

(*Continued*)

Table 1. (*Continued*)

Presence of radioactive minerals	a		Adoption of specific protection for staff members; reduction of exposition time.
Clogging	a	**	Application of soil conditioning agents; reduction of water use.
High temperatures	a		Strengthening of cooling system.
Impact on water resource	c	**	Implementation of specific measures Type 2 and/or 4.

* Not relevant for the present work (excavation in rock with overburden always higher than 80 m).
** Not relevant for mining methods excavations.
*** Not relevant for TBM excavations.

2.3 Risk management

The methodology of risk management ends the overall process involving also the execution phase and is developed in three phases:

1. Choice of the mitigation measures to be expected and of their extent
2. Check of the mitigation measures based on the monitoring data collected during construction
3. In case of exceedance of the threshold values, introduction of changes or alternatives to bring the risk level below the acceptable threshold.

In particular, in the design phase it was envisaged to manage and control the risk of the construction process (phases 2 and 3 of the previous list) through key performance indicators (KPI) defined during the design and periodically verified in dedicated meetings providing also for possible countermeasures in case of exceedance of the threshold values. The implementation of this process during construction is described in the following Chapter 3.

A cyclic risk management process, specifically reiterating for the entire duration of the project, receiving reviewed and regularly updated inputs, was defined to monitor and manage all possible risk sources. This choice derives from the consideration that, especially when dealing with geotechnical aspects, the degree of progressive knowledge and the possible variations in design solutions and implementation technologies require a constant update of assessment, analysis and risk management procedure.

In this phase, the contribution of all professionals involved in the design and construction process (designers, consultants, geologists, construction managers, surveyors, contractors, etc.) is essential, because the parties engaged directly only in the construction works are often too optimistic or tend to underestimate the consequences of the risks, since they are unknowingly influenced by the progress of the works and oriented mainly to their completion, thus failing to capture with complete objectivity the aspects that may represent a source of risk.

2.3.1 Monitoring and Key Performance Indicators (KPI)

The key indicators are the values closely related to the significant variables revealing the safety of the excavation. If it is true that, in general, all the data gathered from the monitoring and surveying system (and further significant data obtained during the excavation), are useful for defining the risk conditions, it was deemed necessary to extrapolate some of them to perform a timely analysis. This does not mean ignoring the others, but to carry out a circular process that includes the following actions:

1. Provide an overall reference framework based on all available data.
2. Perform a quick analysis of the essential data (KPI).
3. Compare the result of the analysis with the framework to confirm or contradict it.
4. Collect additional data, if necessary.
5. Update the frame of reference based on all available data.
6. Repeat the process.

The essential indicators on which the assessment of the degree of risk was based, are those that are most easily and quickly acquired (given their significance to the effects of the analysis) and consequently observable on a more frequent basis. These are summarized in Table 2.

2.3.2 *Thresholds*

Attention Threshold has been defined as the value of any of the KPIs, at which the following actions will have to be implemented:

- Mining methods excavation: all the means and materials must be set up to allow a timely passage to an excavation section heavier than the one currently applied, according to the sequence T2/Rb, T3, T4, T5, T6; see Table 3 for a summary of the support measures of the different type sections.
- TBM excavation: execution of additional surveys.

On the other hand, Alarm Threshold has been defined as the value of any of the KPIs, at which the following actions will have to be implemented:

- Mining methods excavation: Excavation Section heavier than the one currently applied must immediately be adopted, according to the same sequence.

Table 2.   KPI and thresholds.

KPI	Attention threshold	Alarm threshold
Ahead or radial drilling, both with and without core sampling; eventual in situ tests on cores. Analysis of tunnel spoil. **	Small amount of: atypical material with reference to geological model, alterated or clayey material.	Great amount of: atypical material with reference to geological model, alterated or clayey material.
Geomechanical classification of front face and sides. *	Traces of asymmetrical, anisotropic, not homogeneous or not uniform behaviour of rock mass. Sign of possible deterioration of rock mass conditions. Geological discontinuities coherent with geological model and with previous excavation results. RMR near attention threshold of current typical section.	Asymmetrical, anisotropic, not homogeneous or not uniform behaviour of rock mass. Clear deterioration of rock mass conditions. Geological discontinuities incoherent with geological model and with previous excavation results. RMR near alarm threshold of current typical section.
Water flow rate; water physical and chemical properties.	Not in pressure water inflow. Temperature coherent with previous excavation results.	In pressure water inflow. Temperature not coherent with previous excavation results.
Convergence and extrusion monitoring. *	1 % of excavation radius. ***	2 % of excavation radius. ***
TBM parameters. **	Sign of anomalies with respect to previous excavation results. Light reduction of specific energy [J/m^3].	Clear anomalies with respect to previous excavation results. Drastic reduction of specific energy [J/m^3].
Stress/strain monitoring in first stage lining * or in concrete segments **	77% of material design strength	100% of material design strength
Acoustic emission monitoring	Acoustic emission not related to incipient fracture phenomena.	Acoustic emission related to incipient fracture phenomena.

* Only for mining methods excavations.
** Only for TBM excavations.
*** Due to rock mass behaviour in previous excavations in Mules Fault Line (face measured extrusions higher than radial convergence - Fuoco et al. 2016), in this area attention and alarm threshold for extrusion were set to 2 % and 4 % of excavation radius, respectively.

Table 3. Support measures for mining methods typical sections.

Typical section	Application	Support measures
T2	RMR > 60, Overburden < 1000 m	Radial bolting (90°); rigid 1st stage lining (shotcrete).
TRb	RMR > 60, Overburden > 1000 m	Radial bolting (210°); rigid 1st stage lining (shotcrete).
T3	RMR = 41 - 60	Radial bolting (120°); rigid 1st stage lining (shotcrete).
T4	RMR < 41 in Mules Fault Line (better lithologies)	Eventual ahead bolting on face/cavity; rigid 1st stage lining (shotcrete, steel ribs).
T5	RMR < 41 in Mules Fault Line (medium lithologies)	Ahead bolting on face/cavity; radial bolting (270°); rigid 1st stage lining (shotcrete, steel ribs).
T6	RMR < 41 in Mules Fault Line (worse lithologies)	Ahead bolting on face/cavity; radial bolting (360°); deformable 1st stage lining (shotcrete, deform. steel ribs).

- TBM excavation: interruption of advancement, execution of additional surveys, implementation of the specific interventions envisaged, possible definition of other specific actions.

The definition of the threshold values of some KPIs was referred to the average or characteristic values of an advancement process within standard limits. These indicators, whose threshold values have been defined in relative terms basing on the analysis of the previous advancement, are the TBM parameters, the measurement of the water flow rate and that of the main characteristics of inrushes, and acoustic emission monitoring. For surveys, analysis of the excavation material and geomechanical classification of the front and of the excavation walls, threshold values were defined in terms of quality. Finally, the other KPIs (convergence and extrusion measurements, stress-strain measurements in coatings) were based on quantitative threshold values.

The KPIs and related threshold values implemented in the project are summarized in Table 2.

2.3.3 Risk levels, consequent actions and mitigation measures

In Table 4 risk levels, related to KPI and thresholds, are put into correlation with: action that must be performed to better understand the situation; mitigation measures.

Table 4. Relation between KPI, risk levels, consequent actions and mitigation measures.

KPI	Risk level	Consequent actions	Mitigation measures ***
All under attention threshold	Low risk	Only systematic monitoring.	None.
At least one over attention threshold.	Medium risk	Systematic monitoring. Non-systematic monitoring: ahead or radial drilling, with and without core sampling.	Preparation to moving to a heavier/specific typical section. *
At least one over alarm threshold.	High risk	Systematic monitoring. Non-systematic monitoring: ahead or radial drilling, with and without core sampling.	Moving to a heavier/specific typical section. * Excavation interruption; implementation of specific measures (Type 1 to 4); definition of other specific measures. **

* Only for mining methods excavations.
** Only for TBM excavations.
*** If hazard "Presence of gas" and "Presence of radioactive minerals" are over thresholds, specific mitigation measures must be implemented, even if all KPI are under attention threshold; for specific mitigation measures see Table 1.

3 RISK MANAGEMENT PROCEDURE IN CONSTRUCTION PHASE

At present (December 2018), six excavation fronts are active in the Lot Mules 2–3. These fronts have approximately reached the following chainages:

- Exploratory Tunnel: chainage km 15+690 (TBM excavation from km 13+085).
- Railway Tunnels – northern stretch: km 46+288 eastern tunnel, km 46+314 western tunnel (mining methods excavations).
- Railway Tunnels – southern stretch: km 51+387 eastern tunnel, km 51+424 western tunnel (mining methods excavations).
- Access tunnel to Emergency Stop: chainage km 1+825 (mining methods excavation).

The risk management process, started during the design phase, has assumed even more importance in the construction phases (dynamic process), during which an appropriate risk management is being pursued through monitoring activities performed along with the on-going construction works. In fact, in a project of such complexity only the daily update of the monitoring measures and the association between these and the actions taken at the different chainages during the excavation advancement, enable the works to be developed under a continuous and careful risk control.

For TBM excavations the form and method of presentation of forecasted monitoring is being defined. For mining methods excavations, the procedure already tried and tested consists in the following steps:

1. Execution of monitoring (according to the reading frequencies prescribed in the design phase) and transmission of the results to Works Supervisor, Designer and Contractor. In particular, data relating to the following KPIs are transferred on a daily basis: surveys, ahead drillings, geomechanical classification of front face and sides, convergence and extrusion monitoring, acoustic emissions monitoring, stress/strain monitoring in first stage and final lining (Figures 3, 4).
2. Data analysis carried out by the Designer and provision to the Works Supervisor of a simplified report (Figure 5) issued on a daily basis for critical areas and in any case within 48 hours from data reception.

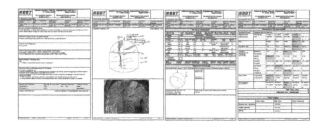

Figure 3.   Example of monitoring reports: geomechanical classification of front face.

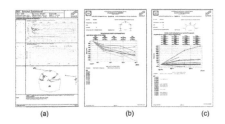

Figure 4.   Examples of monitoring reports: (a) convergence; (b) steel ribs deformation; (c) radial bolts deformation.

Railway tunnel – Northern advancement – Eastern pipe

Chainage [km]	Equipment	Notes	Critical issues	Recommended actions
46+496 - 46+693 (0-300m from current front)	CO	Stable.	x	A
46+585 (90m from current front)	CE	Of the 10 bars installed, 8 are functioning. Of these, one on the left side (= West) has exceeded the yield strength, even if it seems to be a spike during; the other bars register deformations below threshold values. Deformations are stabilising.	x	B
46+863 (270m from current front)	CE	Of the 10 bars installed, 8 are working. Of these, three on the left side (= West), one in the crown and one on the right side (= East); all have exceeded the yield strength, as already happened for the other 5 bars currently not working. The deformations are growing, but stabilising.	x	C / D (if necessary)

Acronym definitions - Equipment:
CO = convergence measurement station
CE = monitored steel rib
ES = extensometer
ESO = extrusion measurements with optical prisms
MP = flat jack test
RD = strain gauge bars in final lining

Acronym definitions - Recommended actions:
A = None
B = Control of the correct functionality of the installed equipment / repetition of measurement
C = Control of the conditions of the first phase lining
D = Implementation of additional confining measures

Figure 5.   Examples of designer daily report.

3. Weekly coordination meeting attended by the Works Supervision Staff, the Contractor and (if necessary) the Designer: discussion of monitoring data and other critical issues that may have occurred during the excavation and choice of modifications or alternatives to be carried out to bring the risk level below the acceptable threshold or to optimize the excavation times.

It is important to highlight that, when the mitigation measures are being chosen, the attendance of all parties involved in the works implementation (Works Supervisor, Contractor and Designer) is essential, so that the degree of residual risk which is left over after the advancement is brought to their attention, and consequently managed and kept under control.

4. Drafting, by the Designer, of a monthly report on the monitoring results regarding all active fronts, including the data within the general context of the works, also in the light of the findings of the previous sections, and providing possible forecasts on future behaviours; at the same time, drafting of a monthly report on the quality of the data received, encompassing as well possible recommendations for improving these data (formats, presentation, reading frequency ...). Delivery of both reports to the Works Supervision Office.

Particularly important was the monitoring of the most difficult section so far crossed, which is the Mules Fault Line (Figure 6). This area is composed by Phyllite, Black Schist, and Mica Schist, often strongly damaged by tectonic actions, alternated to some levels of more competent rock masses (Gneiss and Quartzite); from a geomechanical point of view, the RMR values vary from 20 - 30 in the most damaged section and up to 30 - 50 in the sections with better characteristics. In this section the behaviour to excavation is characterized by front extrusions exceeding the convergences, presumably because of the geo-structural layout of the rock mass having a schistosity perpendicular to the advancement axis.

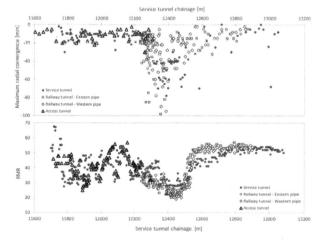

Figure 6.   Maximum radial convergence and RMR in Mules Fault Line, projected on Exploratory Tunnel chainage. All data derived from Lot Mules 2–3 monitoring system, except Exploratory Tunnel convergence and RMR for chainages lower than km 12+461, taken from (Fuoco et al. 2016).

In the Mules Fault Line, the convergences were always kept below the attention threshold values (max 3.5 - 5.0 cm), with the exception of the Exploratory Tunnel stretches km 12+350 - km 12+520, in which the convergences reached values of 8.0 - 10 cm, close to the alarm threshold. In the same section the behaviour of the rock mass turned out to be pushing and crumbling; it was possible to observe considerable stresses on the first phase linings and material was released from the front. In this section, the presence of an efficient risk management system, supported by continuous monitoring activities, proved to be particularly valuable, as it allowed the geological-geomechanical forecast to be updated while advancing (especially for what the advancement of the Exploratory Tunnel is concerned); this allowed to foresee the most critical sections according to the results of previous excavations and to take the appropriate countermeasures.

4 CONCLUSIONS

Risk management is a global and cyclic process, which affected the design but mostly the construction phase of the works at hand.

First of all, the hazards were identified evaluating their possible occurrence and consequences; the risks (risk assessment) were recognized, then organized in a register depending on probability of occurrence and impact; the possible mitigation actions (risk analysis) were then defined relying on this list. Afterwards, a risk management process was implemented based on key performance indicators (KPI), identified during the design phase and periodically verified in terms of thresholds, which were specified a priori or depending on previous advancements.

The above described procedures have led to a risk management process based on monitoring and observation of the excavation behaviour with the contribution of all the parties involved (Works Supervisor, Contractor and Designer). In this way, it was possible to up-date the geomechanical-geological forecast while advancing; that was particularly convenient for the major tectonized zone so far crossed, the Mules Fault Line, in which it was possible to identify in advance the most critical sections, thus providing suitable adjustments/integrations of the excavation sections.

REFERENCES

AFTES 2009. A code of practice for risk management of tunnel works. *Tunnel et ouvrages souterrains.*
Barla G., Lombardi A., Malucelli G., Martinotti, P., Perello, P. Pizzarotti, E.M., Skuk, S., Zurlo, R. et. al. 2010. Problemi di stabilità del fronte durante lo scavo del Cunicolo Esplorativo Aica-Mules della Galleria di Base del Brennero. *Proc, XI ciclo di conferenze di meccanica e ingegneria delle rocce.* Torino.
Bellini, S. & Pizzarotti, E.M. 2015. BBT: approfondimenti sulla caratterizzione geomeccanica degli ammassi rocciosi a seguito dello scavo del Cunicolo Esplorativo. *Expotunnel 2015.* Milano.
Fuoco, S., Zurlo, R., Marini D., Pigorini A. 2016. Tunnel excavation solution in highly tectonized zones. *Proceedings of World Tunneling Congress 2016.* San Francisco.
Gattinoni, P., Pizzarotti, E.M., Scesi L. 2015. Geomechanical characterization of fault rocks in tunnelling: The Brenner Base Tunnel. (Northern Italy) *Tunnelling and Underground Space Technilogy.*
Pizzarotti, E.M. & Skuk, S. 2008. Tunnel di Base del Brennero: modellazione dello stato tensionale dell'ammasso roccioso. *Gallerie e grandi opera sotterranee.*

Tunnels and Underground Cities: Engineering and Innovation meet Archaeology,
Architecture and Art, Volume 7: Long and deep tunnels – Peila, Viggiani & Celestino (Eds)
© 2020 Taylor & Francis Group, London, ISBN 978-0-367-46872-9

Evaluation of excavation method for silo-shaped deep large waste repository underground cavern

K. You
The University of Suwon, Hwaseong-si, Korea

C. Park & J. Park
ESCO Consultant & Engineers Co., Ltd., Seoul, Korea

ABSTRACT: The major concern over large underground silo-shaped caverns, consisted of domes and body shafts, is to select the proper excavation method because they are excavated sequentially by subdividing the large cross section considering safety, constructability and stability of working face. In this study, the excavation method and behavior characteristics of a large dome-shaped cavern, which is the upper part of an underground silo, were evaluated by 3-D numerical analyses in which each excavation sequence was simulated along with 3-D geometry. Bench type and pie type were finally proposed as possible excavation methods and compared each other in terms of construction sequence, support system, constructability, and stability, etc. The results showed that the pie type excavation sequence is more applicable than that of the bench type excavation. In the pie type excavation sequence less stress concentration was observed compared to the bench type excavation sequence. The orders of tunnel displacement and shotcrete stress were also comparatively small. In addition, extensive 3-D numerical analyses were performed considering full excavation modeling for the dome and body shaft of a silo. The results showed that construction with the proposed excavation method was safe and feasible.

1 INTRODUCTION

Typically, underground spaces such as roads, railways, subways and utility tunnels are are constructed as small to medium sized tunnel shapes. As the construction of large underground space is currently limited to a few special facilities such as oil storage facility, LNG storage, agricultural and fishery storage, etc., continuous efforts in planning, designing, and construction technology of a large scale underground space are necessary.

The first phase of Wolsong low and intermediate level radioactive waste (LILW) repository facilities are consisted of 6 large silo-typed caverns as shown Figure 1. The silos are located deeper than 80 meters below sea-level at EL.(-)80m~(-)130m with cylindrical shape. The size of silo is about 50 m in height. The top of silo was planned as dome shape with a diameter of about 30 m and a height of about 16 m. The low part of silo, defined as body shaft, was planned as cylindrical shape about 24m in diameter and 34 m in height. These 6 large underground caverns have been evaluated to be a very high level of difficulty in construction due to the insufficient construction experience and unfavorable geological conditions. In the design and construction of large underground caverns, the selection of excavation method is very important for the purpose of ensuring the stability of the underground excavations because the size of caverns is 10 to 20 times larger than that of the conventional traffic and utility tunnels. This paper discusses some cases of excavation methods considering subdivision of a large cross section and securing of construction safety. By performing extensive 3-D numerical analyses which simulate each excavation step, the behavior characteristics of an excavation method are evaluated.

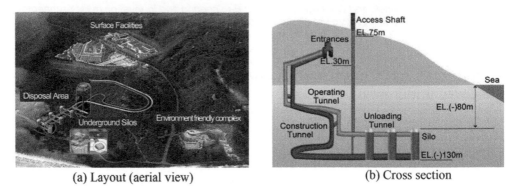

| (a) Layout (aerial view) | (b) Cross section |

Figure 1. Layout of the Wolsong LILW Disposal Center (KORAD, 2011).

2   GEOLOGICAL CONDITION

The site mainly consists of cretaceous sedimentary rocks, tertiary plutonic rocks and intrusive rocks as shown in Figure 2 (Seo et al., 2011). The cretaceous sedimentary rocks, which are the predominantly alternating strata of mudstone, siltstone, and sandstones, belong to the Ulsan formation which is correlated to the upper formation of the Gyeongsang sedimentary basin in southeastern of the Korean Peninsula (Park et al., 2009). The main region in which the silo will be located is mainly composed of granodiorite.

The results of the site characterization were used as a basis for its 3-D numerical analysis as well as information for the construction. Due to the complex geological setting in the project area, proper ground characterization was considered to be the key to success of the LILW repository construction. Detailed in-situ boring investigation was conducted and high-resolution seismic survey exploration and site in-situ tests such as resistivity survey, refraction survey, elastic wave tomography survey were conducted as well. In order to make the silos design feasible, the rock mass at project site was classified considering rock type, rock condition and orientation of discontinuities, etc. The example of geologically predicted cross section

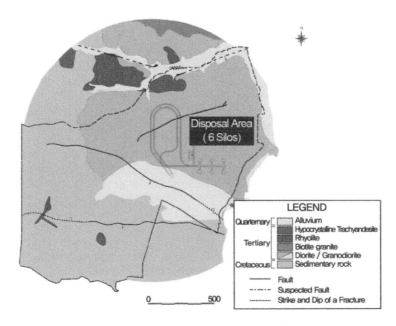

Figure 2. Geological map on the local scale of the Site (Seo et al., 2011).

(a) Cross section with rock grade

(b) Three-dimensional geological modeling

Figure 3. Geological modeling for the site of interest (KTA, 2011).

Table 1. Summary of geotechnical parameters for ground type (KTA, 2011).

Rock Grade	Unit weight (kN/m³)	Cohesion (MPa)	Friction angle (°)	Compliance (MPa)	Poisson ratio
I	2.7	9.6	45	33,000	0.18
II	2.6	4.9	42	19,000	0.21
III	2.5	2.4	38	9,000	0.25
IV	2.3	1.0	32	2,700	0.28
V	2.2	0.2	30	2,000	0.32

and three-dimensional geological structures are as shown in Figure 3. After characterizing the rock types for the site based on the given information, the engineering design parameters of the representative rock grade were determined based on the results of a series of laboratory tests as well as available empirical relations as shown in Table 1.

3 PARAMETRIC STUDY OF EXCAVATION MEHTOD

3.1 Evaluation of excavation method subdividing cross section

In case of full face excavation of large underground cavern, the constructability with time required for mucking and the installation of support should be considered. Therefore, in order to secure constructability and reliability of large section of the tunnel and underground cavern, the subdividing excavation of the cross section is required. In general, most cases of large underground caverns were constructed in good ground conditions and excavation sequences followed a pilot tunnel and multi-subdivided cross section excavation (JTA, 1998). In the tunnel type excavation, there are many cases where the multistage bench excavation method is applied.

Unlike the underground tunnels which are being excavated longitudinally with the same section, the major concern of large dome-shaped underground caverns is the complicated construction sequence resulting from the different 3-D geometry and its behavior with the application of a different excavation method (Park and You, 2015). Generally the construction of dome-shaped underground cavern is based on the sequential excavation subdividing the cross section considering the safety, constructability and the stability of working face. The applicable excavation methods such as multi bench cut, side advance drift cut, and central advance cut with pilot tunnel were selected through the case evaluation of excavation sequences and qualitative 3-D numerical analysis were performed as shown in Table 2.

The advance excavation length and thickness of shotcrete were applied as 2.0m and 250mm respectively. The sensitivity analysis results by perspective of displacement and supporting material stress are summarized in Figure 4. It was analyzed that the case of central advance

Table 2.   Qualitative interpretation of excavation conditions.

CASE/modeling	CASE 1	CASE 2	CASE 3	CASE 4
	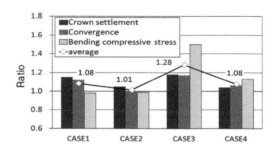			
Crown settlement (mm)	28.4	26.0	29.1	26.1
Convergence (mm)	82.6	73.9	85.0	78.8
Compressive stress (MPa)	3.25	3.26	4.95	3.65

Figure 4.   Increase and decrease ratio by excavation case.

cut with pilot tunnel excavation was relatively advantageous in the aspect of stability. Due to the sensitivity of the subdivided cross section and excavation sequences of the dome-shaped large cavern, the 3-D excavation effect considering 3-D geometry and behavior should be considered.

3.2   Evaluation of dome-shaped cavern excavation method

In the excavation of large underground cavern, subdivided section excavation method is generally applied in order to minimize the occurrence of excessive displacement and stress concentration. Therefore, it is important to select excavation method which has better construction efficiency focusing on the stability and the workspace for construction equipment and installation of the support material such as rock bolts and shotcrete should be considered. In order to evaluate excavation method of dome-shaped cavern, two representative excavation methods, which are defined as bench type and pie type excavations were selected considering such factors as the construction sequence, supports, constructability, stability, etc. as shown in Table 3.

Both eases are alternatives of excavation methods which can improve the constructability and construction time reducing with reference to the existing case study. The bench type excavation sequence has advantages in minimizing temporary support by securing the stability of upper crown excavation area and saving cost of construction. The proposed pie type excavation sequence can improve stability and constructability of working face by advance excavation with tunnel shape and minimum cases of subdivided cross section. It means that same shape of subdivided cross section is repeatedly excavated In order to evaluate excavation method considering stability and constructability of the construction sequence, three-dimensional numerical analysis was conducted for full modeling of large dome shape as shown in Figure 5. Flac-3D (Itasca, 2009) was used in the analysis by applying Mohr-Coulomb model

The ground properties of the representative rock grade were estimated from the results of laboratory tests and field tests as shown in Table 1 and the properties of support materials

Table 3. Outline of dome-shaped excavation method.

Type	Bench type excavation method	Pie type excavation method
Cross section		
Plan		

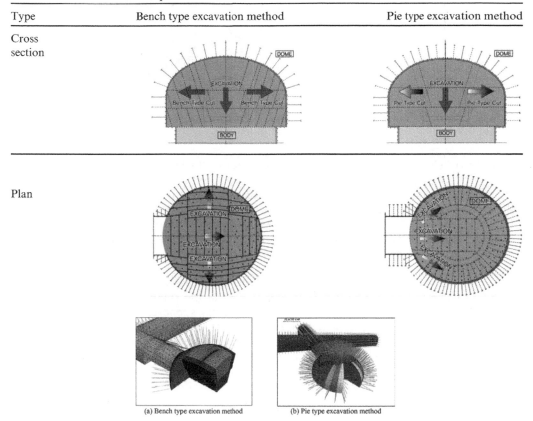

(a) Bench type excavation method　　(b) Pie type excavation method

Figure 5. Numerical analysis modeling.

were applied as 1.0m of advance excavation length, 350 mm thickness of shotcrete, 7.0 length of rock bolt and 20.0m length of cable bolt. The stability of the excavation method was examined by analyzing crown settlement, convergence and shotcrete bending compressive stress of the main support material which was generated by each excavation sequences. The analysis results are summarized in Table 4 and Figure 6.

The crown settlement and convergence by bench type excavation were observed to be 8.4mm and 13.8 mm respectively. The shotcrete bending moment observed 10.42 MPa. In case of pie type excavation, settlement and convergence were estimated to be 5.1 mm, 19.5 mm and shotcrete bending moment was predicted to be 4.08MPa. The stress concentration considerably decreased in pie type excavation compared with the bench type excavation. Therefore, pie type excavation method was considered to be relatively advantageous for ensuring the stability of the dome-shaped cavern excavation, since decrease of stress concentration, crown settlement and convergence compared with the bench type excavation method. The proposed pie type excavation method in this study is advantageous for not only securing stability but also workability because the same shape of subdivided cross section is repeatedly excavated.

Table 4. Result of stability analysis.

Type	Crown settlement (mm)	Convergence(mm)	Compressive stress(MPa)
Bench type	8.4	13.8	10.42
Pie type	5.1	19.5	4.08

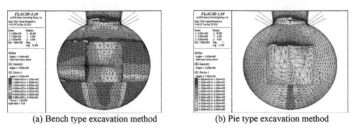

| (a) Bench type excavation method | (b) Pie type excavation method |

Figure 6.   Bending compressive stress distribution.

4   STABILITY EVALUATION OF FULL EXCAVATION OF SILO

For the evaluation of the stability of a silo-shaped underground cavern, body shaft excavation which is consequently excavated after the upper dome excavation should be considered in the analyses. The extensive 3-D numerical analysis was performed considering full excavation modeling of the dome and body shaft of the silo. In this case the stress behavior and displacement of the full body of the silo cavern have to be analyzed considering construction sequences with the 3-D geometry of the dome and body shaft of the silo. The 3-D numerical modeling of full shape of the silo with construction sequences is summarized in Figure 7.

The 3-D numerical analysis was successfully used in the stability assessment of the full shape of silos with emphasis on the ground and primary supports to excavation sequences. The results are summarized in Table 5 and Figure 8 for the selected excavation steps so that the calculated deformation as well as support stress/load were analyzed. The results showed that construction using the proposed excavation method is safe and feasible.

Dome excavation
and support
(pie type excavation)

Body excavation
and support

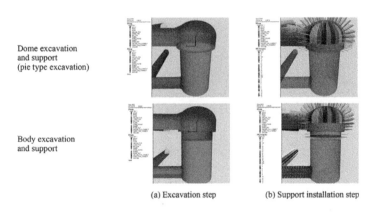

| (a) Excavation step | (b) Support installation step |

Figure 7.   Sequence of numerical analysis summary.

Table 5.   Results of analysis.

			Upper part of dome	Lower part of dome	Full excavation to body shaft
Maximum displacement (mm)	Crown		3.98	3.86	4.02
	Side wall	Dome	2.51	9.27	12.51
		Body	-	-	14.47
Shotcrete bending compressive stress (MPa)	Bending compression	Dome	4.50	4.97	5.67
		Body	-	-	5.81
Axial force(kN)	Rock bolt		62.99	88.52	94.12
	Cable bolt		148.66	151.60	155.07

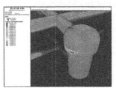
(a) Displacement distribution

(b) Shotcrete compressive stress distribution

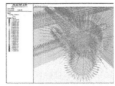
(e) Rock bolt axial force distribution

Figure 8.   Contour plot of 3-D numerical analysis results.

5   CONCLUSION

In this study, the excavation method and behavior characteristics of large dome-shaped and full body of an underground silo were evaluated by 3-D numerical analyses. The comparison and stability analysis of the bench type excavation method and the pie type excavation method were carried out considering the case study, constructability and construction period and so on. The pie-type excavation sequence was introduced as an advantageous excavation method of large dome-shaped underground cavern and stability assessment was performed. The 3-D numerical analyses showed that construction using the proposed excavation method is safe and feasible.

1. The excavation sequences with subdivided cross sections considering stability and constructability is applied when excavating large underground cavern, it should be determined in consideration of the section size, face independence natural ground supporting capacity, etc.
2. It was investigated that most of the silo type cavern excavations was carried out at the center drift subdivided excavation from the case study review on large scale underground cavern excavation methods
3. The pie type excavation sequence showed less stress concentration, and the displacement with support material stress are occurs relatively small amount compared to the bench type excavation sequence.
4. It was found out that the sensitivity of size and number of subdivided cross section in excavation was quite large and consequently the pie type excavation is relatively advantageous as compared with the bench type excavation. The excavation method introduced in this paper is not only advantageous for ensuring safety but also the constructability. In the actual construction, underground cavern for LILW disposal have been completed safely by applying pie type excavation method

REFERENCES

KTA(Korean Tunnelling and Underground Space Association). 2011. *Design of Low and Intermediate Level Nuclear Waste Repository.*
KORAD(Korea Radioactive Waste Agency). 2018. *Brochure of Korea Radioactive Waste Agency.*
Park C. M. & You K. H. 2015. A study on the excavation method of a large dome shape cavern using 3D numerical analysis. KTA 2015 Fall Symposium. *Korean Tunnelling and Underground Space Association*: 170-171.
Park. J.B., Jung, H., Lee, E.Y., Kim, G.Y., Kim, K.S., Koh, Y.K., Park, K.W., Choeng, J.H., Jeong, C. W., Choi, J.S., and Kim, K.D. 2009. Wolsong low and intermediate level radioactive waste disposal center: Progress and challenges, Special Issue in Celebration of the 40th Anniversary of the Korean Nuclear Society. *Nuclear Engineering and Technology.* 41(4): 477-492.
Seo K. W., Baik K. H., and Kim S. J. 2011. Project of the First Wolsong Low-and Intermediate-Level Radioactive Waste Underground Repository in Korea. *Magazine of Korean Tunnelling and Underground Space Association* 13(5): 16-23.
JTA(Japan Tunnelling Association). 1998. Construction of a large underground cavern. Subcommittee on Large-scale Underground Cavern. *Tunnel and Underground.* 29(8): 73-79.
Itasca Consulting Group, Inc. 2009, *Fast Lagrangian Analysis of Continua in 3 Dimensions (Version 5.0).* User Manual, Minnesota, USA.

Tunnels and Underground Cities: Engineering and Innovation meet Archaeology, Architecture and Art, Volume 7: Long and deep tunnels – Peila, Viggiani & Celestino (Eds)
© 2020 Taylor & Francis Group, London, ISBN 978-0-367-46872-9

Real behavior of shotcrete primary lining in squeezing rock mass: The experience at the Brenner Base Tunnel

L. Ziller
SWS Engineering Spa, Trento, Italy

S. Fuoco
BBT-SE, Bolzano, Italy

ABSTRACT: In conventional tunneling, shotcrete is often one of the main support elements and its failure can potentially lead to a collapse. Therefore, in geotechnical monitoring, the definition of the shotcrete stress is essential, even more in squeezing rock mass conditions, where the large deformations that occur can generate relevant stresses in the lining and reach the limit of its resistance capability. This paper presents a "two-step procedure" for the evaluation of shotcrete stress of the primary lining based on the convergence measurements. Although the method can provide only a first estimation of the shotcrete stress, it has proven to be very effective when applied to the case study of the Periadriatic fault at the Brenner Base Tunnel. Very interesting results regarding the shotcrete behavior in squeezing rock mass conditions have been obtained.

1 INTRODUCTION

In conventional tunneling absolute 3D displacement monitoring (i.e. convergences reading) is the most common monitoring method due to its simplicity and its high information content. Optical targets are installed on the tunnel lining and their positions are measured at defined time lapses with a total station.

In order to assess the excavation stability, usually, the measured displacements are compared to threshold values defined during the design phase and if they are lower, the excavation is considered stable and safe. The threshold values are often defined according to experience and literature data as a certain percentage of the tunnel radius.

However, it would be very helpful to obtain at least a first estimation of the stress intensity in the shotcrete based on the convergence readings in order to have immediate information about the utilization level of the lining.

Important issues to overcome are the intrinsic characteristics of shotcrete, which has highly time-dependent properties. In the initial phase it can tolerate significant deformations, compared to standard concrete, without developing considerable stress levels.

This paper aims to show a possible procedure for the determination of the shotcrete stress starting from the convergence data, which is composed of the following two steps:

• evaluation of the lining axial strain with a simple method, which takes advantage of the usual tunnel shape which is designed combining circles and arches;
• calculation of the stresses in the shotcrete using an advance constitutive law which is able to consider the peculiar characteristic of this material.

In the second part of the paper, the application of the method on the monitoring data of a sector of the Brenner Base Tunnel (BBT) exploratory tunnel will be presented. The data are related to the final part of the Periadriatic fault which is characterized by weak rock mass in combination with a significant overburden. In these conditions the high convergences generate stresses in the lining up to the limit of its resistance capability.

Applying the above-mentioned procedure in these squeezing rock mass conditions interesting results have been obtained with respect to the behavior of the shotcrete that was used as primary lining.

2 METHOD STATEMENT

Shotcrete, in combination with bolts and steel ribs, represents the main support element in conventional tunnels where instability phenomena of the cavity are expected. In these cases, the knowledge of the stresses intensity in the shotcrete is very important for the assessment of the lining behavior.

One possibility is to measure it directly, i.e. installing pressure cells or strain gauges in the lining. Due to operative reasons, this type of instrument is installed only in particular locations where high stress levels are expected. On the other hand, the typical distance between two different convergence sections varies in a range between 5 and 20 m, depending on the rock mass conditions.

For this reason, the possibility of evaluating the stress intensity in the shotcrete from the convergence measurement is very important. The process necessarily involves two steps:

- evaluation of the lining strains
- calculation of the shotcrete stress, taking into account its time dependent properties.

Different authors have investigated the topic in the past. The most popular approaches to the problem are the flow rate method and the hybrid method (OeGG Monitoring Handbook 2014).

The deformations are usually calculated imposing the measured displacements to a network of splines that simulate the tunnel lining. In this process, some assumptions on boundary conditions and the displacements field are needed. Assumptions on the actual thickness of the lining are required as well. Both axial and flexural deformations are estimated.

A different approach, proposed by (Rokahr 1997), will be followed in this paper with the goal of keeping the process simpler. Only the axial strain will be evaluated. The main shortcoming doing so is that flexural deformation and consequently the bending moment cannot be calculated. On the other hand, the application of this method does not require any arbitrary assumptions on boundary conditions and on the trend of displacement between the measuring points.

In each case the flexural behaviour of the lining needs to be evaluated, in order to establish if it plays a relevant role. For the case that will be presented following (a deep tunnel in squeezing condition) this component is not so important since the lining works in very close conditions to pure compression. This statement is supported by the fact that the pre-design of tunnel lining in this condition is often performed, with very good results, with the Convergence Confinement Method which considers only the axial component of the lining behaviour.

Regarding the shotcrete constitutive law, both the flow rate and the hybrid method use models which require special tests for the calibration of the parameters used for modelling shotcrete such as calorimeter tests and adiabatic tests.

Also, in this case a different approach has been followed, taking advantage of a novel constitutive shotcrete model proposed recently by (Schädlich & Schweiger 2014).

As a concluding remark of this chapter, it is important to note that the essential conditions for the application of all these methods that use the convergences as input data, are good quality measures and the installation of the measuring sections as close to the face as possible immediately after shotcrete application.

3 EVALUATION OF LINING AXIAL STRAIN

Lining strains are evaluated by using a simple method proposed by (Rokahr 1997) which exploits the typical shape of a tunnel section.

The coordinates of the measuring points that compose a monitoring section are defined for every measure campaign. The lining is then divided into sectors which are defined by three

LEFT   CROWN   RIGHT

Figure 1.   Typical arches layout for a measuring section composed by 5 targets.

adjacent monitoring points. This is because three points geometrically determine an arch. For every measurement, the length of each section can be calculated.

Considering the typical layout of a monitoring section, which is composed by five points, three different arches can be determined as represented in Figure 1.

Lining deformation for each sector is defined as following:

$$\varepsilon_t = \frac{l_t - l_0}{l_0} \tag{1}$$

where:
ε_t is the lining axial deformation at time t
l_t is the length of the arch at time t
l_0 is the length of the arch at the zero reading.

Two sources of error can be identified in this approach:

- the approximation of the actual lining shape with perfect arches;
- the limited accuracy of the convergence readings.

In order to quantify the amount of error given by the first source, a series of numerical analyses with the software Plaxis 2D were performed. The tunnel excavation was modelled with 2D plain strain models. The lining was modelled with a plate element and the tunnel progress was modelled with subsequent stress releases. For each stage of the analysis the actual axial deformation of the lining was compared to the deformation calculated with the arches method applied on imaginary convergence targets placed at the inner boundary of the lining. It has been found that, in general, the arches method provides a slightly overestimation of the actual deformation with an error in most of the cases lower than 10–15% of the exact value.

Regarding the second source of error, the following considerations can be done. The current state of the art of convergence reading can ensure an accuracy of +/- 1 mm in the determination of the position of the points in relation to the neighboring measuring sections. This is often a contractual requirement as well and should be considered as an upper boundary of the error value. For a tunnel with a diameter of about 7 m (as the one presented in chapter 5), the typical length of an arch passing through three measuring points is greater than 5 m. This means that the deformation of an arch is calculated with a precision of +/- 0.2 ‰ (0.001/5.000).

4   SHOTCRETE CONSTITUTIVE MODEL

In order to properly model the shotcrete behavior, the constitutive law proposed by (Schädlich & Schweiger 2014) has been selected. It can account for time dependent strength and stiffness, strain hardening/softening in tension and compression, creep and shrinkage. Since 2014, it has been implemented in the finite element software Plaxis as one of the possible material models.

The shotcrete model counts 28 parameters in total. However, most of them are required to model the post-peak and the tensile behaviour of the material. For our purpose, which is modelling the shotcrete behaviour in compression, in a pre-peak condition, only the 10 parameters reported in Table 1 are significant. All of them can be determined by uniaxial compression tests at different shotcrete ages and one shrinkage test.

The parameters for the analysis presented in the following chapter have been assumed according to the technical literature suggested range of parameters and considering the results of an experimental program presented by (Neuner et al. 2017). The study is focused on the parameters calibration of three different advanced shotcrete models, among which the Schädlich model. The different calibration tests have been performed on shotcrete sampled at one of the Austrian construction sites of the Brenner Base Tunnel. It has not been possible to directly use the data obtained by Neuner due to the different strength class used for the experimental program (C25/30) from the one use at the Mules site (C30/37).

In practice, shotcrete stresses can be calculated by modelling fictitious uniaxial compressive tests with Plaxis 2D imposing the strain paths determined with the arches method. Due to the material time dependent properties it is important to link in the simulation each deformation level to its corresponding time. The time 0 in the simulation is assumed coincident with the zero reading of the measuring section which was performed immediately after shotcrete installation.

A short summary of the main characteristics of the constitutive model is provided following.

4.1 Hardening law

Strain hardening in compression follows a quadratic function up to the peak strength. It is the same approach followed by EN-1992-1-1. Due to the time dependent parameters involved, the curve changes over time.

4.2 Time dependent stiffness and strength

Two types of functions to model the time dependent strength are implemented in the model:

- according to the recommendations by CEB-FIB model code
- according to the early strength classes of EN 14487-1.

In our case the second approach has been considered, mean value of J2 class have been assumed according on the actual shotcrete quality applied on site.

4.3 Creep and shrinkage

Creep strains ε_{cr} increase linearly with stress σ and are related to elastic strains via the creep factor ϕ^{cr}.

$$\varepsilon^{cr}(t) = \phi^{cr} \cdot \sigma \cdot D^{-1} \frac{t - t_0}{t + t_{50}{}^{cr}} \tag{2}$$

Table 1. Shotcrete model relevant parameters and values considered in the calculation.

Parameter	Unit	Value	Description
E_{28}	GPa	25	Young's modulus after 28 days
v	-	0.2	Poisson ratio
$f_{c,28}$	MPa	38	Uniaxial compressive strength after 28 days
E_1/E_{28}	-	0.6	Ratio of Young's modulus after 1 day and 28 days
$f_{c,1}/f_{c,28}$	-	J2	Ratio of f_c after 1 day and 28 days
$\varepsilon_{cp}{}^{p}$	-	- 0.001	Plastic peak strain in uniaxial compression
ϕ^{cr}	-	2.5	Ratio between final creep strain and elastic strain
$t_{50}{}^{cr}$	days	3	Time for development 50% of creep strain
$\varepsilon_{\infty}{}^{shr}$	-	- 0.001	Final shrinkage strain
$t_{50}{}^{shr}$	days	64	Time for development 50% of shrinkage strain

where D is the linear elastic stiffness matrix. The evolution of creep with time t is governed by the start of loading t_0 and the parameter t_{50}^{cr}. For shotcrete utilization higher than 45% of f_c, nonlinear creep effects are accounted for according to EN-1992-1-1.

Shrinkage strains are calculated as

$$\varepsilon^{shr}(t) = \varepsilon_\infty^{shr} \cdot \frac{t}{t + t_{50}^{shr}} \tag{3}$$

with ε_∞^{shr} being the final shrinkage strain and t_{50}^{shr} the time when 50% of shrinkage has occurred.

5 CASE STUDY: PERIADRIADIATIC FAULT AT THE BRENNER BASE TUNNEL

This chapter presents the application of the method on the monitoring data of the Brenner Base Tunnel (BBT) exploratory tunnel in the Periadriatic fault stretch. The development of the method has been driven to improve the understanding of the lining behavior during its construction. Indeed, it was quite clear that the support was working close to its capacity limit (Fuoco et al. 2017) and a quantification of the utilization degree would have been a very valuable information.

5.1 THE BRENNER BASE TUNNEL

The Brenner Base Tunnel is a flat railway tunnel between Austria and Italy. It runs from Innsbruck to Fortezza. Including the Innsbruck railway bypass the entire tunnel system through the Alps will be 64 km long becoming the longest underground rail link in the world.

It consists in a system of tunnels that includes two one-track tubes, a service/exploratory tunnel running mostly parallel to the two main tunnels but 12 m below them, cross-passes between the two main tubes placed every 333 m and 3 emergency stops located roughly 20 km apart.

Overburden is, on average, between 900 and 1000 m, with the highest points about 1800 m at the border between Italy and Austria (Zurlo et al. 2013). The excavation will go through all the geological formations that compose the eastern Alpine Area.

5.2 THE PERIADRIATIC FAULT

One of the most challenging part of the project is constituted by the Periadriatic fault, which is located on the Italian side of the tunnel, close to the Mauls adit. This heavily tectonized zone is generated by the collision between the European and the African tectonic plates. The fault is long some hundreds of meters while the physical overburden varies between 450 and 700 meters.

This study is focused on the last approximately 200 m of the fault which it turned out to be the most difficult. The rock mass is composed by highly tectonized micaschist and fillite. The RMR index, detected during the tunnel construction, was between 25 and 35.

5.3 THE EXPLORATORY TUNNEL CONSTRUCTION

The exploratory tunnel was excavated in advance compared to the main tunnels to obtain additional data to adjust and optimize their consolidation and support systems.

The excavation diameter was equal to about 7 m. Despite the extremely challenging conditions, the exploratory tunnel has been successfully built without any severe problems by using the Italian approach to conventional excavation. This approach consists in applying a shotcrete lining reinforced with steel arches and in pre-confining the rock mass at the tunnel face and around the cavity with rock bolts.

In the sector of interest, two different types of primary lining have been installed:

- Excavation section named "C-T4 bis P" (up to chainage 12360): it was composed by 15 cm of shotcrete reinforced with lattice girders with a spacing of 1.50/2.40
- Excavation section named "C-R1 P" (after chainage 12360): it was composed by 30 cm of shotcrete reinforced with coupled steel arches type IPE180 with a spacing of 1.00.

A secondary shotcrete lining with a concrete invert was installed at 2/3 tunnel diameter from the face.

A considerable amount of monitoring instruments was installed to check the lining behavior such as convergence sections, strain gauges on steel arches and pressure cells embedded in the shotcrete layer.

5.4 CONVERGENCE MEASUREMENTS

More than 30 convergence sections have been installed in the stretch of interest. Each measuring section was composed by 5 optical targets that were evenly distributed on the tunnel lining. They were installed from 1.00 to 1.50 m from the face immediately after the shotcrete installation. Measurements were taken daily for the first days and became more infrequent after about 10 days when the distance between the tunnel face and the convergence section was about 1.5 – 2.0 tunnel diameters.

Figure 2 shows the typical output of the convergence data for measuring section (MS) 12 +440. Vertical and horizontal displacement are plotted against time.

The displacement velocity is higher in the first days and tends to stabilize when the face gets farther.

Figure 3 shows the final radial convergence for the different sections. For each monitoring section, the mean (average of the 5 targets) and the maximum radial convergence corresponding to the last measurement before the installation of the secondary lining are reported.

The maximum radial convergences are equal to about 60 mm, which is slightly lower than 2 % of the tunnel radius. The targets on the sidewalls detected, in most of the cases, greater displacements than the ones applied on the crown area. The anisotropy of the displacement pattern is probably due to both the anisotropy of the rock mass and to the geometry of the

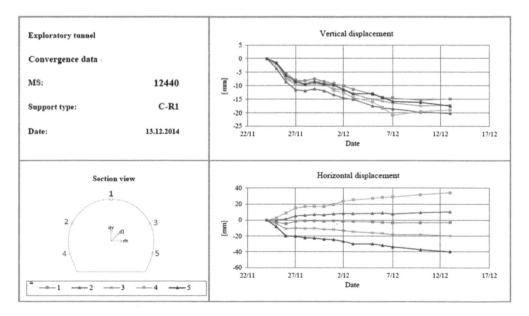

Figure 2.   MS 12+440: convergence measurements.

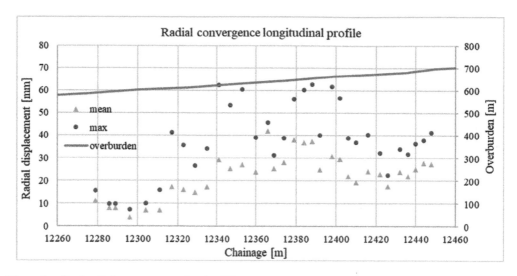

Figure 3. final radial convergences for the different measuring sections.

tunnel lining. In fact, open linings, namely linings that do not have an invert, show an aniso-tropic behavior also when subjected to a hydrostatic state of stress since their stiffness depends on the loading direction.

5.5 LINING DEFORMATION

The results of the application of the arches method are presented hereafter.

Figure 4 shows the development of the lining axial deformation over time for MS 12+440 while Figure 5 shows the lining axial displacement for the different measuring sections before the installation of the secondary lining. The three different values (left, crown, right) refer to the three arches that ae determined with 5 measuring points as shown in Figure 1.

In most of the cases, the greater deformation is measured in the crown area. This is where an open lining reacts in a stiffer manner.

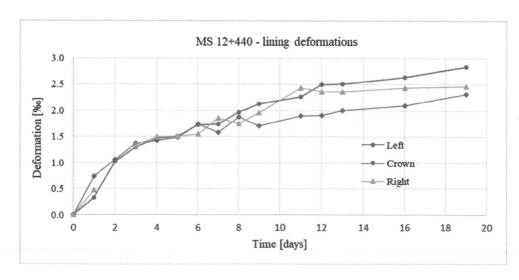

Figure 4. MS 12+440: lining axial deformations over time.

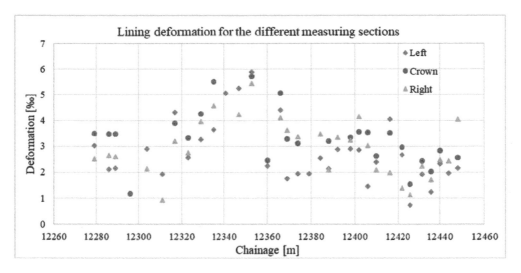

Figure 5.   lining axial deformation for the different measuring sections.

Looking at the results what immediately stands out is that the level of deformation that affects the shotcrete shell is considerably greater than the one that can be expected for a "standard" concrete. It becomes clear that to determine the shotcrete stress it is necessary to adopt a constitutive law that can model in a proper manner its peculiar behavior.

In this part of exploratory tunnel, 5 different sections of strain gauges applied on steel arches were installed. Each section was composed by 3 couples of strain gauges. The maximum deformation detected in this way is equal to about 1.8 ‰. In all the other cases lower values were measured.

This difference can probably be explained considering that the steel arches are not perfectly in contact with the rock mass and in the initial phase they are not loaded by it. This should be further investigated, for example applying measuring targets on steel arches and on shotcrete closely spaced.

5.6   SHOTCRETE STRESS

The estimated stresses derived applying the advance constitutive law on the deformations calculated with the arches method are presented hereafter.

Figure 6 shows the stress variation over time of MS 12+440. After a quick increase of stress in the first days the trend tents to stabilize and reaches an equilibrium.

When measured for a longer period, shotcrete stresses usually decrease due to creep and shrinkage. In our case this was not detected because the secondary lining was installed very close to the face.

Figure 7 shows the maximum stress level estimated from the convergence readings for the different sections along the alignment. The stress corresponding to the maximum deformation of each section is plotted. Two reference thresholds have been defined as well, namely the f_{ck} (characteristic compressive strength) and the f_{cd} (design compressive strength) of the shotcrete.

In this sector there were also 4 pressure cells measuring stations. Unexpectedly the stress level detected by the pressure cells was in all cases much lower the predicted one. The reason is not clear yet and should be surely investigated in the future. In fact, it this quite strange that the stress in the shotcrete is so low when there were multiple signs that the lining was working close to its capacity limit.

On the other hand, the stress estimated from the convergence readings confirms that the shotcrete utilization was very high. The factor of safety of the shotcrete lining was in fact very

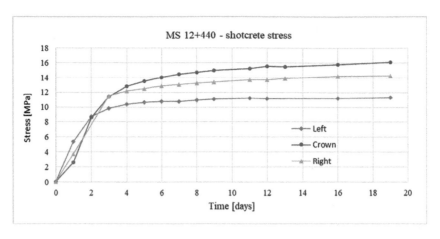

Figure 6.   MS 12+440: shotcrete stress over time.

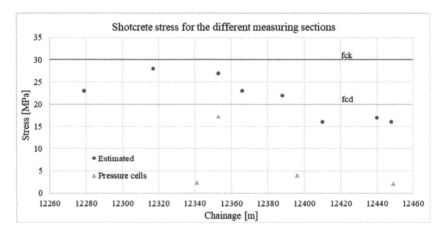

Figure 7.   shotcrete stress for the different measuring sections.

close to one. However, it has to be noted that this condition was temporary because the secondary was installed very close to the face.

It is important to underline that such a high optimization of the support has been possible only thanks to a state of the art application of the observation approach and a very good cooperation of all the parties involved in the project, namely the owner, the contractor and the site supervision.

Figure 7 shows clearly the advantages of the proposed method. Comparing actual and limit stress levels is much easier than comparing displacement for the definition of the safety factor of the lining.

5.7   CONSIDERATIONS ON RESULTS

It is quite interesting that the change to a stiffer support at chainage 12360 (from C-T4 bis P to C-R1 P) almost immediately reduced the amount of deformation. The same trend can be detected also in the shotcrete stress graph. The reduction of deformation is directly linked to the increased stiffness and capacity of the support, which can apply a greater confining stress to the rock mass.

More precisely, the support reaction p_{sh} can be evaluated according to the following equation (considering shotcrete only):

$$p_{sh} = \frac{1}{2} \cdot f_{sh} \cdot \left(1 - \frac{(R - s_{sh})^2}{R^2}\right) \qquad (4)$$

where f_{sh} is the shotcrete stress, s_{sh} is the shotcrete thickness and R is the tunnel radius.

For MS 12+353, one of the last with the "light" support the reaction, which for equilibrium is equal to the rock mass pressure is equal to about 1100 kPa. For MS 12+366, the first with the "heavy" support, the reaction is equal to almost 1500 kPa.

The last consideration concerns the design of shotcrete lining in squeezing conditions. It is common practice, when performing structural calculation of primary lining, to use the "fictitious stiffness" concept in order to model in a simple manner the peculiar characteristic of shotcrete. A reduced Young's modulus is used in the analysis to take into account low initial stiffness, creep and shrinkage. Its value can be directly determined dividing the final stress by the corresponding deformation for each section. The mean value of the "hypothetic stiffness" is equal to 6.3 GPa which is in very good agreement with the 7 GPa proposed by (Pöttler 1990).

6 CONCLUSIONS

This paper has provided a possible method for the evaluation of the shotcrete lining stress based on the convergence readings. The method is very simple and can be used to obtain a first estimation of the lining utilization level in the everyday practice of geotechnical monitoring.

The application of the method on the case study of the Periadriatic fault at the Brenner Base Tunnel has proven that shotcrete can tolerate significant deformations thank to its intrinsic characteristics. The estimated axial deformation of the lining was up to 6 ‰ and greater than 2 ‰ for most of the measuring sections. As was expected the shotcrete lining was working close to its maximum capacity and was highly optimized.

Further investigation should be performed to determine the reason of the not good match between the estimated values by the application of the proposed method and the data of the other monitoring instruments (pressure cells and strain gauges).

REFERENCES

CEB-Fip model code 1990. Design Code – Comite Euro-international du Beton. London: Thomas Telford.

EN-1992-1-1 2004. Eurocode 2: Desing of concrete structures. European Committee for Standardization.

EN 14487-1 (2006). Sprayed concrete – Part 1: Definitions, specifications and conformity. European Committee for Standardization.

Fuoco, S.; Zurlo, R.; Lanconelli, M. 2017. Tunnel deformation limits and interaction with cavity support: the experience inside the exploratory tunnel of the Brenner Base Tunnel. *Proceedings of the World Tunnel Congress 2017*. Bergen.

Neuner, M. et al. 2017. Time-dependent material properties of shotcrete: experimental and numerical studies. *Materials*, 10, 1067.

OeGG - Austrian Society for Geomechanics. 2014. Geotechnical Monitoring in Conventional Tunnelling. Handbook.

Pöttler, R. 1990. Time-dependent rock-shotcrete interaction - a numerical shortcut. *Computers and Geotechnics* 9: 149–169.

Rokahr, R.; Zachow, R. 1997. Ein neues Verfahren zur täglichen Kontrolle der AuslastungeinerSpritzbetonschale. *Felsbau*, 15(6): 430–434.

Schädlich, B.; Schweiger, H.F. 2014. A new constitutive model for shotcrete. In Proceedings of the 8th European Conference on Numerical Methods in Geotechnical Engineering, Delft, The Netherlands, 18–20 June 2014; CRC Press Taylor & Francis: Leiden, The Netherlands, 103–108.

Zurlo, R.; Rea G.; Roccia, M. 2013. Galleria di Base del Brennero. Descrizione dell'opera ed avanzamento attraverso la faglia Periadriatica. *Proceedings on the Italian Tunnel Society Congress*. Bologna: 628–643.

Tunnels and Underground Cities: Engineering and Innovation meet Archaeology,
Architecture and Art, Volume 7: Long and deep tunnels – Peila, Viggiani & Celestino (Eds)
© 2020 Taylor & Francis Group, London, ISBN 978-0-367-46872-9

Tunnel costs related to the quality of the rock mass

R. Zurlo, S. Fuoco, M. Loffredo & A. Marottoli
Brenner Basistunnel BBT-SE, Bolzano-Fortezza, Italy

ABSTRACT: The past decades have shown important progress in the planning and construction of underground works. Tunnel excavation methods have evolved in a similar manner, from
systems that used partial section excavation, to full-bore driving with complete control of the
rock mass deformation process by limiting the size of the plasticized areas around the cavity and
using partial section excavation only in exceptional cases. These technical and technological developments are matched by the difficulty of determining construction costs, as these are linked to the
conditions of the rock mass to be excavated and further impacted by logistics, which is an
extremely significant issue when dealing with long, deep tunnels. The present report shows the
costs resulting from various types of rock mass, comparing them to the cost of excavations.

1 INTRODUCTION

The Brenner Base Tunnel (BBT) is a complex system of tunnels and other engineering works
designed to pass under the Alps from Fortezza in Italy and Innsbruck in Austria along the
new European Scandinavian-Mediterranean Corridor.

The project includes two main single-track tunnels (with a maximum diameter of 0.5 m)
and an exploratory tunnel (maximum diameter 6.6 m) with lies 12 m below the main rail tunnels. The distance between the two main tunnels varies from 40 to 70 m. For operational and
safety reasons, the main tunnels are connected by cross passages at 333 m intervals. The
system also includes three underground emergency stops and connecting tunnels with the
existing Innsbruck bypass (Zurlo et al. 2013).

The main tunnels are just over 55 km long. With the Innsbruck bypass, which is another 9
km in length, the Brenner Base Tunnel will be the longest railway link in the world, once it is
completed in 2027.

2 EXPECTED GEOLOGICAL FORMATIONS (FUOCO ET AL. 2017)

The excavation will drive through basically all the geological formations in the Austroalpine
area, mainly metamorphic rock with a significant amount of plutonic rock mass.

The maximum overburden is about 1,850 m.

The tunnel will intersect the so-called "Periadriatic Line" which is the border between the
European and the African tectonic plate. Due to heavy tectonic activity, the rock mass in this
area is heavily fractured and has very poor geomechanical properties.

The main characteristics of the intact rock are those of the main geological formations
along the tunnel route and are summarized in Figure 1 and in Table 1 (Fuoco et al. 2016).

The diversity of the geological formations to be crossed during excavation work, the presence of significant tectonic areas such as the Periadriatic Fault Line, combined with the length
of the route and the deadline for the start of operations led the authorities to decide that the
excavation was to be carried out with a TBM rather than the traditional demolition machines
or with the drill and blast method. More specifically, drill and blast was used in the sections
prepared for the TBM during the period required to assemble and supply the TBMs

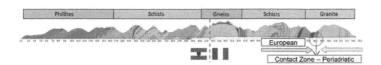

Figure 1. The longitudinal geological profile between Innsbruck and Fortezza.

Table 1. Characteristics of the main geological formations along the tunnel route (Intact Rock).

Lithology	Expected (%)	γ [kN/m^3]	σ_{ci} (Mpa)	E (GPa)
Granite	12	26	89–131	29–34
Tonalite	2	27	45	20–31
Orthogneiss	2	28	115	50
Schists	41	27	50	40
Central Gneiss	12	27	105	45
Phyllites	22	27	33	23
Various	9	23–27	different	different

themselves, and to cross geologically critical areas where there was an overly high risk that the TBMs would become stuck, e.g. in the aforementioned Periadriatic Fault, or, again, when the construction program prescribed the parallel excavation of different sections of the tunnel with construction schedules compatible with both systems.

At the time of writing, in the Italian part of the tunnel and with reference to the construction lot known as Mules 2–3, which is the most important lot as it includes the construction of the tunnels that will reach the border, over 5.5 km of tunnel have been excavated by blasting. The excavation took place through rock mass of varying quality, which required the use of different stabilization systems based on the quality of the rock mass or on the overburden, meaning the stress/load parameters before excavation that influenced the reactions of the rock mass during construction.

Based on the experience gained from the construction of these stretches, an attempt was made to verify the connection between construction costs and rock mass quality, in the present case represented here by the RMR quality index (Bieniawski 1989).

It should be noted that the cost-quality ratio of the rock mass also depends on the approach used for the construction of the tunnel. In this specific case, we refer to the excavation method normally used in Italy.

3   THE ITALIAN TUNNEL PLANING AND CONSTRUCTION METHOD (FUOCO ET AL. 2014)

This method includes three main phases.

First of all, the classification of the rock mass using geological and geotechnical data and defining the various areas where the rock mass reacts in a homogeneous manner.

Secondly, for each homogeneous sector identified, a consolidation and support system must be defined that is appropriate for the rock mass in question and is then summarized in the "section types". The consolidation system is mainly based on the stability conditions at the rock face and on the idea of a full-bore tunnel excavation. If the excavation front is not stable, it will be stabilized with artificial means. The use of these phases allows a certain industrialization of the excavation via a cyclical process.

The third phase is represented by the feedback during the construction phase obtained from a monitoring system to be installed during the excavation which analyses the actual reactions of the rock mass to the excavation. If the response is different than expected, the section type is modified based on the above mentioned guidelines.

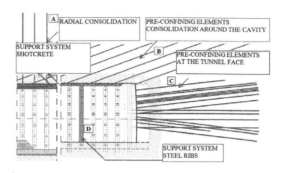

Figure 2.   Italian Tunnel Excavation Method.

Over the past decades, the evolution of excavation techniques and the new technologies available, not to mention a far greater attention to safety, have increasingly industrialized tunnel excavation.

When working in poor-quality rock mass, the industrialization of the process starts with the consolidation of the rock face or - which is even better - of the rock mass just beyond the rock face, known as the advance rock core.

The first to introduce this method in Italy and in Europe was Lunardi (Lunardi 1988). The definition of excavation behaviour is linked to the ratio between the rock mass load resistance and the stress generated by the tunnel excavation and the subsequent rock mass deformations. Based on the rock face stability conditions, Lunardi (Lunardi 1988) identified three different behavioural classes for the rock face:

– behaviour of a stable rock face;
– behaviour of a rock face that is only stable over the "short term"
– behaviour with a unstable rock face.

In the first case, the reaction of the surrounding rock mass is fundamentally elastic. In the second case, non-negligible areas of rock mass plasticity form around the tunnels and the rock face. In the third case, significant areas of rock mass plasticity are present in front of the rock face and surrounding the cavity.

A so called "section type" was defined for each of these classes in order to allow excavation whilst controlling deformations and possible conditions of instability. Consolidation of the rock mass and the support system for the cavity are closely linked to the expected behaviour of the rock mass. Specifically:

If the stress and load at the rock face and surrounding the cavity are close to the limit bearing capacity (almost elastic deformation), the consolidation measures aim to ensure excavation safety (typically, rock bolts with a certain thickness of shotcrete);

If the rock face can be defined as stable over the short term, the consolidation system will aim to reduce the areas of rock mass plasticity in front of the rock face, in order to lessen the effect of the disturbed material surrounding the rock face on the rock mass surrounding the cavity and at a distance from the rock face itself;

If the rock face is unstable, the consolidation system will aim to improve the characteristics of the rock mass or of the surrounding ground so as to continue excavating without encountering unstable conditions.

Applying this concept, a cyclical process can be organized for the consolidation, support and excavation of the ground or the rock mass. This level of stability can be ensured by using different combination of consolidation elements.

This cyclical process allows for swift reactions if the rock mass characteristics change, merely modifying the section type or the number of consolidation elements, based on the characteristics of the excavated rock and/or the reaction of the rock mass to the excavation which are assessed by an appropriate monitoring system.

4 COSTS AS CORRELATED TO ROCK MASS QUALITY

The analysis of the cost/rock mass quality connection summarized as follows is based on the excavation of the tunnels planned as part of the construction lot known as "Mules 2–3".

This lot includes a series of tunnels built by blasting within rock masses of varying types which, with reference to RMR quality index, range from class II (RMR between 60 and 80) to class IV>V (RMR between 20 and 40), and tunnels built by Double Shield TBM.

In order to define the properties that indicate the quality of the rock mass, a particularly significant parameter appears to be the RMR quality index obtained during the excavation of the exploratory tunnel in the previous construction lot, which included excavation works in the rock formations involved in driving the tunnels of the Mules 2–3 lot.

The estimated value of the RMR recorded along the stretch of exploratory tunnel built during the previous construction lot is shown in Figure 3.

Within the most tectonically altered stretch, pre-consolidation and support systems were used that have allowed monitoring of the deformation of the cavity and which were subsequently applied to the excavation stretches in the Mules 2+3 lot which is the subject of the comparative analysis on the costs of the excavated stretches.

The choice of the consolidation system took place on the basis of the above mentioned Italian tunnel planning and construction method.

The almost orthogonal orientation of the geological structure (referred to the tunnel axis) tends to cause a prevailingly longitudinal main deformation of the rock mass (a tendency confirmed during the excavation of the first metres of schist rock mass). A diagram of this expected reaction can be seen in Figure 4.

Based on this observation, two types of primary support were adopted: an "ordinary" radial support system using anchor bolts and shotcrete and a "pre-confining" system with self-drilling rock bolts placed both around the cavity and on the rock face, combined with shotcrete and steel ribs around the cavity.

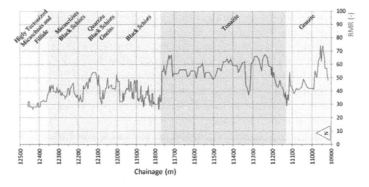

Figure 3. RMR along the excavation route of the exploratory tunnel in the previous construction lot (Fuoco et al. 2017).

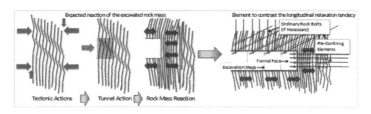

Figure 4. Diagram of the main expected reaction with elements contrasting the relaxation tendency of the rock mass (Fuoco et al. 2016).

The "ordinary" system was to applied to relatively high-quality rock mass where an almost elastic response was expected, whereas the "pre-confining" system should be applied in case of highly de-structured rock mass. In this last case, the concept includes the installation of structural elements to confine the material before excavation, thus improving rock mass resistance and increasing control of longitudinal rock mass relaxation (Figure 4). The system further includes the addition of radial anchors in the stretches where the reaction of the inclined elements was not enough to maintain convergence within the defined limit values.

The number and type of consolidation measures planned and subsequently installed depended on the quality of the rock mass encountered and the actual conditions of the rock mass under excavation.

5 THE CONSOLIDATION AND SUPPORT SYSTEMS AS COMPARED TO ROCK MASS QUALITY

The type of consolidation and support measures planned during excavation define the excavation "section types". The use of these section types, as already mentioned, is a function of rock mass quality represented by the RMR index and by the actual reactions of the rock mass to the excavation activity.

5.1 The section types used for excavation in the Mules 2–3 construction lot

The section types defined by the design are a function of rock mass quality.

The reactions of the rock mass to excavation work are determined by a series of factors, besides the size of the excavation, most especially:

The geo-mechanical properties of the rock mass
The presence of water
The initial stress/load state

All these factors are derived from interpretations and considerations based on specialized, specific studies on the topic. The project, in Italy, defines once and for all the applicability of an excavation section in relation to the expected geo-mechanical profile. In spite of this, the excavation phase is still encumbered with considerable unknowns concerning the reactions of the rock mass which must therefore be constantly monitored. It is extremely important, then, to assess the quality of the rock mass during excavation, for example using the Rock Mass Rating index as is done here, in order to adapt the section type to the actual geological/geo-mechanical conditions encountered.

In other words, the section types defined in the project are applied based on rock mass quality, as represented by the RMR quality index estimated during the measurements at the individual rock faces.

Below are the various excavation section types defined in the planning for the main tunnels with the indication for applicability based on the different types of rock mass. In any case, the consolidation, support and confinement measures set forth for each type on section can be, if properly recalibrated for different excavation geometries, used for the other tunnels considered in this study.

"T2" Excavation Section Type (to be applied to class I or II rock mass RMR>60).

- Excavation with blasting, 5 cm over-excavation, maximum volley length 4.50 m;
- Fibre-reinforced sprayed grout, 30/37, both in the surrounding area (5+10 cm) and at the rock face (5 cm);
- four+five SuperSwellex Pm16 anchor bolts, 3.00 m in length, transv. interval 1.80 m and longitudinal interval 1.50 m.

"T3" Excavation Section Type (to be applied to class III-41≤RMR≤60)

- Excavation with blasting, 5cm over-excavation, maximum volley length 3.00 m;

- Fibre-reinforced sprayed grout, 30/37, both in the surrounding area (5+10cm) and at the rock face (5 cm);
- five+six SuperSwellex Pm24 anchor bolts, 4.50 m in length, transv. interval 1.80 m and longitudinal interval 1.50 m.

"TRb" Excavation Section Type (to be applied to class I or II rock mass (RMR>60) and overburden >1000m)

- Excavation with blasting, 5 cm over-excavation, maximum volley length 1.50m.
- Fibre-reinforced sprayed grout, CFSpC 30/37, both in the surrounding area (5+10cm) and at the rock face (5cm).
- 7/8 anchor bolts, SuperSwellex Pm16, yield strength Ny>140kN, length 3.00m, p = 1.80 m transv x 1.50m long.

"T4" Excavation Section Type (To be applied to class IV-RMR<41 rock mass, mainly gneiss, paragneiss and quarzite)

- Excavation with TBM and/or blasting, 10cm over-excavation, maximum drive or volley length 1.50m.
- Fibre-reinforced sprayed grout, 30/37, both in the surrounding area (5+25cm) and at the rock face (5cm).
- Double 2 IPN180 ribs in S355JR steel, distance from 0.75 to 1.50m.
- 14 R51N self-drilling rods for consolidation around the cavity with yield strength Ny≥630MPa, length 12.00m, transversal interval = 0.75m and longitudinal interval 3.00m, cemented in.
- Stabilization of the rock face (if needed) with 32 R38N self-drilling rods, with yield strength Ny≥400MPa, length 15.00m, overlap 6.00m, cemented in;
- Localized stabilization of the rock face (if needed) with SuperSwellex Pm24 rock bolts, yield strength Ny>200kN, length 5.50m, overlap 2.50m.

"T5" Excavation Section Type (To be applied to class IV-RMR<41 rock mass, mainly schists and cataclastic rock)

- Mechanized excavation, 10cm over-excavation, maximum drive length length 1.50m.
- Fibre-reinforced sprayed grout, 30/37, both in the surrounding area (5+25cm) and at the rock face (5cm).
- Double 2 IPN180 ribs in S355JR steel, distance from 0.75 to 1.50m.
- From 18 to 22 R51N self-drilling rods with yield strength Ny≥630MPa, length 12.00m, transv. interval 0.75m longitudinal interval 3.00m, cemented in.
- Stabilization of the rock face with 34 to 56 R51N self-drilling rods, with yield strength Ny≥630MPa, length 15.00 m, overlap 6.00m, cemented in.
- 14 Dywidag SNØ28 radial anchor bolts in ST670/800 steel, expansion shell; yield strength Ny>413 kN, head strength Ny>200 kN, length 5.50 m, transv. interval 1.50 m long. interval from 0.75 to 1.5 m depending on the interaxis of the ribs. Cemented in.

"C-MS" Excavation Section Type (To be applied to any class of rock mass, in case of TBM excavation)

- Excavation with TBM, 10cm over-excavation, maximum drift length 1.50m;
- Installation of a 30-cm thick lining ring reinforced with 180 kg/mc of steel

6   SECTION TYPE AS COMPARED TO ROCK MASS QUALITY

The study was based on an analysis of excavation costs and RMR values in the stretches of tunnel that have been excavated so far by blasting and by TBM in the Mules 2–3 construction lot, noting that while in the conventional method the cost is a function of the RMR value, with a TBM excavation the cost is the same for whatever RMR value.

The results of the study are described as follows.

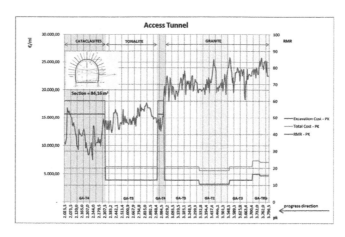

Figure 5. The Access Tunnel to the Trens Emergency Stop.

In the following figures, the term "excavation cost" includes all the costs referred to the excavation, consolidation and primary support system. The "total cost" also includes other costs, as railway systems, cooling system, ventilation, etc.).

The excavation inside the access tunnel to the Trens stop was carried out in an area with high overburden (> 1,000 m), and through a slightly fractured rock mass (Damage zone of the Pustertal valley fault) with elastic properties, where rock burst phenomena occurred requiring the adoption of additional consolidation and support measures.

As can be seen in the graph in Figure 5, the RMR was Class II for 90% of the excavated stretch.

At chainage 2+984, the excavation intercepted the "Core zone" of the Pustertal valley fault, which is 35 m long and characterized by tectonically altered, slightly foliated tonalite and fine-grained cataclastic rock; the graph in Figure 5, shows the drift through the fault core zone and as of chainage 2+984, the RMR index suddenly fell compared to previously recorded values (class III).

In the Exploratory Tunnel (Figure 6) the initial excavation (61 m - from chainage 12+460 to chainage 12+521) drove through tectonically highly altered rock with significant anisotropic behaviour and a 1:10 ratio between the maximum radial convergence (in cm) and the maximum extrusions (in decimeters). The excavation section used was CT5.

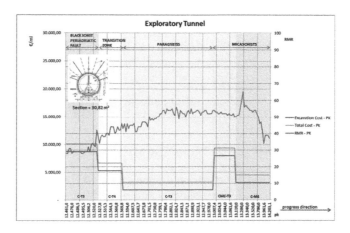

Figure 6. The Exploratory Tunnel.

From chainage location 12+521 rock mass quality began to improve, allowing the use of excavation section CT4, and as of chainage location 12+569 the rock mass featured an even better response to excavation; the presence of more solid material, also confirmed by data gained from the installed geotechnical instruments, allowed the use of a lighter excavation section, i.e. CT 3 instead of CT4.

The assembly chamber for the TBM was constructed from chainage location 13+000 to chainage 13+066; the excavation section went from roughly 30 sqm to about 140 sqm and a T3 section type was used.

The TBM excavation started at chainage 13+085 using a DS-TBM with a 6.70 m diameter; while excavation was ongoing, the prefab ring was set which acts as both a first-phase and final lining.

In the graph in Figure 7, in order to have a proper comparison we considered only the cost of the excavation without the cost of the lining with prefab rings.

During the excavation of the main tunnels northwards a significant fault zone with numerous discontinuities, fragmented rocks and altered rock mass often characterized by a terrigenous matrix, was intercepted; it was particularly squeezing and unstable during excavation.

The measures taken during the construction were very important as they allowed the adaptation of confinement, support and consolidation measures through careful and constant monitoring of the geological/geomechanical data.

In this area, excavation section T5 was used.

It should be noted that in the western main tunnel (Figure 7), localised detensioning led to a sudden fall of the RMR and, as a consequence of this, to an increase in costs. Once this detensioned stretch had been passed, the RMR stabilized at lower values than in the previous section.

For the eastern main tunnel (Figure 8), it should be pointed out that constant monitoring of the rock mass allowed for targeted measures and, starting with the typical T5 section, the adaptation of consolidation measures to the actual response of the rock mass.

Southwards, the main tunnels were excavated through elastic rock mass (Brixner granite) with minor convergences and equally slight plasticized zones around the cavity.

The analysis carried out during excavation works did not highlight significant differences as compared with the definitions made in the executive and detail planning; thus it was not necessary to adjust the consolidation measures in function of the real behaviour of the rock mass after excavation.

The cross-section applied is the GL-TRb-Ter (used to prevent effect of rock burst phenomena).

The analyses in the following Figures (9 and 10) highlight the fact that in both tunnels the rock mass rating RMR almost always varied within the same rock mass class (class II) and that costs were not affected by this variability.

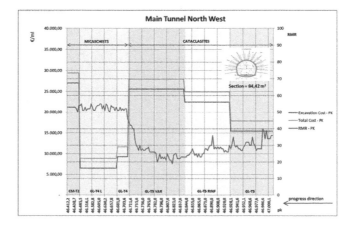

Figure 7.   West main tunnel - North (GLO-N).

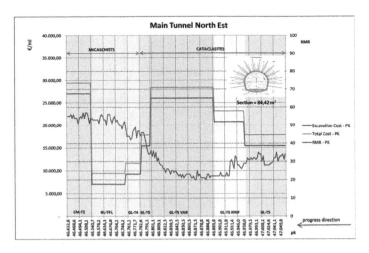

Figure 8.   Main east-north tunnel (GLE-N).

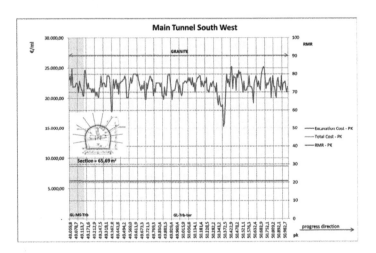

Figure 9.   Main south-west tunnel (GLO-S).

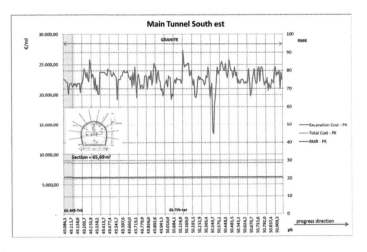

Figure 10.   Main south-east tunnel (GLE-S).

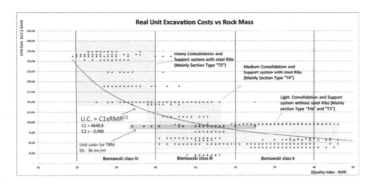

Figure 11.   Real Unit Costs VS Rock Mass Quality.

7   CONCLUSIONS

The final analysis concerned the normalization of costs in function of one cubic meter of excavation. As pointed out in Figure 11, for each tunnel the costs per unit tend to diminish in function of the RMR value. The cost distribution shows the RMR values above which (except for specific situations) costs will vary significantly.

This shows a clear correlation of the RMR with the construction costs, in the sense that a decrease in the RMR value corresponds to a reduction of these costs.

This should be considered as a general rule, whereas in other, varying, local conditions other factors (overburden, presence of water, local instability) can impact the costs for the tunnel in a more or less significant way than the RMR value.

Indeed, looking at Figure 5, one can see that in the first stretch, excavated in a rock mass with high overburdens, costs are influenced, besides by the RMR value, by the measures necessary to contrast rock burst phenomena.

The same applies to Figures 7 and 8, where in local unstable conditions costs were impacted by other factors in a preponderant way as compared with the RMR.

REFERENCES

Bieniawski, Z.T. 1989. *Engineering Rock Mass Classification*. New York: John Wiley and Sons.
Zurlo, R., Rea, G. & Roccia, M. 2013. Galleria di base del Brennero. Descrizione dell'opera ed avanzamento attraverso la faglia Periadriatica. *The Italian Tunneling Society Congress*. Bologna: 628–643.
Fuoco, S., Cucino, P., Schiavinato, L., Oss, A. & Pigorini, A. 2014. The Italian approach for the design and the excavation of conventional tunnels: the case of the Fabriano tunnel. *World Tunneling Congress 2017*. Foz do Iguaçu (Brazil).
Fuoco, S., Zurlo, R., Marini, D. & Pigorini, A. 2016. Tunnel Excavation Solution in Highly Tectonized Zones, Excavation through the Contact between Two Continental Plates. *World Tunneling Congress 2016*. San Francisco (USA).
Fuoco, S., Zurlo, R. & Lanconelli, M. 2017. Tunnel deformation limits and interaction with cavity support: The experience inside the exploratory tunnel of the Brenner Base Tunnel. *World Tunneling Congress 2017*. Bergen (Norway).
Lunardi, P. 1988. ADECO-RS Analisi delle deformazioni controllate nelle rocce e nei suoli e costruzione di gallerie secondo il metodo basato sull'analisi delle deformazioni controllate nelle rocce e nei suoli. *Design and construction of tunnels*. Bergamo.

Tunnels and Underground Cities: Engineering and Innovation meet Archaeology,
Architecture and Art, Volume 7: Long and deep tunnels – Peila, Viggiani & Celestino (Eds)
© 2020 Taylor & Francis Group, London, ISBN 978-0-367-46872-9

The role of construction logistics for long and deep tunnels: The model of the Mules 2-3 construction lot for the Brenner Base Tunnel

R. Zurlo, R. Di Bella & E. Rughetti
Brenner Basistunnel BBT-SE, Bolzano-Fortezza, Italy

ABSTRACT: The excavation of deep, long tunnels is still an important challenge today. High overburdens result in a significant lack of knowledge. The analysis of engineering issues linked to these geological unknown quantities often does not focus on work organization and logistics, matters which are often left to the contractor, without considering that rock face supply and spoil removal issues have an impact on compliance with the estimated timetable and costs for the work, not to mention the environmental impact the works can have on the area in which they are being carried out. This article describes how BBT SE, starting with the planning phase, chose certain logistic solutions for the Mules 2-3 construction lot for the Brenner Base Tunnel. These choices were developed by the contractor after the tendering phase, implementing the system described in this paper which allowed excavation work to proceed in compliance with the planned timetable.

1 INTRODUCTION

The Brenner Base Tunnel, with its access routes, is one of the most important cross-border transport infrastructure projects in Europe. It is a central and crucial part of the Scandinavian-Mediterranean Corridor, which is part of the TEN-T trans-European transport network, on which Europe has concentrated funding for future investment programmes.

The Brenner Pass has always been the most used north-south link in the European Union, over which up to 40 million tonnes of goods are transported annually. However, the steep slopes and the winding curves of the existing railway route do not make for an efficient and modern railway infrastructure, thus jeopardising the possibility of measures to facilitate the transfer of freight from road to rail. The only suitable way to achieve a substantial reduction in slope, thus allowing the transit of longer, larger and faster freight trains, and at the same time reducing travel times, is for the line to run under and through the base of the Alpine mountain crowned by the Brenner Pass, hence the name "base tunnel".

The new railway line runs from Innsbruck (Austria) to Fortezza (Italy), with a maximum gradient of 6,7 ‰ and a length of approximately 20 km less than the existing line. The project consists of a system of two single-track tunnels, each approximately 8 m in diameter, intended exclusively for the transport of goods and passengers by rail. The entire tunnel will be 55 km long (about 24 km in Italy) and will be connected to the existing bypass near Innsbruck, thus reaching a total length of 64 km, which makes it the longest underground railway link in the world.

The two tunnels are connected every 333m by transversal tunnels (bypasses). Along the entire route there will be three emergency stops located at a distance of about 20 km from each other; these will be equipped with emergency systems and with systems for operational management and maintenance; they can all be reached from the outside by a paved tunnel through which vehicles can be driven. An "exploratory tunnel" is built midway between and approximately 12 m below the main tubes, to optimize prospecting of the rock mass and thereby proceed to the following phases of planning and building the tunnel system on the basis of the actual geological conditions present. The location of this lower tunnel allows it to be used for important logistics operations during the tunnel construction, to transport spoil and supply construction materials, and subsequently during the operational phase to drain tunnel waters.

The work is carried out by means of construction lots. Currently, the following construction sites are active in Italy and in Austria: Aica, Mules, Isarco river underpass, Fortezza railway station (in Italy); Wolf, Padastertal valley, Ahrental valley, Ampass, Tulfes, Innsbruck railway station (in Austria). Out of a total of 230 km to be excavated, including main tunnels, exploratory tunnel and logistics or service tunnels, to date about 88 km have already been excavated.

2 THE "MULES 2-3" CONSTRUCTION LOT

The Mules 2-3 construction lot comprises the main part of the Brenner Base Tunnel in Italy. It is located between the State border, to the north (km 32+088 ca) and the adjacent lot "Isarco river underpass", to the south (km 54+015 ca), and extends for approx. 22 km, including a system of tunnels sections and cross passages that cross different geological formations with significantly diversified geomechanical characteristics.

The study of the geomechanical aspects of the rock mass to be crossed, carried out during the design phase, made it possible to define the sections to be excavated with the traditional method and those to be excavated with a TBM. Specifically, the construction lot includes the construction of about 44 km of main tunnels (partly excavated with a TBM (28.5 km), about 15 km of exploratory tunnel (partly driven by drill and blast (about 0.6 km and partly with a TBM (about 14 km)), a 3.8 km long access tunnel to the Trens emergency stop to be excavated using drill and blast and a series of cross passages (excavated by drill and blast).

A construction site of such characteristics and dimensions has inevitably required the development of detailed design solutions that take into account the logistics aspects associated with the large quantities of excavated material to be moved (approx. 4 million cubic metres) and brought to a disposal site, as well as the need to supply the construction materials necessary for the construction, with particular reference to aggregates for projected concrete, aggregates for concrete of prefabricated tunnel linings and shotcrete linings, the prefabricated linings themselves, and the transport of concrete and steel.

Areas near the most suitable access portal have been carefully identified in order to guarantee, in line with the work schedule, the maximum efficiency in the use of the disposal sites and, more in general, of the construction site areas, which are both key elements of the logistics solutions to be implemented.

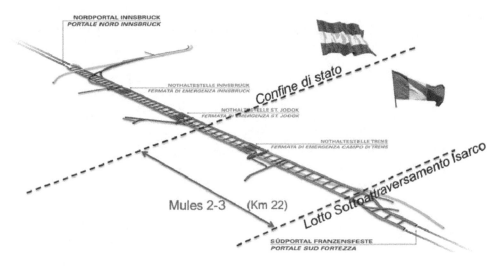

Figure 1. The "Mules 2-3" construction lot.

3 THE CRITERIA AND LOGISTICAL SOLUTIONS ADOPTED IN THE MULES 2-3 CONSTRUCTION LOT

When planning the construction logistics and the materials procurement, particular attention was directed to optimising the construction process and, at the same time, reducing the impact of vehicle traffic on the external environment, avoiding - as far as possible - the crossing of the inhabited centres located near the construction site areas. This is also in compliance with the requirements imposed by the CIPE (Interministerial Committee for Economic Planning) during the authorisation phase of the project.

3.1 *Construction sites*

The two macro-areas serving the Mules 2-3 construction lot, as identified in the design phase, are the area of Mules and the area of Hinterrigger.

The Hinterrigger site, which extends over approximately 22 ha and is located 13 km south of the main BBT access window, is mainly the exclusive recipient of class B+C spoil (i.e. not reusable for the production of concrete) from the tunnels. In the final configuration, about 4.0 million cubic meters of spoil coming from the excavation of the Brenner Base Tunnel in Italy will be disposed of at the Hinterrigger site. During the construction phase, an industrial tubbing prefab plant as well as a stone-crushing and concrete production plant directly connected to the latter will also be built and put into service at Hinterrigger, as illustrated in paragraph 3.4 below.

The Mules construction site covers an area of approximately 3 ha and, being located near the access tunnel, it therefore automatically assumes the role of main site area. On the construction site there are offices, workshops, structures and plants for emergency management, and temporary disposal areas for spoil.

At the Mules construction site there is also a crushing plant for "class A" aggregates used for the production of concrete for the tunnel lining and for the production of shotcrete to stabilize the walls and the rock face.

Figure 2.   The Mules construction site.

3.2 Spoil removal via conveyor belts

The conveyor belt system allows the easy and efficient transport of excavated material from the rock faces and to the temporary/final storage areas, from/to the crushing and concrete production plants in the various construction site areas. The system is gradually developed and expanded as the work progresses.

All the spoil from each single volley/drift using drill and blast/TBM flows into the "logistics node" for the management of the material that is located at the point of intersection of the main tunnels with the exploratory tunnel and the Mules access tunnel.

The logistics node is central to the entire construction process of Mules 2-3 construction lot because all the material flow coming from the excavation activities passes through it and must be sorted towards the disposal sites; the node is organised to link the two levels of the tunnel system: the lowest level in which the exploratory tunnel is located and the level on which the main tubes are excavated. Within this area, support systems for the rock faces and conveyor belts have been installed (and will be implemented over time). By the time the mechanized excavations to the north of the main tunnels start, the belt system will be fully operational, reaching its final configuration; in that phase the system will be well structured and the flow of material will also reach peaks of 2100 ton/h; therefore it will be necessary to have plants and logistics infrastructures capable of processing large quantities of material of different types, which will follow different paths depending on its final use.

In the logistics node there will be transfer devices for each incoming conveyor belt; these are switching systems that can shift the material, according to its classification, onto different belts that cross the whole logistics node and either reach the well that connects the two levels of the tunnel (from where it will be conveyed, through hoppers, through the Aica service tunnel to an aggregates selection system at the Hinterrigger disposal site) or continue on its way to Mules, crossing other belts which were built specially.

In summary, all the belts arriving from the main tunnels will unload the spoil according to classification and final destination, respectively on one of the following three belts: for class A material to the Hinterrigger disposal site; for class B+C material to the Hinterrigger disposal site; for class A material to the Mules disposal site. The following is an overview of the transport of the material, as a function of the excavation methodology:

The system is therefore very well-structured, having to guarantee the removal of spoil from several excavation fronts (up to 6 at the same time) excavated with different methods. The spoil management is even more complicated if we consider the need for great flexibility in

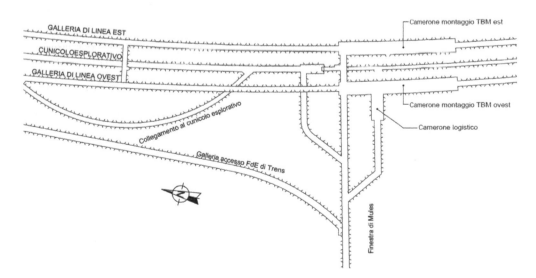

Figure 3. Location of the 'logistics node' for material sorting.

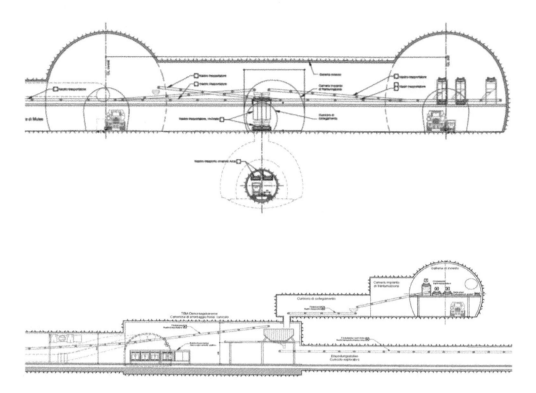

Figure 4. Location of the 'logistics node' for material sorting.

Spoil removal operations in the north and in the south main tunnels, the access tunnel and the exploratory tunnel, excavated with conventional methods	The excavated material is transported from the rock face to the mobile crushing plant located inside the eastern cavern by means of a dumper. The plant crushes the material and continuously unloads it onto the conveyor belts in the logistics node from where it is transported to the destination site (Mules or Hinterrigger)
Spoil removal operations in the north and in the south main tunnels, the access tunnel and the exploratory tunnel, excavated with TBMs	Main tunnels: the spoil is crushed by a primary crushing plant on the TBM and is transferred to conveyor belts travelling in the direction of the logistics node, and from there to the disposal site (Mules or Hinterrigger). Each rock face is provided with an independent conveyor belt system that is progressively extended as the work progresses.
	Exploratory tunnel: the spoil is crushed in a primary crushing plant located in the back-up of the TBM and then unloaded on the conveyor belts through specific hoppers. The conveyor belt that follows the progress of the TBM transports the spoil to the cavern placed along the axis of the tunnel at approx. chainage km 10.5 and from there the spoil, depending on its use, can be transported to Hinterrigger or Mules.

adapting both to the quality of the material to be deposited to the dedicated sites and to the availability of disposal sites.

When fully operational, there will be 26 conveyor belts, with various support structures (hanging, suspended, on the ground), load capacities (200, 350, 500, 600, 1000, 1600 t/h) and belt widths (800, 1000 mm). The belts have variable lengths, appropriate to the context in which they are inserted and always arranged with the aim of optimizing the spaces and paths

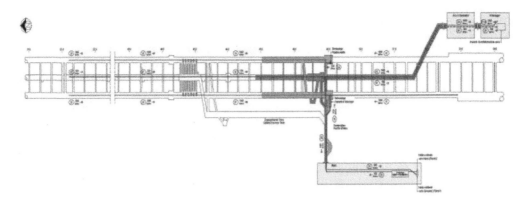

Figure 5. Schematization of the conveyor belts system - final construction phase.

available; the system, when fully operational and in its final configuration, will reach a total length of 78 km.

3.3   The transport system of building materials

The logistics solutions developed for the transport of building materials took into account the time schedule of the works according to the construction program, the requirements linked to the different excavation methods, as well as the need to comply with the authorisation requirements that limit the transit of construction site vehicles on the state roads that pass through residential areas located near the construction sites.

Materials from outside are brought to the Mules construction site via a direct link to the Brenner Highway A22; there is a dedicated motorway entrance/exit that includes two one-way ramps, both on the northern carriageway, where transit is regulated by specific access management systems (through the recognition of authorised license plates) and toll collection.

The material entering Mules mainly consists of steel reinforcements and special elements such as ribs, anchors or drilling material, concrete or additives for the concrete plant.

As far as the supply material for the Unterplattner and Hinterrigger sites is concerned, the existing A22 motorway toll booth in Bressanone Nord, which is less than one km from the entrances to the destination areas, is used. In this case, the types of material include steel reinforcements, concrete, parts for lining elements, spare parts for the train workshop and for the TBM, gasoline, oils and additives for TBMs and dedicated equipment.

The supply of all the material needed for the various rock faces is mainly carried out underground, using the following methods: for the sections excavated with conventional methods wheeled vehicles travel from the temporary storage areas and/or from the concrete plant to their destination, as a function of the various activities provided for in the construction program; for the sections excavated with the TBM, rail transport systems are used, i.e. a system of shuttle trains. Each shuttle train has been designed and sized to transport as many lining elements as possible to the rock face, in order to maximise the transport phase and thus limit rail traffic as much as possible, in the interest of safety. In addition, trains are used for the transport of workers.

The trains which supply the lining elements start from Hinterrigger with lining elements, pea-gravel, backfill mortar for the area behind the inverts and arrive at the logistic node in the cavern of the exploratory tunnel after about 13 km.

They then continue straight ahead if they are going to the TBM of the exploratory tunnel, or they turn left through the connecting tunnel if they are travelling to the TBMs of the main tunnels.

The railway system has been optimised to ensure the best organization and has been properly studied and sized for this purpose; there will be a single track line in the section of exploratory tunnel between Mules and the State border, with bypasses every 2 km, and a double track line in the service tunnel that connects the infrastructure to Hinterrigger. In the sections of the main tunnels to the north, the railway is double-tracked with switches that connect the tracks every 2 km.

The main train station, i.e. the area where the main maintenance of the railway system as a whole is carried out, is located on the Unterplattner site, located further north than the Hinterrigger disposal site and connected to it by means of a 350m service tunnel built within the previous construction lots.

All the transport infrastructures used for the spoil removal operations, including the tracks, comply with the safety and environmental protection regulations in force; the gauge is 900 mm and the slopes of the railway infrastructure are always compatible with the mechanical characteristics of the shuttle trains and with the weight transported with each combination of loads. Train traffic is controlled from a centre at the Mules construction site and is regulated by a controllable signalling system (light signals); in addition, all trains are equipped with a special communication link. The locomotives used for the works in the tunnels are diesel-hydraulic and in Unterplattner, in the TBM maintenance workshops, there is a fuel station.

In the tunnel sections where there are rail tracks, no other vehicles pass through, therefore all passenger and material transport is carried out by trains; for train sizing and the assessment of the number of trains necessary for the supply of construction materials, parameters such as the running speed of the trains both under load and empty, the maximum length of the route to be travelled and the daily production of the TBMs were taken into account. In order to avoid any interruption of the works when waiting for the supply trains, further determining factors have also been taken into account, such as the waiting times at switches and priorities in traffic, and specific loading/unloading time in each work area. When fully operational, the system will include about 15 shuttle trains, also taking into account stand-by trains to cope with potential issues with the supply system.

3.4   Concrete production: description of the concrete plants in Mules 2-3 construction site

Concrete plants have a fundamental role in the logistics system of all large sites: the number, the potential and especially the position within the site are aspects that cannot be neglected if one wants to optimize their use and be competitive in terms of unit production costs. In the case of the Brenner Base Tunnel, the systems present on site have been sized and defined in terms of number and potential, taking into account a number of factors, among which their proximity to the distribution points and the good characteristics of the spoil, which allow them to be completely reused as concrete aggregates, are certainly particularly significant.

There are two concrete plants on the Mules 2-3 construction lot, one exclusively for the production of prefab tunnel linings which is located inside the special plant at Hinterrigger, and one "underground", at the end of the Mules access tunnel, and precisely in the logistics cavern.

Thus, the production is carried out in two concrete plants only where approximately 1 million cubic metres of concrete, needed to make the prefab lining elements and the shotcrete tunnel lining within the Mules 2-3 construction lot during the entire construction phase, are produced. The determining factors that influenced the logistical choices linked to the concrete were proximity between the production site and the installation site and automation, provided that the environment in which they are located allows it and considering the systems for supplying raw materials and other elements (water, cement, additives) that make up the various mix designs foreseen in the project.

In the specific case of Mules, considering the large quantities of concrete and reinforced concrete to be produced and taking into account the limited space available, particular attention is paid to the synchronisation of the various execution phases; the entire cycle of "excavation - extraction and transport to the disposal site of the aggregates - crushing - return to the concrete mixing plant - production of concrete - transport to destination" must be carried out without interruptions, always guaranteeing a quality product that corresponds to the characteristics provided for in the specifications.

Figure 6.   Hinterrigger tubbing prefab plant.

The concrete production at Hinterrigger takes place in the tubbing prefab plant; in this case again, the criterion that was decisive for the logistical process linked to the identification of the most suitable place in which to build the plant was the minimization of the waiting time associated with the transport of concrete to the formworks where the "steam curing" of the tubbings takes place. The site produces the tubbings to feed the 3 double shield TBMs which from spring 2019 will start advancing towards the border (the TBM of the exploratory tunnel began excavating in May 2018); during peak production the prefabrication process must guarantee the production of 250 pieces per day, i.e. a total of 860 cubic meters of concrete.

The imposing concrete plant for prefab elements consists of 4 mixers with 2 cubic metres of concrete, 4 silos for a total of 700 tonnes of stored cement and 2 aggregate storage plants with 2000 cubic metres of aggregates. The aggregate storage plants are fed by a well-structured crushing plant that screens, crushes and selects the aggregates necessary for the preparation of the concrete.

3.5   Ventilation system in tunnels and site logistics: mutual dependencies and interferences

Although the ventilation system cannot properly be considered part of the logistics system, it is an aspect that, in tunnelling projects, must be carefully studied and that cannot be neglected in planning; in the case of deep and long tunnels without many access tunnels connected to the outside, this becomes even more important.

The working environment must comply with the health and safety standards imposed by the regulations on underground work; on the other hand, building materials must be supplied to the construction site and excavation and disposal activities must proceed according to the construction plan. Consequently, the entire logistics system in the construction phase must continually take the ventilation system into consideration.

It is of fundamental importance to guarantee the adequate supply of air from the outside in terms of volumetric flow rate, but another important parameter has also to be considered: air flow speed. In fact, the speed of the air flow must always be between 0.5 m/s and 5 m/s to prevent potential stratification phenomena of natural harmful gases (if present) and at the same time to avoid having overly high recirculation values that are not compatible with a working environment.

The non-derogable constraints on air demand and flow speed therefore also determine another parameter of the entire piping: the size of the ventilation pipes. The size of the pipes, considering the limited space available in the free section of the tunnel, is the interface parameter between the ventilation system and underground construction logistics. The ventilation pipes and their dimensions must not interfere with the shape of the construction vehicles that pass continuously through the tunnel (dumper/shuttle train/excavators), nor must they compromise the rigid layout of the conveyor system.

Therefore, only careful synergic work between the planning of the works and the design of the ventilation system can guarantee the quality of the final result, while adhering to the respect of the schedule and containing costs.

In the Mules 2-3 construction lot for the Brenner Base Tunnel this approach was central both to the executive planning and to develop the construction planning: the careful dislocation along the infrastructure of elements such as the delivery plenum, of any compensation chambers, the use of different types of pipes (rigid, flexible), etc. has ensured, currently ensures and will ensure, even at full capacity, adherence to the schedule set in the construction program. The ventilation system designed and installed collects air from outside through a 40 m high well located in the ventilation cavern (located near the Mules access tunnel). The ventilation chamber houses the electromechanical elements of the ventilation central plant. From here the pipes branch out throughout the tunnel system; the ventilation system is designed to continuously supply fresh air to the rock faces and then ensure its outflow by exploiting the pressure difference created with the external environment.

It is a dynamic system which can be adapted to the planned and different work phases and that can ensure high air quality standards even when, in particular and temporary operating conditions at the construction site, the default path of the piping is modified.

4   CONCLUSIONS

Using underground space to build national and international railways infrastructures like the Brenner Base Tunnel is a wise and appropriate use of this space, considering the benefits in terms of reduction of environmental impacts, increased safety compared to road transport, efficiency, economics, etc.

The construction phase of such infrastructure can, as well, be carried out with a wise and caring use of the resources of the area in which it is being built with proper attention to logistics, spoil recycling and minimization of external impacts.

The example of the Mules 2-3 construction lot of the Brenner Base Tunnel project shows that care in choosing construction and disposal sites allows a "self-feeding" effect for the construction sites themselves, with the only exception being the supply of certain raw materials.

The construction materials supply logistics, the production of prefab elements on the site itself, the transport of spoil exclusively on mile-long conveyor belts, the final and temporary disposal of spoil, are the key logistics elements that not only characterize the construction phase of this construction lot, but ultimately underline the virtuous choices made in the construction of the Brenner Base Tunnel.

REFERENCES

Zurlo, R., Fuoco, S. & Di Bella, R.G., 2017. Wise usage of underground space for transport infrastructure: the model of the Brenner Base Tunnel between Italy and Austria - construction and operation of the longest railway tunnel in the world. In *Livre des résumés – L'espace souterrain notre richesse The value is underground; Congrès international de l'AFTES, Paris, 13–15 November 2017*. Paris (France): AFTES

Zurlo, R., Fuoco, S., Marottoli, A. & Rughetti, E. 2017. Impact of the Rock Mass Quality on the cost of the Conventional Excavation: The experience acquired during the execution of the Brenner Base Tunnel. Proceedings of World Tunneling Congress 2018. Dubai (United Arab Emirates)

Tunnels and Underground Cities: Engineering and Innovation meet Archaeology,
Architecture and Art, Volume 7: Long and deep tunnels – Peila, Viggiani & Celestino (Eds)
© 2020 Taylor & Francis Group, London, ISBN 978-0-367-46872-9

Author Index